Advanced Engineering Mathematics

A. C. BAJPAI

L. R. MUSTOE

D. WALKER

Department of Engineering Mathematics
Loughborough University of Technology

JOHN WILEY & SONS

London · New York · Sydney · Toronto

Library of Congress Cataloging in Publication Data:

Bajpai, Avi C.
Advanced engineering mathematics.

Bibliography: p.
Includes index.
1. Engineering mathematics. I. Mustoe, L.R., joint author. II. Walker, Dennis, joint author. III. Title.
TA330.B33 510'.2'462 77-2198

ISBN 0 471 99521 5 (Cloth)
ISBN 0 471 99520 7 (Paper)

Printed in Great Britain

Preface

This book is the second in a series of books written specifically for engineers and scientists and is aimed at second year undergraduate science and engineering students in universities, polytechnics and colleges in all parts of the world. It would also be useful for students preparing for the Council of Engineering Institutions examinations in mathematics at Part 2 standard. The authors have received a favourable response, both from the United Kingdom and abroad, to their first book *Engineering Mathematics* which followed a new 'integrated' approach. This has encouraged the authors to continue with this approach in the present volume. However, students who have covered the earlier material from an alternative approach will still benefit from using this book. We reproduce without apology part of the preface to *Engineering Mathematics* since we believe the extracts are equally relevant to this work.

The basic concept of the book is that it should provide a motivation for the student. Thus, wherever possible, a topic is introduced by considering a real example and formulating the mathematical model for the problem; its solution is considered by both analytical and numerical techniques. In this way, it is hoped to integrate the two approaches, whereas most previous texts have regarded the analytical and numerical methods as separate entities. As a consequence, students have failed to realise the possibilities of the different methods or that on occasions a combination of both analytical and numerical techniques is needed. Indeed, in most practical cases met by the engineer and scientist the desired answer is a set of numbers; even if the solution can be obtained completely analytically, the final process is to obtain discrete values from the analytical expression.

The authors believe that some proofs are necessary where basic principles are involved. However, in other cases where it is thought that the proof is too difficult for students at this stage, it has been omitted or only outlined. For the numerical techniques, the approach has been to form a heuristically derived algorithm to illustrate how it is used and a formal justification is given only in the simpler cases.

A knowledge of computer programming is assumed since in several parts of the text it is suggested that the reader write and run his own programs in order to obtain a deeper understanding of certain techniques.

Throughout the text there is a generous supply of worked examples which illustrate both the theory and its application. Supplementary problems are provided at the end of each section. Some harder problems have been included for which the student may have to refer to the appropriate book in the bibliography or seek help from his teacher.

A debt of gratitude to the following is acknowledged with pleasure:

Staff and students of Loughborough University of Technology and other institutions who have participated in the development of this text.

John Wiley and Sons for their help and cooperation.

The University of London and the Council of Engineering Institutions for permission to use questions from their past examination papers. (These are denoted by L.U. and C.E.I. respectively.)

Mrs. G. Anthony for drawing the diagrams.

Mrs. Barbara Bell for her great patience and cooperation in typing the entire manuscript and *Gordon Bell* for his help in the general administration.

Contents

Chapter One

Linear Algebra

1.1 INTRODUCTION

Much of the application of mathematics to engineering problems involves linear models. Consider a system with two inputs and one output. If the output of the two inputs combined is equal to the sum of the separate outputs of each input *and* if magnifying an input by a factor α results in the magnification of the output by the same factor, α, the system is said to be a **linear system.**

Great advantages occur if a system is linear. The output from a combination of inputs can be predicted if the outputs from each separate input are known. In this way, a great simplification of the analysis of the system is possible.

But not all systems are linear. We are then faced with the choice of attempting a non-linear analysis or making a linear approximation to the system. We have the worrying problem of whether the powerful methods of linear analysis that we can bring to bear outweigh the possible crudeness of approximation.

However, many of the theoretical models which are widespread in engineering *are* linear and these models have served engineers well. In this chapter we intend to show how linear algebra can act as a unifying link between many branches of mathematics. Sometimes we shall be able to apply our theory directly to practical problems; sometimes we shall not, and you will have to accept the theory as a stepping-stone to further theory which does have direct practical application. Your understanding of some techniques will be enriched by a deeper appreciation of the underlying theory. Having said that, we must point out that we do not dwell too long on abstract theory; references are given in the bibliography to books which take the study of such theory further.

1.2 VECTOR SPACES

Consider the differential equation $$\frac{d^2y}{dx^2} - 4\frac{dy}{dx} + 3y = 0 \qquad (1.1)$$

We find its general solution by forming the auxiliary equation $m^2 - 4m + 3 = 0$ which has roots $m = 3$ and $m = 1$ and we state that two basic solutions of (1.1) are e^{3x} and e^x. We then claim that the **general solution** of (1.1) is found by taking a **linear combination** of the basic solutions, viz.

$$y = Ae^{3x} + Be^x \qquad (1.2)$$

where A and B are real constants. What do we mean by a *general* solution? We mean that *any* solution of (1.1) can be expressed in the form (1.2). Now it is easy to show that (1.2) is *a* solution of (1.1), but its generality is more difficult to establish. For the moment, note that we could represent (1.2) as a pair of coordinates (A, B). Now the particular values of A and B will be determined by the initial conditions attached to

(1.1). Each **particular** solution corresponds to a particular coordinate pair (A, B) which represents in some sense a point in some plane; but, what plane? Also note that if $y_1(x)$ is a solution of (1.1) then so is $\alpha y_1(x)$, where α is any scalar. Further, if $y_2(x)$ is another solution of (1.1) then so is the sum $y_1(x) + y_2(x)$.

Let us now turn to an apparently different example. Consider all polynomial expressions

$$a + bx + cx^2 + dx^3 \tag{1.3}$$

where a, b, c and d are real constants. Now if we multiply the polynomial (1.3) by a scalar, we obtain a polynomial of the same form; likewise, if we add together two such polynomials we again obtain one of the same form. Then the set of polynomials (1.3) is said to be **closed under scalar multiplication** and **closed under addition.** Furthermore we can represent a polynomial of the form (1.3) as the coordinate quartet (a, b, c, d). The structure underlying both examples is seen to be similar.

We can push the ideas further. Suppose we take two solutions of (1.1) represented as (A_1, B_1) and (A_2, B_2). Then their sum may be represented as $(A_1 + A_2, B_1 + B_2)$. For example, if $y_1(x) = 3e^{3x} - e^x \equiv (3, -1)$ and $y_2(x) = e^{3x} + 7e^x \equiv (1, 7)$ then $y_1(x) + y_2(x) = 4e^{3x} + 6e^x \equiv (4, 6) = (3, -1) + (1, 7)$. Similarly, the sum of the polynomials $2 + x - 2x^2 + x^3 \equiv (2, 1, -2, 1)$ and $3 - x^2 + 3x^3 \equiv (3, 0, -1, 3)$ is the polynomial $5 + x - 3x^2 + 4x^3 \equiv (5, 1, -3, 4) = (2, 1, -2, 1) + (3, 0, -1, 3)$. And multiplication by a scalar can also be represented in this vector fashion. It seems as though the underlying structure is essentially of a *vector* nature.

Axioms of a vector space

We shall find it helpful to produce an algebraic structure which extends the ideas of geometric vectors and which embodies the ideas of closure under addition and scalar multiplication. We shall state the axioms for a set of vectors V over a field of real numbers R. Let $\mathbf{v}_1, \mathbf{v}_2, \mathbf{v}_3$ ..... be elements of a set V and α, β any real numbers.

(i) To every pair of vectors $\mathbf{v}_1, \mathbf{v}_2 \in V$ there is a unique vector $(\mathbf{v}_1 + \mathbf{v}_2) \in V$.

(ii) Addition is associative: $\mathbf{v}_1 + (\mathbf{v}_2 + \mathbf{v}_3) = (\mathbf{v}_1 + \mathbf{v}_2) + \mathbf{v}_3$ for any three vectors $\mathbf{v}_1, \mathbf{v}_2, \mathbf{v}_3 \in V$.

(iii) Addition is commutative: $\mathbf{v}_1 + \mathbf{v}_2 = \mathbf{v}_2 + \mathbf{v}_1$.

(iv) There is a zero vector, $\mathbf{0} \in V$ such that $\mathbf{v}_1 + \mathbf{0} = \mathbf{v}_1$ for all $\mathbf{v}_1 \in V$.

(v) For each vector $\mathbf{v}_1 \in V$ there is a unique vector $-\mathbf{v}_1 \in V$ such that $\mathbf{v}_1 + (-\mathbf{v}_1) = \mathbf{0}$.

(vi) To each vector $\mathbf{v}_1 \in V$ and to each scalar $\alpha \in R$ there is a unique product $\alpha\mathbf{v}_1 \in V$.

(vii) $(\alpha\beta)\mathbf{v}_1 = \alpha(\beta\mathbf{v}_1)$.

(viii) $1\,\mathbf{v}_1 = \mathbf{v}_1$.

(ix) $\alpha(\mathbf{v}_1 + \mathbf{v}_2) = \alpha\mathbf{v}_1 + \alpha\mathbf{v}_2$

(x) $(\alpha + \beta)\mathbf{v}_1 = \alpha\mathbf{v}_1 + \beta\mathbf{v}_1$

(ix)–(x): Distributive laws.

Subspaces

If we take a subset of vectors from a vector space then they may form a new vector space by themselves; in such a case we call the subset a **subspace** of V. In order to check whether a subset of vectors does form a subspace it can be shown to be sufficient to demonstrate that the subset is closed under addition and scalar multiplication. For example, if we consider the subset of solutions of (1.1) for which $y(0) = 0$ then the condition on A and B of (1.2) is that $A + B = 0$. Hence, we are dealing with solutions of the form $y = A(e^{3x} - e^x)$. Now we can show that these solutions are closed under addition and scalar multiplication in one fell swoop: we need merely to show that if $y_1(x)$ and $y_2(x)$ are two such solutions, then so is $[\alpha y_1(x) + \beta y_2(x)]$ for any scalars α and β (i.e. any **linear combination of** y_1 and y_2). Let $y_1 = A_1(e^{3x} - e^x)$ and $y_2 = A_2(e^{3x} - e^x)$ and consider $[\alpha y_1(x) + \beta y_2(x)] = \alpha A_1(e^{3x} - e^x) + \beta A_2(e^{3x} - e^x)$; this can be shown to be a solution of the equation (1.1) which has value zero when $x = 0$. Hence the subset of solutions for which $y(0) = 0$ is a subspace.

Example 1

Show that the subset of polynomials of degree $\leqslant 3$ for which the x term is absent forms a subspace.

In vector terms, the subset consists of all vectors of the form $(a, 0, c, d)$. Consider two vectors: $(a_1, 0, c_1, d_1)$ and $(a_2, 0, c_2, d_2)$; the linear combination $\alpha(a_1, 0, c_1, d_1) + \beta(a_2, 0, c_2, d_2) = (\alpha a_1 + \beta a_2, 0, \alpha c_1 + \beta c_2, \alpha d_1 + \beta d_2)$ satisfies the membership criterion of a zero second coordinate. The subset is closed under addition and scalar multiplication and hence forms a subspace.

Example 2

Show that the subset of polynomials of degree $\leqslant 3$ for which the constant term is 3 do *not* form a subspace.

In vector terms, a typical member of the subset is $(3, b, c, d)$. It should be clear that adding two such members produces a polynomial which violates the membership criterion, since it will have a first component of 6. The subset is not closed under addition and nor, as you can show, is it closed under multiplication by scalars other than 1. The subset fails to be a subspace.

See what you can deduce about the subset of solutions of (1.1) for which $y(0) = 1$. What do you think the result means?

Linear Independence

Suppose we look back at the solution (1.2) of equation (1.1). We chose to express the set of solutions as linear combination of two special solutions, viz. e^{3x} and e^x. We needed both of these special solutions to provide an adequate description of the set. When all the vectors in a vector space can be expressed as linear combinations of a subset of vectors, the subset is said to **span** the space. For example, the polynomials $1, x, x^2$ and x^3 span the space of polynomials expressed in (1.3). The unit vectors **i**, **j**, **k** span the space of geometric vectors with three components since any vector (x, y, z) can be written as $x\mathbf{i} + y\mathbf{j} + z\mathbf{k}$.

It may equally well be argued that the functions $3e^{3x} - e^x$ and $2e^{3x} + e^x$ span the space (1.2). Consider a typical solution $4e^{3x} + 3e^x$; this can be written as

$\alpha(3e^{3x} - e^x) + \beta(2e^{3x} + e^x) = (3\alpha + 2\beta)e^{3x} + (-\alpha + \beta)e^x$, provided that $3\alpha + 2\beta = 4$ and $-\alpha + \beta = 3$ i.e. in this case $\alpha = -2/5$ and $\beta = 13/5$. (Try to prove this result for a general solution $Ae^{3x} + Be^x$.) Another subset of solutions which span the space is $\{e^{3x}, e^x, 3e^{3x} - e^x\}$; if we again consider $4e^{3x} + 3e^x$ you should show that it can be written as $1 \times e^{3x} + 4 \times e^x + 1 \times (3e^{3x} - e^x)$. The trouble is that it can also be written as $-2 \times e^{3x} + 5 \times e^x + 2 \times (3e^{3x} - e^x)$ and, indeed, as $(4 - 3\gamma)e^{3x} + (3 + \gamma)e^x + \gamma(3e^{3x} - e^x)$ where γ is any scalar. What has happened is that we no longer have a *unique* representation. Why should this be? One reason might be that we have too many vectors in this new spanning set. The function $3e^{3x} - e^x$ is *itself* a linear combination of e^{3x} and e^x and is in effect redundant. However, neither e^{3x} nor e^x can be expressed as a multiple (special case of a linear combination) of each other. [We should add that we could single out any of the three functions for redundancy: check this.] A set of vectors $\{\mathbf{v}_1, \mathbf{v}_2, \mathbf{v}_3, \ldots\ldots\ldots, \mathbf{v}_n\}$ is said to be **linearly dependent** if and only if there is a set of real numbers $\{\alpha_1, \alpha_2, \alpha_3, \ldots\ldots\ldots, \alpha_n\}$ *not all zero* such that

$$\alpha_1\mathbf{v}_1 + \alpha_2\mathbf{v}_2 + \alpha_3\mathbf{v}_3 + \ldots\ldots\ldots + \alpha_n\mathbf{v}_n = \mathbf{0} \tag{1.4}$$

Note that (1.4) can be rearranged to give a formula for any one of the vectors in terms of the rest; in fact, as a linear combination of the rest.

A set of vectors which is not linearly dependent is **linearly independent**; the condition (1.4) implies that in this case that $\alpha_1 = \alpha_2 = \alpha_3 = \ldots.. = \alpha_n = 0$.

Example 1

The set of polynomials $\{1, x + 2, x^2 + x, x^3\}$ is linearly independent. We write the functions in vector form as (1, 0, 0, 0), (2, 1, 0, 0), (0, 1, 1, 0), (0, 0, 0, 1). Consider the equation $\alpha(1, 0, 0, 0) + \beta(2, 1, 0, 0) + \gamma(0, 1, 1, 0) + \delta(0, 0, 0, 1) = \mathbf{0} = (0, 0, 0, 0)$. Comparing components we obtain the equations $\alpha + 2\beta = 0$, $\beta + \gamma = 0$, $\gamma = 0$, $\delta = 0$. Hence $\alpha = \beta = \gamma = \delta = 0$ and therefore the set is linearly independent.

Example 2

The set of polynomials $\{1, x + 2, x^2 + x, x^3, x^3 - x\}$ is linearly dependent. Prove this.

Example 3

The set of polynomials $\{x + 2, x^3 - 2x, x^3 + 4\}$ is linearly dependent, since (resorting to vector notation) $\alpha(2, 1, 0, 0) + \beta(0, -2, 0, 1) + \gamma(4, 0, 0, 1) = (0, 0, 0, 0) \Rightarrow 2\alpha + 4\gamma = 0$, $\alpha - 2\beta = 0$, $\beta + \gamma = 0$ and these equations have infinitely many solutions of the form $(\alpha, \beta, \gamma) = (-2c, -c, c)$ for any value c.

We can show that a subset of a linearly independent set of vectors is also linearly independent and that a linearly dependent set will remain linearly dependent if one or more vectors from V are added to it.

Basis of a Vector Space

We have seen that we can produce subsets of vectors from a vector space which span the space. If we restrict our attention to those spanning sets which are linearly independent we can summarise the information about the vector space by listing the vectors in a linearly independent spanning set. Since the set spans the space we can produce any vector in the space by a linear combination of the elements of the set. Since the set is linearly independent, we cannot afford to lose any vectors from it or else it will cease to span the space. In other words a spanning set which is linearly independent contains *precisely* the right amount of information to describe the space. Such a set is called a **basis** for the vector space.

Example 1

We can show that the set $\{(1, 0, 0, 0), (0, 1, 0, 0), (0, 0, 1, 0), (0, 0, 0, 1)\}$ is linearly independent, and furthermore it spans the space of vectors with four components. Hence it forms a basis for the space. So does the set $\{(1, 0, 0, 0), (2, 1, 0, 0), (0, 1, 1, 0), (0, 0, 0, 1)\}$ and so too does the set $\{(1, 1, 0, 0), (0, 1, 1, 0), (0, 0, 1, 1), (0, 0, 0, 1)\}$ as you can verify.

Example 2

For the space of solutions of $\frac{d^2y}{dx^2} - 4\frac{dy}{dx} + 3y = 0$, each of the following sets form a basis: $\{(1, 0), (0, 1)\}$, $\{(1, 1), (0, 1)\}$, $\{(1, 1), (1, 0)\}$. There is one thing that *all* bases for a given space have in common: they contain the same number of vectors. The number of vectors in a basis for a given space is called the **dimension** of the space. It is important to realise that any set of n linearly independent vectors will be a basis for a space of dimension n.

Note that where a possible basis for a space consists of the unit vectors of appropriate size, then any other basis can be found by replacing one or more of the unit vectors by a combination of them.

Example 3

The set of polynomials of degree $\leqslant 3$ with no term in x form a vector space. A possible basis is $\{(1, 0, 0, 0), (0, 0, 1, 0), (0, 0, 0, 1)\}$. It should be fairly clear that the set is linearly independent and any polynomial $a + cx^2 + dx^3$ can be written as $(a, 0, c, d) = a(1, 0, 0, 0) + c(0, 0, 1, 0) + d(0, 0, 0, 1)$. Hence the given set spans the space and is a basis for it. Since the basis consists of three vectors, the dimension of this space is 3. Notice that we have taken a subset of the four unit vectors which comprise a basis for the space of all polynomials of degree $\leqslant 3$. In imposing the one restriction that the x term is absent we have reduced the dimension of the space by 1. When we express a vector of the space as a linear combination of the vectors in the basis, the coefficients (in this example, a, b, c and d) are called the **coordinates** of the vector *relative* to the particular basis. We stress the word 'relative' because had we chosen a different basis we should almost certainly have needed different coordinates for the vectors under consideration.

Orthonormal Bases

One advantage of using a basis consisting of the unit vectors is that they are **mutually orthogonal**, i.e. the scalar product of any two of them is zero. Furthermore, they are unit vectors so that the scalar product of any of them with itself is unity; we say that the vectors are **normalised.** A set of normalised, mutually orthogonal vectors is called an **orthonormal** set. If we introduce the **Kronecker delta** symbol δ_{ij} which follows the rules that $\delta_{ij} = 0$ for $i \neq j$ and $\delta_{ij} = 1$, $i = j$, then we may succinctly define an orthonormal set of vectors $\{\mathbf{v}_1, \mathbf{v}_2, ..., ..., \mathbf{v}_n\}$ by requiring that $\mathbf{v}_i \cdot \mathbf{v}_j = \delta_{ij}$.

An important result is that an **orthonormal set is linearly independent.** This follows by stating that $a_1\mathbf{v}_1 + a_2\mathbf{v}_2 + ... + a_r\mathbf{v}_r + ... + a_n\mathbf{v}_n = 0$ and taking the scalar product of both sides with $\mathbf{v}_r$. All the terms on the left-hand side bar one disappear and we are left with the equation $a_r\mathbf{v}_r \,.\, \mathbf{v}_r = 0$. But $\mathbf{v}_r \,.\, \mathbf{v}_r = 1$ and hence $a_r = 0$. By letting $r = 1, 2, 3, ..., n$ we see that all a_r are zero which proves the linear independence.

In geometry we are used to Cartesian axes being mutually orthogonal and this allows a simple representation of vectors relative to these axes. Also, if for example we project a point in space (3 dimensional) onto the $x - y$ plane then the coordinates of the point reduce in a simple way. Thus, the point $(3, 4, -2)$ when projected onto the $x - y$ plane, becomes in the new axes the point $(3, 4)$.

Gram-Schmidt Orthogonalisation Process

This method takes a linearly independent set of vectors from a space V and produces an orthonormal set of vectors from V by a well-defined ritual. In particular it will convert a basis for V into an orthonormal basis. The idea is to start with a linearly independent set $\{\mathbf{v}_1, \mathbf{v}_2, ..., \mathbf{v}_s\}$ and produce an orthonormal set $\{\xi_1, \xi_2, ..., \xi_s\}$ where each $\xi_k = \sum_{i=1}^{k} a_{ik}\, \mathbf{v}_i$.

We know that $\mathbf{v}_1 \neq \mathbf{0}$ and hence $|\mathbf{v}_1| > 0$; therefore we can sensibly define $\xi_1 = \mathbf{v}_1/|\mathbf{v}_1|$ so that $|\xi_1| = 1$.

We now assume that we have found the first r vectors of the orthonormal set and show how to find ξ_{r+1}. We assume that the set $\{\xi_1, \xi_2, ..., \xi_r\}$ is orthonormal and that each of the vectors ξ_k in it is a linear combination of the set $\{\mathbf{v}_1, \mathbf{v}_2, ..., \mathbf{v}_k\}$. Now consider

$$\mathbf{v}'_{r+1} = \mathbf{v}_{r+1} - (\xi_1 \,.\, \mathbf{v}_{r+1})\xi_1 - (\xi_2 \,.\, \mathbf{v}_{r+1}\xi_2 - .... - (\xi_r \,.\, \mathbf{v}_{r+1})\xi_r \qquad (1.5)$$

Let us now take the scalar product of (1.5) in turn with $\xi_1, \xi_2, ..., \xi_r$. First, with ξ_1 we have $\xi_1 \,.\, \mathbf{v}'_{r+1} = \xi_1 \,.\, \mathbf{v}_{r+1} - (\xi_1 \,.\, \mathbf{v}_{r+1})(\xi_1 \,.\, \xi_1) - (\xi_2 \,.\, \mathbf{v}_{r+1})(\xi_1 \,.\, \xi_2) - .... - (\xi_r \,.\, \mathbf{v}_{r+1})(\xi_1 \,.\, \xi_r)$.

Since $\xi_1 \,.\, \xi_2 = \xi_1 \,.\, \xi_3 = ... = \xi_1 \,.\, \xi_r = 0$ and $\xi_1 \,.\, \xi_1 = 1$ (by the properties of an orthonormal set) it follows that $\xi_1 \,.\, \mathbf{v}'_{r+1} = \xi_1 \,.\, \mathbf{v}_{r+1} - (\xi_1 \,.\, \mathbf{v}_{r+1})1 = 0$.

Similarly we can show that $\xi_2 \,.\, \mathbf{v}'_{r+1} = 0, \xi_3 \,.\, \mathbf{v}'_{r+1} = 0, ..., \xi_r \,.\, \mathbf{v}'_{r+1} = 0$. Therefore the vector $\mathbf{v}'_{r+1}$ is orthogonal to the vectors $\xi_1, \xi_2, ..., \xi_r$ and hence $\{\xi_1, \xi_2, ..., \xi_r, \mathbf{v}'_{r+1}\}$ is a mutually orthogonal set. Is $\mathbf{v}'_{r+1}$ of unit length? We can ensure that the additional vector is of unit length by choosing it to be $\xi_{r+1} = \mathbf{v}'_{r+1}/|\mathbf{v}'_{r+1}|$. Hence $\{\xi_1, \xi_2, ..., \xi_r, \xi_{r+1}\}$ is an orthonormal set. In addition, $\mathbf{v}'_{r+1}$ (and therefore ξ_{r+1}) must be a linear combination of the vectors $\mathbf{v}_1, \mathbf{v}_2, ..., \mathbf{v}_{r+1}$; this follows from (1.5) and because we assumed that $\xi_1, \xi_2, ..., \xi_r$ were linear combinations of the vectors $\mathbf{v}_1, \mathbf{v}_2, ..., \mathbf{v}_r$.

If we proceed to replace each $\mathbf{v}_k$ by ξ_k as prescribed above, we shall obtain an orthormal set $\{\xi_1, \xi_2, ..., \xi_s\}$.

Example

We start with the set $\{\mathbf{v}_1, \mathbf{v}_2, \mathbf{v}_3\} = \{(1, 0, 1), (1, -2, 1), (0, 1, 1)\}$ which we can show is a basis for the space of vectors with three components. (Note that we merely need to demonstrate linear independence.)

Then $|\mathbf{v}_1| = \sqrt{1^2 + 0^2 + 1^2} = \sqrt{2}$ and $\xi_1 = \dfrac{1}{\sqrt{2}}(1, 0, 1)$

Now $\mathbf{v}_2' = \mathbf{v}_2 - (\xi_1 \cdot \mathbf{v}_2)\xi_1 = (1, -2, 1) - \left[\dfrac{1}{\sqrt{2}}(1, 0, 1).(1, -2, 1)\right]\dfrac{1}{\sqrt{2}}(1, 0, 1)$

$$= (1, -2, 1) - \frac{1}{2}(2)(1, 0, 1) = (0, -2, 0)$$

Hence $\xi_2 = \mathbf{v}_2'/|\mathbf{v}_2'| = \dfrac{1}{2}(0, -2, 0) = (0, -1, 0)$

Then $\mathbf{v}_3' = \mathbf{v}_3 - (\xi_1 \cdot \mathbf{v}_3)\xi_1 - (\xi_2 \cdot \mathbf{v}_3)\xi_2$

$$= (0, 1, 1) - \left[\frac{1}{\sqrt{2}}(1, 0, 1) . (0, 1, 1)\right]\frac{1}{\sqrt{2}}(1, 0, 1) -$$

$$[(0, -1, 0). (0, 1, 1)]\,(0, -1, 0)$$

$$= (0, 1, 1) - \frac{1}{2}(1, 0, 1) - (-1)(0, -1, 0) = (-\tfrac{1}{2}, 0, \tfrac{1}{2})$$

Finally, $\xi_3 = \dfrac{1}{\sqrt{2}}(-1, 0, 1)$.

We can easily check that $\xi_i \cdot \xi_j = \delta_{ij}$ and hence that $\{\xi_1, \xi_2, \xi_3\}$ is an orthonormal set. Furthermore these three vectors are linearly independent. We have therefore constructed an orthonormal basis.

Check what happens if we write the vectors of the original set in a different order.

You can see an application of this technique in Rosenbrock's algorithm for optimisation (page 109).

Problems

1. Show that the vectors $(1, 2, 2, 1)$, $(3, 4, 4, 3)$ and $(1, 0, 0, 1)$ are linearly dependent and span a vector space of dimension 2; find a basis for this space. Repeat for the vectors $(1, 1, 1, 0)$, $(4, 3, 2, -1)$, $(2, 1, 0, -1)$ and $(4, 2, 0, -2)$.

2. Show that the vector space spanned by the vectors $(1, 2, 1)$, $(1, 2, 3)$ and $(3, 6, 5)$ is identical with that spanned by $(1, 2, 5)$ and $(0, 0, 1)$.

3. Find the dimension of the vector space spanned by each of the following sets of vectors; choose a basis for each space.

   (a) $(1, 2, 3, 4, 5)$, $(5, 4, 3, 2, 1)$ and $(1, 1, 1, 1, 1)$ (b) $(1, 1, 0, -1)$, $(1, 2, 3, 4)$ and $(2, 3, 3, 3)$ (c) $(1, 1, 1, 1)$, $(3, 4, 5, 6)$, $(1, 2, 3, 4)$ and $(1, 0, -1, -2)$.

4. Which of the following sets are subspaces of the vector space of which a typical member is (a, b, c, d), all components being real numbers.

(i) all vectors with $a = b = c = d$ (ii) all vectors with $a = 2$ (iii) all vectors with $a = b$ and $c = 2d$ (iv) all vectors with integer components (v) all vectors with $d = 0$.

5. Find a vector orthogonal to (i) $(1, -2, -2)$ and $(2, -1, 2)$ (ii) $(1, 2, 1)$ and $(2, 1, 2)$ (iii) $(1, 2, 1)$ and $(2, 1, 4)$.

6. Construct an orthonormal basis for the space of vectors with three real components $V_3(R)$ given the basis

   (i) $(2, 1, 3)$, $(1, 2, 3)$, $(1, 1, 1)$ (ii) $(1, -1, 0)$, $(2, -1, -2)$, $(1, -1, -2)$ (iii) $(1, 0, 1)$, $(1, 3, 1)$, $(3, 2, 1)$ (iv) $(2, -1, 0)$, $(4, -1, 0)$, $(4, 0, -1)$.

7. Find an orthonormal basis for $V_3(R)$ given

   (i) $(1, 1, -1)$ and $(2, 1, 0)$ and (ii) $(7, -1, -1)$.

8. Given the basis $\{(1, 0, 1, 0), (1, 1, 0, 0), (0, 1, 1, 1), (0, 1, 1, 0)\}$, use the Gram Schmidt process to obtain an orthonormal basis.

9. Let V be a subspace of U spanned by $\{(0, 1, 1, 0), (0, 5, -3, -2), (-3, -3, 5, -7)\}$. Find an orthonormal basis for V.

1.3 LINEAR TRANSFORMATIONS

In this section we shall be concerned with mappings from one vector space to another. As the chapter unfolds we shall see that these ideas have applications in the solution of ordinary differential equations, the solution of simultaneous linear equations and so on.

The particular class of mappings which interests us is that class which preserves linear structure. Let T be a transformation mapping vector from a space V, known as the **domain**, to another space U, known as the **co-domain**, in such a way that a vector $\mathbf{u}$ from U is associated uniquely with a vector $\mathbf{v}$ from V i.e. $T(\mathbf{u}) = \mathbf{v}$. If we impose the requirement that addition is preserved i.e.

$$T(\mathbf{u}_1 + \mathbf{u}_2) = T(\mathbf{u}_1) + T(\mathbf{u}_2) \tag{1.6a}$$

for all $\mathbf{u}_1, \mathbf{u}_2 \in V$, and if we impose the requirement that multiplication by a scalar is preserved i.e.

$$T(\alpha\mathbf{u}) = \alpha T(\mathbf{u}) \tag{1.6b}$$

then T is called a **linear transformation.**

Note that both conditions can be summarised by the single condition

$$T(\alpha\mathbf{u}_1 + \beta\mathbf{u}_2) = \alpha T(\mathbf{u}_1) + \beta T(\mathbf{u}_2) \tag{1.7}$$

Example 1

Let U and V both be the space of polynomials of degree $\leqslant 3$ and let T be defined by the rule that a polynomial is mapped onto its derivative. Resorting to a more customary notation, we denote elements of U by $p(x)$ so that

$$T[p(x)] = \frac{d}{dx}p(x) = p'(x)$$

Now $$\frac{d}{dx}\left[\alpha p_1(x) + \beta p_2(x)\right] = \alpha p_1'(x) + \beta p_2'(x) \text{ so that}$$

$$T(\alpha\mathbf{u}_1 + \beta\mathbf{u}_2) = \alpha T(\mathbf{u}_1) + \beta T(\mathbf{u}_2)$$

which proves the linearity of T.

One point to note is that the polynomials $p'(x)$ can never be of degree greater than 2. In other words the polynomial $a + bx + cx^2 + dx^3 \equiv (a, b, c, d)$ is mapped onto the polynomial $(b + 2cx + 3dx^2) \equiv (b, 2c, 3d, 0)$. This indicates that we have lost something by the mapping operation.

Example 2

Suppose we extend the above ideas and consider U comprising all functions which can be differentiated twice. Let the mapping T associate the function $f(x)$ from U with the function $\left(\frac{\mathrm{d}^2}{\mathrm{d}x^2} - 4\frac{\mathrm{d}}{\mathrm{d}x} + 3\right)f(x)$. It is left to you to show that T is a linear transformation. Has anything been lost this time?

Example 3

Consider U = all points (x, y) in the plane. Then let T be the operation rotating the plane through an anticlockwise angle θ about the origin so that

$$T[(x, y)] = (x', y') \text{ where } \begin{cases} x' = x\cos\theta - y\sin\theta \\ y' = x\sin\theta + y\cos\theta \end{cases} \tag{1.8}$$

Now $T[\alpha(x_1, y_1) + \beta(x_2, y_2)] = T(\alpha x_1 + \beta x_2,\ \alpha y_1 + \beta y_2)$

$$= [(\alpha x_1 + \beta x_2)\cos\theta - (\alpha y_1 + \beta y_2)\sin\theta,\ (\alpha x_1 + \beta x_2)\sin\theta + (\alpha y_1 + \beta y_2)\cos\theta]$$

$$= (\alpha x_1\cos\theta - \alpha y_1\sin\theta + \beta x_2\cos\theta - \beta y_2\sin\theta,\ \alpha x_1\sin\theta + \alpha y_1\cos\theta + \beta x_2\sin\theta + \beta y_2\cos\theta)$$

$$= \alpha(x_1\cos\theta - y_1\sin\theta,\ x_1\sin\theta + y_1\cos\theta) + \beta(x_2\cos\theta - y_2\sin\theta,\ x_2\sin\theta + y_2\cos\theta)$$

$$= \alpha T(x_1, y_1) + \beta T(x_2, y_2)$$

and linearity is proved. However, in this example, it appears that points have merely changed the values of their coordinates and nothing has been lost.

Let us now quote a useful result.

The subset of vectors of V which are the images under T of vectors from U, i.e. the set of all vectors $T(\mathbf{u})$ where $\mathbf{u}$ is a vector from U, forms a subspace of V. This is called the **image space** of T and is sometimes denoted by $T(U)$. If it is a subspace, its dimension must be less than or equal to the dimension of V. Under what circumstances is equality achieved? The dimension of the image space is called the **rank** of the linear transformation.

The kernel of a linear transformation

Suppose we return to Example 2. We know that there are solutions of the differential equation $\frac{\mathrm{d}^2y}{\mathrm{d}x^2} - 4\frac{\mathrm{d}y}{\mathrm{d}x} + 3y = 0$ i.e. that there are several functions $f(x)$ mapped to the polynomial 0 by the linear transformation $\left(\frac{\mathrm{d}^2}{\mathrm{d}x^2} - 4\frac{\mathrm{d}}{\mathrm{d}x} + 3\right)$. On the other hand, from Example 3 we find that only the point $(0, 0)$ is mapped to the point $(0, 0)$.

The mapping $T[(x, y, z)] = (x, y, 0)$ which projects points in three dimensions onto the x - y plane can easily be shown to be a linear transformation (show it). The image space is the set of all vectors with three components with the last one zero; you can show that it has dimension 2. All points $(0, 0, z)$ are projected onto the point $(0, 0, 0)$. Notice that the basis vectors $(1, 0, 0)$, $(0, 1, 0)$, $(0, 0, 1)$ are mapped onto the vectors $(1, 0, 0)$, $(0, 1, 0)$ and $(0, 0, 0)$. The first two of these form a basis for the image space; the third is redundant.

The subset of vectors in the domain which map onto the zero vector in the co-domain is called the **kernel** or the **null-space** of the linear transformation. We can show that the kernel is itself a vector space and that, moreover, the dimension of the kernel is the dimension of the space V less the dimension of the image space. We call the dimension of the kernel the **nullity** of the linear transformation and we may state the result that if the linear transformation T maps vectors from U into vectors from V then

$$\text{rank of } T + \text{nullity of } T = \text{dimension of } U \tag{1.9}$$

For the projection mapping, the dimension of U is 3, the nullity is 1 since the vector $(0, 0, 1)$ is a basis for the kernel, and the rank is 2.

Let us look back at our three other examples. In Example 1, the domain U had dimension 4 and the image space had dimension 3. The kernel consisted of all polynomials with zero derivative, i.e. all constants, which may be represented $(a, 0, 0, 0)$; the dimension of this space is 1. In Example 2 it is not possible to state the dimension of the domain U, but we have already seen that the dimension of the kernel is 2. In Example 3 only the point $(0, 0)$ is mapped to the point $(0, 0)$ and the kernel, which merely consists of the zero vector, has zero dimension; the dimensions of the domain and the image space are both equal to 2.

Properties of the kernel

Let T map vectors from space U to vectors from space V, and let the kernel of T be written K.

(i) If $\mathbf{u}_0 \in U$ and $\mathbf{k} \in K$ then $\mathbf{u}_0 + \mathbf{k}$ is mapped to the same vector as $\mathbf{u}_0$. This follows since $T(\mathbf{u}_0 + \mathbf{k}) = T(\mathbf{u}_0) + T(\mathbf{k}) = T(\mathbf{u}_0) + \mathbf{0} = T(\mathbf{u}_0)$.

(ii) If $\mathbf{u}_0$ maps to the vector $\mathbf{v}_0$, then all vectors which map to $\mathbf{v}_0$ are of the form $\mathbf{u}_0 + \mathbf{k}$, where $\mathbf{k}$ is some vector from K.

This second property is important since it allows us to produce general solutions to linear ordinary differential equations and simultaneous linear algebraic equations. Before we demonstrate this, consider the operation of differentiating polynomials of degree $\leqslant 3$. We have seen that this is an example of a linear transformation. The kernel is the set of constant polynomials. Therefore, the problem of integration of polynomials of degree <3 can be seen as aiming to find the polynomial which maps to a given polynomial under the operation of differentiation, i.e. we seek $p(x)$ such that $p'(x) = p_1(x)$, which is given. If we can find any polynomial $q(x)$ whose derivative is $p_1(x)$, then by property (ii) we know that *all* solutions to the problem are of the form $q(x) + c$ where c is a constant.

With regard to the solution of the non-homogeneous differential equation

$$\frac{d^2y}{dx^2} - 4\frac{dy}{dx} + 3y = g(x) \tag{1.10}$$

where $g(x)$ is a given function, we can extend our results to show that its most general solution may be written in the form $y = y_{PI} + y_{CF}$, where y_{PI} is *any* solution of (1.10) and y_{CF} is the general solution of the associated homogeneous equation

$$\frac{d^2y}{dx^2} - 4\frac{dy}{dx} + 3y = 0$$

We have not yet dealt with the important class of problems expressible as a set of simultaneous linear algebraic equations. We consolidate ideas with an example from that context.

Example

We require to find the solutions of the equations

$$\left.\begin{aligned} 3x + 2y - z &= 4 \\ x + y + z &= 3\tfrac{1}{2} \end{aligned}\right\} \qquad (1.11)$$

Here we have three unknowns restricted by two equations and therefore we must expect an infinite set of solutions. We take $x = 0$ to find one *particular* solution . You can check that this is $(x, y, z) = (0, 2\frac{1}{2}, 1)$. How can we find other solutions? If we now think in terms of vector spaces, the equations are imposing a mapping which associates (x, y, z) with $(3x + 2y - z, x + y + z)$ and we seek those vectors which are mapped to $(4, 3\frac{1}{2})$. We can show that the mapping is a linear transformation and therefore by property (ii), we need only now find the kernel to obtain the general solution. The kernel is the set of (x, y, z) which map to $(0, 0)$ and which therefore satisfy the equations

$$\left.\begin{aligned} 3x + 2y - z &= 0 \\ x + y + z &= 0 \end{aligned}\right\} \qquad (1.12)$$

To obtain *all* solutions we give z, say, a general value λ and solve the resulting equations in terms of λ. You can check that we obtain $(x,y,z) = (3\lambda , -4\lambda , \lambda)$. By varying λ we obtain all elements in the kernel. The general solution of (1.11) can then be expressed as the sum of the particular solution and the general solution of (1.12) viz. $(x,y,z) = (3\lambda, 2\frac{1}{2} - 4\lambda, 1 + \lambda)$; check for yourself that this claimed solution satisfies (1.11).

You might be tempted to ask what would have happened if we had taken a different particular solution, for example, $(-3, 6\frac{1}{2}, 0)$. We should then have expressed the solution set as $(x,y,z) = (-3 + 3\mu, 6\frac{1}{2} - 4\mu, \mu)$ where μ can be varied. Any specific solution will require different values of μ; for example the solution $(x,y,z) = (1, \frac{7}{6}, \frac{4}{3})$ requires $\lambda = \frac{1}{3}$ or $\mu = \frac{4}{3}$. If the equations (1.11) are replaced by ones with different right-hand sides then the kernel will remain unaltered, and all that is needed is a new particular solution.

Matrices and Linear Transformations

We generalise to m equations in n unknowns. The equations can be regarded as defining a linear transformation from the space of vectors with n real components, denoted R^n, to the vector space R^m. It can be shown that *any* linear transformation mapping R^n to R^m can be defined by a set of m linear equations in n unknowns.

Such a set of equations can be summarised as $\mathbf{Ax} = \mathbf{b}$ where

$$\mathbf{A} = \begin{bmatrix} a_{11} & a_{12} & ..... & a_{1n} \\ a_{21} & a_{22} & ..... & a_{2n} \\ a_{m1} & a_{m2} & ..... & a_{mn} \end{bmatrix}, \quad \mathbf{x} = \begin{bmatrix} x_1 \\ x_2 \\ x_n \end{bmatrix} \quad \text{and} \quad \mathbf{b} = \begin{bmatrix} b_1 \\ b_2 \\ b_m \end{bmatrix} \tag{1.13}$$

A is called the **associated matrix** of the linear transformation.

Suppose we take the vectors $\mathbf{e}_1 = (1, 0, 0, ..., 0)$, $\mathbf{e}_2 = (0, 1, 0, ..., 0)$, ..., $\mathbf{e}_n = (0, 0, 0, ..., 1)$ as a basis for R^n. Then let $\mathbf{x}$ be any vector of R^n so that $\mathbf{x} = x_1\mathbf{e}_1 + x_2\mathbf{e}_2 + ... + x_n\mathbf{e}_n$, i.e. with respect to this basis, $\mathbf{x} = (x_1, x_2, ..., x_n)$. It follows that $T(\mathbf{x}) = x_1 T(\mathbf{e}_1) + x_2 T(\mathbf{e}_2) + ... + x_n T(\mathbf{e}_n)$.

Suppose $T(\mathbf{e}_1) = (a_{11}, a_{21}, a_{31}, ..., a_{m1})$, $T(\mathbf{e}_2) = (a_{12}, a_{22}, a_{32}, ..., a_{m2})$ and so on; then

$$T(\mathbf{x}) = (a_{11}x_1 + a_{12}x_2 + ... + a_{1n}x_n, a_{21}x_1 + a_{22}x_2 + ... + a_{2n}x_n, ..., a_{m1}x_1 + a_{m2}x_2 + ... + a_{mn}x_n)$$

$$= (b_1, b_2, ..., b_m)$$

Hence the transformation can be represented by equations (1.13).

Notice that the columns of the associated matrix are merely $T(\mathbf{e}_1)$, $T(\mathbf{e}_2)$, ..., $T(\mathbf{e}_n)$. We can follow one transformation T_1 mapping vectors from R^n to vectors from R^m by a second transformation T_2 mapping vectors from R^m to vectors from R^p. The combined transformation is denoted by $T_2 \circ T_1$ so that if $\mathbf{x}$ is a vector from R^n, $(T_2 \circ T_1)(\mathbf{x}) \equiv T_2\left[T_1(\mathbf{x})\right]$.

Example 1

Consider T_1 and T_2 both mapping vectors from R^2 to vectors from R^2, defined by the sets of equations

$$\left.\begin{aligned} x_1' &= a_{11}x_1 + a_{12}x_2 \\ x_2' &= a_{21}x_1 + a_{22}x_2 \end{aligned}\right\} \quad \text{and} \quad \left.\begin{aligned} x_1'' &= a_{11}' x_1' + a_{12}' x_2' \\ x_2'' &= a_{21}' x_1' + a_{22}' x_2' \end{aligned}\right\} \quad \text{respectively.}$$

Substituting for x_1', x_2' from the first set into the second set, we obtain

$$\left.\begin{aligned} x_1'' &= (a_{11}' a_{11} + a_{12}' a_{21})x_1 + (a_{11}' a_{12} + a_{12}' a_{22})x_2 \\ x_2'' &= (a_{21}' a_{11} + a_{22}' a_{21})x_1 + (a_{21}' a_{12} + a_{22}' a_{22})x_2 \end{aligned}\right\}$$

The matrix of this combined transformation is

$$\begin{bmatrix} a_{11}' a_{11} + a_{12}' a_{21} & a_{11}' a_{12} + a_{12}' a_{22} \\ a_{21}' a_{11} + a_{22}' a_{21} & a_{21}' a_{12} + a_{22}' a_{22} \end{bmatrix} = \begin{bmatrix} a_{11}' & a_{12}' \\ a_{21}' & a_{22}' \end{bmatrix} \begin{bmatrix} a_{11} & a_{12} \\ a_{21} & a_{22} \end{bmatrix}$$

In this instance the *associated matrix of the combined transformation is the product of the associated matrices of the separate transformations in reverse order of application.* The result applies quite generally.

Example 2

The matrix of the anti-clockwise rotation of the coordinate axes through an angle α is

$\begin{bmatrix} \cos\alpha & -\sin\alpha \\ \sin\alpha & \cos\alpha \end{bmatrix}$. Suppose this is followed by a second rotation through an angle β, which can be represented by the matrix $\begin{bmatrix} \cos\beta & -\sin\beta \\ \sin\beta & \cos\beta \end{bmatrix}$. The combined transformation has matrix $\begin{bmatrix} \cos\beta & -\sin\beta \\ \sin\beta & \cos\beta \end{bmatrix}\begin{bmatrix} \cos\alpha & -\sin\alpha \\ \sin\alpha & \cos\alpha \end{bmatrix}$

$$= \begin{bmatrix} \cos\beta\cos\alpha - \sin\beta\sin\alpha & -\cos\beta\sin\alpha - \sin\beta\cos\alpha \\ \sin\beta\cos\alpha + \cos\beta\sin\alpha & -\sin\beta\sin\alpha + \cos\beta\cos\alpha \end{bmatrix} = \begin{bmatrix} \cos[\beta+\alpha] & -\sin[\beta+\alpha] \\ \sin[\beta+\alpha] & \cos[\beta+\alpha] \end{bmatrix}$$

This suggests that the combined transformation is a rotation through an angle $\beta + \alpha$, as we should have expected. Note that, *in this instance*, the end product is independent of the order of the separate transformation.

Problems

1. Find a linear transformation which maps $(1, 0, 0)$ to $(1, 2, 3)$; $(0, 1, 0)$ to $(3, 1, 2)$ and $(0, 0, 1)$ to $(2, 1, 3)$.

2. Find the images under the transformation of Problem 1 of $(1, 1, 1)$, $(3, -1, 4)$, $(4, 0, 5)$.

3. Which of the following are linear transformations?

   (i) $\sigma(x_1, x_2) = (x_1 + 2,\ 2x_2)$ (ii) $\sigma(x_1, x_2) = (x_1 - x_2,\ x_2)$
   (iii) $\sigma(x_1, x_2) = (x_1\cos 2\alpha - x_2\sin 2\alpha,\ x_1\sin 2\alpha + x_2\cos 2\alpha)$
   (iv) $\sigma(x_1, x_2, x_3) = (2x_1,\ x_2 + x_3,\ 3x_3,\ x_1 + x_2)$.

   A linear transformation maps $(1, 0, 1)$ to $(2, 3, -1)$; $(1, -1, 1)$ to $(3, 0, -2)$ and $(1, 2, -1)$ to $(-2, 7, -1)$. Find the images under this transformation of $(1, 0, 0)$, $(0, 1, 0)$, $(0, 0, 1)$.

4. Find vectors which span the kernel of the transformations whose matrices are

   (i) $\begin{bmatrix} 1 & 1 & 0 \\ 2 & 3 & 1 \\ -2 & 3 & 5 \end{bmatrix}$ (ii) $\begin{bmatrix} 1 & 1 & 3 \\ 1 & 2 & 4 \\ 1 & 1 & 3 \end{bmatrix}$ (iii) $\begin{bmatrix} 1 & 2 & 3 \\ 2 & 4 & 6 \\ 3 & 6 & 9 \end{bmatrix}$

5. Find the rank and nullity of the linear transformations whose associated matrices are

   (i) $\begin{bmatrix} 1 & 2 & 3 \\ 1 & 3 & 5 \end{bmatrix}$ (ii) $\begin{bmatrix} 1 & 3 & 7 \\ 2 & 7 & 16 \end{bmatrix}$ (iii) $\begin{bmatrix} 1 & 2 & 4 & -1 \\ 1 & 3 & 5 & -2 \end{bmatrix}$

   (iv) $\begin{bmatrix} 0 & 1 & -1 \\ -1 & 0 & 1 \\ 1 & -1 & 0 \end{bmatrix}$ (v) $\begin{bmatrix} 0 & 1 & 2 \\ 1 & 2 & 3 \\ 2 & 3 & 4 \end{bmatrix}$ (vi) $\begin{bmatrix} 1 & 1 & 1 & 1 \\ 1 & 2 & 3 & 4 \\ 0 & 1 & 2 & 3 \end{bmatrix}$

6. If $\sigma_1(x_1, x_2) = (x_2, -x_1)$ and $\sigma_2(x_1, x_2) = (x_1, -x_2)$ find $\sigma_1\sigma_2$ and $\sigma_2\sigma_1$.

7. Find the kernel of the transformation $\sigma(x_1, x_2, x_3, x_4) = (3x_1 - 2x_2 - x_3 - 4x_4,\ x_1 + x_2 - 2x_3 - 3x_4)$.

8. Find the solutions of the equations

   (i) $\begin{cases} x_1 + x_2 + x_3 = 3 \\ 2x_1 + 5x_2 - 2x_3 = 3 \end{cases}$ (ii) $\begin{cases} x_1 - 2x_2 + 3x_3 = 0 \\ 2x_1 + 5x_2 + 6x_3 = 0 \end{cases}$

9. Find the dimension of the vector space of solutions of the following systems of linear equations. Find a basis for this space of solutions.

(i) $x + y - z = 0$ (ii) $\begin{cases} 2x + y - z = 0 \\ y + z = 0 \end{cases}$ (iii) $\begin{cases} 2x - 3y + z = 0 \\ x + y - z = 0 \end{cases}$

(iv) $\begin{cases} x + y + z = 0 \\ x - y = 0 \\ y + z = 0 \end{cases}$ (v) $\begin{cases} 2x - 3y + z = 0 \\ x + y - z = 0 \\ 3x + 4y = 0 \\ 5x + y + z = 0 \end{cases}$ (vi) $\begin{cases} x + y + z = 0 \\ 2x + 2y + 2z = 0 \end{cases}$

10. Find the dimension of the set of solutions of the following systems. Find a basis of the space of solutions of the associated homogeneous system and find one solution of the non-homogeneous system. Hence produce general solutions.

(i) $2x + 3y - z = 1$ (ii) $\begin{cases} 2x - y + z = 1 \\ 2x + y + z = 1 \end{cases}$ (iii) $\begin{cases} -x + 4y + z = 2 \\ 3x + y - z = 0 \end{cases}$

(iv) $\begin{cases} x - y + z = 1 \\ 2x - 3y + z = 0 \\ x + y - z = 5 \end{cases}$

1.4 THE SOLUTION OF SIMULTANEOUS LINEAR ALGEBRAIC EQUATIONS

A system of m simultaneous linear equations in n unknowns may have a unique solution, an infinite number of solutions, or no solution at all. Let us investigate the underlying reasons. We denote the system by $\mathbf{Ax} = \mathbf{b}$. If a solution does exist then we first find *one* solution of the system and then find the *general* solution of the system $\mathbf{Ax} = \mathbf{0}$; the sum of these is the general solution of the full system. If the kernel of $\mathbf{A}$ comprises only the zero vector then the solution of the full system will be unique. We denote the columns of $\mathbf{A}$ as $\mathbf{a}_1, \mathbf{a}_2, ..., \mathbf{a}_n$; in other words, we are regarding the matrix as an aggregation of column vectors. We shall choose the usual basis for R^n, viz. $\mathbf{e}_1, \mathbf{e}_2, ..., \mathbf{e}_n$. It follows that if $\mathbf{x} = x_1\mathbf{e}_1 + x_2\mathbf{e}_2 + ... + x_n\mathbf{e}_n$ then $\mathbf{Ax} = x_1\mathbf{a}_1 + x_2\mathbf{a}_2 + ... + x_n\mathbf{a}_n$. For example, if we have the system of equations

$$\left.\begin{aligned} x_1 + 2x_2 &= 1 \\ 4x_1 + x_2 &= 3 \end{aligned}\right\} \tag{1.14}$$

then this may be written as $\mathbf{Ax} \equiv x_1 \begin{bmatrix} 1 \\ 4 \end{bmatrix} + x_2 \begin{bmatrix} 2 \\ 1 \end{bmatrix} = \begin{bmatrix} 1 \\ 3 \end{bmatrix} \equiv \mathbf{b}$

Existence of a solution

Any vector $\mathbf{Ax}$ must be a linear combination of the column vectors $\mathbf{a}_1, \mathbf{a}_2, ..., \mathbf{a}_n$. If a solution of $\mathbf{Ax} = \mathbf{b}$ is to exist, $\mathbf{b}$ must itself be a linear combination of $\mathbf{a}_1, \mathbf{a}_2, ..., \mathbf{a}_n$.

In the example considered, $\mathbf{b}$ is a linear combination of $\begin{bmatrix} 1 \\ 4 \end{bmatrix}$ and $\begin{bmatrix} 2 \\ 1 \end{bmatrix}$ since we can take $x_1 = 5/7$ and $x_2 = 1/7$; this is of course the solution of the system. If we consider the system

$$\left.\begin{aligned} x_1 + 2x_2 &= 1 \\ 4x_1 + 8x_2 &= 3 \end{aligned}\right\} \tag{1.15}$$

we require that $\begin{bmatrix}1\\3\end{bmatrix}$ is a linear combination of $\begin{bmatrix}1\\4\end{bmatrix}$ and $\begin{bmatrix}2\\8\end{bmatrix}$; since $\begin{bmatrix}2\\8\end{bmatrix}$ is a multiple of $\begin{bmatrix}1\\4\end{bmatrix}$ we merely require that $\begin{bmatrix}1\\3\end{bmatrix}$ is a multiple of $\begin{bmatrix}1\\4\end{bmatrix}$. It is *not* and therefore the system has no solutions.

How can we extend these ideas? We first define the **rank** of a matrix **as the maximum** number of linearly independent vectors from the set of its column vectors . In the system (1.14), the rank of $\mathbf{A}$ was 2 (the maximum possible for a matrix with two columns) but in the system (1.15) the rank of $\mathbf{A}$ was 1.

Since $\mathbf{b}$ has to be a linear combination of $\mathbf{a}_1, \mathbf{a}_2, \ldots, \mathbf{a}_n$ for a solution to exist, then the sets $\{\mathbf{a}_1, \mathbf{a}_2, \ldots, \mathbf{a}_n\}$ and $\{\mathbf{a}_1, \mathbf{a}_2, \ldots, \mathbf{a}_n, \mathbf{b}\}$ have the same number of linearly independent vectors. Put another way, the matrices $\mathbf{A} = (\mathbf{a}_1, \mathbf{a}_2, \ldots, \mathbf{a}_n)$ and $(\mathbf{A}|\mathbf{b}) = (\mathbf{a}_1, \mathbf{a}_2, \ldots, \mathbf{a}_n, \mathbf{b})$ have the same rank. The matrix $(\mathbf{A}|\mathbf{b})$, which is formed by adding the column $\mathbf{b}$ to the matrix $\mathbf{A}$ is called the **augmented matrix.**

Example 1

The system (1.14) has augmented matrix $\left[\begin{array}{cc|c}1 & 2 & 1\\4 & 1 & 3\end{array}\right]$ which has rank 2. Note that if column rank and row rank are equal, the augmented matrix cannot have rank >2.

Example 2

The system (1.15) with augmented matrix $\left[\begin{array}{cc|c}1 & 2 & 1\\4 & 8 & 3\end{array}\right]$ has no solution and the augmented matrix has rank 2 which is greater than the rank of $\mathbf{A}$. You can check the rank from a consideration of row vectors.

Uniqueness of a solution

Now we have said earlier that the rank of a linear transformation + the dimension of its kernel = the dimension of the domain space. If the rank of the linear transformation equals the dimension of its domain space then the kernel contains only the zero vector **0**. With reference to the matrix of the transformation, if its rank equals the number of columns then there is a unique solution of the system of equations of which it is the associated matrix.

Then if a system of m equations in n unknowns can be written as $\mathbf{Ax} = \mathbf{b}$, (where $\mathbf{A}$ is an m by n matrix), *provided* a solution can be found, it will be a unique solution if and only if the rank of $\mathbf{A}$ is n. In the special case where $m = n$ then a unique solution *will* exist if and only if the rank of $\mathbf{A} = n$.

Figure 1.1 shows a helpful flow chart for the process.

But suppose that the rank of $\mathbf{A}$ is less than n and more than one solution exists. We quote the general result that if the rank of $\mathbf{A}$ is equal to $r < n$ then all solutions of $\mathbf{Ax} = \mathbf{b}$ can be expressed in terms of $(n - r)$ independent parameters.

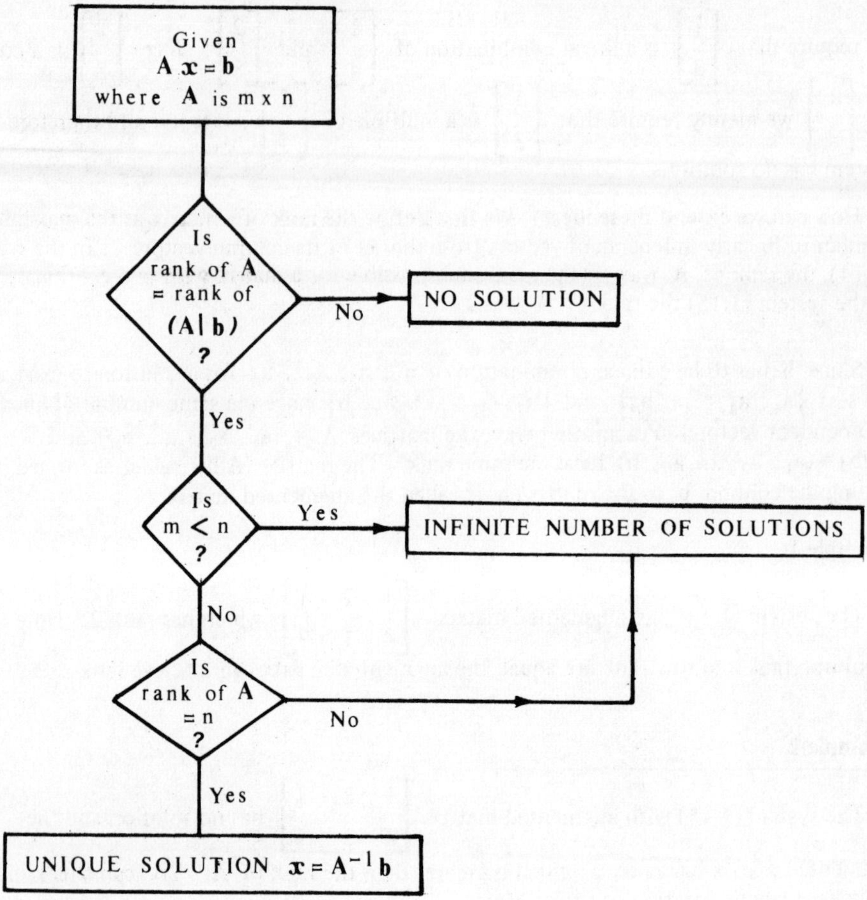

Figure 1.1

Example

Consider the system of equations:

$$\left.\begin{aligned} 3x_1 + 2x_2 - x_3 + 4x_4 &= 4 \\ 4x_1 + 3x_2 - x_3 + 5x_4 &= 4 \\ -2x_1 - x_2 + 2x_3 - 2x_4 &= -3 \\ 3x_1 + 2x_2 + 2x_3 + 7x_4 &= 7 \end{aligned}\right\} \qquad (1.16)$$

which has augmented matrix $\left[\begin{array}{cccc|c} 3 & 2 & -1 & 4 & 4 \\ 4 & 3 & -1 & 5 & 4 \\ 2 & 1 & 2 & -2 & -3 \\ 3 & 2 & 2 & 7 & 7 \end{array}\right]$ Performing a Gauss-Jordan elimination we finally produce $\left[\begin{array}{cccc|c} 1 & 0 & 0 & 3 & 5 \\ 0 & 1 & 0 & -2 & -5 \\ 0 & 0 & 1 & 1 & 1 \\ 0 & 0 & 0 & 0 & 0 \end{array}\right]$. This has a 3 x 3 identity submatrix in its top left-hand corner. The row of zeros indicates that there is no unique solution. Now the rank of $\mathbf{A}$ = rank of $(\mathbf{A}|\mathbf{b}) = 3 < 4$. We should expect to express all solutions in terms of $4 - 3 = 1$ independent parameter, for example x_4.

We have $x_1 = 5 - 3x_4$, $x_2 = -5 + 2x_4$, $x_3 = 1 - x_4$. Hence $(x_1, x_2, x_3, x_4) = (5, -5, 1, 0) + x_4(-3, 2, -1, 1)$ and you can check that $(-3, 2, -1, 1)$ is a basis for the kernel of the transformation whose matrix is $\mathbf{A}$.

The Gauss-Jordan procedure can be regarded as a sequence of pre-multiplications by certain elementary matrices. It can be shown that the rank of the matrix at each stage of the procedure is the same as that of its predecessor.

Application: Dimensional Analysis

In a problem on fluid flow, it is believed from experimental evidence that the velocity v depends upon fluid density ρ, a typical length ℓ, gravitational acceleration g and viscosity μ. Let us assume a relationship of the form

$$v = K\,\rho^{\alpha}\,\ell^{\beta}\,g^{\gamma}\,\mu^{\delta} \tag{1.17}$$

where K, α, β, γ and δ are dimensionless constants.

If we write the variables in terms of mass M, length L and time T we obtain

$$M^0LT^{-1} = (ML^{-3})^{\alpha}\,(L)^{\beta}\,(LT^{-2})^{\gamma}\,(ML^{-1}T^{-1})^{\delta} \tag{1.18}$$

since K is dimensionless. Equating powers of M, L, T respectively we find

$$\alpha + \delta = 0. \quad -3\alpha + \beta + \gamma - \delta = 1 \quad \text{and} \quad -2\gamma - \delta = -1.$$

In matrix form we have $\begin{bmatrix} 1 & 0 & 0 & 1 \\ -3 & 1 & 1 & -1 \\ 0 & 0 & -2 & -1 \end{bmatrix}\begin{bmatrix} \alpha \\ \beta \\ \gamma \\ \delta \end{bmatrix} = \begin{bmatrix} 0 \\ 1 \\ -1 \end{bmatrix}$. The augmented

matrix is $\left[\begin{array}{cccc|c} 1 & 0 & 0 & 1 & 0 \\ -3 & 1 & 1 & -1 & 1 \\ 0 & 0 & -2 & -1 & -1 \end{array}\right]$. This can be reduced to the matrix

$\left[\begin{array}{cccc|c} 1 & 0 & 0 & 1 & 0 \\ 0 & 1 & 0 & \frac{3}{2} & \frac{1}{2} \\ 0 & 0 & 1 & \frac{1}{2} & \frac{1}{2} \end{array}\right]$ by row operations. We see that the augmented matrix has the same rank as that of **A**, viz. 3, but this is less than the number of columns and therefore there is an infinite number of solutions. We now rewrite (1.18) as

$$(ML^{-3})^a\ (L)^b\ (LT^{-2})^c\ (ML^{-1}\ T^{-1})^d\ (LT^{-1})^e = M^0\ L^0\ T^0$$

since K is a dimensionless constant.

The equations can be written $\begin{bmatrix} 1 & 0 & 0 & 1 & 0 \\ -3 & 1 & 1 & -1 & 1 \\ 0 & 0 & -2 & -1 & -1 \end{bmatrix} \begin{bmatrix} a \\ b \\ c \\ d \\ e \end{bmatrix} = \begin{bmatrix} 0 \\ 0 \\ 0 \end{bmatrix}$

As above, this can be reduced to $\begin{bmatrix} 1 & 0 & 0 & 1 & 0 \\ 0 & 1 & 0 & \frac{3}{2} & \frac{1}{2} \\ 0 & 0 & 1 & \frac{1}{2} & \frac{1}{2} \end{bmatrix}$

We have the equations $a + d = 0,\ b + \frac{3}{2}\ d + \frac{1}{2}\ e = 0,\ c + \frac{1}{2}\ d + \frac{1}{2}\ e = 0.$

Now we can form a number of dimensionless combinations of variables since there is no unique solution. Let us decide to find some dimensionless groups involving the velocity v. This means that we want to keep e involved. If we put our solutions in terms of c and d we obtain $a = -d$, $2b + e = -3d$, $e = -2c - d$; we can do this since the rank of our coefficient matrix is 3 and the number of columns is 5.

If we take $c = 0$, $d = -1$ we find $a = 1$, $b = 1$, $e = 1$ and so $\frac{v\rho\ell}{\mu}$ is a dimensionless group. It is well known and called the **Reynold's number.**

If we take $c = 1$, $d = 0$, we obtain $a = 0$, $b = 1$, $e = -2$ and therefore $\ell g/v^2$ is a dimensionless group; the inverse $v^2/\ell g$ is called the **Froude number.** Other dimensionless groups can be formed, but are of less interest.

Problems

1. Determine the rank of the matrix $\mathbf{A} = \begin{bmatrix} 1 & 2 & 3 \\ 2 & 4 & 7 \\ -1 & -2 & -2 \end{bmatrix}$ and reduce **A** to normal form by a sequence of elementary transformations. (C.E.I.)

2. Find the rank of the following matrices

$$\begin{bmatrix} 1 & 5 & -7 \\ 2 & 3 & 1 \end{bmatrix},\ \begin{bmatrix} 2 & 1 & 1 \\ 0 & 1 & -1 \end{bmatrix},\ \begin{bmatrix} 2 & 1 & 3 \\ 7 & 2 & 0 \end{bmatrix},\ \begin{bmatrix} -1 & 2 & -2 \\ 3 & 4 & -5 \end{bmatrix},\ \begin{bmatrix} 1 & 2 & -3 \\ -1 & 2 & 3 \\ 4 & 8 & 12 \\ 0 & 0 & 0 \end{bmatrix}$$

3. Prove that **A** and $\mathbf{A}^T$ have the same rank.

4. A system of m linear equations in n unknowns is said to be *consistent* if it has one or more solutions; otherwise it is *inconsistent.* Show that the equations $x_1 + 2x_2 = 3$, $x_2 - x_3 = 2$, $x_1 + x_2 + x_3 = 1$ are consistent, but that if the third equation is replaced by $x_1 + x_2 + x_3 = 2$ the new set is inconsistent. Interpret the two sets of equations geometrically.

5. Prove that if $\mathbf{A}$ and $\mathbf{B}$ are $m \times n$ matrices, then the rank of $\mathbf{A}+\mathbf{B}$ is less than or equal to the sum of the ranks of $\mathbf{A}$ and of $\mathbf{B}$.

6. A matrix is said to be *regular by rows* if its row-vectors are linearly independent and *regular by columns* if its column-vectors are linearly independent. A matrix $\mathbf{A}$ which is regular either by rows or by columns is called a *regular* matrix. Show that an $m \times n$ matrix has rank r if and only if it contains at least one regular $k \times k$ submatrix for each value of k, $1 \leqslant k \leqslant r$, but none for $k > r$. A submatrix of $\mathbf{A}$ is obtained by deleting complete rows and/or complete columns.

7. The frictional force F on a body immersed in a flowing fluid stream is believed to depend on a particular dimension of the body l, the velocity v, the density ρ and the viscosity μ of the fluid. Show that the quantity $F/(\rho v^2 \ell)$ can be expressed as a function of the reciprocal of the Reynolds' number.

8. The heat lost by an immersed body, q, is believed to depend upon the fluid velocity relative to the body, v; a dimension of the body, ℓ; fluid thermal capacity/unit volume, ρC_p; the thermal conductivity of the fluid, K; the excess temperature of the body over the fluid at large distance, ΔT; and the kinematic viscosity of the fluid, $\nu = \mu/\rho$. Express q in terms of the dimensionless numbers $(\rho C_p v \ell /K)$ and $(\mu C_p/K)$.

1.5 SCHEMES FOR SOLUTION OF LINEAR EQUATIONS

In this section we shall examine a number of variations on Gaussian Elimination which are specially designed to take advantage of particular features of the matrix.

Tridiagonal Matrices

In problems in which finite difference approximations are used to solve differential equations, matrices occur with non-zero elements confined to the leading diagonal, the first sub-diagonal and the first super-diagonal. For a $m \times n$ matrix $\mathbf{A}$ this means that $a_{ij} = 0$ if $i > j+1$ or if $i < j-1$. An example of such a matrix is

$$\mathbf{A} = \begin{bmatrix} 1 & 4 & 0 & 0 & 0 \\ -2 & 2 & 2 & 0 & 0 \\ 0 & 6 & 1 & 3 & 0 \\ 0 & 0 & 4 & 4 & 2 \\ 0 & 0 & 0 & 2 & 7 \end{bmatrix}$$

We are able to develop an efficient algorithm for Gauss Elimination when the coefficient matrix is tri-diagonal. The popular notation for a tri-diagonal matrix is not that used for general matrices. The notation is

$$\begin{bmatrix} b_1 & c_1 & 0 & 0 & & & \bigcirc \\ a_2 & b_2 & c_2 & 0 & & & \\ 0 & a_3 & b_3 & c_3 & \ddots & & \\ & & & \ddots & a_{n-1} & b_{n-1} & c_{n-1} \\ & \bigcirc & & & 0 & a_n & b_n \end{bmatrix}$$

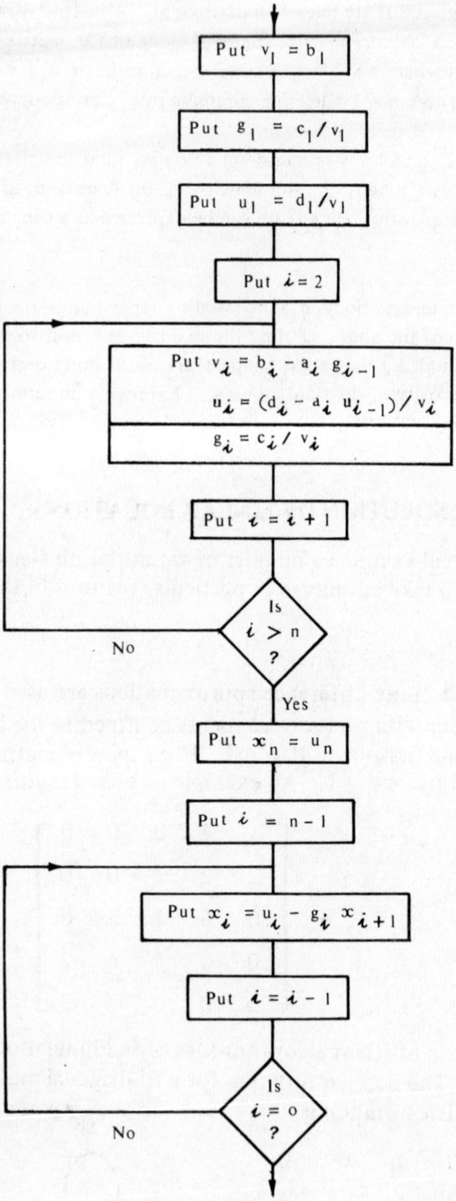

Put v1 = b1
Put g1 = c1/v1
Put u1 = d1/v1
Put i = 2
Put vi = bi − ai gi−1
ui = (di − ai ui−1)/vi
gi = ci / vi
Put i = i + 1
Is i > n ?
No
Yes
Put x n = u n
Put i = n − 1
Put xi = ui − gi xi+1
Put i = i − 1
Is i = o ?
No

Figure 1.2

Consider a simple case. We wish to solve the equations

$$\begin{bmatrix} b_1 & c_1 & 0 & 0 \\ a_2 & b_2 & c_2 & 0 \\ 0 & a_3 & b_3 & c_3 \\ 0 & 0 & a_4 & b_4 \end{bmatrix} \begin{bmatrix} x_1 \\ x_2 \\ x_3 \\ x_4 \end{bmatrix} = \begin{bmatrix} d_1 \\ d_2 \\ d_3 \\ d_4 \end{bmatrix}$$

Eliminating x_1 from the first two equations we obtain an equation connecting x_2 and x_3 only. This is used with the original third equation to eliminate x_2 and produce an equation connecting x_3 and x_4. This last equation is used with the fourth equation of the original set to find a value for x_4. By back-substituting we obtain the values for x_3, x_2 and x_1.

We quote the Thomas algorithm for this kind of matrix in Figure 1.2.

Notice that we have not carried out a check for possible divisions by zero; modify the flow chart to cope. Note also that g_n is calculated but not used.

Triangular decomposition

A matrix which is either upper-triangular (zeros below the leading diagonal) or lower-triangular (zeros above the leading diagonal) is relatively easy to invert.

We first assume that a square matrix **A** can be written as the product of a lower-triangular matrix **L** and an upper-triangular matrix **U** in the order

$$\mathbf{A} = \mathbf{LU} \tag{1.19}$$

This is usually possible when **A** is non-singular.

Example

Let $\mathbf{A} = \begin{bmatrix} 1 & 5 & 3 \\ 3 & 19 & 17 \\ 8 & 36 & 25 \end{bmatrix}$ and suppose that $\mathbf{A} = \mathbf{LU}$ where $\mathbf{L} = \begin{bmatrix} l_{11} & 0 & 0 \\ l_{21} & l_{22} & 0 \\ l_{31} & l_{32} & l_{33} \end{bmatrix}$

and $\mathbf{U} = \begin{bmatrix} u_{11} & u_{12} & u_{13} \\ 0 & u_{22} & u_{23} \\ 0 & 0 & u_{33} \end{bmatrix}$. It follows that

$$\mathbf{LU} = \begin{bmatrix} l_{11}\,u_{11} & l_{11}\,u_{12} & l_{11}\,u_{13} \\ l_{21}\,u_{11} & l_{21}\,u_{12} + l_{22}\,u_{22} & l_{21}\,u_{13} + l_{22}\,u_{23} \\ l_{31}\,u_{11} & l_{31}\,u_{12} + l_{32}\,u_{22} & l_{31}\,u_{13} + l_{32}\,u_{23} + l_{33}\,u_{33} \end{bmatrix} = \mathbf{A}$$

We can set about solving these equations for l_{ij}, u_{ij} in a systematic fashion. However, there are 6 unknowns l_{ij} snd 6 unknown u_{ij} with only 9 equations governing them. We can put the uncertainty in the products $l_{11}\,u_{11}$, $l_{22}\,u_{22}$ and $l_{33}\,u_{33}$. As long as we specify the values of these three products we can choose the values of the components freely. One method takes $u_{ii} = 1$; this is the **Crout** decomposition. A second method takes $l_{ii} = 1$; this is the **Doolittle** method. A third popular variant puts $l_{ii} = u_{ii}$. We shall work with the Crout method.

Therefore $LU = \begin{bmatrix} l_{11} & l_{11}\,u_{12} & l_{11}\,u_{13} \\ l_{21} & l_{21}\,u_{12} + l_{22} & l_{21}\,u_{13} + l_{22}\,u_{23} \\ l_{31} & l_{31}\,u_{12} + l_{32} & l_{31}\,u_{13} + l_{32}\,u_{23} + l_{33} \end{bmatrix} = A$

We can immediately see that $l_{11} = 1$, $l_{21} = 3$, $l_{31} = 8$.

Then $u_{12} = 5/l_{11} = 5$, $u_{13} = 3/l_{11} = 3$, $l_{22} = 19 - l_{21}\,u_{12} = 4$, $l_{32} = 36 - l_{31}\,u_{12} = -4$. Further $u_{23} = (17 - l_{21}\,u_{13})/l_{22} = 2$ and, finally, $l_{33} = 25 - l_{31}\,u_{13} - l_{32}\,u_{23} = 9$.

Hence $A = \begin{bmatrix} 1 & 0 & 0 \\ 3 & 4 & 0 \\ 8 & -4 & 9 \end{bmatrix} \begin{bmatrix} 1 & 5 & 3 \\ 0 & 1 & 2 \\ 0 & 0 & 1 \end{bmatrix}$

You try and repeat the process using Doolittle's method.

General Crout Algorithm

Let A be an $n \times n$ matrix with elements a_{ij}. Then the steps are

i) $l_{i1} = a_{i1}$, $i = 1$, n

ii) $u_{1j} = a_{1j}/l_{11}$, $j = 2$, n

iii) $l_{ik} = a_{ik} - \sum_{m=1}^{k-1} l_{im}\,u_{mk}$, $i = k$, n

iv) $u_{kj} = \left(a_{kj} - \sum_{m=1}^{k-1} l_{km}\,u_{mj} \right) \Big/ l_{kk}$, $j = k+1$, n

$\left. \phantom{\begin{matrix} \\ \\ \\ \\ \\ \end{matrix}} \right\}$ Repeat for $k = 2$, n

Try and write the equations out in full for a 4×4 matrix. Can the algorithm be written with fewer steps? Produce a flow chart for the algorithm.

For a 4×4 matrix the elements are evaluated in the following order.

$l_{11}, l_{21}, l_{31}, l_{41}, u_{12}, u_{13}, u_{14}, l_{22}, l_{32}, l_{42}, u_{23}, u_{24}, l_{33}, l_{43}, u_{34}, l_{44}.$

This is shown schematically below and the advantage in being able to store L and U in the space occupied by A is seen; since we know $u_{ii} = 1$ we do not need to store them.

$$\begin{bmatrix} l_{11} & u_{12} & u_{13} & u_{14} \\ l_{21} & l_{22} & u_{23} & u_{24} \\ l_{31} & l_{32} & l_{33} & u_{34} \\ l_{41} & l_{42} & l_{43} & l_{44} \end{bmatrix}$$

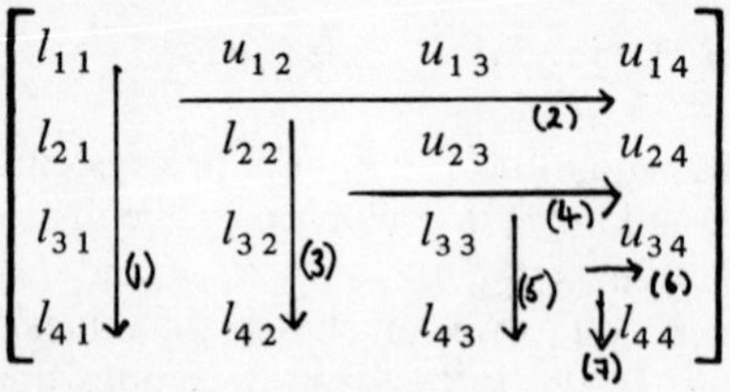

If some $l_{ii} = 0$ then either the decomposition is not possible or, if A is singular, there is an infinite number of possible decompositions.

Solution of linear equations

To solve the system $\mathbf{Ax} = \mathbf{b}$ we first factorise $\mathbf{A}$ into $\mathbf{LU}$ so that $(\mathbf{LU})\mathbf{x} = \mathbf{b}$ i.e. $\mathbf{L}(\mathbf{Ux}) = \mathbf{b}$. If we write

$$\mathbf{Ux} = \mathbf{y} \tag{1.20a}$$

then we need to solve

$$\mathbf{Ly} = \mathbf{b} \tag{1.20b}$$

Having found $\mathbf{y}$ we then use (1.20a) to find $\mathbf{x}$. Each of these two eliminations is easier than the general Gauss Elimination. We work through an example using the decomposition already obtained.

Suppose we wish to solve the system

$$\left.\begin{aligned} x_1 + 5x_2 + 3x_3 &= 22 \\ 3x_1 + 19x_2 + 17x_3 &= 94 \\ 8x_1 + 36x_2 + 25x_3 &= 166 \end{aligned}\right\} \tag{1.21}$$

Then we have the matrix of coefficients as $\mathbf{A}$ of page 21. First we solve $\mathbf{Ly} = \mathbf{b}$ for $\mathbf{y}$ i.e. we solve

$$\begin{bmatrix} 1 & 0 & 0 \\ 3 & 4 & 0 \\ 8 & -4 & 9 \end{bmatrix} \begin{bmatrix} y_1 \\ y_2 \\ y_3 \end{bmatrix} = \begin{bmatrix} 22 \\ 94 \\ 166 \end{bmatrix}$$

Clearly $y_1 = 22$ then $y_2 = 7$ and $y_3 = 2$. We next solve $\mathbf{Ux} = \mathbf{y}$, i.e.

$$\begin{bmatrix} 1 & 5 & 3 \\ 0 & 1 & 2 \\ 0 & 0 & 1 \end{bmatrix} \begin{bmatrix} x_1 \\ x_2 \\ x_3 \end{bmatrix} = \begin{bmatrix} 22 \\ 7 \\ 2 \end{bmatrix}$$

and find $x_3 = 2$, $x_2 = 3$, $x_1 = 1$. We can check that this is the solution of the original equations.

We now work through an example involving 4 equations in 4 unknowns.

Example

Solve

$$\mathbf{Ax} = \begin{bmatrix} 4 & 16 & 8 & 0 \\ 2 & 5 & -14 & -15 \\ 1 & 2 & -6 & 2 \\ 2 & 13 & 35 & 34 \end{bmatrix} \begin{bmatrix} x_1 \\ x_2 \\ x_3 \\ x_4 \end{bmatrix} = \begin{bmatrix} 44 \\ -32 \\ 3 \\ 131 \end{bmatrix} = \mathbf{b}$$

We factorise $\mathbf{A}$ into $\mathbf{LU}$ where

$$\mathbf{L} = \begin{bmatrix} 4 & 0 & 0 & 0 \\ 2 & -3 & 0 & 0 \\ 1 & -2 & 4 & 0 \\ 2 & 5 & 1 & 6 \end{bmatrix} \quad \text{and} \quad \mathbf{U} = \begin{bmatrix} 1 & 4 & 2 & 0 \\ 0 & 1 & 6 & 5 \\ 0 & 0 & 1 & 3 \\ 0 & 0 & 0 & 1 \end{bmatrix}$$

You should check that you can obtain this factorisation.

Then we solve $\mathbf{Ly} = \mathbf{b}$ i.e. $\begin{bmatrix} 4 & 0 & 0 & 0 \\ 2 & -3 & 0 & 0 \\ 1 & -2 & 4 & 0 \\ 2 & 5 & 1 & 6 \end{bmatrix} \begin{bmatrix} y_1 \\ y_2 \\ y_3 \\ y_4 \end{bmatrix} = \begin{bmatrix} 44 \\ -32 \\ 3 \\ 131 \end{bmatrix}$

to obtain $y_1 = 11,\ y_2 = 18,\ y_3 = 7,\ y_4 = 2.$

Finally, we solve $\mathbf{Ux} = \mathbf{y}$ i.e.

$$\begin{bmatrix} 1 & 4 & 2 & 0 \\ 0 & 1 & 6 & 5 \\ 0 & 0 & 1 & 3 \\ 0 & 0 & 0 & 1 \end{bmatrix} \begin{bmatrix} x_1 \\ x_2 \\ x_3 \\ x_4 \end{bmatrix} = \begin{bmatrix} 11 \\ 18 \\ 7 \\ 2 \end{bmatrix}$$

to obtain $x_4 = 2,\ x_3 = 1,\ x_2 = 2,\ x_1 = 1.$

This as we can check is the exact answer, which we should have expected since we were working with exact arithmetic. In practice, we should modify the scheme to take account of round-off by a partial pivotting technique.

General Algorithm

Having found $\mathbf{L}$ and $\mathbf{U}$ we may summarise the solutions for $\mathbf{y}$ and $\mathbf{x}$ by the equations

$$\left.\begin{aligned} y_i &= (b_i - \sum_{j=1}^{i-1} l_{ij}\, b_j)/l_{ii}, \qquad i = 1, 2, \ldots, n \\ x_i &= y_i - \sum_{j=i+1}^{n} u_{ij}\, x_j \qquad j = n, (n-1), \ldots, 1 \end{aligned}\right\} \tag{1.22}$$

It is again left to you to produce a flow chart for the process.

Choleski's Scheme

When the matrix $\mathbf{A}$ is real and symmetric, and positive definite† we can choose the diagonal elements of $\mathbf{L}$ so that $\mathbf{L}$ is real and $\mathbf{U} = \mathbf{L}^T$, the transpose of $\mathbf{L}$. The requirement of positive definiteness for $\mathbf{A}$ is for the purpose of greater numerical stability.

---

† $\mathbf{A}$ is **positive definite** if and only if $\mathbf{x}^T\mathbf{A}\mathbf{x} > 0$ for all $\mathbf{x} \neq \mathbf{0}$.

$$\text{If } \mathbf{A} = \begin{bmatrix} a_{11} & a_{21} & \cdots\cdots & a_{n1} \\ a_{21} & a_{22} & \cdots\cdots & a_{n2} \\ \vdots & \vdots & & \vdots \\ a_{n1} & a_{n2} & \cdots\cdots & a_{nn} \end{bmatrix} = \begin{bmatrix} l_{11} & 0 & \cdots\cdots & 0 \\ l_{21} & l_{22} & \cdots\cdots & 0 \\ \vdots & \vdots & & \vdots \\ l_{n1} & l_{n2} & \cdots\cdots & l_{nn} \end{bmatrix} \begin{bmatrix} l_{11} & l_{21} & \cdots\cdots & l_{n1} \\ 0 & l_{22} & \cdots\cdots & l_{n2} \\ \vdots & \vdots & & \vdots \\ 0 & 0 & \cdots\cdots & l_{nn} \end{bmatrix}$$

$$= \mathbf{LU}$$

it follows that $a_{ii} = \sum_{k=1}^{i} l^2_{ik}$ and $a_{ij} = \sum_{k=1}^{i} l_{ik}\, l_{jk}$. Therefore

$$\left.\begin{aligned} l_{ii} &= \sqrt{a_{ii} - \sum_{k=1}^{i-1} l^2_{ik}}, && 1 \leqslant i \leqslant n \\ l_{ji} &= \Big(a_{ij} - \sum_{k=1}^{i-1} l_{ik}\, l_{jk}\Big)/l_{ii}, && 1 \leqslant i < j \end{aligned}\right\} \quad (1.23)$$

The order of determination of the elements follows the Crout pattern.

To solve a set of equations $\mathbf{Ax} = \mathbf{b}$ where $\mathbf{A}$ is real, symmetric and positive definite, we first solve $\mathbf{Ly} = \mathbf{b}$ for $\mathbf{y}$ and then solve $\mathbf{L}^T\mathbf{x} = \mathbf{y}$ for $\mathbf{x}$. The process uses the formulae

$$\left.\begin{aligned} y_i &= \Big(b_i - \sum_{k=1}^{i-1} l_{ik}\, y_k\Big)/l_{ii}, && i = 1, 2, \ldots\ldots, n \\ x_i &= \Big(y_i - \sum_{j=i+1}^{n} l_{ij}\, x_j\Big)/l_{ii}, && i = n, (n-1), \ldots\ldots, 1 \end{aligned}\right\} \quad (1.24)$$

Example

We require to solve the equations

$$\mathbf{Ax} = \begin{bmatrix} 1 & 2 & 3 \\ 2 & 20 & 26 \\ 3 & 26 & 70 \end{bmatrix} \begin{bmatrix} x_1 \\ x_2 \\ x_3 \end{bmatrix} = \begin{bmatrix} 7 \\ 50 \\ 102 \end{bmatrix} = \mathbf{b}$$

First we perform the decomposition $\mathbf{A} = \begin{bmatrix} 1 & 0 & 0 \\ 2 & 4 & 0 \\ 3 & 5 & 6 \end{bmatrix} \begin{bmatrix} 1 & 2 & 3 \\ 0 & 4 & 5 \\ 0 & 0 & 6 \end{bmatrix} = \mathbf{LL}^T$

then we solve $\begin{bmatrix} 1 & 0 & 0 \\ 2 & 4 & 0 \\ 3 & 5 & 6 \end{bmatrix} \begin{bmatrix} y_1 \\ y_2 \\ y_3 \end{bmatrix} = \begin{bmatrix} 7 \\ 50 \\ 102 \end{bmatrix}$ to obtain $y_1 = 7,\ y_2 = 9,\ y_3 = 6$.

Finally we solve $\begin{bmatrix} 1 & 2 & 3 \\ 0 & 4 & 5 \\ 0 & 0 & 6 \end{bmatrix} \begin{bmatrix} x_1 \\ x_2 \\ x_3 \end{bmatrix} = \begin{bmatrix} 7 \\ 9 \\ 6 \end{bmatrix}$ to obtain the required solution, viz.

$x_3 = 1, \ x_2 = 1, \ x_1 = 2.$

Problems

1. Write a computer program to solve a set of linear equations based on the Thomas algorithm. Use suitable test data.

2. Write a computer program to decompose a matrix by the Crout method. Then incorporate routines to invert a matrix so decomposed and to solve a set of linear equations.

3. Repeat Problem 2 for the Doolittle method and for Choleski's method.

4. Consider the following matrices, where only non-zero elements are shown:

$$\mathbf{A} = \begin{bmatrix} 1 & & \\ a_2 & 1 & \\ & a_3 & 1 \end{bmatrix} \qquad \mathbf{B} = \begin{bmatrix} b_1 & c_1 & \\ & b_2 & c_2 \\ & & b_3 \end{bmatrix}$$

Show that $\mathbf{E} = \mathbf{AB}$ has a special form and give the formulae for its elements. Deduce the general result when **A** and **B** are matrices of the same form but of larger order.

If a matrix **E** having the above structure is given, describe briefly how the matrix factors **A** and **B** can be determined, and how they can be used to solve equations of the form $\mathbf{Ex} = \mathbf{d}$ when **d** is given.

Illustrate this in the following case:

$$\mathbf{E} = \begin{bmatrix} 4 & 2 & & \\ 2 & 5 & 2 & \\ & 2 & 5 & 2 \\ & & 2 & 5 \end{bmatrix} \qquad \mathbf{d} = \begin{bmatrix} 6 \\ 1 \\ 7 \\ 12 \end{bmatrix}$$

(C.E.I.)

5. Define upper (**U**) and lower (**L**) triangular matrices of order 3. Evaluate matrices **U** and **L** explicitly if a matrix $\mathbf{M} = \mathbf{LU}$, where

$$\mathbf{M} = \begin{bmatrix} 2 & -2 & 3 \\ 1 & -4 & 7 \\ -1 & 1 & 2 \end{bmatrix}$$

and the diagonal elements in **U** are all unity. Deduce the value of the determinant of **M**. (L.U.)

6. Apply the method of factorising **A** to determine the solution of the equations:

(i)
$$\begin{aligned} 4x_1 - x_2 + 3x_3 &= 15 \\ 8x_1 - x_2 + 11x_3 &= 43 \\ -12x_1 + 7x_2 + 18x_3 &= 28 \end{aligned}$$

(ii)
$$\begin{aligned} x_1 + 4x_2 + 7x_3 &= 7 \\ 2x_1 + 11x_2 + 5x_3 &= 8 \\ 8x_1 + 6x_2 + 9x_3 &= 9 \end{aligned}$$

(iii)
$$\begin{aligned} 4x_1 + 2x_2 + x_3 + x_4 &= 25 \\ 8x_1 + 6x_2 + 4x_3 + 3x_4 &= 61 \\ 4x_1 - 2x_2 + 3x_3 + x_4 &= 17 \\ 8x_1 + 8x_2 + 12x_3 + 9x_4 &= 79 \end{aligned}$$

(iv)

$$\mathbf{a} = \begin{bmatrix} 2 & 1 & 6 & 3 \\ 4 & -1 & 14 & 8 \\ 8 & -5 & 26 & 19 \\ 10 & -1 & 30 & 25 \end{bmatrix}, \quad \mathbf{b} = \begin{bmatrix} 1 \\ -4 \\ 1 \\ 23 \end{bmatrix}, \quad \mathbf{x} = \begin{bmatrix} x_1 \\ x_2 \\ x_3 \\ x_4 \end{bmatrix}$$

(v)

$$\begin{aligned} x + 3y + 6z &= 17 \\ 2x + 8y + 16z &= 42 \\ 5x + 21y + 45z &= 91 \end{aligned}$$

Show any necessary checks. (L.U.)

7. Show that the inverse of $\mathbf{A} = \mathbf{LU}$ can be written $\mathbf{A}^{-1} = \mathbf{U}^{-1}\mathbf{L}^{-1}$, if it exists. Hence find the inverse of $\mathbf{A} = \begin{bmatrix} 4 & 6 & 8 \\ 6 & 10 & 17 \\ 8 & 17 & 25 \end{bmatrix}$ and hence solve the set of equations $\mathbf{Ax} = \mathbf{b}$ where $\mathbf{b} = (18, 33, 50)^T$.

8. (a) Use the method of Choleski to solve the simultaneous equations

$$\begin{aligned} 8x_1 + x_2 \qquad\qquad &= 4 \\ 2x_1 - \tfrac{3}{4}x_2 + x_3 \qquad &= 1 \\ x_2 + 2x_3 + x_4 &= 2 \\ x_3 - 3x_4 &= -1 \end{aligned}$$

(b) Repeat using the Thomas algorithm.

(c) For this problem, show that $(\mathbf{L}^T)^{-1} = (\mathbf{L}^{-1})^T$.

9. Express the Gauss-Seidel method of solution of simultaneous linear equations as an iterative method in terms of an upper triangular matrix $\mathbf{U}$ and a lower triangular matrix $\mathbf{L}$. (Use the example below to illustrate the formation of these matrices.) If the (column) matrix $\mathbf{e}^{(n)}$ stands for the difference between the exact solution and the nth iteration show that $\mathbf{e}^{(n+1)} = (\mathbf{I} - \mathbf{L})^{-1}\,\mathbf{U}\mathbf{e}^{(n)}$.

Give the explicit forms for $\mathbf{L}$, $\mathbf{U}$ corresponding to the equations

$$\begin{aligned} 25x + 2y + z &= 70 \\ 2x + 10y + z &= 60 \\ x + y + 4z &= 40 \end{aligned}$$

Starting from the matrix $\mathbf{x}^{(0)} = (1, 1, 1)$ find $\mathbf{x}^{(2)}$. (L.U.)

1.6 PARTITIONED MATRICES

In some problems in structural analysis the equations relating imposed forces and resulting deflections can be cast in matrix form. The so-called **flexibility matrix** of coefficients can often be split or **partitioned** into rectangular blocks. The advantage of partitioning is that we can treat the blocks as individual elements and this makes certain operations, for example inversion, more straightforward. In this section, we indicate briefly some of the basic ideas using simple examples for ease of understanding.

(i) Addition

Let $\mathbf{A} = \left[\begin{array}{cc|c} 1 & 2 & 5 \\ 3 & 4 & 6 \\ \hline 7 & 8 & 9 \end{array}\right] = \left[\begin{array}{c|c} \mathbf{A}_1 & \mathbf{A}_2 \\ \hline \mathbf{A}_3 & \mathbf{A}_4 \end{array}\right]$ and $\mathbf{B} = \left[\begin{array}{cc|c} 6 & 7 & 8 \\ 4 & 5 & 9 \\ \hline 1 & 2 & 3 \end{array}\right] = \left[\begin{array}{c|c} \mathbf{B}_1 & \mathbf{B}_2 \\ \hline \mathbf{B}_3 & \mathbf{B}_4 \end{array}\right]$

Then $\mathbf{A} + \mathbf{B} = \left[\begin{array}{c|c} \mathbf{A}_1 + \mathbf{B}_1 & \mathbf{A}_2 + \mathbf{B}_2 \\ \hline \mathbf{A}_3 + \mathbf{B}_3 & \mathbf{A}_4 + \mathbf{B}_4 \end{array}\right] = \left[\begin{array}{cc|c} 7 & 9 & 13 \\ 7 & 9 & 15 \\ \hline 8 & 10 & 12 \end{array}\right]$

(ii) Multiplication

Provided the partitioning has produced compatible blocks we can carry out multiplication in two stages.

Let **A** and **B** be as above. Then

$$\mathbf{AB} = \left[\begin{array}{c|c} \mathbf{A_1 B_1 + A_2 B_3} & \mathbf{A_1 B_2 + A_2 B_4} \\ \hline \mathbf{A_3 B_1 + A_4 B_3} & \mathbf{A_3 B_2 + A_4 B_4} \end{array}\right]$$

$$= \left[\begin{array}{c|c} \begin{bmatrix}1 & 2\\3 & 4\end{bmatrix} \times \begin{bmatrix}6 & 7\\4 & 5\end{bmatrix} + \begin{bmatrix}5\\6\end{bmatrix} \times [1, 2] & \begin{bmatrix}1 & 2\\3 & 4\end{bmatrix} \times \begin{bmatrix}8\\9\end{bmatrix} + \begin{bmatrix}5\\8\end{bmatrix} \times [3] \\ \hline [7, 8] \times \begin{bmatrix}6 & 7\\4 & 5\end{bmatrix} + [9] \times [1, 2] & [7, 8] \times \begin{bmatrix}8\\9\end{bmatrix} + [9] \times [3] \end{array}\right]$$

$$= \left[\begin{array}{cc|c} 19 & 27 & 41 \\ 40 & 53 & 78 \\ \hline 83 & 107 & 155 \end{array}\right]$$

Let $\mathbf{C} = \left[\begin{array}{c|cc} 6 & 3 & 4 \\ \hline 3 & 5 & 2 \\ 1 & 2 & 3 \end{array}\right] = \left[\begin{array}{c|c} \mathbf{C_1} & \mathbf{C_2} \\ \hline \mathbf{C_3} & \mathbf{C_4} \end{array}\right]$. Then, in theory,

$$\mathbf{AC} = \left[\begin{array}{c|c} \mathbf{A_1} & \mathbf{A_2} \\ \hline \mathbf{A_3} & \mathbf{A_4} \end{array}\right] \left[\begin{array}{c|c} \mathbf{C_1} & \mathbf{C_2} \\ \hline \mathbf{C_3} & \mathbf{C_4} \end{array}\right] = \left[\begin{array}{c|c} \mathbf{A_1 C_1 + A_2 C_3} & \mathbf{A_1 C_2 + A_2 C_4} \\ \hline \mathbf{A_3 C_1 + A_4 C_3} & \mathbf{A_3 C_2 + A_4 C_4} \end{array}\right]$$

However, for example, $\mathbf{A_1}$ is 2 x 2 and $\mathbf{C_1}$ is 1 x 2 so that multiplication is not possible.

(iii) Inverse

We consider the special case of a 4 x 4 matrix partitioned into four equal blocks of 2 x 2 submatrices.

Let $\mathbf{A} = \left[\begin{array}{cc|cc} 4 & 2 & 1 & 2 \\ 2 & 2 & 2 & 3 \\ \hline 1 & 2 & 1 & 0 \\ 0 & 1 & 2 & 1 \end{array}\right] = \left[\begin{array}{c|c} \mathbf{A_1} & \mathbf{A_2} \\ \hline \mathbf{A_3} & \mathbf{A_4} \end{array}\right]$ and let $\mathbf{C} = \left[\begin{array}{c|c} \mathbf{C_1} & \mathbf{C_2} \\ \hline \mathbf{C_3} & \mathbf{C_4} \end{array}\right] = \mathbf{A}^{-1}$

so that $\mathbf{AC} = \left[\begin{array}{c|c} \mathbf{A_1 C_1 + A_2 C_3} & \mathbf{A_1 C_2 + A_2 C_4} \\ \hline \mathbf{A_3 C_1 + A_4 C_3} & \mathbf{A_3 C_2 + A_4 C_4} \end{array}\right] = \mathbf{I} = \left[\begin{array}{cc|cc} 1 & 0 & 0 & 0 \\ 0 & 1 & 0 & 0 \\ \hline 0 & 0 & 1 & 0 \\ 0 & 0 & 0 & 1 \end{array}\right]$

It follows that
$$\left.\begin{aligned} \mathbf{A}_1\,\mathbf{C}_1 + \mathbf{A}_2\,\mathbf{C}_3 &= \mathbf{I} && \text{(a)}\\ \mathbf{A}_1\,\mathbf{C}_2 + \mathbf{A}_2\,\mathbf{C}_4 &= \mathbf{0} && \text{(b)}\\ \mathbf{A}_3\,\mathbf{C}_1 + \mathbf{A}_4\,\mathbf{C}_3 &= \mathbf{0} && \text{(c)}\\ \mathbf{A}_3\,\mathbf{C}_2 + \mathbf{A}_4\,\mathbf{C}_4 &= \mathbf{I} && \text{(d)} \end{aligned}\right\} \quad (1.25)$$

If we substitute $\mathbf{C}_3 = -\mathbf{A}_4^{-1}\,\mathbf{A}_3\,\mathbf{C}_1$ from (1.25c) into (1.25a) we obtain $\mathbf{A}_1\,\mathbf{C}_1 - \mathbf{A}_2\,\mathbf{A}_4^{-1}\,\mathbf{A}_3\,\mathbf{C}_1 = \mathbf{I}$ and hence $\mathbf{C}_1 = (\mathbf{A}_1 - \mathbf{A}_2\,\mathbf{A}_4^{-1}\,\mathbf{A}_3)^{-1}$.
We therefore may determine $\mathbf{C}_1$ and $\mathbf{C}_3$. Similarly, we obtain $\mathbf{C}_2 = -\mathbf{A}_1^{-1}\,\mathbf{A}_2\,\mathbf{C}_4$, $\mathbf{C}_4 = (\mathbf{A}_4 - \mathbf{A}_3\,\mathbf{A}_1^{-1}\,\mathbf{A}_2)^{-1}$ and hence determine $\mathbf{C}_4$ and $\mathbf{C}_2$.
Of course $\mathbf{A}_1^{-1}$ and $\mathbf{A}_4^{-1}$ are found in the usual way for 2×2 matrices.

Example

$$\mathbf{A} = \left[\begin{array}{cc|cc} 4 & 2 & 1 & 2\\ 2 & 2 & 2 & 3\\ \hline 1 & 2 & 1 & 0\\ 0 & 1 & 2 & 1 \end{array}\right]$$

Then $$\mathbf{C}_1 = \left(\begin{bmatrix}4 & 2\\ 2 & 2\end{bmatrix} - \begin{bmatrix}1 & 2\\ 2 & 3\end{bmatrix}\begin{bmatrix}1 & 0\\ 2 & 1\end{bmatrix}^{-1}\begin{bmatrix}1 & 2\\ 0 & 1\end{bmatrix}\right)^{-1} = \left(\begin{bmatrix}7 & 6\\ 6 & 7\end{bmatrix}\right)^{-1} = \frac{1}{13}\left(\begin{bmatrix}7 & -6\\ -6 & 7\end{bmatrix}\right)$$

$$\mathbf{C}_3 = -\begin{bmatrix}1 & 0\\ 2 & 1\end{bmatrix}^{-1}\begin{bmatrix}1 & 2\\ 0 & 1\end{bmatrix}\frac{1}{13}\begin{bmatrix}7 & -6\\ -6 & 7\end{bmatrix} = -\frac{1}{13}\begin{bmatrix}-5 & 8\\ 4 & -9\end{bmatrix}$$

$$\mathbf{C}_4 = \left(\begin{bmatrix}1 & 0\\ 2 & 1\end{bmatrix} - \begin{bmatrix}1 & 2\\ 0 & 1\end{bmatrix}\begin{bmatrix}4 & 2\\ 2 & 2\end{bmatrix}^{-1}\begin{bmatrix}1 & 2\\ 2 & 3\end{bmatrix}\right)^{-1} = \frac{1}{13}\begin{bmatrix}-4 & 14\\ -2 & -6\end{bmatrix}$$

$$\mathbf{C}_2 = -\begin{bmatrix}4 & 2\\ 2 & 2\end{bmatrix}^{-1}\begin{bmatrix}1 & 2\\ 2 & 3\end{bmatrix}\frac{1}{13}\begin{bmatrix}-4 & 14\\ -2 & -6\end{bmatrix} = -\frac{1}{13}\begin{bmatrix}3 & -4\\ -10 & 9\end{bmatrix}$$

Finally, $$\mathbf{A}^{-1} = \begin{bmatrix} \frac{7}{13} & -\frac{6}{13} & -\frac{3}{13} & \frac{4}{13}\\ -\frac{6}{13} & \frac{7}{13} & \frac{10}{13} & -\frac{9}{13}\\ \frac{5}{13} & -\frac{8}{13} & -\frac{4}{13} & \frac{14}{13}\\ -\frac{4}{13} & \frac{9}{13} & -\frac{2}{13} & -\frac{6}{13} \end{bmatrix}$$ as can be verified directly.

Problems

1. Prove for the matrix $\mathbf{A} = \left[\begin{array}{cc|cc} 2 & 1 & 3 & 4 \\ 5 & 3 & 2 & 3 \\ \hline 2 & 1 & 2 & 3 \\ 9 & 5 & 1 & 2 \end{array}\right] = \left[\begin{array}{c|c} \mathbf{A}_1 & \mathbf{A}_2 \\ \hline \mathbf{A}_3 & \mathbf{A}_4 \end{array}\right]$ that $\det \mathbf{A} \neq \det \begin{bmatrix} \det \mathbf{A}_1 & \det \mathbf{A}_2 \\ \det \mathbf{A}_3 & \det \mathbf{A}_4 \end{bmatrix}$

2. Find the products of the partitioned matrices

(i) $\left[\begin{array}{cc} 1 & 2 \\ 3 & 4 \\ \hline 5 & 6 \end{array}\right] \begin{bmatrix} 9 & 8 \\ 7 & 6 \end{bmatrix}$ (ii) $\left[\begin{array}{cc|c} 1 & 0 & 5 \\ 0 & 1 & 6 \\ \hline 1 & 0 & 7 \\ 0 & 1 & 8 \end{array}\right] \left[\begin{array}{cc} 2 & 0 \\ 0 & 2 \\ \hline 9 & 10 \end{array}\right]$ (iii) $\left[\begin{array}{ccc} 1 & 2 & 3 \\ 4 & 5 & 6 \\ \hline 7 & 8 & 9 \end{array}\right] \begin{bmatrix} 3 \\ 2 \\ 1 \end{bmatrix}$

(iv) $\left[\begin{array}{cc} a & b \\ \hline c & d \end{array}\right] \left[\begin{array}{c} x \\ \hline y \end{array}\right]$ (v) $\left[\begin{array}{cc|c} a & h & g \\ h & b & f \\ \hline g & f & c \end{array}\right] \left[\begin{array}{c|cc} a & h & g \\ h & b & f \\ \hline g & f & c \end{array}\right]$ (vi) $\left[\begin{array}{c|c} \mathbf{A}_1 & \mathbf{A}_2 \\ \hline \mathbf{A}_3 & \mathbf{A}_4 \end{array}\right] \left[\begin{array}{c|c} \mathbf{I} & \mathbf{0} \\ \hline \mathbf{0} & \mathbf{I} \end{array}\right]$

(vii) $\left[\begin{array}{c|c} \mathbf{A}_1 & \mathbf{A}_2 \\ \hline \mathbf{A}_3 & \mathbf{A}_4 \end{array}\right] \left[\begin{array}{c|c} \mathbf{I} & -\mathbf{A}_1^{-1}\mathbf{A}_2 \\ \hline \mathbf{0} & \mathbf{I} \end{array}\right]$

3. Prove that if the submatrices are compatible for multiplication,

(i) $\left[\begin{array}{c|c} \mathbf{A}_1 & \mathbf{I} \\ \hline \mathbf{I} & \mathbf{A}_1^{-1} \end{array}\right] \left[\begin{array}{c|c} \mathbf{A}_1^{-1} & \mathbf{I} \\ \hline \mathbf{I} & \mathbf{A}_1 \end{array}\right] = 2\left[\begin{array}{c|c} \mathbf{I} & \mathbf{A}_1 \\ \hline \mathbf{A}_1^{-1} & \mathbf{I} \end{array}\right]$

(ii) $\left[\begin{array}{c|c} \mathbf{A}_1 & \mathbf{0} \\ \hline \mathbf{I} & \mathbf{A}_1^{-1} \end{array}\right] \left[\begin{array}{c|c} \mathbf{A}_1^{-1} & \mathbf{I} \\ \hline \mathbf{0} & \mathbf{A}_1 \end{array}\right] = \left[\begin{array}{c|c} \mathbf{I} & \mathbf{A}_1 \\ \hline \mathbf{A}_1^{-1} & 2\mathbf{I} \end{array}\right]$

4. Prove that $\det \left[\begin{array}{c|c} \mathbf{A}_1 & \mathbf{A}_2 \\ \hline \mathbf{A}_3 & \mathbf{A}_4 \end{array}\right] = \det\,[(\mathbf{A}_1\mathbf{A}_4 - \mathbf{A}_3\mathbf{A}_2)]$

5. By partitioning, find the inverses of the following matrices and verify your results.

(i) $\begin{bmatrix} 2 & 4 & 3 & 2 \\ 3 & 6 & 5 & 2 \\ 2 & 5 & 2 & -3 \\ 4 & 5 & 14 & 14 \end{bmatrix}$ (ii) $\begin{bmatrix} 1 & 2 & 3 & 1 \\ 1 & 3 & 3 & 2 \\ 2 & 4 & 3 & 3 \\ 1 & 1 & 1 & 1 \end{bmatrix}$ (iii) $\begin{bmatrix} 1 & -2 & 1 & 0 \\ 1 & -2 & 2 & -3 \\ 0 & 1 & -1 & 1 \\ -2 & 3 & -2 & 3 \end{bmatrix}$

(iv) $\begin{bmatrix} 2 & 1 & -1 & 2 \\ 1 & 3 & 2 & -3 \\ -1 & 2 & 1 & -1 \\ 2 & -3 & -1 & 4 \end{bmatrix}$ (v) $\begin{bmatrix} \cos\alpha & \sin\alpha & \cos\beta & \sin\beta \\ -\sin\alpha & \cos\alpha & -\sin\beta & \cos\beta \\ 0 & 0 & \cos\gamma & \sin\gamma \\ 0 & 0 & -\sin\gamma & \cos\gamma \end{bmatrix}$

Chapter Two

Eigenvalue Problems

2.1 INTRODUCTION

An important set of problems is the class of **boundary value problems**. In these, the values of the dependent variable and its derivative are specified at end-points, or more generally on the boundaries of the domain of the problem. For example, we may be interested in the deflected profile of a beam simply-supported at its ends, where the boundary conditions are the specified values of the deflection and slope of the beam at its two ends. A second example would be the propagation of heat in a slab with insulated faces, where the temperature along the four edges was specified.

This example is one of a **continuous system** and, as we shall see in Chapter 8, the appropriate mathematical model is a partial differential equation. We shall be concerned in this chapter, as in the previous one, with problems where the system is **discrete**. We shall assume that the system possesses a finite number > 1 of degrees of freedom . Alternative names for these systems are **distributed-parameter** and **lumped-parameter** respectively. In the latter class, we make such assumptions as shafts having negligible masses. The reason for these names will become clear as we study the examples in this chapter and in Chapter 8.

A special class of problems is known as the class of **characteristic-value** or **eigenvalue problems.** The main contexts in which we meet such cases is in the study of vibration problems or in certain problems in structures. The distinction that these problems share is that a solution can exist for only a set of values of a parameter of the problem: the eigenvalues. As examples, a system may be able to vibrate only at certain frequencies or a structure may only be in equilibrium in certain modes of deflection.

Vibration problems

Let us pursue a little further the problem of vibrating systems. A system which is subject to an imposed periodic force will vibrate in a way which partly reflects the nature of the imposed force and partly reflects something inherent in the system: the so-called **natural vibration.** The natural frequencies of oscillation are the eigenvalues and the descriptions of the possible vibrations (possibly in terms of displacements of the mass-centres of the system) are the normal modes. If the frequency of the imposed periodic force is close to that of one of these normal modes then the amplitude of the resulting oscillations may build up in an alarming way. It is important therefore for a designer to know the natural frequencies of the system under consideration. The kinds of problem that he wishes to avoid are typified by the collapse of the Ferrybridge cooling towers, the collapse of the Tacoma Narrows bridge, the flutter of aircraft wings and the transverse vibrations of turbine shafts.

2.2 CASE STUDY: VIBRATION OF A TWO-STOREY FRAME

In Figure 2.1(a) we see a schematic representation of a two-storey steel frame. We wish to study the effect of external vibration on this structure.

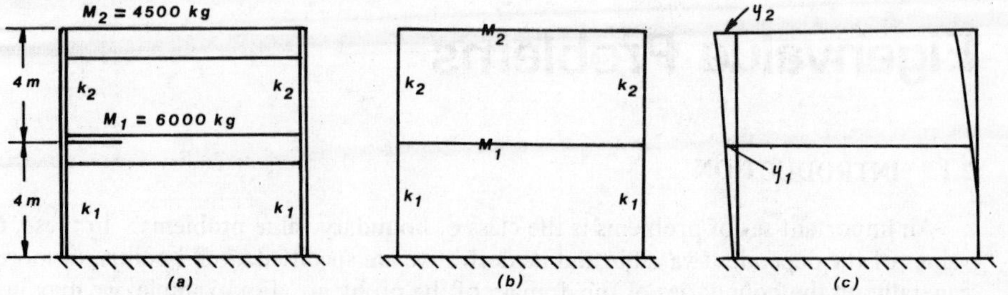

Figure 2.1

The horizontal girders are very stiff compared with the columns of stiffnesses k_1 and k_2, measured in N/m. All the joints are supposed rigid and the masses of the columns may be neglected with regard to the masses carried by the girders. In this physical model we have *lumped* the masses into the girders and we have *lumped* the stiffness into the columns. We shall construct a mathematical model with general masses M_1 and M_2 and stiffness k_1 and k_2. In Figure 2.1(b) we see the frame in its undeflected form with an applied periodic force and in Figure 2.1(c) we present the deflected profile. (We are *assuming* that the frame vibrates in its plane and that the girders remain horizontal during the motion.) We have chosen as coordinates of the system the horizontal deflections of the girders in the plane of the frame. It remains to be seen whether this obvious choice of coordinates is the most suitable for the problem.

The equations of motion are as follows. For the upper girder, whose deflection relative to the lower girder is $(y_2 - y_1)$, we obtain

$$M_2 \ddot{y}_2 = -k_2(y_2 - y_1) \tag{2.1a}$$

and for the lower girder to which a force $F_1 \sin \Omega t$ is applied

$$M_1 \ddot{y}_1 = -k_1 y_1 + k_2(y_2 - y_1) + F_1 \sin \Omega t \tag{2.1b}$$

For the moment we shall consider the free oscillations of the system and we may put $F_1 = 0$ and rearrange the above equations. This produces

$$M_1 \ddot{y}_1 + k_1 y_1 - k_2(y_2 - y_1) = 0 \tag{2.2a}$$

and

$$M_2 \ddot{y}_2 + k_2(y_2 - y_1) = 0 \tag{2.2b}$$

It is reasonable to suppose that the time function is the same for all components of the system. If we let this function be denoted by $f(t)$ and the maximum amplitudes of the two girders be a_1 and a_2 respectively, then the equations (2.2) become

$$M_1 a_1 f''(t) + \left[k_1 a_1 - k_2(a_2 - a_1)\right] f(t) = 0 \tag{2.3a}$$

and

$$M_2 a_2 f''(t) + k_2(a_2 - a_1) f(t) = 0 \tag{2.3b}$$

Either equation then yields the result $f''(t)/f(t) =$ constant. It therefore makes sense to choose the constant the same for each equation, and since we are expecting a periodic behaviour we shall choose the constant to be $-\omega^2$ so that

$$f(t) = A \sin \omega t + B \cos \omega t = C \sin(\omega t + \alpha)$$

where A, B, C and α are suitable constants. By selecting the initial time sensibly we may take $\alpha = 0$; also we take $C = 1$. Substituting the expression $f(t) = \sin \omega t$ into equations (2.3) we obtain, on cancelling the factor $\sin \omega t$ from each term, the results

$$-\omega^2 M_1 a_1 + k_1 a_1 - k_2(a_2 - a_1) = 0 \qquad (2.4a)$$

$$-\omega^2 M_2 a_2 + k_2(a_2 - a_1) = 0 \qquad (2.4b)$$

We have two simultaneous linear equations in the three unknowns ω^2, a_1 and a_2. Suppose we decide to make two unknown quantities, namely ω^2 and the ratio (a_2/a_1). We accomplish this by dividing both equations by a_1; then we have

$$-\omega^2 M_1 + (k_1 + k_2) - k_2(a_2/a_1) = 0 \qquad (2.5a)$$

and

$$\left[-\omega^2 M_2 + k_2\right](a_2/a_1) - k_2 = 0 \qquad (2.5b)$$

We may then obtain the relationships

$$a_2/a_1 = \frac{-\omega^2 M_1 + (k_1 + k_2)}{k_2} = \frac{k_2}{-\omega^2 M_2 + k_2} \qquad (2.6)$$

Cross-multiplying the last two of these gives the **characteristic equation** for the system

$$\omega^4 M_1 M_2 - \omega^2 [M_2(k_1 + k_2) + M_1 k_2] + k_1 k_2 = 0 \qquad (2.7)$$

The roots of this equation are called the **characteristic roots** or **eigenvalues** of the system and will give rise to the natural frequencies at which the system will vibrate. It is more usual to obtain the characteristic equation by rewriting equations (2.4) in terms of a_1 and a_2, expressing these equations in matrix form and then equating the determinant of the matrix to zero.

In this problem we should obtain the equations

$$(-\omega^2 M_1 + k_1 + k_2)a_1 - k_2 a_2 = 0 \quad \text{and} \quad -k_2 a_1 + (-\omega^2 M_2 + k_2)a_2 = 0$$

The determinant that results is

$$\begin{vmatrix} -\omega^2 M_1 + k_1 + k_2 & -k_2 \\ -k_2 & -\omega^2 M_2 + k_2 \end{vmatrix} \qquad (2.8)$$

and equating this to zero produces the characteristic equation (2.7). Let us now take the specific case where $k_1 = k_2 = 1330$ kN/m and the masses M_1 and M_2 have the values shown in Figure 2.1(a). Then the characteristic equation becomes $-0.270\omega^4 + 199.5\omega^2 - 17689 = 0$. This has roots $\omega^2 = 103$ or 635 and hence the natural frequencies are $\omega_1 = 10.15$ rads/sec and $\omega_2 = 25.2$ rads/sec.

Modes of vibration

The first mode of vibration corresponds to the lower frequency ω_1. From equation (2.6) we find $a_2/a_1 = \dfrac{1330 \times 10^3}{-103 \times 4500 + 1330 \times 10^3} = 1.54$. In this mode the upper girder moves in the same direction at each instant as does the lower girder, but moves further by a factor just larger than 1.5.

The second mode of vibration corresponds to ω_2 and is associated with a ratio $a_2/a_1 = -0.87$. In this mode the girders move in opposite senses, and the lower girder moves further. The modes are depicted in Figures 2.2(a) and (b) respectively.

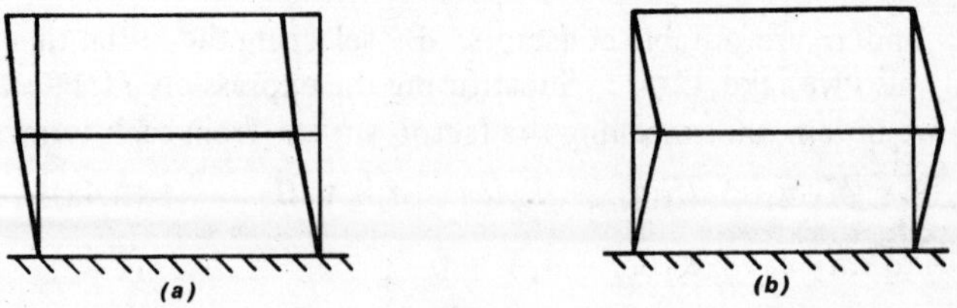

Figure 2.2

Fixed Vibration

If we now consider a horizontal periodic force $F_1 \sin \Omega t$ applied to the lower girder, the governing equations of motion may be reduced by assuming the time function to be $\sin \Omega t$. This will lead us to the *particular integral* part of the solution as opposed to the *complementary function* part obtained as the free vibration. The equations to be solved are

$$-\Omega^2 M_1 a_1 + k_1 a_1 - k_2(a_2 - a_1) = F_1 \text{ and } -\Omega^2 M_2 a_2 + k_2(a_2 - a_1) = 0.$$

We then obtain $a_1 = \dfrac{F_1(-\Omega^2 M_2 + k_2)}{(-\Omega^2 M_1 + k_1 + k_2)(-\Omega^2 M_2 + k_2) - k_2^2}$

and $$a_2 = \frac{F_1 k_2}{(-\Omega^2 M_1 + k_1 + k_2)(-\Omega^2 M_2 + k_2) - k_2^2}$$

Note that when $\Omega = \omega_1$ or ω_2 the denominators are zero (c.f. 2.8) and *resonance* occurs.

Further note that when $\Omega = \sqrt{k_2/M_2}$, $a_1 = 0$ and $a_2 = -F_1/k_2$. To consider a typical case we may take $F_1 = 500$ N. Try and write a computer program to calculate the displacements y_1 and y_2 at several specified times.

Matrix formulation

We now return to equations (2.1). If we let $\mathbf{y} = \begin{bmatrix} y_1 \\ y_2 \end{bmatrix}$ then we may symbolise $\begin{bmatrix} \ddot{y}_1 \\ \ddot{y}_2 \end{bmatrix}$ as $\ddot{\mathbf{y}}$; further the vector $\begin{bmatrix} F_1 \sin \Omega t \\ 0 \end{bmatrix}$ may be written $\mathbf{F}_1$ and $\begin{bmatrix} M_1 \\ M_2 \end{bmatrix}$ as $\mathbf{M}$. The matrix formulation of equations (2.1) is therefore

$$(M_1, M_2) \begin{bmatrix} \ddot{y}_1 \\ \ddot{y}_2 \end{bmatrix} = \begin{bmatrix} -(k_1 + k_2) & k_2 \\ k_2 & -k_2 \end{bmatrix} \begin{bmatrix} y_1 \\ y_2 \end{bmatrix} + \begin{bmatrix} F_1 \sin \Omega t \\ 0 \end{bmatrix}$$

or $\quad \mathbf{M}^T \ddot{\mathbf{y}} = \mathbf{A}\mathbf{y} + \mathbf{F}_1.$

A more useful matrix formulation is found by considering equations (2.4). If we put $M_2 = M_1 = M$ and $k_1 = k_2 = k$ (although the general principles will hold true whatever the ratios of M_2 to M_1 and of k_2 to k_1) then let $\lambda = M\omega^2/k$. Then equations (2.4) may be written as

$$2a_1 - a_2 = \lambda a_1 \qquad \text{and} \qquad -a_1 + a_2 = \lambda a_2$$

i.e. $$\begin{bmatrix} 2 & -1 \\ -1 & 1 \end{bmatrix}\begin{bmatrix} a_1 \\ a_2 \end{bmatrix} = \lambda \begin{bmatrix} a_1 \\ a_2 \end{bmatrix} \qquad (2.9a)$$

or $$\mathbf{Ba} = \lambda \mathbf{a} \qquad (2.9b)$$

where $\mathbf{a} = \begin{bmatrix} a_1 \\ a_2 \end{bmatrix}$ and $\mathbf{B} = \begin{bmatrix} 2 & -1 \\ -1 & 1 \end{bmatrix}$

Alternatively we may write (2.9) as

$$\begin{bmatrix} 2-\lambda & -1 \\ -1 & 1-\lambda \end{bmatrix}\begin{bmatrix} a_1 \\ a_2 \end{bmatrix} = \begin{bmatrix} 0 \\ 0 \end{bmatrix} \qquad (2.10a)$$

or $$(\mathbf{B} - \lambda\mathbf{I})\mathbf{a} = \mathbf{0} \qquad (2.10b)$$

The solution of these equations is found by solving $|\mathbf{B} - \lambda\mathbf{I}| = 0$.

Now consider the determinant $\begin{vmatrix} -\omega^2 M_1 + k_1 + k_2 & -k_2 \\ -k_2 & -\omega^2 M_2 + k_2 \end{vmatrix}$ (2.8)

In the case $M_1 = M_2 = M$ and $k_1 = k_2 = k$, this determinant reduces to

$$\begin{vmatrix} -\omega^2 M + 2k & -k \\ -k & -\omega^2 M + k \end{vmatrix} = k^2 \begin{vmatrix} -\lambda + 2 & -1 \\ -1 & -\lambda + 1 \end{vmatrix}$$

The solutions of the equation derived by equating (2.8) to zero are those of $\begin{vmatrix} 2-\lambda & -1 \\ -1 & 1-\lambda \end{vmatrix} = 0$. This of course is the requirement $|\mathbf{B} - \lambda\mathbf{I}| = 0$.

Definition

The values of λ for which the equations $\mathbf{Ba} = \lambda\mathbf{a}$ have other than the zero solution $\mathbf{a} = \mathbf{0}$ are called the **eigenvalues** of the matrix **B**. To each eigenvalue there is a non-zero solution **a** called the associated **eigenvector**.

In our example we find the eigenvalues by solving the equation

$$\begin{vmatrix} 2-\lambda & -1 \\ -1 & 1-\lambda \end{vmatrix} = 0, \text{ i.e. } (2-\lambda)(1-\lambda) - 1 = 0 \text{ or } \lambda^2 - 3\lambda + 1 = 0$$

The solutions are $\lambda_1 = \dfrac{3-\sqrt{5}}{2} = 0.382$ (3 d.p.) and $\lambda_2 = \dfrac{3+\sqrt{5}}{2} = 2.618$ (3 d.p.)

For $\lambda_1 = 0.382$, we find the eigenvector by solving equations (2.10a), which become

$$1.618a_1 - a_2 = 0 \qquad -a_1 + 0.618a_2 = 0$$

If we write $\lambda_1 = (3-\sqrt{5})/2$ then the equations become

$$\tfrac{1}{2}(1+\sqrt{5})a_1 - a_2 = 0 \qquad -a_1 + \tfrac{1}{2}(\sqrt{5}-1)a_2 = 0$$

Either of these equations produces $a_2/a_1 = \tfrac{1}{2}(1+\sqrt{5}) = 1.618$ (3 d.p.). We see that instead of two independent equations we have really got the same equation repeated.

We should have expected this since the only way that equation (2.9b) would have other than the zero solution was if the matrix **B** had some kind of degeneracy.

We need hardly have bothered with matrices in this simple example, but problems with more than two degrees of freedom really demand matrix methods for efficient solution. In the next section we shall consider some systems with three degrees of freedom.

Problems

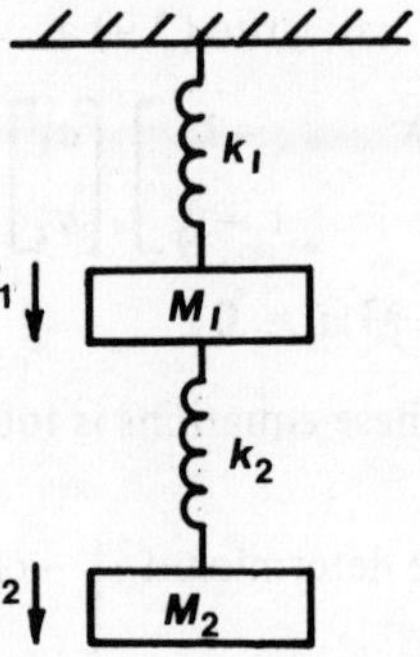

1. For the system of two spring masses as shown, write down the equations of motion in terms of y_1 and y_2, the displacements of the masses from their equilibrium positions. Cast these in matrix form $\ddot{\mathbf{y}} = \mathbf{A}\mathbf{y}$ and produce an eigenvalue problem by writing $\mathbf{y} = \mathbf{x}\, e^{i\omega t}$. Find the eigenvalues and associated eigenvectors of **A** in the case where $k_1 = 3$, $k_2 = 2$, $M_1 = M_2 = 1$. If, initially, $y_1 = 2$, $y_2 = 4$, $\dot{y}_1 = -4\sqrt{6}$, $\dot{y}_2 = 2\sqrt{6}$ find the solutions for $y_1(t)$, $y_2(t)$.

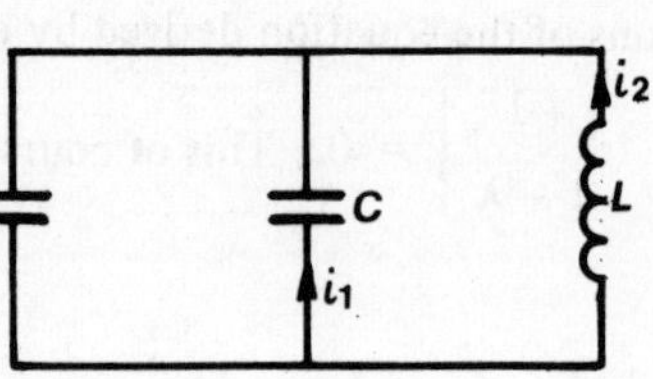

2. Let the currents in the circuit at resonance be given by
$i_1 = a_1 \sin(\omega t + \phi)$,
$i_2 = a_2 \sin(\omega t + \phi)$. Show that the resonant frequencies and modes may be found by solving the eigenvalue problem:

$$3y_1 - y_2 = \lambda y_1, \quad y_1 + y_2 = \lambda y_2$$

where $\lambda = \omega^2 LC$, $y_1 = x_1 + x_2$ and $y_2 = x_1 + 2x_2$.

3.

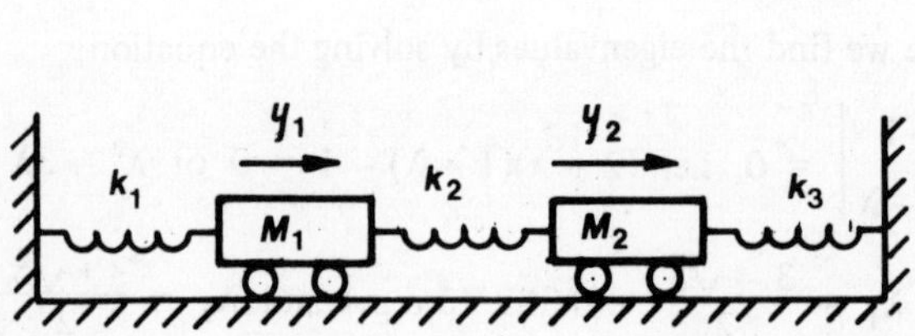

For the system shown, repeat the general strategy of Problem 1. Work the special case $k_1 = k_2 = k_3 = 1$, $M_1 = M_2 = 1$ with initial conditions $y_1 = 1$, $y_2 = \dot{y}_1 = \dot{y}_2 = 0$.

4. Repeat Problem 1 for the conditions

(i) $M_1 = M_2 = 1$, $k_1 = 1.5$, $k_2 = 1$, $y_1(0) = 0 = y_2(0)$, $\dot{y}_1(0) = 4$, $\dot{y}_2(0) = 1$

(ii) $M_1 = M_2 = 1$, $k_1 = 9$, $k_2 = 6$, $y_1(0) = -1$, $y_2(0) = 1$, $\dot{y}_1(0) = 0 = \dot{y}_2(0)$.

2.3 ALGEBRAIC DETERMINATION OF EIGENVALUES

In this section we concentrate on systems with 3 degrees of freedom; these give rise to 3 x 3 matrices. The system shown in Figure 2.3 represents a physical model of a compressor for a jet engine; each disk corresponds to a set of rotor blades. We have ignored any damping effects either between blades and machine housing or between blades. There will be an oscillatory motion in addition to any rigid body rotational motion.

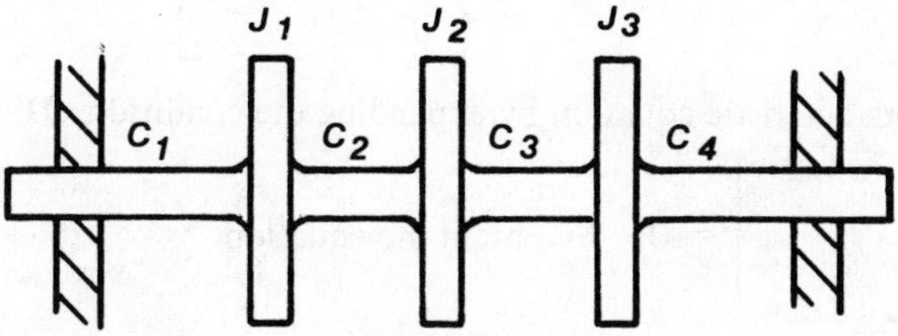

Figure 2.3

We consider the case where the central shaft is free to rotate at either end, the torsional stiffnesses C_i are equal and the moments of inertia J_i of the disks are equal. We shall choose as our coordinates θ_1, θ_2 and θ_3, the angular displacements of the disks in the oscillatory motion. We may write the equations of motion for the oscillatory motion of the disks as follows

$$\left.\begin{aligned} J\ddot{\theta}_1 &= -C(\theta_1 - \theta_2) \\ J\ddot{\theta}_2 &= C(\theta_1 - \theta_2) - C(\theta_2 - \theta_3) \\ J\ddot{\theta}_3 &= C(\theta_2 - \theta_3) \end{aligned}\right\} \qquad (2.11)$$

For an oscillatory motion we may expect $\ddot{\theta}_i = -\omega^2 \theta_i$; substituting for $\ddot{\theta}_i$, writing $\lambda = J\omega^2/C$ and rearranging the equations we obtain

$$\left.\begin{aligned} \theta_1 - \theta_2 &= \lambda\theta_1 \\ -\theta_1 + 2\theta_2 - \theta_3 &= \lambda\theta_2 \\ -\theta_2 + \theta_3 &= \lambda\theta_3 \end{aligned}\right\} \qquad (2.12)$$

In matrix form, equations (2.12) can be written

$$\begin{bmatrix} 1 & -1 & 0 \\ -1 & 2 & -1 \\ 0 & -1 & 1 \end{bmatrix} \begin{bmatrix} \theta_1 \\ \theta_2 \\ \theta_3 \end{bmatrix} = \lambda \begin{bmatrix} \theta_1 \\ \theta_2 \\ \theta_3 \end{bmatrix} \qquad (2.13)$$

or $\qquad \mathbf{A}\boldsymbol{\theta} = \lambda\boldsymbol{\theta}$

Alternatively, we may write the equations as

$$\begin{bmatrix} 1-\lambda & -1 & 0 \\ -1 & 2-\lambda & -1 \\ 0 & -1 & 1-\lambda \end{bmatrix} \begin{bmatrix} \theta_1 \\ \theta_2 \\ \theta_3 \end{bmatrix} = \begin{bmatrix} 0 \\ 0 \\ 0 \end{bmatrix} \qquad (2.14)$$

or $\qquad (\mathbf{A} - \lambda\mathbf{I})\boldsymbol{\theta} = \mathbf{0}$

These equations have a non-zero solution if $|\mathbf{A} - \lambda\mathbf{I}| = 0$. This last condition becomes $(1-\lambda)\left[(2-\lambda)(1-\lambda)-1\right]-(1-\lambda) = 0$ or $(1-\lambda)\lambda(3-\lambda) = 0$. The eigenvalues of matrix **A** are therefore 0, 1 and 3.

Example

Find the eigenvalues of the matrix $\mathbf{D} = \begin{bmatrix} 4 & 2 & -2 \\ 1 & 3 & 1 \\ -1 & -1 & 5 \end{bmatrix}$

We first form the characteristic equation by expanding the condition $|\mathbf{D} - \lambda\mathbf{I}| = 0$, i.e.

$$\begin{vmatrix} 4-\lambda & 2 & -2 \\ 1 & 3-\lambda & 1 \\ -1 & -1 & 5-\lambda \end{vmatrix} = 0.$$ We obtain the equation

$$(4-\lambda)\left[(3-\lambda)(5-\lambda)+1\right] - 2\left[1(5-\lambda)+1\right] - 2\left[-1+(3-\lambda)\right] = 0$$

which factorises into $(4-\lambda)(2-\lambda)(6-\lambda) = 0$. Now let us be honest; we deliberately chose an example which would factorise relatively easily and, of course, we will not always be so lucky.

Determination of Eigenvectors

We shall now find eigenvectors corresponding to each eigenvalue of a matrix. Returning to equations (2.12), we shall substitute $\lambda = 0$, 1 and 3 in turn.

$\underline{\lambda = 0}$. The equations become, with $\lambda = 0$, when written out in full

$$\theta_1 - \theta_2 = 0 \qquad -\theta_1 + 2\theta_2 - \theta_3 = 0 \qquad -\theta_2 + \theta_3 = 0$$

If we add the first and third of these equations together we obtain $\theta_1 - 2\theta_2 + \theta_3 = 0$ which is effectively the same as the second equation of the set. We will not therefore be able to obtain a unique solution (as we must have expected) but will have to content ourselves with merely obtaining the ratio $\theta_1 : \theta_2 : \theta_3$. From the first equation we see that $\theta_1 = \theta_2$ and from the third we find $\theta_2 = \theta_3$. Hence the ratio $\theta_1 : \theta_2 : \theta_3$ is 1 : 1 : 1. Typical eigenvectors satisfying this one condition are (1, 1, 1), (–8, –8, –8), $(1/\sqrt{3}, 1/\sqrt{3}, 1/\sqrt{3})$, and (12.3, 12.3, 12.3). For simplicity's sake we shall choose the simple form (1, 1, 1).

$\underline{\lambda = 1}$. The equations become

$$-\theta_2 = 0 \qquad -\theta_1 + \theta_2 - \theta_3 = 0 \qquad -\theta_2 = 0$$

Again we see that we have only got two independent equations. If we substitute $\theta_2 = 0$ into the second we obtain $\theta_3 = -\theta_1$. Therefore the ratio $\theta_1 : \theta_2 : \theta_3$ is 1 : 0 : –1. Therefore typical eigenvectors are (1, 0, –1), (–3, 0, 3), $(1/\sqrt{2}, 0, -1/\sqrt{2})$. We shall choose the first of these to represent its class.

$\underline{\lambda = 3}$. Now the equations can be written as

$$-2\theta_1 - \theta_2 = 0 \qquad -\theta_1 - \theta_2 - \theta_3 = 0 \qquad -\theta_2 - 2\theta_3 = 0$$

Once more, the second equation is redundant (why?) so that we need only consider the first and last, which give $\theta_1 : \theta_2 : \theta_3 = 1 : -2 : 1$. Two typical eigenvectors are (1, –2, 1) and $(1/\sqrt{6}, -2/\sqrt{6}, 1/\sqrt{6})$.

Interpretation of eigenvectors

We have already remarked that the eigenvectors represent the modes of vibration. Let us investigate the situation further. Corresponding to $\lambda = 0$ we have the eigenvector $(1, 1, 1)$. In other words there is no relative motion of the disks and the disks and the shaft rotate freely in the bearings as a single rigid body. In the case $\lambda = 1$, we chose the eigenvector $(1, 0, -1)$. In this mode the central disk remains motionless whilst the outer two disks rotate with equal amplitudes in opposite senses. In the mode corresponding to $\lambda = 3$ we chose as eigenvector $(1, -2, 1)$ and this may be interpreted as the outer two disks rotating in the same sense, keeping pace with each other whereas the centre disk rotates in the opposite sense with twice the amplitude. In Figure 2.4 we depict these modes; diagrams (a), (b) and (c) schematically represent the modes, while diagrams (d), (e) and (f) attempt to show the motion by considering the positions of markers on each disk at a certain time, given that these markers were initially in a line.

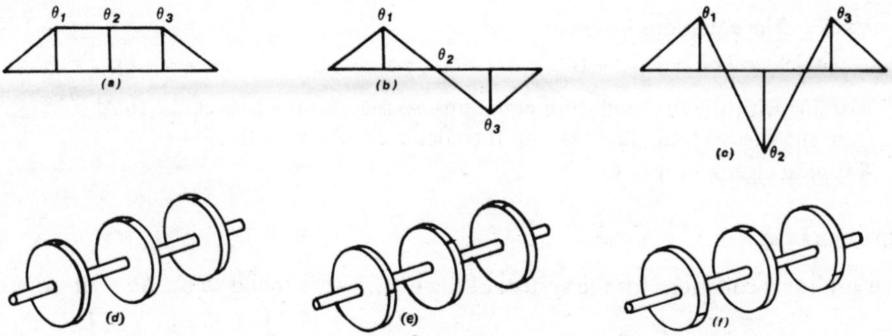

Figure 2.4

In general, the free oscillatory motion will be a combination of these normal modes. Hence in general,

$$\theta_i = A_i \cos \omega_1 t + B_i \sin \omega_1 t + E_i \cos \omega_2 t + F_i \sin \omega_2 t + G_i \cos \omega_3 t + H_i \sin \omega_3 t$$

where ω_1, ω_2 and ω_3 are the three frequencies. Now $\lambda_1 = 0$ and hence $\omega_1 = 0$; $\lambda_2 = 1$ and so $\omega_2 = \sqrt{C/J}$ and $\lambda_3 = 3$ so that $\omega_3 = \sqrt{3C/J}$. Therefore we may write

$$\theta_i = A_i + B_i t + E_i \cos \sqrt{C/J}\, t + F_i \sin \sqrt{C/J}\, t + G_i \cos \sqrt{3C/J}\, t + H_i \sin \sqrt{3C/J}\, t$$

where A_i, B_i, E_i, F_i, G_i and H_i are constants to be determined by the initial displacements and velocities of each disk. (Explain the presence of the B_i term.)

Example

For the matrix $\mathbf{D} = \begin{bmatrix} 4 & 2 & -2 \\ 1 & 3 & 1 \\ -1 & -1 & 5 \end{bmatrix}$ we found eigenvalues 2, 4, 6.

(i) $\lambda = 2$ We solve the equations $(\mathbf{D} - \lambda\mathbf{I})\mathbf{x} = \mathbf{0}$ i.e.

$$\begin{bmatrix} 2 & 2 & -2 \\ 1 & 1 & 1 \\ -1 & -1 & 3 \end{bmatrix}\begin{bmatrix} x_1 \\ x_2 \\ x_3 \end{bmatrix} = \begin{bmatrix} 0 \\ 0 \\ 0 \end{bmatrix}$$

The first row of $(\mathbf{D} - \lambda\mathbf{I})$ is equal to the third row subtracted from the second. Let us concentrate on rows 2 and 3 and work with the equations

$$x_1 + x_2 + x_3 = 0 \qquad -x_1 - x_2 + 3x_3 = 0$$

Adding these equations produces $4x_3 = 0$ and we are left with the single equation $x_1 + x_2 = 0$. A typical eigenvector is $(1, -1, 0)$.

(ii) $\lambda = 4$ The equations to be solved are

$$2x_2 - 2x_3 = 0 \qquad x_1 - x_2 + x_3 = 0 \qquad -x_1 - x_2 + x_3 = 0$$

We therefore obtain $x_2 = x_3$ and hence, $x_1 = 0$. A typical eigenvector is $(0, 1, 1)$.

(iii), $\lambda = 6$ The equations become

$$-2x_1 + 2x_2 - 2x_3 = 0 \qquad x_1 - 3x_2 + x_3 = 0 \qquad -x_1 - x_2 - x_3 = 0$$

Working with the first and third equations we may subtract twice the third from the first to find $4x_2 = 0$ and then deduce $x_1 + x_3 = 0$. A typical eigenvector is $(1, 0, -1)$.

Comparison of systems

The governing equations for the system of Figure 2.3 were found to be $\mathbf{A}\boldsymbol{\theta} = \lambda\boldsymbol{\theta}$ where $\boldsymbol{\theta}$ was the vector of angular displacements, $\mathbf{A}$ was the matrix $\begin{bmatrix} 1 & -1 & 0 \\ -1 & 2 & -1 \\ 0 & -1 & 1 \end{bmatrix}$ and $\lambda = J\omega^2/C$.

In this system the parameters were moments of inertia J and torsional stiffness C. The coordinates were angles θ_i and the critical angular frequencies ω_i had to be determined. We now examine two related systems.

(i) Linear train (Figure 2.5(a))
If we choose as coordinates x_1, x_2 and x_3 – the displacements of the masses from their equilibrium positions – then the governing equation of motion can be written as $\mathbf{A}\mathbf{x} = \lambda\mathbf{x}$ where $\mathbf{x}$ is the vector of linear displacements (x_1, x_2, x_3), $\lambda = M\omega^2/k$ and $\mathbf{A}$ is as before. The system parameters are masses M and spring stiffness k. This time it is frequencies ω which are critical.

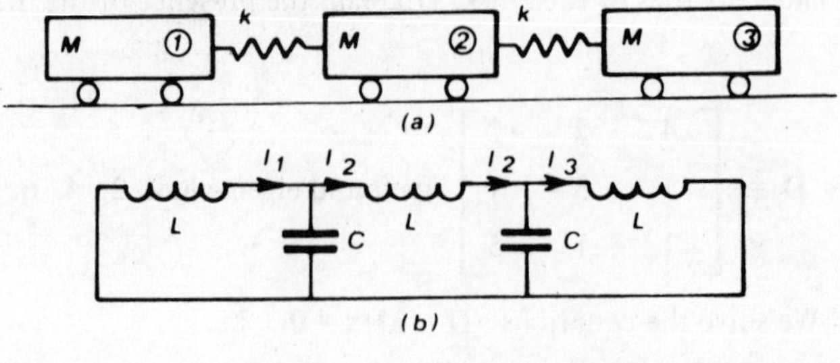

(ii) Electrical Network

For the electrical circuit shown in Figure 2.5(b) we may write the governing equation $\mathbf{AI} = \lambda\mathbf{I}$ where $d^2I_1/dt^2 = -\omega^2 I_1$ etc., $\lambda = \omega^2 CL$ and $\mathbf{I} = (I_1, I_2, I_3)$. The system parameters are inductances L and capacitances C.

All the systems were lumped-parameter approximations of physical situations.

Problems

1. Find the eigenvalues and eigenvectors of the following matrices.

(i) $\begin{bmatrix} 4 & 2 \\ -1 & 1 \end{bmatrix}$ (ii) $\begin{bmatrix} 1 & 0 & 0 \\ 1 & 2 & 0 \\ 2 & -2 & 3 \end{bmatrix}$ (iii) $\begin{bmatrix} 2 & 2 & 1 \\ 1 & 3 & 1 \\ 1 & 2 & 2 \end{bmatrix}$ (iv) $\begin{bmatrix} 1 & 0 & -1 \\ 1 & 2 & 1 \\ 2 & 2 & 3 \end{bmatrix}$

(v) $\begin{bmatrix} 2 & 2 & 0 \\ 2 & 2 & 0 \\ 0 & 0 & 1 \end{bmatrix}$ (vi) $\begin{bmatrix} 2 & 1 & 1 \\ 1 & 2 & 1 \\ 0 & 0 & 1 \end{bmatrix}$ (vii) $\begin{bmatrix} 1 & -1 & -1 \\ 1 & -1 & 0 \\ 1 & 0 & 1 \end{bmatrix}$ (viii) $\begin{bmatrix} -2 & -8 & -12 \\ 1 & 4 & 4 \\ 0 & 0 & 1 \end{bmatrix}$

(ix) $\begin{bmatrix} 1 & 1 & -2 \\ -1 & 2 & 1 \\ 0 & 1 & -1 \end{bmatrix}$ (x) $\begin{bmatrix} 3 & 2 & 2 & -4 \\ 2 & 3 & 2 & -1 \\ 1 & 1 & 2 & -1 \\ 2 & 2 & 2 & -1 \end{bmatrix}$

2. (i) Indicate briefly some physical examples of eigenvalue problems.

(ii) Construct the matrix **A** whose eigenvalues are 0, 1, 3, with corresponding eigenvectors

$$\begin{bmatrix} 1 \\ 0 \\ 1 \end{bmatrix}, \quad \begin{bmatrix} 0 \\ 3 \\ 0 \end{bmatrix}, \quad \begin{bmatrix} 1 \\ 1 \\ 2 \end{bmatrix}.$$

Verify your method by determining the eigenvalues and eigenvectors of the matrix **A**. (C.E.I.)

3. Find the eigenvalues and eigenvectors of the system

$$2x_1 - 2x_2 + 3x_3 - \lambda x_1 = 0; \quad x_1 + x_2 + x_3 - \lambda x_2 = 0; \quad x_1 + 3x_2 - x_3 - \lambda x_3 = 0$$

2.4 FURTHER RESULTS ON EIGENVALUES

We now develop some results concerning eigenvalues and eigenvectors which will be useful in later work.

Miscellaneous results

(i) It is obvious that **A** must be square in order that it should possess eigenvalues.

(ii) If the matrix **A** is singular then $|\mathbf{A}| = 0$ and $\lambda = 0$ is one eigenvalue of **A**.

(iii) If **A** is a lower triangular matrix, then the eigenvalues of **A** are simply the diagonal elements; a similar result holds for an upper triangular matrix.

(iv) The eigenvalues of $\mathbf{A}^T$ are those of **A**: this follows by considering $|\mathbf{A}^T - \lambda\mathbf{I}|$.

(v) The product of the eigenvalues of **A** is $|\mathbf{A}|$.

(vi) The sum of the eigenvalues of $\mathbf{A}$ is the *trace* of $\mathbf{A}$, i.e. the sum of the diagonal elements.

(These last two results form useful checks on numerical estimates of the eigenvalues of a matrix.)

Eigenvalues of a symmetric matrix

Let $\mathbf{A}$ be a **Hermitian** matrix so that $\overline{\mathbf{A}} = \mathbf{A}^T$, where $\overline{\mathbf{A}}$ is the matrix formed by replacing each element of $\mathbf{A}$ by its complex conjugate. Then let λ be an eigenvalue of $\mathbf{A}$ and $\mathbf{x}$ the corresponding eigenvector so that $\mathbf{Ax} = \lambda\mathbf{x}$. Taking transposes we obtain $\mathbf{x}^T\mathbf{A} = \lambda\mathbf{x}^T$. We then take complex conjugates to produce the equation $\overline{\mathbf{x}}^T\overline{\mathbf{A}} = \overline{\lambda}\overline{\mathbf{x}}^T$. Then we take the scalar product of both sides of this equation with $\mathbf{x}$ to give $\overline{\mathbf{x}}^T\overline{\mathbf{A}}\mathbf{x} = \overline{\lambda}\overline{\mathbf{x}}^T\mathbf{x}$. But $\mathbf{Ax} = \lambda\mathbf{x}$ so that $\overline{\mathbf{x}}^T\,\overline{\mathbf{A}}^T\,\mathbf{x} = \overline{\mathbf{x}}^T\,\mathbf{Ax} = \lambda\,\overline{\mathbf{x}}^T\,\mathbf{x}$. Hence $\lambda\,\overline{\mathbf{x}}^T\,\mathbf{x} = \overline{\lambda}\,\overline{\mathbf{x}}^T\,\mathbf{x}$. Now $\overline{\mathbf{x}}^T\,\mathbf{x} \neq 0$ unless $\mathbf{x} = \mathbf{0}$ which is not a valid eigenvector. Therefore $\lambda = \overline{\lambda}$, i.e. the eigenvalues are real.

Similar matrices

Let $\mathbf{A}$ and $\mathbf{B}$ be two matrices related by the equation $\mathbf{B} = \mathbf{P}^{-1}\,\mathbf{AP}$ where $\mathbf{P}$ is a non-singular matrix, then $\mathbf{A}$ and $\mathbf{B}$ are said to be **similar.** Further let $\mathbf{Ax} = \lambda\mathbf{x}$ where λ and $\mathbf{x}$ are an eigenvalue of $\mathbf{A}$ and its associated eigenvector.

$$\begin{aligned}\text{Now } |\mathbf{B} - \lambda\mathbf{I}| &= |\mathbf{P}^{-1}\mathbf{AP} - \lambda\mathbf{I}| = |\mathbf{P}^{-1}\mathbf{AP} - \lambda\mathbf{P}^{-1}\mathbf{IP}| \\ &= |\mathbf{P}^{-1}(\mathbf{A} - \lambda\mathbf{I})\mathbf{P}| = |\mathbf{P}^{-1}|\,.\,|\mathbf{A} - \lambda\mathbf{I}|\,.\,|\mathbf{P}| = \frac{1}{|\mathbf{P}|}.|\mathbf{A} - \lambda\mathbf{I}|\,.\,|\mathbf{P}| \\ &= |\mathbf{A} - \lambda\mathbf{I}|.\end{aligned}$$

It therefore follows that two similar matrices have the same eigenvalues. If we can find a simpler matrix $\mathbf{B}$, especially a diagonal matrix, which is similar to $\mathbf{A}$, then the determination of the eigenvalues could become easier.

Example

We saw that the eigenvalues of $\mathbf{A} = \begin{bmatrix} 1 & -1 & 0 \\ -1 & 2 & -1 \\ 0 & -1 & 1 \end{bmatrix}$ were 0, 1, and 3. Let

$\mathbf{P} = \begin{bmatrix} 1 & 1 & 1 \\ 1 & 0 & -2 \\ 1 & -1 & 1 \end{bmatrix}$; we can show that $\mathbf{P}^{-1} = \frac{1}{6}\begin{bmatrix} 2 & 2 & 2 \\ 3 & 0 & -3 \\ 1 & -2 & 1 \end{bmatrix}$ and that

$\mathbf{P}^{-1}\,\mathbf{AP} = \begin{bmatrix} 0 & 0 & 0 \\ 0 & 1 & 0 \\ 0 & 0 & 3 \end{bmatrix} = \mathbf{B}$. From result (iii), the eigenvalues of $\mathbf{B}$ are 0, 1 and 3.

Powers of a matrix

If a matrix $\mathbf{A}$ has an eigenvalue λ then $\mathbf{A}^m$ has an eigenvalue λ^m for integer m. The proof starts by considering $\mathbf{Ax} = \lambda\mathbf{x}$. Then $\mathbf{A}^2\mathbf{x} = \mathbf{A}\lambda\mathbf{x} = \lambda\mathbf{Ax} = \lambda^2\mathbf{x}$ and the result for positive integers m follows by induction.

Note that if $\mathbf{A}$ is non-singular then the eigenvalues of $\mathbf{A}^{-1}$ are the reciprocals of the eigenvalues of $\mathbf{A}$.

In all cases, corresponding eigenvalues share the same eigenvector.

Example

The eigenvalues of $\mathbf{A} = \begin{bmatrix} -5 & 2 \\ 2 & -2 \end{bmatrix}$ are $\lambda = -6$ and -1. For $\lambda = -6$ we have an eigenvector $\begin{bmatrix} -2 \\ 1 \end{bmatrix}$ and for $\lambda = -1$ we have an eigenvector $\begin{bmatrix} 1 \\ 2 \end{bmatrix}$.

Now $\mathbf{A}^2 = \begin{bmatrix} 29 & -14 \\ -14 & 8 \end{bmatrix}$ (note $\mathbf{A}^2$ is also symmetric).

$|\mathbf{A}^2 - \lambda\mathbf{I}| = \begin{vmatrix} 29-\lambda & -14 \\ -14 & 8-\lambda \end{vmatrix} = 0$ produces $232 - 37\lambda + \lambda^2 - 196 = 0$

i.e. $36 - 37\lambda + \lambda^2 = 0$ which has roots $\lambda = 1 = (-1)^2$ and $\lambda = 36 = (-6)^2$.

Now $(\mathbf{A}^2 - \mathbf{I})\mathbf{x} = 0$ becomes $28x_1 - 14x_2 = 0 \quad -14x_1 + 7x_2 = 0$

so that a typical eigenvector is $\begin{bmatrix} 1 \\ 2 \end{bmatrix}$. Similarly, we can show that a typical eigenvector corresponding to $\lambda = 36$ is $\begin{bmatrix} -2 \\ 1 \end{bmatrix}$. Note that these eigenvectors are **orthogonal** to each other, i.e. their scalar product is zero. We now prove the result generally.

Theorem

If $\mathbf{x}_1$ and $\mathbf{x}_2$ are two eigenvectors of a symmetric matrix $\mathbf{A}$ corresponding to distinct eigenvalues λ_1, λ_2 respectively then $\mathbf{x}_1 \cdot \mathbf{x}_2 = 0$.

Proof

Let $\mathbf{A}\mathbf{x}_1 = \lambda_1\mathbf{x}_1$ and $\mathbf{A}\mathbf{x}_2 = \lambda_2\mathbf{x}_2$. Then $\mathbf{x}_1^T\mathbf{A}\mathbf{x}_2 = \lambda_2\mathbf{x}_1^T\mathbf{x}_2$

i.e. $\mathbf{x}_1^T\mathbf{A}^T\mathbf{x}_2 = \lambda_2\mathbf{x}_1^T\mathbf{x}_2$ (since $\mathbf{A}$ is symmetric) i.e. $(\mathbf{A}\mathbf{x}_1)^T\mathbf{x}_2 = \lambda_2\mathbf{x}_1^T\mathbf{x}_2$

i.e. $(\lambda_1\mathbf{x}_1)^T\mathbf{x}_2 = \lambda_2\mathbf{x}_1^T\mathbf{x}_2$ or $\lambda_1\mathbf{x}_1^T\mathbf{x}_2 = \lambda_2\mathbf{x}_1^T\mathbf{x}_2$

Now if $\lambda_1 \neq \lambda_2$ it follows that $\mathbf{x}_1^T\mathbf{x}_2 = 0$ as required.

Equal eigenvalues

Consider the matrix $\mathbf{C} = \begin{bmatrix} 1 & 0 & -2 \\ 0 & 0 & 1 \\ -2 & 0 & 4 \end{bmatrix}$

Then $|\mathbf{C} - \lambda\mathbf{I}| = \begin{vmatrix} 1-\lambda & 0 & -2 \\ 0 & -\lambda & 1 \\ -2 & 0 & 4-\lambda \end{vmatrix} = -(1-\lambda)\lambda(4-\lambda) - 2(-2\lambda) = -\lambda^3 + 5\lambda^2$.

The eigenvalues are $\lambda = 0$ (twice) and $\lambda = 5$.

For $\lambda = 5$ we solve $-4x_1 - 2x_3 = 0 \quad -5x_2 + x_3 = 0 \quad -2x_1 - x_3 = 0$ to obtain a typical eigenvector $\begin{bmatrix} 5 \\ -2 \\ -10 \end{bmatrix}$.

For $\lambda = 0$ we have the equations $x_1 - 2x_3 = 0 \quad x_3 = 0 \quad -2x_1 + 4x_3 = 0$ and we obtain a typical eigenvector $\begin{bmatrix} 0 \\ 1 \\ 0 \end{bmatrix}$.

Now consider the matrix $\mathbf{D} = \begin{bmatrix} 1 & 0 & -2 \\ 0 & 0 & 0 \\ -2 & 0 & 4 \end{bmatrix}$. It should be clear that the eigenvalues of $\mathbf{D}$ are the same as those of $\mathbf{C}$. However for $\lambda = 5$ we now have to solve the equations

$$-4x_1 - 2x_3 = 0 \qquad -5x_2 = 0 \qquad -2x_1 - x_3 = 0$$

Hence a typical eigenvector is $\begin{bmatrix} 1 \\ 0 \\ -2 \end{bmatrix}$. For $\lambda = 0$ we have the equations $x_1 - 2x_3 = 0$, $-2x_1 + 4x_3 = 0$. We now have only one equation for three unknowns: this leaves us two degrees of freedom. Any vector of the form $\begin{bmatrix} 2\alpha \\ \beta \\ \alpha \end{bmatrix}$ will do for the eigenvector. We shall choose two vectors from this system so that as far as possible we copy the unit vectors $\begin{bmatrix} 1 \\ 0 \\ 0 \end{bmatrix}$, $\begin{bmatrix} 0 \\ 1 \\ 0 \end{bmatrix}$ and $\begin{bmatrix} 0 \\ 0 \\ 1 \end{bmatrix}$. Now the eigenvector $\begin{bmatrix} 1 \\ 0 \\ -2 \end{bmatrix}$ is close to the first unit vector.

As it happens $\begin{bmatrix} 0 \\ 1 \\ 0 \end{bmatrix}$ will do as a second eigenvector. Now we cannot quite copy $\begin{bmatrix} 0 \\ 0 \\ 1 \end{bmatrix}$ but $\alpha = 1$ and $\beta = 0$ gives a third eigenvector $\begin{bmatrix} 2 \\ 0 \\ 1 \end{bmatrix}$. Note that the three eigenvectors $\begin{bmatrix} 1 \\ 0 \\ -2 \end{bmatrix}$, $\begin{bmatrix} 0 \\ 1 \\ 0 \end{bmatrix}$, $\begin{bmatrix} 2 \\ 0 \\ 1 \end{bmatrix}$ are **mutually orthogonal.**

Why do you think that in the case of matrix $\mathbf{C}$ there was only one degree of freedom for the set of eigenvectors associated with the repeated eigenvalue $\lambda = 0$, whereas in the case of matrix $\mathbf{D}$ the eigenvalue $\lambda = 0$ has a class of eigenvectors with two degrees of freedom?

Transformation of coordinates

In Figure 2.6 we show a unit square before and after a shear transformation applied in the x - direction.

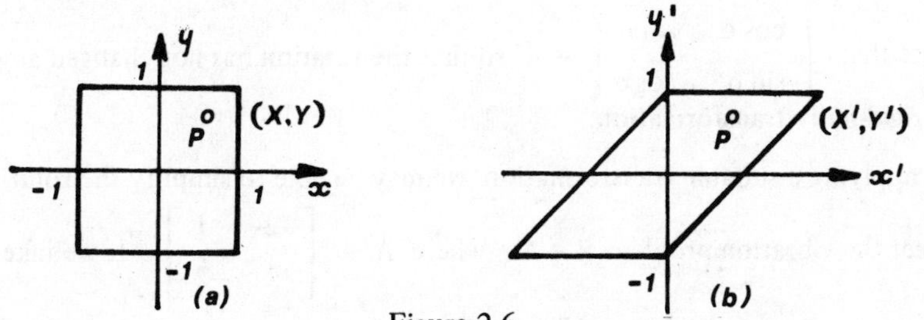

Figure 2.6

The point P has original coordinates (X, Y) and new coordinates (X', Y'). These coordinates are related by $\begin{bmatrix} X' \\ Y' \end{bmatrix} = \begin{bmatrix} 1 & 1 \\ 0 & 1 \end{bmatrix} \begin{bmatrix} X \\ Y \end{bmatrix}$. Now the eigenvalues of the matrix $\begin{bmatrix} 1 & 1 \\ 0 & 1 \end{bmatrix}$ are $\lambda = 1$ (twice). A typical eigenvalue, found by solving $y = 0$ is $\begin{bmatrix} 1 \\ 0 \end{bmatrix}$. The interpretation of this result is that points on the y - axis are the only points for which the vectors connecting them to the origin remain unchanged in direction. These points keep the same distance from the origin $(\lambda = 1)$.

Now consider the matrix $\mathbf{D} = \begin{bmatrix} 2 & 0 \\ 0 & 1/3 \end{bmatrix}$. You should be able to show that this represents the transformation shown in Figure 2.7.

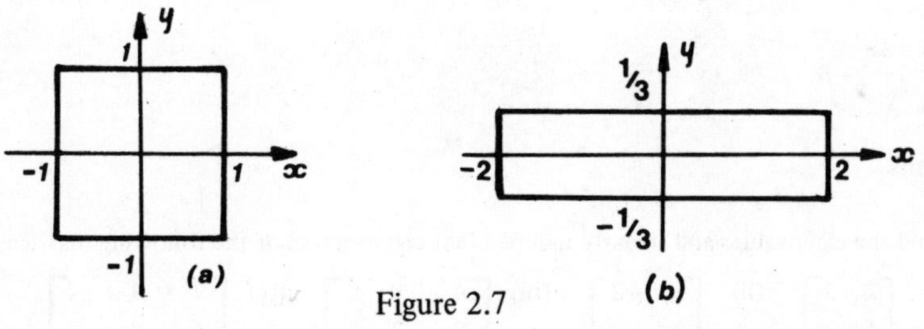

Figure 2.7

The eigenvalues of $\mathbf{D}$ are $\lambda = 2$ and $\lambda = 1/3$. Corresponding to $\lambda = 2$ we have an eigenvector $\begin{bmatrix} 1 \\ 0 \end{bmatrix}$ and corresponding to $\lambda = 1/3$ we have $\begin{bmatrix} 0 \\ 1 \end{bmatrix}$. Geometrically, this means that points on the x - axis have their distance from the origin doubled, whereas points on the y - axis are brought closer to the origin (by a factor of 3). It is interesting to note that $|\mathbf{D}| = 2/3$ which is the ratio of the new area to the area of the square.

Note that the orthogonal matrix $\begin{bmatrix} \cos\alpha & \sin\alpha \\ -\sin\alpha & \cos\alpha \end{bmatrix}$ represents a rotation about the origin through an angle α. The eigenvalues are given by

$\cos^2\alpha - 2\cos\alpha\lambda + \lambda^2 + \sin^2\alpha = 0$ i.e. $\lambda = \cos\alpha \pm \sqrt{\cos^2\alpha - 1} = \cos\alpha \pm i\sin\alpha = e^{i\alpha}$ or $e^{-i\alpha}$

Both eigenvalues are of modulus 1, so that points stay the same distance from the origin after the transformation as they were before it.

Note that $\begin{vmatrix} \cos\alpha & \sin\alpha \\ -\sin\alpha & \cos\alpha \end{vmatrix} = 1$ so that the rotation has not changed any areas: it is a **rigid-body transformation.**

By applying a suitable transformation, we may be able to simplify the solution.

Consider the vibration problem $\ddot{\mathbf{x}} = \mathbf{Ax}$ where $\mathbf{A} = \begin{bmatrix} -2 & 1 \\ 1 & -2 \end{bmatrix}$. If we take

$\mathbf{P} = \begin{bmatrix} 1 & -1 \\ 1 & 1 \end{bmatrix}$ then $\mathbf{P}^{-1} = \tfrac{1}{2}\begin{bmatrix} 1 & 1 \\ -1 & 1 \end{bmatrix}$. (Note that $\mathbf{P}^{-1}$ is a multiple of $\mathbf{P}^{\mathrm{T}}$.)

It follows that $\mathbf{P}^{-1}\mathbf{AP} = \begin{bmatrix} -1 & 0 \\ 0 & -3 \end{bmatrix}$. If we then transform to coordinates $\mathbf{y} = \mathbf{P}^{-1}\mathbf{x}$

the problem becomes $\ddot{\mathbf{y}} = \mathbf{P}^{-1}\ddot{\mathbf{x}} = \mathbf{P}^{-1}\mathbf{Ax} = (\mathbf{P}^{-1}\mathbf{AP})\mathbf{y}$ i.e. $\begin{bmatrix} \ddot{y}_1 \\ \ddot{y}_2 \end{bmatrix} = \begin{bmatrix} -1 & 0 \\ 0 & -3 \end{bmatrix}\mathbf{y}$ which has solutions $\ddot{y}_1 = -y_1$, $\ddot{y}_2 = -3y_2$.

The solution of these equations is $y_1 = A\cos t + B\sin t$ and $y_2 = C\cos\sqrt{3}\,t + D\sin\sqrt{3}\,t$.
In terms of the old coordinates we have, by using $\mathbf{x} = \mathbf{Py}$

$$x_1 = A\cos t + B\sin t - (C\cos\sqrt{3}\,t + D\sin\sqrt{3}\,t)$$
$$x_2 = A\cos t + B\sin t + (C\cos\sqrt{3}\,t + D\sin\sqrt{3}\,t)$$

which are both linear combinations of the normal modes. The **normal** or **natural coordinates** y_1 and y_2 reflect the normal modes of vibration.

Problems

1. Find the eigenvalues and linearly independent eigenvectors of the following matrices.

(i) $\begin{bmatrix} 2 & 5 \\ 4 & 3 \end{bmatrix}$ (ii) $\begin{bmatrix} 3 & -2 \\ 2 & 3 \end{bmatrix}$ (iii) $\begin{bmatrix} 4 & 0 & 0 \\ 9 & -2 & 0 \\ 0 & 8 & 7 \end{bmatrix}$ (iv) $\begin{bmatrix} 7 & 4 & -4 \\ 4 & -8 & -1 \\ 4 & -1 & -8 \end{bmatrix}$

2. Find the eigenvalues and corresponding eigenvectors of $\begin{bmatrix} 2 & 2 & 0 \\ 2 & 2 & 0 \\ 0 & 0 & 1 \end{bmatrix}$. Verify that the eigenvectors are mutually orthogonal and comment on the reason for this.

3. If $\mathbf{C} = \begin{bmatrix} 1 & 0 & 1 \\ 0 & 2 & 2 \\ 1 & 2 & 3 \end{bmatrix}$ and $\mathbf{x}$ is a column vector, show that the equation $\mathbf{Cx} = \lambda\mathbf{x}$ has solutions in which $\mathbf{x} \neq \mathbf{0}$, for each of three different values of λ. Find the corresponding solutions, and if these are denoted by $\mathbf{x}_1, \mathbf{x}_2, \mathbf{x}_3$, show that $\mathbf{x}_1^T\mathbf{x}_2 = \mathbf{x}_2^T\mathbf{x}_3 = \mathbf{x}_3^T\mathbf{x}_1 = \mathbf{0}$. (L.U.)

4. Find the characteristic roots of the matrix $\begin{bmatrix} 2 & 1 & 0 \\ 1 & 2 & 1 \\ 0 & 1 & 2 \end{bmatrix}$. Find also a characteristic vector for each root and verify that these vectors are mutually perpendicular.

5. Find the eigenvalues of the matrix $\begin{bmatrix} 3 & 2 & 3 \\ -1 & 0 & -3 \\ 1 & -2 & 1 \end{bmatrix}$ and the eigenvector corresponding to the smaller of the two positive eigenvalues.

6. Show that the latent vectors of a real symmetric matrix, corresponding to distinct latent roots, are orthogonal.

   The 3 x 3 symmetric matrix $\mathbf{A}$ has latent roots 3, 6, −9. The vector (−2, 2, 1) corresponds to the root 3 and the vector (1, 2, −2) corresponds to the root −9. Find a latent vector corresponding to the root 6 and calculate $\mathbf{A}$ from the relation $\mathbf{P} = \mathbf{MAM}^{-1}$ where $\mathbf{M}$ is a matrix having the latent vectors as its columns and $\mathbf{P}$ is the diagonal matrix of latent roots. (L.U.)

7. Given the matrices $\mathbf{A}_1 = \begin{bmatrix} 1 & -3 & 3 \\ 3 & -5 & 3 \\ 6 & -6 & 4 \end{bmatrix}$ and $\mathbf{A}_2 = \begin{bmatrix} -3 & 1 & -1 \\ -7 & 5 & -1 \\ -6 & 6 & -2 \end{bmatrix}$ show that $\mathbf{A}_1$ and $\mathbf{A}_2$ have the same characteristic polynomial but different eigenvectors. Show which of these matrices can be diagonalised by a transformation of the form $\mathbf{P}^{-1}\mathbf{AP} = \mathbf{D}$. (C.E.I.)

8. The matrix $\mathbf{B} = \begin{bmatrix} 1 & 0 & 0 \\ 0 & 1 & 0 \\ 1 & 1 & 2 \end{bmatrix}$ has characteristic equation $(\lambda - 1)^2(\lambda - 2) = 0$; find three linearly independent eigenvectors for $\mathbf{B}$.

9. The Cayley - Hamilton Theorem states that any square matrix satisfies its own characteristic polynomial. Verify for $\begin{bmatrix} 2 & 3 & -5 \\ -3 & 1 & 2 \\ 1 & -3 & 4 \end{bmatrix}$.

10. Repeat Problem 9 for the matrices of Problem 1, Section 2.4.

11. Use the Cayley - Hamilton Theorem to find $\mathbf{A}^3$, $\mathbf{A}^4$ and $\mathbf{A}^{-1}$ for the matrix $\mathbf{A}$ given by

   (i) $\begin{bmatrix} 1 & 2 \\ 1 & 1 \end{bmatrix}$ (ii) $\begin{bmatrix} 1 & 1 & 2 \\ 3 & 1 & 1 \\ 2 & 3 & 1 \end{bmatrix}$ (iii) $\begin{bmatrix} 3 & 1 & -2 \\ 2 & 4 & -4 \\ 2 & 1 & -1 \end{bmatrix}$

12. A system of vibrating particles satisfies the differential equations:

$$\frac{d^2x}{dt^2} = -5x - y - z; \quad \frac{d^2y}{dt^2} = -x - 3y - z; \quad \frac{d^2z}{dt^2} = -x - y - 3z$$

Using a trial solution of the form

$x = x_0 \sin \omega t$; $y = y_0 \sin \omega t$; $z = z_0 \sin \omega t$

show that ω^2 is an eigenvalue and (x_0, y_0, z_0) an eigenvector of a matrix **A**. Hence find three independent solutions to the equations.

13. Let the distinct eigenvalues of an $n \times n$ matrix **A** be $\lambda_1, \lambda_2, \ldots, \lambda_n$. If there exists a non-singular matrix **P** such that $\mathbf{C} = \mathbf{P}^{-1}\mathbf{AP}$ is a matrix whose only non-zero elements are its diagonal ones which are the eigenvalues of **A**, then we may define matrices $\mathbf{B}_i$ whose only non-zero element is a 1, corresponding to λ_i in **C**. Further, let $\mathbf{E}_i = \mathbf{PB}_i\mathbf{P}^{-1}$; then we may write $\mathbf{A} = \lambda_1\mathbf{E}_1 + \lambda_2\mathbf{E}_2 + \ldots + \lambda_n\mathbf{E}_n$; this is called the *spectral decomposition of* **A**. (The definition can be modified if the eigenvectors are not all distinct.) Obtain this decomposition for matrices (i), (iv) and (v) of Problem 1, Section 2.3.

2.5 QUADRATIC FORMS AND THEIR REDUCTION

An expression such as $3x_1^2 + 4x_1x_2 - x_2^2$ is known as a **quadratic form**; in general, a quadratic form in $x_1, x_2, \ldots, x_n$ is a polynomial in these variables where every term is of the form $a_{ij}x_ix_j$ where the a_{ij} are constants. Quadratic forms occur in many branches of mathematics as the following examples show.

Kinetic and potential energies. With reference to the linear train of Figure 2.5(a), the kinetic energy of the system is $T = \frac{1}{2}M(\dot{x}_1^2 + \dot{x}_2^2 + \dot{x}_3^2)$ and the potential energy of the system is

$$V = \tfrac{1}{2}k(x_2 - x_1)^2 + \tfrac{1}{2}k(x_3 - x_2)^2 = \tfrac{1}{2}kx_1^2 + kx_2^2 + \tfrac{1}{2}kx_3^2 - kx_1x_2 - kx_2x_3.$$

We see that V is a quadratic form in x_1, x_2 and x_3 while T is a quadratic form in $\dot{x}_1$, $\dot{x}_2$ and $\dot{x}_3$.

Equation of a conic. The general equation of a conic is

$$a_{11}x_1^2 + 2a_{12}x_1x_2 + a_{22}x_2^2 + b_1x_1 + b_2x_2 + c_2 = 0$$

The last three terms may be removed by translating the origin of coordinates to the centre of the conic, which will then have the form

$$a_{11}x_1^2 + 2a_{12}x_1x_2 + a_{22}x_2^2 = 0$$

The nature of the conic will be determined by the nature of this quadratic form.

Approximation of a function near a stationary point. Near a stationary point a function $f(x, y)$ may be approximated by

$$f(x_0 + h, y_0 + k) = f(x_0, y_0) + \tfrac{1}{2}h^2 f_{xx}(x_0, y_0) + hkf_{xy}(x_0, y_0) + \tfrac{1}{2}k^2 f_{yy}(x_0, y_0)$$

The nature of the stationary point may be determined by

$$\delta f \equiv f(x_0 + h, y_0 + k) - f(x_0, y_0) = \tfrac{1}{2}h^2 f_{xx}(x_0, y_0) + hkf_{xy}(x_0, y_0) + \tfrac{1}{2}k^2 f_{yy}(x_0, y_0).$$

Since the second partial derivatives are evaluated at (x_0, y_0), δf is given approximately by a quadratic form in h and k. If this quadratic form is **positive definite** (i.e. it takes zero value if h and k are both zero, otherwise it takes positive values), then the stationary point is a local minimum (why?).

Associated symmetric matrix

We can abstract the essential information about a quadratic form into an associated matrix. For example the quadratic form

$$3x_1^2 + 4x_1x_2 - x_2^2 \quad \text{may be written as} \quad (x_1, x_2)\begin{bmatrix} 3 & 2 \\ 2 & -1 \end{bmatrix}\begin{bmatrix} x_1 \\ x_2 \end{bmatrix}$$

as you can verify by multiplication.

However, the representation is not unique; for example, the representations

$$(x_1, x_2)\begin{bmatrix} 3 & 4 \\ 0 & -1 \end{bmatrix}\begin{bmatrix} x_1 \\ x_2 \end{bmatrix}, \quad (x_1, x_2)\begin{bmatrix} 3 & 0 \\ 4 & -1 \end{bmatrix}\begin{bmatrix} x_1 \\ x_2 \end{bmatrix} \quad \text{and} \quad (x_1, x_2)\begin{bmatrix} 3 & 3 \\ 1 & -1 \end{bmatrix}\begin{bmatrix} x_1 \\ x_2 \end{bmatrix}$$

are equally valid. We usually choose that matrix which is symmetric. In this example, the **associated symmetric matrix** is $\begin{bmatrix} 3 & 2 \\ 2 & -1 \end{bmatrix}$.

Let us consider an example in three variables. To settle on the associated symmetric matrix we notice that the diagonal elements are the coefficients of the squared terms and the off-diagonal elements share equally the coefficients of the cross terms. For example, the elements in the positions Row 2, Column 1 and Row 1, Column 2 are each equal to half the coefficient of $x_1 x_2$. Bear in mind that the coefficients may be negative.

The quadratic form $x_1^2 - 2x_2^2 + 3x_3^2 - 6x_1x_2 - 2x_1x_3 + 4x_2x_3$ has associated symmetric matrix $\begin{bmatrix} 1 & -3 & -1 \\ -3 & -2 & 2 \\ -1 & 2 & 3 \end{bmatrix}$.

Finally, we quote an example in four variables; notice there is no term in $u_2 u_4$ nor in u_3^2. The quadratic form $2u_1^2 + 3u_2^2 - u_4^2 - u_1u_2 + 5u_1u_3 + \frac{1}{2}u_1u_4 - 8u_2u_3 + 2u_3u_4$ has associated symmetric matrix $\begin{bmatrix} 2 & -\frac{1}{2} & 2\frac{1}{2} & \frac{1}{4} \\ -\frac{1}{2} & 3 & -4 & 0 \\ 2\frac{1}{2} & -4 & 0 & 1 \\ \frac{1}{4} & 0 & 1 & -1 \end{bmatrix}$. The general formulation of a quadratic form is $\mathbf{x}^T\mathbf{A}\mathbf{x}$, where $\mathbf{x}$ is the vector of variables and $\mathbf{A}$ is the associated matrix.

Nature of a quadratic form

We shall be interested in determining the nature of a quadratic form. The two particular forms in which we are especially interested are positive definite and negative definite forms. The formal definitions are as follows; it is assumed obvious that any quadratic form takes a zero value if all its constituent variables are zero.

The form $Q = \mathbf{x}^T\mathbf{A}\mathbf{x}$ is **positive definite** if $Q > 0$ for all $\mathbf{x} \neq \mathbf{0}$.

The form $Q = \mathbf{x}^T\mathbf{A}\mathbf{x}$ is **negative definite** if $Q < 0$ for all $\mathbf{x} \neq \mathbf{0}$.

Now it is easy to see that the form $3x_1^2 + 4x_2^2 + x_3^2$ is positive definite, that the form $-x_1^2 - 2x_2^2 - 5x_3^2$ is negative definite and that the form $2x_1^2 - x_2^2 + 3x_3^2$ is neither since it can take both positive and negative values (for example $x_1 = 1$, $x_2 = 2$, $x_3 = 1$ gives a positive value whereas $x_1 = 1$, $x_2 = 14$, $x_3 = 1$ gives a negative value.) The reason we are able to determine their nature so easily is that each of these forms is comprised only of squared terms, i.e. the associated matrix is **diagonal**. Now if we could somehow reduce a quadratic form to a sum of squares or, equivalently, find a diagonal matrix related to the original symmetric matrix in such a way that the nature of the quadratic form is unchanged then we shall be able to determine that nature easily.

First we try an ad hoc approach. Consider $Q_1 = x_1^2 + 3x_1x_2 + 5x_2^2$. Now let us try and write Q_1 as a perfect square. It is unlikely that this will be possible, so that we shall have to be content with the compromise of a square term plus a remainder. Hence $Q_1 = (x_1 + \frac{3}{2}x_2)^2 + \frac{11}{4}x_2^2$ and we see that it is positive definite. The form $Q_2 = x_1^2 + 3x_1x_2 + 2x_2^2$ can be written as $(x_1 + \frac{3}{2}x_2)^2 - \frac{1}{4}x_2^2$ and is hence capable of taking both positive and negative values. Finally, the form $Q_3 = x_1^2 + 3x_1x_2 + \frac{9}{4}x_2^2$ can be written exactly as $(x_1 + \frac{3}{2}x_2)^2$; this can take zero values if $x_1 = (-3/2)x_2$ as well as if $x_1 = x_2 = 0$ otherwise it takes positive values. This form is called **positive semi-definite**.

You might well argue that we could equally well have started the process from the 'right-hand end' of each quadratic form and obtained a different result. Suppose we write

$$Q_2 = 2x_2^2 + 3x_1x_2 + x_1^2 = 2(x_2 + \frac{3}{4}x_1)^2 - \frac{1}{8}x_1^2$$

The important point is that the **nature** of the form is revealed to be the same as before. Check Q_1 and Q_3. Finally, consider the three associated matrices $\mathbf{A}_1 = \begin{bmatrix} 1 & 3/2 \\ 3/2 & 5 \end{bmatrix}$ $\mathbf{A}_2 = \begin{bmatrix} 1 & 3/2 \\ 3/2 & 2 \end{bmatrix}$ and $\mathbf{A}_3 = \begin{bmatrix} 1 & 3/2 \\ 3/2 & 9/4 \end{bmatrix}$. We find $|\mathbf{A}_1| = 5 - 9/4 > 0$, $|\mathbf{A}_2| < 0$ and $|\mathbf{A}_3| = 0$. Here then are suitable tests we can employ. Yet consider $Q_4 = -x_1^2 - 3x_1x_2 - 5x_2^2$ which is clearly negative definite. From $\mathbf{A}_4$ we have

$$|\mathbf{A}_4| = \begin{vmatrix} -1 & -3/2 \\ -3/2 & -5 \end{vmatrix} = 5 - 9/4 > 0.$$ The distinction which we can use is that

$a_{11} > 0$ for a positive definite form whereas $a_{11} < 0$ for a negative definite form.

We can extend these ideas to more than two variables; see Problem 7. However, we look for a more suitable method of determining the nature of the quadratic form for more than two variables.

Reduction of a quadratic form via eigenvalues

We shall now present a method which will reduce a quadratic form to a sum of squares by employing the eigenvalues of the associated matrix. First, we collect some results which will prove useful. We have already seen that a real symmetric matrix has eigenvectors which are orthogonal, provided they correspond to distinct eigenvalues.

Theorem

Let $\mathbf{A}$ and $\mathbf{P}$ be $n \times n$ matrices and let $\mathbf{P}^{-1}$ exist. Then $\mathbf{D} = \mathbf{P}^{-1}\mathbf{AP}$ is a diagonal matrix if and only if the columns of $\mathbf{P}$ are the eigenvectors of $\mathbf{A}$ corresponding to the eigenvalues $\lambda_1, \lambda_2, \ldots, \lambda_n$. $\mathbf{D}$ is then the matrix

$$\begin{bmatrix} \lambda_1 & & & & 0 \\ & \lambda_2 & & & \\ & & \cdot & & \\ & & & \cdot & \\ 0 & & & & \lambda_n \end{bmatrix}$$

Proof

If the columns of $\mathbf{P}$, $p_1, p_2, \ldots, p_n$ are the eigenvectors of $\mathbf{A}$ corresponding to eigenvalues $\lambda_1, \lambda_2, \ldots, \lambda_n$ then

$$\mathbf{AP} = (\mathbf{A}p_1, \mathbf{A}p_2, \ldots, \mathbf{A}p_n) = (\lambda_1 p_1, \lambda_2 p_2, \ldots, \lambda_n p_n)$$

$$= \begin{bmatrix} \lambda_1 p_{11} & \lambda_2 p_{12} & \cdots & \lambda_n p_{1n} \\ \lambda_1 p_{21} & \lambda_2 p_{22} & \cdots & \lambda_n p_{2n} \\ \cdot & \cdot & & \cdot \\ \cdot & \cdot & & \cdot \\ \lambda_1 p_{n1} & \lambda_2 p_{n2} & \cdots & \lambda_n p_{nn} \end{bmatrix} = \begin{bmatrix} p_{11} & p_{12} & \cdots & p_{1n} \\ p_{21} & p_{22} & \cdots & p_{2n} \\ \cdot & \cdot & & \cdot \\ \cdot & \cdot & & \cdot \\ p_{n1} & p_{n2} & \cdots & p_{nn} \end{bmatrix} \begin{bmatrix} \lambda_1 & & & 0 \\ & \lambda_2 & & \\ & & \cdot & \\ & & & \cdot \\ 0 & & & \lambda_n \end{bmatrix}$$

$$= \mathbf{PD}$$

Hence $\mathbf{D} = \mathbf{P}^{-1}\mathbf{AP}$.

The argument can be reversed, completing the proof.

Now when $\mathbf{A}$ has n linearly independent eigenvectors it follows that the columns of $\mathbf{P}$ are linearly independent and hence $\mathbf{P}^{-1}$ exists.

We have already seen that the eigenvalues of an orthogonal matrix are of modulus 1 and, since an orthogonal matrix represents rotation about an axis, we would expect to be able to multiply a matrix representing a conic by a suitable orthogonal matrix to realign the coordinate axes along the axes of symmetry of the conic.

We now quote some properties of an orthogonal matrix.

Let $\mathbf{C}$ be an orthogonal matrix. Then the following properties hold.

(i) $\mathbf{C}^{-1} = \mathbf{C}^{\mathrm{T}}$.

(ii) The columns of $\mathbf{C}$ are unit vectors which are mutually orthogonal.

(iii) $\mathbf{C}^{\mathrm{T}}$ is also an orthogonal matrix.

(iv) $|\mathbf{C}| = 1$.

Note that since the distinct eigenvalues of a real symmetric matrix $\mathbf{A}$ are associated with eigenvectors which are mutually orthogonal then the matrix $\mathbf{P}$ which has these eigenvectors as its columns will be orthogonal. Hence we may use the theorem to see that $\mathbf{P}^{-1}\mathbf{AP}$ will be diagonal, with the eigenvalues of $\mathbf{A}$ as its diagonal elements.

Example

The matrix $\mathbf{A} = \begin{bmatrix} -2 & 1 \\ 1 & -2 \end{bmatrix}$ is real and symmetric.

The eigenvalues of $\mathbf{A}$ are $\lambda = -1$ and -3; corresponding to $\lambda = -1$ we have an eigenvector $\begin{bmatrix} 1 \\ 1 \end{bmatrix}$ and corersponding to $\lambda = -3$ we have an eigenvector $\begin{bmatrix} -1 \\ 1 \end{bmatrix}$. Unit eigenvectors are $\begin{bmatrix} 1/\sqrt{2} \\ 1/\sqrt{2} \end{bmatrix}$ and $\begin{bmatrix} -1/\sqrt{2} \\ 1/\sqrt{2} \end{bmatrix}$. Consider $\mathbf{P} = \begin{bmatrix} 1/\sqrt{2} & -1/\sqrt{2} \\ 1/\sqrt{2} & 1/\sqrt{2} \end{bmatrix}$; it is orthogonal and, therefore, $\mathbf{P}^{-1} = \mathbf{P}^{T} = \begin{bmatrix} 1/\sqrt{2} & 1/\sqrt{2} \\ -1/\sqrt{2} & 1/\sqrt{2} \end{bmatrix}$. You can show directly that $\mathbf{P}^{-1}\mathbf{AP} = \begin{bmatrix} -1 & 0 \\ 0 & -3 \end{bmatrix}$.

Now $\mathbf{A}$ is the matrix associated with the quadratic form $Q = -2x_1^2 + 2x_1x_2 - 2x_2^2$. Consider then the transformation $\mathbf{y} = \mathbf{P}^{-1}\mathbf{x}$.

i.e. $$\begin{bmatrix} y_1 \\ y_2 \end{bmatrix} = \begin{bmatrix} 1/\sqrt{2} & 1/\sqrt{2} \\ -1/\sqrt{2} & 1/\sqrt{2} \end{bmatrix} \begin{bmatrix} x_1 \\ x_2 \end{bmatrix}$$

If we express Q in terms of y_1 and y_2 we use $\mathbf{x} = \mathbf{Py}$ so that $x_1 = \frac{1}{\sqrt{2}} y_1 - \frac{1}{\sqrt{2}} y_2$ and $x_2 = \frac{1}{\sqrt{2}} y_1 + \frac{1}{\sqrt{2}} y_2$. Then

$$\begin{aligned} Q &= -2(\tfrac{1}{2} y_1^2 - y_1 y_2 + \tfrac{1}{2} y_2^2) + 2(\tfrac{1}{2} y_1^2 - \tfrac{1}{2} y_2^2) - 2(\tfrac{1}{2} y_1^2 + y_1 y_2 + \tfrac{1}{2} y_2^2) \\ &= -y_1^2 - 3y_2^2 = \lambda_1 y_1^2 + \lambda_2 y_2^2 \end{aligned}$$

This last result holds generally.

If $\mathbf{A}$ is a real symmetric matrix associated with a quadratic form Q in the variables $x_1, x_2, \ldots, x_n$ then we can find an orthogonal matrix $\mathbf{P}$ such that $\mathbf{P}^{-1}\mathbf{AP}$ is a diagonal matrix. The coordinate transformation $y = \mathbf{P}^{-1}\mathbf{x}$ allows Q to be expressed in terms of $y_1, y_2, \ldots, y_n$. Furthermore, if all the eigenvalues of $\mathbf{A}$ are positive the reduced quadratic form is $Q = z_1^2 + z_2^2 + \ldots + z_n^2$. This result follows, because we can apply a transformation $\mathbf{P}$ as demonstrated in the last example to reduce Q to $\lambda_1 y_1^2 + \lambda_2 y_2^2 + \ldots + \lambda_n y_n^2$. A further transformation $y_1 = (1/\sqrt{\lambda_1})z_1$, $y_2 = (1/\sqrt{\lambda_2})z_2$ etc. will produce $Q = z_1^2 + z_2^2 + \ldots + z_n^2$. The second transformation represents contractions in each of the coordinate directions y_i.

Example

The quadratic form $Q = 2x_1^2 - x_2^2 - x_3^2 - 4x_1x_2 + 4x_1x_3 + 8x_2x_3$ has associated matrix $\begin{bmatrix} 2 & -2 & 2 \\ -2 & -1 & 4 \\ 2 & 4 & -1 \end{bmatrix}$. The eigenvalues are given by $\begin{vmatrix} 2-\lambda & -2 & 2 \\ -2 & -1-\lambda & 4 \\ 2 & 4 & -1-\lambda \end{vmatrix} = 0$ and are $\lambda = 3$ (twice) and $\lambda = -6$. For $\lambda = -6$ we need to solve the equations $8x_1 - 2x_2 + 2x_3 = 0$, $-2x_1 + 5x_2 + 4x_3 = 0$, $2x_1 + 4x_2 + 5x_3 = 0$. We obtain as eigenvectors $(\alpha, 2\alpha, -2\alpha)$. We take the unit vector $(1/3, 2/3, -2/3)$. For $\lambda = 3$ we solve $-x_1 - 2x_2 + 2x_3 = 0$ (the other equations are really the same). We shall choose two eigenvectors which are orthogonal to $(1/3, 2/3, -2/3)$ and orthogonal to each other. Suitable vectors are $(2/3, 1/3, 2/3)$ and $(-2/3, 2/3, 1/3)$. Verify that these three vectors are mutually orthogonal.

Let $\mathbf{P} = 1/3 \begin{bmatrix} 1 & -2 & 2 \\ 2 & 2 & 1 \\ -2 & 1 & 2 \end{bmatrix}$. Then $\mathbf{P}$ is orthogonal. Furthermore,

$\mathbf{P}^{-1}\mathbf{AP} = \begin{bmatrix} -6 & 0 & 0 \\ 0 & 3 & 0 \\ 0 & 0 & 3 \end{bmatrix}$. Now consider $\begin{bmatrix} x_1 \\ x_2 \\ x_3 \end{bmatrix} = \mathbf{P} \begin{bmatrix} y_1 \\ y_2 \\ y_3 \end{bmatrix}$. You can check that this information allows us to write $Q = -6y_1{}^2 + 3y_2{}^2 + 3y_3{}^2$.

Theorem

A quadratic form is positive definite if and only if all the eigenvalues of the associated symmetric matrix are positive.

Proof

We know that the transformation $\mathbf{x} = \mathbf{Py}$, as used above will transform Q to the expression $\sum_{i=1}^{n} \lambda_i y_i^2$. Now if all $\lambda_i > 0$, then $Q > 0$ unless $y_i = 0$ for all i i.e. $x_i = 0$ for all i. Conversely, suppose an eigenvalue, say λ_1, is $\leqslant 0$. Then by choosing $y_1 = 1$ and all other $y_i = 0$, Q becomes $\leqslant 0$ for a non-zero $\mathbf{y}$ and hence for a non-zero $\mathbf{x}$, violating the positive definite quality.

Example 1

For the quadratic form of the last example one eigenvector is negative and so the form is not positive definite.

Example 2

The quadratic form $Q = x_1{}^2 + 2x_2{}^2 + 7x_3{}^2 - 2x_1 x_2 + 4x_1 x_3 - 2x_2 x_3$ has associated symmetric matrix $\begin{bmatrix} 1 & -1 & 2 \\ -1 & 2 & -1 \\ 2 & -1 & 7 \end{bmatrix}$ which has eigenvalues satisfying $\lambda^3 - 10\lambda^2 + 17\lambda - 2 = 0$. To find the eigenvalues would require numerical techniques. However we may write

$$\begin{aligned} Q &= (x_1 - x_2 + 2x_3)^2 + 2x_2{}^2 + 7x_3{}^2 - 2x_2 x_3 - x_2{}^2 - 4x_3{}^2 + 4x_2 x_3 \\ &= (x_1 - x_2 + 2x_3)^2 + x_2{}^2 + 3x_3{}^2 + 2x_2 x_3 \\ &= (x_1 - x_2 + 2x_3)^2 + (x_2 + x_3)^2 + 2x_3{}^2 \end{aligned}$$

Therefore we see that Q is positive definite. Note, it does *not* follow that the eigenvalues are 1, 1 and 2 – the coefficients of these squared terms, since the factorisation of Q is not unique and we cannot be sure that our factorisation coincides with that produced via eigenvalues.

Example 3

The inertia matrix of a three-dimensional mass system is $\mathbf{I} = \begin{bmatrix} I_{xx} & -I_{xy} & -I_{xz} \\ -I_{xy} & I_{yy} & -I_{yz} \\ -I_{xz} & -I_{yz} & I_{zz} \end{bmatrix}$.

For a particular mass system the inertia matrix is $\begin{bmatrix} 4 & -1 & -1 \\ -1 & 4 & -1 \\ -1 & -1 & 4 \end{bmatrix}$. Eigenvalues are $\lambda = 2$ and 5 (twice). For $\lambda = 2$ we obtain the unit eigenvector $\begin{bmatrix} 1/\sqrt{3} \\ 1/\sqrt{3} \\ 1/\sqrt{3} \end{bmatrix}$ and for $\lambda = 5$ we choose $\begin{bmatrix} 1/\sqrt{2} \\ -1/\sqrt{2} \\ 0 \end{bmatrix}$ and $\begin{bmatrix} 1/\sqrt{6} \\ -2/\sqrt{6} \\ 1/\sqrt{6} \end{bmatrix}$ to give a mutually orthogonal set.

A set of principal axes for the mass system is given by the eigenvectors and if we rotate the coordinate axes to coincide with these directions the products of inertia disappear and we are left with principal moments of inertia 2, 5 and 5.

Problems

1. Write out in full the quadratic form in x_1, x_2, x_3 whose matrix is

(i) $\begin{bmatrix} 2 & -3 & 1 \\ -3 & 2 & 4 \\ 1 & 4 & -5 \end{bmatrix}$ (ii) $\begin{bmatrix} 3 & 0 & 2 \\ 0 & 4 & -5 \\ 2 & -5 & 6 \end{bmatrix}$ (iii) $\begin{bmatrix} 1 & -2 & 4 \\ -2 & 0 & 3 \\ 4 & 3 & 2 \end{bmatrix}$

2. Write the following quadratic forms in matrix notation:

(i) $2x_1^2 - 6x_1x_2 + x_3^2$ (ii) $3x_1^2 - 8x_1x_2 + 4x_2^2 + 5x_1x_3$

(iii) $x_1^2 - 2x_2^2 - 3x_3^2 + 4x_1x_2 + 6x_1x_3 - 8x_2x_3$

3. Reduce the following quadratic forms to sums of squares

(i) $\begin{bmatrix} 1 & -1 & 0 \\ -1 & 2 & -1 \\ 0 & -1 & 2 \end{bmatrix}$ (ii) $\begin{bmatrix} 1 & 0 & -2 \\ 0 & 0 & 1 \\ -2 & 1 & 3 \end{bmatrix}$ (iii) $\begin{bmatrix} 4 & -4 & 2 \\ -4 & 3 & -3 \\ 2 & -3 & 1 \end{bmatrix}$

(iv) $\begin{bmatrix} 1 & -2 & -1 \\ 2 & 4 & 2 \\ -1 & 2 & 3 \end{bmatrix}$ (v) $\begin{bmatrix} 0 & 2 & 0 & 1 \\ 2 & 4 & 3 & 1 \\ 0 & 3 & 1 & 1 \\ 1 & 1 & 1 & 1 \end{bmatrix}$ (vi) $\begin{bmatrix} 1 & 2 & 3 & 1 \\ 2 & 4 & 6 & 2 \\ 3 & 6 & 9 & 3 \\ 1 & 2 & 3 & 1 \end{bmatrix}$

4. Given that one characteristic root of the matrix $\mathbf{A} = \begin{bmatrix} 5 & -1 & -1 \\ 1 & 3 & 1 \\ -2 & 2 & 4 \end{bmatrix}$ is 2, find the other two, and characteristic vectors corresponding to each of the three roots. Hence find a matrix **H** such that $\mathbf{H}^{-1}\mathbf{AH}$ is diagonal, verifying that your choice is correct.

5. Find the eigenvalues and eigenvectors of the matrix $\mathbf{B} = \begin{bmatrix} 4 & -3 \\ -3 & 4 \end{bmatrix}$. Find a matrix **P** such that $\mathbf{P}^{-1}\mathbf{BP}$ is diagonal and hence show that $u(x, y) \equiv 4x^2 - 6xy + 4y^2$ is positive definite. Show that $\mathbf{B}^2 - 8\mathbf{B} + 7\mathbf{I} = \mathbf{0}$, where $\mathbf{0}$ is the appropriate zero matrix, and hence find $\mathbf{B}^{-1}$.

6. Find the eigenvalues and three linearly independent eigenvectors of the matrix **A**, where $\mathbf{A} = \begin{bmatrix} -1 & 2 & 2 \\ 2 & 2 & -1 \\ 2 & -1 & 2 \end{bmatrix}$. Find the matrix **H** which diagonalises **A**. Is **A** positive definite, negative definite, or indefinite?

7. Verify that the quadratic form $5x_1^2 + 5x_2^2 + 2x_3^2 + 8x_1x_2 + 4x_1x_3 + 4x_2x_3$ is positive definite.

8. Find the eigenvalues and eigenvectors for the matrix $\mathbf{A} = \begin{bmatrix} 5 & 1 & 1 \\ 1 & 3 & 1 \\ 1 & 1 & 3 \end{bmatrix}$. Hence construct a matrix **H** such that $\mathbf{H}^{-1}\mathbf{AH}$ is a diagonal matrix and verify your result.

9. Find three linearly independent characteristic vectors of the matrix $\mathbf{A} = \begin{bmatrix} -3 & 0 & 6 \\ 0 & 3 & 6 \\ 6 & 6 & 0 \end{bmatrix}$ and hence determine a non-singular matrix **P** such that the matrix $\mathbf{P}^{-1}\mathbf{AP}$ is diagonal, verifying the result.

10. Express the quadratic form $4x_1^2 + 3x_2^2 + 4x_3^2 - 4x_1x_2 - 6x_1x_3 + 2x_2x_3$ in matrix form $\mathbf{X'AX}$ where **A** is a symmetric matrix and $\mathbf{X'}$ is the transpose of **X**. Test whether the given quadratic form is positive definite.

2.6 BOUNDARY-VALUE PROBLEMS

We consider as a first example the bending of a uniform strut of length l under an axial compressive force P at either end. In Figure 2.8 we demonstrate an approximate method of solution. We aim to estimate the deflections at five equally spaced internal points and use these as a guide to the deflected profile. The strut is fixed at either end.

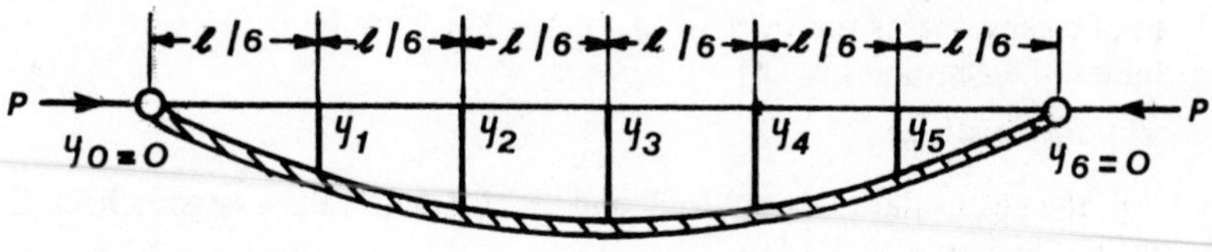

Figure 2.8

The governing differential equation is the bending moment equation

$$EI\frac{d^2y}{dx^2} = -Py \qquad (2.15)$$

where EI is the (constant) flexural rigidity of the strut and y is the deflection of a point on the strut which is distant x from the left-hand end of the strut. Since both ends are fixed $y(0) = 0 = y(l)$;in our approximate method this implies that $y_0 = 0 = y_6$. Since the problem is symmetrical, it follows that $y_1 = y_5$ and $y_2 = y_4$ and we need only find y_1, y_2 and y_3.

We approximate $\frac{d^2y}{dx^2}$ at the point x_r (where the deflection is y_r) by the central difference formula $\frac{y_{r+1} - 2y_r + y_{r-1}}{(l/6)^2}$ which has error $0(l^2/36)$. At x_2, therefore, the differential equation may be approximated by the finite difference equation $EI\left(\frac{y_3 - 2y_2 + y_1}{l^2/36}\right) = -Py_2$ or $y_3 - 2y_2 + y_1 = \frac{-Pl^2}{36EI}y_2$. Now at x_1 the finite difference approximation to $\frac{d^2y}{dx^2}$ is $\frac{y_2 - 2y_1 + y_0}{l^2/36}$; but $y_0 = 0$ and the finite difference equation at x_1 reduces to $y_2 - 2y_1 = \frac{-Pl^2}{36EI}y_1$. If we put $\lambda = Pl^2/(36EI)$, you should be able to see that we can rearrange the three finite difference equations to

$$\left.\begin{aligned} 2y_1 - y_2 &= \lambda y_1 \\ -y_1 + 2y_2 - y_3 &= \lambda y_2 \\ -2y_2 + 2y_3 &= \lambda y_3 \end{aligned}\right\} \qquad (2.16a)$$

In matrix form, these equations may be written

$$\mathbf{Ay} = \lambda\mathbf{y} \qquad (2.16b)$$

where $\mathbf{y} = \begin{bmatrix} y_1 \\ y_2 \\ y_3 \end{bmatrix}$ and $\mathbf{A}$ is the matrix $\begin{bmatrix} 2 & -1 & 0 \\ -1 & 2 & -1 \\ 0 & -2 & 2 \end{bmatrix}$

We have reduced the differential equation to a problem of solving linear simultaneous equations. It always was an eigenvalue problem even if cast in a different light.

Obviously, one solution of (2.16b) is $\mathbf{y} = \mathbf{0}$ which corresponds to no deflection of the strut; this is a possible equilibrium profile. The other possible equilibrium profiles must correspond to the eigenvectors of matrix $\mathbf{A}$, since these are the only non-zero solutions of (2.16).

Now you should be able to show that the eigenvalues of $\mathbf{A}$ are $(2 - \sqrt{3})$, 2 and $(2 + \sqrt{3})$. Remember the useful checks

(i) sum of eigenvalues is the trace of $\mathbf{A} = 2 + 2 + 2 = 6$
(ii) product of eigenvalues is $|\mathbf{A}| = 2$.

Both checks are satisfied.

To 2 d.p. the eigenvalues are 0.27, 2 and 3.73 (these values satisfy check (i) exactly but not exactly check (ii)).

We therefore have three critical values of P: $\frac{9.72EI}{l^2}$, $\frac{72EI}{l^2}$, and $\frac{134.28EI}{l^2}$. The interpretation of these results is that the strut can assume a symmetrical curved profile under an axial load P and remain in equilibrium only if P has one of the three values quoted. It will remain undeflected under any load P (unless this exceeds the compressive strength of the strut). What do these deflected profiles look like? We need to find the eigenvectors of **A**. An eigenvector corresponding to $\lambda = 2 - \sqrt{3}$ is $\begin{bmatrix} 1 \\ \sqrt{3} \\ 2 \end{bmatrix}$; an eigenvector for $\lambda = 2$ is $\begin{bmatrix} 1 \\ 0 \\ -1 \end{bmatrix}$; an eigenvector corresponding to $\lambda = 2 + \sqrt{3}$ is $\begin{bmatrix} 1 \\ -\sqrt{3} \\ 2 \end{bmatrix}$. Sketches of the three profiles are given in Figure 2.9

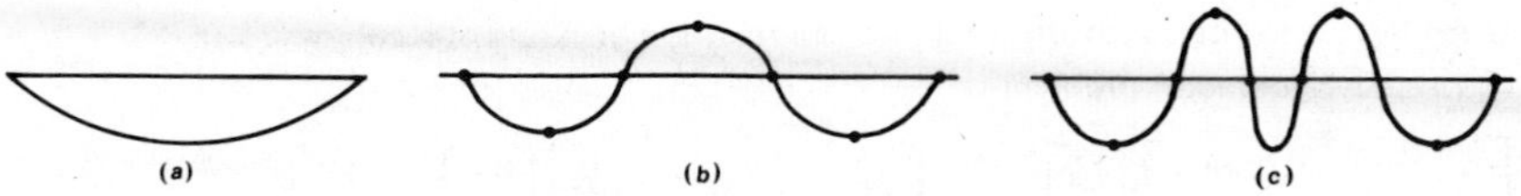

Figure 2.9

However, we might have chosen other eigenvectors from each class; certainly we could have reversed all the signs to obtain profiles which were mirror images of the ones shown. A more serious problem is that we may only determine eigenvectors within a scalar multiple. For profile (a) we could have $\begin{bmatrix} y_1 \\ y_2 \\ y_3 \end{bmatrix} = \begin{bmatrix} 1 \\ \sqrt{3} \\ 2 \end{bmatrix}, \begin{bmatrix} 2 \\ 2\sqrt{3} \\ 4 \end{bmatrix}, \begin{bmatrix} 10 \\ 10\sqrt{3} \\ 20 \end{bmatrix}$ and so on. The actual values of y_1, y_2 and y_3 will not result from this analysis.

In practice, a strut may not be homogenous and uniform and the compressive forces P may not lie exactly along the central axis. We may then model the situation by adding a transverse force F. Let us study the effect of applying an increasing force P. If P is less than the first critical load then the strut will bend into the first mode shown; if P exceeds this value then the strut will buckle to the second mode and deform progressively in this mode until the second critical value of P is reached. By this time, the assumptions on which the mathematical model was constructed will be violated. Of course, if there were no transverse force F there would be no deflected profile.

You should by now be asking yourself what would have happened if the original finite difference approximation were improved, e.g. if we had divided the strut into 12 equal parts. We should then have had 11 internal points and, employing symmetry, 6 equations to solve in 6 unknowns. This would give rise to a 6 x 6 matrix with 6 eigenvalues. In theory there would be 6 critical loads P.

The more inquisitive will also be asking why we do not solve the differential equation by the usual methods. We may restate the equation as $\frac{d^2y}{dx^2} + \frac{P}{EI}y = 0$. This has a general solution $y = A\cos\sqrt{\frac{P}{EI}}\,x + B\sin\sqrt{\frac{P}{EI}}\,x$. Applying the boundary conditions $y(0) = 0 = y(l)$ produces the particular solution $y = B_n \sin\frac{n\pi x}{l}$ where $\sqrt{\frac{P}{EI}} = \frac{n\pi}{l}$ for $n = 1, 2, 3, \ldots$. (The case $n = 0$ gives the undeflected profile $y \equiv 0$.) There is an infinite number of theoretical critical loads given by $P = \frac{n^2\pi^2 EI}{l^2}$. The one of most practical interest is the smallest one where $n = 1$. Then $P = \pi^2 EI/l^2$. Compare this with the approximation we produced of $9.72EI/l^2$. Would we expect the approximation to improve as the number of internal points increases? For the case of 11 internal points we went through a lot of tedious algebra (involving much low cunning based on our experience) to find that the approximation had improved only to $9.79EI/l^2$. It was hardly worthwhile; in such a situation we might well ask if a numerical technique for finding eigenvalues would not have been better. We take up this idea in the next sections.

Problems

1. Using the finite-difference approach, we can treat the buckling of a column as an eigenvalue-eigenvector problem.

$$\begin{bmatrix} 7 & -4 & 1 & 0 \\ -4 & 6 & -4 & 1 \\ 1 & -4 & 5 & -2 \\ 0 & 1 & -2 & 1 \end{bmatrix} \begin{bmatrix} w_b \\ w_c \\ w_d \\ w_e \end{bmatrix} = \begin{bmatrix} 2k & -k & 0 & 0 \\ -k & 2k & -k & 0 \\ 0 & -k & 2k & -k \\ 0 & 0 & -k & k \end{bmatrix} \begin{bmatrix} w_b \\ w_c \\ w_d \\ w_e \end{bmatrix}$$

Find the buckling load $P = 16kEI/l^2$, where the k's are the eigenvalues, and relative values of w_b, w_c, w_d and w_e are represented by the eigenvectors.

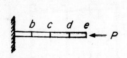

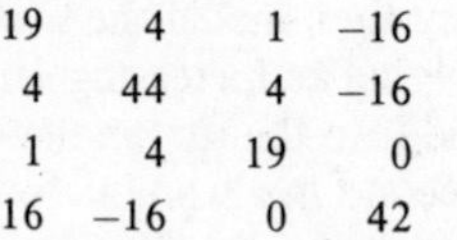

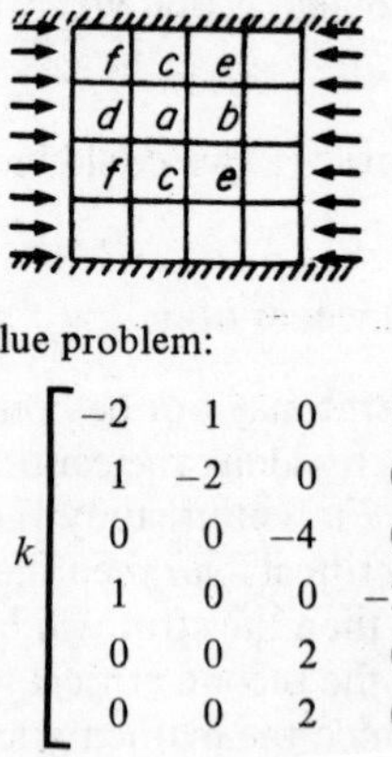

2. The buckling of a plate can be treated as an eigenvalue problem:

$$\begin{bmatrix} 20 & -8 & -16 & -8 & 4 & 4 \\ -8 & 19 & 4 & 1 & -16 & 0 \\ -16 & 4 & 44 & 4 & -16 & -16 \\ -8 & 1 & 4 & 19 & 0 & -16 \\ 4 & -16 & -16 & 0 & 42 & 2 \\ 4 & 0 & -16 & -16 & 2 & 42 \end{bmatrix} \begin{bmatrix} w_a \\ w_b \\ w_c \\ w_d \\ w_e \\ w_f \end{bmatrix} = k \begin{bmatrix} -2 & 1 & 0 & 1 & 0 & 0 \\ 1 & -2 & 0 & 0 & 0 & 0 \\ 0 & 0 & -4 & 0 & 2 & 2 \\ 1 & 0 & 0 & -2 & 0 & 0 \\ 0 & 0 & 2 & 0 & -2 & 0 \\ 0 & 0 & 2 & 0 & 0 & -2 \end{bmatrix} \begin{bmatrix} w_a \\ w_b \\ w_c \\ w_d \\ w_e \\ w_f \end{bmatrix}$$

Find the eigenvalues in terms of k and the eigenvectors.

3. Determine the frequency ω and the modes of the system for which the equation of motion is

$$\begin{bmatrix} y_1 \\ y_2 \\ y_3 \end{bmatrix} = \frac{\omega^2 \delta_{11}}{27} \begin{bmatrix} 27 & 14 & 4 \\ 14 & 8 & 2.5 \\ 4 & 2.5 & 1 \end{bmatrix} \times \begin{bmatrix} M_1 & 0 & 0 \\ 0 & M_2 & 0 \\ 0 & 0 & M_3 \end{bmatrix} \begin{bmatrix} y_1 \\ y_2 \\ y_3 \end{bmatrix},$$

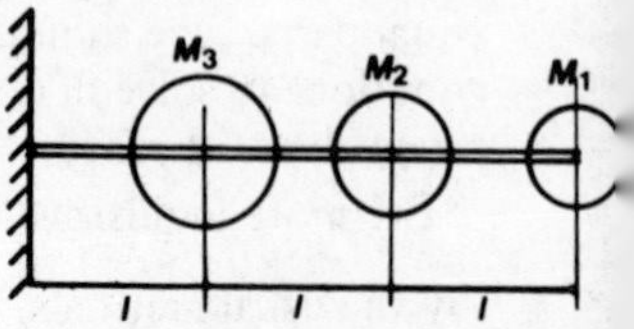

where $M_1 = M_2 = 2$, $M_3 = 3$ and δ_{11} is a constant.

2.7 ITERATIVE METHOD FOR FINDING THE EIGENVALUE OF LARGEST MODULUS

We have so far chosen very simple physical systems. The number of degrees of freedom has been small giving rise to at most a 3 x 3 matrix and we have been careful to make masses and stiffnesses equal. We shall continue to illustrate our ideas with 3 x 3 matrices because these are small enough to allow easy computation and presentation of results, yet are large enough to prevent trivial cases dominating our conclusions.

Consider the system of Figure 2.10. It is effectively that of Figure 2.3, but the masses of the disks are no longer equal and the stiffnesses of different parts of the shaft differ; further, the left-hand end is fixed so that rigid body rotation is no longer possible.

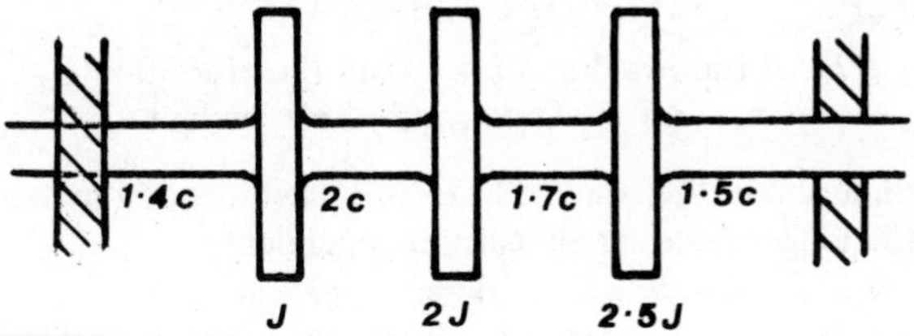

Figure 2.10

For consistency with previous notation, we let x_1, x_2, x_3 be the angles turned through by the disks (reading from left to right). The equations of motion are

$$\left.\begin{aligned} J\ddot{x}_1 &= -1.4CJx_1 - 2C(x_1 - x_2) \\ 2J\ddot{x}_2 &= 2C(x_1 - x_2) - 1.7C(x_2 - x_3) \\ 2.5J\ddot{x}_3 &= 1.7C(x_2 - x_3) \end{aligned}\right\} \tag{2.17}$$

Applying the ideas of Section 2.3, we put $\ddot{x}_i = -\omega^2 x_i$ and write $\lambda = J\omega^2/C$ to obtain the equations in matrix form as

$$\left.\begin{bmatrix} 3.4 & -2 & 0 \\ -2 & 3.7 & -1.7 \\ 0 & -1.7 & 1.7 \end{bmatrix}\begin{bmatrix} x_1 \\ x_2 \\ x_3 \end{bmatrix} = \begin{bmatrix} 1 & 0 & 0 \\ 0 & 2 & 0 \\ 0 & 0 & 2.5 \end{bmatrix}\begin{bmatrix} x_1 \\ x_2 \\ x_3 \end{bmatrix}\right\} \tag{2.18}$$

$$\mathbf{Ax} = \lambda\mathbf{Bx} \tag{2.19}$$

Make sure that you can derive this result yourself.

Although there are certain standard programs available which will handle the eigenvalue problem in the form (2.19) it is easy in this case to form $\mathbf{B}^{-1} = \begin{bmatrix} 1 & 0 & 0 \\ 0 & 0.5 & 0 \\ 0 & 0 & 0.4 \end{bmatrix}$ and then multiply (2.19) by $\mathbf{B}^{-1}$ to produce the equation $\mathbf{Hx} = \lambda\mathbf{x}$ where $\mathbf{H} = \mathbf{B}^{-1}\mathbf{A}$. We have now recovered the more usual form of the eigenvalue problem. In our case we find $\mathbf{H} = \begin{bmatrix} 3.4 & -2 & 0 \\ -1 & 1.85 & -0.85 \\ 0 & -0.68 & 0.68 \end{bmatrix}$. If we were to form the characteristic equation for $\mathbf{H}$ we should have to resort to numerical methods of root-finding. We seek a different approach.

Algebraic foundation of the iterative method

We first state a very useful theorem.

If $\mathbf{A}$ is a real $n \times n$ matrix and λ_1 and λ_2 are two distinct eigenvalues with corresponding eigenvectors $\mathbf{x}_1$ and $\mathbf{x}_2$ then $\mathbf{x}_1$ and $\mathbf{x}_2$ are linearly independent.

The proof follows by assuming a relationship $c_1\mathbf{x}_1 + c_2\mathbf{x}_2 = \mathbf{0}$ with c_1, c_2 scalars.

Then $\quad \mathbf{0} = \mathbf{A}(c_1\mathbf{x}_1 + c_2\mathbf{x}_2) = c_1\mathbf{A}\mathbf{x}_1 + c_2\mathbf{A}\mathbf{x}_2 = c_1\lambda_1\mathbf{x}_1 + c_2\lambda_2\mathbf{x}_2$

But $\quad \lambda_1(c_1\mathbf{x}_1 + c_2\mathbf{x}_2) = \lambda\mathbf{0} = \mathbf{0}$

and hence $\quad \mathbf{0} = c_2\lambda_2\mathbf{x}_2 - \lambda_1 c_2\mathbf{x}_2 = (\lambda_2 - \lambda_1)c_2\mathbf{x}_2$

Since $\mathbf{x}_2 \neq 0$ and $\lambda_2 \neq \lambda_1$ it follows that $c_2 = 0$ and therefore that $c_1\mathbf{x}_1 = \mathbf{0}$ which implies that $c_1 = 0$. Hence $\mathbf{x}_1$ and $\mathbf{x}_2$ are linearly independent.

We generalise the theorem to the result that the eigenvectors of a matrix corresponding to **distinct** eigenvalues are **linearly independent.**

If, therefore, an $n \times n$ matrix $\mathbf{A}$ has n distinct eigenvalues $\lambda_1, \ldots, \lambda_n$ with corresponding eigenvectors $\mathbf{x}_1, \ldots, \mathbf{x}_n$ then these eigenvectors form a basis for the vector space R^n.

Therefore, any vector $\mathbf{y}_0 \in R^n$ can be expressed as a linear combination of these:

$$\mathbf{y}_0 = d_1\mathbf{x}_1 + d_2\mathbf{x}_2 + \ldots + d_n\mathbf{x}_n \tag{2.20}$$

Now consider,
$$\begin{aligned}\mathbf{y}_1 &= \mathbf{A}\mathbf{y}_0 = \mathbf{A}d_1\mathbf{x}_1 + \mathbf{A}d_2\mathbf{x}_2 + \ldots + \mathbf{A}d_n\mathbf{x}_n \\ &= d_1\mathbf{A}\mathbf{x}_1 + d_2\mathbf{A}\mathbf{x}_2 + \ldots + d_n\mathbf{A}\mathbf{x}_n \\ &= \lambda_1 d_1\mathbf{x}_1 + \lambda_2 d_2\mathbf{x}_2 + \ldots + \lambda_n d_n\mathbf{x}_n\end{aligned}$$

Further
$$\begin{aligned}\mathbf{y}_2 &= \mathbf{A}\mathbf{y}_1 = \lambda_1^2 d_1\mathbf{x}_1 + \lambda_2^2 d_2\mathbf{x}_2 + \ldots + \lambda_n^2 d_n\mathbf{x}_n \\ \mathbf{y}_3 &= \mathbf{A}\mathbf{y}_2 = \lambda_1^3 d_1\mathbf{x}_1 + \lambda_2^3 d_2\mathbf{x}_2 + \ldots + \lambda_n^3 d_n\mathbf{x}_n\end{aligned}$$

and, in general, $\mathbf{y}_r = \mathbf{A}\mathbf{y}_{r-1} = \lambda^r_1 d_1\mathbf{x}_1 + \lambda_2^r d_2\mathbf{x}_2 + \ldots + \lambda^r_n d_n\mathbf{x}_n$

You may well be asking what use is this sequence of vectors $\mathbf{y}_0, \mathbf{y}_1, \mathbf{y}_2, \ldots, \mathbf{y}_r$.
If the eigenvalue of largest modulus is λ_1, then the ratios $\left|\frac{\lambda_2}{\lambda_1}\right|, \left|\frac{\lambda_3}{\lambda_1}\right|, \ldots, \left|\frac{\lambda_n}{\lambda_1}\right|$ are all less than 1. Furthermore, the ratios $\left|\frac{\lambda_2}{\lambda_1}\right|^2, \left|\frac{\lambda_3}{\lambda_1}\right|^2, \ldots, \left|\frac{\lambda_n}{\lambda_1}\right|^2$ are all progressively less than 1.

If we write
$$\mathbf{y}_1 = \lambda_1\left[d_1\mathbf{x}_1 + \left|\frac{\lambda_2}{\lambda_1}\right| d_2\mathbf{x}_2 + \ldots + d_n\left|\frac{\lambda_n}{\lambda_1}\right| \mathbf{x}_n\right]$$

$$\mathbf{y}_2 = \lambda_1^2\left[d_1\mathbf{x}_1 + \left|\frac{\lambda_2}{\lambda_1}\right|^2 d_2\mathbf{x}_2 + \ldots + d_n\left|\frac{\lambda_n}{\lambda_1}\right|^2 \mathbf{x}_n\right]$$

and, in general

$$\mathbf{y}_r = \lambda_1^r\left[d_1\mathbf{x}_1 + \left|\frac{\lambda_2}{\lambda_1}\right|^r d_2\mathbf{x}_2 + \ldots + d_n\left|\frac{\lambda_n}{\lambda_1}\right|^r \mathbf{x}_n\right] \tag{2.21}$$

For sufficiently large r we might expect the ratios $\left|\frac{\lambda_2}{\lambda_1}\right|^r, \ldots, \left|\frac{\lambda_n}{\lambda_1}\right|^r$ to be small enough to allow $\mathbf{y}_r \simeq \lambda_1^r d_1 \mathbf{x}_1$ to be a reasonable approximation to the truth.

If this is so, we shall be able to compare $\mathbf{y}_r$ and $\mathbf{y}_{r+1}$. The ratio $\left|\dfrac{\mathbf{y}_{r+1}}{\mathbf{y}_r}\right|$ becomes progressively closer to λ_1 as r becomes larger.

All we have to do is to form the sequence $\{\mathbf{y}_0, \mathbf{A}\mathbf{y}_0, \mathbf{A}^2\mathbf{y}_0, \mathbf{A}^3\mathbf{y}_0, \ldots\}$. We then form a sequence of ratios of successive terms. The limit of this sequence is the eigenvalue of largest modulus.

Before we apply the method we shall add one modification which will allow us the luxury of also finding the eigenvector that corresponds to this eigenvalue.

It helps for the moment to assume we are dealing with a real symmetric matrix with distinct eigenvalues. The corresponding eigenvectors are orthogonal and, in effect, represent a rotated set of Cartesian axes. As we proceed down the sequence $\{\mathbf{y}_0, \mathbf{y}_1, \mathbf{y}_2, \mathbf{y}_3, \ldots\}$ we effectively produce a sequence of vectors whose directions become ever more closely aligned to the axis represented by $\mathbf{x}_1$.

The choice of $\mathbf{y}_0$ is to some extent arbitrary. Let us arrange that one component (which for the purpose of this discussion we may assume is the first) is equal to 1. Then $\mathbf{y}_1 = \mathbf{A}\mathbf{y}_0$ can be written as $b_1\mathbf{y}_1'$ where the first component is also equal to 1. For example, if $\mathbf{y}_0 = \begin{bmatrix}1\\2\\3\end{bmatrix}$ and $\mathbf{y}_1 = \begin{bmatrix}4\\10\\12\end{bmatrix}$ then $b_1 = 4$ and $\mathbf{y}_1' = \begin{bmatrix}1\\2.5\\3\end{bmatrix}$. Thus we can see how closely $\mathbf{y}_0$ and $\mathbf{y}_1'$ agree, component for component. Later on in the sequence, when $\mathbf{y}_r \simeq \lambda_1^r d_1 \mathbf{x}_1$ and $\mathbf{y}_{r+1} \simeq \lambda_1^{r+1} d_1 \mathbf{x}_1$ we should expect that the modified vectors would agree quite closely. If we have modified the $\mathbf{y}_r'$ so that the sequence we really obtain is $\mathbf{y}_0, \mathbf{y}_1', \mathbf{y}_2', \ldots$ then we shall have the relationship $\mathbf{A}\mathbf{y}_r' = \mathbf{y}_{r+1} = b_r \mathbf{y}_{r+1}'$; b_r is an approximation to λ_1.

If you study the following two examples, the technique should become clearer.

Example 1

The matrix $\mathbf{A} = \begin{bmatrix}1 & -1 & 0\\-1 & 2 & -1\\0 & -1 & 1\end{bmatrix}$ has eigenvalues 0, 1, 3. Let us guess that $\mathbf{y}_0 = \begin{bmatrix}1\\-1\\2\end{bmatrix}$. Why? We know that an eigenvector corresponding to $\lambda = 3$ is $\begin{bmatrix}1\\-2\\1\end{bmatrix}$ and for purposes of demonstration we want to be neither too close to the end result nor too far away. In general, the further away our starting vector the more iterations we require to reach a reasonably accurate approximation.

Then $$\mathbf{y}_1 = \begin{bmatrix}1 & -1 & 0\\-1 & 2 & -1\\0 & -1 & 1\end{bmatrix}\begin{bmatrix}1\\-1\\2\end{bmatrix} = \begin{bmatrix}2\\-5\\3\end{bmatrix} = 2\begin{bmatrix}1\\-2.5\\1.5\end{bmatrix} = b_1\mathbf{y}_1'$$

We performed the last step so that the first components of the input $\mathbf{y}_0$ and output $\mathbf{y}_1'$ are equal; we can compare the second and third components. Repeating the step, we obtain

$$\mathbf{y}_2 = \begin{bmatrix} 1 & -1 & 0 \\ -1 & 2 & -1 \\ 0 & -1 & 1 \end{bmatrix} \begin{bmatrix} 1 \\ -2.5 \\ 1.5 \end{bmatrix} = \begin{bmatrix} 3.5 \\ -7.5 \\ 4 \end{bmatrix} = 3.5 \begin{bmatrix} 1 \\ -2.143 \\ 1.143 \end{bmatrix} = b_2 \mathbf{y}_2'$$

and $$\mathbf{y}_3 = \begin{bmatrix} 1 & -1 & 0 \\ -1 & 2 & -1 \\ 0 & -1 & 1 \end{bmatrix} \begin{bmatrix} 1 \\ -2.143 \\ 1.143 \end{bmatrix} = \begin{bmatrix} 3.143 \\ -6.429 \\ 3.286 \end{bmatrix} = 3.143 \begin{bmatrix} 1 \\ -2.045 \\ 1.045 \end{bmatrix} = b_3 \mathbf{y}_3'$$

We continue the calculations in Table 2.1

Table 2.1

r	$\mathbf{y}_r$	$\mathbf{y}_{r+1}$	b_{r+1}	$\mathbf{y}'_{r+1}$
0	(1, –1, 2)	(2, –5, –3)	2	(1, –2.5, 1.5)
1	(1, –2.5, 1.5)	(3.5, –7.5, 4)	3.5	(1, –2.143, 1.143)
2	(1, –2.143, 1.143)	(3.143, –6.429, 3.286)	3.143	(1, –2.045, 1.045)
3	(1, –2.045, 1.045)	(3.045, –6.135, 3.090)	3.045	(1, –2.015, 1.015)
4	(1, –2.015, 1.015)	(3.015, –6.045, 3.030)	3.015	(1, –2.005, 1.005)
5	(1, –2.005, 1.005)	(3.005, –6.015, 3.010)	3.005	(1, –2.002, 1.002)
6	(1, –2,002, 1.002)	(3.002, –6.006, 3.004)	3.002	(1, –2.001, 1.001)
7	(1, –2.001, 1.001)	(3.001, –6.003, 3.002)	3.001	(1, –2.000, 1.000)
8	(1, –2.000, 1.000)	(3.000, –6.000, 3.000)	3.000	(1, –2.000, 1.000)

We can see that the method has converged fairly quickly to the eigenvalue 3 to 3 d.p. But, before we pass the technique with flying colours, consider a second example.

Example 2

The matrix $\mathbf{D} = \begin{bmatrix} 4 & 2 & -2 \\ 1 & 3 & 1 \\ -1 & -1 & 5 \end{bmatrix}$ has eigenvalues 2, 4, 6. Let us make an initial guess of $\mathbf{y}_0 = \begin{bmatrix} 1 \\ 1 \\ 1 \end{bmatrix}$ for the eigenvector associated with the eigenvalue of largest modulus.

Then $$\mathbf{y}_1 = \begin{bmatrix} 4 & 2 & -2 \\ 1 & 3 & 1 \\ -1 & -1 & 5 \end{bmatrix} \begin{bmatrix} 1 \\ 1 \\ 1 \end{bmatrix} = \begin{bmatrix} 4 \\ 5 \\ 3 \end{bmatrix} = 4 \begin{bmatrix} 1 \\ 1.25 \\ 0.75 \end{bmatrix} = b_1 \mathbf{y}_1'$$

Further $\mathbf{y}_2 = \begin{bmatrix} 4 & 2 & -2 \\ 1 & 3 & 1 \\ -1 & -1 & 5 \end{bmatrix} \begin{bmatrix} 1 \\ 1.25 \\ 0.75 \end{bmatrix} = \begin{bmatrix} 5 \\ 5.5 \\ 1.5 \end{bmatrix} = 5 \begin{bmatrix} 1 \\ 1.1 \\ 0.3 \end{bmatrix} = b_2 \mathbf{y}'_2$

The iterations are continued in Table 2.2.

Table 2.2

r	$\mathbf{y}_r$	$\mathbf{y}_{r+1}$	b_{r+1}	$\mathbf{y}'_{r+1}$
0	(1, 1, 1)	(4, 5, 3)	4	(1, 1.25, 0.75)
1	(1, 1.25, 0.75)	(5, 5.5, 1.5)	5	(1, 1.1, 0.3)
2	(1, 1.1, 0.3)	(5.6, 4.6, –0.6)	5.6	(1, 0.821, –0.107)
3	(1, 0.821, –0.107)	(5.856, 3.356, –2.356)	5.856	(1, 0.573, –0.402)
4	(1, 0.573, –0.402)	(5.950, 2.317, –3.583)	5.950	(1, 0.389, –0.602)
5	(1, 0.389, –0.602)	(5.982, 1.565, –4.399)	5.982	(1, 0.262, –0.735)
6	(1, 0.262, –0.735)	(5.995, 1.051, –4.937)	5.995	(1, 0.175, –0.826)
7	(1, 0.175, – 0.826)	(5.998, 0.701, –5.295)	5.998	(1, 0.117, –0.883)
8	(1, 0.117, –0.883)	(6.000, 0.668, –5.532)	6.000	(1, 0.111, –0.922)
9	(1, 0.111, –0.922)	(6.066, 0.411, –5.721)	6.066	(1, 0.068, –0.943)
10	(1, 0.068, –0.943)	(6.022, 0.261, –5.783)	6.022	(1, 0.043, –0.960)

You can see that this time convergence is much slower. Is this because of a poor initial approximation or some other factor? Looking back at formula (2.21) we might expect the rate of convergence to be governed by the ratios $\left|\frac{\lambda_2}{\lambda_1}\right|$ and $\left|\frac{\lambda_3}{\lambda_1}\right|$. In the previous example, these ratios were 1/3 and 0, whereas in this example the ratios are 2/3 and 1/3.

The worst initial approximation we could take is one at 'right angles' to the eigenvector sought, since the sequence of approximating vectors has 'further to go'. In Example 1, the initial approximation was by no means perpendicular to the required eigenvector, but in Example 2 the initial approximation $\begin{bmatrix} 1 \\ 1 \\ 1 \end{bmatrix}$ was perpendicular to the eigenvector $\begin{bmatrix} 1 \\ 0 \\ -1 \end{bmatrix}$.

Example 2 does show that the iteration is really on the eigenvector and *not* on the eigenvalue. We obtained an estimate of the largest eigenvalue correct to 3 d.p. as early as the 8th iteration but the eigenvector was nowhere in sight.

We now apply the iterative technique to the matrix **H** we derived at the beginning

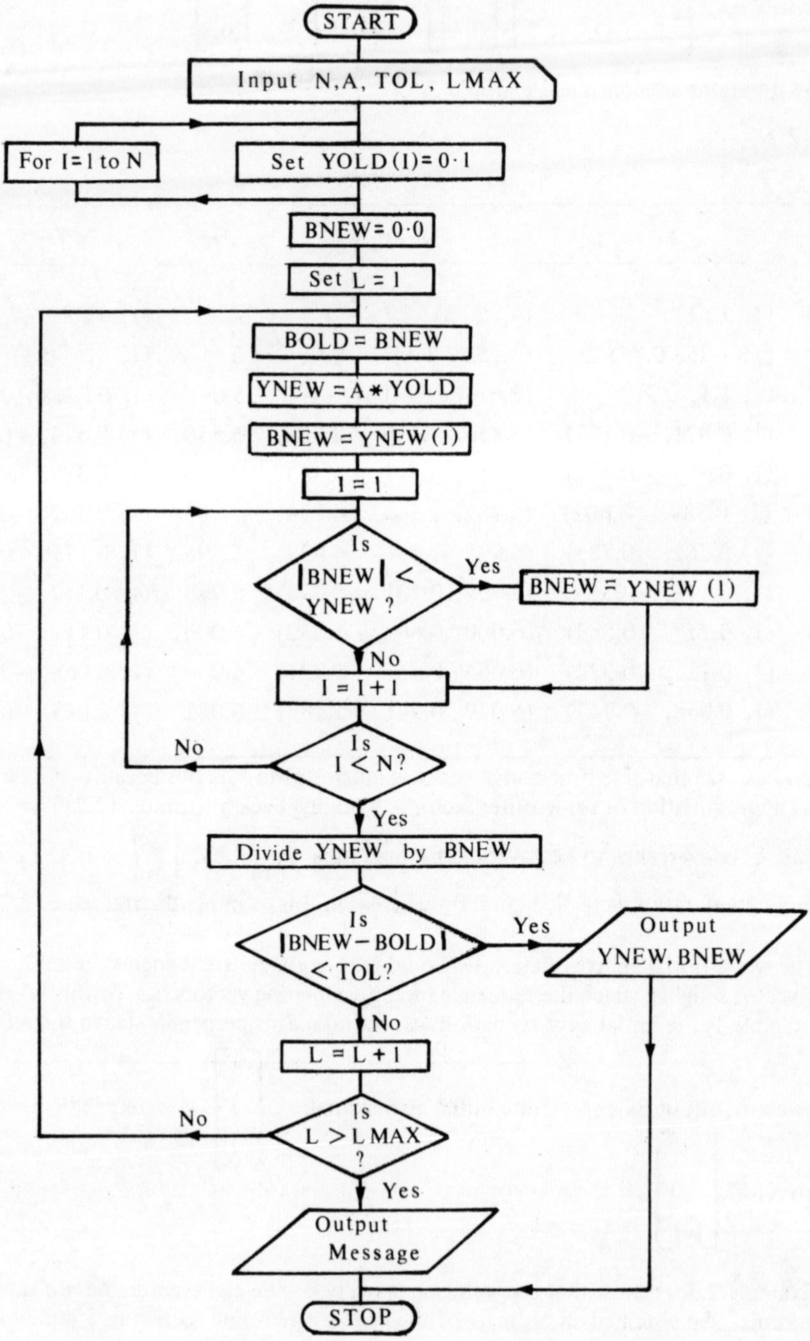
START
Input N,A, TOL, LMAX
For I=1 to N
Set YOLD(I)=0·1
BNEW=0·0
Set L=1
BOLD=BNEW
YNEW=A∗YOLD
BNEW=YNEW(1)
I=1
Is |BNEW| < YNEW ?
Yes
BNEW=YNEW (1)
No
I=I+1
Is I < N?
No
Yes
Divide YNEW by BNEW
Is |BNEW−BOLD| < TOL?
Yes
Output YNEW, BNEW
No
L=L+1
Is L > LMAX ?
No
Yes
Output Message
STOP

Figure 2.11

of this section.

A flow chart for this method is given in Figure 2.11.

The results are: largest eigenvalue = 4.281 (3 d.p.) with corresponding eigenvector (0.913, –0.402, 0.076). Notice that the eigenvector is normalised to unit length.

Breakdown of the iterative technique

The process may break down in certain circumstances.

By a fluke, we may choose as our starting approximation an eigenvector corresponding to an eigenvalue of smaller modulus.

For example, if in Example 1 we start with $\mathbf{y}_0 = \begin{bmatrix} 1 \\ 0 \\ -1 \end{bmatrix}$ we should find $b_1 = 1$,

$$\mathbf{y}_1 = \begin{bmatrix} 1 \\ 0 \\ -1 \end{bmatrix}.$$

Trouble arises if $|\lambda_1| = |\lambda_2|$ which will be the case if $\lambda_2 = -\lambda_1$ or if λ_1 and λ_2 are complex conjugates.

Consider the matrix $\mathbf{C} = \begin{bmatrix} 1 & 0 & 0 \\ -2 & 2 & -1 \\ 0 & 1 & 2 \end{bmatrix}$. If we start with $\mathbf{y}_0 = (0, 1, 1)$ we find that the iterations do not tend toward any limit.

In fact the eigenvalues of $\mathbf{C}$ are 1, 2 – i, 2 + i and the eigenvectors which correspond to them are $\begin{bmatrix} 1 \\ 1 \\ -1 \end{bmatrix}$, $\begin{bmatrix} 0 \\ 1 \\ i \end{bmatrix}$ and $\begin{bmatrix} 0 \\ 1 \\ -i \end{bmatrix}$ respectively.

Problems

1. Find by the iterative method the largest eigenvalue and the associated eigenvector for the matrices of Problem 1, (i), (iv), (v), (vi), (viii), (x) in Section 2.3.

2. Given that the largest eigenvalue of a matrix $\mathbf{A}$ is real and simple (i.e., is not a multiple root) describe an iterative method of finding its value.

   Write a program to carry out the determination of largest eigenvalue of the matrix below, using the iterative method. $\begin{bmatrix} 1 & 1 & 1 \\ 1 & 2 & 3 \\ 1 & 3 & 6 \end{bmatrix}$. (L.U.)

3. If $\mathbf{C} = \begin{bmatrix} 5 & 2 & 0 \\ 2 & 10 & 2 \\ 0 & 2 & 10 \end{bmatrix}$ apply an iterative method to find the dominant eigenvalue and eigenvector (work in fractions). Use an initial approximation of $\begin{bmatrix} 1 \\ 2 \\ 3 \end{bmatrix}$ for the eigenvector.

What factors determine the rate of convergence of this method? Explain *briefly* how you would find the two other eigenvalues of **C**.

4. The matrix $\mathbf{A} = \begin{bmatrix} b & c & 0 & 0 \\ a & b & c & 0 \\ 0 & a & b & c \\ 0 & 0 & a & b \end{bmatrix}$ where $\begin{matrix} a = -0.539 \\ b = \ \ 1.173 \\ c = -0.634 \end{matrix}$ are given. Show that its characteristic equation is $(\lambda - b)^4 - 3ca(\lambda - b)^2 + c^2a^2 = 0$ and deduce the roots $\lambda = b \pm \sqrt{\tfrac{1}{2}ca(3 \pm \sqrt{5})}$.

Determine also the eigenvector **x** corresponding to the largest eigenvalue and verify that the equation $\mathbf{Ax} = \lambda\mathbf{x}$ is satisfied.

5. If $\mathbf{B} = \begin{bmatrix} 10.0 & 2.0 & 0.0 \\ 2.0 & 10.4 & 2.0 \\ 0.0 & 2.0 & 6.4 \end{bmatrix}$ find its dominant eigenvalue and eigenvector working to 2 d.p.

6. Show that if the latent roots of a matrix **A** are real and distinct the iteration process $\mathbf{z}_{i+1} = \mathbf{Ay}_i$; $\mathbf{y}_{i+1} = \mathbf{z}_{i+1}/k_{i+1}$ where k_{i+1} is the numerically largest element of $\mathbf{z}_{i+1}$ and $\mathbf{y}_0$ is an arbitrary vector, can be used to obtain the dominant latent root and latent vector of **A**.

Determine the dominant latent root of the matrix $\mathbf{A} = \begin{bmatrix} 9 & 10 & 8 \\ 10 & 5 & -1 \\ 8 & -1 & 3 \end{bmatrix}$ to the nearest whole number.

7. Verify the Cayley - Hamilton theorem for the matrix $\mathbf{A} = \begin{bmatrix} -1 & 2 & 4 \\ 2 & 2 & -2 \\ 4 & -2 & -1 \end{bmatrix}$. Show also that in this case **A** satisfies an equation of lower degree than its characteristic equation.
Hence, or otherwise, determine the inverse of **A**.
Further, apply direct iteration $\mathbf{x}^{(k+1)} = \dfrac{1}{\lambda_{k+1}} \mathbf{Ax}^{(k)}$, $\mathbf{x}^{(0)} = (-1, 0, 1)$, to obtain approximate values for the dominant eigenvalue (-6) and eigenvector $(-1, \tfrac{1}{2}, 1)$ of **A**. (C.E.I.)

8. Specify clearly as an algorithm *or* in the form of a brief flow diagram the iterative method suggested by $\mathbf{x}_{k+1} = \dfrac{1}{\lambda_{k+1}} \mathbf{Ax}_k$ for determining the dominant eigenvalue and eigenvector of the system $\mathbf{Ax} = \lambda\mathbf{x}$.
In a three-dimensional stress problem we have the matrix $\begin{bmatrix} \sigma_x & \tau_{xy} & \tau_{zx} \\ \tau_{xy} & \sigma_y & \tau_{yz} \\ \tau_{zx} & \tau_{yz} & \sigma_z \end{bmatrix}$.

Using the above method, or otherwise, write a program in a specified code to determine the largest principal stress (eigenvalue) and its associated direction ratios (eigenvector.) (C.E.I.)

2.8 DETERMINATION OF OTHER EIGENVALUES

In practical situations there is often more interest in the lowest eigenvalue; sometimes it is important to know *all* the eigenvalues. In this section we examine some ways by which the other eigenvalues may be found.

Lowest Eigenvalue

Let $\mathbf{A}$ be a non-singular matrix, λ one of its eigenvalues and $\mathbf{x}$ a corresponding eigenvector, so that $\mathbf{Ax} = \lambda\mathbf{x}$. Then, premultiplication by $\mathbf{A}^{-1}$ produces $\mathbf{A}^{-1}\mathbf{Ax} = \lambda\mathbf{A}^{-1}\mathbf{x}$ i.e. $\mathbf{x} = \lambda\mathbf{A}^{-1}\mathbf{x}$. Dividing both sides by λ (which we may do since $\mathbf{A}$ being non-singular implies that $\lambda \neq 0$) we obtain $(1/\lambda)\mathbf{x} = \mathbf{A}^{-1}\mathbf{x}$. This last equation may be written as $\mathbf{A}^{-1}\mathbf{x} = (1/\lambda)\mathbf{x}$ which demonstrates that $1/\lambda$ is an eigenvalue of $\mathbf{A}^{-1}$ and $\mathbf{x}$ is a corresponding eigenvector.

Example

The matrix $\mathbf{D} = \begin{bmatrix} 4 & 2 & -2 \\ 1 & 3 & 1 \\ -1 & -1 & 5 \end{bmatrix}$ had eigenvalues 2, 4 and 6. (page 38)

Corresponding eigenvectors were $\begin{bmatrix} 1 \\ -1 \\ 0 \end{bmatrix}$, $\begin{bmatrix} 0 \\ 1 \\ 1 \end{bmatrix}$ and $\begin{bmatrix} 1 \\ 0 \\ -1 \end{bmatrix}$ respectively. Therefore the matrix $\mathbf{D}^{-1} = \dfrac{1}{24}\begin{bmatrix} 8 & -4 & 4 \\ -3 & 9 & -3 \\ 1 & 1 & 5 \end{bmatrix}$ will have eigenvalues $\dfrac{1}{2}, \dfrac{1}{4}, \dfrac{1}{6}$ with corresponding eigenvectors $\begin{bmatrix} 1 \\ -1 \\ 0 \end{bmatrix}$, $\begin{bmatrix} 0 \\ 1 \\ 1 \end{bmatrix}$ and $\begin{bmatrix} 1 \\ 0 \\ -1 \end{bmatrix}$ respectively. A procedure to find the lowest eigenvalue of a matrix then is as follows. Invert the matrix and find the eigenvalue of largest modulus of the inverse by the iterative technique developed in the last section. The reciprocal of the result is the eigenvalue we seek and the eigenvector obtained by the iterations will serve as an eigenvector for our purposes.

You apply the iterative technique to $\mathbf{D}^{-1}$.

The rate of convergence is governed by the rules described in the last section.

However, the explicit determination of the inverse matrix is a lengthy process. Then instead of finding $\mathbf{y}_r$ from the equation $\mathbf{y}_{r+1} = \mathbf{A}^{-1}\mathbf{y}_r$, we solve the set of equations $\mathbf{Ay}_{r+1} = \mathbf{y}_r$ by Gauss Elimination.

Example

We seek to find the lowest eigenvalue of matrix $\mathbf{D}$ as above. We shall find $\mathbf{y}_r$ via

$$\mathbf{Dy}_r = \mathbf{y}'_{r-1} \qquad (2.22)$$

If we start with an initial guess $\mathbf{y}_0 = \begin{bmatrix} 1 \\ 1 \\ 1 \end{bmatrix}$ then $\begin{bmatrix} 4 & 2 & -2 \\ 1 & 3 & 1 \\ -1 & -1 & 5 \end{bmatrix} \begin{bmatrix} y_1 \\ y_2 \\ y_3 \end{bmatrix} = \begin{bmatrix} 1 \\ 1 \\ 1 \end{bmatrix}$

Gaussian elimination with partial pivotting produes the equations

$$\begin{bmatrix} 4 & 2 & -2 \\ 0 & 2.5 & 1.5 \\ 0 & 0 & 4.8 \end{bmatrix} \begin{bmatrix} y_1 \\ y_2 \\ y_3 \end{bmatrix} = \begin{bmatrix} 1 \\ 3/4 \\ 1.4 \end{bmatrix}$$

Hence $y_3 = 7/24, \; y_2 = 3/24, \; y_1 = 8/24$

so that $\mathbf{y}_1 = (8/24, \; 3/24, \; 7/24) = 8/24(1, \; 3/8, \; 7/8) = b_1\mathbf{y}'_1$.

Now we need to solve the equations $\begin{bmatrix} 4 & 2 & -2 \\ 1 & 3 & 1 \\ -1 & -1 & 5 \end{bmatrix} \begin{bmatrix} y_1 \\ y_2 \\ y_3 \end{bmatrix} = \begin{bmatrix} 1 \\ 3/8 \\ 7/8 \end{bmatrix}$

Since we shall be carrying out the same sequence of operations each time that we use the iterative formula, we can establish the row operations in the first place and then simply apply these rules to the sequence of vectors $\mathbf{y}'_1, \mathbf{y}'_2, \ldots$. You continue the computation of the lowest eigenvalue of $\mathbf{D}$. When producing $\mathbf{y}'_r$ from $\mathbf{y}_r$ we divide by the component of largest modulus, which is denoted b_r.

Eigenvalue nearest a prescribed value

Suppose we know that there is an eigenvalue approximately equal to some number p. Let the matrix $\mathbf{A}$ have eigenvalues $\lambda_1, \lambda_2, \ldots, \lambda_n$ with corresponding eigenvectors $\mathbf{x}_1, \mathbf{x}_2, \ldots, \mathbf{x}_n$. Then if $\mathbf{A}\mathbf{x}_1 = \lambda_1\mathbf{x}_1$ it follows that $(\mathbf{A} - p\mathbf{I})\mathbf{x}_1 = (\lambda_1 - p)\mathbf{x}_1$. $\mathbf{A}\mathbf{x}_2 = \lambda_2\mathbf{x}_2$ implies that $(\mathbf{A} - p\mathbf{I})\mathbf{x}_2 = (\lambda_2 - p)\mathbf{x}_2$ and so on. The eigenvalues of the matrix $(\mathbf{A} - p\mathbf{I})$ are $(\lambda_1 - p), (\lambda_2 - p), \ldots, (\lambda_n - p)$; the eigenvector corresponding to $(\lambda_i - p)$ is the same as that which corresponds to λ_i. For example if a matrix $\mathbf{D}$ has eigenvalues 2, 4, 6 and we seek the middle of these, then the matrix $(\mathbf{D} - 3.5\mathbf{I})$ has eigenvalues -1.5, 0.5 and 2.5. The iteration procedure to find the eigenvalue of least modulus will produce the value $\lambda_2 = 0.5$ with associated eigenvector $\mathbf{x}_2$. Then we add back 3.5 to λ_2 to obtain the required value 4.0; $\mathbf{x}_2$ is the required eigenvector. This method does require knowledge of the approximate location of the eigenvalues.

Acceleration of convergence

We may use the above technique to accelerate the convergence of the iterative method. For example if the eigenvalues of a matrix $\mathbf{A}$ are 5, 4, 3, 2, 1 then the iterative method applied to $\mathbf{A}$ will converge at a rate which depends upon the speed at which $(4/5)^r \to 0$. If we form $(\mathbf{A} - 2.5\mathbf{I})$ then the appropriate eigenvalues are 2.5, 1.5, 0.5, -0.5 and -1.5 and the rate of convergence is determined by the speed at which $((1.5)/(2.5))^r \to 0$. Note that if we form $(\mathbf{A} - 3.5\mathbf{I})$ so that the eigenvalues are now 1.5, 0.5, -0.5, -1.5 and -2.5 the rate of convergence depends upon $(0.5/1.5)^r$.

Subdominant eigenvalues – the method of deflation

A popular method of determining eigenvalues of a matrix other than those of largest or smallest modulus is the method of **deflation** of the matrix. In effect this eliminates a previously determined mode from the system and works with a smaller matrix.

Let λ_1 be the eigenvalue of largest modulus of a $n \times n$ matrix $\mathbf{A}$ and let $\mathbf{x}_1$ be the corresponding eigenvector; the other eigenvalues are $\lambda_2, \lambda_3, \ldots, \lambda_n$. Let us suppose the largest component of $\mathbf{x}_1$ is the first one. We partition $\mathbf{A}$ as $\mathbf{A} = \left[\begin{array}{c}\mathbf{a}_1\\ \hline \mathbf{B}\end{array}\right]$ where $\mathbf{a}_1$ is the first row of $\mathbf{A}$. It follows that $\mathbf{B}$ is the $(n-1) \times n$ matrix obtained by deleting the first row of $\mathbf{A}$. Note that if we carry out our customary practice of dividing approximate eigenvectors by their largest element, then the first element of $\mathbf{x}_1$ will be 1.

Now consider the matrix $\quad \mathbf{A}_1 = \mathbf{A} - \mathbf{x}_1\mathbf{a}_1 \qquad (2.23)$

It can be shown that this matrix has eigenvalues $0, \lambda_2, \lambda_3, \ldots, \lambda_n$.

Example

Let $\mathbf{A} = \begin{bmatrix} 4 & 2 & -2 \\ 1 & 3 & 1 \\ -1 & -1 & 5 \end{bmatrix}$ which has eigenvalues 2, 4, 6. Then suppose we have already found the largest eigenvalue by iteration together with its associated eigenvector $\begin{bmatrix} 1 \\ 0 \\ -1 \end{bmatrix}$.

We now form $\mathbf{A}_1 = \mathbf{A} - \mathbf{x}_1\mathbf{a}_1 = \begin{bmatrix} 4 & 2 & -2 \\ 1 & 3 & 1 \\ -1 & -1 & 5 \end{bmatrix} - \begin{bmatrix} 1 \\ 0 \\ -1 \end{bmatrix} (4 \quad 2 \quad -2)$

$= \begin{bmatrix} 4 & 2 & -2 \\ 1 & 3 & 1 \\ -1 & -1 & 5 \end{bmatrix} - \begin{bmatrix} 4 & 2 & -2 \\ 0 & 0 & 0 \\ -4 & -2 & 2 \end{bmatrix} = \begin{bmatrix} 0 & 0 & 0 \\ 1 & 3 & 1 \\ 3 & 1 & 3 \end{bmatrix}$. Now theory predicts that $\mathbf{A}_1$ has eigenvalues 4 and 2. Notice that the top row is zero. This was expected.

If we were using the traditional algebra approach we would solve

$\begin{vmatrix} 0-\lambda & 0 & 0 \\ 1 & 3-\lambda & 1 \\ 3 & 1 & 3-\lambda \end{vmatrix} = 0$. Because of the two zeros on the top row this reduces to

$-\lambda \begin{vmatrix} 3-\lambda & 1 \\ 1 & 3-\lambda \end{vmatrix} = 0$, showing that there is a zero eigenvalue. To determine other eigenvalues of $\mathbf{A}_1$ we may effectively remove its first row and first column and work with matrix $\mathbf{B}_1 = \begin{bmatrix} 3 & 1 \\ 1 & 3 \end{bmatrix}$ which has 4 as the eigenvalue of largest modulus. The associated eigenvector is $\mathbf{x}_2^* = \begin{bmatrix} 1 \\ 1 \end{bmatrix}$. Then a second eigenvalue of $\mathbf{A}$ is 4. We may repeat this procedure to find the remaining eigenvalue. We form $\mathbf{A}_2 = \mathbf{A}_1 - \mathbf{x}_2^* \mathbf{a}_2$ $= \begin{bmatrix} 3 & 1 \\ 1 & 3 \end{bmatrix} - \begin{bmatrix} 1 \\ 1 \end{bmatrix} \begin{bmatrix} 3 & 1 \end{bmatrix} = \begin{bmatrix} 0 & 0 \\ 2 & 2 \end{bmatrix}$. Ignoring the first row and first column of $\mathbf{A}_2$ we are left with the element 2: the third eigenvalue of $\mathbf{A}$.

The determination of the eigenvectors related to these eigenvalues is in general not as simple as is often claimed, especially if these eigenvectors have a zero element.

Problems

1. Find by inverse iteration the smallest eigenvalue and associated eigenvector of the matrices of Section 2.3, Problem 1, (iv), (x), Problem 3; Section 2.4, Problem 1 (iii).

2. If we have an approximate eigenvector $\mathbf{x}$ and corresponding eigenvalue $\lambda^{(n)}$ we may obtain an improved estimate of the eigenvalue by means of the *Rayleigh Quotient* $\lambda^{(n+1)} = \dfrac{\mathbf{x}^T \mathbf{A}\mathbf{x}}{\mathbf{x} \cdot \mathbf{x}}$

   Given the approximate eigenvector $\mathbf{x} = (0.731, 0.233, 1)^T$ for matrix $\mathbf{A} = \begin{bmatrix} 2 & 1 & 3 \\ 1 & -1 & 1 \\ 3 & 1 & 4 \end{bmatrix}$

   obtain an improved estimate of λ.

3. Find all the eigenvalues and eigenvectors of the matrices.

   (i) $\begin{bmatrix} 2 & -1 & 0 \\ -1 & 2 & -1 \\ 0 & -1 & 2 \end{bmatrix}$ (ii) $\begin{bmatrix} 1 & 1 & 1 & 1 \\ 1 & 2 & 3 & 4 \\ 1 & 3 & 6 & 10 \\ 1 & 4 & 10 & 20 \end{bmatrix}$

4. The moment of inertia matrix of a 3-dimensional rigid body is $mr^2 \begin{bmatrix} 10.000 & 0.134 & -0.866 \\ 0.134 & 6.500 & -1.000 \\ -0.866 & -1.000 & 7.500 \end{bmatrix}$

   the eigenvalues being the principal moments of inertia and the eigenvectors the directions of principal axes.

   Determine by iteration (from the characteristic equation) correct to 3 d.p. the numerically smallest eigenvalue and hence find its associated eigenvector. How can the remaining solutions be obtained?

5. Find by direct iteration correct to 2d.p. the dominant eigenvalue and eigenvector of the matrix

   $\mathbf{A} = \begin{bmatrix} 9 & 10 & 8 \\ 10 & 5 & -1 \\ 8 & -1 & 3 \end{bmatrix}$, taking $\mathbf{x}^{(0)} = \begin{bmatrix} 5 \\ 3 \\ 2 \end{bmatrix}$

   Then using the fact that if λ is an eigenvalue of $\mathbf{A}$, $\lambda - a$ is an eigenvalue of $\mathbf{A} - a\mathbf{I}$; similarly obtain a second eigenvalue and associated eigenvector.

6. The equations of motion of a particular 3 mass system lead to the eigenvalue formulation

   $\mathbf{Ax} = \lambda \mathbf{Bx}$, where $\mathbf{A} = \begin{bmatrix} 2 & -1 & 0 \\ -1 & 2 & -1 \\ 0 & -1 & 2 \end{bmatrix}$ and $\mathbf{B} = \begin{bmatrix} 1 & 0 & 0 \\ 0 & 2 & 0 \\ 0 & 0 & 3 \end{bmatrix}$.

Form the characteristic equation and hence determine the eigenvalues and eigenvectors (fundamental frequencies and modes of vibration).

Expressing $\mathbf{B} = \mathbf{G}^2$ where $\mathbf{G}$ is a diagonal matrix, form the matrix $\mathbf{Q} = \mathbf{G}^{-1}\mathbf{A}\mathbf{G}^{-1}$ and verify that the eigenvalues of $\mathbf{Q}$ are exactly those obtained above while each eigenvector of $\mathbf{Q}$ is $\mathbf{G}$ times the corresponding eigenvector of $\mathbf{Ax} = \lambda\mathbf{Bx}$.

7. Find by direct iteration the dominant eigenvalue and eigenvector of the matrix $\mathbf{A} = \begin{bmatrix} 9 & 10 & 8 \\ 10 & 5 & -1 \\ 8 & -1 & 3 \end{bmatrix}$ taking $\mathbf{x}_0 = \begin{bmatrix} 1 \\ 1 \\ 0 \end{bmatrix}$ as a starting vector and working to one decimal place with 3 iterations.

Using the fact that if λ is an eigenvalue of $\mathbf{A}$, then $\lambda - p$ is an eigenvalue of $\mathbf{A} - p\mathbf{I}$ where $\mathbf{I}$ is the unit matrix, obtain a second eigenvalue and associated eigenvector of $\mathbf{A}$, again working to one decimal place and starting with $\begin{bmatrix} -1 \\ 1 \\ 1 \end{bmatrix}$. (Only two iterations are required.)

8. The largest eigenvalue, λ_1, of the matrix $\mathbf{C} = \begin{bmatrix} 5 & -4 & 1 \\ -4 & 6 & -4 \\ 1 & -4 & 7 \end{bmatrix}$ is sought. Starting with $\mathbf{x}^{(0)} = (1, 1, 1)$ verify that successive iterations yield; $\lambda_1^{(1)} = 2$; $\mathbf{x}^{(1)} = (1, -1, 2)$; $\lambda_1^{(2)} = 11$; $\mathbf{x}^{(2)} = (1, -1.63, 1.72)$; $\lambda_1^{(3)} = 13.24$; $\mathbf{x}^{(3)} = (1, -1.56, 1.48)$; $\lambda^{(8)} = 12.265$; $\mathbf{x}^{(8)} = (1, -1.485, 1.325)$.

Use the Rayleigh Quotient to produce a better estimate of λ_1 from $\lambda_1^{(1)}$.

2.9 TRANSFORMATION METHODS FOR SYMMETRIC MATRICES

The eigenvalues of a diagonal matrix are simply the diagonal elements themselves. There are other kinds of matrix whose eigenvalues are easy to determine; for example, tri-diagonal matrices. We present methods for symmetric matrices and indicate where these methods may be modified to apply to non-symmetric matrices. We describe one method in detail and mention the main features of two others. Whilst these methods have been superseded by more powerful techniques, it is important that you understand the basic ideas to give you a clue as to how the more recent methods perform.

Jacobi's method

Let $\mathbf{A}$ be a symmetric matrix. Then suppose we find an orthogonal matrix $\mathbf{P}$ (i.e. $\mathbf{P}^{-1} = \mathbf{P}^T$) such that $\mathbf{B} = \mathbf{P}^T\mathbf{A}\mathbf{P}$ is diagonal. The eigenvalues of $\mathbf{B}$ are the same as those of $\mathbf{A}$. The eigenvectors of $\mathbf{A}$ are the columns of $\mathbf{P}$. This result follows from the more general result that if $\mathbf{Q}$ is any orthogonal matrix and $\mathbf{C} = \mathbf{Q}^T\mathbf{A}\mathbf{Q}$, then $\mathbf{C}$ and $\mathbf{A}$ have the same eigenvalues. For $|\mathbf{Q}^T\mathbf{A}\mathbf{Q} - \lambda\mathbf{I}| = |\mathbf{Q}^{-1}\mathbf{A}\mathbf{Q} - \lambda\mathbf{Q}^{-1}\mathbf{I}\mathbf{Q}| = |\mathbf{Q}^{-1}| \,.\, |\mathbf{A} - \lambda\mathbf{I}|\; |\mathbf{Q}|$ $= |\mathbf{A} - \lambda\mathbf{I}|$

Now a simple orthogonal 3 x 3 matrix is $\begin{bmatrix} \cos\theta & \sin\theta & 0 \\ -\sin\theta & \cos\theta & 0 \\ 0 & 0 & 1 \end{bmatrix}$ which represents a rotation clockwise through an angle θ about the z - axis (and, hence, in the x - y plane). However, for large matrices **A** the direct determination of the matrix **P** is not easy. What we do is to obtain an approximation to **P** of the form $\mathbf{P} = \mathbf{P}_1\mathbf{P}_2\mathbf{P}_3 \ldots \mathbf{P}_r$ where each $\mathbf{P}_i$ is an orthogonal matrix corresponding to a rotation in some plane. We say 'approximation' since to obtain equality we should need an infinite number of $\mathbf{P}_i$. The trouble is that the sensible way to proceed from matrix **A** is to systematically reduce off-diagonal elements to zero; we need only bother with half of these and use symmetry to nullify the others. However, later $\mathbf{P}_i$ will change elements of **A** previously set to zero. If we decide to eliminate the largest off-diagonal element at each stage, then we may perform an iterative process, stopping when all off-diagonal elements are less in modulus than some predetermined small tolerance. The diagonal elements of the current matrix are taken as approximations to the eigenvalues of **A**. A drawback is that we do not know how fast the method will converge, or even whether it will converge at all. In addition, terminating the iteration may introduce damaging errors in the approximations. The iteration proceeds $\mathbf{A}$, $\mathbf{P}_1^T\mathbf{A}\mathbf{P}$, $\mathbf{P}_2^T\mathbf{P}_1^T\mathbf{A}\mathbf{P}_1\mathbf{P}_2$, $\mathbf{P}_3^T\mathbf{P}_2^T\mathbf{P}_1^T\mathbf{A}\mathbf{P}_1\mathbf{P}_2\mathbf{P}_3$, ...†. If we decide to eliminate the element a_{ij} (and also a_{ji}) then we choose the orthogonal matrix **P** where $p_{kk} = 1$, $k \neq i$ or j, $p_{ii} = \cos\theta = p_{jj}$, $p_{ij} = -\sin\theta$, $p_{ji} = \sin\theta$; all other $p_{lm} = 0$. A suitable value for θ has to be found.

$$\text{Consider} \begin{bmatrix} \cos\theta & -\sin\theta & 0 \\ \sin\theta & \cos\theta & 0 \\ 0 & 0 & 1 \end{bmatrix} \begin{bmatrix} a_{11} & a_{12} & a_{13} \\ a_{12} & a_{22} & a_{23} \\ a_{13} & a_{23} & a_{33} \end{bmatrix} \begin{bmatrix} \cos\theta & \sin\theta & 0 \\ -\sin\theta & \cos\theta & 0 \\ 0 & 0 & 1 \end{bmatrix}$$

$$= \begin{bmatrix} a_{11}\cos^2\theta - a_{12}\sin 2\theta + a_{22}\sin^2\theta & \tfrac{1}{2}(a_{11} - a_{12})\sin 2\theta + a_{12}\cos 2\theta & a_{13}\cos\theta - a_{23}\sin\theta \\ \tfrac{1}{2}(a_{11} - a_{22})\sin 2\theta + a_{12}\cos 2\theta & a_{11}\sin^2\theta + a_{12}\sin 2\theta + a_{22}\cos^2\theta & a_{13}\sin\theta + a_{23}\cos\theta \\ a_{13}\cos\theta - a_{23}\sin\theta & a_{13}\sin\theta + a_{23}\cos\theta & a_{33} \end{bmatrix}$$

Notice that the matrix remains symmetric after the transformation and that one element (not in the 1 and 2 rows nor in the 1 and 2 columns) is unaltered. Now, the elements in the positions (1, 3) and (2, 3) are altered and so are, of course, those in positions (3, 1) and (3, 2) whereas, the intention was to reduce the elements a_{12} and a_{13} to zero. This aim will be achieved if we choose θ so that $\tfrac{1}{2}(a_{11} - a_{22})\sin 2\theta + a_{12}\cos 2\theta = 0$ i.e. $\tan 2\theta = -2a_{12}/(a_{11} - a_{12})$.

In general, to eliminate a_{ij} we should choose $\tan 2\theta = -2a_{ij}/(a_{ii} - a_{jj})$.

We really want the values of $\cos\theta$ and $\sin\theta$. These may be found by the rules††

$$\sin 2\theta = \frac{-2a_{ij}\,\text{sign}\,(a_{ii} - a_{jj})}{\sqrt{4a_{ij}^2 + (a_{ii} - a_{jj})^2}} \qquad \sin\theta = \frac{\sin 2\theta}{\sqrt{2(1 + (1 - \sin^2 2\theta)^{1/2})}}$$

$$\cos\theta = \sin 2\theta/(2\sin\theta)$$

† Note that $(\mathbf{P}_1\mathbf{P}_2)^T = \mathbf{P}_2^T\mathbf{P}_1^T = \mathbf{P}_2^{-1}\mathbf{P}_1^{-1} = (\mathbf{P}_1\mathbf{P}_2)^{-1}$ so that $(\mathbf{P}_1\mathbf{P}_2)$ is orthogonal. Similar results hold for $(\mathbf{P}_1\mathbf{P}_2\mathbf{P}_3)$ etc.

†† sign $x = 1, 0$ or -1 according as to whether $x > 0$, $= 0$ or < 0.

Givens' Method

This method was designed to preserve zero off-diagonal elements once they had been created zero. If the rotation is carried out in the (i, j) plane then θ is chosen so that $a_{i-1,j}$ is made zero. It can be shown that this condition reduces to $\tan\theta = a_{i-1,j}/a_{i-1,i}$. The elements which are reduced to zero are, in order, (1, 3), (1, 4), (1, 5), ..., $(1, n)$, (2, 4), (2, 5), ..., $(2, n)$, ...,$(n-2, n)$. Two differences from Jacobi's method should be noted. First, the process is a direct method and ceases after all these elements have been reduced to zero, i.e. after $\frac{1}{2}(n-1)(n-2)$ rotations; whereas Jacobi's method was an *iterative* method. Second, the end product is *not* a diagonal matrix; it is a **tri-diagonal matrix**, i.e. it has zeros except in the principal diagonal and the two adjacent diagonals – one on either side. An advantage of this method is that as the method proceeds, the number of operations at each rotation decreases because of the increasing number of zeros. It has been claimed that the reduction to tridiagonal form is 20 times faster than the comparable Jacobi reduction to diagonal form.

Once we have obtained the tridiagonal form we need to find its eigenvalues, which are, of course, the same as the eigenvalues of the original matrix. We consider this problem shortly.

Householder's Method

This method followed that of Givens by a few years. Its advantage lies in its ability to introduce at a stroke all the zeros possible on any one row. The reduction is again to tridiagonal form and takes only $(n-2)$ rotations; however, the computation in a single transformation is more complicated. Even so, there is a distinct saving in computer time over Givens' method and, of course, the fewer the operations, the less the propagation of error.

Non-symmetric matrices

The Givens and Householder methods can be applied to non-symmetric matrices. In such cases, the reduction is not to tridiagonal form but to 'almost triangular' or **Hessenberg** form.

An Upper Hessenberg matrix $\mathbf{H}$ is one where $h_{ij} = 0$ for $i > j + 1$. It has a sub-diagonal of (possible) non-zero elements more than an upper triangular matrix.

Examples

$$\begin{bmatrix} 1 & 2 & 3 & 4 & 5 \\ 6 & 7 & 6 & 5 & 4 \\ 0 & 3 & 2 & 1 & 2 \\ 0 & 0 & 3 & 2 & 1 \\ 0 & 0 & 0 & 2 & 3 \end{bmatrix}, \quad \begin{bmatrix} 8 & 3 & 2 & 6 \\ 7 & 2 & 1 & 5 \\ 0 & 1 & 0 & 4 \\ 0 & 0 & -1 & 3 \end{bmatrix}.$$

It can be shown that the determination of the eigenvalues and eigenvectors of a Hessenberg matrix is markedly easier than for the original matrix. The reduction in the number of independent elements from n^2 in $\mathbf{A}$ to $\frac{1}{2}(n-1)(n-2)$ in $\mathbf{H}$ is apparently small, but allows a powerful algorithm to operate.

Eigenvalues of a symmetric tridiagonal matrix

Let **C** be a symmetric tridiagonal matrix derived from a symmetric matrix **A**. Then its characteristic equation is given by $|\mathbf{C} - \lambda\mathbf{I}| = 0$ i.e.

$$\begin{vmatrix} c_{11} - \lambda & c_{12} & & & \\ c_{12} & c_{22} - \lambda & c_{23} & & \\ & c_{23} & c_{33} - \lambda & c_{34} & \\ & & & \ddots & \\ & & & & c_{n-1,\,n} \\ & & & c_{n-1,\,n} & c_{nn} - \lambda \end{vmatrix} = 0 \quad \text{or}$$

$$p_n(\lambda) = \begin{vmatrix} \alpha_1 - \lambda & \beta_2 & & & \\ \beta_2 & \alpha_2 - \lambda & \beta_3 & & \\ & \beta_3 & \alpha_3 - \lambda & \beta_4 & \\ & & & \ddots & \\ & & & & \beta_n \\ & & & \beta_n & \alpha_n - \lambda \end{vmatrix} = 0$$

If we expand the determinant by its last column we obtain the recurrence relationship $p_n(\lambda) = (\alpha_n - \lambda)p_{n-1}(\lambda) - \beta^2_n\, p_{n-2}(\lambda)$. Then we may use the recurrence relationship

$$p_r(\lambda) = (\alpha_r - \lambda)p_{r-1}(\lambda) - \beta^2_r\, p_{r-2}(\lambda) \qquad (2.24)$$

together with initial values $p_1(\lambda) = \alpha_1 - \lambda$ and $p_0(\lambda) = 1$ (for convenience) to find the eigenvalues as follows.

(i) Calculate $p_r(\lambda)$ for $r = 2, 3, \ldots\ldots, n$ by using (2.24)

(ii) Count the number of sign changes in the sequence $\{p_r(\lambda)\}$ for several values of λ and tabulate the results. If a p_r is zero regard its sign as opposite to that of p_{r-1} †

(iii) Locate the eigenvalues approximately; the number of sign changes equals the number of eigenvalues whose values are $\leqslant$ the current value of λ.

(iv) Improve the approximate locations by the method of bisection.

Example

$$\mathbf{C} = \begin{bmatrix} 1 & -1 & 0 & 0 \\ -1 & 2 & -1 & 0 \\ 0 & -1 & 2 & -1 \\ 0 & 0 & -1 & 1 \end{bmatrix} \qquad \mathbf{C} - \lambda\mathbf{I} = \begin{bmatrix} 1-\lambda & -1 & 0 & 0 \\ -1 & 2-\lambda & -1 & 0 \\ 0 & -1 & 2-\lambda & -1 \\ 0 & 0 & -1 & 1-\lambda \end{bmatrix}$$

† If no β_r is zero there are no two consecutive zero p_r and hence no multiple eigenvalues. If any $\beta_r = 0$ then the problem can be reduced to the solution of several smaller tridiagonal problems.

Now $p_0(\lambda) = 1; \quad p_1(\lambda) = 1 - \lambda; \quad p_2(\lambda) = (2-\lambda)(1-\lambda) - 1(1);$

$$p_3(\lambda) = (2-\lambda)(1-3\lambda+\lambda^2) - 1(1-\lambda) = 1 - 6\lambda + 5\lambda^2 - \lambda^3;$$

$$p_4(\lambda) = (1-\lambda)(1-6\lambda+5\lambda^2-\lambda^3) - 1(1-3\lambda+\lambda^2) = -4\lambda + 10\lambda^2 - 6\lambda^3 + \lambda^4.$$

We now try several values of λ

λ	−4	0	4	10	1	2	3
p_0	1	1	1	1	1	1	1
p_1	5	1	−3	−9	0	−1	−2
p_2	29	1	5	71	−1	−1	1
p_3	169	1	−7	−559	−1	1	1
p_4	816	0	16	4960	1	0	−3
sign changes	0	1	4	4	2	3	3

Initially we found one root between $\lambda = -4$ and 0 and three between $\lambda = 0$ and $\lambda = 4$. Further investigation reveals one root between 0 and 1, one between 1 and 2 and one between 3 and 4.

2.10 FURTHER NOTES ON COMPUTATION

Two methods which have found favour with writers of standard computer programs are the **LR and QR algorithms.** The *LR algorithm* in its simple form decomposes a matrix $\mathbf{A}$ into the product of a lower and an upper triangular matrix: $\mathbf{A} = \mathbf{A}_1 = \mathbf{L}_1\mathbf{R}_1$. Then we form $\mathbf{A}_2 = \mathbf{R}_1\mathbf{L}_1$ and decompose $\mathbf{A}_2$ as $\mathbf{L}_2\mathbf{R}_2$; in general, $\mathbf{A}_s = \mathbf{L}_s\mathbf{R}_s$ and $\mathbf{R}_s\mathbf{L}_s = \mathbf{A}_{s+1}$. If the eigenvalues of $\mathbf{A}$ are distinct in modulus then the sequence $\{\mathbf{A}_s\}$ tends to an upper triangular matrix which will usually have as its elements, the eigenvalues of $\mathbf{A}$ in order of decreasing modulus. Coincident eigenvalues do not usually present a problem, but unequal eigenvalues which share the same modulus may cause a modification of the basic result. Drawbacks of this method are that the factorisation $\mathbf{A}_s = \mathbf{L}_s\mathbf{R}_s$ may not exist for some s and even where it does it may be numerically unstable. Wilkinson, an eminent worker in the field of eigenvalue problems, reports that with large matrices convergence may require over 100 iterations, with a steady loss of accuracy as succeeding iterations occur.

The *QR algorithm* overcomes these difficulties. This time the decomposition is into the product of a unitary matrix† $\mathbf{Q}$ and an upper triangular matrix $\mathbf{R}$. Then $\mathbf{A}_1 = \mathbf{A}$ and $\mathbf{A}_s = \mathbf{Q}_s\mathbf{R}_s$, $\mathbf{R}_s\mathbf{Q}_s = \mathbf{A}_{s+1}$ for $s = 1, 2, \ldots\ldots$ Note that

$$\mathbf{A}_{s+1} = \mathbf{Q}_s^*\mathbf{A}_s\mathbf{Q}_s = \mathbf{Q}_s^*\mathbf{Q}_{s-1}^* \ldots \mathbf{Q}_2^*\mathbf{Q}_1^*\mathbf{A}\mathbf{Q}_1\mathbf{Q}_2 \ldots \mathbf{Q}_{s-1}\mathbf{Q}_s$$

$$= (\mathbf{Q}_1\mathbf{Q}_2 \ldots \mathbf{Q}_{s-1}\mathbf{Q}_s)^* \mathbf{A}(\mathbf{Q}_1\mathbf{Q}_2 \ldots \mathbf{Q}_{s-1}\mathbf{Q}_s)$$

† $\mathbf{Q}$ is a **unitary** matrix if $(\overline{\mathbf{Q}})^T = \mathbf{Q}$. We denote $(\overline{\mathbf{Q}})^T$ by $\mathbf{Q}^*$.

showing that the process is equivalent to performing a set of unitary transformations, which is known to be numerically quite stable. We can show that any $n \times n$ matrix **A** can be decomposed as $\mathbf{A} = \mathbf{QR}$ where **R** is an upper triangular matrix with real, non-negative diagonal elements and **Q** is a unitary matrix, which is unique if $\mathbf{A}^{-1}$ exists.

If a non-singular matrix **A** has eigenvalues of different modulus then under the **QR** transformation the elements below the main diagonal tend to zero, those above this diagonal tend to values with fixed modulus and those on the main diagonal tend to the eigenvalues of **A**. If **A** has some eigenvalues sharing the same modulus then there is a slight modification to this statement.

Since the amount of computation for one iteration is proportional to n^3 for a general matrix, it is usual to first transform the matrix to Hessenberg form by Householder's method (or, better still, a tridiagonal matrix if **A** is symmetric) and then apply the **QR** algorithm; the amount of computation is reduced to being proportional to n^2 at each iteration.

Balancing

If the matrix **A** has elements which vary considerably in size it is advisable to transform the matrix into a more *balanced* one so that maximum accuracy is achieved in calculating eigenvalues and eigenvectors. By interchanging rows and columns it is possible to push the smallest elements in the top left-hand corner of the matrix which should be done if the original matrix **A** is real and symmetric: the resulting matrix will also be symmetric. If **A** is not symmetric or if it contains complex elements then the object of balancing is to ensure that row i and column i, regarded as vectors, have the magnitudes which are of the same order. In either situation the eigenvalues of original and resulting matrices are the same. There are sometimes standard computer routines available for balancing.

Chapter Three

Optimization I – Linear Programming

3.1 INTRODUCTION

Many engineering problems are concerned with optimizing the performance of a system. In this chapter we look at a special class of such problems, namely those in which we optimize a linear **objective function** subject to **linear constraints.** These problems are named **linear programming** problems. Although the restriction of linearity may seem severe, there are many problems in which cost is to be minimised or profit maximised where this technique is applicable.

We shall first look at two linear programming problems which can be solved graphically, in order to study the main features of this class of optimization problems. Then we shall find an algebraic solution and develop the **simplex** algorithm. Finally, we shall mention some of the modifications to our basic algorithm which will enable a wider range of linear problems to be tackled. We stress that we are merely providing you with an *introduction* to linear programming.

3.2 GRAPHICAL SOLUTION

Example 1

A company markets two mixes of concrete, X and Y. The company is capable of producing up to 10 truckloads per hour of X or up to 5 truckloads per hour of Y. The trucks available can carry up to 6 loads per hour of X and up to 10 loads per hour of Y, because of the different distances that have to be covered. However, no more than 7 loads per hour can be handled at the loading phase of the operation. The profit on a truckload of mix X is £4 and on a truckload of mix Y is £6.

We may express the problem algebraically. Let the number of truckloads of mixes X and Y that are made per hour be x and y respectively. Clearly, our aim is to maximise the profit $4x + 6y$; however there are certain constraints on the values that x and y can take. First of all, the limitation of production means that $\frac{x}{10} + \frac{y}{5} \leqslant 1$; note that $x/10$ is the fraction of an hour occupied in producing mix X and $y/5$ is the fraction of an hour occupied in producing mix Y. Similarly, the limitation on delivery implies that $\frac{x}{6} + \frac{y}{10} \leqslant 1$; finally, the limitation on loading implies that $x + y \leqslant 7$. Perhaps 'finally' was a little hasty: we ought to add for completeness a restriction which, however obvious, must appear explicitly – that x and y are to be non-negative, i.e. $x \geqslant 0$, $y \geqslant 0$. To summarise, we wish to maximise the linear **objective function:**

$$P = 4x + 6y \tag{3.1}$$

subject to the constraints

$$x + 2y \leqslant 10; \quad 5x + 3y \leqslant 30; \quad x + y \leqslant 7 \tag{3.2}$$

and

$$x \geqslant 0 \quad y \geqslant 0 \tag{3.3}$$

We can represent the combination x and y by a point (x, y) in the Cartesian plane. The constraints (3.3) restrict the possible points to the first quadrant and it is relatively straightforward to see that the **region of feasible solutions**, i.e. those points which also satisfy the constraints (3.2) is the shaded region in Figure 3.1(a). If we regard each constraint as specifying a set of points then the feasible region is the intersection of these sets.

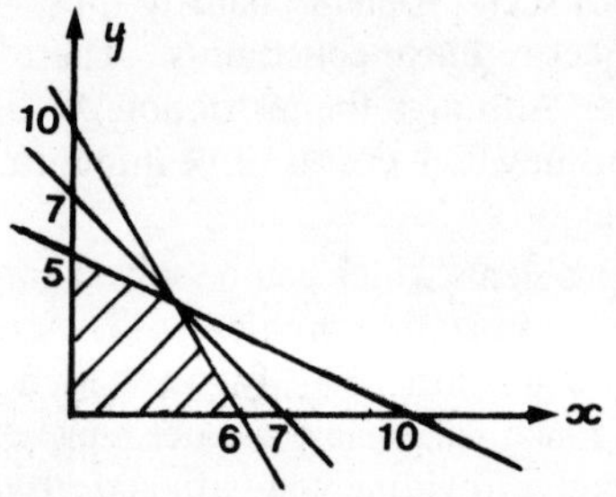

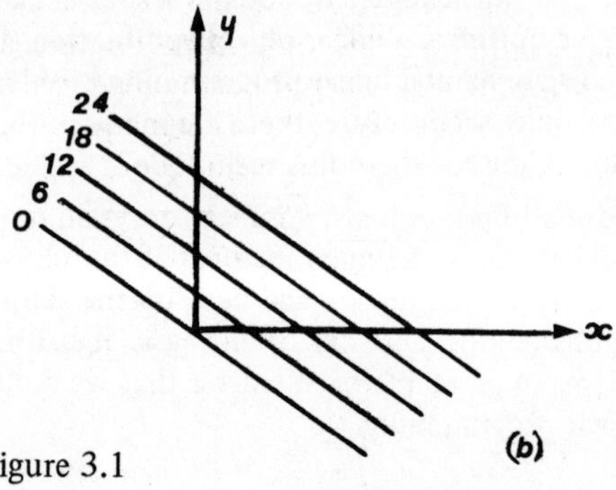

Figure 3.1

In Figure 3.1(b) we have drawn five lines of the form $4x + 6y = P$ where the appropriate value of P is attached to each line. Along each of these **iso-profit** lines any combination of x and y gives rise to the same profit, P. We see that the further away from the origin we draw one of these lines, the greater the profit it represents.

We require a point (or points) which lies in the feasible region and which lies on the furthest possible iso-profit line from the origin. It is easiest to see where this optimum point (or **optimal solution**) occurs by superimposing the basic features of Figures 3.1; this is done in Figure 3.2.

We have included a broken line which passes through vertex B of the feasible region. Any iso-profit lines further from the origin will have no contact with the feasible region, whereas one nearer the origin will have contact but will provide a lower profit. The point B represents the optimal solution. To find the corresponding values of x and y we may construct the features of Figure 3.2 on squared paper or solve the equations of sides AB and BC

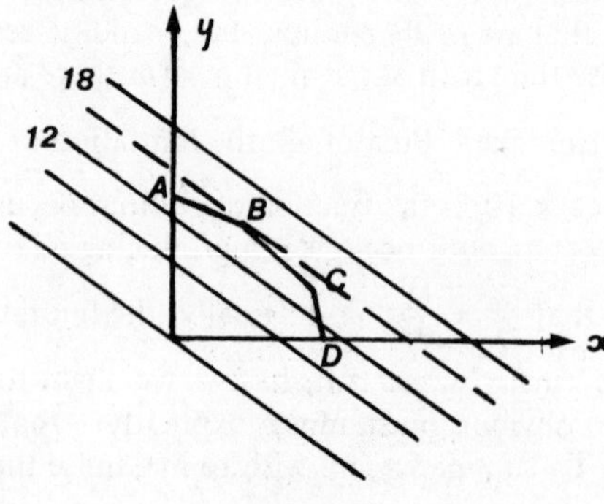

Figure 3.2

simultaneously. These equations are $x + 2y = 10$ and $x + y = 7$, respectively. It is straightforward to obtain the solution $x = 4$, $y = 3$ and therefore the company's policy should be to produce 4 truckloads of mix X and 3 truckloads of mix Y thereby providing an hourly profit of £34. Notice that the loading time is fully used and production is fully maintained; however, the delivery time is not totally used and some waiting occurs.

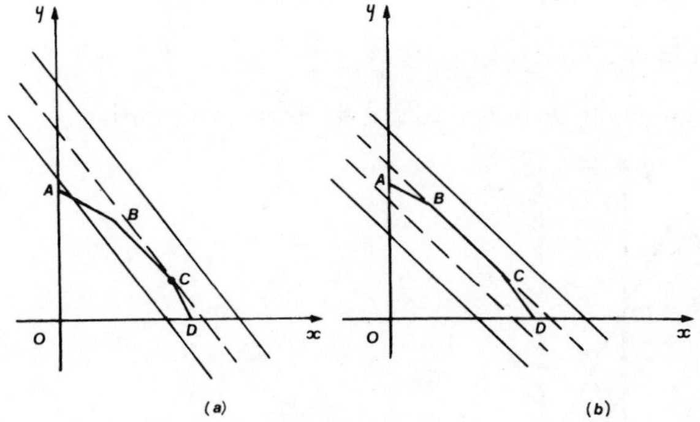

Figure 3.3

Having obtained the optimum solution, we may carry out a **sensitivity analysis** and determine the effect on the optimum of altering constraints or the coefficients in the profit expression. We shall content ourselves with considering the effect of two further profit expressions. In Figure 3.3(a) we use the expression $P = 4x + 3y$ and in Figure 3.3(b) we use the expression $P = 4x + 4y$. In Figure 3.3(a) the optimum point is now C which has coordinates (4.5, 2.5). In Figure 3.3(b), the iso-profit lines are parallel to the side BC: any point on that side will give the same profit; in order to determine a unique optimum point, we must bring other considerations into account. We may well question the appearance of non-integer coordinates for point C: as long as the working day is an even number of hours, we shall be able to produce complete truckloads of each mix. You should examine our model, detailed in equations (3.1) to (3.3) from a practical standpoint in this and other respects. It is interesting to note before passing to a second example, that in all three cases considered the maximum profit point was at a vertex of the feasible region.

Example 2

Figure 3.4 shows a three-membered rigid frame; it is loaded as shown. We require to find the minimum frame weight; the plastic-moment capacities of the beam and the columns are called x and y respectively. There are three possible collapse mechanisms: beam mechanism, sway mechanism and combined mechanism. The mechanism constraints can be obtained using the principles of virtual work. We may state the problem algebraically as follows:

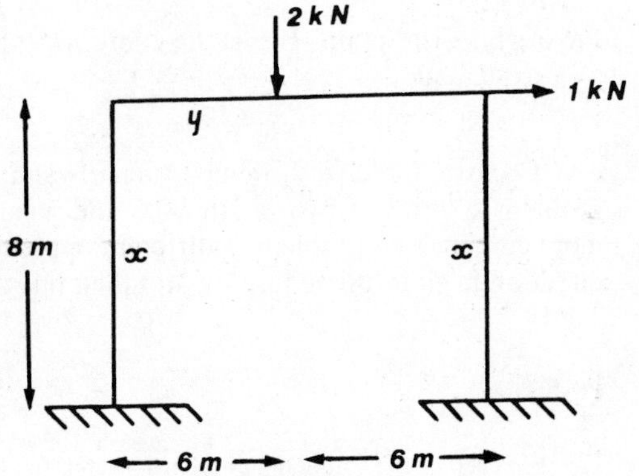

Figure 3.4

Minimise $16x + 12y$ (3.4)
subject to the constraints

beam mechanism	$4y \geqslant 12; \quad 2x + 2y \geqslant 12$	(3.5a)
sway mechanism	$2x + 2y \geqslant 8; \quad 4x \geqslant 8$	(3.5b)
combined mechanism	$4x + 2y \geqslant 20; \quad 2x + 4y \geqslant 20$	(3.5c)
In addition, we have	$x \geqslant 0, \quad y \geqslant 0$	(3.6)

Figure 3.5 shows the main features of the graphical solution

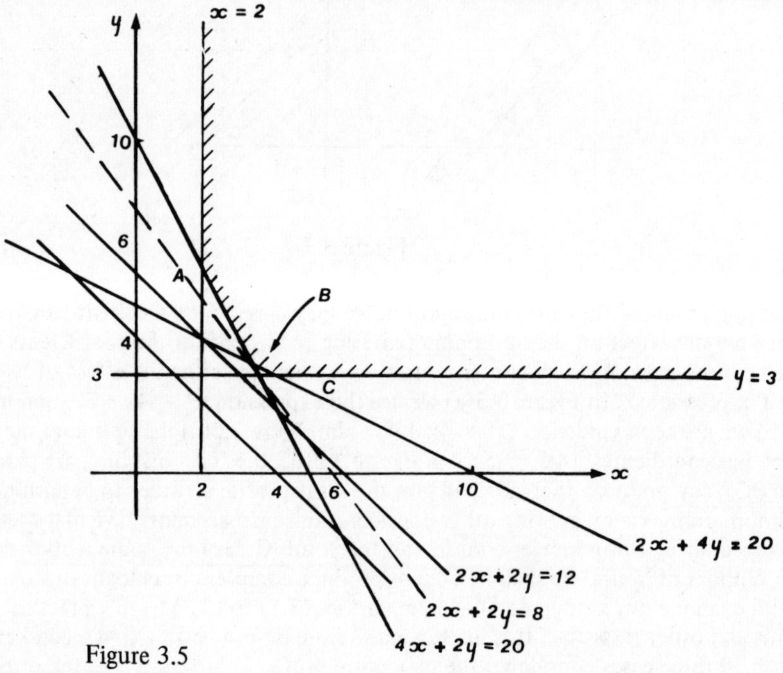

Figure 3.5

The broken line indicates that the optimum point is B (note again this occurs at a vertex); the shaded feasible region is not bounded. This time we require the **iso-cost** line nearest the origin. The point B has the coordinates $\left(\frac{10}{3}, \frac{10}{3}\right)$ and yields a minimum frame weight of $\frac{280}{3}$.

Of course, we have chosen particularly simple examples where we can express the problem in terms of two variables x and y and hence obtain a graphical solution. In more complicated problems a different approach is needed. We shall head in the general direction in an intuitive fashion and then put the method on a reasonably rigorous footing.

Problems

1. For each of the following problems, sketch the feasible region, and a few profit lines and hence find the optimal solution: specify the optimal value of the objective function and the corresponding values of the basic variables. If the solution does not possess integer coordinates

choose that point with integer coordinates which is optimal. In all cases you may assume $x \geq 0$, $y \geq 0$, unless otherwise stated.

(a) Maximise $x + y$, subject to $x + 3y \leqslant 9$, $10x + 7y \leqslant 44$.

(b) Repeat (a) for the objective functions $x + 2y$, $x + 3y$, $2x + y$.

(c) Maximise $2x + 3y$ subject to $x + 2y \leqslant 82$, $8x + 5y \leqslant 370$.

(d) Maximise and minimise $2x + y$ subject to $6x + y \leqslant 6$, $40x + 11y \leqslant 99$, $4x + 3y \leqslant 18$, $x \leqslant 1\frac{1}{2}$, $y \geqslant 2$.

(e) Maximise $3x + y$ subject to $15x + 4y \geqslant 120$, $20x + 3y \leqslant 120$, $y \leqslant 14$.

(f) Maximise $2x + y$ and minimise $x + 2y$ subject to $3 \leqslant y \leqslant 15$, $5x + 2y \leqslant 50$, $6x + y \geqslant 15$, $x + 2y \geqslant 10$.

2. Sketch the feasible regions for the following problems. Again, x, $y \geqslant 0$ unless stated otherwise.

(a) Constraints are $4x_1 + 3x_2 \leqslant 6$, $3x_1 + 4x_2 \leqslant 6$, $3x_1 + 5x_2 \leqslant 7\frac{1}{2}$. Note the redundancy of one of the constraints. The problem is said to be DEGENERATE.

(b) Constraints are $5x_1 + 2x_2 \leqslant 10$, $x_1 + 3x_2 \leqslant 6$, $2x_1 + 5x_2 \leqslant 10$. Comment on the region. What happens if we wish to maximise $4x_1 + 3x_2$?

(c) Constraints are $3x_1 + 2x_2 \leqslant 8$, $3x_1 + 2x_2 \geqslant 10$.

(d) Constraints are $3x_1 - x_2 \leqslant -4$, $x_1 - x_2 \geqslant 1$.

(e) Constraints are $5x_1 + 4x_2 \leqslant 10$, $x_1 + x_2 \geqslant 1.5$, $4x_1 + x_2 \geqslant 2$, $7x_1 + 10x_2 \leqslant 17.5$, $x_1 + 4x_2 \geqslant 2$.

(f) Constraints are $3x_1 - 2x_2 \geqslant -2$, $x_1 - 2x_2 \geqslant -6$. What happens if we try to maximise $5x_1 + 4x_2$?

3.3 THE SIMPLEX ALGORITHM

We shall examine Example 1 of the last section with a view to developing an algorithm capable of handling more than two variables. The clues are that the maximum points in the three cases considered occurred at vertices of the feasible region and that at each side of this polygonal region either one of the $\leqslant$ constraints is an equality or one of the original problem variables is zero: see Figure 3.6

Consider the inequality $x + 2y \leqslant 10$ which is represented by AB. If we introduce a variable r which represents unused production time then the constraint can be written $x + 2y + r = 10$ where we impose the restriction $r \geqslant 0$. We can similarly introduce variables s and t, satisfying $x + y + s = 7$, $5x + 3y + t = 30$, $s \geqslant 0$, $t \geqslant 0$. The variables r, s and t are called **slack variables** and we can see from Figure 3.6 that the feasible region has as many sides as

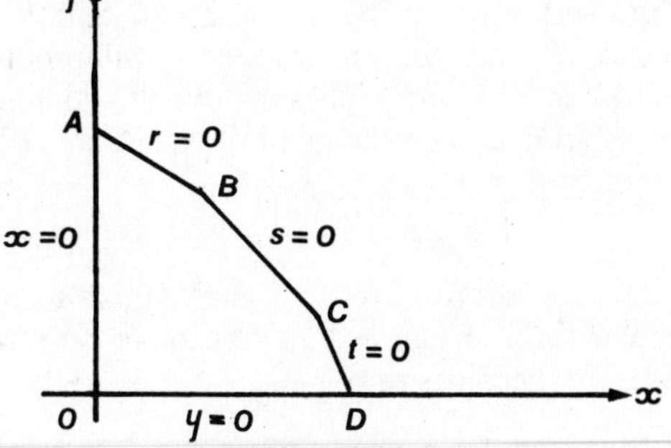

Figure 3.6

there are variables and that each side represents the restriction that one of the variables is zero. Hence, a vertex, where an optimum is to be found, represents the restriction that two of the variables are zero. It might seem that we have made the problem more complicated by increasing the number of variables from two to five, but we are now able to express the three constraints as equalities as long as we remember that we must impose the restriction that all variables are positive.

The second feature to note is that, since the optimum appears to be found at a vertex, all we need do is to evaluate the objective function at each vertex and see where it takes its least value. In practice it will be quicker to start at 0 and proceed round the vertices, either clockwise or anti-clockwise until the maximum is reached; we detect this by finding a smaller value of the objective function at the next vertex. (If two adjacent vertices provide the same maximum value, then all points on the side joining them will provide the same value.) If the maximum value occurs at B, it will be more efficient to proceed clockwise from 0, whereas, if the maximum occurs at C, an anti-clockwise journey will be the shorter. We need, therefore a means of checking which direction will be the quicker; this will be especially important in problems with several variables.

One final point: the feasible region can be defined as that region where all the five variables are non-negative simultaneously.

Development of the algorithm

Let us now summarise the equations we have obtained

$$4x + 6y = P \qquad (3.1)$$

$$x + 2y + r = 10 \qquad (3.7a)$$

$$x + y + s = 7 \qquad (3.7b)$$

$$5x + 3y + t = 30 \qquad (3.7c)$$

We also have the restrictions

$$x \geqslant 0, \quad y \geqslant 0, \quad r \geqslant 0, \quad s \geqslant 0, \quad t \geqslant 0 \qquad (3.8)$$

We represent the steps of the simplex algorithm in flow-chart form in Figure 3.7. Let us follow them through with our example and, when we have seen the procedures involved, we shall represent them in matrix form to make use of our linear equations structure.

We start from 0 where our zero variables are x and y. The values of other variables at this vertex are $r = 10$, $s = 7$, $t = 30$ and $P = 0$. Now increasing either x or y will increase P, but a unit increase in y will produce a more rapid increase in P than a unit increase in x. This implies that we should hold x at zero and increase y until one of the constraints is on the point of violation. First we express (3.7a) to (3.7c) in terms of y, noting that $x = 0$:

$$r = 10 - 2y \qquad s = 7 - y \qquad t = 30 - 3y$$

As y increases from zero, so r, s and t decrease. The danger signals are when they become zero. When $y = 5$, r becomes zero, when $y = 7$, s becomes zero and when $y = 10$, t becomes zero.

Note that, for example, when $y = 7$, r is negative. Therefore we must take the least of these values for y, viz. 5, and our new vertex, A, is where $x = r = 0$. The next

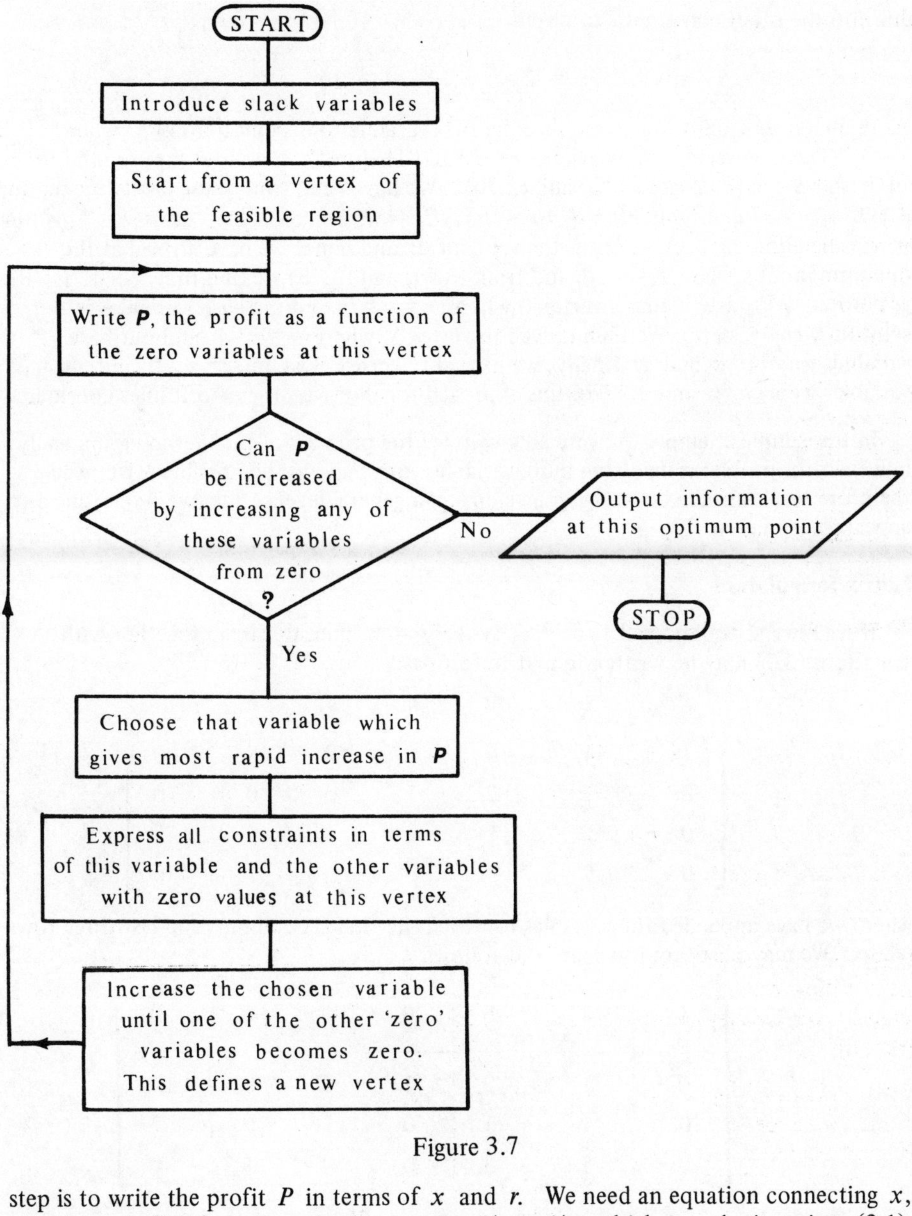

Figure 3.7

step is to write the profit P in terms of x and r. We need an equation connecting x, y and r, viz (3.7a), and we obtain $y = 5 - \frac{1}{2}x - \frac{1}{2}r$ which we substitute into (3.1) to get

$$30 + x - 3r = P \qquad (3.9)$$

Now an increase in x *will* produce an increase in P, whereas an increase in r *will not*. There is no choice to be made this time. We express (3.7a) to (3.7c) in terms of x, noting that r is zero at the vertex. We have

$$y = 5 - \frac{1}{2}x \qquad s = 7 - y - x \qquad t = 30 - 3y - 5x$$

We have expressed y in terms of x from the first constraint and we must substitute this into the other constraints to obtain

$$y = 5 - \frac{1}{2}x \qquad s = 2 - \frac{1}{2}x \qquad t = 15 - \frac{7}{2}x$$

As x increases from zero we see that the first variable to become zero is s, when $x = 4$. The new vertex, B, is where $r = s = 0$. The profit P is now expressed in terms of r and s via (3.9) and (3.7a) and (3.7b). We have first, from these last two equations, $x + 2s - r = 4$ and then $30 + (4 + r - 2s) - 3r = P$ i.e. $34 - 2r - 2s = P$. This time, increasing either r or s *reduces* the value of P and hence we have arrived at the optimum point. Here $r = s = 0$ and from (3.7a) and (3.7b) we find that $x = 4$, $y = 3$, as before. We started from a vertex 0 where $x = y = 0$ and the **basic variables** in our solution were r, s, t. We then moved to vertex A where $x = r = 0$ and our basic variables were y, s and t; finally, we moved to vertex B where $r = s = 0$ and our basic variables were x, y and t. Try this approach for the two other profit lines considered.

In this simple example we were able to trace the progress of the method graphically; however, for problems involving more variables, this may not be possible. We need therefore to obtain the basic algebraic steps on a general level so that we do not need to appeal to geometry.

Matrix formulation

If we rewrite equation (3.1) as $P - 4x - 6y = 0$ then this form together with equations (3.7) may be written in matrix form as

$$\begin{array}{c} \\ \\ r \\ s \\ t \end{array}
\begin{array}{c} \begin{array}{ccccccc} P & x & y & r & s & t & \end{array} \\ \left[\begin{array}{ccccccc} 1 & -4 & -6 & 0 & 0 & 0 & 0 \\ 0 & 1 & 2 & 1 & 0 & 0 & 10 \\ 0 & 1 & 1 & 0 & 1 & 0 & 7 \\ 0 & 5 & 3 & 0 & 0 & 1 & 30 \end{array}\right] \end{array}$$

where we have appended the variables to which the first six columns and last three rows relate. We may **partition** this matrix as follows

$$\left[\begin{array}{c|cc|ccc|c} 1 & -4 & -6 & 0 & 0 & 0 & 0 \\ \hline 0 & 1 & 2 & 1 & 0 & 0 & 10 \\ 0 & 1 & 1 & 0 & 1 & 0 & 7 \\ 0 & 5 & 3 & 0 & 0 & 1 & 30 \end{array}\right]$$

In practice, we tend to omit the first column since this gives us no useful information as regards the solution of the problem. The reduced matrix is known as the **Simplex Tableau.**

$$\left[\begin{array}{cc|ccc|c} -4 & -6 & 0 & 0 & 0 & 0 \\ \hline 1 & 2 & 1 & 0 & 0 & 10 \\ 1 & 1 & 0 & 1 & 0 & 7 \\ 5 & 3 & 0 & 0 & 1 & 30 \end{array}\right]$$

The first step is to decide which of the non-basic variables to convert to a basic variable. Scanning the top row, we notice that both non-zero elements are negative and we choose the more negative of these. This concentrates our attention on the second column. To select the variable which will be replaced as a basic variable, we look down the second column and divide the number in that position in each of the rows into the element in the last column of that row. We select the least resulting number. In this example the second row gives $\frac{10}{2} = 5$, the third gives $\frac{7}{1} = 7$ and the fourth yields $\frac{30}{3} = 10$. The least of these numbers came from the second row and we circle the element in row 2 column 2 (i.e. the rth row and the yth column); this is to indicate that y is to take over from r as a basic variable. The circled element is the **PIVOT**.

$$\left[\begin{array}{ccccc|c} -4 & -6 & 0 & 0 & 0 & 0 \\ 1 & \textcircled{2} & 1 & 0 & 0 & 10 \\ 1 & 1 & 0 & 1 & 0 & 7 \\ 5 & 3 & 0 & 0 & 1 & 30 \end{array}\right]$$

We divide the row which contains the pivot by the value of the pivot.

$$\left[\begin{array}{ccccc|c} -4 & -6 & 0 & 0 & 0 & 0 \\ 1/2 & 1 & 1/2 & 0 & 0 & 5 \\ 1 & 1 & 0 & 1 & 0 & 7 \\ 5 & 3 & 0 & 0 & 1 & 30 \end{array}\right] \quad \begin{array}{l} \text{Row 1} + 6\ \text{Row 2} \\ \\ \text{Row 3} - \text{Row 2} \\ \text{Row 4} - 3\ \text{Row 2} \end{array}$$

Then we subtract multiples of this row from the other rows so that the entries in their y column become zero.

$$\left[\begin{array}{ccccc|c} -1 & 0 & 3 & 0 & 0 & 30 \\ 1/2 & 1 & 1/2 & 0 & 0 & 5 \\ 1/2 & 0 & -1/2 & 1 & 0 & 2 \\ 7/2 & 0 & -3/2 & 0 & 1 & 15 \end{array}\right]$$

From this tableau we find that, at the new vertex, the basic variables are y, s and t (1 in the appropriate column with zero entries) and the profit, P, is 30 (top right element). Furthermore, the values of y, s and t at the vertex are 5, 2, 15 respectively (last column) and, of course x and r are both zero. Now we repeat the procedure.

We scan the top row and find only one negative entry and we concentrate on the first column. As we move down, the division process gives $5\Big/\frac{1}{2} = 10$ from row 2, $2\Big/\frac{1}{2} = 4$ from row 3 and $15\Big/\frac{7}{2} = \frac{30}{7}$ from row 4. The least of these being 4, we select row 3 and circle the **pivot** $\frac{1}{2}$ in that row

$$\left[\begin{array}{ccccc|c} -1 & 0 & 3 & 0 & 0 & 30 \\ 1/2 & 1 & 1/2 & 0 & 0 & 5 \\ \textcircled{1/2} & 0 & -1/2 & 1 & 0 & 2 \\ 7/2 & 0 & -3/2 & 0 & 1 & 15 \end{array}\right]$$

First we divide the third row by the value of the pivot.

$$\left[\begin{array}{ccccc|c} -1 & 0 & 3 & 0 & 0 & 30 \\ 1/2 & 1 & 1/2 & 0 & 0 & 5 \\ 1 & 0 & -1 & 2 & 0 & 4 \\ 7/2 & 0 & -3/2 & 0 & 1 & 15 \end{array}\right] \begin{array}{l} \text{Row 1 + Row 3} \\ \text{Row 2} - 1/2 \text{ Row 3} \\ \\ \text{Row 4} - 7/2 \text{ Row 3} \end{array}$$

We subtract suitable multiples of row 3 from the other rows to reduce the entries in their first column to zero.

$$\left[\begin{array}{ccccc|c} 0 & 0 & 2 & 2 & 0 & 34 \\ 0 & 1 & 1 & -1 & 0 & 3 \\ 1 & 0 & -1 & 2 & 0 & 4 \\ 0 & 0 & 4/2 & -7 & 1 & 1 \end{array}\right]$$

From this tableau we see that the values of the basic variables at this new vertex are $y = 3$, $x = 4$, $t = 1$ and the value of P is 34. Scanning the top row, we see that both entries are positive and this is the clue that no further improvement in the solution can be made.

Problems

1. Solve the Problems 1(a), (b) and (c) of Section 3.2 by the Simplex method.

2. Solve the following problem by the simplex method.

   Maximise $5x_1 + 2x_2$ subject to $2x_1 + x_2 \leqslant 9$, $x_1 - 2x_2 \leqslant 2$, $-3x_1 + 2x_2 \leqslant 3$; $x_i \geqslant 0$. Interpret your result graphically.

3.4 GENERAL FORMULATION OF THE SIMPLEX ALGORITHM

Let us now state the ideas behind linear programming in a linear algebra context.

The problems we have been studying can be formulated as:– maximise

$$z = \sum_{j=1}^{n} c_j x_j \text{ subject to } \sum_{j=1}^{n} a_{ij} x_j = b_i \quad i = 1, 2, \ldots, m \text{ where } x_j \geqslant 0 \quad j = 1, 2, \ldots, n$$

In other words, we have n variables and m constraints which means that there are m slack variables and $(n - m)$ original basic variables. In matrix form we have

$$\text{maximise} \quad z = \mathbf{c}^{\mathrm{T}}\mathbf{x} \quad \text{subject to} \quad \mathbf{A}\mathbf{x} = \mathbf{b} \quad \text{and} \quad \mathbf{x} \geqslant \mathbf{0}$$

(Notice that we have expressed the inequality as an equality via slack variables.)

A **feasible solution** to the problem is a vector $\mathbf{x} = (x_1, x_2, \ldots\ldots, x_n)$ which satisfies all the constraints; a **basic feasible solution** is a feasible solution which contains at most m positive x_j. If the number of positive x_j is *exactly* m, then the basic feasible solution is called **non-degenerate.**

We now quote some results that we shall need.

(i) If there exists a set of $k(<m)$ linearly independent columns of $\mathbf{A}$, denoted by $\mathbf{a}_1, \mathbf{a}_2, \ldots\ldots, \mathbf{a}_k$ such that $x_1\mathbf{a}_1 + x_2\mathbf{a}_2 + \ldots\ldots + x_k\mathbf{a}_k = \mathbf{b}$ with all $x_j \geqslant 0$ then the point $\mathbf{x} = (x_1, x_2, \ldots\ldots, x_k, 0, \ldots\ldots, 0)$ is an extreme point of the set of feasible solutions, X.†

(ii) Each extreme point of X has m linearly independent vectors from the set of columns of $\mathbf{A}$ associated with it.

(iii) If there is to be at least *one* solution of the equations $\mathbf{Ax} = \mathbf{b}$ then we know that if the rank of $\mathbf{A}$ is m then there is at least one subset of m vectors from the set $\mathbf{a}_1, \mathbf{a}_2, \ldots\ldots, \mathbf{a}_n$ which is linearly independent.

Note that we can summarise the set of feasible solutions X as $\{\mathbf{x}|\mathbf{Ax} = \mathbf{b},\ \mathbf{x} \geqslant \mathbf{0}\}$. Since we seek a solution from the finite subset of feasible solutions corresponding to extreme points we might be tempted to think that the simplest procedure is to examine the profit function z at each of them. Consider though the problem where $n = 200$, $m = 100$, which is not unduly large as linear programming problems go. The number of possible sets of m linearly independent vectors that can be chosen from a total of n is ${}^nC_m$ and in our example is ${}^{200}C_{100} = \dfrac{200!}{100!\ 100!}$; this number is so large that it would be impossible to examine each possibility. We want a method that will select only a very small number of extreme points in order to obtain the optimal solution.

Now let a matrix $\mathbf{B}$ consist of m linearly independent columns of $\mathbf{A}$; we call $\mathbf{B}$ the **basis matrix.** Any other column of $\mathbf{A}$ can be written as a linear combination of the columns of $\mathbf{B}$. For convenience, we assume that the linearly independent columns are $\mathbf{a}_1, \mathbf{a}_2, \ldots\ldots, \mathbf{a}_m$.

If we consider the matrix $\mathbf{A}$ as being partitioned into a basis matrix $\mathbf{B}$ and a non-basis matrix $\mathbf{N}$, viz. $\mathbf{A} = (\mathbf{B}, \mathbf{N})$ and $\mathbf{x}$ partitioned as $\mathbf{x}_B$ and $\mathbf{x}_N$ then

$\mathbf{Ax} = (\mathbf{B}, \mathbf{N})\left(\dfrac{\mathbf{x}_B}{\mathbf{x}_N}\right) = \mathbf{b}$ and hence $\mathbf{Bx}_B + \mathbf{Nx}_N = \mathbf{b}$ but since $\mathbf{x}_N = \mathbf{0}$ we have $\mathbf{Bx}_B = \mathbf{b}$ and therefore $\mathbf{x}_B = \mathbf{B}^{-1}\mathbf{b}$. We call the variables which constitute $\mathbf{x}_B$ **basic variables.**

The vector $\mathbf{c}^{\mathrm{T}}$ can be partitioned as $\left(\mathbf{c}_B^{\mathrm{T}}, \mathbf{c}_N^{\mathrm{T}}\right)$ and

$$z = \mathbf{c}^{\mathrm{T}}\mathbf{x} = \left(\mathbf{c}_B^{\mathrm{T}}, \mathbf{c}_N^{\mathrm{T}}\right)\left(\dfrac{\mathbf{x}_B}{\mathbf{x}_N}\right) = \mathbf{c}_B^{\mathrm{T}}\mathbf{x}_B + \mathbf{c}_N^{\mathrm{T}}\mathbf{x}_N = \mathbf{c}_B^{\mathrm{T}}\mathbf{x}_B\,.$$

Any basic feasible solution to $\mathbf{Ax} = \mathbf{b}$ can be written as $\sum_{i=1}^{m} \mathbf{x}_{Bi}\mathbf{a}_i = \mathbf{b}$ where we have for convenience taken the first m columns of $\mathbf{A}$ as a basis. If we seek a new basic feasible solution we must remove one vector (or more) from $\mathbf{B}$ and replace each one with a vector from $\mathbf{N}$. The simplex method replaces one vector at a time; this ensures that we move from one vertex to an adjacent vertex.

---

† For example a triangle is a convex set of points in the plane, the extreme points are the vertices.

Example in three variables

So far we have considered only a maximisation problem in two variables, which we were able to accomplish graphically. We now consider an example in three variables.

A manufacturer of plastic products makes three kinds of articles X, Y and Z, each of which requires three stages in its manufacture: moulding, painting and assembling. These operations for one article of X take 4, 2 and 3 hours respectively. The corresponding times for one article of Y are 3, 2 and 2 hours, respectively and for one article of Z are 5, 2 and 3 hours, respectively. There are available 400 hours of moulding time, 200 hours of painting time and 300 hours of assembling time. The profit per article of X is £4, per article of Y is £4 and per article of Z is £5. How many articles of each type should be manufactured so as to maximise the total profit?

We let the number of articles of each type manufactured be x_1, x_2 and x_3 respectively. Then we formulate the problem as follows.

Maximise $z = 4x_1 + 4x_2 + 5x_3$ subject to $4x_1 + 3x_2 + 5x_3 \leqslant 400$, $2x_1 + 2x_2 + 2x_3 \leqslant 200$, $3x_1 + 2x_2 + 3x_3 \leqslant 300$ with $x_1, x_2, x_3 \geqslant 0$.

We show schematically the feasible region in Figure 3.8. The three dashed triangles represent the three planes where the constraints achieve equality signs. The solid lines represent the edges of the feasible region. The vertices of this region are marked.

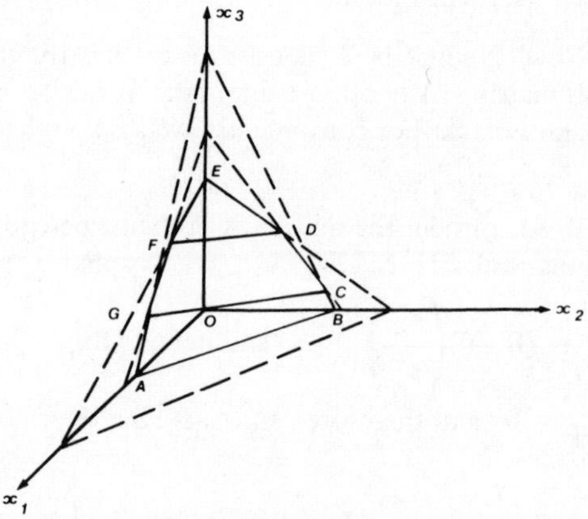

Figure 3.8

To employ the simplex algorithm we introduce slack variables r, s and t, all positive, so that

$$4x_1 + 3x_2 + 5x_3 + r = 400$$
$$2x_1 + 2x_2 + 2x_3 + s = 200$$
$$3x_1 + 2x_2 + 3x_3 + t = 300$$

The simplex tableau becomes as follows. We indicate the main tableau in the development of the solution. Pivots are circled. The solution starts at vertex 0.

$$\left[\begin{array}{rrrrrr|r} -4 & -4 & -5 & 0 & 0 & 0 & 0 \\ 4 & 3 & \textcircled{5} & 1 & 0 & 0 & 400 \\ 2 & 2 & 2 & 0 & 1 & 0 & 200 \\ 3 & 2 & 3 & 0 & 0 & 1 & 300 \end{array}\right]$$

$$\left[\begin{array}{rrrrrr|r} -4 & -4 & -5 & 0 & 0 & 0 & 0 \\ 4/5 & 3/5 & 1 & 1/5 & 0 & 0 & 80 \\ 2 & 2 & 2 & 0 & 1 & 0 & 200 \\ 3 & 2 & 3 & 0 & 0 & 1 & 300 \end{array}\right]
\begin{array}{l} \text{Row 1 + 5 Row 2} \\ \\ \text{Row 3} - \text{2 Row 2} \\ \text{Row 4} - \text{3 Row 2} \end{array}$$

$$\left[\begin{array}{cccccc|c} 0 & -1 & 0 & 1 & 0 & 0 & 400 \\ 4/5 & 3/5 & 1 & 1/5 & 0 & 0 & 80 \\ 2/5 & \textcircled{4/5} & 0 & -2/5 & 1 & 0 & 40 \\ 3/5 & 1/5 & 0 & -3/5 & 0 & 1 & 60 \end{array}\right]$$

$$\left[\begin{array}{cccccc|c} 0 & -1 & 0 & 1 & 0 & 0 & 400 \\ 4/5 & 3/5 & 1 & 1/5 & 0 & 0 & 80 \\ 1/2 & 1 & 0 & -1/2 & 5/4 & 0 & 50 \\ 3/5 & 1/5 & 0 & -3/5 & 0 & 1 & 60 \end{array}\right] \quad \begin{array}{l} \text{Row 1 + Row 3} \\ \text{Row 2 – 3/5 Row 3} \\ \\ \text{Row 4 – 1/5 Row 3} \end{array}$$

$$\left[\begin{array}{cccccc|c} 1/2 & 0 & 0 & 1/2 & 5/4 & 0 & 450 \\ 1/2 & 0 & 1 & 1/2 & -3/4 & 0 & 50 \\ 1/2 & 1 & 0 & -1/2 & 5/4 & 0 & 50 \\ 1/2 & 0 & 0 & -1/2 & -1/4 & 1 & 50 \end{array}\right]$$

Let us now outline the steps taken. In the original profit formula, increasing x_3 gave the most rapid return and this governed the choice of the third column of the matrix. The pivot was found to be 5 and row 2 was divided by this value; other entries in column 3 were reduced to zero by row operations and the third tableau represents the situation at the new vertex, which is E. At this vertex the profit is £400 and the values of the basic variables x_3, s and t are 80, 40 and 60 respectively. There being only one negative entry in row 1 of this tableau, we are restricted to column 2 where we find the pivot as 4/5. We divide row 3 by this pivot and then reduce other entries in column 2 to zero by row operations. We are left with the fifth tableau which represents the state of affairs at the new vertex, D. At this vertex the profit is £450 and the values of the basic variables x_2, x_3 and t are 50, 50 and 50 respectively (x_1, r and s are zero). It is perhaps worth remarking that in proceeding from E to D we have maintained zero wastage r, reduced wastage s to zero and decreased wastage t.

In the final solution then, the manufacturer should produce 50 articles each of X and Z, none of Y and make a profit of £450. He will fully use the time available for moulding and painting but will have 50 hours of assembly time idle.

Problems

1. (a) Maximise $3x_1 + x_2 + 2x_3$ subject to $x_1 - 2x_2 - x_3 \leqslant 10$, $2x_1 + x_2 + 2x_3 \leqslant 12$, $x_1 - x_2 + x_3 \leqslant 5$; $x_i \geqslant 0$.

   (b) Maximise $3x_1 + 2x_2 + x_3$ subject to $4x_1 + x_2 + x_3 \leqslant 8$, $3x_1 + 3x_2 + 2x_3 \leqslant 9$; $x_i \geqslant 0$.

   (c) Maximise $x_1 + 3x_2 + 2x_3$ subject to $2x_1 + x_3 \leqslant 10$, $2x_1 + x_2 \leqslant 12$, $x_1 + 2x_2 + 3x_3 \leqslant 30$; $x_i \geqslant 0$.

   (d) Maximise $5x_1 + 3x_2 + 2x_3$ subject to $x_1 + 2x_2 + 3x_3 \leqslant 25$, $2x_1 + 3x_2 + x_3 \leqslant 6$, $3x_1 + x_2 + 2x_3 \leqslant 20$; $x_i \geqslant 0$.

   (e) Maximise $4x_1 + 5x_2 + x_3$ subject to $2x_1 + x_2 + x_3 \leqslant 100$, $2x_1 + 3x_2 \leqslant 120$, $x_1 + 2x_3 \leqslant 80$.

2. A manufacturer has three products which each require the same three stages in their manufacture. Product X requires 4 hours in stage I, 3 hours in stage II and 1 hour in stage III per article. The

corresponding times for an article of Y are 1, 2 and 3 hours and for an article of Z are 3, 2 and 4 hours. However, in a given week, there is a limited amount of time available for each stage and these are 39 hours for stage I, 34 hours for stage II and 28 hours for stage III. The profit on an article of X is £200, of Y is £300 and of Z is £400. Find how many articles of each product should be made.

3. Using the simplex method, solve the following problems.

   (a) Maximise $2x_1 + x_2 + 6x_3 + 2x_4$, subject to $2x_1 + 3x_2 + x_3 + 7x_4 \leqslant 13$, $5x_1 + x_2 + 3x_3 + 2x_4 \leqslant 5$, $3x_1 + 5x_2 + 2x_3 + x_4 \leqslant 4$; $x_i \geqslant 0$.

   (b) Maximise $x_1 + x_2 + 2x_3 + x_4$ subject to $x_2 + x_3 + x_4 \leqslant 2$, $x_1 + 2x_2 + x_4 \leqslant 4$, $2x_1 - x_2 - x_3 + x_4 \leqslant 3$, $-x_1 + x_2 + x_3 - 2x_4 \leqslant 3$; $x_i \geqslant 0$.

4. A factory is required to produce 100 tons per week of an alloy containing 40% lead, 30% zinc, 30% tin. Raw materials available are unlimited supplies of pure lead at £90 per ton, zinc at £120 per ton, tin at £150 per ton, and limited quantities of two types of scrap alloy. Scrap alloy A contains 20% lead, 70% zinc, 10% tin; scrap alloy B contains 10% lead, 50% zinc, 40% tin. One week there is available 90 tons of A at £30 per ton and 110 tons of B at £33 per ton. (Note that prices are fictitious.)
   How much of each of the raw materials should the factory use in the week to minimise the cost, based on the available data?

5. A rolling mill produces three products, A, B and C. Before going into the rolling mill the products must be heated in a soaking pit which can heat product A at the rate of 1200 tons/shift, product B at 600 tons/shift and C at 400 tons/shift. The rolling mill proper can roll A at the rate of 600 tons/shift, B at 750 tons/shift and C at 1000 tons/shift. The output of A and B together must exceed 500 tons/shift. The profits per ton on A, B and C are in the ratio 1:3:4. Use the simplex method to derive the production plan for one shift which gives the maximum profit within the operating constraints.

3.5 OTHER FEATURES OF LINEAR PROGRAMMING

We have not yet considered Example 2 of Section 3.2 in the light of the simplex method. We can transform it into an example on maximisation by seeking to maximise $-16x - 12y$. An alternative method is the following application of the principle of **duality**.

Duality

Consider the two problems below

	I		II
Maximise	$4x_1 + 6x_2$	Minimise	$10y_1 + 7y_2$
subject to	$x_1 + 2x_2 \leqslant 10$	subject to	$y_1 + y_2 \geqslant 4$
	$x_1 + x_2 \leqslant 7$		$2y_1 + y_2 \geqslant 6$
	$x_1, \ x_2 \geqslant 0$		$y_1, \ y_2 \geqslant 0$

In Figure 3.9(a) we present the feasible region for problem I together with the profit line passing through the optimal vertex. In Figure 3.9(b) we similarly represent the features of problem II.

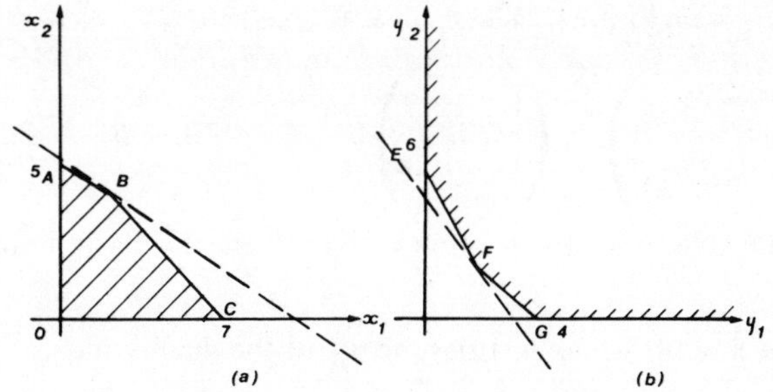

Figure 3.9

The optimal point of problem I is B, at which $x_1 = 4$, $x_2 = 3$ and the optimum value of the objective function is 34.

The optimum point of problem II is F, at which $y_1 = y_2 = 2$ and the optimum value of the objective function is 34. (Is this a coincidence?)

You may also have noticed a tie-up between the coefficients of the two problems. We can see this more clearly if we remember the general formulation for problem I:

$$\text{Maximise} \quad z = \sum_{j=1}^{n} c_j x_j$$

$$\text{subject to} \quad \sum_{j=1}^{n} a_{ij} x_j = b_i \qquad i = 1, 2, \ldots\ldots, m$$

$$\text{where} \quad x_j \geqslant 0 \qquad j = 1, 2, \ldots\ldots, n$$

If we now remove the m slack variables from consideration we have the statement:

$$\text{Maximise} \quad z = \sum_{j=1}^{n-m} c_j x_j$$

$$\text{subject to} \quad \sum_{j=1}^{n-m} a_{ij} x_j \leqslant b_i \qquad i = 1, 2, \ldots\ldots, m$$

$$\text{where} \quad x_j \geqslant 0 \qquad j = 1, 2, \ldots\ldots, (n - m)$$

The associated problem,

$$\text{Minimise} \quad z = \sum_{i=1}^{m} b_i y_i$$

$$\text{subject to} \quad \sum_{i=1}^{m} a_{ji} y_i \geqslant c_j \qquad j = 1, 2, \ldots\ldots, (n - m)$$

$$\text{where} \quad y_i \geqslant 0 \qquad i = 1, 2, \ldots\ldots, m$$

is called the **dual** of the first problem.

In problem I, $c_1 = 4,\ c_2 = 6,\ n = 4,\ m = 2$

$$\begin{pmatrix} a_{11} & a_{22} \\ a_{21} & a_{22} \end{pmatrix} = \begin{pmatrix} 1 & 2 \\ 1 & 1 \end{pmatrix}, \quad b_1 = 10, \quad b_2 = 7$$

and you should be able to see the relation with problem II. Each problem is, in fact, the dual of the other.

We now quote a result which justifies the use of the duality idea.

If a linear programming problem and its dual are both feasible then they both possess optimal solutions; the value of the objective function is the same in both cases. However, if one of the problems is not feasible, neither of them has an optimal solution.

Before we proceed further, we quote two more examples of duals to fix ideas.

Examples

(1) Problem I

Maximise $8x + 30y + 7z$

subject to

$$\begin{aligned} x + 5y + 2z &\leqslant 10 \\ 4x + 6y + z &\leqslant 15 \\ x,\ y,\ z &\geqslant 0 \end{aligned}$$

Problem II

Minimise $10u + 15v$

subject to

$$\begin{aligned} u + 4v &\geqslant 8 \\ 5u + 6v &\geqslant 30 \\ 2u + v &\geqslant 7 \\ u,\ v &\geqslant 0 \end{aligned}$$

(2) Problem I

Minimise $50x + 12y + 30z$

subject to

$$\begin{aligned} 3x + 2y + 5z &\geqslant 2 \\ 4x + y + 3z &\geqslant 3 \\ 5x + 2y + z &\geqslant 5 \\ x,\ y,\ z &\geqslant 0 \end{aligned}$$

Problem II

Maximise $2u + 3v + 5w$

subject to

$$\begin{aligned} 3u + 4v + 5w &\leqslant 50 \\ 2u + v + 2w &\leqslant 12 \\ 5u + 3v + w &\leqslant 30 \\ u,\ v,\ w &\geqslant 0 \end{aligned}$$

Let us solve a simple minimisation problem via its dual.

Example

Minimise $9u + 4v + 9w$ subject to

$$\begin{aligned} 3u + v + w &\geqslant 5 \\ u + v + 3w &\geqslant 4 \\ u,\ v,\ w &\geqslant 0 \end{aligned}$$

We first convert this problem into its dual.

Maximise $5x + 4y$ subject to

$$\begin{aligned} 3x + y &\leqslant 9 \\ x + y &\leqslant 4 \\ x + 3y &\leqslant 9 \\ x,\ y &\geqslant 0 \end{aligned}$$

We shall now solve this problem by the simplex method.

The appropriate simplex tableau is

$$\begin{bmatrix} -5 & -4 & 0 & 0 & 0 & 0 \\ \textcircled{3} & 1 & 1 & 0 & 0 & 9 \\ 1 & 1 & 0 & 1 & 0 & 4 \\ 1 & 3 & 0 & 0 & 1 & 9 \end{bmatrix}$$

The following tableaux are presented without comment:

$$\begin{bmatrix} -5 & -4 & 0 & 0 & 0 & 0 \\ 1 & 1/3 & 1/3 & 0 & 0 & 3 \\ 1 & 1 & 0 & 1 & 0 & 4 \\ 1 & 3 & 0 & 0 & 1 & 9 \end{bmatrix} \quad \begin{matrix} \text{Row 1 + 5 Row 2} \\ \\ \text{Row 3 – Row 2} \\ \text{Row 4 – Row 2} \end{matrix}$$

$$\begin{bmatrix} 0 & -7/3 & 5/3 & 0 & 0 & 15 \\ 1 & 1/3 & 1/3 & 0 & 0 & 3 \\ 0 & \textcircled{2/3} & -1/3 & 1 & 0 & 1 \\ 0 & 8/3 & -1/3 & 0 & 1 & 6 \end{bmatrix}$$

$$\begin{bmatrix} 0 & -7/3 & 5/3 & 0 & 0 & 15 \\ 1 & 1/3 & 1/3 & 0 & 0 & 3 \\ 0 & 1 & -1/2 & 3/2 & 0 & 3/2 \\ 0 & 8/3 & -1/3 & 0 & 1 & 6 \end{bmatrix} \quad \begin{matrix} \text{Row 1 + 7/3 Row 3} \\ \text{Row 2 – 1/3 Row 3} \\ \\ \text{Row 4 – 8/3 Row 3} \end{matrix}$$

$$\begin{bmatrix} 0 & 0 & 1/2 & 7/2 & 0 & 18\tfrac{1}{2} \\ 1 & 0 & 1/2 & -1/2 & 0 & 5/2 \\ 0 & 1 & -1/2 & 3/2 & 0 & 3/2 \\ 0 & 0 & 1 & -4 & 1 & 2 \end{bmatrix}$$

We have arrived at the optimum solution. We have a maximum value of 18½ and the values of the variables involved are $x = 5/2$, $y = 3/2$, $r = s = 0$, $t = 2$. Interpreting this in terms of our original problem, we have a minimum value of 18½, the values of the variables involved are $u = 1/2$, $v = 7/2$, $w = 0$ (top row of the last tableau), both slack variables being zero.

Now we return to Example 2 of Section 3.2. We first formulate the dual.

$$\begin{aligned} \text{Maximise} \quad & 12u_1 + 12u_2 + 8u_3 + 8u_4 + 20u_5 + 20u_6 \\ \text{subject to} \quad & 0u_1 + 2u_2 + 2u_3 + 4u_4 + 4u_5 + 2u_6 \leqslant 16 \\ & 4u_1 + 2u_2 + 2u_3 + 0u_4 + 2u_5 + 4u_6 \leqslant 12 \\ & u_1, u_2, u_3, u_4, u_5, u_6 \geqslant 0 \end{aligned}$$

You can show that the solution to the dual problem is that the maximum is $\frac{280}{3}$. It follows that the minimum frame weight is $\frac{280}{3}$ and the corresponding values of x and y are both $\frac{10}{3}$.

It is perhaps worth noting that corresponding to the variables u_5 and u_6 we must have two constraints in the primal problem being satisfied by equality, i.e. $4x + 2y = 20$, $2x + 4y = 20$; solving these simultaneously yields the optimum solution (see Figure 3.5).

Modification to the simplex algorithm

Although the simplex algorithm we have represented is the basic algorithm, it is not necessarily the most efficient one. Revised algorithms have been developed and programmed for digital computers. We want to mention some of the modifications we may need to employ in order to cater for other kinds of feasible region. You will find it helpful to refer to the problems at the end of Section3.2 (and their solutions).

Problem 3.2 2(a) provided a **degenerate problem.** The constraint $x_1 + 3x_2 \leqslant 6$ was redundant. This will lead to a basic variable becoming zero. If we arrive at a point where more constraints are meeting than we expect to be the case, we need to proceed along the one which is not redundant and the algorithm must be modified to take account of this.

We have not yet considered a mixture of constraints where we have both $\leqslant$ and $\geqslant$ signs. Consider, for example, the problem of maximising $3x_1 + 2x_2$ subject to $3x_1 + 5x_2 \leqslant 30$, $3x_1 + x_2 \leqslant 6$, $7x_1 + 5x_2 \geqslant 16$; $x_1, x_2 \geqslant 0$. A useful technique is seen in the following development. First, convert the inequalities into equations. For the third constraint we need to *subtract* the variable t, which we call a **surplus variable.**

The constraints then become

$$3x_1 + 5x_2 + r = 30 \qquad (3,10a)$$

$$3x_1 + x_2 + s = 6 \qquad (3.10b)$$

$$7x_1 + 5x_2 - t = 16 \qquad (3.10c)$$

However, we are faced with the problem of where to start: an initial basic feasible solution. We cannot choose the origin $x_1 = x_2 = 0$ for this would provide a negative value of t. We resort to the idea of introducing an **artificial variable** a_1, so that the last constraint becomes $7x_1 + 5x_2 + a_1 - t = 16$. We choose a new objective function $z = 3x_1 + 2x_2 - a_1 M$ where M is a large positive constant. This objective function, subject to constraints (3.10a), (3.10b) and (3.10c) form a new linear programming problem. We have, in effect, added a new dimension to our original problem. The initial basic feasible solution is $r = 30$, $s = 6$, $t = 0$, $a_1 = 16$, $x_1 = x_2 = 0$. We apply the simplex algorithm, bearing in mind that M is very large. When we reach a stage that a_1 is a non-basic variable (i.e. $a_1 = 0$ at the vertex in question) we leave the a_1 column blank and put $M = 0$ for succeeding tableaux. We have now got an initial basic feasible solution to our original problem and we proceed with the solution as though we had just commenced.

Another problem arises if the solution must contain integer coordinates. Then it is not always clear how to move from the theoretical optimal solution to one with integer coordinates; it is all too possible to step outside the feasible region

Problems

1. For each of the following problems, state the dual. Solve both graphically.

   (a) Maximise $3x_1 + 4x_2$ subject to $2x_1 + 2x_2 \leqslant 11$, $x_1 + 3x_2 \leqslant 14$; $x_i \geqslant 0$.

   (b) Minimise $-6x_1 - 3x_2$ subject to $2x_1 + 4x_2 \leqslant 4$, $3x_1 - 3x_2 \geqslant 2$, $x_i \geqslant 0$.

2. Solve the duals of the example of Problem 1,Section 3.2

3. Attempt to use the simplex method to maximise $x + 2y$ subject to $x - y \leqslant 1$, $y - 2x \leqslant 2$; $x, y \geqslant 0$. Use a sketch to explain the failure.

4. (i) Minimise $x_1 + x_2 + x_3 + x_4 + x_5$ Subject to

$$\begin{aligned} 2x_1 + x_2 + x_3 + x_4 - x_5 &= 4 \\ -x_1 + x_2 + 3x_3 - 2x_4 + x_5 &= 4 \end{aligned}$$

all variables to be non-negative.

Is the solution unique? Is the solution of the dual problem unique? Give reasons for your answer.

(ii) Consider the following problem:

Maximise $13x_1 + 3x_2 + 16x_3 + 8x_4 + 4x_5$

Subject to

$$\begin{aligned} 5x_1 + 12x_3 - 5x_4 + 11x_5 &\leqslant 86 \\ 3x_1 + 13x_3 - 10x_4 + 7x_5 &\leqslant 54 \\ 8x_1 + 3x_2 + 7x_3 + 8x_4 + 4x_5 &\leqslant 77 \end{aligned}$$

all $x_i \geqslant 0$.

A solution $x_1 = 4$, $x_2 = 7$, $x_3 = x_4 = 0$, $x_5 = 6$ is proposed. Is it (a) feasible, (b) optimal, (c) an extreme point of the feasible region? (C.E.I.)

5. Solve the following problems by the Simplex method: hence draw conclusions about the solutions of the dual problems: state the latter:

Maximise $5x_1 + x_2$

Subject to $x_1 + 2x_2 \leqslant 4$, $2x_1 + 3x_2 \leqslant 3$

all x_i non-negative

Maximise $5x_1 + x_2$

Subject to $x_1 - 2x_2 \leqslant 2$, $x_1 - 3x_2 \leqslant 3$

all x_i non-negative

Maximise $3x_1 + x_2$

Subject to $x_1 - x_2 \leqslant 1$, $-x_1 + x_2 \leqslant -2$

all x_i non-negative

Solve the first problem also with the additional condition that all variables must have integer values. (C.E.I.)

Chapter Four

Optimization II — Non-Linear Problems

4.1 INTRODUCTION

We first remind ourselves of some basic results concerning functions of one variable viz. $f(x)$. There is a **local minimum** of $f(x)$ at $x = a$ if $f(a+h) > f(a)$ for small values of h (small enough not to stray into regions where other features of $f(x)$ occur) and a similar definition applies to a **local maximum**. At a **point of inflection** the quantity $\left[f(a+h_1) - f(a)\right] . \left[f(a-h_2) - f(a)\right] < 0$, where h_1 and h_2 are small positive numbers. A **stationary point** is one at which $f'(x) = 0$. Let one of these stationary points be a. If the first non-vanishing derivative at $x = a$ is of *even* order, then we have a local minimum if the value of this derivative is positive and a local maximum if the value is negative. However, if the first non-vanishing derivative at $x = a$ is of *odd* order, we have a point of inflection.

In optimization probelms we are often concerned with **global maxima** or **global minima.** Sometimes we restrict the values of x to an interval, and sometimes we place no restriction on the values of x.

We have sketched in Figure 4.1 a function which, in the range $a \leqslant x \leqslant b$ exhibits a global maximum (b) and global minimum (x_1), local maxima (x_6 and a), a local minimum (x_4) and two points of inflection (x_2 & x_5). In addition, we have a **valley** (x_7) and a **ridge** (x_3): these are points where $f''(x) = 0$, but unlike a point of inflection the second derivative does not change sign at the point.

The function $f(x)$ has a global maximum in the interval $a \leqslant x \leqslant b$ at one of the following three places

(i) where $f'(x) = 0$

(ii) at a boundary

(iii) where $f'(x)$ has a discontinuity.

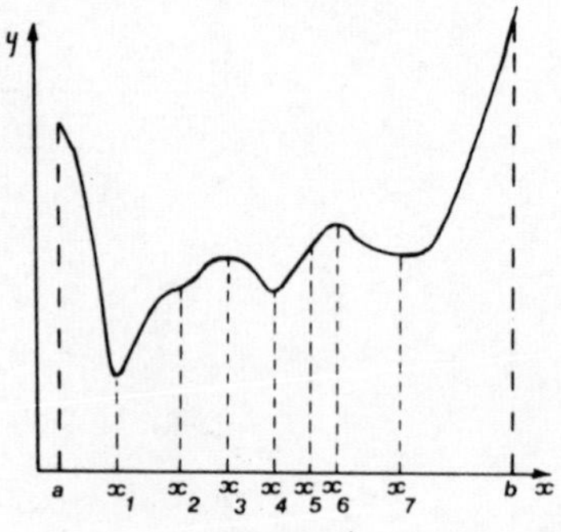

Figure 4.1

Consider Figure 4.2; it depicts the function $f(x) = (x^3 - 2x^2)^{1/5}$ in the range $-1 \leqslant x \leqslant 5$.

We can isolate the global minimum (*least value*) of $f(x)$ by examining the value of the function at each of the three classes of place listed above.

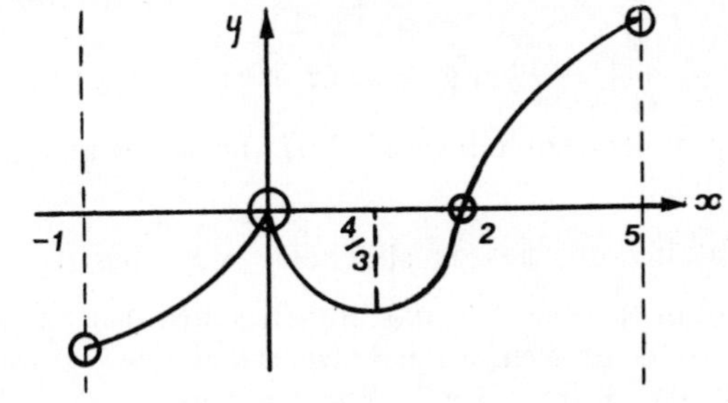

Figure 4.2

(i) $f'(x) = \dfrac{1}{5}\dfrac{(3x^2 - 4x)}{(x^3 - 2x^2)^{4/5}}$. Therefore $f'(x) = 0$ where $3x^2 - 4x = 0$
i.e. $x(3x - 4) = 0$ i.e. $x = 0$ or $x = \dfrac{4}{3}$

(ii) The boundaries are $x = -1$ and $x = 5$

(iii) $f'(x)$ possibly has a discontinuity at $x^3 - 2x^2 = 0$ i.e. $x = 0$ or $x = 2$. We therefore have a set of values of x to examine, viz. $\left\{-1, 0, \frac{4}{3}, 2, 5\right\}$
$f(-1) = (-3)^{1/5} = -(3)^{1/5}$, $f(0) = 0$, $f(\frac{4}{3}) = (-32/27)^{1/5} = -(32/27)^{1/5}$,
$f(2) = 0$, $f(5) = (75)^{1/5}$
In this instance the global minimum occurs at $x = -1$.

A further pair of definitions is helpful. A function $f(x)$ is called **convex** over some interval of x if, for any two values x_1 and x_2 in the interval and for all λ which satisfy $0 \leqslant \lambda \leqslant 1$, we have the inequality

$$f\left[\lambda x_2 + [1-\lambda]x_1\right] \leqslant \lambda f(x_2) + (1-\lambda)f(x_1) \qquad (4.1)$$

We illustrate the idea in Figure 4.3(a). Note that values of a convex function are always overestimated by *linear interpolation.*

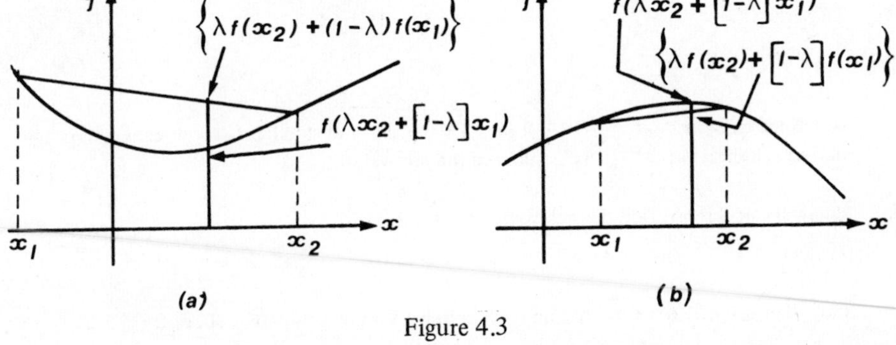

Figure 4.3

Similarly, a function $f(x)$ is called **concave** over some interval of x if, for any two values x_1 and x_2 in the interval and for all λ which satisfy $0 \leqslant \lambda \leqslant 1$, we have the inequality

$$f[\lambda x_2 + [1-\lambda]x_1] \geqslant \lambda f(x_2) + (1-\lambda)f(x_1) \tag{4.2}$$

This concept is depicted in Figure 4.3(b), and we see that linear interpolation provides an underestimate.

Note that, in both cases, equality holds at x_1 and x_2.

One word of caution. We have blithely stated that we must find the values of x for which $f'(x) = 0$; however, it is not always that easy. Consider for example $f(x) = 5x^6 - 4x^4 + 3x^2 - x + 7$. The derivative vanishes where $f'(x) = 30x^5 - 16x^3 + 6x - 1 = 0$. To solve this equation requires techniques such as that of Newton-Raphson.

If we know that a function is convex in a closed interval we can straight away use the result that a local minimum of the function in that interval is also the global minimum there. Similarly if a function is concave in a closed interval, a local maximum of the function in that interval is also the global maximum there. We complement these results with the following: (i) the global maximum of a convex function $f(x)$ in a closed interval is taken at one or both boundaries of the interval; (ii) the global minimum of a concave function $f(x)$ in a closed interval is taken at one or both boundaries of the interval.

Example

Find the least value of $f(x) = x^2 + 3x + 2$ in the interval $[-2, 4]$.

It should first be shown that the function is convex. (Verify this for yourselves.)

Since $f(x)$ is a convex function, it is convex in the closed interval $[-2, 4]$. Now $f'(x) = 2x + 3$ and hence $f'(x) = 0$ when $x = -\frac{3}{2}$. This local minimum is therefore the global minimum and the value of the function at the point in question is

$$\frac{9}{4} - \frac{9}{2} + 2 \quad \text{i.e.} \quad -\frac{1}{4}.$$

Problems

1. Sketch the curve $x^3 - 6x^2 + 9x + 4$ in $-1 \leqslant x \leqslant 5$. Indicate local and global maxima and minima. Repeat for $x^3 - 18x^2 + 96x$ in the interval $(0, 9)$.

2. Why will calculus not find the minimum of

   (i) $f(x) = |x|$ (ii) $f(x) = x^2 - 2x + 2;\ x \geqslant 2$.

3. Show that $(ax + b)/(cx + d)$ has no local extrema for any values of a, b, c, d.

4.2 SEARCH TECHNIQUES IN ONE VARIABLE

Sometimes a mathematical relationship between variables cannot be found and in order to obtain values of $f(x)$ we have to carry out an experiment or a series of calculations. This means that we want to keep the number of evaluations to a minimum. The strategy will be to start from a *base point* and select where next to make an evaluation of $f(x)$; depending on whether we get a value of $f(x)$ nearer to the optimum (e.g. a lower value if we seek a minimum) we make a decision as to where next to choose x for a further evaluation. The values of x that we select will form a sequence $\{x_i\}$. The problem then arises as to how we detect convergence; this can be a complex problem and there is a danger in using too simple a criterion for convergence.

In addition to the sequence $\{x_i\}$, we have a companion sequence $\{f(x_i)\}$. Therefore, we have two possible criteria: for seeking a least value of $f(x)$ these are

$$|f(x_i) - f(x_{i+1})| < \epsilon \tag{4.3a}$$

and

$$|x_i - x_{i+1}| < \epsilon^* \tag{4.3b}$$

where ϵ and ϵ^* are prescribed positive values. Consider Figure 4.4; the first diagram i.e.4.4(a) represents a shallow valley where large changes in x give rise to only small changes in $f(x)$ and the criterion (4.3b) would be the more useful. Figure 4.4(b) shows a steep-sided valley and small changes in x now cause large changes in $f(x)$; here the criterion (4.3a) would be the better choice.

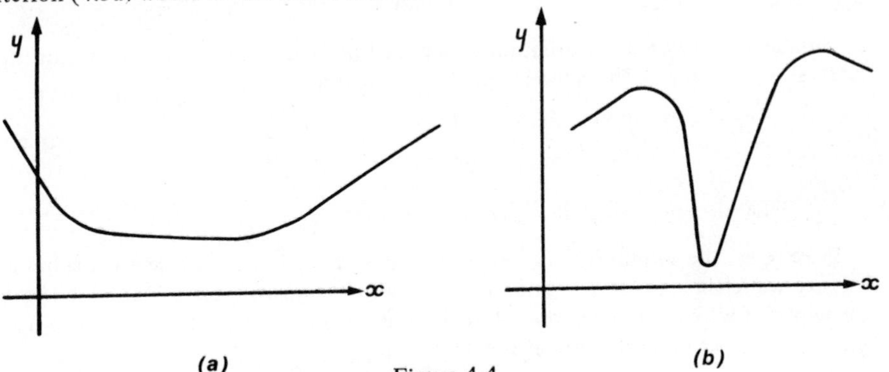

Figure 4.4

A possible ploy would be to require both criteria to hold for each of a given number of iterations. Whilst not completely safe, it should nevertheless be reasonably safe; indeed it may, with some problems, be too safe.

Search techniques fall into two main categories: those which specify an interval in which the optimum value lies and which seek to reduce successively the interval and those which specify the position of the optimum point by a point which approximates to it. We shall consider examples of each class; the **Fibonacci Search** for the former, the **Davies, Swann and Campey algorithm** and **Powell's algorithm** for the latter.

We first assume that the function in question is unimodal, i.e. there is only one stationary value in the interval of interest. We further assume that the stationary value x^* is a minimum.

[Note: if we wish to find the maximum of a function we can consider the equivalent problem of minimising $-f(x)$.]

Search strategy

We recall that if a function $f(x)$ is unimodal in an interval (a, b) then it is necessary to evaluate the function at two internal points before the location of the stationary value can be confined to a subinterval of (a, b). Try and prove this result. Suppose we let the current interval be (a, b) then we evaluate the function at the internal points x_1 and x_2 where $a < x_1 < x_2 < b$.

Consider Figure 4.5(a). We have evaluated the function $f(x)$ at two internal points x_1 and x_2. Note that $b - x_2 = x_1 - a$. The result $f(x_1) > f(x_2)$ leads us to reject the interval $a \leqslant x \leqslant x_1$.

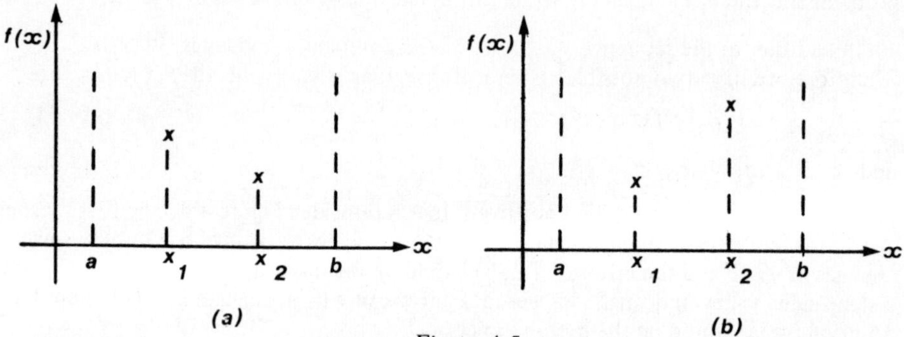

Figure 4.5

Similarly, if we obtain a situation as depicted in Figure 4.5(b), we shall reject the interval $x_2 < x \leqslant b$. This suggests the following rule:

$$\left.\begin{array}{lll}\text{If} \quad x_1 < x_2 & \text{then} & f(x_1) > f(x_2) \Rightarrow x^* > x_1 \\ & \text{and} & f(x_1) < f(x_2) \Rightarrow x^* < x_2\end{array}\right\} \qquad (4.4)$$

What do you deduce if $f(x_1) = f(x_2)$?

If we reject the subinterval (a, x_1) we relabel x_1 as a; if we reject the subinterval (x_2, b) we relabel x_2 as b. In either event we have a new interval (a, b) and it would be helpful if we were to keep the internal point (x_2 or x_1) so that we needed just one further function evaluation at the next stage.

Fibonacci search

We seek an algorithm which gives the largest ratio of initial to final interval length for a fixed number of function evaluations. For simplicity let us take the final interval as being of length 1 and find L_n, the length of the largest interval which can be reduced to 1 after n function evaluations.

First we find an upper bound for L_n, the length of the interval (a, b). As before we evaluate the function at the internal points x_1 and x_2. If the minimum lies in (a, x_1) then we have only $(n - 2)$ evaluations left to refine our interval of uncertainty and hence $x_1 - a \leqslant L_{n-2}$. On the other hand, if the minimum lies in (x_1, b) we already have one point, x_2, in this subinterval and therefore the interval (x_1, b) can be refined by $(n - 1)$ evaluations provided we use x_2 in our strategy; hence $b - x_1 \leqslant L_{n-1}$. Adding these inequalities, we obtain $(a, b) = L_n \leqslant L_{n-2} + L_{n-1}$. Further, for consistency we must define L_0 and L_1 to be equal to 1, since we need *two* function evaluations to reduce the interval of uncertainty.

Suppose we could find a sequence of numbers F_n such that

$$F_n = F_{n-1} + F_{n-2} \tag{4.5}$$

with $\quad F_1 = F_2 = 1$

Then we would have the ideal optimum for the interval reduction, (L_n would have achieved its upper bound).

The sequence of numbers governed by (4.5) are the **Fibonacci numbers**; the first few are

1, 1, 2, 3, 5, 8, 13, 21, 34, 55, ...

In Problem 2(a) at the end of this section we ask you to derive a formula for F_n.

The **Fibonacci strategy** is as follows. If the case illustrated in Figure 4.5(a) occurs then we choose an internal point x_3 as far to the left of b as x_2 is to the right of x_1 so that $b - x_3 = x_2 - x_1$; see Figure 4.6(a). Conversely, if the case of Figure 4.5(b) holds, then we choose x_3 as far to the right of a as x_1 is to the left of x_2 i.e. $x_3 - a = x_2 - x_1$; see Figure 4.6(b). Notice that, in either case, we only require one further function evaluation at each stage after the initial one, since one of the four points at any stage remains an intermediate point at the next stage.

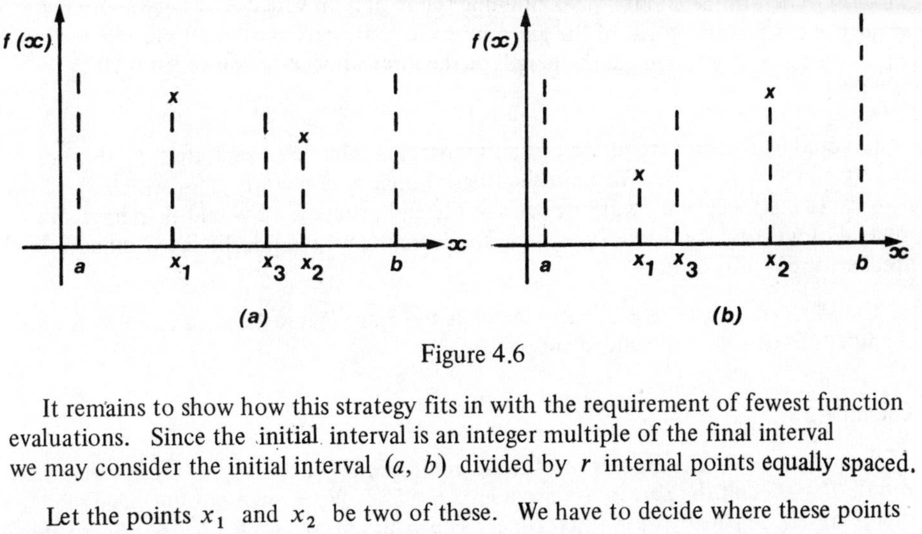

Figure 4.6

It remains to show how this strategy fits in with the requirement of fewest function evaluations. Since the initial interval is an integer multiple of the final interval we may consider the initial interval (a, b) divided by r internal points equally spaced.

Let the points x_1 and x_2 be two of these. We have to decide where these points will be.

Once this decision is made then the above strategy will select the remaining points. The set $\{x_1, x_2, x_3, \ldots\ldots, x_s\}$ will be a subset of the internal points.

In Figure 4.7 we depict successive stages in the reduction of the interval of uncertainty. An interval of length 13 ($= F_7$) is divided into two intervals of length 5 and one of length 3 (Figure 4.7(a)). Note that $x_1 - a = 5 = b - x_2$ and that $x_1 - a = 5 = F_5$ whilst $b - x_1 = 8 = F_6$. Suppose we consistently discard the right-hand sub-interval†. Then a point is chosen $F_4 = 3$ units from the left-hand end of the remaining interval to be placed symmetrically with the remaining internal point; Figure 4.7(b)

† The results derived hold quite generally.

At the stage of diagram (e) we need one further function evaluation to locate the minimum. We place this, not at the same place as x_5 as theory would suggest, but slightly to the left as shown in Example 2.

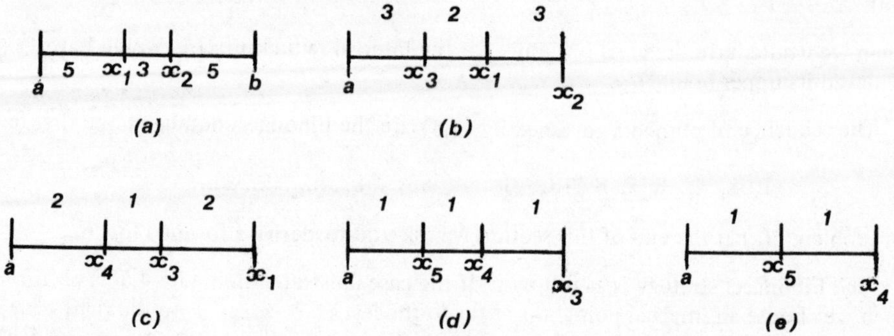

Figure 4.7

Example 1

Suppose $f(x)$ has a minimum in the interval $(a, b) = (0, 10)$. Further suppose that we wish to determine an interval containing the minimum which is of length $d = 0.2$ at most, i.e. which is $\leqslant 2\%$ of the given interval. The pertinent question is how many stages are necessary to reduce the length of the interval of uncertainty from 10 to $\leqslant 0.2$?

It would be fortuitous if there *was* a Fibonacci number F_N such that $F_N d$ was *exactly* equal to $(b - a)$. We want the first Fibonacci number F_N for which $F_N d > (b - a)$ *because*, working back to the end product, we would then have an interval of uncertainty $< d$. In our current example, we require the first Fibonacci number > 50; this is $F_8 = 55$.

The Flow chart for the Fibonacci search is shown in Figure 4.8; it should be read in conjunction with the following example.

Example 2

We programmed a computer to find the minimum of $f(x) = x^2 - 7x + 4$ given that $A = 0$, $B = 10$ and the percentage accuracy, $P = 5\%$. We printed out the value of N, such that $F_N > 100/P$ for the first time. The values of A, L, R, B are printed above the corresponding values of $f(A)$, $f(L)$, $f(R)$ and $f(B)$ for each of the $(N - 2)$ stages. We present the results in Table 4.1 and work through the first 2 stages. (Via calculus we find that the minimum value of $f(x)$ occurs at $x = 3.5$.)

At stage 1 we notice that $f(L) < f(R)$; therefore we reject the interval $[R, B]$. We then relabel B at 6.19048, R at 3.80952 and calculate L as 2.38095; we note that $6.19048 - 3.80952 = 2.38096$, but this discrepancy occurs primarily because of round-off in printing the numbers. At stage 2 we see that $f(R) < f(L)$ and we discard the interval $[A, L]$, relabel A at 2.38095, L at 3.80952 and calculate R as 4.7619 which is as far from 6.19048 as 3.80952 is from 2.38095. At stage 6, both numbers L and R should be 3.33333, which is equidistant from 2.85714 and 3.80952; in fact we took the other internal point as $3.80952 - 0.495 \times (3.80952 - 2.85714)$. Notice that *all* the other values of L and R are original internal points (as closely as round-off allows). At stage 6 we conclude that the interval of uncertainty is (3.33333, 3.80952),

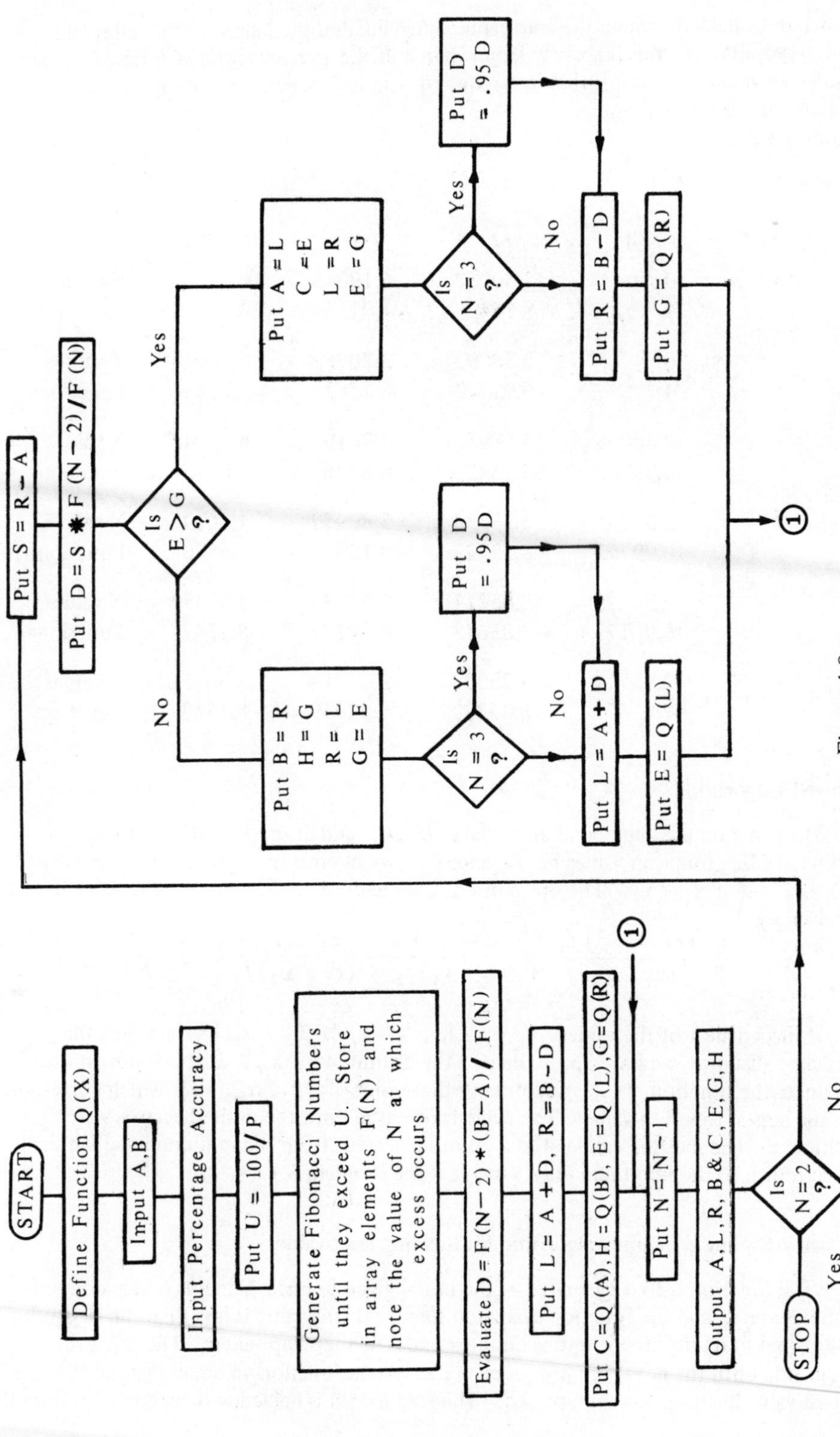
START
Define Function Q(X)
Input A,B
Input Percentage Accuracy
Put U = 100/P
Generate Fibonacci Numbers until they exceed U. Store in array elements F(N) and note the value of N at which excess occurs
Evaluate D = F(N − 2) * (B − A) / F(N)
Put L = A + D, R = B − D
Put C = Q (A), H = Q(B), E = Q (L), G = Q(R)
1
Put N = N − 1
Output A, L, R, B & C, E, G, H
Is N = 2 ?
Yes
No
STOP
Put S = R − A
Put D = S * F (N − 2) / F (N)
Is E > G ?
No
Yes
Put B = R H = G R = L G = E
Is N = 3 ?
Yes
No
Put D = .95 D
Put L = A + D
Put E = Q (L)
Put A = L C = E L = R E = G
Is N = 3 ?
Yes
No
Put D = .95 D
Put R = B − D
Put G = Q (R)
1

Figure 4.8

which does indeed include the true value 3.5. Further the length of this interval is 0.47619; although this is slightly larger than half the previous interval because of the displaced value L) it is less than the required length 0.5, since we used $F_7 = 21$ instead of 19 internal points.

Table 4.1

The Value of N is 7.

STAGE	A	L	R	B	
1	0	3.80952	6.19048	10	X-values
	4	–8.1542	–1.01134	34	Function
2	0	2.38095	3.80952	6.19048	X-values
	4	–6.99773	–8.1542	–1.01134	Function
3	2.38095	3.80952	4.7619	6.19048	X-values
	–6.99773	–8.1542	–6.6576	–1.01134	Function
4	2.38095	3.33333	3.80952	4.7619	X-values
	–6.99773	–8.22222	–8.1542	–6.6576	Function
5	2.38095	2.85714	3.33333	3.80952	X-values
	–6.99773	–7.83673	–8.22222	–8.1542	Function
6	2.85714	3.33333	3.35714	3.80952	X-values
	–7.83673	–8.22222	–8.22959	–8.1542	Function

Powell's algorithm

The function is evaluated at an initial point x_1 and at $x_2 = x_1 + d$; let the corresponding function values be f_1 and f_2. We choose $x_3 = x_1 + 2d$ if $f_1 > f_2$ and $x_3 = x_1 - d$ if $f_1 < f_2$. The optimum of the quadratic fitted through the three points is given by

$$x_m = \frac{1}{2}\,\frac{(x_2^2 - x_3^2)\,f_1 + (x_3^2 - x_1^2)\,f_2 + (x_1^2 - x_2^2)\,f_3}{(x_2 - x_3)\,f_1 + (x_3 - x_1)\,f_2 + (x_1 - x_2)\,f_3}$$

If the smallest of the values $|x_1 - x_m|$, $|x_2 - x_m|$, $|x_3 - x_m|$ is less than the required distance, we have approximated the optimum by x_m. If this is not so then we evaluate the function at x_m and discard that point of $\{x_1, x_2, x_3\}$ which corresponds to the largest function value. The cycle is repeated until the desired accuracy is achieved. The method can lead to a point far distant from the minimum and we shall then return to the minimum only slowly. (See Problem 3.)

Davies, Swann & Campey algorithm for locating least values

The function is first evaluated at the initial given point. A step is taken along the line of search and the function evaluated there. If this value is less than the first value (or equal to it) the step-length is doubled and a further step taken. The process is repeated until the next step produces an increase in function value, indicating that the least value location has been passed. The step length is halved and a step taken from the previous (last 'successful') point. This will give four equally spaced points at which the function has been evaluated, one of which gives rise to the least value so far encountered. One of the extreme two points of the quartet is rejected, that one which is further from the 'best minimum' point. The remaining three points can be labelled $x_1 = x_2 - d$, x_2, $x_3 = x_2 + d$. Let the corresponding function values be f_1, f_2, f_3. A quadratic curve is fitted to these three points. This quadratic will have a minimum at $x_2 + d_m$ where $d_m = d(f_1 - f_3)/[2(f_1 - 2f_2 + f_3)]$. This concludes a stage of the algorithm. A new stage begins with a reduced step length and starts from either x_2 or $x_2 + d_m$, whichever corresponds to a smaller function value. See Figure 4.9.

If the very first step gives rise to an increase in $f(x)$, we reverse the direction of search.

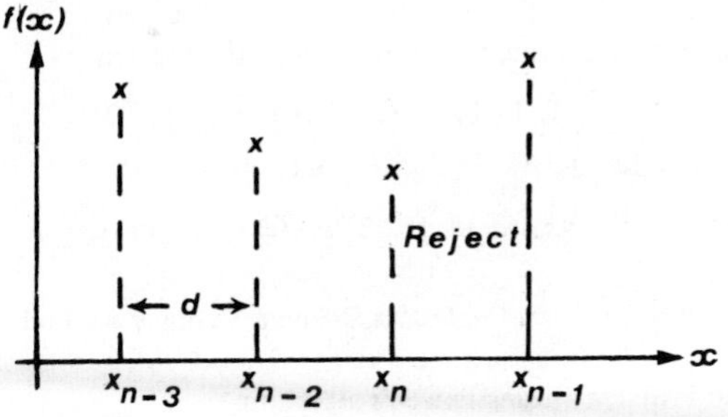

Figure 4.9

Problems

1. (a) Uniform search evaluates $f(x)$ at points equally spaced in the original interval of uncertainty. Show that the fraction of the original interval within which the optimum lies after N evaluations is given by $F = 2/(N + 1)$.

   (b) Uniform dichotomous search divides the original interval into $(N/2) + 1$ intervals of width $L/[(N/2) + 1]$. At each of the $(N/2)$ points, two function evaluations are performed, one on either side of the points, separated by a small distance 2ϵ. Find the fraction F for this search method.

   (c) Sequential dichotomous search commences by evaluating the function at two points distance ϵ on either side of the mid-point of the interval of uncertainty. Almost half this interval can now be rejected. The next two evaluations are performed close to the mid-point of the new interval of uncertainty, and so on. Show that $F = (1/2)^{N/2}$.

   (d) Show that for the Fibonacci search, $F = 1/F_N$.

2. (a) Assuming that the recurrence relation $F_n = F_{n-1} + F_{n-2}$ has a solution of the form $F_n = k^n$, derive the result $F_n = \frac{1}{\sqrt{5}}\left\{\left[\frac{1+\sqrt{5}}{2}\right]^{n+1} - \left[\frac{1-\sqrt{5}}{2}\right]^{n+1}\right\}$. Show that for large values of n, the successive reductions in interval are asymptotically approximately $\frac{1}{2}(1 + \sqrt{5}) = \tau$. **Golden Section** search has a constant interval reduction of τ. Hence $F = 1/\tau^N$. Show that the ratio of interval reduction by Fibonacci search to that by Golden Section search $F_N/\tau^{N-1} \simeq 1.17$.

   (b) Since the ratio of the whole interval at any stage to the larger sub-interval is equal to the ratio of the larger sub-interval to the smaller one, show that this result leads to the equation $\tau^2 = \tau + 1$. Solve this equation.

   (c) Repeat Example 2 page 102 by Golden Section search.

(d) Write a computer program for (c).

(e) Write a program to evaluate the first 100 Fibonacci numbers.

3. A second disadvantage of Powell's method is that x_m will correspond to a maximum if $\dfrac{(x_2 - x_3)f_1 + (x_3 - x_1)f_2 + (x_1 - x_2)f_3}{(x_1 - x_2)(x_2 - x_3)(x_3 - x_1)} \geqslant 0$. Prove this statement. [Note: if either disadvantage arises, we take a maximum permissible step in the direction of decreasing function values and the new point replaces one of x_1, x_2 or x_3 as in the main statement of the algorithm.]

4. Minimise the following functions by Fibonacci search.

(a) $|\cos x|$ over $[-1, 1]$; (b) $\cos x$ over $[0, 2]$;

(c) $e^{3x} - 3x + 1 - (x + 1) \log_e (x + 1)$ in $[0, 4]$;

(d) $x^4 - 2x^2 - 4x + 3$ in $[1, 2]$; (e) $(x - 1)^3 \log_e x$.

5. Repeat Problem 4 via the Davies, Swann & Campey method.

6. Repeat Problem 4 using Powell's algorithm.

7. Draw a flow diagram to represent in more detail the algorithm outlined below for locating by direct search the absolute maximum of a function $f(x)$ in the range $a \leqslant x \leqslant b$.

(a) Select an interval of search h.

(b) Compute points $x_i = a + ih$, $f_i = f(x_i)$, $i = 1, 2, ...$

(c) At each stage in (b) examine whether f_i is greater than the previous maximum. Write a program to determine by the above method the absolute maximum of $f(x) = e^x \sin 6x + \log_e (1 + \cos x)$ in the range $0 \leqslant x \leqslant \pi/3$. (C.E.I.)

8. A function $f(x)$ can be evaluated for any given value of x in the range $a \leqslant x \leqslant b$ but its derivatives are not available. Derive an algorithm for finding a maximum from the following principles.

Take steps of length h from the point a. When a maximum turning point is located, take smaller steps from a suitable point and repeat. Stop when the step length is reduced below a given value, or if no turning point is detected in the first search. (C.E.I.)

4.3 FUNCTIONS OF SEVERAL VARIABLES. DIRECT SEARCH METHODS

As soon as we move into more than one dimension, the problem of finding greatest or least values of a function becomes much more complicated. We shall try and give you an insight into the problem by adopting a parallel approach: we shall give examples of two variables, thus allowing geometrical interpretation, side by side with formulae relating to the general case of n dimensions. In the general case, the point $(x_1, x_2, x_3, ..., x_n)$ is conveniently referenced as $\mathbf{x}$ and we shall employ vector notation extensively. It will be helpful to plot *contours* of a function of two variables, a contour being a curve connecting points where the function values are equal. In Figure 4.10 we depict a set of contours which illustrate some of the main features to be encountered.

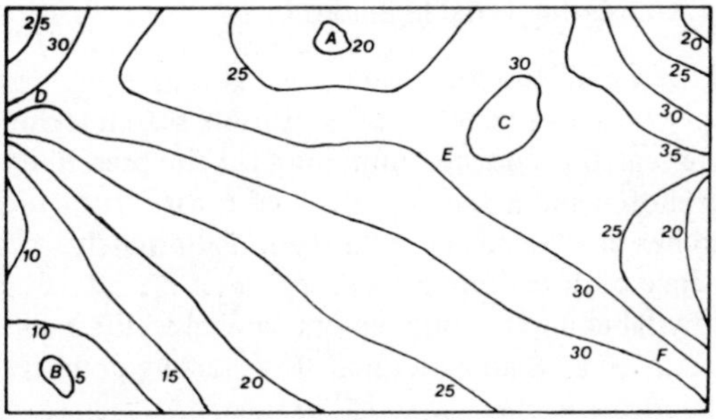

Figure 4.10

Points A and B represent local minima, point C represents a local maximum and points D, E and F represent possible **saddle points**. (Geometric illustrations of these types of features are shown in Figure 4.16, Section 4.4.). Can you detect any other features?

Suppose we consider the problem of cutting down uncertainty. In Figure 4.11(a) we see an **interval of uncertainty** of 10% of the original interval. If we now consider a function of two variables, each of which is restricted to a 10% interval of uncertainty, we see that the **area of uncertainty** in evaluating $f(x, y)$ is 1%. However, the situation is really the reverse: we specify the area of uncertainty as a percentage of the original area and then determine the uncertainty in each variable. Since we aim for 10% uncertainty in this discussion we are led to a percentage uncertainty of $(100/\sqrt{10}) \simeq 31.62\%$. This is depicted in Figure 4.11(b); here we see how much less precisely the values of x_1 and x_2 are located. If we consider a function of three variables, the same 10% 'volume' of accuracy would mean a percentage uncertainty in each variable of 46.5%. If we take the case of 50 variables, the percentage uncertainty in each variable is 91%. We therefore are forced to seek regions of uncertainty which are very small fractions of the original region in which the problem was defined.

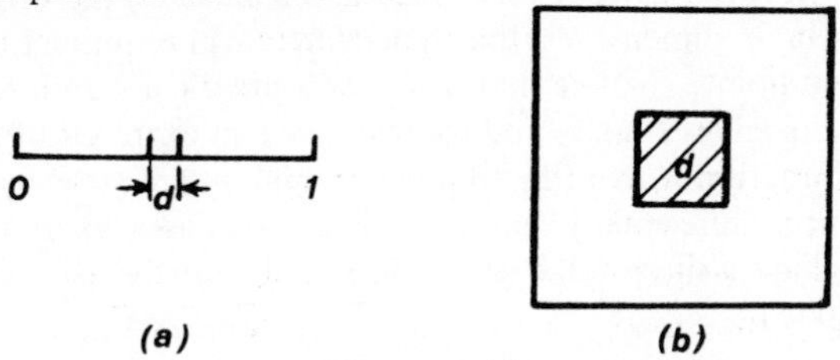

Figure 4.11

Methods of simultaneous search

We could set up a uniform grid over the problem region and evaluate at each node. We could, but we would be ill-advised to do so; with 10 variables and each variable divided into 10 sub-intervals we would need 10^{10} evaluations. Alternatively, we could generate points in the region by selecting sets of random numbers.

However, we shall study methods of sequential search, where information already accumulated is used to help select new points.

Sequential search - Alternating variable method

In this method the variables are examined one at a time and a *univariate search* made along the line of the chosen variable. Any univariate search technique may be employed. When the univariate search produces a minimum (to the prescribed accuracy) the next variable in order is chosen and a search performed for it. If there are only two variables involved, then searches take place in each direction alternately; if there are three variables, the sequence of directions is $\{x_1, x_2, x_3, x_1, x_2, x_3, x_1, ...\}$. In vector notation, we start by labelling the unit vectors parallel to the axes $\mathbf{e}_i$, $i = 1, ..., n$ and denote d_i, $i = 1, ..., n$ as scaling factors. The starting point is labelled $\mathbf{x}_0$; searches are made in the directions $\mathbf{e}_1, \mathbf{e}_2, ..., \mathbf{e}_n$ in turn. At each stage we move to a new point given by $\mathbf{x} = \mathbf{x}_s + d_i \mathbf{e}_i$, where $\mathbf{x}_s$ is the starting point.

The method is very simple but, naturally, has its drawbacks. It is at its best when the function under consideration is the analogue of a sphere in n - dimensions, called a **hypersphere.**

Consider Figure 4.12(a). Here we show contours of $f(x, y) = x^2 + y^2$.

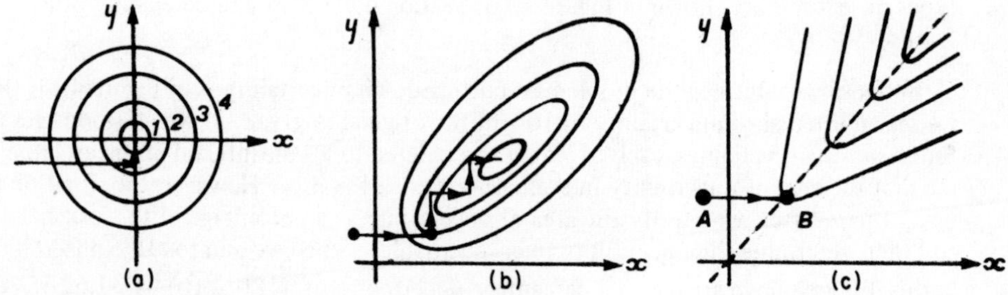

Figure 4.12

We start our search from $\mathbf{x}_0 = (-3.5, -1.6)$ and move along the line of $\mathbf{e}_1$, keeping y fixed. A search procedure will reveal the minimum in this direction to be at $\mathbf{x} = (0, -1.6)$. Keeping $\mathbf{x}$ fixed, we move along the line of $\mathbf{e}_2$ and locate the minimum at (0, 0). We have achieved our aim in a set of two searches. (We could have been lucky and started on either axis, or very lucky and started on the origin.) There is a general result that in n dimensions, the hypersphere will require at *most* n searches to locate the optimum point. Notice that if we use a practical search technique, we shall not locate minimum values exactly and we may need an extra set of searches. Figure 4.12(b) depicts a function where the variables x and y are strongly inter-related; here, the minimum point requires many searches. There are cases where the method fails: in Figure 4.12(c) we have a sharp ridge which rises to the north-east. Once at the point B we shall remain there.

Produce a flow chart for this method.

The method of Hooke and Jeeves. Pattern search

There have been several methods devised which comprise two major stages, an *exploratory* stage and a *pattern* stage. We now examine one of the simplest methods, due to Hooke and Jeeves. The algorithm is designed to seek the best direction of search and move fast in that direction.

First of all, a base camp is established at $\mathbf{x}^{(0)}$. For this first stage of the expedition, this point is also the advanced camp $\mathbf{x}^{(1)}$, from which explorations are made. Then a search party is sent out. They move to a new point by increasing x_1 by a fixed amount d and evaluate $f(\mathbf{x})$ at this point. If it is less than $f(\mathbf{x}^{(0)})$, progress has been made; if it is not, then we decrease x by d and test again; if this still produces no decrease, we stay at x_1. From whichever point we are at, we now repeat the process on x_2 and, in fact, on all the other variables in turn. When all variables are thus dealt with, the exploratory stage is complete. It is deemed a success if the value of the function at the latest point, $\mathbf{x}^*$, is less than $f(\mathbf{x}^{(0)})$. Otherwise it is a failure. If successful, the base camp is moved to $\mathbf{x}^*$, and the advanced camp is moved to $\mathbf{x}^{(1)} = \mathbf{x}^* + (\mathbf{x}^* - \mathbf{x}^{(0)})$; this means that the advanced camp is as far from the new base camp as this latter is from the old base camp and in the direction of old base camp to new base camp. [In computer programs we set $\mathbf{x}^{(1)}$ first and then $\mathbf{x}^{(0)} = \mathbf{x}^*$; why?]. A new search is carried out from the new advanced camp and the process repeated. If an exploratory search fails there are two basic reasons in general, although as we shall see, only one applies at the initial stage, but when writing a computer program, we must cater for the general case. For clarity, suppose we are at a stage other than the first and the exploratory search has failed. Assuming that the advanced camp from which we made our latest search does not coincide with the base camp then we move the advanced camp back to coincide with the base camp and try again. If the step just mentioned has been carried out to no avail then we reduce the step size and begin again. When the step size is decreased below a given level, we stop and hope we have reached the minimum.

A useful ploy is to start the method from a different point and compare results. In Figure 4.13 we present a flow chart for the method. The unit vectors in the chosen directions of search, which we usually assume are parallel to the axes, are denoted by $\mathbf{e}_1, \mathbf{e}_2, \ldots\ldots, \mathbf{e}_n$. $\mathbf{x}^{(\mathrm{I})}$ represents that point in the exploratory stage at which we find ourselves *before* moving in the direction of $\mathbf{e}_\mathrm{I}$.

The flow chart can be modified to output results at each stage to enable us to monitor progress.

Rosenbrock's Method

This method modifies the alternating variable method. A step of a given length is taken in each of the coordinate directions in turn. If a success is encountered in a direction insofar as a lower function value is achieved, then the step length in that direction is increased for the next search in that direction. However, if a failure is encountered that step is decreased and the next search in that direction takes place in the opposite sense. When a complete cycle is accomplished, a new cycle takes place with the revised step lengths. The next direction is then searched. When in each direction, a failure has been recorded after a success, the time has come to realign the axes to take account of the local geometry of the function surface.

The idea behind the realignment is to retain the direction of most progress and select directions orthogonal to it. Suppose that we have an objective function of three variables and that from the previous base point we have advanced d_1, d_2, and d_3 in the directions $\mathbf{x}_1$, $\mathbf{x}_2$, $\mathbf{x}_3$. We choose $\mathbf{q}_1 = d_1\mathbf{x}_1 + d_2\mathbf{x}_2 + d_3\mathbf{x}_3$, $\mathbf{q}_2 = d_1\mathbf{x}_1 + d_2\mathbf{x}_2$, $\mathbf{q}_3 = d_1\mathbf{x}_1$ and we must then select a set of three orthogonal directions which we do by means of the **Gram-Schmidt orthogonalisation process**. (Refer back to Section 1.2 for details.) We now advance from the latest point along these directions. This method will also handle constrained functions. In Figure 4.14 we show the algorithm schematically in two dimensions. Try and produce a flow chart for the method.

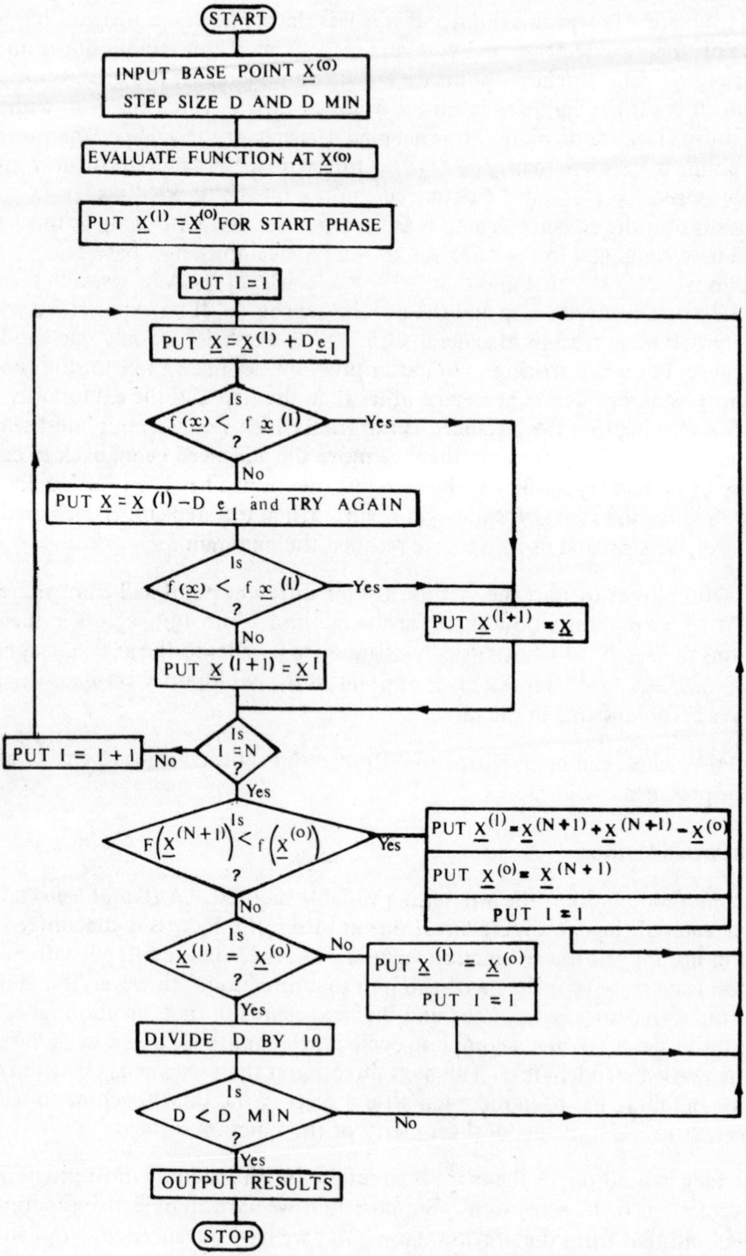
START
INPUT BASE POINT X(0)
STEP SIZE D AND D MIN
EVALUATE FUNCTION AT X(0)
PUT X(1) = X(0) FOR START PHASE
PUT I = 1
PUT X = X(1) + D e_I
Is f(x) < f x(1) ?
Yes
No
PUT X = X(1) − D e_I and TRY AGAIN
Is f(x) < f x(1) ?
Yes
No
PUT X(I+1) = X
PUT X(I+1) = X I
Is I = N ?
PUT I = I + 1
No
Yes
Is F(X(N+1)) < f(X(0)) ?
Yes
PUT X(1) = X(N+1) + X(N+1) − X(0)
PUT X(0) = X(N+1)
PUT I = 1
No
Is X(1) = X(0) ?
No
PUT X(1) = X(0)
PUT I = 1
Yes
DIVIDE D BY 10
Is D < D MIN ?
No
Yes
OUTPUT RESULTS
STOP

Figure 4.13

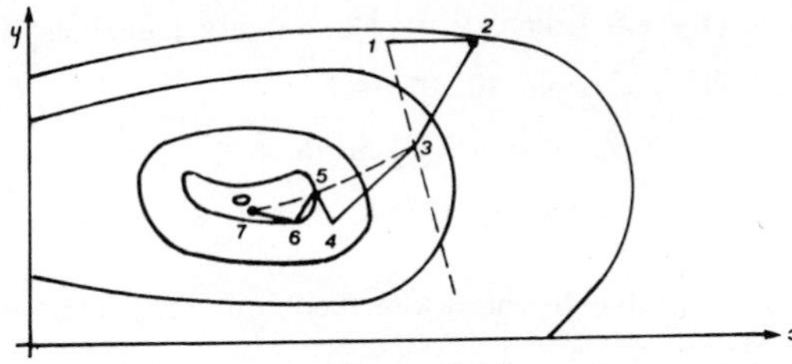

Figure 4.14

The simplex method of Nelder and Mead

The **simplex method** is named thus because in this branch of mathematics a **simplex** in n - dimensions is defined as that entity with $(n + 1)$ vertices. In two variables (or two dimensions) the simplex has three vertices and is a triangle. In three dimensions, the simplex has four vertices and is a tetrahedron. A basic simplex method sets up a regular simplex and evaluates the objective function at each of the $(n + 1)$ vertices. The point at which the objective function has its largest value is reflected in the centroid of the simplex; the new point takes over from its reflection and a new simplex is defined. Near a minimum, one point will remain as a vertex of several successive simplexes; this is a pointer to halving the distances of the other vertices from this particular vertex and repeating the search procedure. In Figure 4.15 points 1, 2, 3 start, then point 1 is reflected to obtain point 4.

Then point 2 is reflected to obtain point 5 and point 4 is reflected to obtain point 6. The search would, in fact, continue.

There is a need for modification since the simplex needs to adjust to the local topography. In narrow valleys with steep sides, we want a much smaller simplex, whereas in flat 'country' the simplex could well expand. The method of Nelder and Mead allows for this possibility.

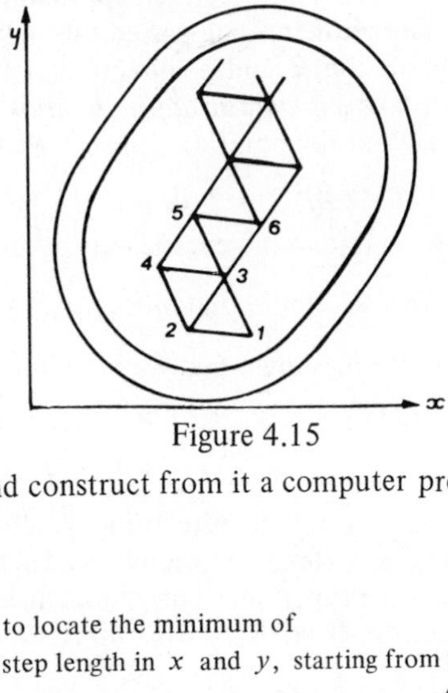

Figure 4.15

Try to produce a flow chart for the method and construct from it a computer program.

Problems

1. Use the method of Hooke and Jeeves in an attempt to locate the minimum of $z = 3x^2 + 2y^2 - 4xy - 3x - 7y + 10$ using a unit step length in x and y, starting from the origin. Check your answer by calculus. (L.U.)

2. Use the method of Hooke and Jeeves in a computer program to find the minimum of the function $f(x, y) = x^3 + y^3 - 12x^2 - 9y^2 + 256$. Try plotting the contours of $f(x, y)$ and mark on these the successive points reached by the method.

3. Minimise the following functions by the method of Hooke and Jeeves:

   (a) $8x^2 - 4xy + 5y^2$; start from $(5, 2)$

   (b) $(x - 1)^2 + (y - 1)^2 + \left(\frac{x}{4} + \frac{y}{4} - 1\right)^2$ from $(0, 0)$

(c) $100(y - x^2)^2 + (1 - x)^2$ from $(0, 0)$ {Rosenbrock's function}

(d) $100(y - x^3)^2 + (1 - x)^2$ from $(0, 0)$

(e) $(x^2 + xy^3 - 9)^2 + (3x^2 y - y^3 - 4)^2$ from $(0, 0.2)$

(f) $(x^2 + y^2 - 1)^2 + x^2 y^2$ from $(0, 1)$, $(0.8, 0.2)$, $(0.8, -0.2)$, $(1.2, 0.4)$

4. Repeat Problem 3 using (i) the Rosenbrock method, (ii) the Simplex method.

4.4 CALCULUS APPROACH TO FUNCTIONS OF SEVERAL VARIABLES

The behaviour of functions of one variable near stationary points can be determined by considering the Taylor's series for the function, centred at the stationary point currently under study. We now examine the analogous result for a function of two variables.

The **tangent plane approximation** is:

$$f(x_0 + h, y_0 + k) \simeq f(x_0, y_0) + h f_x(x_0, y_0) + k f_y(x_0, y_0) \tag{4.6}$$

A stationary point occurs at (x_0, y_0) when the tangent plane is horizontal there, i.e.

$$f_x(x_0, y_0) = 0 = f_y(x_0, y_0) \tag{4.7}$$

We need a more refined approximation than (4.6); this ensured that the plane given by the right-hand side passed through the point $(x_0, y_0, f(x_0, y_0))$ with the correct slopes in the x and y directions. In order to get the curvature correctly matched we must take a *quadratic approximation* whose second derivatives agree with those of $f(x, y)$ at the point (x_0, y_0). We take the general quadratic function in two variables:

$$q(x, y) = \alpha + \beta_1 x + \beta_2 y + \gamma_1 x^2 + \gamma_2 xy + \gamma_3 y^2$$

Now we obtain the coefficients by matching the 1st and 2nd derivatives at (x_0, y_0).

We also require that $q(x_0, y_0) = f(x_0, y_0)$.

It can be shown that we have, on setting $x = x_0 + h$, $y = y_0 + k$ the approximation

$$f(x_0 + h, y_0 + k) \simeq f(x_0, y_0) + h f_x(x_0, y_0) + k f_y(x_0, y_0) + \tfrac{1}{2}[h^2 f_{xx}(x_0, y_0) + 2hk f_{xy}(x_0, y_0) + k^2 f_{yy}(x_0, y_0)] \tag{4.8}$$

As we were able to extend the quadratic approximation for a function of one variable, so we can extend equation (4.8). Taylor's Theorem for a function of two variables states that if the function is continuous in some closed domain D of the x - y plane and possesses partial derivatives up to order $(n + 1)$ in D, then

$$f(x_0 + h, y_0 + k) = f(x_0, y_0) + h f_x(x_0, y_0) + k f_y(x_0, y_0) + \frac{1}{2!}[h^2 f_{xx}(x_0, y_0) + 2hk f_{xy}(x_0, y_0) + k^2 f_{yy}(x_0, y_0)] + \frac{1}{3!}\left[h\frac{\partial}{\partial x} + k\frac{\partial}{\partial y}\right]_0^3 f(x_0, y_0) + \ldots + \frac{1}{n!}\left[h\frac{\partial}{\partial x} + k\frac{\partial}{\partial y}\right]_0^n f(x, y) + \frac{1}{(n+1)!}\left[h\frac{\partial}{\partial x} + k\frac{\partial}{\partial y}\right]_0^{n+1} f(\xi, \eta) \tag{4.9}$$

where for example,

$$\left[h\frac{\partial}{\partial x}+k\frac{\partial}{\partial y}\right]_0^3 f(x, y) \equiv \quad h^3\frac{\partial^3 f(x, y)}{\partial x^3}+3h^2k\frac{\partial^3 f(x, y)}{(\partial x^2)\partial y}$$

$$+3hk^2\frac{\partial^3 f(x, y)}{\partial x\ \partial y^2}+k^3\frac{\partial^3 f(x, y)}{\partial y^3}$$

evaluated at (x_0, y_0) and where (ξ, η) is a point in the interior of D.

(It should be clear how to extend this result for a function of three or more variables.) If we add the condition that

$$\lim_{n\to\infty}\left[\frac{1}{(n+1)!}\left[h\frac{\partial}{\partial x}+k\frac{\partial}{\partial y}\right]_0^{n+1} f(\xi, \eta)\right] = 0 \text{ for all } (x, y) \text{ in } D$$

then we can obtain the infinite series expansion. Note that it is convenient to write $f(x_1, x_2, x_3, \ldots.., x_n)$ as $f(\mathbf{x})$ and therefore we may generalise the quadratic approximation to

$$f(\mathbf{x}) \simeq f(\mathbf{x}_0)+(\mathbf{x}-\mathbf{x}_0)\,.\,\nabla_0 f + \tfrac{1}{2}(\mathbf{x}-\mathbf{x}_0)^T H_0(\mathbf{x}-\mathbf{x}_0) \tag{4.10}$$

We shall quote the meanings of the new symbols for two variables

$$\mathbf{x}_0 \equiv (x_0, y_0) \text{ and } \mathbf{x} \equiv (x, y) = (x_0+h, y_0+k)$$

$\nabla f \equiv \left(\dfrac{\partial f}{\partial x}, \dfrac{\partial f}{\partial y}\right)$ and is called the **gradient vector** of f, pronounced 'grad f'

$H \equiv \begin{bmatrix} \dfrac{\partial^2 f}{\partial x^2} & \dfrac{\partial^2 f}{\partial x\ \partial y} \\ \dfrac{\partial^2 f}{\partial y\ \partial x} & \dfrac{\partial^2 f}{\partial y^2} \end{bmatrix}$ is called the **Hessian matrix** of f (it is the matrix of second partial derivatives and is symmetrical);

the suffix $_0$ means that the derivatives are evaluated at (x_0, y_0).

Maximum and Minimum values

We quote a theorem due to Weierstrass.

If a function $f(\mathbf{x})$ is continuous in a closed domain D then it has a greatest and a least value either in D or on the boundary of D.

We know that a necessary condition for $f(x, y)$ to have a local maximum or a local minimum at (x_0, y_0) is that the tangent plane should be horizontal there, i.e. $\nabla f|(x_0, y_0) = \mathbf{0}$. Therefore we may write (4.10) as

$$\delta f \equiv f(\mathbf{x}) - f(\mathbf{x}_0) = \tfrac{1}{2}(\mathbf{x}-\mathbf{x}_0)^T H_0(\mathbf{x}-\mathbf{x}_0)$$

We shall develop the appropriate conditions for two variables and quote the generalisation. The resulting quadratic form $(h, k)^T H_0(h, k)$ can be symbolised as $Ah^2 + 2Bhk + Ck^2$ where A, B and C are constants. If the point (x_0, y_0) is a local maximum then δf will be negative for all values of h and k [small] i.e. the quadratic form is **negative definite.** Correspondingly, if the point (x_0, y_0) is a local minimum then δf will be positive for all small values of h and k i.e. the quadratic form is **positive definite.** Some indefinite forms will be geometrically described as **saddle points,** so that passing over the point (x_0, y_0) in some directions gives the impression of it being a local maximum and in other directions, the impression created is that of a local minimum. See Figure 4.16.

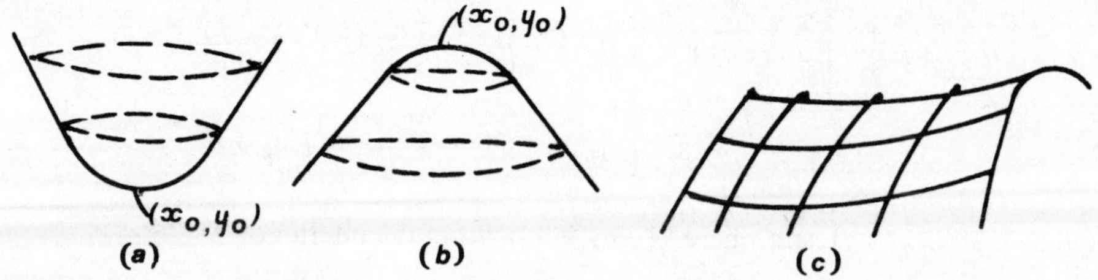

Figure 4.16

Other more esoteric features exist. If the quadratic form is identically zero, then we should consider third-order terms.

The conditions for classifying stationary points are:

for a local maximum, $f_{xx} < 0$ and $f_{xx}f_{yy} - (f_{xy})^2 > 0$ (4.11a)

for a local minimum, $f_{xx} > 0$ and $f_{xx}f_{yy} - (f_{xy})^2 > 0$ (4.11b)

If $f_{xx}f_{yy} - (f_{xy})^2 < 0$, we designate this a saddle point.

We note that for two variables $D = f_{xx}f_{yy} - (f_{xy})^2$ is the determinant of the Hessian matrix. [Why could we substitute $f_{yy} > 0$ for $f_{xx} > 0$ in the conditions for a local minimum and similarly for a local maximum?]

Example 1

Consider $f(x, y) = x^3 + y^3 - 12x^2 - 9y^2 + 256$; $f_x = 3x^2 - 24x$; $f_y = 3y^2 - 18y$; $f_{xx} = 6x - 24$; $f_{xy} = 0$; $f_{yy} = 6y - 18$.

We have stationary points where $f_x = 0$ and $f_y = 0$, i.e. where $3x(x - 8) = 0$ *and* $3y(y - 6) = 0$. This gives us four possibilities: (0, 0), (8, 0), (0, 6), (8, 6). To examine the nature of these stationary points we are well advised to draw up a table.

Point	(0, 0)	(8, 0)	(0, 6)	(8, 6)
f_{xx}	–24	24	–24	24
f_{yy}	–18	–18	18	18
f_{xy}	0	0	0	0
D	>0	<0	<0	>0
	Local Maximum	Saddle Point	Saddle Point	Local Minimum

Example 2

$f(x, y) = x^2y - 4x^2 - 2y^2 + 16y$; $f_x = 2xy - 8x$; $f_y = x^2 - 4y + 16$;
$f_{xx} = 2y - 8$; $f_{xy} = 2x$; $f_{yy} = -4$

Stationary points occur when $2x(y - 4) = 0$ and $x^2 = 4y - 16$. Then $x = 0$ and hence $y = 4$ is one possibility; if $y = 4$, $x^2 = 0$. This means we can find only one stationary point at (0, 4); there $f_{xx} = 0$, $f_{xy} = 0$ and $f_{yy} = -4$. This gives $D = 0$. We need further investigation. The quadratic approximation is

$$f(h, 4+k) \simeq f(0, 4) + hf_x(0, 4) + kf_y(0, 4) + \frac{1}{2!}\left[h^2 f_{xx}(0, 4) + 2hkf_{xy}(0, 4) + k^2 f_{yy}(0, 4)\right]$$

i.e. $\delta f \equiv f(h, 4+k) - f(0, 4) \simeq \frac{1}{2}k^2(-4) = -2k^2$

Thus the value of δf is always negative, which implies $f(x, y)$ has a local minimum at (0, 4). Try to draw the contours of $f(x, y)$.

Example 3

A rectangular tank is to be constructed to contain ½ cu.m. of hot liquid; in order to minimise heat loss we must so construct the tank that the surface area is least. In addition, we require the base to be of double thickness. What dimensions do we choose?

Let the lengths of three mutually perpendicular edges be x, y, z metres. Then the volume restriction implies

$$xyz = ½ \qquad (4.12)$$

The surface area is given by

$$S = 3xy + 2xz + 2yz \qquad (4.13)$$

It is symmetrical with respect to x and y but not z.

If we substitute for z from (4.12) into (4.13) we obtain

$$S = 3xy + (x+y)/xy = 3xy + (1/y) + (1/x) \qquad (4.14)$$

To locate the stationary points, we find S_x and S_y and equate to zero.

$$S_x = 3y - 1/x^2 = 0 = 3x - 1/y^2 = S_y$$

Hence we obtain $3x^2y - 1 = 0$ and $3xy^2 - 1 = 0$ which produce, on subtraction, $3x^2y - 3xy^2 = 0$, i.e. $3xy(x-y) = 0$. We reject the possibilities $x = 0$ and $y = 0$ since they do not represent a physical solution to our problem; these, in theory, give a maximum surface area. The remaining possibility is $y = x$ which leads to $x = y = \sqrt[3]{(1/3)}$, $z = ½\sqrt[3]{9}$ or to 3 s.f. $x = y = 0.693$, $z = 1.05$ and a minimum surface area of $S_{min} = \sqrt[3]{3} + 2\sqrt[3]{3} = 4.33\text{m}^2$.

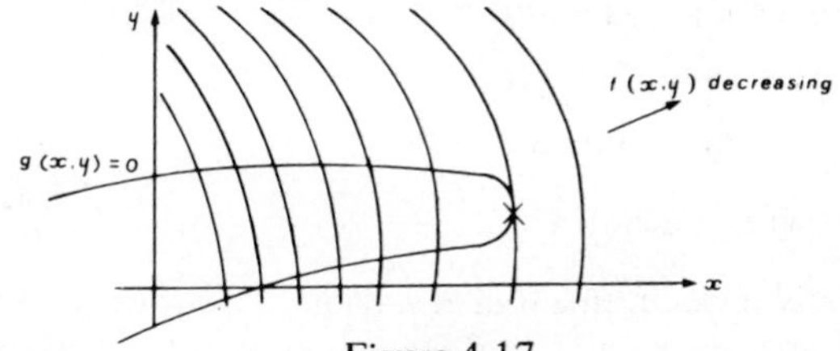

Figure 4.17

Constrained Problems

Example 3 contained a constraint, viz. the volume of the tank was a fixed quantity; we overcame the constraint by making a substitution to remove one of the variables. In this subsection we now develop a technique—the method of **Lagrange multipliers** which effectively delays the substitution. We approach the technique from a geometrical viewpoint for a function of two variables. Figure 4.17 shows a set of contours of the function $f(x, y)$ and the curve representing the constraint $g(x, y) = 0$. The point marked with a cross is the minimum value of $f(x, y)$ subject to the constraint $g(x, y) = 0$. This point lies on a particular contour $f(x, y) = \text{constant}$ and it should be clear that this contour and the constraint curve have a common tangent at this point.

We can represent the constraint curve in parametric form, with the parameter chosen as s, the length along the curve from a given point. Hence $x = x(s)$ and $y = y(s)$ so that the objective function can be written in terms of s, as follows

$$f(x, y) \equiv f[x(s), y(s)] = F(s)^{\dagger} \tag{4.15}$$

By the chain rule

$$F'(s) = f_x \frac{dx}{ds} + f_y \frac{dy}{ds} \tag{4.16}$$

and the maximum and minimum points of $f(x, y)$ on the constraint curve can be found by solving $F'(s) = 0$. Now (4.16) can be written as $F'(s) = (\nabla f) \cdot \mathbf{t}$ where $\mathbf{t} = \frac{dx}{ds}\mathbf{i} + \frac{dy}{ds}\mathbf{j}$. The condition $F'(s) = 0 \Rightarrow \nabla f \cdot \mathbf{t} = 0$ and therefore ∇f is normal to $\mathbf{t}$. However, $\mathbf{t}$ is a tangent to the constraint curve. (Why?) Therefore ∇f must be normal to this curve at a maximum or minimum point. But the vector ∇g is also normal to the constraint curve. Because we have two vectors normal to a given curve at a given point, it follows that one is a scalar multiple of the other i.e.

$$\nabla f + \lambda \nabla g = 0 \tag{4.17}$$

where λ is a scalar. Hence, to find the maxima and minima of $f(x, y)$ subject to $g(x, y) = 0$ we must solve the equations

$$f_x + \lambda g_x = 0; \quad f_y + \lambda g_y = 0; \quad g(x, y) = 0 \tag{4.18}$$

If there are three original variables, then the procedure is as below.

Example 1

Consider the rectangular tank problem of page 115 where we seek to minimise $f(x, y, z) = 3xy + 2yz + 2xz$ subject to $g(x, y, z) = xyz - \frac{1}{2} = 0$.

It helps to form $F(x, y, z) \equiv 3xy + 2yz + 2zx + \lambda(xyz - \frac{1}{2})$

$$\equiv f(x, y, z) + \lambda g(x, y, z)$$

Then $F_x = f_x + \lambda g_x = 3y + 2z + \lambda yz = 0$ (4.19a)

$$F_y = f_y + \lambda g_y = 3x + 2z + \lambda xz = 0 \tag{4.19b}$$

$$F_z = f_z + \lambda g_z = 2y + 2x + \lambda xy = 0 \tag{4.19c}$$

and $g(x, y, z) = xyz - \frac{1}{2} = 0$ (4.19d)

We have apparently introduced a fourth variable λ, but we have four equations to solve for the three variables x, y, z and the extra variable λ. We use the first three equations to express x, y and z in terms of λ and hence the ratio $x : y : z$ and we use the fourth equation to find the actual values of these variables at any stationary point. We have already remarked on the symmetry between x and y and we can make use of this (we ought to add that the quick solution of such equations depends on the relative skill and experienced insight of the solver). We subtract (4.19b) from (4.19a) to obtain

$$3(y - x) + \lambda z(y - x) = 0 \text{ i.e. } (y - x)[3 + \lambda z] = 0$$

We have two alternatives to consider and we substitute the possibility $y = x$ into

---

† Note that since the expression in s will be different from that in x and y we use a different symbol for it.

(4.19c) to obtain $4x + \lambda x^2 = 0$. This yields $x = 0$ (rejected – why?) or $\lambda x = -4$. The case $\lambda x = -4$ gives $x = y = -4/\lambda$ and substitution into either (4.19a) or (4.19b) yields $-(12/\lambda) + 2z - 4z = 0$ i.e. $z = -6/\lambda$.
We must now return to the condition $3 + \lambda z = 0$.
If we substitute into (4.19a) we obtain

$$3y + 2z - 3y = 0 \text{ i.e. } z = 0 \text{ which we reject.}$$

We have as our only possibility $x : y : z = 2 : 2 : 3$.
Substituting in (4.19d) we obtain $x = y = \sqrt[3]{(1/3)}$, $z = \tfrac{1}{2}\sqrt[3]{9}$.
We argue for a minimum on physical grounds.

Example 2

Find the least value of the function $f(x, y, z) = x^2 + y^2 + z^2$ along the line of intersection of the planes $2x + y - z = 2$ and $2x - y + z = -2$. We seek to minimise $f(x, y, z) = x^2 + y^2 + z^2$ subject to $g(x, y, z) = 2x + y - z - 2 = 0$ and $h(x, y, z) = 2x - y + z + 2 = 0$. We form
$F(x, y, z) \equiv f(x, y, z) + \lambda g(x, y, z) + \mu h(x, y, z)$ where λ and μ are scalars. Then

$$F_x = 2x + 2\lambda + 2\mu = 0 \tag{4.20a}$$

$$F_y = 2y + \lambda - \mu = 0 \tag{4.20b}$$

$$F_z = 2z - \lambda + \mu = 0 \tag{4.20c}$$

$$g(x, y, z) = 2x + y - z - 2 = 0 \tag{4.20d}$$

$$h(x, y, z) = 2x - y + z + 2 = 0 \tag{4.20e}$$

We solve the first three equations to obtain x, y, z in terms of λ, μ. Hence $x = -\lambda - \mu$ and $y = -(\lambda/2) + (\mu/2)$ and $z = (\lambda/2) - (\mu/2)$. Substitution in the last two equations produces $\lambda = -1$, $\mu = 1$ and hence $x = 0$, $y = 1$, $z = -1$ giving a minimum value for $f(x, y, z)$ of 2. Can you see geometrically that this means that the nearest distance from the origin of any point on the line of intersection is 2? Note why we can claim the stationary point is a minimum and note further that if we seek such least distances, it is better to work with the square of the distance.

Although it is not discussed here, it should be noted that the method of Lagrange multipliers can logically be extended to cover the case of an objective function having three variables, subject to one constraint.

Problems

1. How do the following functions behave at the origin

   (i) $x^2 + xy + y^2 + x^3 + x^2y + y^3$ (ii) $x^2 + 2xy + y^2 + x^4 + x^2y^2 + y^4$?

2. Show that $z = x^3 + y^3 - 3\lambda(x + y) + 6xy$ has four stationary values where $\lambda > 3$. Show that when $\lambda = 4$, z has one minimum value which is $28 - 20\sqrt{5}$. (L.U.)

3. Find the absolute maximum and minimum values of $f(x, y) = 5x + 2y$ in the region $x^2 + 4y^2 \leqslant 1$.

4. Show that $z = x^2 + y^2 - 4xy^2$ has one stationary point and find its nature. (L.U.)

5. Find the stationary values of $f(x,y) = x^3 + ay^2 - 6axy$, where $a > 0$ and determine whether they are maxima, minima or saddle points. (L.U.)

6. Show that the sum of the squares of the distances of the point (x_0, y_0) from the vertices of a triangle in the $x - y$ plane is least when (x_0, y_0) is the centroid of the triangle. (L.U.)

7. Find the local maximum of the function $x^3 + y^3 + z^3 + 47.3$ subject to $x^2 - y^2 + z^2 = 1$. (L.U.)

8. Find that point on the sphere $x^2 + y^2 + z^2 = 1$ furthest from the point $(1, 3, 2)$. (L.U.)

9. A rectangular block with edges x, y, z is cut from a sphere of radius b so that the volume xyz has a maximum value. Find x, y, z and show that the maximum value is $8b^3/3\sqrt{3}$. (L.U.)

10. State conditions sufficient to ensure that a function $f(x,y)$ of the two independent variables x, y should have a minimum value at the point (a, b).

A closed capsule is to be constructed of material of negligible thickness to have a constant volume V. It comprises a hollow right circular cylinder of fixed radius R but variable length z, surmounted at the ends by two right circular cones of base radius R and variable heights x and y. Show that the surface area S of the capsule may be written in the form

$$S = \frac{2V}{R} - \frac{2}{3}\pi R(x + y) + \pi R(x^2 + R^2)^{1/2} + \pi R(y^2 + R^2)^{1/2}.$$

If $V = 2\pi R^3\sqrt{5}/3$ show that S assumes a minimum value when $x = y = z = 2R/\sqrt{5}$. (L.U.)

11. (i) Show that the function $\dfrac{x^2 + y^2 + 1}{(x + y + 1)^2}$ is a minimum at the point $(1, 1)$.

(ii) Use Lagrange's method of undetermined multipliers to prove that a closed tank, in the form of a circular cylinder with plane ends, has maximum volume for given surface area when the length is equal to the diameter of the ends. (L.U.)

12. Using Lagrange multipliers find a point (x, y, z) on the unit sphere $x^2 + y^2 + z^2 = 1$ which minimizes the function $x + y^2 + yz + 2z^2$. (L.U.)

13. (i) The lengths a, b, c vary so that $a + b + c = 3k$, where k is constant. Find the maximum volume enclosed by the ellipsoid $\dfrac{x^2}{a^2} + \dfrac{y^2}{b^2} + \dfrac{z^2}{c^2} = 1$.

(ii) Find the greatest value of the function $(x^2 - y^2)\exp\left[-x^2 - 2y^2\right]$. (L.U.)

14. (i) If $z = x^4 + y^4 - 2(x - y)^2$ show that z has minimum values at $(\sqrt{2}, -\sqrt{2})$ and $(-\sqrt{2}, \sqrt{2})$.

(ii) Positive numbers x, y and z are subject to the condition $x + y + z$ = constant. Prove that their product is greatest when they are equal. (L.U.)

15. (i) Prove that the function $xy(3 - x - y)$ has a maximum when $x = y = 1$.

(ii) Use Lagrange's method of undetermined multipliers to find the shortest distance from the origin to the plane $2x - y + z = 12$. Check your result by the methods of coordinate geometry. (L.U.)

16. Describe briefly how the method of Lagrange's multipliers can be used in the determination of extrema of a function $f(x,y,z)$ subject to the constraint $g(x,y,z) = 0$.

A solid has the form of a right circular cylinder of radius R and height H surmounted by a right circular cone of height h whose base coincides with the upper end of the cylinder. If the solid is to have a given constant volume, use the above method to show that when its surface area is a minimum, $H : R : h = (\sqrt{5}+1) : \sqrt{5} : 2$. (C.E.I.)

17. A closed right circular cylindrical storage tank is designed to hold 1000 m^3 of hot liquid. It is proposed to minimize the heat loss by making the surface area as small as possible.

    Use the method of Lagrange multipliers to obtain the optimum dimensions of the tank. (C.E.I.)

18. Use the method of Lagrange's multipliers to find the minimum of the function $f(x,y,z,t) = x^2 + y^2 + z^2 + t^2$ subject to the conditions $x + y - z + 2t = 2$, $2x - y + z + 3t = 3$. (C.E.I.)

4.5 METHODS USING THE GRADIENT OF A FUNCTION

The previous search techniques that we have mentioned used only the properties of continuous functions. We now examine methods which call upon further information from a function, namely the derivatives of the function. The simplest method works on the principle of choosing the quickest way down.

Method of steepest descent

In this technique we effectively fit a tangent plane to the function at specified points. In each tangent plane we find the line of steepest slope, and proceed down this line either by a fixed step size or until we find the lowest point (which can be accomplished by a one-variable search). At this lowest point we find the new direction of steepest descent and repeat the procedure. Let us first see how to choose this direction analytically.

We consider a small change $(\delta x_1, \delta x_2, \ldots\ldots, \delta x_n)$ from the present point. The first approximation to the change in the value of the function is given by

$$df \simeq \sum_{j=1}^{n} f_{x_j}\, \delta x_j$$

where the derivatives are evaluated at the present point. Of all possible changes of a given magnitude

$$D = \sqrt{\sum_{j=1}^{n} [\delta x_j]^2}$$

we seek that one which causes greatest change in $f(\mathbf{x})$. We work the idea through for two independent variables; x_1 and x_2. We want to maximise

$$df = \frac{\partial f}{\partial x_1} \,.\, \delta x_1 + \frac{\partial f}{\partial x_2} \,.\, \delta x_2 \quad \text{subject to } D^2 - (\delta x_1)^2 - (\delta x_2)^2 = 0$$

This time, our variables are δx_1 and δx_2; if this confuses you, call them l and m, for example. We use the method of Lagrange multipliers and form

$$F(\delta x_1, \delta x_2, \lambda) = \frac{\partial f}{\partial x_1}\,\delta x_1 + \frac{\partial f}{\partial x_2}\,\delta x_2 + \lambda\,[D^2 - (\delta x_1)^2 - (\delta x_2)^2]$$

The equations to be solved are

$$\frac{\partial f}{\partial x_1} - 2\lambda\,\delta x_1 = 0 \quad \frac{\partial f}{\partial x_2} - 2\lambda\,\delta x_2 = 0 \quad \text{and} \quad (\delta x_1)^2 + (\delta x_2)^2 = D^2$$

Therefore $\dfrac{1}{2\lambda} = \dfrac{\delta x_1}{\partial f/\partial x_1} = \dfrac{\delta x_2}{\partial f/\partial x_2}$ and this tells us to choose the direction of change so that the components of the change are proportional to the partial derivatives of $f(\mathbf{x})$ i.e. we proceed along the gradient vector, ∇f. The constant of proportionality, λ, will be chosen to be negative so that we make a *descent.*

Note that the gradient vector is orthogonal to the contours. Suppose we move from our present point along the contour which passes through it by a change $\mathbf{d} = (d_1, d_2)$.

Then $df = 0 = \dfrac{\partial f}{\partial x_1}\, d_1 + \dfrac{\partial f}{\partial x_2}\, d_2$; but $\nabla f . \mathbf{d} = \dfrac{\partial f}{\partial x_1} . d_1 + \dfrac{\partial f}{\partial x_2} . d_2 = 0$, hence orthogonality is proved.

In general, the method of Lagrange multipliers shows that the direction of local steepest descent is given by ∇f and the orthogonality property still holds.

A variant on the method is to form the **normalised gradient vector** $\dfrac{\nabla f}{|\nabla f|}$ and step a fixed distance, repeating this until no reduction in function value is obtained. Then the step size is reduced and the method restarted from the previous best point.

At each minimum point, the direction of steepest descent is orthogonal to the most recent search direction. For functions of two variables, only two directions are used and this procedure is akin to an alternating variable search. Methods of speeding up the process have been suggested but the trend is towards employing different techniques.

Example

$$f(x, y) = x^2 + 4y^2 + 2xy$$

The contours of the function are shown in Figure 4.18.

We start at the point $P_0 = (2, 1)$. Now $f_x = 2x + 2y$ and $f_y = 2x + 8y$, and at $(2, 1)$ the values are 6 and 12 respectively. Hence we find that the local direction of steepest descent is given by the unit vector $\left(\dfrac{1}{\sqrt{5}}, \dfrac{2}{\sqrt{5}}\right)$. [Why?]. We have to proceed along the line through $(2, 1)$ with gradient 2, i.e. along the line $y = 2x - 3$, until we come to the lowest point. In practice we would start with a given step length and reduce the

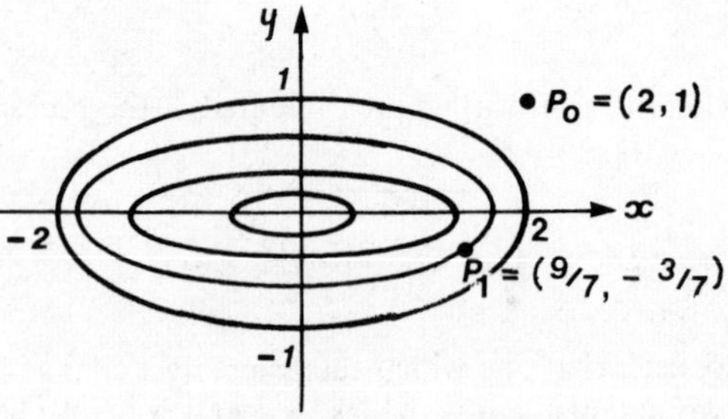

Figure 4.18

step length when necessary until it obtained a critical size; the point then reached would be an approximation to the point we shall obtain by calculus.

We seek the minimum of $x^2 + 4y^2 + 2xy$ subject to $y = 2x - 3$. Using the method of Lagrange multipliers, we minimise $F(x, y, \lambda) = x^2 + 4y^2 + 2xy + \lambda(y - 2x + 3)$. We solve $2x + 2y - 2\lambda = 0$ and $2x + 8y + \lambda = 0$.

Eliminating λ, we obtain $6x + 18y = 0$, i.e. $y = -x/3$ and hence we obtain the point $P_1 = (9/7, -3/7)$. At this point $f_x = 12/7$ and $f_y = -6/7$.

The local direction of steepest descent is given by the unit vector $\left(\frac{2}{\sqrt{5}}, \frac{-1}{\sqrt{5}}\right)$

and the cycle repeated. Further results are shown in Table 4.2. You can see the approach to the minimum at (0, 0).

Table 4.2

Stage	Start Point	f_x	f_y	Unit Vector	End point
0	$(2, 1)$	6	12	$\left(\frac{1}{\sqrt{5}}, \frac{2}{\sqrt{5}}\right)$	$\left(\frac{9}{7}, \frac{-3}{7}\right)$
1	$\left(\frac{9}{7}, \frac{-3}{7}\right)$	$\frac{12}{7}$	$\frac{-6}{7}$	$\left(\frac{2}{\sqrt{5}}, \frac{-1}{\sqrt{5}}\right)$	$\left(\frac{6}{28}, \frac{3}{28}\right)$
2	$\left(\frac{6}{28}, \frac{3}{28}\right)$	$\frac{9}{14}$	$\frac{18}{14}$	$\left(\frac{1}{\sqrt{5}}, \frac{2}{\sqrt{5}}\right)$	$\left(\frac{27}{196}, \frac{-9}{196}\right)$
3	$\left(\frac{27}{196}, \frac{-9}{196}\right)$	$\frac{18}{98}$	$\frac{-9}{98}$	$\left(\frac{2}{\sqrt{5}}, \frac{-1}{\sqrt{5}}\right)$	$\left(\frac{18}{784}, \frac{9}{784}\right)$
4	$\left(\frac{18}{784}, \frac{9}{784}\right)$	$\frac{27}{392}$	$\frac{54}{392}$	$\left(\frac{1}{\sqrt{5}}, \frac{2}{\sqrt{5}}\right)$	$\left(\frac{81}{5488}, \frac{-27}{5488}\right)$

If a direct search method is used at each stage, progress may be slower.

The efficiency of this method is not high and is related to the scales that we use. It works best when the contours are circular (or the analogue in higher dimensions) and the closer the contours are to being circular, the more rapid the convergence. Consider the function $z = x^2 + 36y^2$, some contours of which are shown in Figure 4.19. Starting from the point P_0, the direction of steepest descent is decidedly *not* towards the lowest point, which is the origin. However, if we make the transformation $v = 6y$ the function becomes $z = x^2 + v^2$ and the direction of steepest descent is *directly* towards the origin. This choice of scale is not simply graphical; if the two variables

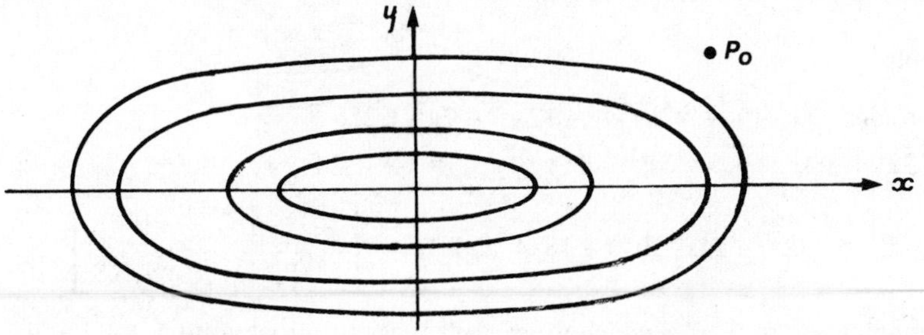

Figure 4.19

represent different physical quantities such as temperature and pressure it may not be possible to achieve a suitable set of contours. Attempts have been made to accelerate the convergence, but since the method of steepest descent is not a practical contender, we shall not discuss them.

Method of Newton-Raphson

Near a local minimum, the contours are nearly elliptical (in two dimensions). Many methods make use of this quadratic property. If we return to the quadratic approximation

$$f(\mathbf{x}) \simeq f(\mathbf{x}_0) + (\mathbf{x} - \mathbf{x}_0) \cdot \nabla f_0 + \tfrac{1}{2}(\mathbf{x} - \mathbf{x}_0)^T H_0 (\mathbf{x} - \mathbf{x}_0) \qquad (4.21)$$

we know that at the minimum, which we assume is $\mathbf{x}_0$, $\nabla f = \mathbf{0}$.

For two variables, we have

$$f(x,y) \simeq f(x_0,y_0) + (x - x_0)f_x(x_0,y_0) + (y - y_0)f_y(x_0,y_0)$$
$$+ \tfrac{1}{2}[(x - x_0)^2 f_{xx}(x_0,y_0) + 2(x - x_0)(y - y_0)f_{xy}(x_0,y_0) + (y - y_0)^2 f_{yy}(x_0,y_0)]$$

Differentiating this with respect to x and with respect to y we obtain

$$f_x(x,y) \simeq 0 + f_x(x_0,y_0) + (x - x_0)f_{xx}(x_0,y_0) + (y - y_0)f_{xy}(x_0,y_0)$$

and $f_y(x,y) \simeq 0 + f_y(x_0,y_0) + (x - x_0)f_{xy}(x_0,y_0) + (y - y_0)f_{yy}(x_0,y_0)$.

But if (x, y) is the local minimum then $f_x(x, y) = 0 = f_y(x, y)$.

Suppose we are at the point (x_0, y_0), seeking the minimum. Then let the minimum point be expressed as $(x_0 + h, y_0 + k)$ where h and k are found approximately by solving the equations

$$\left.\begin{aligned} 0 &= f_x(x_0, y_0) + hf_{xx}(x_0, y_0) + kf_{xy}(x_0, y_0) \\ 0 &= f_y(x_0, y_0) + hf_{xy}(x_0, y_0) + kf_{yy}(x_0, y_0) \end{aligned}\right\} \qquad (4.22)$$

We generalise to the case of a function of n variables; let $\mathbf{l}_0$ be the vector of step sizes $(h_1, h_2, \ldots, h_n)$. The equations to be solved are $\nabla f_0 = -\mathbf{H}_0\mathbf{l}_0$ where $\mathbf{H}$ is the Hessian matrix. The solution is obtained by matrix inversion as $\mathbf{l}_0 = -\mathbf{H}_0^{-1} \nabla f_0$ and we move to the point $\mathbf{x}_0 + \mathbf{l}_0$. For a quadratic function this will give an exact minimum in one move. For a function near the minimum, we set up an iterative scheme.

$$\mathbf{x}_{i+1} = \mathbf{x}_i - \mathbf{H}_i^{-1} \nabla f_i$$

where $\mathbf{H}_i$ and ∇f_i are evaluated at $\mathbf{x}_i$.

Note the analogy to the Newton-Raphson method for locating roots of non-linear equations in one variable.

Example

Consider $f(x, y) = x^3 + y^3 - 12x^2 - 9y^2 + 256$

$f_x = 3x^2 - 24x$; $f_y = 3y^2 - 18y$; $f_{xx} = 6x - 24$; $f_{xy} = 0$; $f_{yy} = 6y - 18$

Hence $\nabla f = (3x^2 - 24x,\ 3y^2 - 18y)$ and $\mathbf{H} = \begin{bmatrix} 6x - 24 & 0 \\ 0 & 6y - 18 \end{bmatrix}$

At the point $\mathbf{x}_0 = (1, 1)$ which we can select as our initial point,

$$\nabla f = (-21, -15) \quad \text{and} \quad \mathbf{H} = \begin{bmatrix} -18 & 0 \\ 0 & -12 \end{bmatrix}$$

Hence $\mathbf{H}^{-1} = \begin{bmatrix} \frac{-1}{18} & 0 \\ 0 & \frac{-1}{12} \end{bmatrix}$ and $\mathbf{l} = \begin{bmatrix} \frac{-1}{18} & 0 \\ 0 & \frac{-1}{12} \end{bmatrix} \begin{bmatrix} -21 \\ -15 \end{bmatrix} = \begin{bmatrix} \frac{7}{6} \\ \frac{5}{4} \end{bmatrix}$

We therefore advance to the point $\mathbf{x}_1 = (1, 1) - \left(\frac{7}{6}, \frac{5}{4}\right) = \left(-\frac{1}{6}, -\frac{1}{4}\right)$.

At $\mathbf{x}_1$, $\nabla f = \left(\frac{147}{36}, \frac{75}{16}\right)$ and $\mathbf{H} = \begin{bmatrix} -25 & 0 \\ 0 & \frac{-39}{2} \end{bmatrix}$. Hence

$$\mathbf{H}^{-1} = \begin{bmatrix} \frac{-1}{25} & 0 \\ 0 & \frac{-2}{39} \end{bmatrix} \quad \text{and} \quad \mathbf{l} = \begin{bmatrix} -0.163 \\ -0.240 \end{bmatrix}.$$

We advance to the point $\mathbf{x}_2 = (-0.167, -0.25) - (-0.163, -0.240) = (-0.004, -0.010)$.

Further iterations on a computer terminal yielded $\mathbf{x}_3 = (-1.39 \times 10^{-6}, -1.54 \times 10^{-5})$ $\mathbf{x}_4 = (0, 4.00 \times 10^{-11})$, $\mathbf{x}_5 = (0, 0)$.

There are certain drawbacks to this method, apart from the fact that it relies upon the function being approximately quadratic. The matrix **H** must be positive definite which will only for certain be true for a convex function. As the method stands, we require $\mathbf{H}^{-1}$ at every step and this may require a numerical inversion which demands $n(n+3)/2$ evaluations; then the solution of $\mathbf{l} = \mathbf{H}^{-1}\nabla f$ may demand a further number of operations of $0(n^3)$. Other difficulties also exist.

Again, modifications have been suggested, including a method due to Davidon. In effect, he approximates $\mathbf{H}^{-1}$ in such a way that the successive approximations are positive definite.

Methods using conjugate directions

These methods use the properties of ellipses (or their n - dimensional analogues). They aim to collect information on the curvature of the objective function, but do not calculate the second derivatives of the function explicitly. In this way, they are as rapid as Newton's method (when it works), but they avoid its drawbacks. Provided the initial point is near the minimum, the contours will be closed and almost elliptical. The origin of the conjugate direction method lies in a simple result concerning the ellipse.

Theorem

The line joining the points of contact of two parallel tangents to an ellipse will pass through the centre of the ellipse. (See Figure 4.20).

In this Figure we show four parallel tangents and the line joining their points of contact P_1, P_2, P_3, P_4. This line passes through Q, the centre of the ellipse. The proof is left to you. The direction of the tangents is said to be **conjugate** to the direction of the line through the centre of the ellipse.

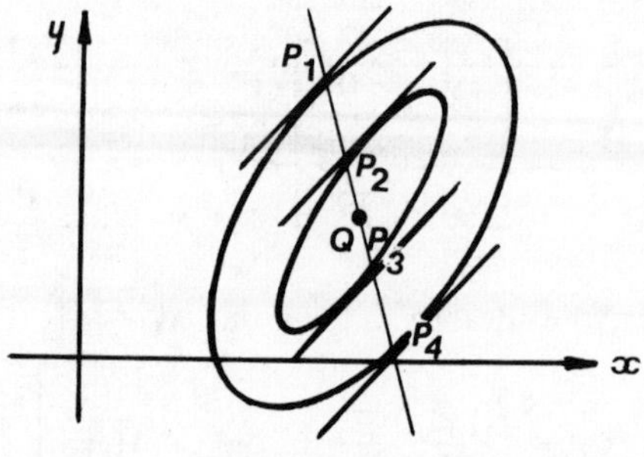

Figure 4.20

In n - dimensions the result may be generalised: if P_1 is the point of contact to a quadric (generalisation of an ellipse) of an $(n-1)$ dimensional tangent subspace and P_2 is the point of contact of a parallel tangent subspace, then the centre of the quadric lies on the line P_1P_2. It is no use trying to appeal to geometry in n - dimensions so we need an algebraic definition of conjugacy. The general quadric is of the form

$$F(\mathbf{x}) = f_0 + \mathbf{x}^T\mathbf{a} + \frac{1}{2}\mathbf{x}^T\mathbf{A}\mathbf{x}$$

(for example $f(x,y) = 4 + 3x + 2y + x^2 + 2xy + 4y^2$ can be written as above with

$$\mathbf{x} = \begin{bmatrix} x \\ y \end{bmatrix}, \quad \mathbf{a} = \begin{bmatrix} 3 \\ 2 \end{bmatrix}, \quad \mathbf{A} = \begin{bmatrix} 1 & 1 \\ 1 & 4 \end{bmatrix}).$$

If two vectors $\mathbf{x}$ and $\mathbf{y}$ are such that $\mathbf{x}^T\mathbf{A}\mathbf{y} = 0$ then $\mathbf{x}$ and $\mathbf{y}$ are defined to be *conjugate* to **A**.

The algebraic definition is a generalisation of the geometric result.

A set of directions $\mathbf{t}_k$, $k = 1, 2, 3, \ldots\ldots, n$ is said to be **mutually conjugate** with respect to a matrix **A** if $\mathbf{t}_i^T\mathbf{A}\mathbf{t}_j = 0$ for all $i, j \in \{0, 1, \ldots\ldots, n\}$ where $i \neq j$.

The great advantage of a method employing conjugate directions is embodied in the following theorem.

If linear searches are performed along a set of n directions which are mutually conjugate with respect to the Hessian matrix of a quadratic function then the minimum will be located in at most n searches.

Example 1

$$f(x, y) = x^2 + 2xy + 4y^2 - 8x - 6y + 16$$

The Hessian matrix is $\begin{bmatrix} 2 & 2 \\ 2 & 8 \end{bmatrix}$. Suppose we start from the point (0, 0) along the x - axis, i.e. in the direction given by (1, 0). We seek the conjugate direction to (1, 0); let this be given by (u, v). Then

$$(1, 0)\begin{bmatrix} 2 & 2 \\ 2 & 8 \end{bmatrix}\begin{bmatrix} u \\ v \end{bmatrix} = 0 \text{ i.e. } (2, 2)\begin{bmatrix} u \\ v \end{bmatrix} = 0 \text{ i.e. } 2u + 2v = 0;$$

hence a vector representing the conjugate direction is $(1, -1)$.

In the direction $(1, 0)$ the function becomes $x^2 - 8x + 16$ and this has a minimum at $(4, 0)$. We now start from $(4, 0)$ along the direction given by $(1, -1)$ i.e. along the line $y = 4 - x$; the function along this line can be written $x^2 + 2x(4-x) + 4(4-x)^2 - 8x - 6(4-x) + 16$ i.e. $3x^2 - 26x + 56$. This has a minimum at $x = \frac{13}{3}$ (and $y = \frac{-1}{3}$). We have located the point $(\frac{13}{3}, \frac{-1}{3})$ after two linear searches. You can check that this is the minimum by partially differentiating $f(x, y)$.

Alternatively, suppose we find the gradient vector at $(0, 0)$. This is $(-8, -6)$. If we choose the tangent at $(0, 0)$ as one direction, i.e. $(6, -8)$ or $(3, -4)$ then the conjugate direction (u, v) is given by

$$(3, -4)\begin{bmatrix} 2 & 2 \\ 2 & 8 \end{bmatrix}\begin{bmatrix} u \\ v \end{bmatrix} = 0 \text{ i.e. } (-2, -26)\begin{bmatrix} u \\ v \end{bmatrix} = 0 \text{ hence } \begin{bmatrix} u \\ v \end{bmatrix} = \begin{bmatrix} 13 \\ -1 \end{bmatrix}$$

In the direction $(6, -8)$ or $y = \frac{-4}{3}x$ the function becomes $\frac{49}{9}x^2 + 16$ which has a minimum at $(0, 0)$. Starting from this point, we move along the line $y = \frac{-1}{13}x$ and find the minimum at $\left(\frac{13}{3}, \frac{-1}{3}\right)$ as before.

General remarks on conjugate directions

We now quote some general results concerning conjugate directions. First, it is important to realise that a positive definite quadratic function has a set of mutually orthogonal principal axes such that if the coordinate axes are rotated to align with the principal axes then in the new coordinate system the quadratic function can be expanded as a sum of squared terms in the new independent variables. This would permit the function to be minimised by n single-variable searches in turn.

Second, if we find a point $\mathbf{x}_1$ at which $(n-1)$ conjugate directions are tangential to the quadratic surface, the minimum of the function lies on the line through $\mathbf{x}_0$ in the direction of the nth conjugate direction.

The final result we need is that if we have r conjugate directions $\mathbf{t}_1, \mathbf{t}_2, \ldots\ldots, \mathbf{t}_r$ where $r < n$ and if $\mathbf{x}_1$ and $\mathbf{x}_2$ are two points obtained by single-variable searches along these directions then the direction $\mathbf{t}_{r+1} = \mathbf{x}_2 - \mathbf{x}_1$ is conjugate to $\mathbf{t}_1, \mathbf{t}_2, \ldots\ldots, \mathbf{t}_r$.

Example 2

$$f(x,y,z) = 3x^2 + 9y^2 + 2z^2 + 8xy + 2yz + 2zx - 10x - 30y + 8z + 6$$

The gradient vector $\nabla f = (6x + 8y + 2z - 10,\ 18y + 8x + 2z - 30,\ 4z + 2y + 2x + 8)$

and the Hessian $\mathbf{H} = \begin{bmatrix} 6 & 8 & 2 \\ 8 & 18 & 2 \\ 2 & 2 & 4 \end{bmatrix}$. At the starting point $\mathbf{x}_0 = (0, 0, 0)$,

$\nabla f = (-10, -30, 8)$. We choose the first of the three conjugate directions, arbitrarily; we shall take it orthogonal to ∇f for example, $(4, 0, 5) = \mathbf{t}_1$.

The second conjugate direction is conjugate to the first, i.e.

$$(u, v, w)\begin{bmatrix} 6 & 8 & 2 \\ 8 & 18 & 2 \\ 2 & 2 & 4 \end{bmatrix}\begin{bmatrix} 4 \\ 0 \\ 5 \end{bmatrix} = 0, \qquad (u, v, w)\begin{bmatrix} 34 \\ 42 \\ 28 \end{bmatrix} = 0; \text{ hence } 34u + 42v + 28w = 0$$

We choose this direction $\mathbf{t}_2$ so that it is tangential to the quadratic surface at (0, 0, 0) i.e. $\mathbf{t}_2 . \nabla f = 0$ there; this implies $-10u - 30v + 8w = 0$.

Hence we may take $\mathbf{t}_2 = (49, -23, -25)$.

The third conjugate direction is conjugate to each of the others, hence

$$(u, v, w)\mathbf{H}\begin{bmatrix} 4 \\ 0 \\ 5 \end{bmatrix} = 0 \quad \text{and} \quad (u, v, w)\mathbf{H}\begin{bmatrix} 49 \\ -23 \\ -25 \end{bmatrix} = 0$$

A solution of these equations is $\mathbf{t}_3 = (0, 4, -6)$.

We can show by partial differentiation that the minimum of $f(x, y, z)$ occurs at the point $(0, 2, -3)$ which lies on the line through $\mathbf{x}_0$ parallel to the direction given by $\mathbf{t}_3$ as suggested in one of our results.

Powell's Method

The algorithm we shall describe here is due to Powell. His method avoids the need to work out derivatives. The method is flowcharted in Figure 4.21.

Problems

1. Find the unit vectors which represent the directions of steepest descent for the function $z = 3x^2 + y^3 - 2x^2y + 10$ at the points (1, 1), (2, 1), (1, 2), (L.U.)

2. Using the steepest descent find the least value of $z = x^2 + xy + y^2 + x^3 + xy^2 + y^3 + 10$, starting from (1, 1). Why can we not reach (0, 0) in one step? (L.U.)

3. Repeat Problem 3, Section 4.3, using any of the methods described in this section.

4. Distinguish between the classical steepest descent method and a typical gradient method for minimisation of a function of several variables.

   Starting from the origin, locate correct to 1 decimal place by one of the above methods the minimum value of $f(x_1, x_2) = 2x_1^2 - 2x_1x_2 + 2x_2^2 - 6x_1 + 6$ in the region $x_1 \geqslant 0$, $x_2 \geqslant 0$, $x_1 + x_2 \leqslant 2$. (C.E.I.)

5. From the starting point (1, 1, 1), take one step of the steepest descent method towards finding the minimum of the function $f(x, y, z) = 2x^2 + 2xy + y^2 + yz + z^2$. (C.E.I.)

6. Describe in detail a gradient method for minimizing the unconstrained function $f(x_1, x_2, \ldots\ldots, x_n)$. Starting from the point $(-1, 2)$, find the value, correct to 2 decimal places, of the function $F = (x^2 + y^2 - 1)^2 + (x + y - 1)^2$ after taking one step using the Newton-Raphson method. (C.E.I.)

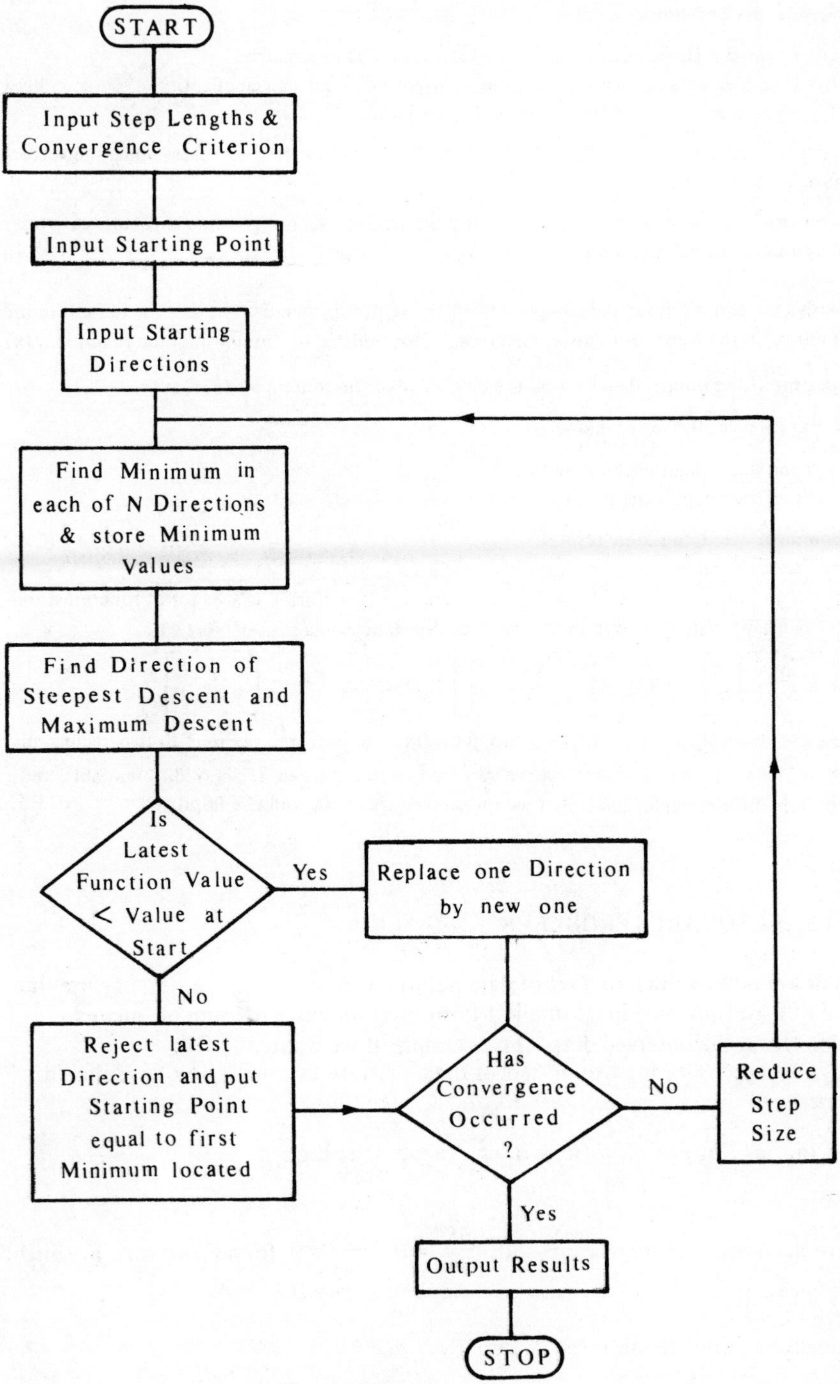
START
Input Step Lengths & Convergence Criterion
Input Starting Point
Input Starting Directions
Find Minimum in each of N Directions & store Minimum Values
Find Direction of Steepest Descent and Maximum Descent
Is Latest Function Value < Value at Start
Yes
Replace one Direction by new one
No
Reject latest Direction and put Starting Point equal to first Minimum located
Has Convergence Occurred ?
No
Reduce Step Size
Yes
Output Results
STOP

Figure 4.21

7. Two stages of the Newton-Raphson algorithm for the minimization of a function of several variables are as follows:

   (i) Form $\mathbf{h} = \mathbf{Hg}$ where $\mathbf{h}$ and $\mathbf{g}$ are vectors and $\mathbf{H}$ is a matrix.
   (ii) If all elements of $\mathbf{h}$ are not greater in magnitude than a given quantity ϵ, stop; otherwise replace $\mathbf{x}$ by $\mathbf{x} - \mathbf{h}$, where $\mathbf{x}$ is also a vector.

   Write a section of program in a stated code to perform these two stages, assuming the vectors have m elements.

   Also write a section of program to compute the gradient vector $\mathbf{g}$ in the particular case when the function to be minimized is $(\sin(x_1 x_2) - x_2)^2 + (e^{2x} - 1)\sin x_2$. (C.E.I.)

8. Derive the equations for the components of the vector ξ where $x + \xi$ is a single step from the point $\mathbf{x}$ in the Newton-Raphson direction for maximising (or minimising) the function $f(\mathbf{x})$.

   Starting at the point $(0, -1)$ find the new value of the function $f(x,y) = (x - y)^2 + \frac{1}{9}(x + y - 10)^2$ after taking

   (a) one step using the above method
   (b) two unit steps using the gradient method.

   Comment briefly on these results. (C.E.I.)

9. The iteration $\mathbf{x}^{(k+1)} = \mathbf{x}^{(k)} + \mathbf{A}\mathbf{f}(\mathbf{x}^{(k)})$ where $\mathbf{x}$ is a vector, $\mathbf{f}$ is a vector of functions, and $\mathbf{A}$ is a constant matrix, converges in certain circumstances to a root of $\mathbf{f}(\mathbf{x}) = 0$.

   If $\mathbf{x}^{(0)} = \begin{bmatrix} 2 \\ 2 \end{bmatrix}$, $\mathbf{f}(\mathbf{x}) = \begin{bmatrix} x^2 + y^2 - 9 \\ x - y^2 + 1 \end{bmatrix}$ and $\mathbf{A} = -\frac{1}{10}\begin{bmatrix} 2 & 2 \\ 1 & -2 \end{bmatrix}$

   use the above algorithm to find a solution of $\mathbf{f}(\mathbf{x}) = 0$ near $\mathbf{x}^{(0)}$ correct to two decimal places. State concisely how a suitable matrix $\mathbf{A}$ could be found when $\mathbf{f}$ is (a) differentiable and (b) non-differentiable, and state how the convergence rate could be improved. (C.E.I.)

4.6 LEAST SQUARES PROBLEMS

To fit a function $\phi(\mathbf{x})$ to a set of data points $\mathbf{x}_i$, $i = 1, \ldots, n$ we may use the method of least squares. In its simplest form this minimises the sum of squared deviations from the observed data. For example, if we wanted to fit $w = a_1 x^2 + a_2 y^3 + a_3 \log z$ to n sets of data quartets (x_i, y_i, z_i, w_i) we should minimise

$$S(a_1, a_2, a_3) = \sum_{i=1}^{n} (w_i - a_1 x_i^2 - a_2 y_i^3 - a_3 \log z_i)^2$$

To solve the normal equations $\dfrac{\partial S}{\partial a_1} = 0$; $\dfrac{\partial S}{\partial a_2} = 0$; $\dfrac{\partial S}{\partial a_3} = 0$ for a_1, a_2 and a_3 might be very difficult.

It might be better to minimise S directly.

Later in this section we look at ways of solving simultaneous non-linear equations. The bonus we pick up in these least squares problems is that we may obtain an approximation to the Hessian of S which does not require second derivatives. We shall derive the results for a simple example and then generalise. For the example we shall choose a function of x and y, partly to make the example simpler to understand and partly to show that functions which are sums of squares need not necessarily have come from a least squares problem. Consider then

$$S(x,y) \equiv [f_1(x,y)]^2 + [f_2(x,y)]^2 = \mathbf{f}^T\mathbf{f}, \text{ where } \mathbf{f} = \left[f_1(x,y),\ f_2(x,y)\right]$$

Now $\dfrac{\partial S}{\partial x} = 2f_1 \cdot \dfrac{\partial f_1}{\partial x} + 2f_2 \cdot \dfrac{\partial f_2}{\partial x}$ and $\dfrac{\partial S}{\partial y} = 2f_1 \cdot \dfrac{\partial f_1}{\partial y} + 2f_2 \cdot \dfrac{\partial f_2}{\partial y}$. Hence

$\nabla S \equiv 2\mathbf{J}^T\mathbf{f}$ where $\left[\mathbf{J}_{ij}\right] = \left[\dfrac{\partial f_i}{\partial x_j}\right]$ – the **Jacobian** matrix.

Further, $\dfrac{\partial^2 S}{\partial x^2} = 2f_1 \dfrac{\partial^2 f_1}{\partial x^2} + 2\left(\dfrac{\partial f_1}{\partial x}\right)^2 + 2f_2 \dfrac{\partial^2 f_2}{\partial x^2} + 2\left(\dfrac{\partial f_2}{\partial x}\right)^2$ with similar results for the other derivatives.

Hence, the Hessian matrix of S is given by $\mathbf{H} = 2\mathbf{J}^T\mathbf{J} + 2 \sum_{i=1}^{2} \mathbf{f}_i \mathbf{K}_i$ where $\mathbf{K}_i$ is the matrix of second derivatives of f_i .

If the functions f_i are nearly linear or are small at the minimum point then the matrices $\mathbf{K}_i$ contain small elements and can be neglected so that $\mathbf{H} \simeq 2\mathbf{J}^T\mathbf{J}$ which requires only the evaluation of first derivatives.

This result has allowed algorithms to be developed for these special cases which are somewhat more rapid. A method due to Powell using only function values updates at each iteration an approximation to the Jacobian. Of course, the methods previously mentioned for general optimisation will still work.

The solution of simultaneous non-linear equations

In this subsection, we seek the solution of $f(x, y) = 0$ and $g(x, y) = 0$. We can extend most of the methods used for one non-linear equation in one variable. We first develop an analogue of **Basic Iteration** and we demonstrate this on the equations $x^2 + 4y^2 = 4$, $e^x + y = 1$; see Figure 4.22.

Graphically, we find approximate solutions at $(-1.3,\ 0.7)$ and $(0.7,\ -0.9)$. We rearrange the equations so that one has x as the subject and the other has y as the subject. Thus we have $x = -\sqrt{4 - 4y^2}$ and $y = 1 - e^x$; we have chosen the negative sign for the square root and this means that we shall aim to find the root with negative x - value. If we want the other root, we shall have to take the positive square root. The iterative cycle we choose is

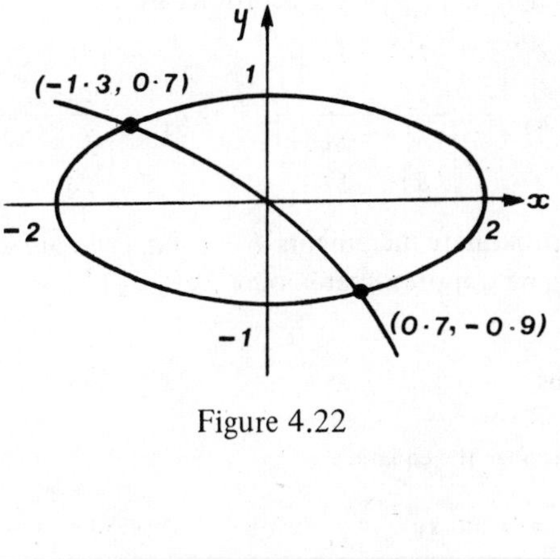

Figure 4.22

$$x_{n+1} = -2\sqrt{1-y_n^2} \qquad y_{n+1} = 1 - e^{x_{n+1}}$$

We start with $y_0 = 0.7$ and develop further approximations as shown below. (Results quoted to 3 d.p.)

y - values	0.7	0.760	0.727	0.747	0.736	0.742	.....	0.740
x - values	−1.4	−1.299	−1.373	−1.331	−1.355	−1.341	.....	−1.346

We leave it to you to find the other root.

We could, of course, rearrange the equations the other way round, i.e. $x = \log(1-y)$, $y = -\sqrt{(4-x^2)/4}$. If we start with $y_0 = -0.9$ we develop the following iterative values.

y - values	−0.9	−0.947	−0.943	.....	−0.943
x - values	0.642	0.666	0.664	.....	0.664

It can be shown that a sufficient condition for convergence is

$$\left|\frac{\partial f}{\partial x}\right| + \left|\frac{\partial g}{\partial x}\right| \leqslant M < 1 \quad \text{and} \quad \left|\frac{\partial f}{\partial y}\right| + \left|\frac{\partial g}{\partial y}\right| \leqslant M < 1$$

The smaller the value of M, the more rapid is the convergence.

Newton-Raphson method

The Newton-Raphson method described on page 122 is now modified because the Hessian matrix **H** is approximated by $2\mathbf{J}^T\mathbf{J}$. Alternatively, we may develop the algorithm directly via the expansions

$$f(x^*,y^*) = 0 = f(x_1,y_1) + f_x(x_1,y_1)(x^*-x_1) + f_y(x_1,y_1)(y^*-y_1) + \ldots..$$
$$g(x^*,y^*) = 0 = g(x_1,y_1) + g_y(x_1,y_1)(x^*-x_1) + g_y(x_1,y_1)(y^*-y_1) + \ldots..$$

where the desired root is (x^*, y^*) and (x_1, y_1) is a nearby point.

If the nearness is sufficient we may neglect terms not shown and solve the equations approximately for the increments $(x^* - x_1)$ and $(y^* - y_1)$. If all derivatives are evaluated at (x_1, y_1) the solutions are

$$\delta x_1 = \frac{\begin{vmatrix} -f & f_y \\ -g & g_y \end{vmatrix}}{\begin{vmatrix} f_x & f_y \\ g_x & g_y \end{vmatrix}}; \qquad \delta y_1 = \frac{\begin{vmatrix} f_x & -f \\ g_x & -g \end{vmatrix}}{\begin{vmatrix} f_x & f_y \\ g_x & g_y \end{vmatrix}}$$

The approximate increments δx_1 and δy_1 are added to x_1 and y_1 respectively to obtain a new approximate point (x_2, y_2).

Problems

1. Rearrange the equations

$$f(x,y) = \sin(xy) - \frac{y}{2\pi} - x = 0 \qquad g(x,y) = \left[1 - \frac{1}{4\pi}\right](e^{2x} - e) + \frac{ey}{\pi} - 2ex = 0$$

in the form

$$x = F(x,y),\ y = G(x,y)$$

and state a sufficient condition for the convergence of a direct iteration method based on this arrangement.

Starting with initial values $x_0 = 0.4$, $y_0 = 3.0$, iterate on the above until the solution $x = \frac{1}{2}$, $y = \pi$ is obtained correct to two significant figures.

Indicate briefly another method, involving minimization or otherwise, for solving the given equations. (C.E.I.)

2. Derive the Newton-Raphson method for solving the non-linear algebraic equations $f_i(x_1, x_2, \ldots, x_n) = 0$, $i = 1, 2, \ldots, n$.

   Use the above to find a solution of the equations

   $$f(x,y) = x^2 + y^2 - 1 = 0, \quad g(x,y) = x^2 + 9y^2 - 4x + 18y + 9 = 0$$

   correct to 2 decimal places in the neighbourhood of the point $(1, -\frac{1}{2})$.

   With the aid of a sketch, or otherwise, deduce a second solution of the given equations. (C.E.I.)

3. Derive the formulae used in solving the simultaneous equations $F(x,y) = 0$, $G(x,y) = 0$ by the method of Newton-Raphson. Describe briefly how this can be of use in finding the minimum of a function of two variables.

   Illustrate this by taking a single step starting from the point $(1, 0)$ for the function $x^2y^2 + 4xy^2 + 2x^2y + 2x^2 + y^2$.

   State briefly any disadvantages of this method. (C.E.I.)

4. Solve by Basic Iteration and Newton-Raphson method the equations $3x^2 + y^3 = 21$, $2x + y^2 + 2 = 0$.

5. Find the solution of the equations $x^2 + xy + y^2 = 12$, $x^2 - \sin y = 0$, starting from the point $(1, 1)$.

6. The cost of transporting a product from a factory to a customer is directly proportional to distance. Where should a factory be situated if customers are located at the points $(5, 4)$, $(2, 3)$, $(80, 40)$, $(25, 65)$? Assume equal demands from each customer.

   If a factory is placed instead at $(10, 10)$, where should a second factory be placed so as to minimise the total cost, assuming that goods are shipped from the nearest factory?

4.7 CONSTRAINED OPTIMIZATION TECHNIQUES

Several optimization techniques are specifically designed to handle the constrained problem; among these are **gradient projection** and **reduced gradient** methods.

Other techniques involve modification of unconstrained methods. The first category of these take constraints explicitly into account at each step. In this category we may distinguish two sub-categories: **direct search** methods and **gradient** methods. Direct search methods sense when the present point is near a constraining boundary and modify accordingly the search directions. Gradient methods are variations on the method of steepest descent and modify, if necessary, the gradient direction so that the new point lies within the feasible region. However, the gradients within the infeasible region may be unreliable. If the constraints are complicated then both types of method in this first category can become inefficient.

The second category comprises those methods which use global **penalty functions.** The essential idea here is to add a sequence of penalty functions to the basic objective function so that the solutions to the sequence of new unconstrained objective functions tends to the solution of the constrained problems. The difficulty here is that the penalty functions may introduce unusual features into the problem. When a function is added which penalise constraint violations it is called an **exterior** penalty function. For constraints $h_i(x) = 0$ we can take penalty functions of the form $r^{-1}[h_i(x)]^2$. A sequence of values of r is chosen which tends to zero as the search proceeds. For constraints of the form $g_i(x) \leqslant 0$, **barrier** functions are chosen which are non-zero in the interior of the feasible region and infinite on its boundaries.

It would be impossible to do justice to the subject of constrained methods without going into the subject fairly extensively. We refer you to the bibliography.

Problem

Find the minimum value of $x^2 + y^2$ subject to $x + y = 6$ (a) by the method of Lagrange multipliers (b) by the created response surface technique, which is as follows. Take

$\phi(x,y,r) = x^2 + y^2 + (1/r)(x + y - 6)^2$

Start with $r = 10$ and determine the minimum point. Now choose a smaller value of r and obtain a new minimum point. Repeat until a sequence of points is obtained which should tend to the required constrained minimum point.

4.8 PRACTICAL USE OF ALGORITHMS

A measure of the efficiency of an algorithm is the number of function evaluations needed to locate the minimum of the objective function to a specified accuracy. Some algorithms require the estimation of derivatives and allowances must be made for this additional calculation.

When setting up a subroutine to specify the objective function, it is important to avoid cancellation errors, especially when we use function values to estimate derivatives.

The scaling of variables is also an important factor. If there is a disparity in the values of the variables of x then x^2 can be swamped by just one of the variables. Care should be taken to ensure that the magnitudes of the variables are roughly on a par. The gradient vector is quite dependent on scaling and therefore steep-sided valleys may cause severe problems if the variables are not well scaled. Of course, if the variables are similar physical quantities for example, lengths, there should be no problem of scale; if, however, the variables are temperature and pressure, this may not be so easy. An advantage of scaling is that the Hessian matrix may be quite close to the unit matrix of the same size.

It is important not to demand too much accuracy for the final solution or else rounding errors may become of comparable magnitude.

We can also carry out a sensitivity analysis on the final solution to see how the optimum is related to small changes in the objective function.

There is no 'best' algorithm for all optimization problems and, if a library of computer routines is available, there is usually a user's guide at hand.

Chapter Five

Ordinary Differential Equations

5.1 INTRODUCTION

In this chapter we first study numerical methods of solving initial-value ordinary differential equations and, by means of difference equations, we consider the stability of these methods. The methods we introduce are examples of **predictor-corrector** methods which comprise two parts: a predictor formula which generates an estimate of a new value for the dependent variable and a corrector formula which aims to improve upon the predicted value.

Second, we examine the analytical features of systems of differential equations related to problems arising in practice.

Finally, we turn our attention to boundary-value problems and consider numerical methods of solution.

We present the numerical methods for solving equations in the form $\frac{dy}{dx} = f(x, y)$. Equations of higher order can be solved in the equivalent form of a system of first order equations. For example, the equation $\frac{d^2y}{dx^2} + P(x)\frac{dy}{dx} + Q(x)y = R(x)$ can be formulated as the system

$$\frac{dy}{dx} = z \qquad \frac{dz}{dx} = R(x) - Q(x).y - P(x).z$$

This is an example of the general formulation $\frac{dy}{dx} = f(x,y,z)$, $\frac{dz}{dx} = g(x,y,z)$.

5.2 ONE-STEP AND MULTISTEP METHODS

The objective in solving numerically an initial-value differential equation $\frac{dy}{dx} = f(x, y)$ with $y(0) = y_0$, is to generate a sequence of points on the solution curve, viz., (x_1, y_1), (x_2, y_2), (x_3, y_3), ..., (x_m, y_m). **One step methods** use only the previously-obtained solution point (x_n, y_n) to generate the next point (x_{n+1}, y_{n+1}).

The most popular one-step methods are the **Runge-Kutta methods**. They are designed to give the same values as a Taylor series truncated to the term in h^p, where h is the (constant) step size $(x_r - x_{r-1})$, and p is the **order** of the method. The advantage they possess is that they require only the evaluation of $f(x, y)$ for certain pairs of values (x, y), and then take a weighted average of these values to approximate the change $y_r - y_{r-1}$; they do not require derivatives.

The simplest such method is **Euler's method** which effectively approximates the solution via Taylor's series up to the term in h. It is therefore a linear approximation. The working formula is

$$y_{n+1} = y_n + hf(x_n, y_n) \tag{5.1}$$

Second-order Runge-Kutta Methods

When we move to second-order methods, we begin to obtain a broader outlook on the problems involved. The assumption made is that the value y_{n+1} can be computed from a formula of the general kind

$$y_{n+1} = y_n + ak_1 + bk_2 \tag{5.2}$$

where $\quad k_1 = hf(x_n, y_n) \quad$ and $\quad k_2 = hf(x_n + \alpha h, y_n + \beta k_1)$

The four unknowns a, b, α and β have to be chosen so that the formula (5.2) is equivalent to a quadratic approximation.

$$y_{n+1} \simeq y_n + hy'(x_n) + \frac{h^2}{2} y''(x_n) \tag{5.3}$$

Now we can expand k_2 by a Taylor's series and when we match formulae (5.2) and (5.3) we obtain the following relationships

$$\left.\begin{aligned} a + b &= 1 \\ a\alpha = b\beta &= \tfrac{1}{2} \end{aligned}\right\} \tag{5.4}$$

We have only three equations connecting four unknowns and, therefore, one dregee of of freedom to play with. There is thus an infinity of possible second-order methods. Choosing $a = \frac{1}{2}$ produces $b = \frac{1}{2}$, $\alpha = \beta = 1$. Formula (5.2), then becomes

$$y_{n+1} = y_n + \frac{h}{2}\left\{f(x_n, y_n) + f(x_n + h, y_n + hf[x_n, y_n])\right\} \tag{5.5}$$

the **Improved Euler Method.**

It can be shown that the local truncation error is given by

$$y_{n+1} - y_n = \frac{h^3}{12}(f_{xx} + 2f.f_{xy} + f^2.f_{yy} - 2f_x f_y - 2f.f_y^2) + 0(h^4)$$

This is so complicated as to render error estimation very tedious. However, we see that the local truncation error is of order h^3 as opposed to Euler's method where this error is of order h^2. This means that the extra evaluation of $f(x, y)$ is compensated by the larger step size that we can use to maintain a required accuracy.

Fourth-order methods

By far the most popular Runge-Kutta methods are those of the fourth order, which implies a local truncation error of order h^5. One algorithm is as follows.

$$\left.\begin{aligned} &k_1 = hf(x_n, y_n), \quad k_2 = hf(x_n + \tfrac{1}{2}h, y_n + \tfrac{1}{2}k_1) \\ &k_3 = hf(x_n + \tfrac{1}{2}h, y_n + \tfrac{1}{2}k_2), \quad k_4 = hf(x_n + h, y_n + k_3) \\ &y_{n+1} = y_n + \frac{1}{6}(k_1 + 2k_2 + 2k_3 + k_4) \end{aligned}\right\} \tag{5.6}$$

We shall not attempt to derive the local truncation error but merely make a few remarks about it. The difficulty is that without a knowledge of the truncation error it is not easy to choose a suitable step-size h. Collatz suggested that if the quantity $|(k_2 - k_3)/(k_1 - k_2)|$ becomes greater that a few hundredths it is wise to decrease h.

Merson derived a modification of this fourth-order method which kept an error of order h^5, but which allowed an error estimate to be made relatively easily. The snag is that $f(x, y)$ has to be evaluated five times. The equations are:

$$\left.\begin{aligned}
&k_1 = hf(x_n, y_n),\ k_2 = hf(x_n + h/3,\ y_n + k_1/3)\\
&k_3 = hf(x_n + h/3,\ y_n + k_1/6 + k_2/6),\ k_4 = hf(x_n + h/2,\ y_n + k_1/8 + 3k_3/8)\\
&k_5 = hf(x_n + h,\ y_n + k_1/2 - 3k_3/2 + 2k_4)\\
&y_{n+1} = y_n + \frac{1}{6}(k_1 + 4k_4 + k_5)
\end{aligned}\right\} \tag{5.7}$$

We may estimate the local truncation error ϵ_T by

$$\epsilon_T \simeq \frac{1}{30}(2k_1 - 9k_3 + 8k_4 - k_5) \tag{5.8}$$

If $f(x, y)$ is of the form $ax + by + c$, where a, b and c are constants, then formula (5.8) is exact.

Suppose the evaluation of $f(x, y)$ is quite involved. Then the extra calculation in (5.7) may become quite expensive in computation time. In addition, the Runge-Kutta methods, being one-step methods, are quite wasteful since, having proceeded from (x_n, y_n) to (x_{n+1}, y_{n+1}), we have to start from scratch to proceed to (x_{n+2}, y_{n+2}); all previous information (except for y_{n+1}) is abandoned.

There is another disadvantage to Runge-Kutta methods. For algebraic convenience we illustrate the point via a second-order method. Consider the equation $dy/dx = -8y$ with the condition $y(0) = 1$; this has the analytical solution $y = e^{-8x}$. The second-order formula (5.5) becomes

$$y_{n+1} = y_n + \frac{h}{2}[-8y_n - 8(y_n - h\,.\,8y_n)] = y_n(1 - 8h + 32h^2)$$

which is equivalent to the first three terms of the expansion $y_n(e^{-8h})$. It follows from the initial condition that $y_{n+1} = (1 - 8h + 32h^2)^{n+1}$.

This result can be obtained from any second-order Runge-Kutta method. Its significance lies in the fact that $(1 - 8h + 32h^2) > 1$ if $h > 0.25$. For step sizes > 0.25, y_{n+1} grows indefinitely larger as n increases. However, the true solution, $y = e^{-8x}$, *decreases* as x, and hence n, increases. To match such behaviour we would have to choose a step size smaller than 0.25. This feature is known as **partial instability**. The phenomenon is *not* restricted to cases where the true solution shows an exponential decay.

Multistep Methods

We attempt to use previously calculated information in making our next step. An example of such a method is the **Adams-Bashforth** formula

$$y_{n+1} = y_n + h\left(1 + \frac{1}{2}\nabla + \frac{5}{12}\nabla^2 + \frac{3}{8}\nabla^3 + \frac{251}{720}\nabla^4 + \ldots\right)f_n \tag{5.9}$$

or its equivalent *ordinate* form (as far as the term in $\nabla^3 f_n$)

$$y_{n+1} = y_n + \frac{h}{24}(55f_n - 59f_{n-1} + 37f_{n-2} - 9f_{n-3}) \tag{5.10}$$

You will notice that this formula requires three values prior to f_n to get the method started. A useful ploy is to use a Runge-Kutta method three times with the initial value to obtain y_0, y_1, y_2 and y_3, and hence f_0, f_1, f_2 and f_3. The Adams-Bashforth formula (5.10) can then be used for further values y_n. We shall look more closely at the formulae (5.9) and (5.10) in the next section. The local truncation error in formula (5.10) is given by $\epsilon_T = \frac{251}{720} y^{(v)}(\xi)$ where $x_n < \xi < x_{n+1}$.

Example

The equation $\frac{dy}{dx} = x - y^2$ is given, together with the initial condition $y(0) = 0$. We require to estimate y for $x = 0.2, 0.4, 0.6, \ldots\ldots, 2.0$.

We can obtain the Maclaurin series for y:

$$y = \frac{1}{2}x^2 - \frac{1}{20}x^5 + \frac{1}{160}x^8 - \ldots\ldots \tag{5.11}$$

We could then use this series to calculate y for $x = 0.2$, 0.4 and 0.6 and then embark with formula (5.10). However, we shall assume that we are using a digital computer throughout and estimate these values via the fourth-order Runge-Kutta method (5.6). Notice that if we use (5.10) we are predetermining the local truncation error; we could alternatively use formula (5.9) with as many terms as a prescribed accuracy required.

Let us work the first step with the Runge-Kutta method.

Here $h = 0.2$ and $f(x, y) = x - y^2$.

Then $k_1 = 0.2(0 - 0^2) = 0$, $k_2 = 0.2(0.1 - 0^2) = 0.02$

$k_3 = 0.2(0.1 - (0 + 0.01)^2) = 0.2(0.0999) = 0.01998$

$k_4 = 0.2(0.2 - (0 + 0.01998)^2) = 0.2(0.19961) = 0.03992$

$y_1 \simeq 0 + \frac{1}{6}(0 + 0.04 + 0.03996 + 0.03992) = \frac{1}{6}(0.11988) = 0.01998$

Table 5.1

x_n	y_n	$f_n = x_n - y_n^2$
0.2	0.01998	0.199600
0.4	0.079484	0.393682
0.6	0.176204	0.568952
0.8	0.304957	0.707001
1.0	0.455680	0.792356
1.2	0.617135	0.819144
1.4	0.778322	0.794215
1.6	0.930876	0.733470
1.8	1.069703	0.655735
2.0	1.193073	

We can compute two further values y_2 and y_3 and thence we obtain $f_0 = 0$, $f_1 = 0.199600$, $f_2 = 0.393682$ and $f_3 = 0.568952$.

Using (5.10) we have $y_4 = 0.304957$.

We continued the application of (5.10) to obtain the remaining values of y_n. Results are tabulated in Table 5.1, to 6 decimal places. How many of these decimal places are justified?

Comparison of approaches

Runge-Kutta methods do not require information from previously calculated data points and are therefore self-starting. This allows an easy change in the step size. However, they require several evaluations of $f(x, y)$ at each step and demand much computing time. Furthermore, they do not easily provide information about the local truncation error.

Multistep methods, on the other hand, *do* require information about previously calculated data points. They can also provide a reasonable estimate of the local truncation error. A drawback is that to change the step size requires turning back to a Runge-Kutta method to provide sufficient starting values.

A point to bear in mind is that when a Runge-Kutta method is used to provide starting values for a multistep method, the former should have a local truncation error of the same order as the multistep method. However, if the truncation error of the multistep method is of order higher than h^5, a fourth-order Runge-Kutta method *can* be used with a step size which is an integer divisor of the step size of the multistep method.

Problems

1. You are given the equation $y' = -2y$ with the condition $y(0) = 1$. We require the value of $y(0.2)$. With a step size $h = 0.1$ apply in turn the formulae (5.1), (5.5), (5.6), (5.7), (5.9). Compare your results with the analytical solution. With a step size $h = 1$ estimate $y(1.0)$ via (5.7). Comment.

2. The response x of a given hydraulic valve subject to sinusoidal input variation is given by

$$\frac{dx}{dt} = \sqrt{2\left\{1 - \frac{x^2}{\sin^2 t}\right\}} \quad \text{with } x = 0 \text{ at } t = 0.$$

Show whether a Taylor series solution of this equation is possible.

Show that $\left.\dfrac{dx}{dt}\right|_{x=0} = \sqrt{\dfrac{2}{3}}$ and hence use the Runge-Kutta fourth order method to obtain a solution at $t = 0.2$. (C.E.I.)

3. Use the method of isoclines to sketch the solution curves of $\dfrac{dy}{dx} = x^2 + y^2$ in the region $-2 < x < 2$, $-2 < y < 2$.

A second-order Runge-Kutta formula in the usual notation is

$k_1 = hf(x, y)$, $k_2 = hf(x + h, y + k_1)$, $y(x + h) = y(x) + ½(k_1 + k_2)$

Illustrate how this is used by taking two steps with $h = 0.2$ starting from (0, 1) for the above differential equation. Compare this with a Taylor series solution and explain any discrepancy. Would you have found a discrepancy if a fourth-order method had been used? (C.E.I.)

4. Solve the following differential equations via the Runge-Kutta second – and fourth – order methods and the Adams-Bashforth methods.

(i)	$y' = x + y + xy + 1$	$y(0) = 2$
(ii)	$y' = x/y$	$y(0) = 1$
(iii)	$y' = 1/(x + y)$	$y(0) = 2$
(iv)	$y' = x^3 + y^2$	$y(0) = 0$

In each case estimate $y(1)$ using various step sizes h. Draw conclusions about the most suitable method for each equation.

5. Write a computer program for the Merson method. Use it on any of the equations in Problem 4.

5.3 PREDICTOR-CORRECTOR METHODS

In this section we study two methods from a general class of predictor-corrector methods. We have chosen the two which we believe to be among the most popular and wide-spread. First, however, we should mention some general points.

Most of the formulae can be derived by integrating the differential equation between two values of x at which tabulated values of y are required.

A **predictor** formula can be based on the integral

$$y(x_{n+1}) - y(x_{n-s}) = \int_{x_{n-s}}^{x_{n+1}} f(x, y)\,dx \tag{5.12a}$$

If we approximate $f(x, y)$ by an interpolating polynomial $P(x)$ (the interpolating points being $x_n, x_{n-1}, x_{n-2}, \ldots x_{n-s+1}, \ldots$) we obtain the relationship

$$y^p_{n+1} = y_{n-s} + \int_{x_{n-s}}^{x_{n+1}} P(x)\,dx \tag{5.12b}$$

where y_{n-s} is an approximation to $y(x_{n-s})$ and y^p_{n+1} is the obtained approximation to $y(x_{n+1})$. This is an explicit formula since x_{n+1} is not used in defining $P(x)$. The integration extends past x_n and therefore we are extrapolating $P(x)$ to x_{n+1}; hence (5.12b) is a predicting formula: it is an **explicit** or **open** formula.

Having obtained a predicted value y^p_{n+1} we may now define a new interpolating polynomial $Q(x)$ which is defined also at x_{n+1}. Then we obtain a **corrector** formula from

$$y_{n+1} - y_{n-r} = \int_{x_{n-r}}^{x_{n+1}} f(x, y)\,dx$$

where r may not be the same as s in (5.12).

The formula is

$$y^c_{n+1} = y_{n-r} + \int_{x_{n-r}}^{x_{n+1}} Q(x)\,\mathrm{d}x \tag{5.13}$$

This is an **implicit** or **closed** formula. (y^c_{n+1} is the corrected value).

We can use the predictor formula once to estimate y_{n+1} and then apply (5.13) as an iterative formula to generate successively better approximations to y_{n+1}. It is usually preferred, however, to use the corrector formula once only and to choose a small enough step size to allow for reasonably accurate representation of y_{n+1}.

Milne-Simpson Method

Starting from $\dfrac{\mathrm{d}y}{\mathrm{d}x} = f(x, y)$ we integrate between x_{n+1} and x_{n-3} as follows.

$$\int_{x_{n-3}}^{x_{n+1}} \frac{\mathrm{d}y}{\mathrm{d}x}\,\mathrm{d}x = \int_{x_{n-3}}^{x_{n+1}} f(x, y)\mathrm{d}x \simeq \int_{x_{n-3}}^{x_{n+1}} P(x)\mathrm{d}x \tag{5.14}$$

where $P(x)$ is a quadratic interpolating polynomial agreeing with $f(x, y)$ at x_n, x_{n-1} and x_{n-2}. Notice that we extrapolate to x_{n+1} and to x_{n-3} which could give rise to a sizeable error. Now the polynomial $P(x)$ is found to be given by

$$\begin{aligned} P(x) \equiv P(x_n + rh) &= f_n + r\nabla f_n + \frac{r(r+1)}{2}\nabla^2 f_n \\ &= ½(r+2)(r+1)f_n - r(r+2)f_{n-1} + ½r(r+1)f_{n-2} \end{aligned}$$

Hence
$$\begin{aligned} \int_{x_{n-3}}^{x_{n+1}} P(x)\mathrm{d}x &= \int_{-3}^{1} P(x_n + rh)\,.\,h\,\mathrm{d}r \\ &= h\int_{-3}^{1} (f_n + r\nabla f_n + \tfrac{1}{2}r(r+1)\nabla^2 f_n)\mathrm{d}r \\ &= h\,(4f_n - 4\nabla f_n + \tfrac{8}{3}\nabla^2 f_n) = h\,(\tfrac{8}{3}f_n - \tfrac{4}{3}f_{n-1} + \tfrac{8}{3}f_{n-2}) \end{aligned}$$

Since $\displaystyle\int_{x_{n-3}}^{x_{n+1}} \frac{\mathrm{d}y}{\mathrm{d}x}\,\mathrm{d}x = y_{n+1} - y_{n-3}$ we establish the **Milne predictor formula**

$$y_{n+1} = y_{n-3} + \frac{4h}{3}(2f_n - f_{n-1} + 2f_{n-2}) \tag{5.15}$$

By recourse to the error in the interpolation polynomial it can be shown that the truncation error is given by

$$\epsilon_T \simeq (28/90)h^5\,y^{(v)}(\xi_1), \text{ where } x_{n-3} < \xi_1 < x_{n+1} \tag{5.16}$$

We may then estimate $f(x_{n+1}, y_{n+1})$.

The corrector formula which completes the Milne-Simpson method is found by integrating the differential equation over the range $[x_{n-1}, x_{n+1}]$. The interpolating polynomial is not now extrapolated; it agrees with $f(x, y)$ at x_{n-1}, x_n and x_{n+1}. The **Simpson corrector formula**, so called because of its resemblance to the Simpson integration rule, is

$$y_{n+1} = y_{n-1} + \frac{h}{3}(f_{n+1} + 4f_n + f_{n-1}) \tag{5.17}$$

We quote the truncation error

$$\xi_T = \frac{1}{90} h^5 y^{(v)}(\xi_2), \text{ where } x_{n-1} < \xi_2 < x_{n+1} \tag{5.18}$$

Example

We apply the method to the differential equation $\dfrac{dy}{dx} = x - y^2$ with the starting values given on page 136. The results are shown in Table 5.2.

Table 5.2

x_n	y_n	$f_n = x_n - y_n^2$	y^p_{n+1}	f^p_{n+1}	y^c_{n+1}
0	0	0			
0.2	0.01998	0.199600			
0.4	0.079484	0.393682			
0.6	0.176204	0.568952	0.304913	0.707028	0.304586
0.8	0.304586	0.707228	0.455411	0.792601	0.455568
1.0	0.455568	0.792458	0.616975	0.819342	0.617679
1.2	0.617679	0.818472	0.778589	0.793800	0.779578
1.4	0.779578	0.792259	0.931509	0.732291	0.932332
1.6	0.932332	0.730757	1.070441	0.654157	1.070874
1.8	1.070874	0.653228	1.193737	0.574992	1.193576
2.0	1.193576				

Adams-Moulton Method

The derivation of the predictor and corrector formulae for this method follow a similar line to that used for the Milne-Simpson method. For the predictor formula we approximate $f(x, y)$ by a cubic agreeing with it at x_{n-3}, x_{n-2}, x_{n-1} and x_n, and we perform the integration over $[x_n, x_{n+1}]$. The formula we obtain is

$$y_{n+1} = y_n + (h/24)(55f_n - 59f_{n-1} + 37f_{n-2} - 9f_{n-3}) \tag{5.10}$$

which is just the Adams-Bashforth predictor formula , obtained earlier.

As we have stated, the truncation error is given by

$$\epsilon_T = \frac{251}{720} y^{(v)}(\xi_1), \text{ where } x_n < \xi_1 < x_{n+1} \tag{5.19}$$

Having predicted y_{n+1}, we may estimate $f_{n+1} \equiv f(x_{n+1}, y_{n+1})$ and then apply the **Adams-Moulton correction formula.** In this case we approximate $f(x, y)$ by a cubic agreeing at x_{n-2}, x_{n-1}, x_n and x_{n+1} and again integrate over $[x_n, x_{n+1}]$. The formula obtained is

$$y_{n+1} = y_n + (h/24)(9f_{n+1} + 19f_n - 5f_{n-1} + f_{n-2}) \tag{5.20}$$

with truncation error given by

$$\epsilon_T = \frac{19}{720} y^{(v)}(\xi_2), \text{ where } x_{n-2} < \xi_2 < x_{n+1} \tag{5.21}$$

We saw that (5.10) could be expressed in finite difference form as

$$y_{n+1} = y_n + h\left(f_n + \frac{1}{2}\nabla f_n + \frac{5}{12}\nabla^2 f_n + \frac{3}{8}\nabla^3 f_n\right)$$

In the same way, (5.20) can be written as

$$y_{n+1} = y_n + h\left(f_{n+1} - \frac{1}{2}\nabla f_{n+1} - \frac{1}{12}\nabla^2 f_{n+1} - \frac{1}{24}\nabla^3 f_{n+1}\right)$$

Example

Consider again the differential equation $\frac{dy}{dx} = x - y^2$; $y(0) = 0$. This time we need four starting values. On page 136 we really did half the stages. Let us assume that we have already calculated y_1, y_2 and y_3 by the Runge-Kutta method. We then estimate y_4 by (5.10) as before to obtain the value 0.304957. Now we are able to estimate $f_4 = x_4 - y_4^2 = 0.707001$.

The corrector formula (5.20) is used to obtain a revised estimate for y_4. It produces $y_4 = 0.304573$.

In Table 5.3 we have carried forward the solution and we show the predicted and corrected values of y_{n+1}.

Table 5.3

x_n	y_n	f_n	y^p_{n+1}	f^p_{n+1}	y^c_{n+1}
0	0	0			
0.2	0.01998	0.199600			
0.4	0.079484	0.393682			
0.6	0.176204	0.568952	0.304957	0.707001	0.304573
0.8	0.304573	0.707235	0.455403	0.792608	0.455572
1.0	0.455572	0.792454	0.616957	0.819364	0.617769
1.2	0.617769	0.818361	0.778620	0.793750	0.779749
1.4	0.779749	0.791992	0.931682	0.731969	0.932551
1.6	0.932551	0.730349	1.070792	0.653404	1.071014
1.8	1.071014	0.652927	1.194004	0.574353	1.192770
2.0	1.192770				

Accuracy and Convergence

It has been stated that we could, if we wish, use the corrector formula repeatedly to obtain successively better estimates of y_{n+1}. Does the corrector formula necessarily converge, and if it does, will it be to the true value $y(x_{n+1})$? Let us examine the problem of convergence by studying a simpler predictor-corrector method. The predictor is given by $y_{n+1} = y_{n-1} + 2hf_n$ and the corrector by

$$y_{n+1}^{(i+1)} = y_n + \frac{h}{2}[f(x_n, y_n) + f(x_{n+1}, y_{n+1}^{(i)})] \tag{5.22}$$

The notation $y_{n+1}^{(i)}$ means the ith approximation to y_{n+1}.

Now, *if* the corrector formula converges we have, in the limit,

$$y_{n+1} = y_n + \frac{h}{2}[f(x_n, y_n) + f(x_{n+1}, y_{n+1})].$$

Therefore $y_{n+1}^{(i+1)} - y_{n+1} = \frac{h}{2}[f(x_{n+1}, y_{n+1}^{(i)}) - f(x_{n+1}, y_{n+1})]$ represents the discrepancy between the $(i+1)$th approximation and the value y_{n+1} to which convergence occurs.

Let us now suppose that $f(x, y)$ and $\partial f/\partial y$ are continuous functions in the interval over which we are solving the differential equation and suppose that $|f_y| < M$ in the neighbourhood of the point (x_{n+1}, y_{n+1}) where M is a positive constant. Then we can show that by application of the mean value theorem.

$$y_{n+1}^{(i+1)} - y_{n+1} = \frac{h}{2}[y_{n+1}^{(i)} - y_{n+1}] f_y(x_{n+1}, y_{n+1}^{*})$$

where y_{n+1}^{*} is a value between $y_{n+1}^{(i)}$ and y_{n+1}. For application of the corrector formula, x_n is fixed and we need only consider variations in y.

Since f_y is bounded in the neighbourhood of (x_{n+1}, y_{n+1}), it follows that

$$|y_{n+1}^{(i+1)} - y_{n+1}| < \frac{h}{2}M\,|y_{n+1}^{(i)} - y_{n+1}| \quad \text{and by induction that}$$

$$|y_{n+1}^{(i+1)} - y_{n+1}| < \left[\frac{h}{2}M\right]^{i+1} |y_{n+1}^{(0)} - y_{n+1}|,$$

where $y_{n+1}^{(0)}$ is the initial approximation to y_{n+1} provided by the predictor.

Hence the sequence $\{y_{n+1}^{(i)}\}$ converges to some value y_{n+1} *if* $h < 2/M$. However, we do *not* know whether this will be the true value $y(x_{n+1})$; the discrepancy between the two is the local error.

It is the practice in the Milne-Simpson and Adams–Moulton methods not to use the corrector formula iteratively. Instead we try to use a step size h which is small enough to render re-corrections unnecessary. We shall merely quote results.

For the Milne-Simpson method we have first the condition that successive recorrections converge:

$$h < 3/|f_y(x_n, y_n)| \tag{5.23a}$$

Then we have the condition that the first corrected value is adequate to N decimal places and will not be altered in the Nth decimal place by further corrections:

$$|y_{n+1}^{c} - y_{n+1}^{p}| < \frac{3}{10^N h\,|f_y(x_n, y_n)|} \tag{5.23b}$$

where y^c_{n+1} and y^p_{n+1} are the corrected and predicted values of y_{n+1} via (5.17) and (5.15). Finally, a consideration of the predictor and corrector formulae would suggest that when the predicted and corrected values of y_{n+1} do not agree, the true value is likely to lie between them and nearer to the corrector value. A working criterion for accuracy to N decimal places is

$$|y^c_{n+1} - y^p_{n+1}| < 29/10^N \tag{5.23c}$$

For the **Adams-Moulton method** the criteria are, respectively,

$$h < \frac{24}{9\,|f_y(x_n, y_n)|} \tag{5.24a}$$

$$|y^c_{n+1} - y^p_{n+1}| < \frac{24}{9\,.\,10^N h\,|f_y(x_n, y_n)|} \tag{5.24b}$$

and $$|y^c_{n+1} - y^p_{n+1}| < \frac{14}{10^N} \tag{5.24c}$$

Example

Let us check the criteria for the Adams-Moulton method on the results obtained on page 141.

x_n	y_n	x_{n+1}	y^p_{n+1}	y^c_{n+1}
1.8	1.071014	2.0	1.194004	1.192770

Since $y' = x - y^2$, $f_y(x, y) = -2y$
Hence $f_y(x_n, y_n) = -2.142028$

Criterion (5.24a) requires $h < 1.2449$, which it is.

Criterion (5.24b) requires $10^N < 24/\left[9|y^c_{n+1} - y^p_{n+1}|(0.2)(2.142028)\right]$
i.e. $N < 3.7$.

Criterion (5.24c) similarly requires $N < 4.05$, so that we can rely on 3 d.p. only.

It should be noted that the criteria do **not** apply in a straight-forward extension to systems of first-order equations. The criteria for such systems are very complicated and beyond our scope.

Flow chart for Adams-Moulton method

In Figure 5.1 we present a flow chart for the Adams-Moulton method with the Runge-Kutta fourth order method used to provide the starting values. Note that we have not made any provision for modifying the step size. You should realise that we cannot change the step size without rendering the predicted value invalid. A simple ploy is to halt the calculations, take the previously determined point (x_n, y_n) as a new starting point and apply the Runge-Kutta method again before re-commencing the use of the predictor and corrector formulae. See whether you can modify the flow chart to allow for this sophistication.

Here we use N as the number of solution points required.

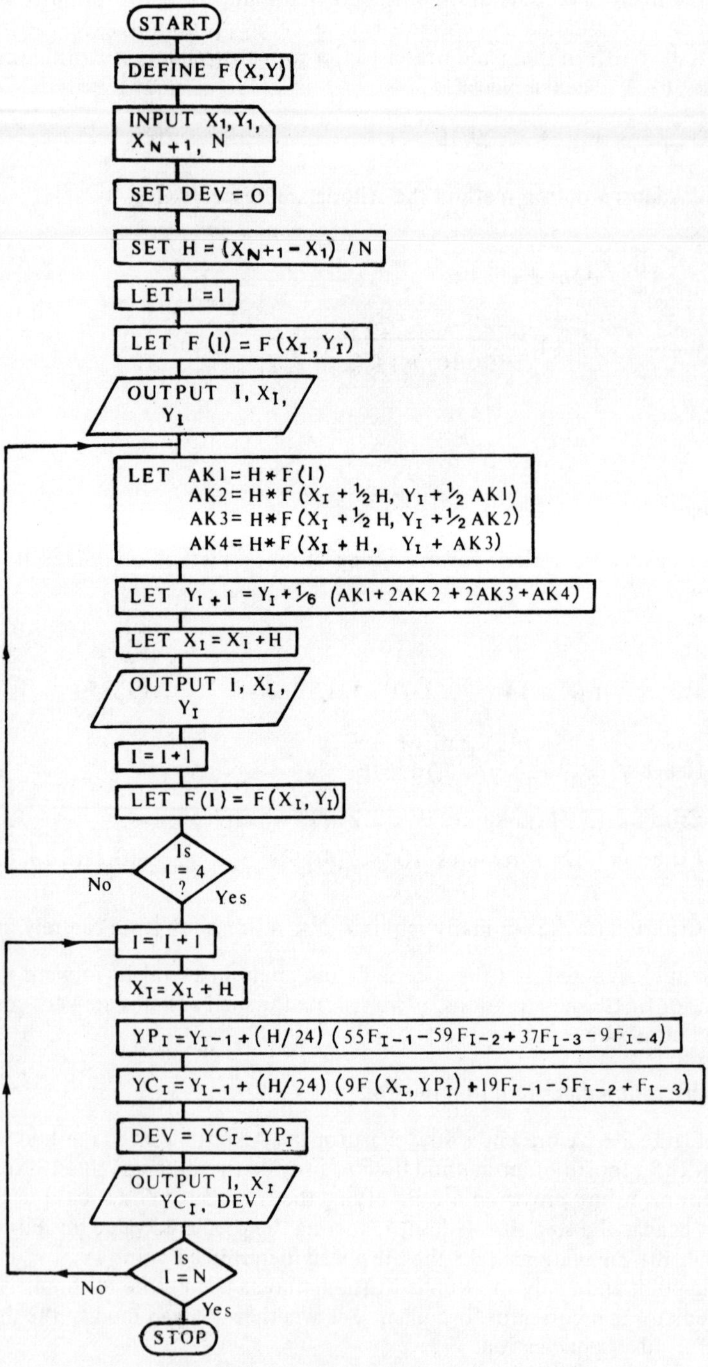
START
DEFINE F(X,Y)
INPUT X1, Y1, XN+1, N
SET DEV = O
SET H = (XN+1 − X1) / N
LET I = 1
LET F(I) = F(XI, YI)
OUTPUT I, XI, YI
LET AK1 = H * F(I)
AK2 = H * F(XI + ½ H, YI + ½ AK1)
AK3 = H * F(XI + ½ H, YI + ½ AK2)
AK4 = H * F(XI + H, YI + AK3)
LET YI+1 = YI + 1/6 (AK1 + 2AK2 + 2AK3 + AK4)
LET XI = XI + H
OUTPUT I, XI, YI
I = I + 1
LET F(I) = F(XI, YI)
Is I = 4 ?
No
Yes
I = I + 1
XI = XI + H
YPI = YI−1 + (H/24) (55FI−1 − 59FI−2 + 37FI−3 − 9FI−4)
YCI = YI−1 + (H/24) (9F(XI, YPI) + 19FI−1 − 5FI−2 + FI−3)
DEV = YCI − YPI
OUTPUT I, XI YCI, DEV
Is I = N
No
Yes
STOP

Figure 5.1

The flow chart can be modified to handle any predictor-corrector formula – try to produce a flow chart for the Milne-Simpson method.

We remark that if we are working with numbers markedly different from 1, a criterion to be used in place of (5.24c) might be

$$\left|y_{n+1}^{c} - y_{n+1}^{p}\right| / \left|y_{n+1}^{c}\right| < \frac{14}{10N}$$

For a general program we might adopt the following check at each step. If the relative error exceeds a prescribed upper bound, halve the step, compute four starting values and return to the predictor-corrector formulae. If the relative error is less than a second prescribed value, indicating that the step size is so small that we are obtaining too much accuracy for our purposes, double the step size, compute four starting values and return to the predictor-corrector formulae.

Problems

1. Use a Taylor series method to obtain values of y to three decimal places at $x = -0.1$, 0.1 and 0.2, if $y'' + (x+1)y^2 = 0$ with $y(0) = 1$ and $y'(0) = 1$. Then use the following predictor-corrector pair of formulae to obtain the value of y at $x = 0.3$.

   Predictor: $\quad y_{n+1} = 2y_{n-1} - y_{n-3} + \dfrac{4h^2}{3}(y_n'' + y_{n-1}'' + y_{n-2}'')$

   Corrector: $\quad y_{n+1} = 2y_n - y_{n-1} + \dfrac{h^2}{12}(y_{n+1}'' + 10y_n'' + y_{n-1}'')$

   Describe *briefly* how such forward integration formulae can be used to solve a boundary value problem. (C.E.I.)

2. By means of the first 3 non-zero terms of a Taylor's series obtain the solution of the differential equation $\dfrac{dz}{dt} = e^{2t} - z^2$ for the values $t = 0.10$, $t = 0.20$, $t = 0.30$, given that $z = 1$ when $t = 0$. Using the Adams-Bashforth method for continuing the solution by means of the backward difference equation $z_{i+1} = z_i + h\,[1 + ½\,\nabla + \frac{5}{12}\nabla^2 + \frac{3}{8}\nabla^3]\left(\dfrac{dz}{dt}\right)_i$ obtain the value of z when $t = 0.4$. (L.U.)

3. The table below gives the starting values for the solution of $\dfrac{dy}{dx} = 1 + y^2$. Use the formula (from Simpson's rule) $y_1 = y_{-1} + 2h\left[1 + \frac{1}{6}\delta^2\right] f_0$ as a 'corrector' formula to obtain the value of y for $x = 0.6$. Give your reasons for your choice of the 'predicted' value from which you commence. Give your answer correct to 3 decimal places.

x	y	$2hf$
0.0	0.000	0.4000
0.2	0.203	0.4164
0.4	0.423	0.4715

(L.U.)

4. A first order differential equation may be solved numerically by using one of the pairs (a) or (b) of predictor and corrector formulae.

(a) $\begin{cases} y_1 = y_0 + h(1 + \frac{1}{2}\nabla + \frac{5}{12}\nabla^2 + \frac{3}{8}\nabla^3 + \frac{251}{720}\nabla^4 + ...)y_0' \\ y_1 = y_0 + h(1 - \frac{1}{2}\nabla - \frac{1}{12}\nabla^2 - \frac{1}{24}\nabla^3 - \frac{19}{720}\nabla^4 + ...)y_1' \end{cases}$

(b) $\begin{cases} y_4 = y_0 + \frac{4}{3}h(2y_1' - y_2' + 2y_3') \\ y_4 = y_2 + \frac{1}{3}h(y_2' + 4y_3' + y_4') \end{cases}$

Prove *one only* of the predictor formulae.

The differential equation $y' = x - \frac{1}{10}y^2$ is to be solved with the initial value $y = 1$ when $x = 0$. Assuming that the following starting values have been obtained, find the value of y at $x = 0.3$.

x	−0.2	−0.1	0.1	0.2
y	1.04068	1.01513	0.99507	1.00013

(L.U.)

5. Obtain, by the Adams-Moulton formulae or otherwise, the next line, beginning $x = 0.5$, of the table giving the start of the numerical solution of $\frac{dy}{dx} = (x + y)^{1/2}$ given the values below. First form a difference table as far as $\nabla^2 f$.

x	0.1	0.2	0.3	0.4
y	0.3071	0.3771	0.4587	0.5511

(L.U.)

Use the Adams-Moulton formulae in finite difference form to extend the table below by one step, in the step-by-step solution of the differential equation $\frac{dy}{dx} = f(x, y) = \sqrt{(x + y)}$.

x	0	0.1	0.2	0.3
y	0.2500	0.3071	0.3771	0.4587

(L.U.)

6. The table below gives the starting values for the numerical solution of the differential equation $\frac{dy}{dx} = 1 + y^2$, with $y = 0$ when $x = 0$. Form a difference table as far as $\nabla^3 y'$. Explain how the two formulae

(a) $y_{n+1} = y_n + h(1 + \frac{1}{2}\nabla + \frac{5}{12}\nabla^2 + \frac{3}{8}\nabla^3)y_n'$

(b) $y_{n+1} = y_n + h(1 - \frac{1}{2}\nabla - \frac{1}{12}\nabla^2 - \frac{1}{24}\nabla^3)y_{n+1}'$

are used to continue the solution, and obtain the value of y at $x = 0.3$.

x	−0.2	−0.1	0.0	0.1	0.2
y	−0.20271	−0.10033	0.00000	0.10033	0.20271

(L.U.)

7. Describe a predictor-corrector method for solving the differential equation $\frac{dy}{dx} = f(x, y)$ with initial condition $y = y_0$ at $x = x_0$.
Solve numerically the differential equation $\frac{dy}{dx} = x - y^2$, $y = 1$ at $x = 0$ for the values of $x = 0.2$ and $x = 0.4$, specifying your results to 3 decimal places. (C.E.I.)

5.4 LINEAR DIFFERENCE EQUATIONS

In studying the stability of numerical methods of solving ordinary differential equations we encounter linear difference equations with constant coefficients. These are equations of the form

$$y_{k+r} + a_{r-1} y_{k+r-1} + \ldots\ldots + a_1 y_{k+1} + a_0 y_k = f(k) \tag{5.25}$$

The theory of linear difference equations is very closely aligned to the corresponding theory of linear differential equations. We shall discuss methods of solution and give examples of their application.

Assuming that $a_0 \neq 0$ then equation (5.25) is said to be of **order** r.

(a) **First-order homogeneous equations**

Consider a typical equation $y_{k+1} = Ay_k$ for A constant.

Let us suppose that we are given the initial condition $y_0 = \alpha$, a constant.

Then $y_1 = Ay_0 = A\alpha$, $y_2 = Ay_1 = A^2\alpha$ and, in general,

$$y_k = A^k\alpha \tag{5.26}$$

This solution satisfies the difference equation and the initial condition; it is therefore *the* solution (we have not *proved* this statement).

Example $y_{n+1} = 3y_n$; $y_0 = 1$. The solution is $y_n = 3^n$.

(b) **First-order non-homogeneous equations**

Consider $y_{k+1} = Ay_k + B$, where A and B are constants.

Then $y_1 = Ay_0 + B$, $y_2 = Ay_1 + B = A^2y_0 + AB + B$,

$y_3 = Ay_2 + B = A^3y_0 + A^2B + AB + B$ and so

$y_k = A^ky_0 + B(A^{k-1} + A^{k-2} + \ldots\ldots + A + 1)$

$$\text{Hence } y_k = \begin{cases} A^ky_0 + B(A^k - 1)/(A-1), & A \neq 1 \\ y_0 + Bk, & A = 1 \end{cases} \tag{5.27}$$

Example $2y_{k+1} = y_k + 4$, $y_0 = 3$.

First we write the equation in the form $y_{k+1} = \frac{1}{2}y_k + 2$ and then apply (5.27) to obtain $y_k = 4 - (\frac{1}{2})^k$.

(c) **Second-order homogeneous equations**

The equations are of the form $y_{k+2} + \alpha y_{k+1} + \beta y_k = 0$ where α and β are constant. We need two initial conditions y_0 and y_1 to determine the solution uniquely. We look for two linearly independent solutions. The technique is to substitute $y_k = m^k$ to obtain the **auxiliary equation** $m^2 + \alpha m + \beta = 0$.

Examples

(i) $y_{k+2} - 3y_{k+1} + 2y_k = 0$; $y_0 = 3/2$, $y_1 = 2$.

Substitution of $y_k = m^k$ produces the auxilary equation $m^2 - 3m + 2 = 0$ which has roots $m = 1$ and $m = 2$.

The general solution is $y_k = A.2^k + B.1^k \equiv A.2^k + B$ where A and B are constants.

Since $y_0 = 3/2$, it follows that $3/2 = 1.A + B$. Similarly, from $y_1 = 2$ we obtain $2 = 2A + B$. Hence $A = ½$, $B = 1$ and $y_k = 2^{k-1} + 1$.

(ii) $y_{k+2} + 6y_{k+1} + 9y_k = 0$; $y_0 = 1$, $y_1 = 2$.

The auxiliary equation is $m^2 + 6m + 9 = 0$ which has a repeated root $m = -3$.

By analogy with differential equations we may quote the general solution as $y_k = (Ak + B)(-3)^k$.

From $y_0 = 1$ and $y_1 = 2$ we find that $1 = (B)1$ and $2 = (A + B)(-3)$ so that $B = 1$, $A = -5/3$ to give the solution $y_k = (1 - 5k/3)(-3)^k$.

(iii) $y_{k+2} - 2y_{k+1} + 2y_k = 0$; $y_0 = 0$, $y_1 = 1$.

The auxiliary equation is $m^2 - 2m + 2 = 0$ which has roots

$$m = 1 + i = \sqrt{2}(\cos \pi/4 + i \sin \pi/4) = \sqrt{2}\, e^{i\pi/4}$$

and $m = 1 - i = \sqrt{2}(\cos \pi/4 - i \sin \pi/4) = \sqrt{2}\, e^{-i\pi/4}$

The general solution is $y_k = (\sqrt{2})^k (A \cos k\pi/4 + B \sin k\pi/4)$ and from the initial conditions, $A = 0$, $B = 1$, so that $y_k = (\sqrt{2})^k \sin k\pi/4$.

(d) **Second-order non-homogeneous equations**

We look for *the* solution in the form: *a* particular solution + *the* complementary solution. To find a particular solution we employ the method of trial solutions.

For the equation $y_{k+2} + \alpha y_{k+1} + \beta y_k = f(k)$ we present in Table 5.4 suitable trial solutions for several forms of $f(k)$.

Table 5.4

$f(k)$	Trial solution
a^k, a constant	Aa^k, A constant
$\sin bk$, $\cos bk$	$A \sin bk + B \cos bk$, A, B constant
k^h	$A_0 + A_1 k + ... + A_h k^h$, A_i constant
$k^h \,.\, a^k$	$a^k(A_0 + A_1 k + ... + A_h k^h)$
$a^k \sin bk$, $a^k \cos bk$	$a^k(A \sin bk + B \cos bk)$
$f_1(k) + f_2(k)$	Trial solution for $f_1(k)$ + trial solution for $f_2(k)$

Example $y_{k+2} - 6y_{k+1} + 9y_k = 2k + 3k^2$

The auxiliary equation is $m^2 - 6m + 9 = 0$ which leads to the complementary solution $y = (Ak + B)3^k$.

We try a particular solution of the form $y_k = (\alpha_0 + \alpha_1 k + \alpha_2 k^2)$

Then $y_{k+1} = \alpha_0 + \alpha_1(k+1) + \alpha_2(k+1)^2$ and $y_{k+2} = \alpha_0 + \alpha_1(k+2) + \alpha_2(k+2)^2$.

Substitution into the difference equation produces

$$\alpha_0 + \alpha_1(k+2) + \alpha_2(k+2)^2 - 6\alpha_0 - 6\alpha_1(k+1) - 6\alpha_2(k+1)^2 + 9\alpha_0 + 9\alpha_1 k + 9\alpha_2 k^2 = 2k + 3k^2.$$

Equating the coefficients of like powers of k we obtain the conditions

$\alpha_2 - 6\alpha_2 + 9\alpha_2 = 3 \Rightarrow \alpha_2 = 3/4$, $\alpha_1 + 4\alpha_2 - 6\alpha_1 - 12\alpha_2 + 9\alpha_1 = 2 \Rightarrow \alpha_1 = 2$,
$\alpha_0 + 2\alpha_1 + 4\alpha_2 - 6\alpha_0 - 6\alpha_1 - 6\alpha_2 + 9\alpha_0 = 0 \Rightarrow \alpha_0 = 19/8$

We obtain a particular solution of $y_k = (19/8) + 2k + (3/4)k^2$

Therefore the general solution is $y_k = (Ak + B)3^k + (19/8) + 2k + (3/4)k^2$

Application to a string insulator problem

In Figure 5.2(a) we show a representation of a string insulator for an electric power line. The $(n-1)$ insulators which form the device are connected by metallic conductors. The top insulator is attached to a tower and hence earthed; the lowest insulator is connected to the line conductor which carries an alternating current.

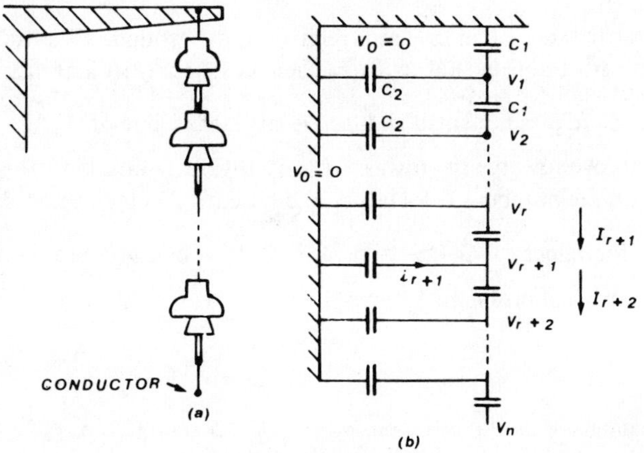

Figure 5.2

Now we assume that all the conductors are identical and all the insulators are identical. The rate at which the charge in one of the conductors changes depends on its capacitance relative to all other conductors (C_1) and to earth (C_2). We consider however its capacitance relative to the two adjacent conductors and to the earth. The equivalent network is shown in Figure 5.2(b). Note that if we denote the potential of the rth conductor by V_r then the current between the rth and $(r+1)^{\text{th}}$ conductors is given by $I_{r+1} = i\omega C_1(V_r - V_{r+1})$; similarly, the current between the $(r+1)^{\text{th}}$ and $(r+2)^{\text{th}}$ conductors is given by $I_{r+2} = i\omega C_1(V_{r+1} - V_{r+2})$. Now the current flowing between the $(r+1)^{\text{th}}$ conductor and the ground is given by $i_{r+1} = -i\omega C_2 V_{r+1}$. From Kirchhoff's first law for circuits it follows that $I_{r+2} = i_{r+1} + I_{r+1}$ so that we obtain

$$i\omega C_1(V_{r+1} - V_{r+2}) = -i\omega C_2 V_{r+1} + i\omega C_1(V_r - V_{r+1})$$

or

$$V_{r+2} - \left(2 + \frac{C_2}{C_1}\right)V_{r+1} + V_r = 0 \tag{5.28}$$

The auxiliary equation is $m^2 - \left(2 + \frac{C_2}{C_1}\right)m + 1 = 0$. This has roots

$$m = 1 + \frac{C_2}{2C_1} \pm \sqrt{\frac{C_2}{C_1} + \frac{C_2^2}{4C_1^2}}$$

It simplifies the algebra if we write $\cosh\mu = 1 + \frac{C_2}{2C_1}$, for the auxiliary equation then becomes $m^2 - 2m\cosh\mu + 1 = 0$ which has roots $m = \cosh\mu \pm \sinh\mu$ i.e. $2e^{\mu}$ and $2e^{-\mu}$. Therefore the general solution of (5.28) may be written as

$$V_r = \alpha\cosh\mu r + \beta\sinh\mu r \tag{5.29}$$

where α and β are constants.

Now equation (5.28) applies for $r = 1, 2, \ldots, (n-2)$. The boundary conditions we impose are that $V_0 = 0$ and $V_n = V$, some given value for the line conductor. Applying (5.29) with $r = 0$ we obtain $0 = \alpha\cosh\mu r$ so that $\alpha = 0$; applying (5.29) with $r = n$ we obtain $V = \beta\sinh\mu n$ and we may finally write

$$V_r = V\sinh\mu r/\sinh\mu n; \quad r = 0, 1, \ldots, n \tag{5.30}$$

where $\cosh\mu = 1 + \frac{C_2}{2C_1}$.

Suppose the capacitance of an insulator relative to the ground, C_2, can be neglected in comparison with the capacitance C_1. Then $\cosh\mu \simeq 1$ so that $\mu \simeq 0$ and $V_r \simeq \frac{V_r}{n}$. If C_2/C_1 is not small and μn is large, the value of $V_{r+1} - V_r$ decreases rapidly as we move towards the tower. (Prove this for yourself.) Therefore there is a sensible limit on the number of insulators to be used. To emphasise this, plot graphs of $\frac{V_r}{V}$ against r for values of $C_2/C_1 = 0, 1/3, 2/3, 1$ and interpret your results in the context of the original problem.

Problems

1. Obtain a solution of the difference equation $y_{n+2} - 2r\cos\theta . y_{n+1} + r^2 y_n = 0$ in the form $y = r^n[A\cos n\theta + B\sin n\theta]$.

   Solve the equation $y_{n+2} - 3y_{n+1} + 9y_n = 13n + 3^n$. (L.U.)

2. (a) Solve the difference equation $y_{n+2} + 6y_{n+1} - 7y_n = 3^n$.

   (b) Show that the equation $y_{n+1} \cdot y_n + 2y_{n+1} + 8y_n + 9 = 0$ can be reduced by the substitution $y_n = (V_{n+1}/V_n) - 2$ to an equation with constant coefficients. Hence solve the given equation when $y_0 = -5$. (L.U.)

3. (a) Solve the difference equation $y_{n+2} - 4y_{n+1} + 4y_n = 3^n$.

   (b) Obtain the solution of the simultaneous difference equations $x_{n+1} = 2x_n + y_n + n$, $y_{n+1} = 2x_n + 3y_n - n$, for which $x_0 = 3$ and $y_0 = 0$. (L.U.)

4. (a) Solve the difference equation $y_{n+2} - 6y_{n+1} + 9y_n = 2^n + n$.

   (b) Show that $y_n = n$ is a solution of the difference equation $n(n+1)y_{n+2} - 5n(n+2)y_{n+1} + 4(n+1)(n+2)y_n = 0$. By substituting $y_n = nu_n$, obtain the general solution of this equation and show that the solution for which $y_1 = 12$ and $y_2 = 60$ is $y_n = \frac{3}{2}n \cdot 4^n + 6n$. (L.U.)

5. Solve the difference equations

   (a) $y_{n+2} - 6y_{n+1} + 13y_n = 2^n$

   (b) $\Delta^2 y_n - 4\Delta y_n + 4y_n = n \cdot 2^n + 3^n$. (L.U.)

6. A long row of trucks, each of mass M, are standing in a line, touching one another, joined by couplings which become tight when the distance between trucks is a. A steady force $P = Mf_0$ is applied to the end truck in a direction moving it away from its neighbour. In the subsequent motion, v_n denotes the common velocity of the moving trucks immediately after the nth truck starts to move. Assuming that frictional resistance can be neglected, show that

   $(n+1)^2 v_{n+1}^2 = n^2 v_n^2 + 2naf_0$ and deduce that $v_{n+1}^2 = \dfrac{n}{n+1} af_0$. (L.U.)

7. A mechanical model of a band-pass filter comprises a series of identical disks each connected to a fixed base by an identical spring. If θ_x is the angle turned through by the xth disk, the equation of its motion is $c(\theta_{x-1} - \theta_x) + c(\theta_{x+1} - \theta_x) - k\theta_x = I\dfrac{d^2\theta_x}{dt^2}$ where c, k and I are constants. Substitute $\theta_x = \Theta_x \sin \omega t$, put $\beta = 1 + \dfrac{k}{2c} - \dfrac{I\omega^2}{2c}$ and discuss the solutions of the resulting difference equation for the cases $-1 \leqslant \beta \leqslant 1$, $\beta < -1$ and $\beta > 1$.

8. Calculate the frequencies and normal modes of vibration of a ten storey building similar to the two storey case discussed in Section 2.2. Compare the difference equation approach and the eigenvalue approach.

5.5 STABILITY OF NUMERICAL PROCEDURES

In Section 5.2 we saw the phenomenon of partial instability. Roughly speaking stability is the phenomenon of the computed solution behaving like the true solution of the differential equation. It is concerned with the way errors propagate through the solution. Much of the study of stability has been confined to the equation $dy/dx = \lambda y$, for a constant λ; this is because the general problem is so very complicated. Let us summarise the basic facts.

Stability may depend upon the differential equation under study, the method used for solution, and the step size, h.

There is the possibility of **inherent instability** where small variations in the initial conditions attached to the problem give rise to large variations in the true solution. Such ill-conditioning must be reflected in the numerical scheme and will occur whatever method is used.

Consider the differential equation $dy/dx = y - 2x$. This has the analytical solution $y = Ae^x + 2x + 2$. Suppose the initial conditions are $y = 2$ when $x = 0$; then $A = 0$ and the particular solution becomes $y = 2x + 2$. The term Ae^x, which would dominate the solution for large values of x, is absent. However, round-off error and truncation error will ensure that e^x will appear in the solution, albeit with a small coefficient. This will mean that for large values of x, this spurious term will swamp the true solution. Since the phenomenon is a feature of the differential equation, the only hope is to reformulate the problem.

Partial Instability occurs when the numerical method concerned represents the true solution badly for step sizes above a certain value.

Certain finite difference schemes are unsuitable for use with some differential equations, no matter how small the step size may be. For other differential equations, the schemes may work quite well. This phenomenon is called **weak instability**.

There are situations in which the method, despite a low truncation error, introduces spurious solutions which increase, in a way which cannot be controlled by the step size, for any differential equation. This is the disastrous property of **strong instability**. This often results because a high-accuracy corrector formula has been employed which results in several spurious solutions, at least one of which grows uncontrollably.

General test for stability

A general test for stability is the following. Find the appropriate difference equation when the numerical method is applied to a particular equation. Let the roots of the difference equation be m_1, m_2, ..., m_k where k is the order of the difference equation. Then the general solution of the difference equation may be written $y_r = a_1 m_1^r + a_2 m_2^r + ... + a_k m_k^r$ where a_i are constants. One of the roots, say m_1, will lead to the exact solution as $h \to 0$. If the roots satisfy $|m_i| < 1$ then the method is said to be **strongly stable.** If any of the extraneous roots satisfy $|m_i| > 1$ then errors will grow exponentially and the method is **unstable.**

Schemes which are weakly stable should not be integrated over long intervals.

In the next section we apply a predictor-corrector method of Hamming to a practical problem. Hamming's method is a compromise between stability and accuracy. The price that is paid for good stability is a fairly high truncation error. The problem involves a *system* of first-order equations arising from a second-order differential equation.

Problems

1. Consider the equation $y' = y - x$ with initial condition $y(0) = 1$. Write down the analytical solution. Let the value of $y(0)$ vary by $\pm 1\%$ and obtain the analytical solutions. Comment on the variation in $y(x)$ for $x = 5$ and deduce what will happen for larger values of x. This is an example of *inherent instability*.

2. Consider the equation $y' = -10y$ with initial condition $y(0) = 1$. The analytical solution is $y = e^{-10x}$. Show that a Runge-Kutta second-order method leads to $y_{m+1} = (1 - 10h + 50h^2)^m$ and show that the expression in parentheses exceeds 1 if $h > 0.2$. What effect will this have on computed values of x? Repeat the study for a Runge-Kutta fourth-order method. This is an example of *partial instability*.

3. Repeat the principles of Problem 2 on the equation $y' = -5y$ with $y(0) = 1$.

4. Use a second-order Runge-Kutta method on the equation $y' = -10y$ with $y(0) = 1$ and a step size of 0.2. Comment.

5. Apply the Milne-Simpson corrector formula to the equation $y' + 40y = ½ + 20x$ with $y(1) = 0.5$ and a step size of 0.1. Tabulate y for $x = 1.0, 1.1, \ldots, 1.5, 2.0, 3.0, 4.0$ and 5.0 and compare with the analytical solution. Apply the formula to the equation $y' + \lambda y = 0$, $\lambda > 0$. You should obtain a second order difference equation. This will have *two* roots, yet it should be an approximation to a *first-order* differential equation. We have introduced an unwanted solution which, in fact, will grow even if the step size $h \to 0$. If, however, $\lambda < 0$ the spurious solution decays exponentially. This shows that the Milne-Simpson method is *weakly unstable* for this equation.

5.6 CASE STUDY - SURGE TANK

In a hydro-electric scheme, water may be conveyed to the turbines by a very long pipe. The very large mass of water will require considerable forces to retard or accelerate it when the demands of the turbines change. There might be a change in demand of output or there might be a sudden failure in the system which cannot be coped with, except by the incorporation of a surge tank near the power station. In Figure 5.3(a) we represent schematically a typical set-up. If the flow to the turbine increases, the level of water in the surge tank falls and hence an acceleration head is

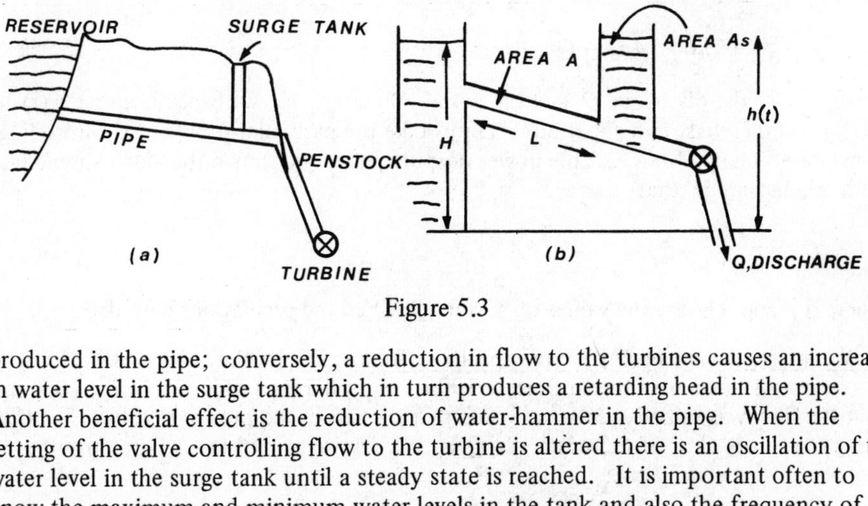

Figure 5.3

produced in the pipe; conversely, a reduction in flow to the turbines causes an increase in water level in the surge tank which in turn produces a retarding head in the pipe. Another beneficial effect is the reduction of water-hammer in the pipe. When the setting of the valve controlling flow to the turbine is altered there is an oscillation of the water level in the surge tank until a steady state is reached. It is important often to know the maximum and minimum water levels in the tank and also the frequency of oscillation; there is the possibility of prolonged oscillations leading to resonance. A further problem is whether the system is in stable equilibrium.

We shall assume the pipe to be rigid and the water to be incompressible. This latter assumption is acceptable if the duration of an oscillation in the tank is much greater than the time taken for a pressure wave to travel along the pipe. Refer to Figure 5.3(b) during the derivation of the governing model equations.

We shall let the elevation of the water surface in the surge tank be given by $h(t) = h_0 + y(t)$ where h_0 is the equilibrium elevation and we let the velocity of water in the pipe be given by $v(t) = v_0 + x(t)$ where v_0 is the velocity corresponding to steady discharge.

We have therefore introduced two new variables $y(t)$ and $x(t)$ which represent respectively departures from equilibrium values of head in the surge tank and velocity of water in the pipe.

The momentum equation is found by applying Newton's second law to the water in the pipe. Assuming that there is an upward velocity $u(t)$ of water in the surge tank, it can be written

$$\frac{L}{g}\frac{dv}{dt} + (h - h_0) + F_p v|v| + F_T u|u| = 0 \tag{5.31}$$

where F_p is the friction coefficient for the pipe and F_T is the so-called 'throttle' friction coefficient. Note that

$$u = \frac{dh}{dt} \tag{5.32}$$

Finally, we have the equation of continuity applied at the base of the surge tank; this is written

$$vA = \frac{dh}{dt} A_s + Q \quad \text{, where } Q \text{ is the discharge.} \tag{5.33}$$

We shall only consider the case where $F_T = 0$. Introducing the variables $y(t)$ and $x(t)$ we may rewrite (5.31) and (5.33) as follows

$$\frac{L}{g}\frac{dx}{dt} = -y - F_p v|v| \tag{5.34}$$

$$A_s \frac{dy}{dt} = Ax + Av_0 - Q \tag{5.35}$$

If we specify the discharge Q and the initial values h_0, v_0 we can determine $x(t)$ and $y(t)$ from these last two equations. Let us take the particular problem of demanding constant efficiency for a variable power output. It can be shown that for a simple tank, with F_T negligible, that

$$Q = \frac{h_0 + y_f}{h_0 + y} Q_f \tag{5.36}$$

where y_f and Q_f are the values of y and Q when the oscillations have decayed.

We now make the problem more specific. We let $L = 2000\text{m}$, $A = 4.909\text{m}^2$, h_0 (the static head over the turbines) $= 100\text{m}$, $F_p = 0.5$ and $A_s = 78.54\text{m}^2$. Suppose that the flow was initially $30\text{m}^3\text{s}^{-1}$ and that the power requirement is altered so that there is a steady flow of $15\text{m}^3\text{s}^{-1}$ ($= Q_f$). We require to find the maximum surge height.

In the final steady state $\frac{dx}{dt} = 0$ so that $y_f = -F_p v_f|v_f|$ and $\frac{dy}{dt} = 0$ so that $A(x_f + v_0) = Av_f = Q_f$.

Therefore $v_f = Q_f/A = 15/4.909 = 3.056\text{ms}^{-1}$

$y_f = -0.5(3.056)|3.056| = -4.67\text{m}$

Initially, $Av_0 = Q_0$ so that $v_0 = 30/4.909 = 6.11\text{ms}^{-1}$; also $y_0 = -F_p v_0|v_0| = -18.67\text{m}$.

During the oscillations $Q = \dfrac{100 - 4.67}{100 + y} \times 15 = \dfrac{15 \times 95.33}{100 + y}$

We now particularise the equations (5.34) and (5.35) to the forms

$$\left.\begin{aligned} \frac{dx}{dt} &= \frac{9.81}{2000}\left(-y - 0.5(x+v_0)|x+v_0|\right) \\ \frac{dy}{dt} &= \frac{1}{78.54}\left(4.909(x+v_0) - \frac{15 \times 95.33}{100+y}\right) \end{aligned}\right\} \tag{5.37}$$

These are special cases of the equations

$$\left.\begin{aligned} \frac{dx}{dt} &= f_1(x, y, t) \\ \frac{dy}{dt} &= f_2(x, y, t) \end{aligned}\right\} \tag{5.38}$$

When f_1 and f_2 are independent of time t, the equations describe an **autonomous system.**

Hamming's method

Hamming's method was designed to choose that corrector formula (from a general class of formulae) which would produce accurate values for polynomials up to degree four and would have a much stronger stability than the Milne-Simpson method. The outline of his method for a single differential equation follows. As usual we consider the equation $dy/dx = f(x, y)$.

Step 1 We have starting values y_0, y_1, y_2, y_3 (and hence we can calculate corresponding values f_0, f_1, f_2, f_3). Usually, y_0 will be the given initial condition and y_1, y_2, y_3 can be found by a fourth-order Runge-Kutta method.

Step 2 y_{n+1} is predicted by the Milne formula

$$y_{n+1}^{p} = y_{n-3} + \frac{4h}{3}(2f_n - f_{n-1} + 2f_{n-2}) \tag{5.15}$$

Step 3 The predicted value is modified by assuming that the local truncation error does not change appreciably on successive intervals. The modification is

$$y_{n+1}^{m} = y_{n+1}^{p} + \frac{112}{121}(y_n^{c} - y_n^{p}) \tag{5.39}$$

Note that, for $n = 3$ there is no corrected value y_n^c and this step is by-passed in that instance.

Step 4 The corrected value for y_{n+1} is obtained from the Hamming formula

$$y_{n+1}^{c} = \frac{9}{8}y_n - \frac{1}{8}y_{n-2} + \frac{3h}{8}(f_{n+1} + 2f_n - f_{n-1}) \tag{5.40}$$

Although we can use the formula iteratively, it is usual to choose the step size such that one application will ensure that the required accuracy is obtained.

Step 5 The truncation error for the corrector formula is calculated from

$$\epsilon_c \simeq \frac{9}{121}(y_{n+1}^{c} - y_{n+1}^{p}) = e_t \tag{5.41}$$

This formula is quoted without derivation.

Step 6 The final value for y_{n+1} is obtained from

$$y_{n+1} = y_{n+1}^{c} - e_t \tag{5.42}$$

The value of f_{n+1} is also calculated.

Step 7 If the local truncation error is acceptably small, then we return to step 2, provided more values of y are required. If the estimate (5.41) lies outside the acceptable limits then we reduce the step size and return to step 2.

Try to produce a flow chart for this method.

Application to a system of differential equations

Suppose we wish to apply Hamming's method to the system of first-order ordinary differential equations

$$\left.\begin{aligned} \frac{dy_1}{dx} &= f_1(x, y_1, y_2, \ldots\ldots, y_n) \\ \frac{dy_2}{dx} &= f_2(x, y_1, y_2, \ldots\ldots, y_n) \\ &\vdots \\ \frac{dy_n}{dx} &= f_n(x, y_1, y_2, \ldots\ldots, y_n) \end{aligned}\right\} \tag{5.43}$$

In essence, we simply perform a step on each equation before proceeding to the next step. The only problem comes in storing the required values efficiently.

Application to the surge problem

We applied Hamming's method to equations (5.37) and using a step size of 10 seconds. Some results are shown in Table 5.5 to 3 d.p. They look reasonable; but how reliable do you think they are?

Table 5.5

t	10	20	30	100	200
x	−0.035	−0.129	−0.267	−1.354	−2.902
y	−7.085	−15.488	−13.924	−3.162	12.219

5.7 PHASE-PLANE DIAGRAMS

There are often engineering systems which are governed by second-order differential equations that do not involve the independent variable *explicitly*. We are able to reduce the governing equation to a first order equation. For example, the system governed by the equation $\ddot{x} + x\dot{x} + x^2 = 0$ can be studied more easily if we put $y = \dot{x} = \frac{dx}{dt}$ so that $\ddot{x} = \dot{y} \equiv \frac{dy}{dt} = \frac{dy}{dx} \cdot \frac{dx}{dt} = y'y$. We then have as governing equation $yy' + xy + x^2 = 0$ which is a first order equation in y. The initial conditions to the original problem should have specified $x(0)$ and $\dot{x}(0)$. For the reduced equation, then, we have a corresponding pair of values (x, y) at the initial time. This allows a solution to be found. If we plot the graph of $y = y(x)$, this is called a **phase-plane diagram.** As t varies, both x and y will change and the point (x, y) will move along the diagram.

Example 1

Consider a system experiencing simple harmonic motion of frequency ω, where the governing equation is $\ddot{x} + \omega^2 x = 0$. Writing $y = \dot{x}$, we obtain $yy' + \omega^2 x = 0$. Integrating, we obtain $\frac{1}{2}y^2 + \frac{1}{2}\omega^2 x^2 = \text{constant}$. Suppose that initially $x = a$ and $y = \dot{x} = 0$. Then we finally obtain the solution as

$$y^2 + \omega^2 x^2 = \omega^2 a^2 \tag{5.44}$$

In Figure 5.4 we sketch the phase-plane diagram.

Notice that above the x - axis $y = \dot{x}$ is positive and therefore x increases with time; this suggests that the imposed arrow marks the progress of the motion.

The initial point is P_1. From the diagram we see that x oscillates between $-a$ and $+a$ and $y = \dot{x}$ between $-a\omega$ and $+a\omega$. The closed contour implies a periodic motion.

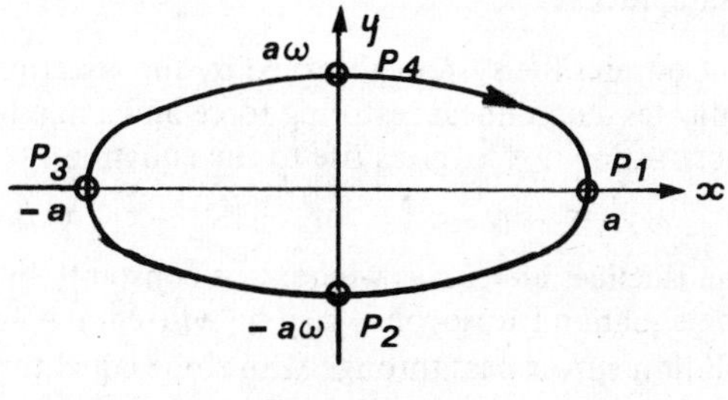

Figure 5.4

Example 2

We now consider a simple pendulum of length l moving in a vacuum. The equation of motion is $\ddot{x} + \omega^2 \sin x = 0$, where $\omega^2 = g/l$ and x is the angle made with the vertical; the equation can be converted to $yy' + \omega^2 \sin x = 0$. Integration produces the equation $\frac{1}{2}y^2 - \omega^2 \cos x = \text{constant}$. Now let $x = 0$ and $y = \dot{x} = u$ at $t = 0$. Then $\frac{1}{2}u^2 - \omega^2 = \text{constant}$ so that

$$y^2 = 2\omega^2 \cos x + (u^2 - 2\omega^2) \tag{5.45}$$

There are three cases to consider

(i) $u^2 > 4\omega^2$ so that $y^2 > 2\omega^2(\cos x + 1)$ and $y^2 > 0$ for all values of x.

(ii) $u^2 = 4\omega^2$ so that $y^2 = 2\omega^2(\cos x + 1) = 4\omega^2 \cos^2 \frac{1}{2}x$. Therefore $y = \pm 2\omega \cos \frac{1}{2}x$, except that $y = +2\omega \cos \frac{1}{2}x$ if $x > 0$ and $y = -2\omega \cos \frac{1}{2}x$ if $x < 0$.

(iii) $u^2 < 4\omega^2$ so that y^2 can be negative for some values of x and therefore the solution is not defined for all x.

It is left to you to show that the phase-plane diagram is as shown in Figure 5.5. Notice that for small initial velocities the motion is periodic whereas, for larger velocities the motion is that of a progressively increasing angle x or a progressively decreasing one.

Try to interpret these cases physically. (diagram not to scale)

We leave you to try and extend this diagram for the general case $x = a$ and $y = \dot{x} = u$ at $t = 0$.

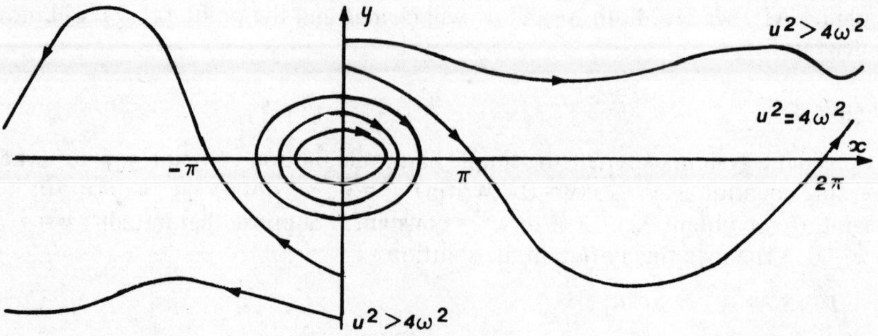

Figure 5.5

Sometimes the phase-plane diagram is not easy to draw and the use of isoclines is often helpful.

Example 3

Consider the system governed by the equation of motion $\ddot{x} + x\dot{x} + x^2 = 0$ which indicates a non-linear restoring force and a non-linear damping term. The usual substitution $y = \dot{x}$ gives rise to the equation

$$yy' = -xy - x^2 \quad \text{or} \quad y' = -x - x^2/y \tag{5.46}$$

The **isoclines** are curves where $y' = \text{constant}$, in this case $-x - x^2/y = C$, C a constant. The equation for isoclines may be written $y = -x^2/(x + C)$. On each isocline, the solution curves pass through with slope equal to C. In Figure 5.6(a) we have drawn the isoclines and we have indicated the local slopes of the solution curves; in Figure 5.6(b) we have indicated some solution curves.

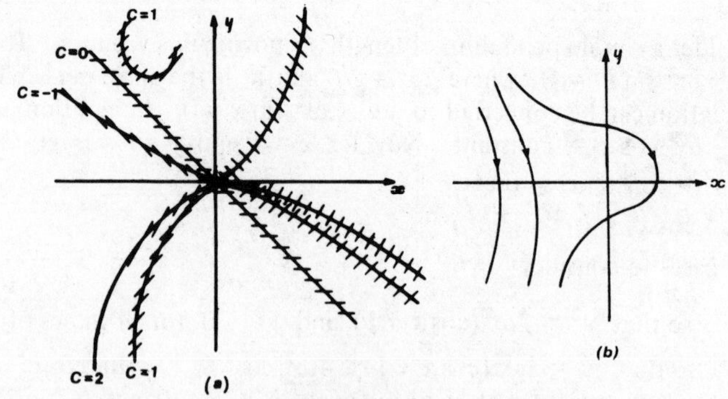

Figure 5.6

We leave you to check the isoclines and subsequent development.

Note that $x = 0$ corresponds to $y' = 0$ from equation (5.46).

Finally we show an example of a **limit cycle** which is a closed contour to which all solution curves tend as $t \to \infty$, no matter what the initial conditions.

Example 4 – Van der Pol's Equation

In Figure 5.7(a) we show a circuit diagram for a simple electronic oscillator. We state, without proof, that a suitable governing equation is

$$\ddot{x} - \epsilon(1 - x^2)\dot{x} + x = 0 \qquad (5.47)$$

where $\epsilon > 0$ is a constant of the circuit and x is a function of the current i. The equation represents a sustained periodic oscillation. For $\epsilon = 1$ we sketched the phase-plane diagram shown in Figure 5.7(b).

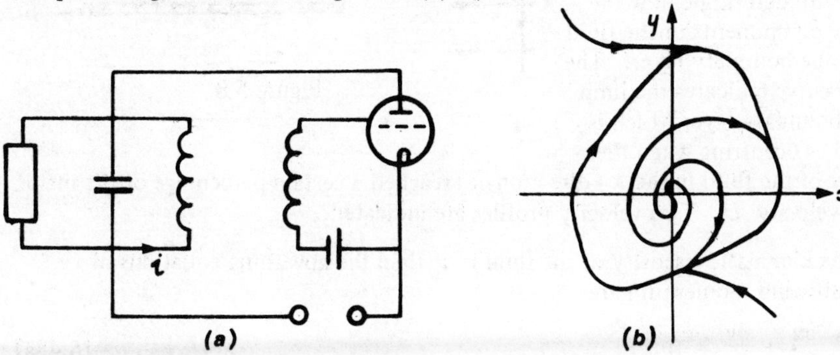

Figure 5.7

Problems

1. There are various ways of investigating the behaviour of systems described by the Van der Pol equation $\dfrac{d^2x}{dt^2} - \epsilon(1 - x^2)\dfrac{dx}{dt} + x = 0.$

   Using the substitution $y = dx/dt$, express the equation in the (phase-plane) form $dy/dx = f(x, y)$, sketch the isoclines $f(x,y) = \lambda$ for $\epsilon = 1$ and $\lambda = 0, \pm 1, \pm 2, \infty$, and hence graphically deduce the nature of the solution. (C.E.I.)

2. Sketch the phase-plane diagram for each of the following

   (i) $\ddot{x} - 9x = 0$ (ii) $\ddot{x} + 2x - 1.5x^2 = 0$

   (iii) $\ddot{x} - 2x + 3x^3 = 0$ (iv) $\ddot{x} + \dot{x}|\dot{x}| + x = 0$

3. Use isoclines to sketch the phase-plane diagram in the cases

   (i) $\ddot{x} + 2x\dot{x} + x^2 = 0$ (ii) $\ddot{x} + \dot{x}^2 - 2x = 0$ (iii) $\ddot{x} + 2\dot{x} + 8x = 0$

5.8 BOUNDARY-VALUE PROBLEMS: SHOOTING METHODS

In Figure 5.8 we show a flat thin plate immersed in a fluid moving with uniform velocity parallel to the plate. Near the plate the viscous forces dominate in the **boundary layer.** It is required to determine how the velocity components of the fluid vary in the boundary layer. The broken curve indicates the limit of the boundary layer which is defined as occurring when the velocity of the fluid in the x - direction has reached a certain percentage of the main stream velocity U. Two velocity profiles are indicated.

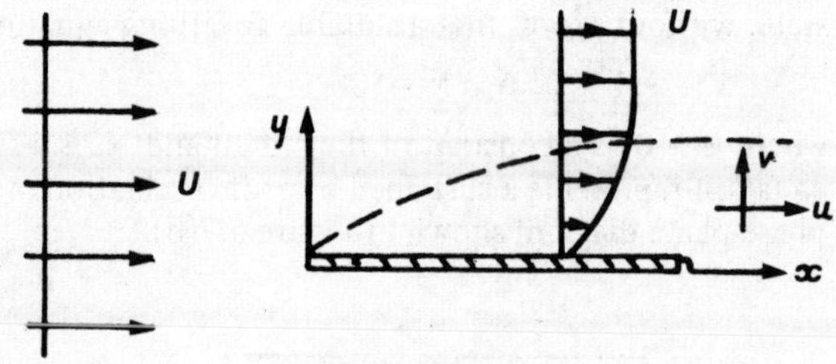

Figure 5.8

If the kinematic viscosity of the fluid is ν_o then the governing equations of continuity and momentum are

$$\frac{\partial u}{\partial x} + \frac{\partial v}{\partial y} = 0 \qquad (5.48a)$$

$$u\,\frac{\partial u}{\partial x} + v\,\frac{\partial v}{\partial y} = \nu_o \frac{\partial^2 u}{\partial y^2} \qquad (5.48b)$$

The boundary conditions are $u = v = 0$ at $y = 0$ and $u \to U$ as $y \to \infty$.

It is customary to introduce the dimensionless coordinate $\eta = y\sqrt{\dfrac{U}{\nu_o x}}$ and a stream function $\psi = \sqrt{\nu_o x\, U}\; f(\eta)$ with $u = \partial\psi/\partial y$ and $v = -\partial\psi/\partial x$. Then we may reduce equations (5.48) to a single ordinary differential equation

$$ff'' + 2f''' = 0 \qquad (5.49)$$

The boundary conditions are $f = 0$ and $f' = 0$ at $\eta = 0$ and $f' \to 1$ as $\eta \to \infty$.

We now replace (5.49) by a set of three first-order equations.

$$\frac{df_1}{d\eta} = f_2, \quad \frac{df_2}{d\eta} = f_3, \quad \frac{df_3}{d\eta} = -\tfrac{1}{2} f_1 f_3 \qquad (5.50)$$

where $f_2 \equiv f'$, and $f_3 \equiv f''$.

The boundary conditions are now written as $f_1 = f_2 = 0$ at $\eta = 0$ and $f_2 \to 1$ as $\eta \to \infty$.

Shooting methods aim to convert a boundary-value problem into an initial-value problem. In this instance, we search for a value of f_3 at $\eta = 0$ which will produce a solution for which $f_2 \to 1$ as $\eta \to \infty$.

One practical point needs to be made here. The condition $f_2 \to 1$ as $\eta \to \infty$ must be approximated by $f_2 = 1$ at $\eta = \eta^*$, where η^* is chosen sufficiently large so that the solution does not alter much for $\eta > \eta^*$.

The technique is to assume a value for f_3 at $\eta = 0$ and proceed with the initial-value problem (via a Runge-Kutta approach, for example) to obtain a solution and a value for f_2 at η^*. Then, a second value of f_3 at $\eta = 0$ is chosen and the procedure repeated. Hopefully, the required value of f_3 at $\eta = 0$ will lie between the two selected and any root-finding method (such as successive bisection) can be used to locate more precisely the required value.

Any second-order differential equation may be decomposed into two simultaneous first-order equations; for example $y'' - 6y' + 4y = 0$ can be written as $z = y'$ and $z' - 6z + 4y = 0$ i.e. as $y' = z$ and $z' = 6z - 4y$. An illustration of how to solve such equations using a Runge-Kutta method is given in Bajpai, Mustoe & Walker, Engineering Mathematics page 578.

Problems

1. Describe how a second order boundary-value problem (second order ordinary differential equation) may be solved over a given interval using the Euler method with trapezoidal correction.

   For the boundary-value problem $yy'' + y'^2 + 1 = 0$, $y(0) = 1$, $y(1) = 2$, solve for y consistent to two decimal places at $x = 0.5$ and $x = 1.0$ using the above method and a trial value $y'(0) = 1$. From the result estimate a new value for $y'(0)$.

   Note: The Euler-trapezoidal formulae applied to the *first order* differential euqation $y' = f(x,y)$ are

   $$y_{r+1}^{(0)} = y_r + hf(x_r, y_r),\quad y_{r+1}^{(n)} = y_r + \tfrac{1}{2}h[f(x_r, y_r) + f(x_{r+1}, y_{r+1}^{(n-1)})]$$ (C.E.I.)

2. The Blasius problem relating to boundary layer fluid flow past a flat plate can be reduced to the dimensionless form $f(\eta)f''(\eta) + 2f'''(\eta) = 0$ with boundary conditions $f(0) = f'(0) = 0$, $f'(\infty) = 1$. Express this as a set of first-order differential equations. Assuming $f''(0) = 0.4$, take one step $\Delta\eta = 1$ in the solution of the resulting initial-value problem, using Runge-Kutta or any equivalent method.

   Indicate briefly how the original problem could be solved using this approach. (C.E.I.)

3. Solve $y'' + xy' - 3y = 4.2x - 3$ with $y(0) = 1$, $y(1) = 2.9$. Use $h = 0.25$ and compare the values obtained with the analytical solution. Verify that this latter is $y = x^3 + 0.9x + 1$.

4. Solve the equation $y'' - yy' = e^x$ with $y(0) = 1$, $y(1) = -1$.

5. The lateral deflection of a pin-ended strut is governed by the equation

   $$\frac{d^2y}{dx^2} + \frac{P}{EI}\left(1 + \left[\frac{dy}{dx}\right]^2\right)^{3/2} y = 0,$$ where P is the axial load and E and I are constants.

   Given that $y(0) = y(L) = 0$ find the deflection at the mid-point, using suitable test data for P, E, I and L.

5.9 BOUNDARY-VALUE PROBLEMS: FINITE DIFFERENCE METHODS

This technique, of substituting finite differences for derivatives, is dealt with more fully in the chapter on partial differential equations. We shall indicate its working here as applied to a particular problem. In Figure 5.9 we depict the steady flow of a highly viscous fluid between two stationary, flat, parallel plates. There is a pressure gradient acting in the x - direction and this gives rise to a parabolic velocity distribution as shown. The flow is everywhere in the x - direction and the maximum velocity U occurs at $y = H$. The general velocity is denoted by u. It can be readily shown that

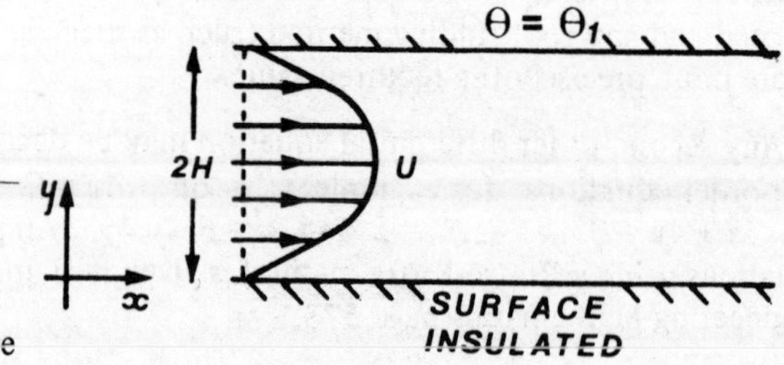

Figure 5.9

$$u = \frac{U}{H^2}\,(2Hy - y^2) \tag{5.51}$$

Now let the upper plate be maintained at a fixed temperature and let the lower plate be thermally insulated. It can be shown that the temperature distribution in the fluid $\theta(y)$ is given by

$$\frac{d^2\theta}{dy^2} = -\frac{\mu}{k}\left(\frac{du}{dy}\right)^2 = -\frac{\mu}{k}\,\frac{4U^2}{H^4}\,(H - y)^2 \tag{5.52}$$

where μ is the dynamic viscosity of the fluid and k is its thermal conductivity. The boundary conditions are

$$\theta = \theta_1 \text{ at } y = 2H, \text{ and } \frac{d\theta}{dy} = 0 \text{ at } y = 0$$

It is usual to introduce dimensionless variables $Y = \dfrac{y}{2H}$, $T = \dfrac{k(\theta - \theta_1)}{\mu U^2}$

so that equation (5.52) becomes

$$\frac{d^2T}{dY^2} = -16(1 - 2Y)^2 \tag{5.53a}$$

and the boundary conditions are

$$T = 0 \text{ at } Y = 1, \quad \frac{dT}{dY} = 0 \text{ at } Y = 0 \tag{5.53b}$$

Now we can, in fact, check the performance of a finite difference technique since we can obtain the analytical solution

$$T = \frac{1}{3}\,[9 - 8Y - (1 - 2Y)^4] \tag{5.54}$$

Note that $T(0) = 8/3$.

(You are justified in asking why we bother to develop a numerical solution to this problem when an analytical solution can be found. We are in a position to check the reliability of a numerical method on this problem with a known answer before using it on examples where analytical solutions are difficult or impracticable to obtain.)

From Taylor's series we are able to obtain finite difference approximations to derivatives, together with an order of magnitude for the truncation error. Suppose that the aim of the numerical solution is to obtain estimates of T at values of Y equally spaced a distance h apart. Let us suppose that $f(x)$ is to be specified at the equally-spaced points 2, 3, 4, ..., $(n-1)$ as shown in Figure 5.10.

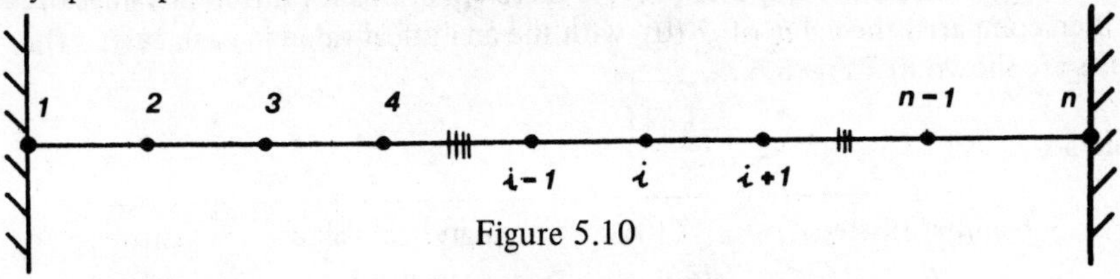

Figure 5.10

Now $f(x_{i+1}) = f(x_i) + hf'(x_i) + (h^2/2!)f''(x_i) + (h^3/3!)f'''(x_i) + \ldots$

$$f(x_{i-1}) = f(x_i) - hf'(x_i) + (h^2/2!)f''(x_i) - (h^3/3!)f'''(x_i) + \ldots$$

so that, on addition, we obtain

$$f(x_{i+1}) + f(x_{i-1}) = 2f(x_i) + h^2 f''(x_i) + 0(h^4)$$

Rearranging this equation, we obtain for $i = 2, 3, \ldots, (n-1)$

$$f''(x_i) = [f(x_{i+1}) - 2f(x_i) + f(x_{i-1})]/h^2 + 0(h^2) \tag{5.55a}$$

In fact, we can show that the error term $0(h^2)$ is actually

$$-\frac{h^2}{12} f^{(iv)}(x_i) \tag{5.55b}$$

Suppose we know that $f(x_n) = 0$. Then

$$f''(x_{n-1}) \simeq \frac{f(x_n) - 2f(x_{n-1}) + f(x_{n-2})}{h^2} = \frac{-2f(x_{n-1}) + f(x_{n-2})}{h^2}$$

If, however, we consider the boundary condition $f'(x_1) = A$ then we could use the approximation $f'(x_1) = \dfrac{f(x_2) - f(x_1)}{h}$; however, as you can show from Taylor's series, the truncation error is $0(h)$. It would therefore be unwise to use the approximate relation $f(x_1) = f(x_2) - Ah$ in (5.55a).

(If $A = 0$ we can invent a fictitious point x_0 and then write $f(x_2) = f(x_0)$ so that $f''(x_1) \simeq [2f(x_2) - 2f(x_1)]/h^2$).

Instead, we use the approximation $f'(x_1) = [-3f(x_1) + 4f(x_2) - f(x_3)]/2h$ which has truncation error $0(h^2)$.

Hence, since $f'(x_1) = 0$, we obtain the condition $-3f(x_1) + 4f(x_2) - f(x_3) = 0$. We have therefore n equations (5.55a) for the n unknowns $f(x_1)$, ..., $f(x_n)$.

For the problem under consideration we may write these equations in matrix form as follows.

$$\begin{bmatrix} -3 & 4 & -1 & & & & \bigcirc \\ 1 & -2 & 1 & & & & \\ & 1 & -2 & 1 & & & \\ & & & \ddots & & & \\ & \bigcirc & & & 1 & -2 & 1 \\ & & & & & 0 & 1 \end{bmatrix} \begin{bmatrix} T(Y_1) \\ T(Y_2) \\ T(Y_3) \\ \vdots \\ T(Y_{n-1}) \\ T(Y_n) \end{bmatrix} = h^2 \begin{bmatrix} 0 \\ \phi_2 \\ \phi_3 \\ \vdots \\ \phi_{n-1} \\ 0 \end{bmatrix} \tag{5.56}$$

where $\phi_j = -16(1-2Y_j)^2$, $j = 2, 3, \ldots, (n-1)$

We programmed the computer to solve these equations for different values of n. We then compared the value of $T(0)$ with the analytical value in each case. The results are shown in Table 5.6

Table 5.6

Number of steps n	$T(0)$ (4 d.p.)	Analytical value (4 d.p.)	| Error | (4 d.p.)
8	2.3125	2.6667	0.3542
16	2.5703	2.6667	0.0964
32	2.6416	2.6667	0.0351
64	2.6603	2.6667	0.0064
128	2.6651	2.6667	0.0016

Now, we have not got a tridiagonal matrix of coefficients, which we should have had if we had used the simpler approximation for $f'(x_1) = 0$. However, the results would have been less accurate.

In Table 5.7 we show some values of T for the case $n = 128$, and compare with the analytical values of T computed from (5.54); both are recorded to 4 d.p.

Table 5.7

Y	0	0.25	0.5	0.75	1.0
$T(Y)$ computed	2.6667	2.3203	1.6862	1.0004	0.0412
$T(Y)$ analytical	2.6667	2.3125	1.6667	0.9792	0

Problems

1. Show that the boundary-value problem $\dfrac{d^2y}{dx^2} = \dfrac{0.06912\,x(x-1)}{I(x)}$, $y(0) = y(1) = 0$, can be approximated at the points $x_1, x_2, \ldots, x_r = x_0 + rh, \ldots$ by the algebraic equations

$$y_{r-1} - 2y_r + y_{r+1} = \frac{0.06912\,h^2 x_r(x_r-1)}{I(x_r)}, \text{ for } r = 1, 2, \ldots \left[\frac{1}{h} - 1\right]$$

Write down these equations given

x	1/6	1/3	1/2	2/3	5/6
$I(x)$	0.01281	0.01327	0.01342	0.01327	0.01281

and taking $h = 1/6$.

Given the initial approximations $(y_1, y_2, y_3, y_4, y_5) = (0.07, 0.12, 0.14, 0.12, 0.07)$, obtain more accurate results by applying the Gauss-Seidel iteration. (The actual number of equations to be solved may be reduced by assuming symmetry of y about $x = ½$.) (C.E.I.)

2. Given $y'' + (x^2 - 1)y = 0$, $y(0) = 0$, $y'(0) = 1$, find $y(0.1)$, $y(0.2)$ from the Taylor series for y, and $y(0.3)$, from the central difference formula $\delta^2 y = (1 + \frac{1}{12}\delta^2 + ...)(h^2 y'')$, retaining four decimal places. (L.U.)

3. Solve the equation $y'' + y = 0$ with $y(0) = 0$, $y(\pi/2) = 1$. Use a step size $h = 0.25$ after normalising the interval.

4. Solve the boundary-value problem $y'' + xy' - xy = 2x$, $y(0) = 1$, $y(1) = 0$ using $h = 0.2$.

5. Solve the boundary-value problem $y'' + xy' + y = 2x$, $y(0) = 1$, $y(1) = 0$.

6. Solve the boundary-value problem $y'' + 2y' + y = x$, $y(0) = y(1) = 0$.

Chapter Six

Special Functions

6.1 INTRODUCTION: A PROBLEM IN HEAT TRANSFER

When a mathematical model is produced there is no guarantee that a solution can be found in the form of a compact formula involving standard functions of the calculus. If for example we cannot find the solution of an ordinary differential equation in such a form, we may either resort to numerical or graphical techniques or express the solution as a power series in the independent variable. The reason that we study functions named *sine*, *cosine*, *logarithm*, *exponential*, etc. is that the behaviours they describe occur very frequently in model solutions. Certain ordinary differential equations occur often in various branches of engineering; their importance has led to the creation and study of **special functions** by means of which we may express their solutions. In this chapter we shall briefly examine some of these functions, after discussing power series solutions of ordinary differential equations.

Cooling Fins

In order to speed up the rate of convective heat transfer from a surface, such as a domestic radiator, thin metal strips or fins are attached to the primary surface and conduct heat away from the main construction unit, thereby increasing the area available for cooling by the surrounding fluid. A first analysis of the effect of a fin can often be made by assuming a one-dimensional flow of heat. In Figure 6.1 we depict a one-dimensional fin with variable cross-section. We shall assume that steady-state conditions prevail.

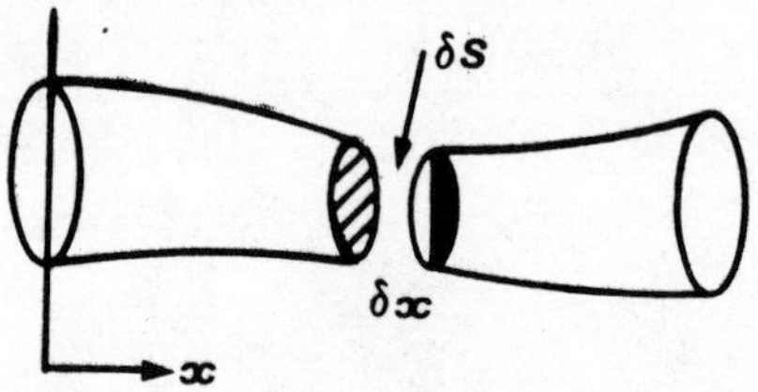

Figure 6.1

$A(x)$ is the cross-section area at a general point and δS represents the surface area of the fin between the planes x and $x+\delta x$. Note that δS is a function of x. Consider the heat flow in this section of the fin.

Let the rate of flow of heat, dQ/dt, be denoted by $R(x)$. Then the equation of heat balance for the section of fin is

$$R(x) = R(x+\delta x) + dQ_c/dt \tag{6.1}$$

where Q_c is the heat convected through the surface element δS. We may expand $R(x+\delta x)$ as a Taylor series to obtain

$$R(x+\delta x) = R(x) + \delta x\, R'(x) + \frac{(\delta x)^2}{2!} R''(x) + \ldots.. \qquad (6.2)$$

Let the temperature of the surrounding fluid be zero and let $\theta(x)$ be the temperature at any point inside the fin. Then

$$\frac{dQ_c}{dt} = h\,\delta S\,\theta \qquad (6.3)$$

where h is the coefficient of heat transfer. Further,

$$R(x) = -kA(x)[d\theta/dx] \qquad (6.4)$$

where k is the thermal conductivity of the fin material.

Substituting (6.2), (6.3) and (6.4) into (6.1) we find that as $\delta x \to 0$ the heat balance equation becomes

$$\frac{1}{A(x)}\frac{d}{dx}\left(-kA(x)\frac{d\theta}{dx}\right) + \frac{h}{A(x)}\frac{dS}{dx}\theta = 0 \qquad (6.5)$$

or, expanding the first term and dividing by $(-k)$,

$$\frac{d^2\theta}{dx^2} + \frac{1}{A(x)}\frac{dA}{dx}\cdot\frac{d\theta}{dx} - \frac{h}{k}\frac{1}{A(x)}\frac{dS}{dx}\theta = 0 \qquad (6.6)$$

Suppose we consider a truncated triangular fin as shown in Figure 6.2

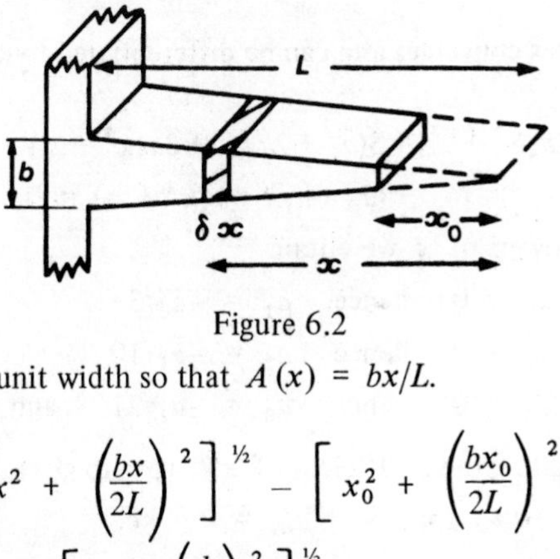

Figure 6.2

Let the fin be of unit width so that $A(x) = bx/L$.
Furthermore,

$$S(x) = 2\left\{\left[x^2 + \left(\frac{bx}{2L}\right)^2\right]^{1/2} - \left[x_0^2 + \left(\frac{bx_0}{2L}\right)^2\right]^{1/2}\right\}$$

$$= 2(x - x_0)\left[1 + \left(\frac{b}{2L}\right)^2\right]^{1/2}$$

Substituting these expressions for $A(x)$ and $S(x)$ into (6.6) we obtain

$$\frac{d^2\theta}{dx^2} + \frac{1}{x}\frac{d\theta}{dx} - \frac{r^2}{x}\theta = 0 \qquad (6.7)$$

where $r^2 = \dfrac{2hL}{kb}\left[1 + \left(\dfrac{b}{2L}\right)^2\right]^{1/2}$. Further, we impose the boundary conditions that the temperature at the base of the fin is given i.e.

$$\theta = \theta_0 \quad \text{at} \quad x = L \qquad (6.8a)$$

and convection occurs at the tip of the fin, i.e.

$$-k\frac{d\theta}{dx} = h_0\,\theta \quad \text{at} \quad x = x_0 \qquad (6.8b)$$

6.2 SERIES SOLUTION OF ORDINARY DIFFERENTIAL EQUATIONS

We shall attempt to solve equation (6.7) by a series method.

It is a particular case of the general form $\frac{dy}{dx} + P(x)\frac{dy}{dx} + Q(x)y = 0$. We have to be careful in our choice of expansion. The straightforward substitution $y = \sum_{m=0}^{\infty} a_m x^m$ may not always give the most general solution and the method of **variation of parameters** may have to be called in.

Example

$$2x\frac{d^2y}{dx^2} + 3\frac{dy}{dx} + y = 0 \tag{6.9}$$

If we cast this in the form $\frac{d^2y}{dx^2} + \frac{3}{2x}\frac{dy}{dx} + \frac{1}{2x}y = 0$ we see that we cannot expand $P(x)$ and $Q(x)$ as power series involving only positive integral powers of x.

However let us nevertheless try the substitution $y = \sum_{m=0}^{\infty} a_m x^m$.

Assuming that this series converges and can be differentiated twice term by term we obtain from (6.9)

$$2x(2a_2 + 6a_3x + 12a_4x^2 + \ldots) + 3(a_1 + 2a_2x + 3a_3x^2 + \ldots) + (a_0 + a_1x + a_2x^2 + \ldots) = 0$$

Equating coefficients of powers of x we obtain

x^0: $3a_1 + a_0 = 0$ hence $a_1 = -a_0/3$

x^1: $4a_2 + 6a_2 + a_1 = 0$ hence $a_2 = -a_1/10$

x^2: $12a_3 + 9a_3 + a_2 = 0$ hence $a_3 = -a_2/21$ and so on.

You can show that equating the coefficient of x^m to zero gives

$2m(m+1)a_{m+1} + 3(m+1)a_{m+1} + a_m = 0$ i.e.

$$\frac{a_{m+1}}{a_m} = \frac{-1}{(2m+3)(m+1)} \quad \text{for all} \quad m \geqslant 0.$$

Hence we may deduce that

$$\frac{a_{m+1}}{a_0} = \left[\frac{-1}{(2m+3)(m+1)}\right]\left[\frac{-1}{(2m+1).m}\right]\ldots\left[\frac{-1}{3.1}\right]$$

$$= (-1)^{m+1}\frac{2^{m+1}}{(2m+3)(2m+2)(2m+1)(2m)\ldots 3.2} = (-1)^{m+1}.2^{m+1}/(2m+3)!$$

It is more customary to work in terms of a_m and therefore $a_m = a_0(-1)^m 2^m/(2m+1)!$ We have a solution of (6.9) in the form

$$y = a_0 \sum_{m=0}^{\infty} (-1)^m 2^m x^m/(2m+1)! \tag{6.10}$$

Now (6.10) contains only one aribtrary constant, a_0. To obtain the general solution of (6.9) we shall have to resort to variation of parameters. First we try and cast (6.10) in terms of elementary functions: *this procedure may not often be possible.*

$$y = a_0 \left[1 - \frac{2x}{3!} + \frac{(2x)^2}{5!} - \frac{(2x)^3}{7!} + \ldots\ldots \right]$$

$$= \frac{a_0}{\sqrt{(2x)}} \left[(2x)^{1/2} - \frac{(2x)^{3/2}}{3!} + \frac{(2x)^{5/2}}{5!} - \frac{(2x)^{7/2}}{7!} + \ldots\ldots \right]$$

$$= \frac{a_0}{\sqrt{(2x)}} \sin \sqrt{(2x)} \qquad (6.11)$$

(We had to use some insight to obtain the result.)

We now put $y = \dfrac{v(x)}{\sqrt{(2x)}} \sin \sqrt{(2x)}$ and differentiate twice to find expressions for $\mathrm{d}y/\mathrm{d}x$ and $\mathrm{d}^2y/\mathrm{d}x^2$. Substituting these three expressions into (6.9) and simplifying, we obtain

$$2x \sin \sqrt{(2x)} \frac{\mathrm{d}^2 v}{\mathrm{d}x^2} + [2\sqrt{2}\sqrt{x} \cos \sqrt{(2x)} + \sin \sqrt{(2x)}] \frac{\mathrm{d}v}{\mathrm{d}x} = 0 \quad (6.12)$$

This equation can be solved by the integrating factor technique, since it is essentially of first order in $\mathrm{d}v/\mathrm{d}x$. The final solution is

$$y = A. \frac{\sin \sqrt{(2x)}}{\sqrt{(2x)}} + B \frac{\cos \sqrt{(2x)}}{\sqrt{(2x)}} \qquad (6.13)$$

where A and B are constants. We have therefore found a second solution

$$\cos \sqrt{(2x)} \,/\, (\sqrt{2x}) = \frac{x^{-1/2}}{\sqrt{2}} \left[1 - \frac{(2x)}{2!} + \frac{(2x)^2}{4!} + \ldots\ldots \right]$$

$$= \frac{1}{\sqrt{2}} \left[x^{-1/2} - x^{1/2} + \frac{x^{3/2}}{3!} - \ldots\ldots \right]$$

and we see that the second solution involves fractional powers of x. The following method of solution allows this possibility.

The Frobenius Method

We seek a solution in the form

$$y = x^c \sum_{n=0}^{\infty} a_n x^n = \sum_{n=0}^{\infty} a_n x^{n+c} \qquad (6.14)$$

where c is not necessarily a positive integer. It is implicit that $a_0 \neq 0$. This allows us the flexibility we require.

Singular points of a differential equation

In the general form

$$\frac{\mathrm{d}^2 y}{\mathrm{d}x^2} + P(x) \frac{\mathrm{d}y}{\mathrm{d}x} + Q(x) \,.\, y = 0 \qquad (6.15)$$

if both $P(x)$ and $Q(x)$ can be expanded as a Taylor series in the neighbourhood of

$x = x_0$ then (6.15) is said to have an **ordinary point** at $x = x_0$. If the expansions are valid over an interval $(x_0 - \alpha,\ x_0 + \alpha)$ then it can be shown that the series solution, obtained as above, is valid for that interval. If either $P(x)$ or $Q(x)$ cannot be expanded as a Taylor series about a point $x = x_1$ then x_1 is said to be a **singular point** for the differential equation.

If x_1 is a singular point and if both $(x - x_1)P(x)$ *and* $(x - x_1)^2 Q(x)$ can be expanded as Taylor series about $x = x_1$ then x_1 is said to be a **regular singular point.** In such a case (6.15) can be written as

$$(x - x_1)^2 \frac{d^2y}{dx^2} + (x - x_1)P_1(x)\frac{dy}{dx} + Q_1(x)y = 0 \tag{6.16}$$

where $P_1(x)$ and $Q_1(x)$ can be expanded as Taylor series about $x = x_1$.

Many practical problems have governing differential equations of this form. Two that we shall study are

Bessel's equation: $$x^2\frac{d^2y}{dx^2} + x\frac{dy}{dx} + (x^2 - \nu^2)y = 0 \tag{6.17}$$

Legendre's equation: $$(1 - x^2)\frac{d^2y}{dx^2} - 2x\frac{dy}{dx} + \nu(\nu + 1)y = 0 \tag{6.18}$$

Bessel's equation has a regular singular point at $x = 0$, while Legendre's equation has one at $x = 1$ and a second at $x = -1$. The latter case is not so obvious, but we can rewrite (6.18) as

$$(1 - x)^2\frac{d^2y}{dx^2} - \frac{2x(1 - x)}{(1 + x)}\frac{dy}{dx} + \frac{\nu(\nu + 1)(1 - x)}{(1 + x)}y = 0$$

so that $$P_1(x) = 2x/(1 + x) \text{ and } Q_1(x) = \nu(\nu + 1)(1 - x)/(1 + x).$$

Both $P_1(x)$ and $Q_1(x)$ can be expanded as Taylor series about $x = 1$ and therefore $x = 1$ is a regular singular point. Similarly, we may show that $x = -1$ is a regular singular point.

In effect, the series expansion of y about a regular singular point is valid for those values of x for which the Taylor series expansions of $P_1(x)$ and $Q_1(x)$ are valid.

Example

Find the singular points in the finite plane of

(a) $$x^2\frac{d^2y}{dx^2} - 4\frac{dy}{dx} + 2y = 0$$

(b) $$x(x - 2)\frac{d^2y}{dx^2} + (x - 2)\frac{dy}{dx} + 6xy = 0$$

(c) $$5\frac{d^2y}{dx^2} + 4x\frac{dy}{dx} + 3y = 0$$

We have stated that we are concerned with the finite plane since we do not wish to deal with points at infinity.

(a) The singular point is at $x = 0$, $P(x) = -4/x^2$ and $Q(x) = 2/x^2$ and therefore it is not regular.

(b) The singular points are at $x = 0$ and $x = 2$; both are regular.

(c) There are no singular points.

The indicial equation

We expect that the Frobenius method will yield two linearly independent solutions, both in the form of infinite series. Apart from the coefficients a_n, the unknown is c. As with the auxiliary equation employed in the method of complementary function where we may have two real roots, one real root or no real roots, so we must expect different possibilities for the values of c. To see how Frobenius' method works, we shall attempt to solve equation (6.7); first we multiply it by x and change the dependent variable to y to obtain

$$x\frac{d^2y}{dx^2} + \frac{dy}{dx} - r^2 y = 0 \tag{6.19}$$

We now seek a solution in the form

$$y = x^c \sum_{n=0}^{\infty} a_n x^n = \sum_{n=0}^{\infty} a_n x^{n+c} \tag{6.14}$$

Therefore

$$\left.\begin{aligned} \frac{dy}{dx} &= \sum_{n=0}^{\infty} a_n . (n+c)x^{n+c-1} \\ \frac{d^2y}{dx^2} &= \sum_{n=0}^{\infty} a_n(n+c)(n+c-1)x^{n+c-2} \end{aligned}\right\} \tag{6.20}$$

Substituting into (6.19) gives

$$\sum_{n=0}^{\infty} a_n (n+c)(n+c-1)x^{n+c-1} + \sum_{n=0}^{\infty} a_n (n+c)x^{n+c-1}$$

$$- r^2 \sum_{n=0}^{\infty} a_n x^{n+c} = 0$$

or

$$\sum_{n=0}^{\infty} a_n (n+c)^2 x^{n+c-1} - r^2 \sum_{n=0}^{\infty} a_n x^{n+c} = 0 \tag{6.21}$$

Now we can expand the series in (6.21). Since the results hold for all x, we can equate the coefficient of each power of x on the left-hand side to zero. The smallest power of x is x^{c-1} and the coefficient of this is

$$a_0 c^2 = 0 \tag{6.22}$$

(There is no contribution from the second summation.)

Equation (6.22) is called the **indicial equation**, and is always found by equating the coefficient of the lowest power of x to zero. Since $a_0 \neq 0$, we have the repeated root $c = 0$. This will lead us only to one solution and we have hit trouble. We shall now look at various possibilities for the indicial equation by selecting four arbitrary examples. Then we shall return to this problem.

Roots of indicial equation differ by a non-integer

Consider the equation

$$3x\frac{d^2y}{dx^2} + 2\frac{dy}{dx} + y = 0 \tag{6.23}$$

If we substitute (6.14) and (6.20) we obtain

$$\sum_{n=0}^{\infty} a_n(n+c)(3[n+c-1]+2)x^{n+c-1} + \sum_{n=0}^{\infty} a_n x^{n+c} = 0$$

Equating the coefficient of the lowest power of x, viz. x^{c-1}, to zero gives the indicial equation $a_0c(3c-1) = 0$. This has two roots $c = 0$ and $c = 1/3$ which differ by a non-integer.

We now equate the coefficient of x^{n+c-1}, $n > 1$ to zero to obtain

$$a_n(n+c)(3n+3c-1) + a_{n-1} = 0 \tag{6.24}$$

Now we take the two values of c in turn.

$c = 0$ The General equation (6.24) now becomes $a_n n(3n-1) + a_{n-1} = 0$. Therefore

$$a_n = \frac{-a_{n-1}}{n(3n-1)} = \frac{(-1)^2}{n(3n-1)}\frac{a_{n-2}}{(n-1)(3n-4)} = \ldots = \frac{(-1)^n a_0}{n!(3n-1)(3n-4)\ldots 5.2}$$

and we have one solution from (6.14)

$$y = a_0\left\{1 - \frac{x}{2.1!} + \frac{x^2}{2.5.2!} - \frac{x^3}{2.5.8.3!} + \ldots\ldots\right\} \tag{6.25}$$

$c = 1/3$ Equation (6.24) is now $a_n(n+1/3)(3n) + a_{n-1} = 0$ so that

$$a_n = \frac{-a_{n-1}}{n(3n+1)} = \frac{(-1)^2 a_{n-2}}{n(3n+1)(n-1)(3n-2)} = \frac{(-1)^n a_0}{n!(3n+1)(3n-2)\ldots 7.4}$$

and we have a second solution

$$y = a_0 x^{1/3}\left\{1 - \frac{x}{4.1!} + \frac{x^2}{4.7.2!} - \frac{x^3}{4.7.10.3!} + \ldots\ldots\right\} \tag{6.26}$$

The general solution is a linear combination of the two, thus

$$y = A\left\{1 - \frac{x}{2.1!} + \frac{x^2}{2.5.2!} - \frac{x^3}{2.5.8.3!} + \ldots\right\} + Bx^{1/3}\left\{1 - \frac{x}{4.1!} + \frac{x^2}{4.7.2!} - \frac{x^3}{4.7.10.3!} + \ldots\right\} \tag{6.27}$$

Roots of indicial equation equal

The general solution for equations of the last category may be written

$$\begin{aligned} y &= Au(x, c_1) + Bu(x, c_2) \\ &= (A+B)u(x, c_1) + B(c_2 - c_1)\left\{\frac{u(x, c_2) - u(x, c_1)}{c_2 - c_1}\right\} \\ &= \alpha u(x, c_1) + \beta\left\{\frac{u(x, c_2) - u(x, c_1)}{c_2 - c_1}\right\} \end{aligned}$$

Now take the limit as $c_2 \to c_1$ to obtain

$$y = \alpha u(x, c_1) + \beta \left.\frac{\partial u}{\partial c}\right|_{c=c_1} \tag{6.28}$$

Example 1

Consider the equation

$$x\frac{d^2y}{dx^2} + \frac{dy}{dx} - xy = 0 \tag{6.29}$$

Substituting for y and its derivatives we obtain

$$\sum_{n=0}^{\infty} a_n (n+c)(n+c-1+1)\, x^{n+c-1} - \sum_{n=0}^{\infty} a_n x^{n+c+1} = 0$$

The indicial equation is $a_0 c^2 = 0$ which has a repeated root $c = 0$.
The coefficient of x^c equated to zero produces $a_1(c+1)^2 = 0$ i.e. $a_1 = 0$.
The coefficient of x^{n+c-1} $(n > 1)$ equated to zero gives $a_n(n+c)^2 - a_{n-2} = 0$ i.e.
$a_n = \dfrac{a_{n-2}}{(n+c)^2}$.
Now $a_3 = a_5 = a_7 = ... = 0$ and $a_2 = a_0/(2+c)^2$, $a_4 = a_2/(4+c)^2$ $= a_0/\left[(2+c)^2 . (4+c)^2\right]$ etc. so that the only solution we can obtain is found by taking $c = 0$ in

$$y = a_0 x^c \left\{1 - \frac{x^2}{(2+c)^2} + \frac{x^4}{(2+c)^2(4+c)^2} - .....\right\} = a_0\, u(x, c) \tag{6.30}$$

The second solution is found from $y = \dfrac{\partial u}{\partial c}(x, c)$ i.e.

$$y = a_0 x^c \log x \left\{1 - \frac{x^2}{(2+c)^2} + \frac{x^4}{(2+c)^2(4+c)^2} - .....\right\}$$
$$+ a_0 x^c \left\{\frac{2x^2}{(2+c)^3} - \frac{2x^4}{(2+c)^3(4+c)^2} - \frac{2x^4}{(2+c)^2(4+c)^3} + ...\right\} \tag{6.31}$$

We now let $c \to 0$ in (6.30) and (6.31), and put $a_0 = 1$, to obtain

$$y_1 = 1 - \frac{x^2}{2^2} + \frac{x^4}{2^2 . 4^3} - ..... \tag{6.32a}$$

and
$$y_2 = \log x \left\{1 - \frac{x^2}{2^2} + \frac{x^4}{2^2 . 4^2} - .....\right\}$$
$$+ \frac{2x^2}{2^3} - \frac{2x^4}{2^3 . 4^2} - \frac{2x^4}{2^2 . 4^3} + ...$$

i.e.
$$y_2 = \log x \left\{1 - \frac{x^2}{2^2} + \frac{x^4}{2^2 . 4^2} - .....\right\} + \frac{x^2}{2^2} - \frac{x^4}{2^2 . 4^2}\left(1 - \frac{2}{4}\right) + ... \tag{6.32b}$$

The general solution of (6.29) is $y = Ay_1 + By_2$.

Example 2

We saw that the fin equation

$$x\frac{d^2y}{dx^2} + \frac{dy}{dx} - r^2y = 0 \tag{6.19}$$

had an indicial equation with repeated root $c = 0$.

Equating the coefficient of x^c to zero produces $a_1(1+c)^2 - r^2 a_0 = 0$ and the general recurrence relationship is $a_n(n+c)^2 - r^2 a_{n-1} = 0$.

Hence one solution is

$$y_1 = 1 + \frac{r^2x}{1^2} + \frac{r^4x^2}{1^2 . 2^2} + ...$$

The second solution is

$$y_2 = \log x\left\{1 + \frac{r^2x}{1^2} + \frac{r^4x^2}{1^2 . 2^2} + ...\right\} - 2\left\{\frac{r^2x}{1^3} + \frac{r^4x^2}{1^3 . 2^3}(2+1) + ...\right\}$$

Roots of indicial equation differing by an integer

Trouble arises in cases where the roots of the indicial equation differ by an integer. The trouble can be of two kinds.

(i) One of the coefficients a_n becomes infinite.
(ii) One of the coefficients a_n becomes indeterminate.

We shall merely outline the way each of these troubles can be overcome.

Example 1

Consider the equation

$$x^2\frac{d^2y}{dx^2} + x\frac{dy}{dx} + (x^2 - 9)y = 0 \tag{6.33}$$

The indicial equation is $a_0(c-3)(c+3) = 0$ of which the roots are $c_1 = -3$, $c_2 = 3$. The equation for the coefficient of x^{c+1} is $a_1(c-2)(c+4) = 0$ and hence $a_1 = 0$. The recurrence relation for the coefficients a_n, $n \geqslant 2$ is $(c+n-3)(c+n+3)a_n + a_{n-2} = 0$. Therefore $a_3 = a_5 = a_7 = ... = 0$.

Further $a_2 = -a_0/(c-1)(c+5)$, $a_4 = -a_2/(c+1)(c+7)$, $a_6 = -a_4/(c+3)(c+9)$ and so on. The trouble is that a_6, and all subsequent 'even' coefficients will have $(c+3)$ in their denominator. This renders the case $c = -3$ invalid. Note that it is the smaller root which causes trouble.

We can, of course, carry on with $c_2 = 3$ to obtain the first solution

$$y = a_0u(x, c_2) = a_0x^3\left\{1 - \frac{x^2}{2.8} + \frac{x^4}{2.4.8.10} - \frac{x^6}{2.4.6.8.10.12} + ...\right\} \tag{6.34}$$

We quote, without proof, that the way to find a second solution of (6.33) is to find $\frac{\partial}{\partial c}\{(c-c_1)u(x, c)\}$ and then evaluate the expression at $c = c_1 = -3$. Now

$$u(x,c) = x^c\left\{1 - \frac{x^2}{(c-1)(c+5)} + \frac{x^4}{(c-1)(c+1)(c+5)(c+7)} - \frac{x^6}{(c-1)(c+1)(c+3)(c+5)(c+7)(c+9)} + \cdots\right\}$$

and $$(c+3)u(x,c) = x^c\left\{1 - \frac{(c+3)x^2}{(c-1)(c+5)} + \frac{(c+3)x^4}{(c-1)(c+1)(c+5)(c+7)} - \frac{x^6}{(c-1)(c+1)(c+5)(c+7)(c+9)} + \ldots\ldots\right\}$$

Notice we have effectively removed $(c+3)$ from the denominators. Then

$$\frac{\partial}{\partial c}\{(c+3)u(x,c)\} = x^c \log x\left\{1 - \frac{(c+3)x^2}{(c-1)(c+5)} + \frac{(c+3)x^4}{(c-1)(c+1)(c+5)(c+7)} - \frac{x^6}{(c-1)(c+1)(c+5)(c+7)(c+9)} + \ldots\ldots\right\}$$

$$+ x^c\left[\frac{-(c+3)x^2}{(c-1)(c+5)}\left\{\frac{1}{(c+3)} - \frac{1}{(c+5)} - \frac{1}{(c-1)}\right\}\right.$$

$$+ \frac{(c+3)x^4}{(c-1)(c+1)(c+5)(c+7)}\left\{\frac{1}{(c+3)} - \frac{1}{(c-1)} - \frac{1}{(c+1)} - \frac{1}{(c+5)} - \frac{1}{(c+7)}\right\}$$

$$\left. - \frac{x^6}{(c-1)(c+1)(c+5)(c+7)(c+9)}\left\{\frac{-1}{(c-1)} - \frac{1}{(c+1)} - \frac{1}{(c+5)} - \frac{1}{(c+7)} - \frac{1}{(c+9)}\right\} + \ldots\right]$$

If we now put $c = -3$ we obtain

$$v(x, c_1) = x^{-3}\log x\left\{1 - \frac{x^6}{(-4)(-2)2.4.6} + \ldots\ldots\right\} + x^{-3}\left\{\frac{-x^2}{(-4).2} + \frac{x^4}{(-4)(-2).2.4} - \frac{x^6}{(-4).(-2)(2.4.6)}\left[-\frac{1}{(-4)} - \frac{1}{(-2)} - \frac{1}{2} - \frac{1}{4} - \frac{1}{6}\right] + \ldots\right\} \quad (6.35)$$

The general solution of (6.33) is a linear combination of expressions (6.34) and (6.35).

Example 2

Consider the equation

$$(1 - x^2)\frac{d^2y}{dx^2} - 4x\frac{dy}{dx} + 4y = 0 \quad (6.36)$$

The indicial equation is $a_0\, c(c-1) = 0$. The coefficient of x^{c-1} is

$$a_1(c+1)c = 0 \quad (6.37)$$

and the general recurrence relation for $n > 2$ reduces to

$$a_n = \frac{a_{n-2}(c+n+2)(c+n-3)}{(c+n)(c+n-1)} \quad (6.38)$$

The roots of the indicial equation are $c = 1$ or $c = 0$. However, if we take $c = 0$ then (6.37) does not give us a value for a_1; a_1 is indeterminate. Let us assume it is finite. Then consider $c = 0$ so that (6.38) becomes $a_n = a_{n-2}(n+2)(n-3)/n(n-1)$

Hence $a_2 = \dfrac{a_0\, 4.(-1)}{2.1}$, $a_4 = \dfrac{a_2 .6.1}{4.3} = \dfrac{a_0 .6.4.1.(-1)}{4.3.2.1}$,

$a_6 = \dfrac{a_4 .8.3}{6.5} = \dfrac{a_0 .8.6.4.3.1.(-1)}{6.5.4.\,3.2.1}$, and $a_3 = a_1 .0$, making $a_5 = a_7 = \ldots\ldots = 0$.

We use a_1 and a_0 as the two arbitrary constants. The general solution is

$$y = a_0\left\{1 - \frac{2x^2}{1} - \frac{3x^4}{3} - \frac{4x^6}{5} - \frac{5x^8}{7} - \cdots\cdots\right\} + a_1 x \qquad (6.39)$$

We could replace a_0 and a_1 by A and B respectively, to be consistent.

Note that the second part of the solution is a finite series and that the power of x which appears is the value of c not yet considered, viz. 1.

To justify the statement that (6.39) is the general solution of (6.36) we could now consider the case $c = 1$. In fact, we obtain merely $y = a_1 x$ which is covered by (6.39).

Convergence of Power Series

We have so far tacitly assumed that the series we derive will converge. Strictly speaking we should test them by, for example, the ratio test to decide over what range of values of x they will be valid expansions. More importantly, we should like to know over what range of x the series are useful for practical computation.

Expansion about other points

So far we have considered examples where $x = 0$ was a regular singular point. If, for example, $x = 1$ is a regular singular point then we could attempt a Frobenius solution about $x = 1$. The easiest way perhaps to accomplish this is to change the independent variable by the transformation $x = 1 - z$.

Furthermore, there may be cases where the Frobenius method fails and then it is sometimes possible to expand effectively about $x = \infty$ by transforming the variables via $x = 1/z$ and hoping that $z = 0$ is a regular singular point of the transformed differential equation. In such a case we would end up with an expansion in descending powers of x.

Problems

1. Use the method of Frobenius to solve the following differential equations.

   (a) $4xy'' + (1 - 4x)y' - 4y = 0$

   (b) $x^2y'' - \frac{1}{2}xy' + \frac{1}{2}(1 - x^2)y = 0$

   (c) $x(1 - x)y'' - 3y' + 2y = 0$

   (d) $4xy'' + (4x + 3)y' + y = 0$ (L.U.)

   (e) $2x^2y'' + x(2x - 1)y' + y = 0$ (L.U.)

   (f) $xy'' + 2y' + xy = 0$ (L.U.)

   (g) $3x(1 - 3x)y'' - 4y' + 4y = 0$ (L.U.)

   (h) $xy'' + (1 + x)y' + 2y = 0$ (L.U.)

2. Find the general series solution of the differential equation $4xy'' + 6y' + y = 0$ and show that it can be expressed in the form $(A\cos\sqrt{x} + B\sin\sqrt{x})/\sqrt{x}$. (C.E.I.)

3. Use the method of Frobenius to obtain the general solution of $2xy'' + y' + y = 0$ up to the term in x^3, and deduce the solution which satisfies the initial conditions $y = -y' = 1$ at $x = 0$. (C.E.I.)

4. What is meant by a regular singular point of a differential equation? Obtain the solution of the differential equation $xy'' - (3+x)y' + 2y = 0$ in the form of a series. Give the general term of the infinite series. (C.E.I.)

5. Show that the differential equation $xy'' + (1+x)y' + 2y = 0$ has a regular singular point at $x = 0$. Hence show that

$$y = 1 - 2x + \frac{3}{2}x^2 - \frac{2}{3}x^3 + \ldots\ldots \text{ is a solution.}$$ (C.E.I.)

6. (i) Classify the singular points in the finite plane for the equation $x^4(x^2+1)(x-1)^2y'' + 4x^3(x-1)y' + (x+1)y = 0.$

(ii) Show that the solution of the differential equation $xy'' + (1-x)y' - y = 0$ can be expressed in the form $y = Ae^x + B\left[e^x \log_e x - \sum_{n=0}^{\infty} \frac{x^n}{n!}\left(1 + \frac{1}{2} + \frac{1}{3} + \ldots\ldots + \frac{1}{n}\right)\right]$ where A and B are arbitrary constants. (C.E.I.)

7. Find a series solution $u(x)$ of the differential equation $xy'' + y' - y = 0$ which is finite for $x = 0$. A second, independent solution of this equation has the form

$y = u(x)\log x + \sum_{r=1}^{\infty} b_r . x^r$. Show that $b_1 + 2a_1 = 0$ and

$b_{r+1}(r+1)^2 - b_r + 2a_{r+1}(r+1) = 0 \quad (r = 1, 2, \ldots\ldots)$. By substituting $b_r = k_r a_r$ show that $k_{r+1} - kr = -2/(r+1)$ and obtain an expression for the second solution. (L.U.)

8. Show that the indicial equation of the differential equation $xy'' + (\lambda - x)y' + 3y = 0$, is $c(c+\lambda-1) = 0$ and show further that $(c+r)(c+r+\lambda-1)a_r = (c+r-4)a_{r-1}$.

Show that one solution is $1 - \frac{3}{\lambda}x + \frac{3}{\lambda(\lambda+1)}x^2 - \frac{1}{\lambda(\lambda+1)(\lambda+2)}x^3$, and obtain the first three terms of the other solution (a) when λ is not an integer, (b) when $\lambda = 2$. (L.U.)

9. By putting $y = u\exp(-\frac{1}{4}x^2)$, transform the differential equation $y'' + (n + \frac{1}{2} - \frac{1}{4}x^2)y = 0$, into $\frac{d^2u}{dx^2} - x\frac{du}{dx} + nu = 0$, and obtain two solutions of the second equation in the form of a series of ascending powers of x.

Show that if n is a positive integer, one of these series terminates. Hence obtain one solution of the equation $y'' + \left[\frac{9}{2} - \frac{x^2}{4}\right]y = 0$, such that $y = 1$ when $x = 0$ and $y \to 0$ as $x \to \infty$. (L.U.)

10. A tubular gas preheater works by drawing cool air through a cylindrical heated tube. For a particular tube the governing equation for the temperature of the air is

$\frac{d^2T}{dx^2} - 7500\frac{dT}{dx} - 3500\frac{T}{x^{1/2}} = 0$ where T is the difference between the temperature of the wall of the tube and the air at a distance x from the inlet. Find the series solution for T *as far as the term in* $x^{5/2}$. (It helps to substitute $x = z^2$.) The second derivative term is due to the thermal conductivity of the air. Can this be neglected to leave a reasonably good approximate model equation?

6.3 THE GAMMA FUNCTION

In this section we look at a function which is useful in describing the Bessel and Legendre functions to which the rest of this chapter is directed.

Basically speaking, the **gamma function** $\Gamma(x)$ extends the idea of factorial from the range of non-negative integers.

For positive values of x we may define

$$\Gamma(x) = \int_0^\infty t^{x-1} e^{-t}\, dt \qquad (6.40)$$

and extend the definition by the additional rule, [which can be shown to be true for $x > 0$ from (6.40)].

$$\Gamma(x+1) = x\,\Gamma(x) \qquad (6.41)$$

We emphasise that $\Gamma(x)$ is not defined for $x = 0, -1, -2, -3, \ldots..$

Since we can show from (6.40) that

$$\Gamma(1) = 1 \qquad (6.42)$$

it follows by induction that

$$\Gamma(n+1) = n! \qquad (6.43)$$

if n is a positive integer, so that we see the relationship with the factorial function.

Example

Tables of $\Gamma(x)$ are usually given for the range $1 < x < 2$ and other values can be found from applying (6.41) enough times.

For example, if we know that to 4 d.p., $\Gamma(1.27) = 0.9126$ then

$$\Gamma(2.27) = 1.27\,\Gamma(1.27) = 1.1590$$

and $$\Gamma(-0.73) = \frac{\Gamma(-0.73+1)}{-0.73} = -\frac{\Gamma(0.27)}{0.73} = -\frac{\Gamma(0.27+1)}{(0.73)(0.27)} = \frac{-\Gamma(1.27)}{(0.73)(0.27)} = 4.6301$$

(4 d.p.)

Further properties

We can show directly that

$$\Gamma(\tfrac{1}{2}) = \sqrt{\pi} \qquad (6.44)$$

Also it can be proved that

$$\Gamma(n+\tfrac{1}{2}) = \frac{1.3.5.\ \ldots..\ (2n-1)\sqrt{\pi}}{2^n} \quad \text{for } n = 1,\ 2,\ 3,\ \ldots.. \qquad (6.45)$$

and that

$$\Gamma(x)\,\Gamma(1-x) = \pi \operatorname{cosec} \pi x \quad \text{for } x \neq 0,\ \pm 1,\ \pm 2,\ \ldots.. \qquad (6.46)$$

In Figure 6.3 we show the graph of $\Gamma(x)$.(diagram not to scale)

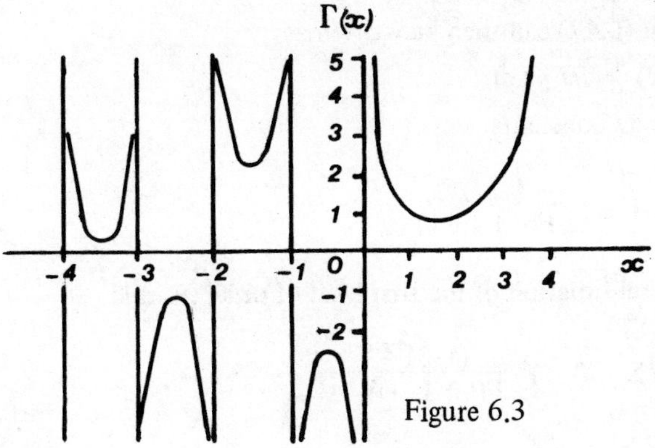

Figure 6.3

Problems

1. Show that $\int_0^{\pi/2} \sin^n x \, dx = \frac{1}{2}\sqrt{\pi}\, \Gamma\left[\frac{n+1}{2}\right] \Big/ \Gamma\left[\frac{n}{2}+1\right]$

2. Evaluate in terms of $\Gamma(1.6)$ the following: $\Gamma(0.6)$, $\Gamma(2.6)$, $\Gamma(12.6)$, $\Gamma(-18.6)$.

3. Find the length of the lemniscate $r^2 = \cos 2\theta$.

4. Find the area bounded by the curve $y^2 = 1 - x^4$.

5. Find an expression for $\int_0^a \sqrt{a^n - x^n}\, dx$ and verify the answer directly in the case $a = 10$, $n = 2$.

6. By means of a suitable substitution, express $\int_0^{\frac{1}{2}\pi} \left[\frac{1}{\sin^2\theta} - \frac{1}{\sin\theta}\right]^{1/3} \cos\theta \, d\theta$ in terms of the Γ function.

6.4 BESSEL FUNCTIONS OF THE FIRST AND SECOND KIND

The subject of Bessel Functions can be approached from many directions. One major application lies in the solution of certain partial differential equations where circular symmetry exists.

Bessel's equation of order ν is

$$x^2 \frac{d^2y}{dx^2} + x\frac{dy}{dx} + (x^2 - \nu^2)y = 0 \qquad (6.47)$$

Note that this is a second-order differential equation; the parameter ν refers to the order of the function which is a solution of (6.47). The equation has a regular singular point at the origin and we may use the Frobenius method to effect a solution. We shall not go into details here, but merely present results. We first consider the case when ν is a non-integer.

The solution of (6.47) can then be written as

$$y = AJ_\nu(x) + BJ_{-\nu}(x) \tag{6.48}$$

where A and B are constants.

$$J_\nu(x) = \sum_{s=0}^{\infty} (-1)^s \frac{(x/2)^{\nu+2s}}{\Gamma(\nu+s+1)\,.\,s!} \tag{6.49a}$$

and is called a **Bessel function of the first kind of order** ν, and

$$J_{-\nu}(x) = \sum_{s=0}^{\infty} (-1)^s \frac{(x/2)^{2s-\nu}}{\Gamma(s+1-\nu)\,.\,s!} \tag{6.49b}$$

When $\nu = 0$, $J_{-0}(x) \equiv J_0(x)$ and we seem to have only one solution

$$J_0(x) = \sum_{s=0}^{\infty} (-1)^s \frac{(x/2)^{2s}}{(s!)^2} \tag{6.50}$$

which is the Bessel function of the first kind of order zero.

A second solution will emerge via the Frobenius method; it is given by

$$Y_0(x) = \frac{2}{\pi}\left[\left\{\log(\tfrac{1}{2}x) + \gamma\right\}J_0(x) - \sum_{s=1}^{\infty}\left\{\frac{(-1)^s (x/2)^{2s}}{(s!)^2}\left(1 + \frac{1}{2} + \ldots\ldots + \frac{1}{s}\right)\right\}\right] \tag{6.51}$$

(γ is the so-called **Euler constant**; $\gamma = \lim_{n\to\infty}\left(1 + \frac{1}{2} + \ldots\ldots + \frac{1}{n} - \log n\right) = 0.5772$ (4 d.p.))

This complicated expression is called the **Bessel function of the second kind of order zero.** The general solution of (6.47) is then

$$y = AJ_0(x) + BY_0(x) \tag{6.52}$$

Now the reason we have labelled these functions $J_\nu(x)$ and $Y_0(x)$ etc. is that the equations of which they are the solution crop up frequently in models of engineering situations. Consequently, just as we give the series $x - x^3/3! + x^5/5! - \ldots\ldots$ the name $\sin x$ we give the series (6.49), (6.50) and (6.51) simple notations. If we can find the main properties of these series for different values of x we shall be able to understand the qualitative nature of the solution of equations like (6.47) as well as calculating approximate values from the series expansions. Since tables of Bessel functions exist we merely need to be able to express solutions in terms of these just as we could do for $\sin x$ and $\cos x$. One case of (6.47) remains to be considered: when ν is a non-zero integer, n. The solution is

$$y = AJ_n(x) + BY_n(x) \tag{6.53}$$

where

$$J_n(x) = \sum_{s=0}^{\infty} \frac{(-1)^s (x/2)^{2s+n}}{(s+n)!\, s!} \tag{6.54}$$

and

$$Y_n(x) = \frac{2}{\pi}\left[\left\{\log(\tfrac{1}{2}x) + \gamma\right\}J_n(x) - \frac{1}{2}\sum_{s=0}^{n-1} \frac{(n-s-1)!}{s!}\left[\frac{x}{2}\right]^{2s-n}\right.$$

$$-\frac{1}{2}\sum_{s=0}^{\infty}\frac{(-1)^s(x/2)^{2s+n}}{s!(n+s)!}\left\{1+\frac{1}{2}+\frac{1}{3}+\ldots+\frac{1}{s}+1+\frac{1}{2}+\ldots+\frac{1}{n+s}\right\}\Bigg] \tag{6.55}$$

We note that the term with $s = 0$ in the last summation is $\frac{(x/2)^n}{n!}\left\{1+\frac{1}{2}+\ldots+\frac{1}{n}\right\}$
All the series converge for all x.

You must not be put off by these formulae. We have written them down simply to show the advantages of developing properties of the Bessel functions and merely dealing with expressions like $J_{1/2}(x)$, $Y_3(x)$ etc. Now we shall produce graphs of some Bessel functions of positive integer order and look at some main features.

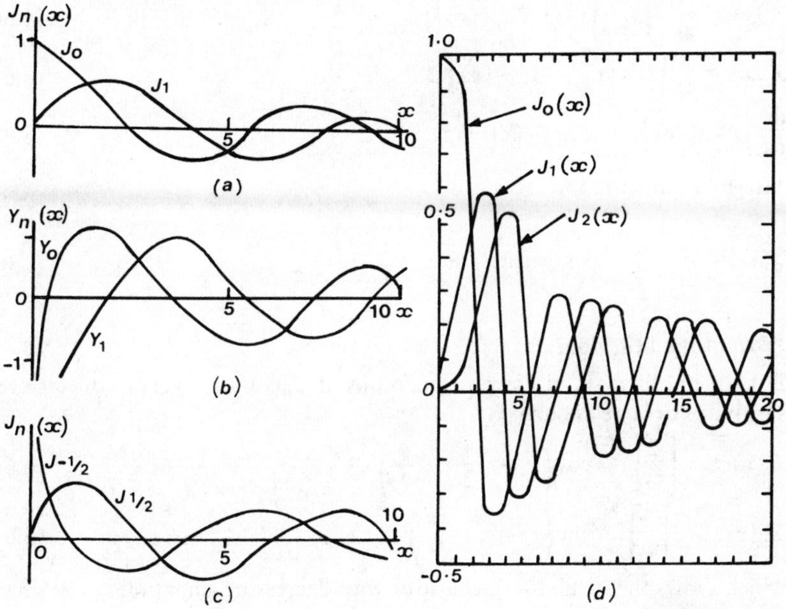

Figure 6.4

In Figure 6.4 we see the graphs of $J_0(x)$ and $Y_0(x)$; note that both are oscillatory with oscillations of decaying amplitude, which is a characteristic of all these Bessel functions. Whereas $J_0(x)$ is finite at $x = 0$, $Y_0(x)$ is not, because of the presence of a logarithm in its definition. From graph (d) we see that between any two consecutive values of x for which $J_n(x) = 0$ there will be one zero of $J_m(x)$ where $m \neq n$. This is called *interlacing of zeros.* The larger n the larger are the values of the corresponding zeros. Note that apart from $J_0(x)$ all $J_n(x)$ are zero at $x = 0$. Further, each $J_n(x)$ oscillates with smaller amplitude than its predecessor. In graph (b) we see a similar set-up for the $Y_n(x)$. In graph (c) we show $J_{1/2}(x)$ and $J_{-1/2}(x)$ which are involved in the solution of the wave equation in spherical polar coordinates. They also have the advantage that they can be expressed as

$$J_{1/2}(x) = \sqrt{2/(\pi x)} \,.\, \sin x \tag{6.56}$$

and

$$J_{-1/2}(x) = \sqrt{2/(\pi x)} \,.\, \cos x \tag{6.57}$$

Properties of Bessel Functions

We now quote some useful formulae which are helpful when equations have to be reduced to a form of Bessel's equation. We shall assume that n is an integer and ν is any real number.

(a) $J_{-n}(x) = (-1)^n J_n(x)$ (6.58)

(b) **Generating Function**

$$e^{\frac{1}{2}x(t-t^{-1})} = \sum_{n=-\infty}^{\infty} J_n(x) \, . \, t^n \tag{6.59}$$

(c) **Recurrence relations**

$$J_\nu(x) = \frac{x}{2\nu}\left\{J_{\nu-1}(x) + J_{\nu+1}(x)\right\} \tag{6.60}$$

$$J'_\nu(x) = \frac{1}{2}\left\{J_{\nu-1}(x) - J_{\nu+1}(x)\right\} \tag{6.61}$$

$$\frac{d}{dx}\left\{x^\nu J_\nu(x)\right\} = x^\nu J_{\nu-1}(x) \tag{6.62}$$

(d) **Integral representation**

$$J_0(x) = \frac{2}{\pi} \int_0^{\pi/2} \cos(x \sin\theta)\, d\theta \tag{6.63}$$

(e) **Asymptotic behaviour**

For large positive x, the series expansions of $J_\nu(x)$ and $Y_\nu(x)$ converge very slowly. For large enough x,

$$J_\nu(x) \sim \left[\frac{2}{\pi x}\right]^{\frac{1}{2}} \cos\left(x - \frac{\nu\pi}{2} - \frac{\pi}{4}\right) \tag{6.64a}$$

$$Y_\nu(x) \sim \left[\frac{2}{\pi x}\right]^{\frac{1}{2}} \sin\left(x - \frac{\nu\pi}{2} - \frac{\pi}{4}\right) \tag{6.64b}$$

which shows the oscillatory behaviour with decreasing amplitude. For small values of x, it may be shown that

$$J_\nu(x) \sim (x/2)^\nu/\Gamma(\nu+1) \tag{6.65}$$

The symbol $\sim$ means asymptotically equivalent to.

(f) **Zeros**

Using exact equivalents of (6.64) we may locate the zeros of $J_n(x)$ by iteration. In Table 6.1 we show the first zeros of $J_0(x)$, $J_1(x)$, $J_2(x)$ to 2 d.p.

Table 6.1

n	1st root	2nd root	3rd root
0	2.40	5.52	8.65
1	3.83	7.02	10.17
2	5.14	8.42	11.62

(g) **Orthogonal property**

If α_1 and α_2 are two unequal positive roots of $J_n(\alpha a) = 0$ then

$$\int_0^a x\,J_n(\alpha_1 x)\,J_n(\alpha_2 x)\,dx = 0 \qquad (6.66)$$

The factor x is said to be a **weighting factor.**

Under the condition that α_r are the roots of $J_n(\alpha a) = 0$ it follows that a function $f(x)$ can be expanded as

$$f(x) = \sum_{r=0}^{\infty} A_r J_n(\alpha_r x) \quad ; \quad \int_0^a x\,J_n^2(\alpha x)\,dx = \tfrac{1}{2} a^2 [J'(\alpha a)]^2 A_r \qquad (6.67)$$

where the A_r are coefficients. This property will prove useful in the solution of certain partial differential equations. We note in passing that there are other conditions under which orthogonal expansion is possible.

Problems

1. Defining the Bessel function $J_n(x)$ by means of the generating function

$\exp\left\{\tfrac{1}{2}x(t - t^{-1})\right\} = \sum_{n=-\infty}^{\infty} t^n J_n(x)$ show that, if n is an integer,

(a) $J_n(x) = (\tfrac{1}{2}x)^n \sum_{r=0}^{\infty} \dfrac{(-x^2/4)^r}{r!\,(n+r)!}$ (b) $J_{-n}(x) = (-1)^n J_n(x)$

(c) $J_{n-1}(x) + J_{n+1}(x) = \dfrac{2n}{x} J_n(x)$ (d) $J_{n-1}(x) - J_{n+1}(x) = 2J_n'(x)$

Deduce that $J_n(x)$ satisfies the differential equation $x^2y'' + xy' + (x^2 - n^2)y = 0$. (L.U.)

2. Show that a solution of Bessel's equation $x^2y'' + xy' + (x^2 - \nu^2)y = 0$ is

$J_\nu(x) = \sum_{r=0}^{\infty} \dfrac{(-1)^r}{r!\,\Gamma(\nu + r + 1)} \left\{\dfrac{x}{2}\right\}^{\nu+2r}$ where ν is a constant. Hence obtain

$J_{1/2}(x) = \left[\dfrac{2}{\pi x}\right]^{1/2} \sin x$. Prove the recurrence relation $\dfrac{\nu}{x} J_\nu(x) - J_\nu'(x) = J_{\nu+1}(x)$ and deduce that $J_0'(x) = -J_1(x)$. (C.E.I.)

3. Show that the following relationships hold.

(i) $J_{1/2}(x) = \sin x\,(2/\pi x)^{1/2}$ (ii) $J_{-1/2}(x) = \cos x\,(2/\pi x)^{1/2}$

(iii) $J_{3/2}(x) = \dfrac{\sin x}{x} - \cos x \quad (2/\pi x)^{1/2}$

4. In the formula $\exp\left\{\tfrac{1}{2}x(t - 1/t)\right\}$ put $t = e^{i\theta}$ and deduce that

$\cos(x \sin\theta) = J_0(x) + 2J_2(x)\cos 2\theta + 2J_4(x)\cos 4\theta + \ldots..$
$\sin(x \sin\theta) = 2[\,J_1(x)\sin\theta + J_3(x)\sin 3\theta + J_5(x)\sin 5\theta + \ldots..]$.

5. (i) Express $J_5(x)$ in terms of $J_0(x)$ and $J_1(x)$.

   (ii) Express $J_{-3/2}(x)$ in terms of $\sin x$ and $\cos x$.

   (iii) Find the first derivative with respect to x of the following: $x^2 J_3(2x)$, $xJ_0(x^2)$, $J_2(x)$.

6. Find the indefinite integrals of the following: $J_0(x)\cos x$, $J_0(x)\sin x$, $J_1(x)\cos x$, $J_1(x)\sin x$, $xJ_0(x)\cos x$, $xJ_0(x)\sin x$, $xJ_1(x)\cos x$, $xJ_1(x)\sin x$, $xJ_0(x)$, $x^2 J_0(x)$, $x^3 J_0(x)$, $J_1(x)$, $xJ_1(x)$.

7. (i) Prove that $\frac{d}{dx}\left[x^n J_n(ax)\right] = ax^n J_{n-1}(ax)$

   (ii) Show that $\left(\frac{1}{x}\frac{d}{dx}\right)^r\left[x^\nu J_\nu(x)\right] = x^{\nu-r} J_{\nu-r}(x)$

   $\left(\frac{1}{x}\frac{d}{dx}\right)^r\left[x^{-\nu} J_\nu(x)\right] = (-1)^r x^{-\nu-r} J_{\nu+r}(x)$ where r is a positive integer.

8. The current in a rectifying valve is $a \exp(b \cos \omega t)$ where a and b are constants. Prove that the mean value of the current over one period is $aJ_0(ib)$. (L.U.)

   (Hint: Use a similar approach to Problem 4 with $t = ie^{i\theta}$.)

9. Show that

   (i) $J_n(x)Y_{n+2}(x) - Y_n(x)J_{n+2}(x) = \dfrac{-4(n+1)}{\pi x^2}$

   (ii) $J_\nu(x)Y_\nu'(x) - Y_\nu(x)J_\nu'(x) = \dfrac{2}{\pi x}$

   (iii) $Y_{1/2}(x) = -\left[\dfrac{2}{\pi x}\right]^{1/2}\cos x$ (iv) $Y_{n+1/2}(x) = (-1)^{n+1} J_{-n-1/2}(x)$

10. Compute the values of $J_0(\tfrac{1}{2})$, $J_1(\tfrac{1}{2})$, $J_2(\tfrac{1}{2})$ directly from the series. Comment on the practical convergence of the series.

11. If the positive roots of the equation $J_0(x) = 0$ are $\alpha_1, \alpha_2, \ldots$ show that for $0 < x < 1$,

$$\sum_{r=1}^{\infty} \frac{J_0(\alpha_r x)}{\alpha_r J_1(\alpha_r)} = \tfrac{1}{2}.$$

6.5 MODIFIED BESSEL FUNCTIONS

Sometimes, when solving Laplace's equation in cylindrical coordinates we obtain the equation

$$x^2 \frac{d^2y}{dx^2} + x\frac{dy}{dx} - (x^2 + \nu^2)y = 0 \tag{6.68}$$

and this is called the modified Bessel equation and for ν non-integral it has a general solution in the form

$$y = AI_\nu(x) + BI_{-\nu}(x) \tag{6.69}$$

where A and B are constants and $I_\nu(x)$ is called the **modified Bessel function of the first kind**, of order ν. It can be shown that

$$I_\nu(x) = (\mathrm{i})^{-\nu} J_\nu(\mathrm{i}x) \tag{6.70}$$

For the case ν integral we have a **modified Bessel function of the second kind,** $K_n(x)$ which has a role relative to $I_n(x)$ which is similar to the role of $Y_n(x)$ relative to $J_n(x)$. The general solution of (6.68) with ν integral is

$$y = AI_\nu(x) + BK_\nu(x) \tag{6.71}$$

Problems

1. Show that $\exp\left[\tfrac{1}{2}x(t + 1/t)\right] = I_0(x) + I_1(x)t + I_2(x)t^2 + \ldots + I_{-1}(x)\left\{\frac{1}{t}\right\} + I_{-2}(x)\left\{\frac{1}{t}\right\}^2 + \ldots$
Hence deduce the properties.

(i) $I_{-n}(x) = I_n(x)$;

(ii) $2I_n'(x) = I_{n-1}(x) + I_{n+1}(x)$;

(iii) $2nI_n(x) = x[I_{n-1}(x) - I_{n+1}(x)]$

(iv) $xI_n'(x) = xI_{n-1}(x) - nI_n(x)$;

(v) $xI_n'(x) = nI_n(x) + xI_{n+1}(x)$;

(vi) $I_0'(x) = I_1(x)$;

(vii) $I_{\frac{1}{2}}(x) = \sinh x\,[2/(\pi x)]^{\frac{1}{2}}$;

(viii) $I_{-\frac{1}{2}}(x) = \cosh x\,[2/(\pi x)]^{\frac{1}{2}}$;

(ix) $I_{-\frac{1}{2}}(x) - I_{\frac{1}{2}}(x) = \frac{2}{\pi} K_{\frac{1}{2}}(x)$

(x) $K_{\frac{1}{2}}(x) = \left[\frac{\pi}{2x}\right]^{\frac{1}{2}} e^{-x}$

6.6 TRANSFORMATIONS OF BESSEL'S EQUATION

Many problems can be modelled by a differential equation which can be transformed to Bessel's equation.

We quote the result that the differential equation

$$x^2\frac{\mathrm{d}^2y}{\mathrm{d}x^2} + (1 - 2\alpha)x\frac{\mathrm{d}y}{\mathrm{d}x} + [\beta^2\gamma^2x^{2\gamma} + (\alpha^2 - \nu^2\gamma^2)]\,y = 0 \tag{6.72}$$

has general solution

$$y = x^\alpha[AJ_\nu(\beta x^\gamma) + BJ_{-\nu}(\beta x^\gamma)],\ \nu \text{ a non-integer} \tag{6.73a}$$

or

$$y = x^\alpha[AJ_n(\beta x^\gamma) + BY_n(\beta x^\gamma)],\ \nu = n, \text{ an integer} \tag{6.73b}$$

Example 1

Consider the thin strut shown in Figure 6.5. We wish to find what length it can be without buckling under its own weight.

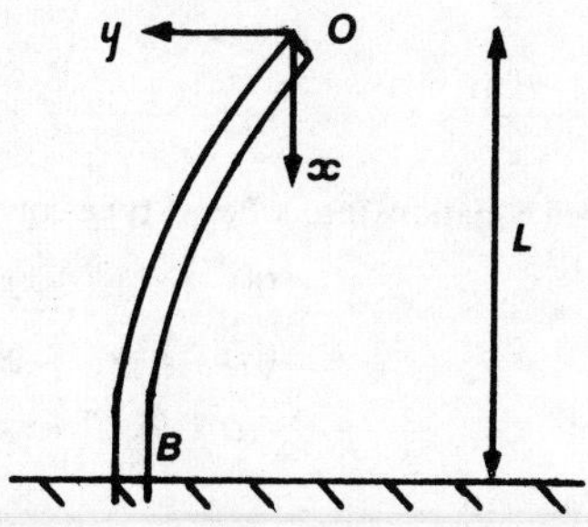

Figure 6.5

It can be shown that the governing equation is

$$\frac{d^2p}{dx^2} + k^2xp = 0 \tag{6.74}$$

where $p = dy/dx$ and $k = w/EI$, w being the weight per unit length. The boundary conditions are

$$p'(0) = 0, \; p'(L) = 0 \tag{6.75}$$

We can either use the results of (6.72) directly or carry out three transformations of variables. We shall take the easy way out. Multiplying (6.74) by x^2 we obtain $x^2 \dfrac{d^2p}{dx^2} + k^2x^3p = 0$ which compares with (6.72) if $1 - 2\alpha = 0$, $\beta^2\gamma^2x^{2\gamma} = k^2x^3$ and $\alpha^2 - \nu^2\gamma^2 = 0$. Therefore $\alpha = \dfrac{1}{2}$, $\gamma = \dfrac{3}{2}$, $\beta = 2k/3$, $\nu = \dfrac{1}{3}$. Therefore, from (6.73a) we have as general solution

$$p = x^{1/2}[AJ_{1/3}(\tfrac{2}{3}kx^{3/2}) + BJ_{-1/3}(\tfrac{2}{3}kx^{3/2})] \tag{6.76}$$

It can be shown after much algebra that the condition $p' = 0$ at $x = 0$ implies that $A = 0$ and the condition $p'(L) = 0$ means that if the column has buckled (in the sense that its vertical equilibrium position is unstable) then $B \neq 0$ and hence $J_{-1/3}(\tfrac{2}{3}kL^{3/2}) = 0$. From tables we find that the smallest positive root of this last equation is approximately $\tfrac{2}{3}kL^{3/2} = 1.866$ and hence $L \simeq 1.986k^{-2/3} = 1.986\left[\dfrac{EI}{w}\right]^{1/3}$ For greater lengths, the column will buckle under its own weight.

Example 2

In the same way, the modified Bessel equation (6.68) may be generalised to give an equation similar to (6.72). It then transpires that the solution of the cooling fin problem may be written as

$$\theta = AI_0(2rx^{1/2}) + BK_0(2rx^{1/2}) \tag{6.77}$$

From the boundary conditions

$$\theta = \theta_0 \text{ at } x = L \tag{6.78a}$$

and

$$-\frac{Kd\theta}{dx} = h_0\theta \text{ at } x = x_0 \tag{6.78b}$$

we may evaluate A and B.

Problems

1. Transform the following seven equations into a Bessel-type equation and solve.

(i) $xy'' + y' + xy = 0$ (ii) $x^2y'' + xy' + (x^2 - 9)y = 0$

(iii) $y'' - (1/x)y' + y = 0$ (iv) $x^2y'' + xy' + 9(x^2 - 1)y = 0$

(v) $y'' + (3/x)y' + y = 0$ (vi) $9xy'' + 9y' + y = 0$

(vii) $y'' + (k^2e^x - n^2)y = 0$.

2. By means of the substitutions $y = x^{1/2}u$, $z = kx$ reduce the equation $y'' + \left[k^2 + \frac{1}{4x^2}\right]y = 0$ to the form $\frac{d^2u}{dz^2} + \frac{1}{z}\frac{du}{dz} + u = 0$. If $x^{-1/2}y$ is finite when $x = 0$ and if $y' = a^{1/2}$ when $x = a$ show that the solution is $y = \dfrac{2ax^{1/2}J_0(kx)}{J_0(ka) - 2kaJ_1(ka)}$. (L.U.)

3. In a problem on the stability of a tapered strut the displacement y satisfies the equation $y'' + \frac{K^2}{4x}y = 0$ where K has to be determined. By writing $y = x^{1/2}u$, $z = Kx^{1/2}$, the equation can be reduced to the form $z^2\frac{d^2u}{dz^2} + z\frac{du}{dz} + (z^2 - 1)u = 0$. If $dy/dx = 0$ at $x = a$ and at $x = l$, show that the equation for K is $J_0(Ka^{1/2})\,Y_0(Kl^{1/2}) = J_0(Kl^{1/2})\,Y_0(Ka^{1/2})$. (L.U.)

4. Show that $J_n(x)/x^n$ is the solution of $y'' + ([1 + 2n]/x)y' + y = 0$ and that $\sqrt{x}\,J_n(kx)$ is a solution of $y'' + (k^2 - (4n^2 - 1)/4x^2)y = 0$. In both cases, n is a positive integer.

5. Two thin-walled metal pipes, joined by circular flanges, carry steam. The exposed surfaces of the flanges lose heat to the surrounding air which is at T_1° C. The conductivity of the flange metal is k and the heat transfer coefficient is h. Let the inner radius of the flanges be a and the outer radius b. Further, let the steam be at T_0° C. The heat balance equation can be written $r\frac{d^2T}{dr^2} + \frac{T}{r} - \frac{2h}{k}r(T - T_1) = 0$ where T is the temperature of the flange metal at radius r from the centre of the flanges. Write down the boundary conditions and find the rate of heat loss from a pipe through the flanges and the proportion of this which escapes from the rim.

6.7 SOLUTION OF PARTIAL DIFFERENTIAL EQUATIONS

We conclude our look at Bessel Functions with an example of their use in solving partial differential equations. We do not go into details of the derivation, merely outlining the main steps.

Consider a long circular cylinder of radius a and of diffusivity constant k. Initially, the cylinder is at constant temperature T_0 and its curved surface is kept at $T = 0$ for all $t > 0$. Find the temperature T as a function of distance from the axis, r, and time t.

We can assume that the temperature is independent of vertical coordinate z and angular coordinate θ so that the governing equation is $\frac{\partial^2 T}{\partial r^2} + \frac{1}{r}\frac{\partial T}{\partial r} = \frac{1}{k}\frac{\partial T}{\partial t}$. By methods discussed in Chapter 8, it can be shown that the general solution of this equation which is relevant to this problem is

$$T = \sum_{n=1}^{\infty} A_n\, e^{-kp_n^2 t}\, J_0(p_n r)$$

where the p_n are the positive roots of $J_0(p_n a) = 0$; this choice ensures that $T = 0$ at $r = a$ as required.

At $t = 0$, $T = T_0$ for $0 < r < a$. It can be shown that

$$A_n = \frac{2}{a^2 J_1^2(p_n a)} \int_0^a r T_0 J_0(p_n r)\, dr$$

By substituting $u = p_n r$ we obtain

$$A_n = \frac{2T_0}{p_n^2 a^2 J_1^2(p_n a)} \int_0^{p_n a} u J_0(u)\, du$$

and using the result that $\int_0^a x^\nu J_{\nu-1}(x)\, dx = a^\nu J_\nu(a)$ we find that

$A_n = \dfrac{2T_0}{p_n a J_1(p_n a)}$. Hence we have the solution

$$T = \frac{2T_0}{a}\left\{\sum_{n=1}^{\infty} e^{-kp_n^2 t} \frac{J_0(p_n r)}{p_n J_1(p_n r)}\right\} \tag{6.79}$$

Problems **(Attempt these after reading Chapter 8.)**

1. If α is a positive root of $J_0(x) = 0$, obtain the following expansions

   (i) $1 = 2 \sum_\alpha \dfrac{J_0(\alpha x)}{\alpha J_1(\alpha)}$ (ii) $x^2 = 2 \sum_\alpha \dfrac{(\alpha^2 - 4)}{\alpha^3 J_1(\alpha)} J_0(\alpha x)$

   where the functions are defined in the range $0 < x < 1$.

2. A finite cylinder of raidus a and length l has one end and the curved surface maintained at a constant temperature (which we may assume to be zero) and the temperature of the other end is kept at $\phi(r)$ where r is distance from the axis of the cylinder. A steady state has been reached and the temperature is finite at all points in the cylinder. Take cylindrical coordinates (r, ϕ, z) with the z - axis along the axis of the cylinder and the origin at the centre of one end. The governing equation for the temperature T is $\dfrac{\partial^2 T}{\partial r^2} + \dfrac{1}{r}\dfrac{\partial T}{\partial r} + \dfrac{\partial^2 T}{\partial z^2} = 0$. Solve the problem in the case $\phi(r) = A$, constant.

3. Let x be the current density at radius r in a wire of circular cross-section through which alternating current flows. If a is the radius of the wire, μ its permeability and ρ its specific resistance, we have the equation $\dfrac{1}{r}\dfrac{\partial}{\partial r}\left[r \dfrac{\partial x}{\partial r}\right] = \dfrac{4\pi\mu}{\rho}\dfrac{\partial x}{\partial t}$. If we let the total current through the wire be $C \cos \omega t$ find x and show that the magnetic intensity H at radius r and time t is given by the real part of $\dfrac{2C}{aI_0'(ka)} I_0'(kr)\, e^{i\omega t}$.

4. A circular membrane of radius a with fixed circumference is made to vibrate. The displacement z satisfies the equation $\dfrac{\partial^2 z}{\partial t^2} = c^2\left\{\dfrac{\partial^2 z}{\partial r^2} + \dfrac{1}{r}\dfrac{\partial z}{\partial r} + \dfrac{1}{r^2}\dfrac{\partial^2 z}{\partial \theta^2}\right\}$. To find the normal modes of vibration we try a solution of the form $z = R(r)\,\Theta(\theta) \cos(\omega t - \epsilon)$. Find the solution which satisfies the condition $z = 0$ at $r = a$.

6.8 AN INTRODUCTION TO LEGENDRE POLYNOMIALS

Any solution of Laplace's equation $\nabla^2 u = 0$ is called a **harmonic function.** In spherical polar coordinates (r, θ, ϕ), Laplace's equation becomes

$$\frac{\partial}{\partial r}\left\{r^2 \frac{\partial u}{\partial r}\right\} + \frac{1}{\sin\theta}\frac{\partial}{\partial\theta}\left(\sin\theta\,\frac{\partial u}{\partial\theta}\right) + \frac{1}{\sin^2\theta}\frac{\partial^2 u}{\partial\phi^2} = 0 \tag{6.80}$$

For problems which have axial symmetry, i.e. are independent of ϕ, we obtain

$$\frac{\partial}{\partial r}\left\{r^2 \frac{\partial u}{\partial r}\right\} + \frac{1}{\sin\theta}\frac{\partial}{\partial\theta}\left(\sin\theta\,\frac{\partial u}{\partial\theta}\right) = 0 \tag{6.81}$$

This equation has solutions $r^n P_n(\mu)$ and $r^{-n-1} P_n(\mu)$ where n is a positive integer $\mu = \cos\theta$ and $P_n(\mu)$ satisfies **Legendre's equation** viz.

$$\frac{\mathrm{d}}{\mathrm{d}\mu}\left\{(1-\mu^2)\frac{\mathrm{d}P_n}{\mathrm{d}\mu}\right\} + n(n+1)P_n = 0 \tag{6.82}$$

The functions $P_n(\mu)$ are called **Legendre polynomials** and their counterparts $r^n P_n(\mu)$ and $r^{-n-1} P_n(\mu)$ are called **spherical harmonics** and are especially useful in solving problems with spherical boundaries. Note that $-1 \leqslant \mu \leqslant 1$. We remark that it can be shown that the Legendre polynomials are linearly independent.

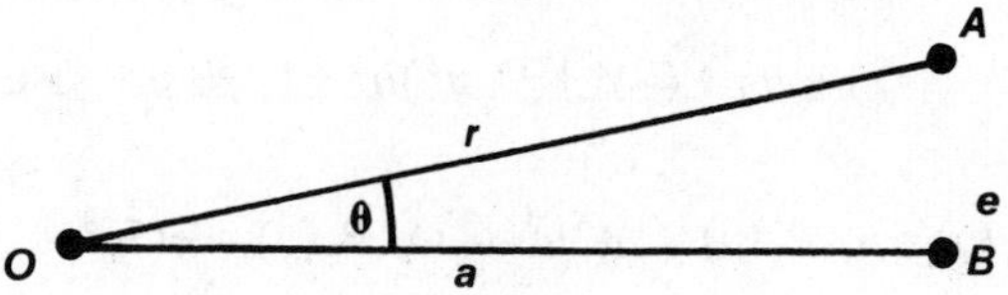

Figure 6.6

Generating Function

In some potential problems it is useful to expand the potential as a series in r or r^{-1}. Figure 6.6 shows a point electric charge e at B which is a distance a from 0. The electrostatic potential V at A due to the charge at B is given by

$$V = e/AB = e\left[r^2 - 2ra\cos\theta + a^2\right]^{-½} = e\left[r^2 - 2ra\mu + a^2\right]^{-½}$$

Hence

$$V = \begin{cases} (e/a)\{(r/a)^2 - 2(r/a)\mu + 1\}^{-½} & 0 \leqslant r \leqslant a \\ (e/r)\{1 - 2(a/r)\mu + (a/r)^2\}^{-½} & a \leqslant r \end{cases}$$

It can be shown that

$$V = \begin{cases} (e/a) \displaystyle\sum_{n=0}^{\infty} P_n(\mu)\left[\frac{r}{a}\right]^n & 0 \leqslant r \leqslant a \\ (e/r) \displaystyle\sum_{n=0}^{\infty} P_n(\mu)\left[\frac{a}{r}\right]^n & a \leqslant r \end{cases} \tag{6.83}$$

This means that we are saying that $P_n(\mu)$ is the coefficient of h^n in the expansion of $\{1 - 2h\mu + h^2\}^{-½}$. (In the case $0 \leqslant r \leqslant a$ we used $h = r/a$ and for $r \geqslant a$ we used $h = a/r$.)

It needs to be shown that this definition of $P_n(\mu)$ allows it to satisfy equation (6.82). In other words the definition

$$\{1 - 2h\mu + h^2\}^{-1/2} = \sum_{n=0}^{\infty} h^n P_n(\mu) \tag{6.84}$$

where $-1 \leqslant \mu \leqslant 1$ and $h \leqslant 1$ and the equation (6.82) are equivalent.

Examples of $P_n(\mu)$

We can obtain expressions for the $P_n(\mu)$ from (6.84). The general formula for $P_n(\mu)$ is

$$P_n(\mu) = \frac{(2n)!\,\mu^n}{2^n(n!)^2}\left\{1 - \frac{n(n-1)}{2(2n-1)}\frac{1}{\mu^2} + \frac{n(n-1)(n-2)(n-3)}{2.4(2n-1)(2n-3)}\frac{1}{\mu^4} + \dots\right\} \tag{6.85}$$

which could have been derived by applying the Frobenius technique to Legendre's equation (6.82). We can, however, obtain the first few polynomials directly from (6.84). For

$$(1 - 2\mu h + h^2)^{-1/2} = 1 + \frac{(-1/2)}{1}(-2\mu h + h^2) + \frac{(-1/2)(-3/2)}{1.2}(-2\mu h + h^2)^2 + \dots$$

$$= 1 + \mu h + (-1/2 + 3/2\,\mu^2)h^2 + (-3/2\,\mu + 5/2\,\mu^3)h^3 + \dots$$

Hence

$$P_0(\mu) = 1,\ P_1(\mu) = \mu,\ P_2(\mu) = 1/2(3\mu^2 - 1),\ P_3(\mu) = 1/2(5\mu^3 - 3\mu)$$

Note that $P_{2n+1}(\mu)$ involves only odd powers of μ, $P_{2n}(\mu)$ involves only even powers. Also $P_n(\mu)$ is a polynomial in μ of degree n.

Example

We can, in fact find $P_n(0)$, $P_n(1)$ and $P_n(-1)$ easily. From (6.84) we obtain

$$\sum_{n=0}^{\infty} P_n(0)h^n = (1 + h^2)^{-1/2} = 1 + \frac{(-1/2)}{1}h^2 + \frac{(-1/2)(-3/2)}{1.2}h^4 + \dots$$

so that $P_n(0) = 0$ for n odd and $P_0(0) = 1$, $P_2(0) = -1/2$, $P_4(0) = 3/8$,

$$P_{2n}(0) = (-1)^n \frac{1.3\dots(2n-1)}{2.4\dots\dots 2n}.$$

In the same way we can show that $P_n(1) = 1$ for all n and $P_n(-1) = (-1)^n$. We could have deduced this last result as a special case of $P_n(-\mu) = (-1)^n P_n(\mu)$ which follows directly from (6.85).

In Figure 6.7(a) we graph P_0, P_1, P_2 and P_3 for $0 < \mu < 1$; in Figure 6.7(b) we show the graphs with θ the horizontal coordinate.

The graph of $P_{2n}(\mu)$ is symmetrical about the vertical axis since it is an even function; $P_{2n+1}(\mu)$ is an odd function.

The expressions for higher order P_n are best found by means of recurrence relations which we are now going to study.

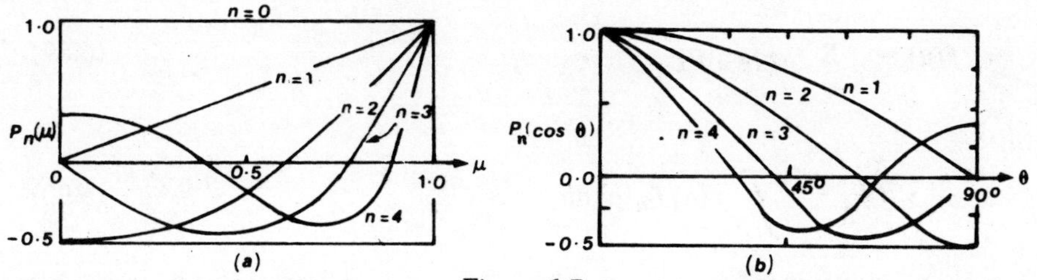

Figure 6.7

Recurrence relations

From

$$\left\{1 - 2h\mu + h^2\right\}^{-1/2} = \sum_{n=0}^{\infty} h^n P_n(\mu) \tag{6.84}$$

we may derive two main recurrence relations for $P_n(\mu)$.

Differentiating with respect to h we obtain

$$(\mu - h)(1 - 2h\mu + h^2)^{-3/2} = \sum_{n=0}^{\infty} nh^{n-1} P_n(\mu)$$

i.e. $$(\mu - h)(1 - 2h\mu + h^2)^{-1/2} = (1 - 2h\mu + h^2) \sum_{n=0}^{\infty} nh^{n-1} P_n(\mu)$$

or $$(\mu - h) \sum_{n=0}^{\infty} h^n P_n(\mu) = (1 - 2h\mu + h^2) \sum_{n=0}^{\infty} nh^{n-1} P_n(\mu)$$

Equating coefficients of h^n we find that

$$\mu P_n - P_{n-1} = (n+1)P_{n+1} - 2n\mu P_n + (n-1)P_{n-1}$$

i.e. $$(2n+1)\mu P_n = (n+1)P_{n+1} + n P_{n-1} \tag{6.86}$$

Similarly, by differentiating (6.84) with respect to μ we can derive

$$\mu P'_n - P'_{n-1} = n P_n \tag{6.87}$$

There are many other relationships that can be derived.

Orthogonality

The Legendre polynomials form an orthogonal set so that

$$\int_{-1}^{1} P_n(\mu) P_m(\mu)\, d\mu = \frac{2}{2n+1} \delta_{mn} \tag{6.88}$$

where δ_{mn} is the Kronecker delta such that $\delta_{mn} = \begin{cases} 0 & n \neq m \\ 1 & n = m \end{cases}$

We can, as a consequence, expand $f(\mu)$ for $-1 \leqslant \mu \leqslant 1$ as

$$f(\mu) = \sum_{n=0}^{\infty} a_n P_n(\mu) \tag{6.89}$$

where

$$a_n = \frac{2n+1}{2} \int_{-1}^{1} f(\mu) P_n(\mu) \, d\mu \tag{6.90}$$

The problem is to evaluate the coefficients a_n. A formula which helps in this task is **Rodrigues' formula**

$$P_n(\mu) = \frac{1}{2^n n!} \frac{d^n}{d\mu^n} \left\{ (\mu^2 - 1)^n \right\}$$

Example of application to Potential Theory

A point charge Q is situated at a point A which is distance h from the origin of a sphere $r = a$ which is constructed from material of dielectric constant K. Find the potentials ϕ_1 outside the sphere and ϕ_2 inside the sphere; see Figure 6.8

Now the potentials ϕ_1 and ϕ_2 must satisfy Laplace's equation and the conditions that at $r = a$, $\phi_1 = \phi_2$ and $\dfrac{\partial \phi_2}{\partial r} = K \dfrac{\partial \phi_1}{\partial r}$; the derivation of these results can be found in a textbook on electrostatics.

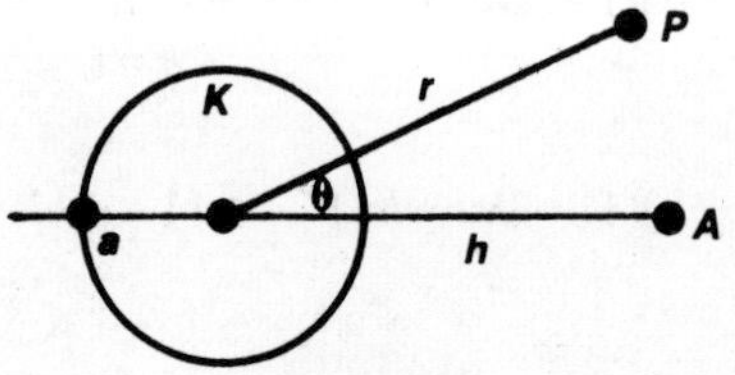

Figure 6.8

Let P be a point outside the sphere. The potential at P due to the charge at A is

$$\frac{Q}{\left\{h^2 + r^2 - 2hr \cos\theta\right\}^{1/2}} \text{ which equals } \sum_{n=0}^{\infty} \frac{Qr^n}{h^{n+1}} P_n(\mu), \; r < h.$$

The contribution to the potential outside the sphere due to the dielectric must $\to 0$ as $r \to \infty$ and therefore has to be of the form $\displaystyle\sum_{n=0}^{\infty} \frac{A_n P_n(\mu)}{r^{n+1}}$ which leads to

$$\phi_2 = \sum_{n=0}^{\infty} \left[\frac{Qr_n}{h^{n+1}} + \frac{A_n}{r^{n+1}} \right] P_n(\mu) \tag{6.91}$$

Inside the sphere the potential must be finite at $r = 0$ and therefore it must be

$$\phi_2 = \sum_{n=0}^{\infty} B_n r^n P_n(\mu) \tag{6.92}$$

If we apply the boundary conditions at $r = a$, we find (as you can verify) that

$$A_n = \frac{Qn(1-K)a^{2n-1}}{[n(K+1)+1]h^{n+1}} \qquad B_n = \frac{Q(2n+1)}{[n(K+1)+1]h^{n+1}} \tag{6.93}$$

Problems

1. Find the expressions for $P_4(x)$, $P_5(x)$ and $P_6(x)$ directly and deduce expresisons for $P_7(x)$, $P_8(x)$ and $P_9(x)$.

2. Prove that $P_5'(x) = 9P_4(x) + 5P_2(x) + P_0(x)$.

3. Using Rodrigues' formula obtain expressions for the polynomials $P_3(x)$, $P_4(x)$.

4. In terms of Legendre polynomials, find expressions for $1, x, x^2, x^3, x^4$. Comment on your results.

5. Show that

(i) $$\int_{-1}^{1} xP_n(x)\,P_{n+1}(x)\,dx = 2(n+1)/\ (2n+1)(2n+3)$$

(ii) $$\int_{-1}^{1} \{P_n(x)\}^2\,dx = \frac{2}{n+1}$$

6. Charges e, $-e$ are placed at points A, B respectively with AB $= 2a$. Polar coordinates are established with pole at O, the mid-point of AB and the angle θ is measured from OA. Show that the electrostatic potential at the point P $= (r, \theta)$ is given by

$$\phi(r,\theta) = \begin{cases} \dfrac{2e}{r}\displaystyle\sum_{n=0}^{\infty}\left\{\frac{a}{r}\right\}^{2n+1} P_{2n+1}(\cos\theta), & (r > a) \\ \dfrac{2e}{a}\displaystyle\sum_{n=0}^{\infty}\left\{\frac{r}{a}\right\}^{2n+1} P_{2n+1}(\cos\theta) & (r < a) \end{cases}$$

7. Show that

(i) $P_n'(x) = (2n-1)P_{n-1}(x) + (2n-5)P_{n-3}(x) + (2n-9)P_{n-5}(x) + \ldots..$

(ii) $xP_n'(x) = nP_n(x) + (2n-3)P_{n-2}(x) + (2n-7)P_{n-4}(x) + \ldots..$

8. Given that the steady state temperature, T, in a solid sphere of radius a is given by $\sin\theta\,\dfrac{\partial}{\partial r}\left\{r^2\,\dfrac{\partial T}{\partial r}\right\} + \dfrac{\partial}{\partial\theta}\left\{\sin\theta\,\dfrac{\partial T}{\partial\theta}\right\} = 0$ find the solution which satisfies the boundary conditions that the top hemispherical surface is maintained at a constant temperature T_a whilst the lower hemispherical surface is kept at a constant temperature T_b.

9. A circular wire of radius a carries a charge of constant line density e and surrounds an earthed spherical conductor of radius c, the centre of the circle coinciding with the centre O of the sphere. Taking O as origin and choosing the polar axis to be perpendicular to the plane of the wire, show that the surface density of charge at the general point of the surface of the conductor is given by

$$\sigma = -\frac{e}{2c}\left\{1 - 5.\frac{1}{2}\left\{\frac{c}{a}\right\}^2 P_2(\cos\theta) + 9.\frac{1.3}{2.4}\left\{\frac{c}{a}\right\}^4 P_4(\cos\theta) + 13.\frac{1.3.5}{2.4.6}\left\{\frac{c}{a}\right\}^6 P_6(\cos\theta)\right\}$$

(L.U.)

10. Two similar rings of radius a lie opposite each other in parallel planes, so spaced that the radius of one ring subtends an angle α at the other. Show that when the rings carry charges e and e' the repulsion between them is

$$\frac{ee'}{a^2} \sum_{n=0}^{\infty} \frac{(-1)^n (2n+1)!}{2^{2n}(n!)^2} (\sin\alpha)^{2n+2} P_{2n+1}(\cos\alpha) . \qquad \text{(L.U.)}$$

6.9 THE ERROR FUNCTION

This function occurs in probability theory, heat conduction and mass diffusion among other areas. It is defined by

$$\text{erf}(x) = \frac{2}{\sqrt{\pi}} \int_0^x e^{-t^2}\, dt \tag{6.94}$$

The factor $2/\sqrt{\pi}$ is chosen so that $\text{erf}(x) \to 1$ as $x \to \infty$. In Figure 6.9 we depict the interpretation of $\text{erf}(x)$.

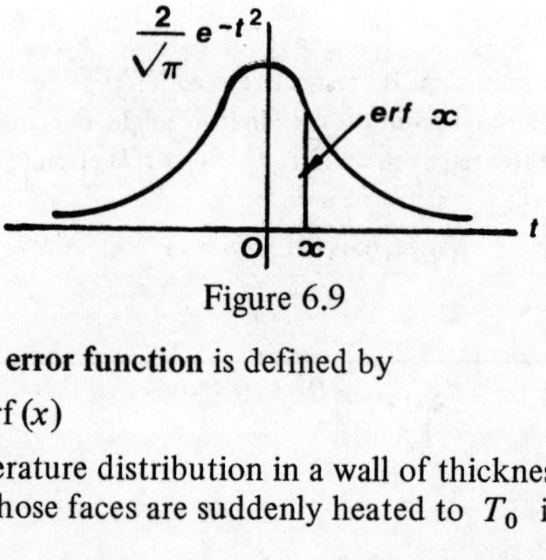

Figure 6.9

The **complementary error function** is defined by

$$\text{erfc}(x) = 1 - \text{erf}(x) \tag{6.95}$$

For example, the temperature distribution in a wall of thickness L with an initial uniform temperature, whose faces are suddenly heated to T_0 is given by

$$T = T_0 \sum_{n=0}^{\infty} (-1)^n \left\{ \text{erfc}\,[(x+nL)/2\sqrt{Kt}\,] + \text{erfc}\,[(nL+L-x)/2\sqrt{Kt}\,] \right\}$$

where K is the conductivity of the wall. We naturally do not expect you to be able to derive this result, but merely to see that the use of erfc allows a closed formula to be produced.

Problems

1. Prove that $\text{erf}(x) = \dfrac{2}{\sqrt{\pi}}\left\{x - \dfrac{x^3}{3} + \dfrac{x^5}{5.2!} - \dfrac{x^7}{7.3!} + \ldots\right\} = 1 - \dfrac{e^{-x^2}}{x\sqrt{\pi}}\left\{1 - \dfrac{1}{2x^2} + \dfrac{1.3}{(2x^2)^2} - \dfrac{1.3.5}{(2x^2)^3} + \ldots\right\}$

What advantages has either formula?

Chapter Seven

Approximation of Functions

7.1 INTRODUCTION

In this chapter we are concerned with the approximation of functions by other functions; the approximating functions are often polynomials, but may be for example sines or cosines as in the case of Fourier series. If our aim is to evaluate our given function at a given point to a certain number of decimal places then, provided the value of the approximating function agrees to the required number of decimal places we shall achieve our aim. We have to decide what constitutes a *good* approximation and whether the approximation we are using is, in fact, good in the sense of our criterion; furthermore, is it the *best* approximation in that sense.

The simplest approximations are via polynomials which arise from truncating infinite power series; they are especially easy to evaluate. We start this chapter by looking at aspects of the Maclaurin series for $\sin x$. Then we consider a different criterion for approximation: the **mini-max approximation.** This leads us to the study of the use of **Chebyshev polynomials.** We then move on to the **least squares criterion** for approximation and hence to **Fourier series.** The chapter concludes by examining the use of **cubic spline fitting.**

7.2 TAYLOR SERIES APPROXIMATION

The Taylor series approximation to a function represents the function near the centre of the expansion with very little error, but the error increases rapidly as we move to a point further from the centre. Suppose we require a smaller maximum error over some interval; the price we must pay is the loss of the very high accuracy at the centre of the interval. To illustrate the nature of the problem, consider the calculation of $\sin x$ by the Maclaurin's series.

$$f(x) = \sin x = x - \frac{x^3}{3!} + \frac{x^5}{5!} - \frac{x^7}{7!} + \frac{x^9}{9!} - \ldots.. \tag{7.1}$$

We can save some calculation by using this series to represent $\sin x$ in the interval $[0, \pi/2]$ and employing trigonometric identities to deal with angles outside this interval. Suppose we decide that we wish to keep the maximum error below 0.00005. Then we require that the remainder term, $R_n = f^{(n)}(\xi)\,\dfrac{x^n}{n!}$ where $0 < \xi < x \leqslant \dfrac{\pi}{2}$, be less than 0.00005 for $0 < x \leqslant \dfrac{\pi}{2}$.

Now $|R_n| \leqslant M\,\dfrac{x^n}{n!}$ where M is the maximum value taken by $|f^{(n)}(x)|$ in the interval $[0, \pi/2]$. Therefore we require that

$$|R_n| \leqslant M \frac{x^n}{n!} < 0.00005$$

But the derivatives of $\sin x$ are $\pm\sin x$, $\pm\cos x$ and these are all less than or equal to 1 in modulus and hence $M = 1$. We then have

$$1 \cdot \frac{x^n}{n!} < 0.00005$$

Clearly x^n increases as x increases towards $\pi/2$ and we therefore seek the smallest value of n so that $\left[\frac{\pi}{2}\right]^n \Big/ n! < 0.00005$. We find that $n = 10$. However since there is no term in x^{10} we take the series (7.1) only as far as but not including the term in x^{11}. This means that we are calculating five terms and possibly getting more accuracy than was required.

Let us only use this number of terms of the series and ask how we might spread out the error more evenly. Since the dominant term far from $x = 0$ is the last term, $x^9/9!$, we could try the effect of reducing the coefficient of this term. Now 9! is 362880 and we consider the effect of changing this number to 370000, i.e. we use the approximation

$$\sin x = x - x^3/6 + x^5/120 - x^7/5040 + x^9/370000 \tag{7.2}$$

Now we programmed a digital computer to evaluate the approximation (7.2) for $x = 0(\pi/18)\pi/2$, i.e. $0^\circ(10^\circ)90^\circ$, and the errors in the results, when compared with the values obtained from the standard routine for sine evaluation provided by the computer software, are shown in Table 7.1. The notation E–11 means 10^{-11} etc.

Table 7.1

x (Degrees)	Approximate Error (7.2)	Approximate Error (7.1)
10	0.363 E–11	0.364 E–11
20	0.368 E–11	0.728 E–11
30	–0.124 E–9	0.291 E–10
40	–0.160 E–8	0.487 E–9
50	–0.998 E–8	0.559 E–8
60	–0.390 E–7	0.413 E–7
70	–0.970 E–7	0.225 E–6
80	–0.968 E–7	0.973 E–6
90	0.455 E–6	0.354 E–5

Economical evaluation of polynomials

In high-precision work, the bug-bear is round-off error. In order to cut down on its effects, we look for ways of evaluating polynomials such as (7.2) which reduce round-off error as much as possible. Furthermore, multiplication on a computer is much more time-consuming than additions or subtractions and we should aim for a method of evaluation which keeps multiplication to a minimum. One way of achieving this aim is to use the idea of **nested multiplication.** For example, the polynomial $ax^3 + bx^2 + cx + d$ requires for its direct evaluation $(3 + 2 + 1) = 6$ multiplications and 3 additions. The equivalent nested form $[(ax + b)x + c]x + d$ requires 3 multiplications and 3 additions. Here we have halved the number of multiplications and not increased the number of additions.

The method of **economical evaluations** may increase the number of additions, but will cut down on multiplication and hence reduce overall running time. Consider the polynomial

$$p(x) = ax^4 + bx^3 + cx^2 + dx + e \tag{7.3}$$

The nested form is $\{[(ax+b)x+c]x+d\}x+e$ which reduces the number of multiplications from 10 to 4. The economical evaluation method would be to express the polynomial as

$$[(\alpha x+\beta)^2 + \alpha x + \gamma]\,[(\alpha x+\beta)^2 + \delta] + \epsilon \tag{7.4}$$

We determine the coefficients as follows: $\alpha = (a)^{1/4}$, $\beta = (b - \alpha^3)/4\alpha^3$, $\delta = 3\beta^2 + 8\beta^3 + (d\alpha - 2c\beta)/\alpha^2$, $\gamma = (c/\alpha^2) - 2\beta - 6\beta^2 - \delta$, $\epsilon = e - \beta^4 - \beta^2(\gamma+\delta) - \gamma\delta$.

Now this may seem to defeat totally the object of the exercise, but you should remember that once we have specified the polynomial by its coefficients we can determine α, β, γ, δ and ϵ and store these in the computer so that we may input any value of x and obtain the value $p(x)$. For the polynomial expressed as (7.4) we need 3 multiplications and 5 additions, a saving of 1 multiplication over the nested form (7.3) at the expense of 1 addition. The method of evaluation, given a value for x, would be to form $z_1 = \alpha x$, $z_2 = (z_1+\beta)^2$ and $z_3 = (z_2 + z_1 + \gamma)\,.\,(z_1+\delta) + \epsilon$. The first stage requires one multiplication, the second requires one addition and one multiplication and the third requires one multiplication and four additions. For the approximation (7.2) we note that x is a common factor and that the remaining powers of x are even. The nested form is

$$x\left\{\left[\left\{(0.00000270x^2 - 0.00019841)x^2 + 0.00833333\right\}x^2 - 0.16666667\right]x^2 + 1\right\} \tag{7.5}$$

We should, of course, evaluate x^2 only once. The economical evaluation is as follows.

$$q = x^2, \quad z_1 = 0.04053600\;q, \quad z_2 = (z_1 - 0.99469973)^2$$

$$p(x) = x[(z_2 + z_1 - 4.02594644)(z_2 + 5.15025875) + 19.6433554] \tag{7.6}$$

Unfortunately, although the economical evaluation technique reduces the number of multiplications it can lead to propagation of round-off error which may make it less accurate than the nesting method. Sometimes a combination of economical evaluation followed by nesting is employed to get the best of both worlds.

In Table 7.2 we show the approximation to $\sin x$ evaluated by formulae (7.5) and (7.6).

Normalisation of a series

We use the series (7.1) in the interval $[-\pi/2, \pi/2]$; however, we can always make an interval of a desired length by choosing a suitable scaling factor. If we measure a new angle $x^1 = 2x/\pi$, i.e. $x = \pi x^1/2$, then $x \in [-\pi/2, \pi/2]$ implies $x^1 \in [-1, 1]$ and vice-versa. This means that the coefficients of the series for $\sin x$ will have to be recalculated. When we have got the function defined in the standard range $[-1, 1]$ we can look quite generally at ways of approximating it.

Table 7.2

x (Degrees)	Error (7.5)	Error (7.6)
10	–0.146 E–10	0.879 E–3
20	–0.156 E–9	0.703 E–2
30	–0.713 E–9	0.237 E–1
40	–0.317 E–8	0.562 E–1
50	–0.136 E–7	0.110
60	–0.474 E–7	0.189
70	–0.118 E–6	0.301
80	–0.150 E–6	0.450
90	0.317E–6	0.641

7.3 MINI-MAX APPROXIMATION

In an effort to spread out the error in the approximation of $f(x)$ by a polynomial $p_n(x)$ of degree n in the interval $[-1, 1]$ we choose the coefficients so that the maximum value of $|f(x) - p_n(x)|$ over $[a, b]$ is smaller than the corresponding quantity for any other polynomial of degree n. Since we are minimising the maximum of some quantity we call the chosen polynomial the **minimax approximation.** Provided that the function $f(x)$ is continuous on the interval $[-1, 1]$ then we can be guaranteed that a unique minimax polynomial exists.

It is difficult in general to find such a polynomial, but we shall aim to find a way round this problem by making use of the following result.

If $f(x)$ is continuous on $[-1, 1]$, $p_n(x)$ is a polynomial of degree n, $\epsilon(x) = p_n(x) - f(x)$, and there are at least $(n+2)$ points x_i such that $-1 \leqslant x_0 \leqslant x_1 \leqslant \ldots\ldots \leqslant x_{n+1} \leqslant 1$ with the properties $\epsilon(x_0) = M$, $\epsilon(x_1) = -M$, $\epsilon(x_2) = M, \ldots$, $\epsilon(x_{n+1}) = (-1)^{n+1}M$, then $p_n(x)$ is the minimax approximation of degree n to the function $f(x)$, in the interval $[-1, 1]$.

It follows that the maximum error of approximation is $|M|$.

Example

Find the minimax approximation of degree 1 to the function $f(x) = x^3 + x^2 + 2x$ in $[-1, 1]$.

We want to find the coefficients of the straight line approximation $p_1(x) = a_0 + a_1 x$. Now the error function

$$\epsilon(x) \equiv p_n(x) - f(x) = a_0 + (a_1 - 2)x - x^2 - x^3 \tag{7.7}$$

We wish to find the three points at which $\epsilon(x)$ will be $\pm M$. Let us guess that two of these will be the end-points -1 and 1 (this turns out in practice to be the case quite often); we shall label the third point x_1 and assume that

$$\left.\begin{aligned} \epsilon(-1) &= a_0 - a_1 + 2 &&= M \\ \epsilon(x_1) &= a_0 + (a_1 - 2)x_1 - x_1^2 - x_1^3 &&= -M \\ \epsilon(1) &= a_0 + a_1 - 4 &&= M \end{aligned}\right\} \tag{7.8}$$

(Since we have not specified whether M is positive or negative, we can make $\epsilon(-1) = M$ without loss of generality.)

Subtracting the first of these equations from the third yields $2a_1 - 6 = 0$ i.e. $a_1 = 3$. We then have $p_1(x) = a_0 + 3x$. The error $\epsilon(x) = a_0 + x - x^2 - x^3$ and this has local turning points where $\epsilon'(x) = 0$, i.e. $1 - 2x - 3x^2 = 0$ i.e. at $x = \frac{1}{3}$ or $x = -1$. We therefore choose $x_1 = \frac{1}{3}$ and find that $\epsilon(-1) = a_0 - 1 = M$, and $\epsilon(\frac{1}{3}) = a_0 + \frac{1}{3} - \frac{1}{9} - \frac{1}{27} = -M$. Therefore $a_0 = \frac{11}{27}$ and $M = \frac{-16}{27}$. This implies that the minimax straight line approximation is $p_1(x) = \frac{11}{27} + 3x$ and the maximum error in the interval $[-1, 1]$ is $\frac{16}{27}$. Draw the function $f(x)$ and its approximation $p_1(x)$ for yourselves. Of course, we have chosen a particularly simple example and, in general, the coefficients of the minimax polynomial are much harder to find.

Chebyshev Polynomials

In order to ease the calculations in the case where $f(x)$ is itself a polynomial we introduce some special polynomials, called **Chebyshev polynomials** which have certain useful properties. Before we describe these polynomials you may well ask why we bother to approximate one polynomial by another. The reason is contained in the following result.

If $f(x)$ is a polynomial of degree n, defined on $[-1, 1]$ then the minimax polynomial approximation of degree $(n-1)$ or less is the polynomial $p(x) = f(x) - \alpha T_n(x)$; here, $T_n(x)$ is the Chebyshev polynomial of degree n and α is a constant equal to the coefficient of the highest power of x in $f(x)$. This means that $p(x)$ is of degree $(n-1)$ at most and hence represents a saving in calculation time.

The **Chebyshev polynomial of degree** n is defined by

$$T_n(\cos\theta) = \cos n\theta \quad \text{or} \quad T_n(x) = \cos n(\cos^{-1} x) \tag{7.9}$$

To evaluate the first few polynomials we use the first definition

$$\left.\begin{aligned} &T_0(\cos\theta) = 1 \text{ i.e. } T_0(x) = 1, \quad T_1(\cos\theta) = \cos\theta \text{ i.e. } T_1(x) = x \\ &T_2(\cos\theta) = \cos 2\theta = 2\cos^2\theta - 1 \text{ i.e. } T_2(x) = 2x^2 - 1 \end{aligned}\right\} \tag{7.10}$$

Similarly we may show that $T_3(x) = 4x^3 - 3x$ and $T_4(x) = 8x^4 - 8x^2 + 1$. However, the higher the value of n, the more tedious the direct evaluation will become. We can obtain the higher degree polynomials by making use of a recurrence relation. Further polynomials are shown in Table 7.3.

Consider the first definition of (7.9) It follows that
$T_{n+1}(x) = T_{n+1}(\cos\theta) = \cos(n+1)\theta = \cos n\theta \cos\theta - \sin n\theta \sin\theta$ and
$T_{n-1}(x) = \cos(n-1)\theta = \cos n\theta \cos\theta + \sin n\theta \sin\theta$.

Hence by addition of these two equations we have
$T_{n+1}(x) + T_{n-1}(x) = 2\cos n\theta \cos\theta = 2T_n(x) \,.\, x$. Therefore we have the recurrence relation

$$T_{n+1}(x) = 2xT_n(x) - T_{n-1}(x) \tag{7.11}$$

This relation, together with the results $T_0(x) = 1$, $T_1(x) = x$ is enough to generate all the Chebyshev polynomials. Check that $T_2(x)$, $T_3(x)$ and $T_4(x)$ obtained this way agree with the stated results.

Table 7.3

$T_0(x) = 1 \quad T_1(x) = x \quad T_2(x) = 2x^2 - 1 \quad T_3(x) = 4x^3 - 3x$

$T_4(x) = 8x^4 - 8x^2 + 1 \quad T_5(x) = 16x^5 - 20x^3 + 5x$

$T_6(x) = 32x^6 - 48x^4 + 18x^2 + 1 \quad T_7(x) = 64x^7 - 112x^5 + 56x^3 - 7x$

$T_8(x) = 128x^8 - 256x^6 + 160x^4 - 32x^2 + 1$

$T_9(x) = 256x^9 - 576x^7 + 432x^5 - 120x^3 + 9x$

Properties of Chebyshev polynomials

1. The even-numbered polynomials contain only even powers of x and are hence even functions. The odd-numbered polynomials are likewise odd functions.
2. The coefficient of x^n in $T_n(x)$ is 2^{n-1}; the coefficients of $T_n(x)$ alternate in sign.
3. $T_n(1) = 1, \quad T_n(-1) = (-1)^n$.
4. If $-1 \leqslant x \leqslant 1$, $-1 \leqslant T_n(x) \leqslant 1$, i.e. we can confine the graph of $T_n(x)$ within the square shown in Figure 7.1. This follows because $|\cos n\theta| < 1$.
5. If we consider any other polynomial of degree n which has the coefficient of 2^{n-1} but may have other coefficients different from those of $T_n(x)$, then it will be > 1 or < -1 for at least one value of x in $[-1, 1]$. This is the **mini-max** property.
6. If we put $T_n(x) = 0$ we obtain n real roots in $[-1, 1]$. We can show that if n is odd, then $x = 0$ is always a root, and for any value of n the non-zero roots are in pairs: to each positive root there is a negative root of equal magnitude.
7. In between two zeros of $T_n(x)$, there is either a local minimum of -1 or, alternately, a local maximum of 1.

In Figure 7.1 we show the first few Chebyshev polynomials. Try and graph the next three.

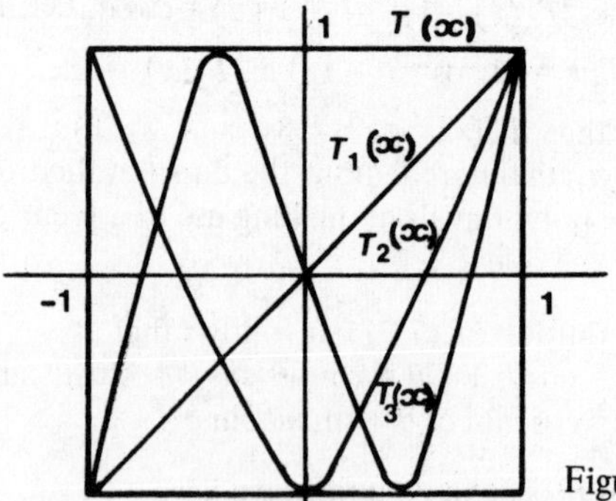

Figure 7.1

(Note, incidentally, that we can change the interval of definition of these polynomials from $[-1, 1]$. For example to change to the interval $[a, b]$ we define new polynomials $T_n^1(x) = T_n(u)$ where $u = \left(\frac{2}{b-a}\right)x - \left(\frac{b+a}{b-a}\right)$ and where $x \in [a, b]$).

Economising a polynomial

In order to make use of Chebyshev polynomials in the economisation of a polynomial, we first express powers of x in terms of the Chebyshev polynomials as linear combinations of them. These expressions, being identities, will hold for any x, but the domain of $T_n(x)$ restricts the range of x. Table 7.4 shows the expressions for the first nine powers of x.

Table 7.4

$$1 = T_0, \quad x = T_1, \quad x^2 = \frac{1}{2}(T_0 + T_2), \quad x^3 = \frac{1}{4}(3T_1 + T_3),$$

$$x^4 = \frac{1}{8}(3T_0 + 4T_2 + T_4), \quad x^5 = \frac{1}{16}(10T_1 + 5T_3 + T_5),$$

$$x^6 = \frac{1}{32}(10T_0 + 15T_2 + 6T_4 + T_6), \quad x^7 = \frac{1}{64}(35T_1 + 21T_3 + 7T_5 + T_7),$$

$$x^8 = \frac{1}{128}(35T_1 + 56T_3 + 28T_5 + 8T_7 + T_9),$$

$$x^9 = \frac{1}{256}(126T_1 + 84T_3 + 36T_5 + 9T_7 + T_9)$$

There are many properties of the coefficients that can be deduced. We mention two. First, the coefficient of $T_n(x)$ in the expansion of x^n is $1/2^{n-1}$. Second, each expression is a weighted average of Chebyshev polynomials.

If we return to the Maclaurin expansion for $\sin x$, we may normalise it to produce

$$\sin\left[\frac{\pi x}{2}\right] = 1.57079633x - 0.64596410x^3 + 0.079692626x^5 - 0.004681754x^7 + 0.000160441x^9 \tag{7.12}$$

Now we substitute for powers of x from Table 7.4 and combine like polynomials to obtain

$$\sin\left[\frac{\pi x}{2}\right] = 1.133649778T_1 - 0.138070634T_3 + 0.004491284T_5 - 0.000067511T_7 + 0.000000627T_9 \tag{7.13}$$

Suppose we now omit the term in $T_9(x)$. What error will we have committed? Notice that so far, apart from round-off error, there is the truncation error already committed by truncating the Maclaurin series. Had we chosen the infinite series for $\sin x$ we should have slightly different coefficients for the Chebyshev polynomials T_1, T_3, T_5, T_7 and T_9. But the contributions from higher powers of x to changes in these coefficients will be of secondary importance.

If we omit the term in $T_9(x)$, then the maximum error we commit is $0.000000627 \,.\, 1$, since $|T_9(x)| < 1$.

Now, to date we have committed two errors, apart from round-off error; the error due to economising, at most 0.000000627 and the error due to truncating the original Maclaurin's series; this latter is at most $(\pi/2)^{11}/11!$ i.e. 0.00000360 (3 s.f.) and outweighs the economising error.

We now convert the expansion (7.13) with the last term removed, back to a polynomial in x, by using Table 7.3. We obtain, after regrouping like terms

$$\sin\left[\frac{\pi x}{2}\right] = 1.570790685x - 0.645888889x^3 + 0.079421881x^5 - 0.004320761x^7 \qquad (7.14)$$

Note that the greater change in coefficients occurs for lower powers of x. We are free to nest this last polynomial.

In Table 7.5 we compute $\sin\left[\frac{\pi x}{2}\right]$ for $x = 0(0.1)1$ by the series (7.12) which is a modification of the series (7.1) truncated onto the terms shown, and by the series (7.14). We compare with the standard computer function. The graphs of the Maclaurin error and the Chebyshev error are shown in Figure 7.2.

Table 7.5 (Results quoted to 7 d.p.)

x	Approximation (7.12)	Approximation (7.14)	Error in (7.12)	Error in (7.14)
1/9	0.1736481	0.1736476	0.465661 E–09	–0.528058 E–06
2/9	0.3420201	0.3420195	0.465660 E–09	–0.565778 E–06
3/9	0.5000000	0.4999999	0.465661 E–09	–0.526196 E–07
4/9	0.6427876	0.6427881	0.186264 E–08	0.528991 E–06
5/9	0.7660444	0.7660449	0.651925 E–08	0.525265 E–06
6/9	0.8660254	0.8660252	0.428408 E–07	–0.135041 E–06
7/9	0.9396928	0.9396922	0.226311 E–06	–0.393949 E–06
8/9	0.9848087	0.9848089	0.976026 E–06	0.123406 E–05
1	1.0000034	1.0000029	0.354461 E–05	0.291503 E–05

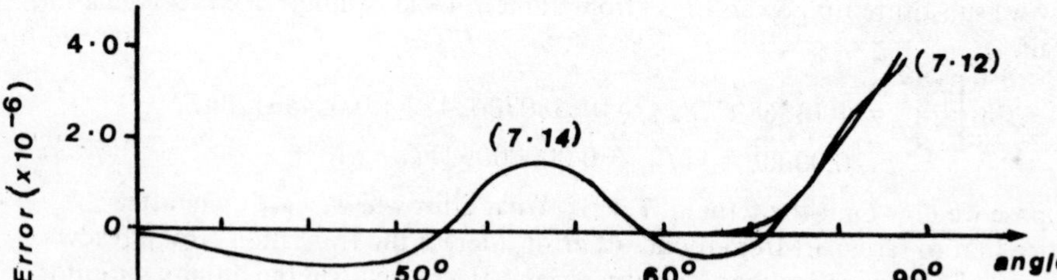

Figure 7.2

Problems

1. Write a program to carry out the calculations for $\sin x$ using (7.12) and (7.14).

2. Prove that

(i) $(1 - x^2)\,T_n''(x) - xT_n'(x) + n^2\,T_n(x) = 0;$

(ii) $$\int_{-1}^{1} \frac{T_m(x)\,T_n(x)\,dx}{\sqrt{(1-x^2)}} = \begin{cases} 0 & \text{if } m \neq n \\ \tfrac{1}{2}\pi & \text{if } m = n \neq 0 \\ \pi & \text{if } m = n = 0 \end{cases}$$

(iii) $$\int_{-1}^{1} \frac{x^2 T_{n-1}(x)\, T_{n+1}(x)\, dx}{\sqrt{(1-x^2)}} = \frac{\pi}{8} \text{ when } n \geqslant 2.$$ (L.U.)

3. Given $\arctan x = x - \frac{x^3}{3} + \frac{x^5}{5} - \frac{x^7}{7} + \frac{x^9}{9} - \ldots$. Plot the error over the interval $[-1, 1]$ when the above series is truncated after the term in x^7. Economize the ninth-degree truncated power series three times (giving a third-degree expression), and plot its error over the interval $[-1, 1]$.

4. The series for $\log_e(1 + x)$ converges slowly for $-1 < x < 1$;
$\log_e(1 + x) = x - \frac{1}{2}x^2 + \frac{1}{3}x^3 - \frac{1}{4}x^4 + \frac{1}{5}x^5 - \ldots$. Since it is an alternating series, its error is less than the first abandoned term. Convert to Chebyshev series, including terms up to $T_5(x)$. What is the maximum error of the truncated Chebyshev series on $[-1, 1]$? How many terms of the Maclaurin series are needed to give a smaller maximum error?

5. Find the best approximation of degree 2 to the following functions and state the maximum error.

   (i) $f(x) = (x + 1)^3$, $-1 \leqslant x \leqslant 1$ (ii) $f(x) = x^2$, $0 \leqslant x \leqslant 1$

   (iii) $f(x) = x^3$, $\frac{1}{2} \leqslant x \leqslant \frac{3}{2}$

6. How would you compute $T_n(x)$ for $n = 39$, $x = 0.2$? Give reasons for your answer.

7. Show that $p_1(x) = \frac{5}{4} - x$ is a good approximation to $f(x) = e^{-x}$, $-1 \leqslant x \leqslant 1$. Show also that there is no linear approximation with a maximum error smaller than 0.1. (Take $e \simeq 2.72$, $e^{-1} \simeq 0.37$, as good enough approximations for this purpose.) Say whether the linear (first-order) Taylor approximation to e^{-x}, about $x = 0$, is a good approximation.

8. Express $f(x) = x^3 + x^2 + 1$, $(x \in [-1, 1])$, in the form of a linear combination of Chebyshev polynomials, and deduce the minimax quadratic approximation to f.
If the first two terms in your expression are $\frac{1}{2}a_0 T_0 + a_1 T_1$, in what error measure is this the best *linear* approximation to f.

9. Write the first three terms of the Maclaurin series for e^x, and reduce down to two terms. Compare with the two-term Maclaurin series. Graph the function e^x for $-1 \leqslant x \leqslant +1$, and compare with the Taylor series of two terms and the reduced series. What is the maximum error in the two series approximations?

7.4 PERIODIC PHENOMENA - AN INTRODUCTION TO FOURIER SERIES

Many phenomena that are studied in engineering are periodic in nature. For example the current and voltage in an alternating current circuit, the displacement, velocity and acceleration of the piston shown in Figure 7.3 and many parameters in a vibrating system are all periodic. To model such situations would suggest that sine and cosine functions, being the most well-known periodic functions, might well be used as the building blocks. We would hope that such modelling would give adequate representation over the whole cycle of periodicity rather than the local nature of the Taylor series representation.

Many wave phenomena are also periodic and it is well known, especially through acoustics, that a wave can in general be decomposed or analysed into several distinct waves of different frequencies. In the terminology of acoustics the wave with the lowest frequency is called the **fundamental** and the others are called **harmonics.** The plotting of amplitude versus frequency is called **spectral analysis.**

Approximation of a function by a trigonometric series

A series of the form,

$$\tfrac{1}{2}a_0 + a_1 \cos x + b_1 \sin x + a_2 \cos 2x + b_2 \sin 2x + a_3 \cos 3x + \ldots.. \tag{7.15}$$

where the a_i and b_i are constants is called an **infinite trigonometric series.** It can also be written in the shortened form

$$\tfrac{1}{2}a_0 + \sum_{k=1}^{\infty} (a_k \cos kx + b_k \sin kx) \tag{7.16}$$

The reason for writing the constant term as $\tfrac{1}{2}a_0$ will become apparent later. Whether the series converges or not will depend on the value of x chosen and the coefficients a_k and b_k. Note that if it converges in any closed interval $[c, c + 2\pi]$, the periodic nature of the cosine and sine functions guarantees convergence for all values of x. Remember that a function $g(x)$ is **periodic** with period c if $g(x + c) = g(x)$ for all values of x. Such a function will have period $2c$, $3c$, $4c$, ... We usually choose the period as the **smallest** value of c for which $g(x + c) = g(x)$ is true for all x.

We shall now address ourselves to the problem of approximating a function $f(x)$ by a finite trigonometric series

$$S_n(x) = \frac{a_0}{2} + \sum_{k=1}^{n} (a_k \cos kx + b_k \sin kx) \tag{7.17}$$

over the interval $(-\pi, \pi)$ and, more generally, the interval $(c, c + 2\pi)$. We shall in the next section extend the ideas to making an approximation over the interval $(c, c + L)$.

Suppose we decide that we shall select the coefficients a_k, b_k by choosing that finite trigonometric series which is the *least squares* best approximation of $f(x)$ in the interval $(-\pi, \pi)$. There may be certain conditions imposed on $f(x)$: these will become apparent as we progress. The least squares criterion is interpreted as minimising

$$I_n = \int_{-\pi}^{\pi} \left\{ f(x) - \tfrac{1}{2}a_0 - \sum_{k=1}^{n} (a_k \cos kx + b_k \sin kx) \right\}^2 dx \tag{7.18}$$

We need only to assume that $f(x)$ is integrable in $[-\pi, \pi]$.

To find the normal equations which will yield a solution for a_k, b_k we need to differentiate I_n partially with respect to a_k, b_k. Therefore

$$\frac{\partial I_n}{\partial a_0} = \frac{-1}{2} \cdot 2 \int_{-\pi}^{\pi} \left\{ f(x) - \frac{a_0}{2} - \sum_{k=1}^{n} (a_k \cos kx + b_k \sin kx) \right\} dx$$

$$= -\left\{ \int_{-\pi}^{\pi} f(x)\,dx - \frac{1}{2}a_0 \int_{-\pi}^{\pi} dx - \sum_{k=1}^{n} \int_{-\pi}^{\pi} (a_k \cos kx + b_k \sin kx)\,dx \right\} \tag{7.19a}$$

For $r \neq 0$

$$\frac{\partial I_n}{\partial a_r} = -2 \int_{-\pi}^{\pi} \left\{ f(x) - \frac{a_0}{2} - \sum_{k=1}^{n} (a_k \cos kx + b_k \sin kx) \right\} . \cos rx\,dx$$

$$= -2 \left[\int_{-\pi}^{\pi} f(x) \cos rx\,dx - \frac{a_0}{2} \int_{-\pi}^{\pi} \cos rx\,dx - \sum_{k=1}^{n} \left\{ \int_{-\pi}^{\pi} (a_k \cos kx \cos rx + b_k \sin kx \cos rx)\,dx \right\} \right] \tag{7.19b}$$

Similarly,

$$\frac{\partial I_n}{\partial b_r} = -2 \left[\int_{-\pi}^{\pi} f(x) \sin rx\,dx - \frac{a_0}{2} \int_{-\pi}^{\pi} \sin rx\,dx - \sum_{k=1}^{n} \left\{ \int_{-\pi}^{\pi} (a_k \cos kx \sin rx + b_k \sin kx \sin rx)\,dx \right\} \right] \tag{7.19c}$$

Now we can effect a considerable simplification by noting the following standard results of integration where r and k are any positive integers. The results still hold if $\int_{-\pi}^{\pi}$ is replaced by $\int_{c}^{c+2\pi}$.

(i) $\int_{-\pi}^{\pi} \cos rx\,dx = 0$ (ii) $\int_{-\pi}^{\pi} \sin rx\,dx = 0$

(iii) $\int_{-\pi}^{\pi} \sin rx \cos kx\,dx = 0, \; k \neq r$ (iv) $\int_{-\pi}^{\pi} \sin rx \cos rx\,dx = 0$

(v) $\int_{-\pi}^{\pi} \sin rx \sin kx\,dx = 0, \; k \neq r$ (vi) $\int_{-\pi}^{\pi} \sin^2 rx\,dx = \pi$

(vii) $\int_{-\pi}^{\pi} \cos rx \cos kx\,dx = 0, \; k \neq r$ (viii) $\int_{-\pi}^{\pi} \cos^2 rx\,dx = \pi$

Therefore, $$\frac{\partial I_n}{\partial a_0} = -\left\{ \int_{-\pi}^{\pi} f(x)\,dx - \pi a_0 - 0 \right\} \tag{7.20a}$$

since all remaining integrals in (7.19a) are zero by (i) and (ii). Likewise

$$\frac{\partial I_n}{\partial a_r} = -2\left\{\int_{-\pi}^{\pi} f(x)\cos rx\,dx - 0 - \int_{-\pi}^{\pi} a_r\cos^2 rx\,dx\right\}$$

$$= -2\left\{\int_{-\pi}^{\pi} f(x)\cos rx\,dx - a_r\,.\,\pi\right\} \quad \text{by (viii)} \tag{7.20b}$$

since all remaining integrals in (7.19b) are zero by (i), (iii), (iv), (vii). Similarly

$$\frac{\partial I_n}{\partial b_r} = -2\left\{\int_{-\pi}^{\pi} f(x)\sin rx\,dx - b_r\,\pi\right\} \tag{7.20c}$$

since all remaining integrals in (7.19c) are zero by (ii), (iii), (iv), (v) and we have used (vi).

Now the least squares criterion requires $\dfrac{\partial I_n}{\partial a_0} = \dfrac{\partial I_n}{\partial a_r} = \dfrac{\partial I_n}{\partial b_r} = 0$. Therefore we obtain the following formulae for the coefficients

$$\left.\begin{aligned} a_0 &= \frac{1}{\pi}\int_{-\pi}^{\pi} f(x)\,dx \\ a_r &= \frac{1}{\pi}\int_{-\pi}^{\pi} f(x)\,.\,\cos rx\,dx; \quad r = 1,\ 2,\ 3,\ \ldots.. \\ b_r &= \frac{1}{\pi}\int_{-\pi}^{\pi} f(x)\,.\,\sin rx\,dx; \quad r = 1,\ 2,\ 3,\ \ldots.. \end{aligned}\right\} \tag{7.21}$$

Now you see the reason for choosing the constant term in the trigonometric series to be $\frac{1}{2}a_0$: it allowed a systematic representation of the coefficients. Also note that a_0 is the average value of $f(x)$ over a cycle.

The coefficients defined above are the **Fourier coefficients** of $f(x)$.

Example

Find the Fourier coefficients for $f(x) = x + 2,\ -\pi \leqslant x \leqslant \pi$

$$a_0 = \frac{1}{\pi}\int_{-\pi}^{\pi}(x+2)\,dx = \frac{1}{\pi}\left[\frac{x^2}{2} + 2x\right]_{-\pi}^{\pi} = 4$$

$$a_r = \frac{1}{\pi}\int_{-\pi}^{\pi}(x+2)\cos rx\,dx = \frac{1}{\pi}\left\{\left[\frac{(x+2)\sin rx}{r}\right]_{-\pi}^{\pi} - \int_{-\pi}^{\pi} 1\,.\,\frac{\sin rx}{r}\,dx\right\}$$

$$= \frac{1}{\pi}\left\{ 0 - 0 + \left[\frac{\cos rx}{r^2}\right]_{-\pi}^{\pi} \right\} = 0$$

$$b_r = \frac{1}{\pi}\int_{-\pi}^{\pi} (x+2)\sin rx \, dx = \frac{1}{\pi}\left\{ \left[\frac{(x+2)(-\cos rx)}{r} \right]_{-\pi}^{\pi} + \int_{-\pi}^{\pi} 1 \, . \, \frac{\cos rx}{r}\, dx \right\}$$

$$= \frac{1}{\pi}\left\{ \frac{(-\pi+2)}{r}\cos r\pi - \frac{(\pi+2)}{r}\cos r\pi + \left[\frac{\sin rx}{r^2}\right]_{-\pi}^{\pi} \right\}$$

$$= \frac{1}{\pi}\left[\frac{-2\pi\cos r\pi}{r}\right] = \frac{-2}{r}\cos r\pi$$

Now $\cos\pi = -1$, $\cos 2\pi = 1$, $\cos 3\pi = -1$ etc. Hence, $b_1 = \frac{2}{1}$, $b_2 = \frac{-2}{2}$, $b_3 = \frac{2}{3}$, $b_4 = \frac{-2}{4}$, $b_5 = \frac{2}{5}$, ...

Infinite trigonometric series

We saw with Taylor's series that under certain conditions, the more terms that were taken, the better the approximation. What is the effect of allowing n to tend to infinity in the trigonometric series approximation? For a fixed value of n, the Fourier coefficients are those which give the least squares best fit.

Let us expand the integrand in the right-hand side of (7.18) and integrate formally the terms which are independent of $f(x)$. Then

$$I_n = \int_{-\pi}^{\pi} \left\{f(x)\right\}^2 dx - a_0 \int_{-\pi}^{\pi} f(x)\,dx - 2\sum_{k=1}^{n}\left\{\int_{-\pi}^{\pi} f(x)(a_k\cos kx + b_k \sin kx)\,dx\right\} + a_0 \sum_{k=1}^{n}\left\{\int_{-\pi}^{\pi}(a_k\cos kx + b_k\sin kx)\,dx\right\} + \frac{a_0^2\,\pi}{2}$$

$$+ \sum_{k=1}^{n}\sum_{l=1}^{n}\left\{\int_{-\pi}^{\pi}(a_k\cos kx + b_k\sin kx)(a_l\cos lx + b_l\sin lx)\,dx\right\}$$

By using the standard integration formulae and formulae (7.21) we obtain after simplification.

$$I_n = \int_{-\pi}^{\pi}\left\{f(x)\right\}^2 dx - \pi\left[\frac{a_0^2}{2} + \sum_{k=1}^{n}(a_k^2 + b_k^2)\right] \qquad (7.22)$$

Now I_n, being the integral of a squared function can never be negative and $\sum_{k=1}^{n}(a_k^2 + b_k^2)$ is an increasing function of n; hence I_n is a monotonically decreasing function of n. Therefore the approximation of $f(x)$ by a finite trigonometric series with Fourier coefficients (finite Fourier series) improves with increasing n.

We make the remark that the Fourier coefficients tend to zero with increasing n, provided $f(x)$ is an integrable function in $[-\pi, \pi]$.

We have not yet proved that the **Fourier series**

$$\tfrac{1}{2} a_0 + \sum_{k=1}^{\infty} (a_k \cos kx + b_k \sin kx) \tag{7.23}$$

with the coefficients defined by (7.21) will converge for any value of x and even if it does converge, whether the result is $f(x)$. All we have so far demanded of $f(x)$ is that it be integrable in $[-\pi, \pi]$. Now we are in a certain difficulty in so far as the terms in (7.23) are periodic with period 2π. How then could we possibly represent a function such as $f(x)$? The answer is to strive only for a representation in the range $[-\pi, \pi]$ and forget about representation outside this range.

Consider Figure 7.3. In the first diagram we show the function $f(x) = x^2$ and in the second diagram we show the function $g(x) = x^2$, $-\pi < x < \pi$, $g(x + 2\pi) = g(x)$.

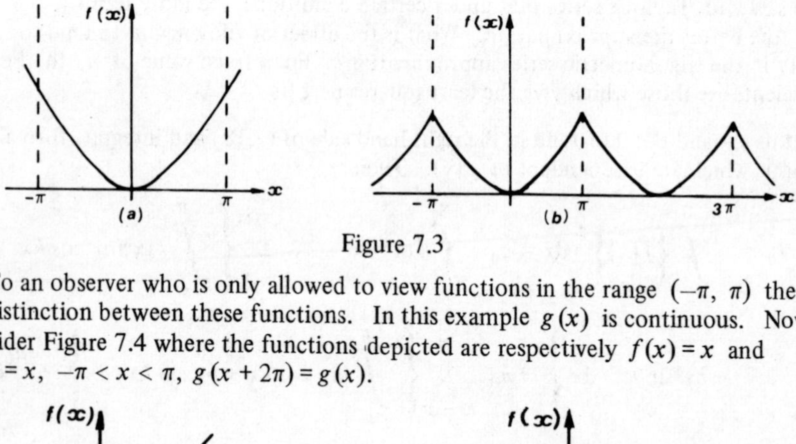

Figure 7.3

To an observer who is only allowed to view functions in the range $(-\pi, \pi)$ there is no distinction between these functions. In this example $g(x)$ is continuous. Now consider Figure 7.4 where the functions depicted are respectively $f(x) = x$ and $g(x) = x$, $-\pi < x < \pi$, $g(x + 2\pi) = g(x)$.

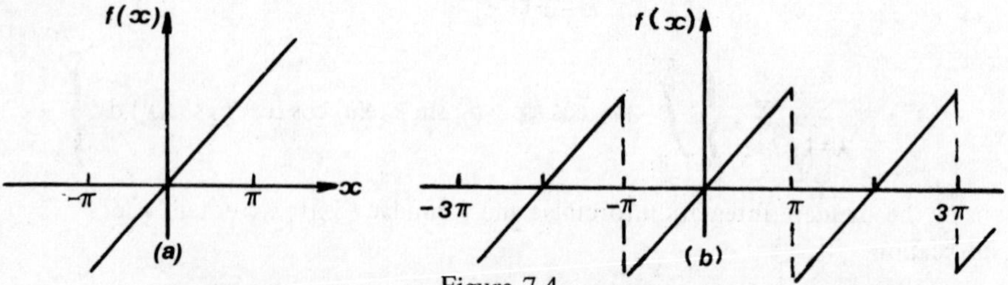

Figure 7.4

Again the functions are indistinguishable in the range $(-\pi, \pi)$ but this time the function $g(x)$ is not continuous. However, at each point of discontinuity, it has a finite jump of -2π. These points are called points of **ordinary discontinuity**.

Dirichlet's Criterion for representation

A well-known sufficient condition for uniform convergence of the Fourier series is due to Dirichlet. It states

If $f(x)$ is a function, defined in $-\pi \leqslant x < \pi$ and extended outside this interval by $f(x+2\pi) = f(x)$, having a finite number of points of ordinary discontinuity in the interval $[-\pi, \pi]$ and having a finite number of local maxima and minima there, then the series (7.23) with the coefficients given by (7.21) converges at every point x_0 of $[-\pi, \pi]$ to the value $\frac{1}{2}[f(x_0+) + f(x_0-)]$.

Note that at points of continuity $f(x_0+) = f(x_0-)$ and therefore convergence is to simply $f(x_0)$, as we should hope ($f(x_0+)$ means that $f(x)$ is evaluated at a value of x slightly greater than x_0). At the points of discontinuity the value converged to is the average of the values on either side. Note that, in comparison with the somewhat restrictive conditions a function has to satisfy to possess a Taylor series, a very wide class of functions do possess Fourier series which represent them. We can, of course, manufacture continuous functions which have infinitely many local maxima and minima in the interval $[-\pi, \pi]$ and hence do not necessarily permit a Fourier series representation, but most of the functions arising in engineering problems do satisfy the Dirichlet criterion.

Further notes on the Fourier coefficients

Note that if the Fourier series approximation including terms up to $\cos nx$ and $\sin nx$ has coefficients $a_0, a_1, \ldots, a_n, b_1, \ldots, b_n$, then to improve the approximation we merely need to calculate extra coefficients a_{n+1}, b_{n+1}; the existing coefficients are unchanged.

A set of continuous functions $\{f_1(x), f_2(x), \ldots, f_r(x), \ldots\}$ which do not vanish identically in the interval $[a, b]$ is said to be **orthogonal** on the interval $[a, b]$ if

$$\int_a^b f_r(x) f_s(x)\,dx = 0, \; r \neq s \quad \text{and} \quad \int_a^b \{f_r(x)\}^2\,dx = d_r^2, \text{ where } d_r \neq 0, \text{ for all } r.$$

If $d_r = 1$ for all r the functions are said to be **orthonormal** on the interval $[a, b]$.

Therefore the functions $\frac{1}{\sqrt{2\pi}}, \frac{\cos x}{\sqrt{\pi}}, \frac{\sin x}{\sqrt{\pi}}, \frac{\cos 2x}{\sqrt{\pi}}, \frac{\sin 2x}{\sqrt{\pi}}, \ldots$ form an orthonormal set on the interval $[-\pi, \pi]$.

The set of orthonormal functions $\phi_1(x), \phi_2(x), \phi_3(x)$ is said to be **complete** if the conditions $\int_a^b \psi(x)\phi_1(x)\,dx = 0$, $\int_a^b \psi(x)\phi_2(x)\,dx = 0$, $\int_a^b \psi(x)\phi_3(x)\,dx = 0$ .... imply that $\psi(x) \equiv 0$ on $[a, b]$.

In general if we assume that a function $f(x)$ which is defined on the interval $[a, b]$ can be expanded in terms of a complete set of orthonormal functions $\phi_1(x), \phi_2(x), \phi_3(x), \ldots$ in the form $f(x) = \sum_{k=1}^{\infty} c_k \phi_k(x)$ then the coefficients can be found by the formulae $a_r = \int_a^b f(x)\phi_r(x)\,dx$. It is now time to analyse some standard wave forms to see the kind of representation that Fourier series provide.

Example 1 – Square Wave

Figure 7.5 shows this wave form, and we now find the Fourier coefficients.

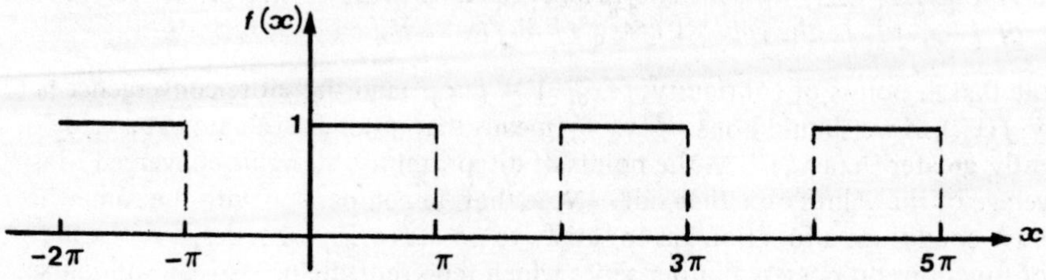

Figure 7.5

We can define this function by $f(x) = \begin{cases} 0, & -\pi < x < 0 \\ 1, & 0 \leqslant x < \pi \end{cases}$ $\quad f(x + 2\pi) = f(x).$

The function is periodic, with period 2π. It arises in practice if a repeated impulse is applied to a mechanical system.

Now the formulae (7.21) for coefficients involve integrations over the interval $[-\pi, \pi]$. We shall split this range into $[-\pi, 0]$ and $[0, \pi]$ to reflect the different definitions of $f(x)$. Hence

$$\pi a_0 = \int_{-\pi}^{\pi} f(x)\,dx = \int_{-\pi}^{0} 0 .\, dx + \int_{0}^{\pi} 1 .\, dx = \pi \text{ and so } a_0 = 1.$$

The fact that $f(0) = 1$ does not alter the value of $\int_{-\pi}^{0} f(x)\,dx$. Further,

$$\pi a_r = \int_{-\pi}^{\pi} f(x) \cos rx \,dx = \int_{-\pi}^{0} 0 .\, dx + \int_{0}^{\pi} \cos rx\,dx = \left[\frac{\sin rx}{r}\right]_0^{\pi} = 0$$

Also, $b_r = \dfrac{1}{\pi r}(1 - \cos r\pi)$. [Verify this].

Now $\cos r\pi = 1$ for even r and -1 for odd r (sketch a graph of $\cos x$) and therefore $b_r = \dfrac{2}{r\pi}$ for odd values of r and $b_r = 0$ for even values of r. Then we may write

$$f(x) = \frac{1}{2} + \sum_{r=1,3,5,..}^{\infty} \frac{2}{r\pi} \sin rx \qquad \text{i.e.}$$

$$f(x) = \frac{1}{2} + \frac{2}{\pi}\left[\frac{\sin x}{1} + \frac{\sin 3x}{3} + \frac{\sin 5x}{5} + \ldots..\right] \tag{7.24}$$

In Figure 7.6 we sketch the first four partial sums (as far as the term in $\sin 5x$) on the graph of $f(x)$. Then, for completeness, we show the partial sum up to the term in $\sin 11x$. The graphs can be extended in either direction by periodicity.

We see that as we take more terms in the series the approximation improves in the following respects. First, the maximum amplitude of the oscillations decreases, meaning that no point is too badly approximated. Second, the frequency of the oscillations increases, meaning that more points (though still only a handful) are exactly represented. Furthermore, the representation stays reasonable over a greater part of the interval $[0, \pi]$ and of the interval $[-\pi, 0]$. However, there is still a poor representation near the points of discontinuity and we shall see that even when we want to represent a continuous function, the necessity to model a periodic function will cause this phenomenon to occur. Note that all partial sum curves pass through the points $(0, \frac{1}{2})$, $(\pi, \frac{1}{2})$, $(-\pi, \frac{1}{2})$.

It is of little use obtaining a result such as (7.24) if we do not get a feel for the approximation by substituting values of x.

Suppose we substitute $x = \pi/2$. The right-hand side of (7.24) becomes

$\frac{1}{2} + \frac{2}{\pi}\left[1 - \frac{1}{3} + \frac{1}{5} \ldots\ldots\ldots\right]$ and since Dirichlet's criterion is met, we are assured that the series in square brackets will converge, in fact to $\pi/4$ (so that the total value is 1). What we are not assured is the rate of convergence of the series; in this instance the spin-off of a series for $\pi/4$ is devalued by the slowness of convergence.

Since we are dealing with a series of sine functions it does not follow that the smaller the value of x, the more rapid the convergence, as was the case with Maclaurin's series. Since the Fourier coefficients are proportional to $1/r$ it means in physical terms that the square wave contains many high-frequency components and if an electronic device will not pass such components then the input will appear in a distorted form as output.

Notice that $a_0 = \frac{1}{2}$ is the average value of $f(x)$ over a cycle of length 2π. Notice further that at $x = n\pi$, for integral n, the right-hand side of (7.24) has all its terms zero except the first and hence predicts the value $\frac{1}{2}$ for $f(x)$.

This value is midway between the values on either side of this discontinuity as predicted by Dirichlet's condition.

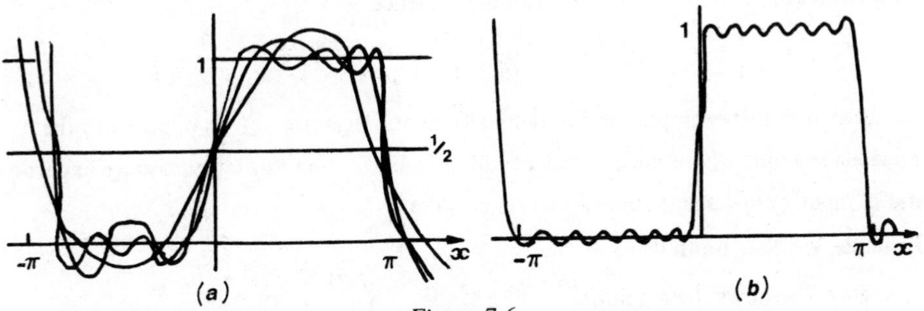

Figure 7.6

Example 2 Rectified cosine wave

Rectifiers are devices which convert alternating currents into direct currents. When the current flows in one direction, the rectifier will conduct, but when the current flows in the opposite direction, the rectifier will not conduct. If we combine two rectifiers we can produce full-wave rectification, in which the 'negative' half-cycles are reversed. A sketch of the resulting waveform is given in Figure 7.7.

The function which represents this waveform is $f(x) = \cos \tfrac{1}{2}x$, $-\pi < x < \pi$ with $f(x + 2\pi) = f(x)$. The function we have chosen is of period 2π. (You can see that the fundamental frequency of this rectified wave is twice that of the original cosine wave which had period 4π.) The term $\tfrac{1}{2}a_0$ is the direct current term since it is not affected by oscillations. It is given by

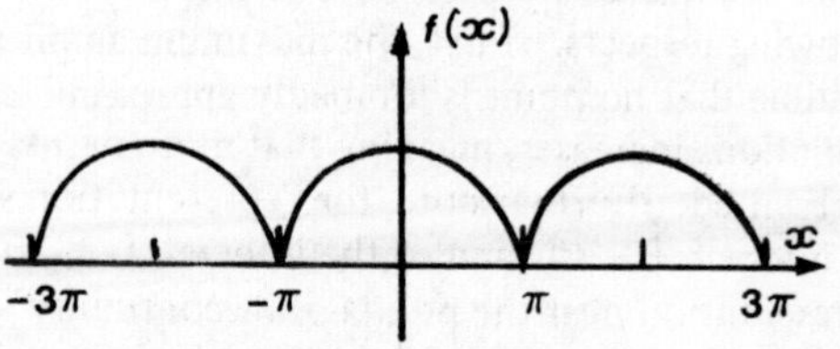

Figure 7.7

$$\pi a_0 = \int_{-\pi}^{\pi} \cos \frac{x}{2}\, dx = \left[2 \sin \frac{x}{2}\right]_{-\pi}^{\pi} = 4. \text{ Therefore } a_0 = 4/\pi.$$

$$\text{Also } \pi a_r = \int_{-\pi}^{\pi} \cos \frac{x}{2} \cos rx\, dx = \tfrac{1}{2}\int_{-\pi}^{\pi} \left\{ \cos (r + \tfrac{1}{2})x + \cos (r - \tfrac{1}{2})x \right\} dx$$

$$= \tfrac{1}{2}\left[\frac{\sin (r + \frac{1}{2})x}{(r + \frac{1}{2})} + \frac{\sin (r - \frac{1}{2})x}{(r - \frac{1}{2})}\right]_{-\pi}^{\pi}$$

$$\text{Hence } a_r = \frac{1}{\pi}\left[\frac{\sin (r + \frac{1}{2})\pi}{(r + \frac{1}{2})} + \frac{\sin (r - \frac{1}{2})\pi}{(r - \frac{1}{2})}\right]$$

Now if r is odd, $\sin (r + \tfrac{1}{2})\pi$ and $\sin (r - \tfrac{1}{2})\pi$ are respectively -1 and 1 and

$$a_r = \frac{1}{\pi}\left\{\frac{-1}{(r + \frac{1}{2})} + \frac{1}{(r - \frac{1}{2})}\right\} = \frac{1}{\pi(r^2 - \frac{1}{4})} = \frac{4}{\pi(4r^2 - 1)}$$

If r is even, $\sin (r + \tfrac{1}{2})\pi$ and $\sin (r - \tfrac{1}{2})\pi$ are respectively 1 and -1. $a_r = \dfrac{-4}{\pi(4r^2 - 1)}$.

Similarly, $\pi b_r = 0$ for all r. [Verify this]. Hence

$$f(x) = \frac{4}{\pi}\left\{\frac{1}{2} + \frac{\cos x}{(4.1^2) - 1} - \frac{\cos 2x}{(4.2^2) - 1} + \frac{\cos 3x}{(4.3^2) - 1} - \cdots\right\}$$

Notice that in this example the Fourier coefficients decay like $\dfrac{1}{r^2}$ as opposed to the previous example where the coefficients decayed like $\dfrac{1}{r}$; in this latter case we expect a useful rapidly convergent approximation to $f(x)$.

Example 3 Saw-tooth wave

Figure 7.8 shows the graph of a waveform frequently occurring in electronics. It is defined by $f(t) = t/\pi$, $-\pi < t < \pi$ with $f(t + 2\pi) = f(t)$. Since we are dealing with a time variable problem we shall use t for the independent variable; do not be put off by this change.

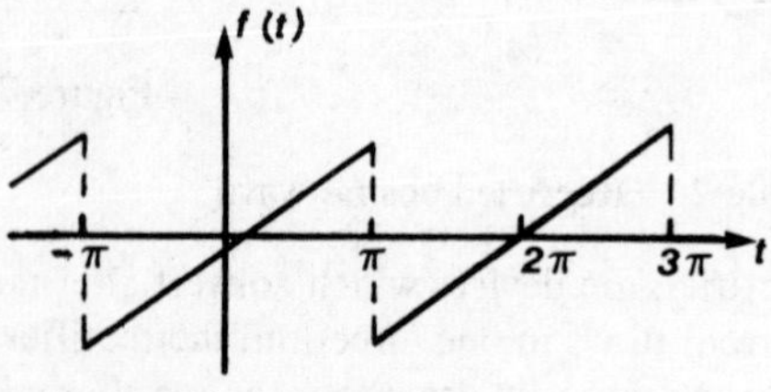

Figure 7.8

Let $f(t) = \frac{1}{2}a_0 + \sum_{k=1}^{\infty} (a_k \cos kt + b_k \sin kt)$. Then

$$\pi a_0 = \int_{-\pi}^{\pi} f(t)\,\mathrm{d}t = \frac{1}{\pi}\int_{-\pi}^{\pi} t\,\mathrm{d}t = 0$$

$$\pi a_r = \frac{1}{\pi}\int_{-\pi}^{\pi} t \cos rt\,\mathrm{d}t = \frac{1}{\pi}\left[\frac{t \sin rt}{r}\right]_{-\pi}^{\pi} - \frac{1}{\pi}\int_{-\pi}^{\pi} 1 \,.\, \frac{\sin rt}{r}\,\mathrm{d}t$$

$$= 0 - 0 - \frac{1}{\pi}\left[\frac{-\cos rt}{r^2}\right]_{-\pi}^{\pi} = 0$$

Hence $a_r = 0$. Likewise $b_r = \dfrac{-2\cos r\pi}{\pi r}$. If r is even, $\cos r\pi = 1$ and $b_r = -2/\pi r$ whereas, if r is odd, $\cos r\pi = -1$ and $b_r = 2/\pi r$. Therefore

$$f(t) = \frac{2}{\pi}\left[\frac{\sin t}{1} - \frac{\sin 2t}{2} + \frac{\sin 3t}{3} - \ldots\right].$$

Problems

1. For each of the functions below, defined in the range $[-\pi, \pi]$ and assumed to have period 2π, find their Fourier series.

   (a) $f(x) = 2,\ -\pi < x < 0;\ f(x) = 1,\ 0 < x < \pi$
   (What is the value of the sum of the series when $x = -\pi, 0, \pi$? Comment.)

   (b) $f(x) = 0,\ -\pi < x < 0;\ f(x) = x,\ 0 < x < \pi$

   (c) $f(x) = x^2,\ -\pi < x < 0;\ f(x) = x^2,\ 0 < x < \pi$

   (d) $f(x) = |\sin x|,\ -\pi < x < \pi$

   (e) $f(x) = -\frac{1}{2},\ -\pi < x < 0;\ f(x) = \frac{1}{2},\ 0 < x < \pi$
   (What is the value of the sum of the series when $x = -\pi, 0, \pi$? Comment.)

   (f) $f(x) = 1,\ -\pi < x < 0;\ f(x) = x,\ 0 < x < \pi$

   (g) $f(x) = 0,\ -\pi < x < 0;\ f(x) = 1,\ 0 < x < \pi/2;\ f(x) = -1,\ \pi/2 < x < \pi$

2. A function $f(x)$ is of period 2π and is defined in the interval $0 \leqslant x \leqslant 2\pi$ in the form $f(x) = \sin x,\ 0 \leqslant x \leqslant \pi;\ f(x) = 0,\ \pi \leqslant x \leqslant 2\pi$. Show that

   $$f(x) = \frac{1}{\pi} + \frac{1}{2}\sin x - \frac{2}{\pi}\sum_{n=1}^{\infty} \frac{\cos 2nx}{4n^2 - 1}.$$

   Deduce the sum of the infinite series $\frac{1}{1.3} - \frac{1}{3.5} + \frac{1}{5.7}\ \ldots$ (L.U.)

3. A function $f(x)$ is defined in the range $(0, \pi)$ as follows: $f(x) = x$ for $0 \leqslant x < \frac{1}{4}\pi$, $f(x) = \frac{1}{4}\pi$ for $\frac{1}{4}\pi \leqslant x < \frac{3}{4}\pi$, $f(x) = \pi - x$ for $3\pi/4 < x < \pi$. If, in addition, $f(\pi + x) = -f(x)$ for all values of x, expand $f(x)$ as a Fourier series, giving the first 3 non-zero terms and the general term. (L.U.)

4. If $f(t)$ is a periodic function of period 2π and $f(t) = t/\pi$ for $0 < t < \pi$, $f(t) = (2\pi - t)/\pi$ for $\pi < t < 2\pi$, show that

$$f(t) = \frac{1}{2} - \frac{4}{\pi^2} \sum_{1}^{\infty} \frac{\cos(2m+1)t}{(2m+1)^2}$$

. Show also, that if ω is not an integer,

$$y = \frac{1}{2\omega^2}(1 - \cos\omega t) - \frac{4}{\pi^2} \sum_{m=1}^{\infty} \frac{\cos(2m+1)t - \cos\omega t}{(2m+1)^2[\omega^2 - (2m+1)^2]}$$

satisfies the equation $d^2y/dt^2 + \omega^2 y = f(t)$ with the initial conditions $y = dy/dt = 0$ when $t = 0$. (L.U.)

5. If $f(x)$ is an even periodic function with period 2π defined by $f(x) = x^2$ for $0 \leqslant x \leqslant \pi$, sketch its graph in the range $-3\pi \leqslant x \leqslant 3\pi$. Show that the Fourier series for $f(x)$ is

$$f(x) = \frac{\pi^2}{3} + 4 \sum_{n=1}^{\infty} \frac{(-1)^n \cos nx}{n^2}$$

Obtain a Fourier sine series, valid in the range $0 < x < \pi$, for the function $f(x)$. (L.U.)

7.5 ODD AND EVEN FUNCTIONS; HALF-RANGE SERIES

The rectified wave of Example 2 of the last section gave rise to a series without sine terms, whereas the saw-tooth wave gave rise to a series containing sine terms only. The rectified waveform is an even function of x and the saw-tooth wave form is an odd function of t. It can easily be shown that for an even function $f(x)$ the Fourier coefficients $b_r = 0$ for all r and as a bonus

$$a_0 = \frac{2}{\pi} \int_0^{\pi} f(x)\,dx, \quad a_r = \frac{2}{\pi} \int_0^{\pi} f(x) \cos rx\,dx$$

If $f(x)$ is an odd function, the Fourier coefficients $a_r = 0$ for all r and $a_0 = 0$. Furthermore

$$b_r = \frac{2}{\pi} \int_0^{\pi} f(x) \sin rx\,dx$$

Check these properties with Examples 2 and 3.

Consequently we should always check at the outset whether a function is even or odd: we could save ourselves much calculation.

To reinforce these ideas, let us look at two variants on the square wave of Example 1 of the last section.

Example 1. Anti-symmetric square wave

In Figure 7.9(a) we sketch the waveform $f(t) = \begin{cases} -1, & -\pi < t < 0 \\ 1, & 0 < t < \pi \end{cases}$ with $f(t + 2\pi) = f(t)$. The function is odd and therefore: $a_0 = a_r = 0$.

Now $b_r = \frac{2}{\pi}\int_0^\pi f(t)\sin rt\,dt$, therefore $\frac{\pi b_r}{2} = \int_0^\pi 1\,.\,\sin rt\,dt = \left[\frac{-\cos rt}{r}\right]_0^\pi$

i.e. $b_r = \frac{2}{r\pi}(1 - \cos r\pi)$. If r is even, $b_r = 0$ whereas if r is odd, $b_r = 4/r\pi$.

Therefore $f(t) = \frac{4}{\pi}\left[\frac{\sin t}{1} + \frac{\sin 3t}{3} + \frac{\sin 5t}{5} + \text{.........}\right]$

Compare this series with the one for Example 1 of the last sub-section and comment on the similarities and differences.

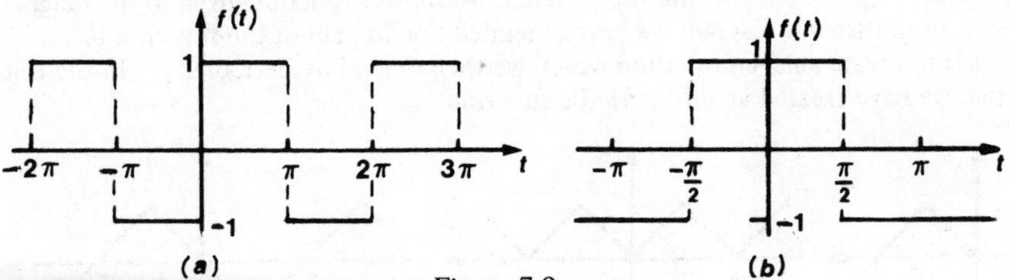

Figure 7.9

Example 2. Symmetric square wave

In Figure 7.9(b) the wave form is sketched whose definition is

$$f(t) = \begin{cases} -1, & -\pi < t < -\frac{\pi}{2} \\ 1, & -\frac{\pi}{2} < t < \frac{\pi}{2} \\ -1, & \frac{\pi}{2} < t < \pi \end{cases} \qquad f(t + 2\pi) = f(t).$$

The function is even and therefore $b_r = 0$

$$a_0 = \frac{2}{\pi}\int_0^\pi f(t)\,dt = \frac{2}{\pi}\int_0^{\pi/2} 1\,dt + \frac{2}{\pi}\int_{\pi/2}^{\pi}(-1)\,.\,dt = 0$$

This we could have expected, since the average value of $f(t)$ over a cycle is 0.

$$a_r = \frac{2}{\pi}\int_0^\pi f(t)\cos rt\,dt = \frac{2}{\pi}\int_0^{\pi/2}\cos rt\,dt + \frac{2}{\pi}\int_{\pi/2}^{\pi}(-\cos rt)\,dt$$

$$= \frac{2}{\pi}\left[\frac{\sin rt}{r}\right]_0^{\pi/2} + \frac{2}{\pi}\left[\frac{-\sin rt}{r}\right]_{\pi/2}^{\pi} = \frac{4\sin(r\pi/2)}{r\pi}$$

If r is even, $a_r = 0$. If $r = 1, 5, 9, \ldots\ldots$, $a_r = 4/r\pi$, whereas if $r = 3, 7, 11, \ldots\ldots$, $a_r = -4/r\pi$. Hence

$$f(t) = \frac{4}{\pi}\left[\frac{\cos t}{1} - \frac{\cos 3t}{3} + \frac{\cos 5t}{5} \text{.........}\right]$$

Note that we should be able to deduce this result from the result of the previous example by substituting $(t + \pi/2)$ for t. Try the substitution.

In this, as well as the last example, the even harmonics were zero. In the next section we investigate the underlying cause.

Half-range series

We can often make use of the oddness or evenness of a function giving rise to a sine series or a cosine series respectively. In the solution of some partial differential equations, the boundary conditions may restrict us to a series which contains only sine terms (see Section 8.4). We shall therefore need to investigate how to manufacture an odd function or an even function, given a function which may be neither.

Consider Figure 7.10. In the first diagram we show a function given in the range $[0, \pi]$. In the second diagram we have extended the domain of the function to $[-\pi, \pi]$ to create an even function which we then extend by periodicity. In the third diagram we have created an odd periodic function.

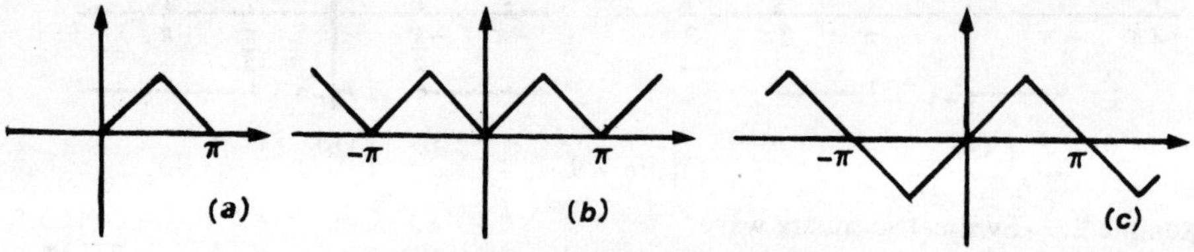

Figure 7.10

Therefore the function defined on $[0, \pi]$ can be represented by either a cosine series or a sine series.

If the function of Figure 7.10(a) be $f(t) = \begin{cases} t & 0 < t < \frac{\pi}{2} \\ \pi - t, & \frac{\pi}{2} < t < \pi \end{cases}$ then for the function of Figure 7.10(b)

$$b_r = 0,\ a_0 = \frac{2}{\pi}\int_0^{\pi/2} t\,dt + \frac{2}{\pi}\int_{\pi/2}^{\pi} (\pi - t)\,dt \quad \text{and}$$

$$a_r = \frac{2}{\pi}\int_0^{\pi/2} t\cos rt\,dt + \frac{2}{\pi}\int_{\pi/2}^{\pi} (\pi - t)\cos rt\,dt$$

This gives rise to the series

$$f(t) = \frac{\pi}{4} - \frac{2}{\pi}\left[\frac{\cos 2t}{1^2} + \frac{\cos 6t}{3^2} + \frac{\cos 10t}{5^2} + \ldots\ldots\ldots\right] \tag{7.25a}$$

For the function of Figure 7.10(c) $a_0 = 0,\ a_r = 0,$ and the series obtained is,

$$f(t) = \frac{4}{\pi}\left[\frac{\sin t}{1^2} - \frac{\sin 3t}{3^2} + \frac{\sin 5t}{5^2} - \ldots\ldots\ldots\right] \tag{7.25b}$$

For purposes of comparison we truncated the two series to the terms shown and estimated $f(t)$ for $t = 0,\ \pi/8,\ \pi/4,\ 3\pi/8,\ \pi/2$. The results are shown in Table 7.6

Table 7.6

t	$f(t)$	Approximation (7.25a)		Approximation (7.25b)	
		Estimate (4 d.p.)	Error	Estimate (4.d.p.)	Error
0	0	0.0526	0.0526	0	0
$\pi/8$	0.3927	0.4033	0.0106	0.4036	0.0109
$\pi/4$	0.7854	0.7854	0	0.7643	0.0211
$3\pi/8$	1.1781	1.1675	−0.0106	1.2110	0.0329
$\pi/2$	1.5708	1.5182	−0.0526	1.4656	−0.1052

Problems

1. Which of the following functions are odd, even or neither?

   (a) $\cos x - \cos 5x$, (b) $\sin x - 5\cos 5x$, (c) $2 + x$, (d) $x/\{(x+1)(x-1)\}$,
   (e) $x \sin x$, (f) $\sin x + \cos x$.

2. Find the Fourier series for the function

   $f(x) = 0,\ -\pi < x < -\tfrac{1}{2}\pi;\quad f(x) = \cos x,\ -\tfrac{1}{2}\pi \leqslant x \leqslant \tfrac{1}{2}\pi;\quad f(x) = 0,\ \tfrac{1}{2}\pi < x < \pi.$

3. Find the Fourier sine series and the Fourier cosine series for each of the following functions and graph the two periodic extensions for each case.

   (a) $f(x) = 2,\ 0 < x < \pi$; (b) $f(x) = x^2,\ 0 < x < \pi$; (c) $f(x) = x + 1,\ 0 < x < \pi$.

4. Given that $f(x) = 1 + \cos x,\ 0 < x < \pi$, find a half-range Fourier sine series for $f(x)$.
   Find the sum of the series when $x = 3\pi/2$ and show that $\frac{\pi}{4} = 1 - \frac{1}{3} + \frac{1}{5} - \frac{1}{7} + \frac{1}{9} - \ldots\ldots$

5. Prove that for $0 < x < \pi$, $\sin x = \frac{2}{\pi} - \frac{4}{\pi}\left[\frac{\cos 2x}{2^2 - 1} + \frac{\cos 4x}{4^2 - 1} + \frac{\cos 6x}{6^2 - 1} + \ldots\right]$.

   Sketch for $-2\pi \leqslant x \leqslant 3\pi$ the graph of the function represented by the right hand side of this equation.

   Find the sum of the series $\frac{1}{1.3} - \frac{1}{3.5} + \frac{1}{5.7} - \frac{1}{7.9} + \ldots$ (L.U.)

6. Show that the Fourier cosine series for the function $f(x) = x$, where $0 \leqslant x \leqslant \pi$, is
   $\frac{\pi}{2} - \frac{4}{\pi}\sum_{0}^{\infty}\frac{\cos(2n+1)x}{(2n+1)^2}$. Express $f(x)$ as a Fourier sine series. Sketch the graph of each of these series for $-2\pi < x < 3\pi$. (L.U.)

7. Obtain the Fourier expansion of an even function $f(x)$ defined by

   $f(x) = x$, for $0 \leqslant x < \pi/3$; $f(x) = 0$, for $\pi/3 \leqslant x \leqslant 2\pi/3$;
   $f(x) = x - 2\pi/3$, for $2\pi/3 < x \leqslant \pi$,

   and simplify the coefficients of the first four terms. (L.U.)

7.6 FURTHER FEATURES OF FOURIER SERIES

In this section we develop some more results for Fourier series. Consider the series representation

$$f(x) = \tfrac{1}{2}a_0 + \sum_{k=1}^{\infty} (a_k \cos kx + b_k \sin kx) \tag{7.16}$$

Suppose we replace x by $(x + \pi)$, then

$$f(x+\pi) = \tfrac{1}{2}a_0 + \sum_{k=1}^{\infty} (a_k \cos kx \cos k\pi + b_k \sin kx \cos k\pi)$$

Now, if k is even $\cos k\pi = 1$ whereas if k is odd, $\cos k\pi = -1$.

Therefore the two terms $a_k \cos kx \cos k\pi$ and $b_k \sin kx \cos k\pi$ are equal to their counterparts in (7.16) when k is even but are the negatives of their counterparts if k is odd.

If we require $f(x + \pi) = f(x)$ then the odd harmonics vanish so that

$$f(x+\pi) = f(x) = \tfrac{1}{2}a_0 + \sum_{k=2,4,6\ldots}^{\infty} (a_k \cos kx + b_k \sin kx) \tag{7.26}$$

In this case $f(x)$ is of period π ; it is also of period 2π.

Example

Consider the function shown in Figure 7.11(a). It is specified either by

$$f(x) = \begin{cases} x\ , & 0 < x < \pi \\ x - \pi, & \pi < x < 2\pi \end{cases} \qquad f(x + 2\pi) = f(x)$$

or by $f(x) = x,\ 0 < x < \pi;\ f(x + \pi) = f(x)$.

We shall adapt the formulae (7.21) to extend over the interval $[0, 2\pi]$ instead of $[-\pi, \pi]$. We find

$$\pi a_0 = \int_0^{2\pi} f(x)\,dx = \int_{\pi}^{2\pi} (x - \pi)\,dx + \int_0^{\pi} x\,dx = \frac{\pi^2}{2} + \frac{\pi^2}{2} = \pi^2$$

Hence $a_0 = \pi$. Similarly, $a_r = 0$ and $b_r = -\dfrac{1}{r}(1 + \cos r\pi)$.

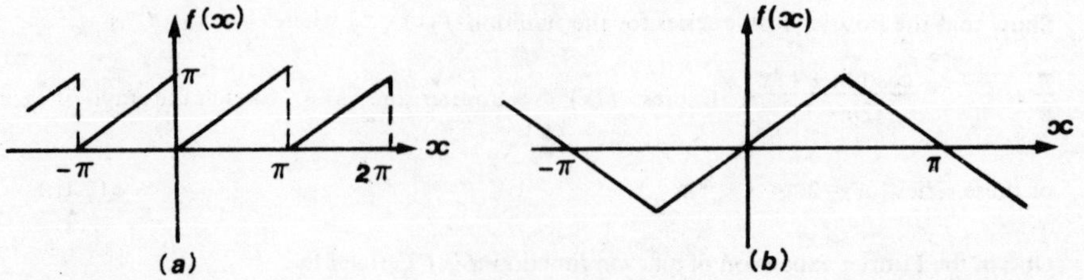

Figure 7.11

Therefore, $b_r = 0$ for r odd and $b_r = -2/r$ for r even. Hence

$$f(x) = \frac{\pi}{2} - 2\left[\frac{\sin 2x}{2} + \frac{\sin 4x}{4} + \frac{\sin 6x}{6} + \dots\right]$$

and the odd harmonics have vanished as promised.

The function graphed in Figure 7.11(b) can be defined by $f(x) = x$, $\frac{-\pi}{2} < x < \frac{\pi}{2}$; $f(x + \pi) = -f(x)$

By the standard method we can derive the Fourier series as

$$f(x) = \frac{4}{\pi}\left[\frac{\sin x}{1} - \frac{\sin 3x}{3} + \frac{\sin 5x}{5} - \dots\right]$$

the even harmonics have vanished.

This function satisfies the condition $f(x - \pi) = -f(x)$.

We could have calculated the non-zero coefficients by $2\int_0^{\pi}$ instead of $\int_{-\pi}^{\pi}$ as indeed we could have done for the previous example.

Just as any function can be written as the sum of an odd and an even function, any function can be written as the sum of a function with odd harmonics only and a function with even harmonics only. Now we can begin to see a way of producing a Fourier series with fewer terms and hence a greater rate of convergence.

Consider the function $f(x) = x$, $0 < x < \pi/2$. We can first extend the function to $f(x) = x$, $-\pi/2 < x < 0$, i.e. making it odd, and then extend it by periodicity to the function shown in Figure 7.11(b). We shall then lose all but the sine terms; of these we shall lose either the odd or the even ones.

General period of a function

Suppose $f(x)$ has a period T and suppose its basic domain of definition is $[a, a + T]$. Then it will be capable of a Fourier series representation under the Dirichlet conditions as

$$f(x) = \tfrac{1}{2}a_0 + \sum_{k=1}^{\infty}\left(a_k \cos\frac{2\pi kx}{T} + b_k \sin\frac{2\pi kx}{T}\right) \tag{7.27}$$

where the coefficients are given by

$$\left.\begin{aligned} a_0 &= \frac{2}{T}\int_a^{a+T} f(x)\,\mathrm{d}x \\ a_r &= \frac{2}{T}\int_a^{a+T} f(x)\cos\frac{2\pi rx}{T}\,\mathrm{d}x \\ b_r &= \frac{2}{T}\int_a^{a+T} f(x)\sin\frac{2\pi rx}{T}\,\mathrm{d}x \end{aligned}\right\} \tag{7.28}$$

Example 1

The function $f(x) = x$ is to be represented by a Fourier sine series in the interval $[0, 4]$.

We first extend the function by defining $f(x) = 8 - x$ in $[4, 8]$ and then create an odd function in $[-8, 8]$. We then have $f(x + 16) = f(x)$; [c.f. Figure 7.11(b)] and we get a Fourier series representation by taking our basic interval as $[-8. 8]$. Since the function is odd, we need only consider sine terms. Then the representation

$$f(x) = \tfrac{1}{2}a_0 + \sum_{k=1}^{\infty} \left(a_k \cos \frac{2\pi kx}{16} + b_k \sin \frac{2\pi kx}{16} \right)$$

has $a_0 = a_r = 0$ and $b_r = \frac{2}{16} \cdot \int_{-8}^{8} f(x) \sin \frac{2\pi rx}{16}\, dx$

$$= \frac{4}{16} \left\{ \int_0^4 x \sin \frac{\pi rx}{8}\, dx + \int_4^8 (8 - x) \sin \frac{\pi rx}{8}\, dx \right\}$$

Hence $b_r = \frac{32}{\pi^2 r^2} \sin \frac{\pi r}{2}$. Clearly, $b_r = 0$ if r is even. It follows that

$$f(x) = \frac{32}{\pi^2} \left\{ \frac{\sin \frac{\pi x}{8}}{1^2} - \frac{\sin \frac{3\pi x}{8}}{3^2} + \frac{\sin \frac{5\pi x}{8}}{5^2} - \cdots\cdots \right\}$$

Suppose we had tried to represent $f(x)$ by a Fourier series based on $[-4, 4]$ simply by making the extension $f(x) = x$ on $[-4, 0]$ to make it odd. Then we should have found

$$f(x) = \sum_{k=1}^{\infty} b_k \sin \frac{2\pi kx}{8} \quad \text{where } b_r = 2 \cdot \frac{2}{8} \int_0^4 x \sin \frac{2\pi rx}{8}\, dx = -\frac{8}{\pi r} \cos r\pi$$

Therefore $f(x) = \frac{8}{\pi} \left[\frac{\sin \pi x/4}{1} - \frac{\sin 2\pi x/4}{2} + \frac{\sin 3\pi x/4}{3} - \cdots\cdots \right]$. Compare the series after four terms each for $x = 0(\tfrac{1}{2})4$.

Example 2

A uniform horizontal beam of length l, freely supported at its ends carries a point load W at a distance a from one end. Show that the deflection is the same as would be produced by a non-uniformly distributed load $w(x)$ per unit length where

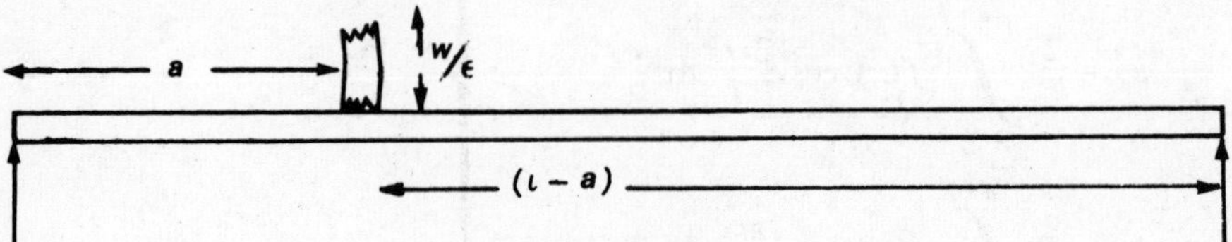

Figure 7.12

$$w(x) = \frac{2W}{l}\left[\sin\frac{\pi a}{l}\,.\,\sin\frac{\pi x}{l} + \sin\frac{2\pi a}{l}\,.\,\sin\frac{2\pi x}{l} + \ldots\ldots\ldots\right]$$

Find the deflected profile.

The approximation we make is that the point load can be replaced by a uniformly distributed load W/ϵ over a small distance ϵ as shown in Figure 7.12.

We seek a sine series representation and therefore the function we choose is

$$f(x) = \begin{cases} 0, & 0 < x < a \\ W/\epsilon, & a < x < a+\epsilon \\ 0, & a+\epsilon < x < l \end{cases} \qquad f(x) \text{ odd and } f(x+2l) = f(x)$$

$$\text{Then } f(x) = \sum_{r=1}^{\infty} b_r \sin\frac{r\pi x}{l}, \qquad b_r = \frac{2}{l}\int_0^l f(x)\sin\frac{r\pi x}{l}\,dx \qquad \text{i.e.}$$

$$b_r = \frac{2W}{l\epsilon}\int_a^{a+\epsilon} \sin\frac{r\pi x}{l}\,dx = \frac{2W}{l\epsilon}\left[-\frac{l}{r\pi}\cos\frac{r\pi x}{l}\right]_a^{a+\epsilon}$$

$$= \frac{2W}{r\pi\epsilon}\left[-\cos\frac{r\pi}{l}(a+\epsilon) + \cos\frac{r\pi}{l}a\right] = \frac{4W}{r\pi\epsilon}\sin\frac{r\pi\,(a+\tfrac{1}{2}\epsilon)}{l}\sin\frac{r\pi\epsilon}{2l}$$

$$\simeq \frac{4W}{r\pi\epsilon}\sin\frac{r\pi a}{l}\,.\,\frac{r\pi\epsilon}{2l} \quad \text{if } \epsilon \text{ is small i.e. } b_r \simeq \frac{2W}{l}\sin\frac{r\pi a}{l}$$

Hence the result follows.

To obtain the deflected profile we use the formula $EI\dfrac{d^4y}{dx^4} = w(x)$. Integrating four times and using the boundary conditions applicable to this problem we obtain for the deflected profile

$$y = \frac{2Wl^3}{\pi^4 EI}\left[\sin\frac{\pi a}{l}\,.\,\sin\frac{\pi x}{l} + \frac{1}{2^4}\sin\frac{2\pi a}{l}\,.\,\sin\frac{2\pi x}{l} + \ldots\ldots\right]$$

Gibbs Phenomenon

In Figure 7.6 we saw how the Fourier series representation of a square wave exhibited the phenomenon of breaking away from the function at points of ordinary discontinuity. Since in practice we only ever take a finite number of terms in the series approximation, it is important to know how many terms to take in order to overcome this poorness of approximation. It is interesting to note that the first observation of this phenomenon was reported by Michelson, who had constructed a machine to analyse numerically a waveform into its harmonics. Gibbs then investigated the phenomenon mathematically.

An infinite Fourier series will converge slowly near a discontinuity. The high-frequency oscillations of the truncated series around the true function value will always be present, but their amplitudes are usually small enough not to cause concern. Near a discontinuity, they become noticeable and unfortunately we cannot eliminate them by taking more terms of the series. We can merely reduce the range of x over which they are noticeable; we cannot reduce the amplitude at the discontinuity.

To investigate the phenomenon, we examine the error involved in approximating the square wave

$$f(x) = \begin{cases} -1, & -\pi < x < 0 \\ 1, & 0 < x < \pi \end{cases} \qquad f(x + 2\pi) = f(x)$$

We have chosen this form of square wave for simplicity of algebra.

Let the partial sum up to the nth harmonic be

$$S_n(x) = \tfrac{1}{2}a_0 + \sum_{k=1}^{n} (a_k \cos kx + b_k \sin kx)$$

It can be shown after much algebra that

$$2\pi S_n(x) = \int_{-x}^{x} \frac{\sin(n+\tfrac{1}{2})\theta}{\sin\theta/2}\, d\theta - \int_{\pi-x}^{\pi+x} \frac{\sin(n+\tfrac{1}{2})\theta}{\sin\theta/2}\, d\theta$$

We have expressed $S_n(x)$ in terms of two integrals, one centred on 0 and the other centred on π: the two points of ordinary discontinuity.

The second integral is evaluated over a region where the size of the integrand is approximately 1. The first integrand has a value at $\theta = 0$ of $(2n+1)$, as we can discover from L'Hopital's rule. Therefore, for values of n which are not small (and this will be the case for the situation we are studying) we can neglect the second integral and use as our approximation

$$S_n(x) \simeq \frac{1}{2\pi} \int_{-x}^{x} \frac{\sin(n+\tfrac{1}{2})\theta}{\sin\theta/2}\, d\theta = \frac{1}{\pi} \int_{0}^{x} \frac{\sin(n+\tfrac{1}{2})\theta}{\sin\theta/2}\, d\theta$$

since the integrand is even. Further, since we are interested in values of θ near the origin we approximate $\sin\theta/2$ by $\theta/2$ and put $m = (n + \tfrac{1}{2})$ to obtain

$$S_n(x) \simeq \frac{2}{\pi} \int_{0}^{x} \frac{\sin m\theta}{\theta}\, d\theta$$

We wrote a computer program and obtained a table of values for $S_n(x)$ from which we produced the graph shown in Figure 7.13(a).

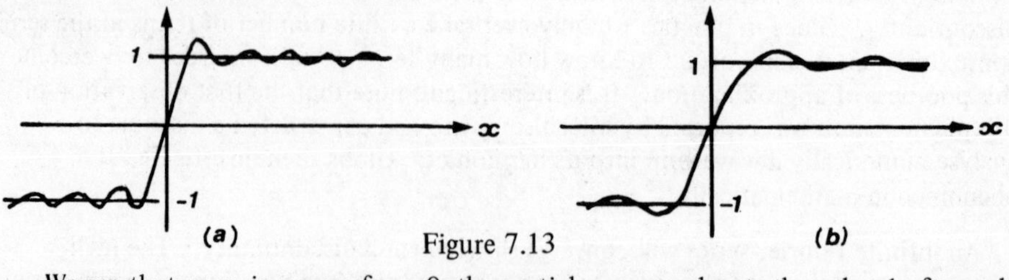

Figure 7.13

We see that as x increases from 0, the partial sum overshoots the value 1 after only a small interval of x. It reaches a maximum value and then oscillates about the value 1 with decaying amplitude; similar remarks apply for negative values of x. It can be shown that this first maximum has a value of about 1.18 irrespective of the value of n. Its position does change, however, moving nearer $x = 0$ as n increases.

Several methods of overcoming this problem have been proposed. The one we mention here is the Lanczos smoothing factor. The basic idea is to replace the approximate value $S_n(x) \simeq \frac{1}{2}a_0 + \sum_{k=1}^{n} (a_k \cos kx + b_k \sin kx)$ by the average of $S_n(x)$ between $(x - \pi/n)$ and $(x + \pi/n)$. It can be shown that this produces the revised approximation

$$\frac{1}{2}a_0 + \sum_{k=1}^{n} \frac{\sin(\pi k/n)}{\pi k/n}(a_k \cos kx + b_k \sin kx) \tag{7.29}$$

We have multiplied the oscillatory terms by the sigma-factors $\sigma_k = \dfrac{\sin(\pi k/n)}{\pi k/n}$

For consistency we can take $\sigma_0 = 1$.

The effect of introducing the sigma factors is shown in Figure 7.13(b) where we see the approximations up to $\cos 20x$ and $\sin 20x$.

The price we pay for reducing the maximum amplitude is that the rate of rise to this first maximum is slowed by half.

Parseval's result

Assume that $f(x)$ has a Fourier series representation (7.16). Parseval's result is that for a suitably integrable function $f(x)$,

$$\frac{1}{2\pi}\int_{-\pi}^{\pi}\left\{f(x)\right\}^2 dx = \tfrac{1}{4}a_0^2 + \tfrac{1}{2}\sum_{k=1}^{\infty}(a_k^2 + b_k^2) \tag{7.30}$$

The left-hand side is the **power** in the wave form given by $f(x)$.

Example 1

Find the Fourier series expansion for the function $f(x)$ defined by $f(x) = \begin{cases} -x, & -\pi < x < 0 \\ x, & 0 < x < \pi \end{cases}$ $f(x + 2\pi) = f(x)$ and deduce the power in the wave form $g(x) = x^2$, $-\pi < x < \pi$, $g(x + 2\pi) = g(x)$.

The first part we shall leave to you (including a sketch of $f(x)$). The Fourier coefficients are $a_0 = \pi$, $b_r = 0$, $a_r = \dfrac{2}{\pi r^2}[\cos r\pi - 1]$. Now $g(x) = [f(x)]^2$ and from Parseval's result

$$\frac{1}{2\pi}\int_{-\pi}^{\pi} x^2\, dx = \frac{1}{4}\cdot\pi^2 + \frac{8}{\pi^2}\sum_{k=1}^{\infty}\frac{1}{(2k-1)^4} \text{ which is the power.}$$

Since the left-hand side can be integrated directly to produce the result $\dfrac{\pi^2}{3}$ we may deduce the result that

$$\frac{\pi^4}{96} = \frac{1}{1^4} + \frac{1}{3^4} + \frac{1}{5^4} + \ldots\ldots\ldots$$

Example 2

The root-mean-square (RMS) value of $f(x)$ of period T is defined by

$$(\text{RMS value})^2 = \frac{1}{T}\int_0^T [f(x)]^2 \, dx$$

Find the RMS value of the wave $e = \sum_{k=1}^{\infty} E_k \sin(kwx + \alpha_k)$

Then using the result that ½ a_0 is the mean value of $f(x)$ over one cycle, denoted by $\bar{f}$, say, we have, from the generalised form of Parseval's result the relationship

$$(\text{RMS value})^2 = \bar{f}^2 + \tfrac{1}{2}\sum_{k=1}^{\infty}(a_k^2 + b_k^2)$$

For the given wave,

$$e = \sum_{k=1}^{\infty}\left\{E_k \sin\alpha_k \cos kwx + E_k \cos\alpha_k \sin kwx\right\}$$

The period $T = 2\pi/w$ and therefore, since $\bar{f} = 0$

$$(\text{RMS value})^2 = \tfrac{1}{2}\sum_{k=1}^{\infty}\left[(E_k \sin\alpha_k)^2 + (E_k \cos\alpha_k)^2\right] = \tfrac{1}{2}\sum_{k=1}^{\infty} E_k^2$$

Integration of Fourier series

Since integration is a smoothing process, we expect that integration of a Fourier series is permissible. We quote the result that if $f(x)$ satisfies the Dirichlet condition then we can integrate its Fourier series term by term over the range $[a, x] \subset [-\pi, \pi]$ to obtain a series which converges to the integral of the original function. That is,

$$\int_a^x f(x)\,dx = \frac{a_0}{2}[x - a] + \sum_{k=1}^{\infty}\frac{1}{k}[b_k(\cos ka - \cos kx) + a_k(\sin kx - \sin ka)]$$

Note that the coefficients have an additional factor of k in the denominator and the new series should converge more rapidly than the original series. However, unless $a_0 = 0$ the new series will not be a *Fourier series* although the series for

$\int_a^x f(x)\,dx - a_0[x - a]/2$ will be.

Example

Find the Fourier series of $f(x) = \begin{cases} -1, & -\pi < x < 0 \\ 1, & 0 < x < \pi \end{cases}$ $\quad f(x + 2\pi) = f(x)$

and deduce the Fourier series for the function $g(x) = |x|$, $-\pi < x < \pi$, $g(x + 2\pi) = g(x)$.

The Fourier series for $f(x)$ can be found to be $\frac{4}{\pi} \sum_{k=1,3,5,...}^{\infty} \frac{\sin kx}{k}$

Here $a_0 = 0$ since $f(x)$ is odd; also $a_r = 0$, $b_r = \frac{4}{r\pi}$ if r is odd and, $b_r = 0$ if r is even. Now

$$g(x) = \int_0^x f(x)\,dx = \sum_{k=1,3,5...}^{\infty} \frac{1}{k} \cdot \frac{4}{\pi} \cdot \frac{1}{k} (1 - \cos kx)$$

$$= \frac{4}{\pi} \sum_{k=1,3,5...}^{\infty} \frac{1}{k^2} - \frac{4}{\pi} \sum_{k=1,3,5...}^{\infty} \frac{\cos kx}{k^2}$$

Now, if we put $x = \pi$, the right hand side becomes $2 \cdot \frac{4}{\pi} \sum_{k=1,3,5...}^{\infty} \frac{1}{k^2}$, since all the cosine terms are -1. Since $g(x)$ is continuous at $x = \pi$ and $g(\pi) = \pi$, it follows that $\frac{8}{\pi} \sum_{k=1,3,5...}^{\infty} \frac{1}{k^2} = \pi$ and therefore $g(x) = \frac{\pi}{2} - \frac{4}{\pi} \sum_{k=1,3,5...}^{\infty} \frac{\cos kx}{k^2}$

Differentiation of Fourier series

Since differentiation tends to destroy smooth properties of a function, we cannot expect the term-by-term differentiation of the Fourier series for $f(x)$ to converge uniformly to $f'(x)$.

Now let

$$f(x) = \tfrac{1}{2} a_0 + \sum_{k=1}^{\infty} (a_k \cos kx + b_k \sin kx),$$

as usual and suppose $f'(x)$ can have a Fourier series representation

$$f'(x) = \tfrac{1}{2} A_0 + \sum_{k=1}^{\infty} (A_k \cos kx + B_k \sin kx).$$

Then

$$A_0 = \frac{1}{\pi} \int_{-\pi}^{\pi} f'(x)\,dx = \frac{1}{\pi} [f(\pi) - f(-\pi)]$$

$$A_r = \frac{1}{\pi} \int_{-\pi}^{\pi} f'(x) \cos rx\,dx = \frac{1}{\pi} [f(\pi) - f(-\pi)] \cos r\pi + rb_r$$

Similarly $B_r = -ra_r$

Term-by-term differentiation yields $\sum_{k=1}^{\infty}(-ka_k \sin kx + kb_k \cos kx)$ and this series is the same as the Fourier series of $f'(x)$ if $f(\pi) = f(-\pi)$ i.e. if the periodic extension of $f(x)$ is continuous at the points $x = n\pi$.

Example 1

Consider the function $f(x) = x$, $-\pi < x < \pi$, $f(x + 2\pi) = f(x)$. Its Fourier series is

$$2\sum_{k=1}^{\infty}(-1)^{k+1}\frac{1}{k}\sin kx$$

Differentiating this series term by term gives the series $2\sum_{k=1}^{\infty}(-1)^{k+1}\cos kx$ which does not converge for any value of x. Note that there is a discontinuity of $f(x)$ at $x = n\pi$.

Example 2

Consider the function $f(x) = |x|$, $-\pi < x < \pi$, $f(x + 2\pi) = f(x)$, which is continuous at $x = n\pi$. Its Fourier series is

$$f(x) = \frac{\pi}{2} - \frac{4}{\pi}\sum_{k=1,3,5\ldots}^{\infty}\frac{1}{k^2}\cos kx.$$

Differentiation of this series term-by-term produces

$$\frac{4}{\pi}\sum_{k=1,3,5\ldots}^{\infty}\frac{1}{k}\sin kx$$

which you can show is the Fourier series for the function

$$g(x) = \begin{cases} -1, & -\pi < x < 0 \\ 1, & 0 < x < \pi \end{cases} \qquad g(x + 2\pi) = g(x)$$

Amplitude spectrum

The plot of the amplitude of the Fourier coefficients against the frequencies of the components they represent is called an **amplitude spectrum.**

For the last example considered, the spectra of $f(x)$ and $g(x)$ are shown in Figure 7.14(a) and (b) respectively.

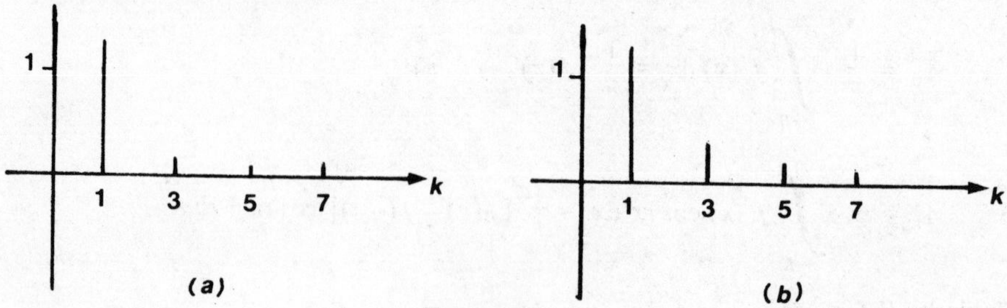

Figure 7.14

Addendum

The sending of signals without distortion is an important function of many electronic devices. The amplification of a waveform can be studied by Fourier analysis. The original wave is analysed into its harmonics and the amplification of each harmonic calculated from the characteristics of the device (it may, for example, distort very high frequency waves). The Fourier analysis will tell us how much of each harmonic is present in the original wave and hence the degree of distortion expected. The range of freedom from such distortion is called the **bandwidth** of the device. We need to know how many harmonics can be 'lost' before the quality of reproduction suffers markedly. The effects of filters can also be studied by first analysing the input wave into its harmonics.

Problems

1. Find the Fourier series for the following functions; sketch their graphs

(a) $f(x) = x,\ -1 < x \leqslant 0;\ f(x) = x + 2,\ 0 < x \leqslant 1$

(b) $f(x) = -1,\ -2 < x \leqslant -1;\ f(x) = x,\ -1 < x < 1;\ f(x) = 1,\ 1 < x < 2$

(c) $f(x) = \frac{2}{3}x,\ 0 < x < \frac{1}{3}\pi;\ f(x) = \frac{1}{3}(\pi - x),\ \frac{\pi}{3} < x < \pi$

(d) $f(x) = \frac{4ax}{l},\ 0 < x < \frac{l}{4};\ f(x) = \frac{4a}{l}(\tfrac{1}{2}l - x),\ \frac{l}{4} < x < \frac{3l}{4};\ f(x) = \frac{4a}{l}(x - l),\ \frac{3l}{4} < x < l$

(e) $f(x) = ax/b,\ 0 < x < b;\ f(x) = a(l - x)/(l - b),\ b < x < l.$

2. Find the Fourier series for the functions whose graphs are shown.

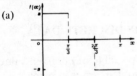

(a)

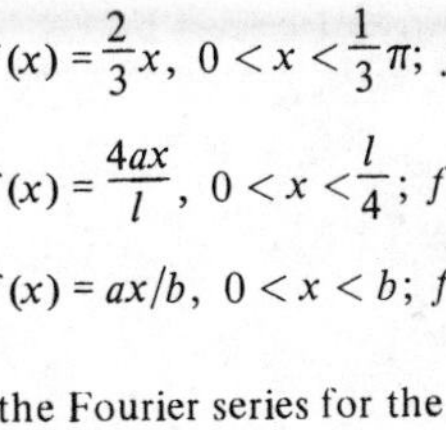

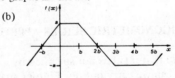

(b)

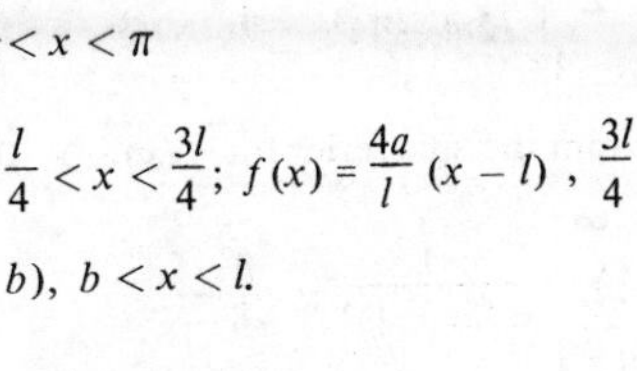

3. Sketch the graph of each of the following functions and find their Fourier series. Assume periodicity outside the quoted range.

(a) $f(x) = \begin{cases} 8, & 0 < x < 2 \\ -8, & 2 < x < 4 \end{cases}$ (b) $f(x) = \begin{cases} -x, & -4 \leqslant x \leqslant 0 \\ x, & 0 \leqslant x \leqslant 4 \end{cases}$

(c) $f(x) = 4x,\ 0 < x < 10$ (d) $f(x) = \begin{cases} 2x, & 0 \leqslant x < 3 \\ 0, & -3 < x < 0 \end{cases}$

4. Prove that if $-\pi < x < \pi$,

$$\cosh ax = \frac{2a}{\pi}\left\{\frac{1}{2a^2} + \sum_{n=1}^{\infty} (-1)^n \frac{1}{(n^2 + a^2)} \cos nx\right\} \sinh a\pi$$

and deduce that, for the same range of values of x,

$$\sinh ax = \frac{2}{\pi}\left\{\sum_{n=1}^{\infty} (-1)^{n-1} \frac{n}{(n^2 + a^2)} \sin nx\right\} \sinh a\pi$$

5. Show that, for $-\pi \leqslant x \leqslant \pi$,

$$x \cos x = -\tfrac{1}{2} \sin x + 2\left\{\frac{2}{1.3} \sin 2x - \frac{3}{2.4} \sin 3x + \frac{4}{3.5} \sin 4x - \ldots\right\}$$

and deduce that, for $-\pi \leqslant x \leqslant \pi$

$$x \sin x = 1 - \tfrac{1}{2} \cos x - 2\left\{\frac{\cos 2x}{1.3} - \frac{\cos 3x}{2.4} + \frac{\cos 4x}{3.5} - \ldots\right\}$$

By differentiating this last result, show that for $0 \leqslant x \leqslant \pi$

$$x = \tfrac{1}{2}\pi - \frac{4}{\pi}\left[\frac{\cos x}{1^2} + \frac{\cos 3x}{3^2} + \frac{\cos 5x}{5^2} + \ldots\right]$$

6. Use Parseval's result and the Fourier series for $f(x) = x$ in $-\pi < x < \pi$ to deduce that

$$\sum_{m=1}^{\infty} \frac{1}{m^2} = \frac{\pi^2}{6}$$

7. (a) From the cosine series for $f(x) = x,\ 0 < x < \pi$ deduce that

$$\sum_{m=1}^{\infty} \frac{1}{(2m-1)^4} = \frac{\pi^4}{96}$$

(b) From the sine series for $f(x) = 1,\ 0 < x < \pi$ deduce that

$$\sum_{m=1}^{\infty} \frac{1}{(2m-1)^2} = \frac{\pi^2}{8}$$

7.7 TRIGONOMETRIC SERIES APPROXIMATION OF DISCRETE DATA

Suppose that $f(x)$ is not specified by a formula but by a table of observed values. How then can we carry out our Fourier analysis? This problem arises in many contexts, an important one being in analysing the motion of the sea by taking sample readings over a period of time.

First consider the problem of fitting a finite trigonometric series to a function $f(x)$ specified by $(2n + 1)$ values: $f(x_0),\ f(x_1),\ f(x_2),\ \ldots,\ f(x_{2n})$. (If the number of data points is $2n$, we in effect omit the last sine term.) We shall merely state the relevant results and leave you either to carry out the proofs yourselves or consult suitable books.

The $(2n + 1)$ points will determine uniquely the coefficients in the series

$$\tfrac{1}{2}a_0 + \sum_{k=1}^{n} \left(a_k \cos \frac{2\pi kx}{(2n+1)} + b_k \sin \frac{2\pi kx}{(2n+1)} \right) \qquad (7.31)$$

We have the following properties of orthogonality, given in the set of equations (7.32) below

$$\sum_{i=0}^{2n} \sin\frac{2\pi j x_i}{(2n+1)} \cos\frac{2\pi k x_i}{(2n+1)} = 0 \tag{7.32}$$

$$\sum_{i=0}^{2n} \sin\frac{2\pi j x_i}{(2n+1)} \sin\frac{2\pi k x_i}{(2n+1)} = \begin{cases} 0 \text{ if } j \neq k \\ (2n+1)/2 \text{ if } j = k \text{ (and neither is zero)} \end{cases}$$

$$\sum_{i=0}^{2n} \cos\frac{2\pi j x_i}{(2n+1)} \cos\frac{2\pi k x_i}{(2n+1)} = \begin{cases} 0 & \text{if } j \neq k \\ (2n+1)/2 & \text{if } j = k \text{ and neither is 0 or } (2n+1) \\ (2n+1) & \text{if } j = k \text{ and is either 0 or } (2n+1) \end{cases}$$

We can then proceed to determine the coefficients by the formulae

$$\left.\begin{aligned} a_k &= \frac{2}{(2n+1)} \sum_{i=0}^{2n} f(x_i) \cos\frac{2\pi k x_i}{(2n+1)}, \quad k = 0, 1, 2, \ldots.., n \\ b_k &= \frac{2}{(2n+1)} \sum_{i=0}^{2n} f(x_i) \sin\frac{2\pi k x_i}{(2n+1)}, \quad k = 1, 2, \quad \ldots.., n \end{aligned}\right\} \tag{7.33}$$

Example

The following data is given. Fit a trigonometric approximation and estimate $f(2.5)$. Note $n = 2$.

x	0	1	2	3	4
$f(x)$	0	2	4	2	0

We find from (7.33) that $a_0 = \frac{2}{5}\left[0 + 2 + 4 + 2 + 0\right] = 3.2$

$$a_1 = \frac{2}{5}\left[0 \times \cos\frac{2\pi.0}{5} + 2 \times \cos\frac{2\pi.1}{5} + 4 \times \cos\frac{2\pi.2}{5} + 2 \times \cos\frac{2\pi.3}{5} + 0 \times \cos\frac{2\pi.4}{5}\right] = -1.6944 \text{ (4 d.p.)}$$

$$a_2 = \frac{2}{5}\left[0 \times \cos\frac{2\pi.2.0}{5} + 2 \times \cos\frac{2\pi.2.1}{5} + 4 \times \cos\frac{2\pi.2.2}{5} + 2 \times \cos\frac{2\pi.2.3}{5} + 0 \times \cos\frac{2\pi.2.4}{5}\right] = 0.0944 \text{ (4 d.p.)}$$

Similarly $b_1 = 1.2310$ (4 d.p.) and $b_2 = -0.2906$ (4 d.p.)

Then $f(x) \simeq 1.6 - 1.6944 \cos\frac{2\pi x}{5} + 0.0944 \cos\frac{4\pi x}{5} + 1.2310 \sin\frac{2\pi x}{5}$

$$-0.2906 \sin\frac{4\pi x}{5}$$

Hence $f(2.5) = 3.389$.

Least squares approximation

Suppose we wish to fit to a set of data a trigonometric series which has fewer coefficients than the number of data points. We shall then have an infinite number of series from which to choose and we select one of these by the method of least squares. For $(2n + 1)$ data points and a series with $(2m + 1)$ coefficients, where $m \leqslant n$ we seek to minimise

$$E = \sum_{i=0}^{2n} \left[f(x_i) - S_m(x_i) \right]^2 \qquad (7.34)$$

where

$$S_m(x) = \tfrac{1}{2}\alpha_0 + \sum_{k=1}^{m} \left[\alpha_k \cos \frac{2\pi kx}{(2n+1)} + \beta_k \sin \frac{2\pi kx}{(2n+1)} \right] \qquad (7.35)$$

The coefficients α_k, β_k which give rise to the minimum value of E are just those in (7.33). In other words, the Fourier coefficients are fixed, irrespective of the value of $m \leqslant n$. If we fit a trigonometric series with more terms we do not alter these earlier coefficients; if we have determined coefficients up to a_4, b_4 and we decide to take two more terms in the series approximation then we need only determine a_5 and b_5, since the other coefficients will remain at their chosen values. The same result held for a fomula specification of $f(x)$ in Section 7.4.

This useful result is a direct consequence of orthogonality, as expressed in (7.32). We do not find this true for least squares polynomial fitting, where the selection of a higher degree polynomial necessitates the determination of all the coefficieints from afresh.

Example

Fit a least squares approximation of the form $f(x) \simeq b_1 \sin x + b_3 \sin 3x + b_5 \sin 5x$ to the data

x	0	$\pi/6$	$\pi/3$	$\pi/2$	$2\pi/3$	$5\pi/6$	π
$f(x)$	0	2.5	4	4.5	4	2.5	0

This is a slight variant on the material discussed in the text, but there are fewer coefficients than data points and the principles still apply , so

$$b_1 = \frac{2}{7}\left[0 \text{ x } \sin\left(\frac{2\pi}{7} . 0\right) + 2.5 \text{ x } \sin\left(\frac{2\pi}{7} \cdot \frac{\pi}{6}\right) + 4 \text{ x } \sin\left(\frac{2\pi}{7} \cdot \frac{\pi}{3}\right) + \ldots\ldots \right]$$

$$= 4.112 \text{ (3 d.p.)}$$

$$b_3 = \frac{2}{7}\left[0 \text{ x } \sin\left(\frac{2\pi.3}{7} . 0\right) + 2.5 \text{ x } \sin\left(\frac{2\pi}{7} . 3 . \frac{\pi}{6}\right) + 4 \text{ x } \sin\left(\frac{2\pi}{7} . 3 . \frac{\pi}{3}\right) + \ldots \right]$$

$$= -0.263 \text{ (3 d.p.)}$$

Similarly $b_5 = -0.235$ (3 d.p.)

Hence, $f(x) \simeq 4.112 \sin x - 0.263 \sin 3x - 0.235 \sin 5x$

Data smoothing

Suppose we have some numerical data which may have some random 'noise' in the sense that it represents the true values of a function with random errors superimposed. If we further suppose that the true function is relatively smooth, whereas the errors are not, then we may make a start towards separating the two; because the true function is smooth, its Fourier coefficients will decay rapidly, whereas the error, being unsmooth, will have slowly decreasing Fourier coefficients. The higher frequency harmonics of the series representing the total function will be almost entirely due to error. We shall, of course, have some error contribution in the lower frequency harmonics.

Example

The data from an experiment is as shown below.

Assume that the first and last values are indeed zero and extend the function as an odd function. Smooth the data using trigonometric series and calculate the root-mean-square error of the given data and the smoothed data, on the assumption that the true function was $f(x) = x(2-x)$ and that random errors were superimposed.

x	0	0.25	0.50	0.75	1.00	1.25	1.50	1.75	2.00
$f(x)$	0.000	0.407	0.823	0.841	1.061	0.946	0.845	0.348	0.000

Since we are manufacturing an odd function we need only compute the coefficients b_k, using the formulae (7.33). Having extended the function to an odd function of period 4, we note that the a_k are zero and the coefficients

$$b_k = \frac{1}{4} \sum_{j=1}^{7} f(x_j) \sin \frac{2\pi k x_j}{8}$$

Table 7.7 shows the smoothed values and the values of $f(x) = x(2-x)$. (All values are quoted to 3 d.p.).

Table 7.7

Given values	0.000	0.407	0.823	0.841	1.061	0.946	0.845	0.348	0.000
Smoothed values	0.000	0.452	0.750	0.910	1.009	1.000	0.777	0.377	0.000
'True' values	0.000	0.438	0.750	0.938	1.000	0.938	0.750	0.438	0.000

The subject of data smoothing is quite extensive and we have only touched on it here. The interested reader may consult 'Applied Analysis' by C. Lanczos (Pitman & Sons, London 1957) for a fuller treatment.

We merely remark that in order to get a series which should converge rapidly (so that a greater separation between true values and noise can be achieved) we first transform $f(x)$ into $g(x) = f(x) - (\alpha + \beta x)$ where α and β can be found from the boundary conditions $g(x_0) = 0 = g(x_{2n})$.

We can then extend $g(x)$ to an odd function. The resulting function will have no discontinuities in either itself or in its derived function. Therefore we determine the coefficients of the sine series in the usual way.

We must be careful, however, not to oversmooth the values and destroy the qualitative nature of the function.

Numerical Harmonic Analysis

We conclude this section with an example on fitting a Fourier series to a table of data by evaluating those terms which make a significant contribution. The data is the observed values of the current I flowing in an electronic device, the observations taken at regular intervals of time. We have chosen the times to be multiples of $\pi/6$ merely for numerical convenience. The data is as follows

t	0	$\pi/6$	$\pi/3$	$\pi/2$	$2\pi/3$	$5\pi/6$	π	$7\pi/6$	$4\pi/3$	$3\pi/2$	$5\pi/3$	$11\pi/6$	2π
$I(t)$	0	4.6	11.0	17.8	21.6	22.8	19.8	9.6	0	0	0	0	0

Now we assume a Fourier series representation

$$I(t) = \tfrac{1}{2}a_0 + \sum_{k=1}^{\infty} (a_k \cos kt + b_k \sin kt)$$

Then $a_0 = \dfrac{1}{\pi}\displaystyle\int_0^{2\pi} I(t)\,dt = 2\left(\text{mean value of } I(t) \text{ over } [0, 2\pi]\right)$

$$a_r = \frac{1}{\pi}\int_0^{2\pi} I(t)\cos rt\,dt = 2\left(\text{mean value of } I(t)\cos rt \text{ over } [0, 2\pi]\right)$$

Also, $b_r = 2\left(\text{mean value of } I(t)\sin rt \text{ over } [0, 2\pi]\right)$

We find mean values by summing the contributions from the twelve points $0(\pi/6)\,11\pi/6$ and divide by 12. However, since we would then multiply the result by 2 we shall in fact obtain the total contribution and divide by 6. We display in Table 7.8 the result of the calculations for a_0, a_1, b_1 and summarise the results for higher coefficients.

Results for coefficients are quoted to 1 d.p.
Hence $a_0 = 17.9$, $a_1 = -8.2$, $b_1 = 9.2$. Similarly, we can find that $a_2 = 0.7$, $b_2 = -2.8$, $a_3 = -1.5$, $b_3 = 0.0$, $a_4 = 0.5$, $b_4 = 0.3$, $a_5 = -0.2$, $b_5 = -0.4$.

In general, there is the result that with r intervals it is worth carrying this kind of analysis only as far as the $\left(\dfrac{r}{2} - 1\right)^{\text{th}}$ harmonic. In this example $r = 12$ and hence we need go no further than the 5th harmonic.

Problems

1. Find a Fourier series, as far as the third harmonic, to represent the periodic function $f(x)$ given by the values below.

x	0°	30°	60°	90°	120°	150°	180°	210°	240°	270°	300°	330°
f	−7.24	−11.32	17.5	18.26	21.72	23.48	16.8	6.68	−13.52	−27.88	−36.28	−28.5

2. Repeat Problem 1 for the f - values below:

(a) 1.0, 2.4, 4.0, 4.8, 3.2, 4.3, 4.3, 3.1, −2.0, −2.0, −0.8, 0.6

(b) 0, 1.0, 1.9, 2.8, 3.2, 3.4, 2.6, 1.8, 1.4, 0.8, 0.5, 0.2

(c) 0, 2.1, 2.8, 3.2, 2.1, 0, 0, 0, 0, 0, 1.7, 1.7

3. Find a Fourier series, as far as the harmonic requested for the data below.

(a)

x	0	0.5	1.0	1.5	2.0	2.5	3.0	3.5	4.0	4.5	(4th)
f	3.0	0.3	−0.3	−1.4	−1.5	−1.0	0.1	1.4	3.5	4.6	

(b)

x	0	0.2	0.4	0.6	0.8	1.0	1.2	1.4	1.6	1.8	2.0	(5th)
f	0	0.8	2.8	5.6	2.4	0	−2.4	−5.6	−2.8	−0.8	0	

Table 7.8

t	$I(t)$	$\cos t$	$I \cos t$	$\sin t$	$I \sin t$
0	0	1	0	0	0
$\pi/6$	4.6	0.867	3.98	0.5	2.30
$\pi/3$	11.0	0.5	5.50	0.867	9.52
$\pi/2$	17.8	0	0	1	17.80
$2\pi/3$	21.6	−0.5	−10.8	0.867	18.70
$5\pi/6$	22.8	−0.867	−19.74	0.5	11.4
π	19.8	−1	−19.8	0	0
$7\pi/6$	9.6	−0.867	−8.33	−0.5	−4.8
$4\pi/3$	0	−0.5	0	−0.867	0
$3\pi/2$	0	0	0	−1	0
$5\pi/3$	0	0.5	0	−0.867	0
$11\pi/6$	0	0.867	0	−0.5	0
Σ	107.2		−49.18		54.92
$\Sigma/6$	17.9		−8.2		9.2

7.8 CUBIC SPLINE FITTING

Suppose that we have a set of data points (x_1, y_1), (x_2, y_2), ..., (x_n, y_n) and we wish to fit a smooth curve to them. To fit a polynomial of degree $(n - 1)$ would seem to be unnecessarily complicating the relationship between x and y and we look for a curve of fairly low degree which still gives a reasonable fit. A technique which is increasing in usage is the mathematical analogue of the draughtsman's method of using a **spline** – which is a flexible strip of material that can be made to pass through each of the data points, progressing smoothly from one interval to the next.

We aim to copy this device by fitting cubic polynomials in the intervals $(x_i,\ y_i)$ to $(x_{i+1},\ y_{i+1})$ and requiring not only that two adjacent cubics meet (so that we have a continuous curve) but that the slopes of the adjacent cubics are equal at the meeting point and their curvatures at the meeting point are equal. In other words, the first derivatives of two adjacent cubics are equal at the meeting point, and their second derivatives are equal there also.

We shall write the equation for the cubic passing through $(x_i,\ y_i)$ and $(x_{i+1},\ y_{i+1})$ as

$$y = \alpha_i (x - x_i)^3 + \beta_i (x - x_i)^2 + \gamma_i (x - x_i) + \delta_i \tag{7.36}$$

You should convince yourself that this does not restrict the cubic in any further way: it is merely a form of equation which will allow ease of algebra.

At the left-hand end of the interval, viz. $(x_i,\ y_i)$,

$$y = y_i = \delta_i \tag{7.37}$$

At the right-hand end, viz. (x_{i+1}, y_{i+1}),

$$y = y_{i+1} = \alpha_i(x_{i+1} - x_i)^3 + \beta_i(x_{i+1} - x_i)^2 + \gamma_i(x_{i+1} - x_i) + \delta_i$$

Writing $h_i = (x_{i+1} - x_i)$ we have

$$y_{i+1} = \alpha_i h_i^3 + \beta_i h_i^2 + \gamma_i h_i + \delta_i \tag{7.38}$$

To ensure that the slopes of two adjacent cubics match at their meeting point we find

$$y' = 3\alpha_i(x - x_i)^2 + 2\beta_i(x - x_i) + \gamma_i \tag{7.39}$$

Similarly, in order that the second derivatives match, we need to find

$$y'' = 6\alpha_i(x - x_i) + 2\beta_i \tag{7.40}$$

It has become customary to simplify the subsequent algebra by writing the equations in terms of the second derivatives of the cubic polynomials at each point; we write $S_i = y''$ for the value at x_i.

Then we have from equation (7.40), $S_i = 6\alpha_i \,.\, 0 + 2\beta_i = 2\beta_i$ and $S_{i+1} = 6\alpha_i h_i + 2\beta_i$. We can solve these equations to obtain

$$\beta_i = \tfrac{1}{2} S_i \tag{7.41}$$

$$\alpha_i = (S_{i+1} - S_i)/6h_i \tag{7.42}$$

We may now substitute these values into (7.38) to obtain for the coefficient γ_i the formula

$$\gamma_i = \frac{(y_{i+1} - y_i)}{h_i} - \frac{h_i(2S_i + S_{i+1})}{6} \tag{7.43}$$

Since we have already a formula for δ_i, viz. (7.38), we can now obtain a fearsome-looking expression for our cubic spline in terms of the given data values and the second derivatives. However, we have not used the requirement that the slopes of the cubic polynomials must agree at each x_i. If we impose this requirement, we find from (7.39) that $y_i' = 3\alpha_i \,.\, 0 + 2\beta_i \,.\, 0 + \gamma_i$. From the previous interval, we have the slope at the right-hand end of the interval $[x_{i-1},\ x_i]$ is

$$y_{i-1}' = 3\alpha_{i-1}(x_i - x_{i-1})^2 + 2\beta_{i-1}(x_i - x_{i-1}) + \gamma_{i-1}$$

i.e. $$y_{i-1}' = 3\alpha_{i-1} h_{i-1}^2 + 2\beta_{i-1} h_{i-1} + \gamma_{i-1}$$

Equating the two expressions for y_i' and substituting for α_i, β_i etc. we obtain

$$\frac{1}{6}\left\{h_{i-1}\,S_{i-1} + 2(h_{i-1} + h_i)S_i + h_i\,S_{i+1}\right\} = \frac{(y_{i+1} - y_i)}{h_i} - \frac{(y_i - y_{i-1})}{h_{i-1}}$$

In the case of equally spaced data, $h_i = h_{i-1} = h$ and we obtain the simplified form

$$\frac{1}{6}(S_{i-1} + 4S_i + S_{i+1}) = \frac{y_{i+1} - 2y_i + y_{i-1}}{h_i^2} = \Delta^2 y_{i-1} \tag{7.44}$$

Whichever is the formula to use, it holds for $i = 2, 3, \ldots, (n-1)$. This means we have got only $(n-2)$ equations for n unknowns. The other two equations are obtained by placing conditions on S_1 and S_n. The choice of conditions influences the kind of fit near the end points x_1 and x_n. A common choice is to require S_1 to be a linear extrapolation from S_3 and S_2 and, similarly, S_n to be a linear extrapolation from S_{n-2} and S_{n-1}. The advantage of this choice over others (e.g. the simpler one of $S_1 = 0 = S_n$) is that if the data lies on a cubic curve this coincides with our cubic spline.

$$\left.\begin{aligned} &\text{We have chosen } \frac{S_2 - S_1}{h_1} = \frac{S_3 - S_2}{h_2} \\ &h_2\,S_1 - (h_1 + h_2)S_2 + h_1\,S_3 = 0 \end{aligned}\right\} \tag{7.45}$$

Similarly,

$$h_{n-1}\,S_{n-2} - (h_{n-2} + h_{n-1})S_{n-1} + h_{n-2}\,S_n = 0$$

In the special case where $h_i = h$ for all i, the above equations reduce to

$$\left.\begin{aligned} S_1 - 2S_2 + S_3 &= 0 \\ S_{n-2} - 2S_{n-1} + S_n &= 0 \end{aligned}\right\} \tag{7.46}$$

We shall continue our discussion with this special case in mind. We then write the system of equations (7.44) and (7.46) in matrix form

$$\begin{bmatrix} 1 & -2 & 1 & 0 & 0 & \cdots & & & 0 \\ 1 & 4 & 1 & 0 & 0 & \cdots & & & 0 \\ 0 & 1 & 4 & 1 & 0 & \cdots & & & 0 \\ 0 & 0 & 1 & 4 & 1 & \cdots & & & 0 \\ \cdots & & & & & & & & \\ 0 & 0 & 0 & 0 & 0 & \cdots & 1 & 4 & 1 \\ 0 & 0 & 0 & 0 & 0 & \cdots & 1 & -2 & 1 \end{bmatrix} \begin{bmatrix} S_1 \\ S_2 \\ S_3 \\ S_4 \\ \vdots \\ S_{n-1} \\ S_n \end{bmatrix} = \frac{6}{h^2} \begin{bmatrix} 0 \\ \Delta^2 y_1 \\ \Delta^2 y_2 \\ \Delta^2 y_3 \\ \vdots \\ \Delta^2 y_{n-2} \\ 0 \end{bmatrix} \tag{7.47}$$

Because the coefficient matrix is tridiagonal we may use the methods of Section 1.5

Example

Table 7.9 shows a set of data points; the second differences have been evaluated for simplicity. We have to fit a cubic spline to the data. Note that $h = 1$.

The system of equations (7.47) gives rise to the solutions: $S_1 = \dfrac{179}{56}$, $S_2 = 1$,

$S_3 = \dfrac{-67}{56}$, $S_4 = \dfrac{53}{14}$, $S_5 = \dfrac{-109}{56}$, $S_6 = 4$, $S_7 = \dfrac{557}{56}$

Table 7.9

x	y	Δy	$\Delta^2 y$
0	0		
		1	
1	1		1
		2	
2	3		0
		2	
3	5		2
		4	
4	9		0
		4	
5	13		4
		8	
6	21		

Because of the choice of end conditions, we shall see that the cubic polynomials in the intervals [0, 1] and [1, 2] are the same; a similar remark applies to the intervals [4, 5] and [5, 6]. In Table 7.10 we display the coefficients for the four different polynomials which make up the spline.

Table 7.10

Interval	[0, 2]	[2, 3]	[3, 4]	[4, 6]
α_i	–41/112	93/112	–107/112	111/112
β_i	179/112	–67/112	53/28	–109/112
γ_i	–13/56	99/56	343/112	223/56
δ_i	0	3	5	9

For example, in the interval [3, 4] the cubic polynomial is

$$y = \frac{-107}{112}(x-3)^3 + \frac{53}{28}(x-3)^2 + \frac{343}{112}(x-3) + 5$$

Had we used the alternative approach of four cubic polynomials obtained by interpolation techniques (so that each of these passes through four adjacent data points) we should have found that at the overlap regions the predicted values of y differ and the slopes at data points are different. Hence the advantages of cubic splines with the smooth overlap regions are clear.

You should produce a flow chart for the process of obtaining a cubic spline fit when the data values are equally spaced. Write and run a computer program using the data of the example above.

Problems

1. Confirm the statement that for a set of data exactly fitted by a cubic, the values of S at the two ends will be linearly related to the adjacent S - values if the spline curve and the cubic polynomial are the same function. If other conditions are imposed on S in the end intervals ($S_1 = S_n = 0$ or $S_1 = S_2$ and $S_n = S_{n-1}$), how should the set of (7.47) be modified? Will this change the portions of the spline curve in intervals other than the first and last?

2. Find the coefficient matrix and the right-hand-side vector for fitting a cubic spline to the following data. Use a linearity condition on the terminal S - values:

x	0.15	0.76	0.89	1.07	1.73	2.11
y	0.3945	0.2989	0.2685	0.2251	0.0893	0.0431

3. Solve the set of equations of Problem 2 (you may wish to utilize a computer program for this), and then determine the constants of the various cubics. The data are the ordinates of the normal probability function. Compare a few interpolated values with tabulated values of the function say at $x = 0.30,\ 0.80,\ 1.50,\ 2.00$.

4. Fit a cubic spline to the exponential function, utilizing values of e^x for $x = 0.0\ (0.5)\ 2.0$, and compare the interpolated with the true values for $x = 0.25,\ 1.25,\ 1.75$

5. Consider the function $f(x) = 1/(1 + 25x^2)$ on $-1 \leqslant x \leqslant 1$. For equally spaced nodes $x_j = -1 + jh,\ j = 0, 1, 2, \ldots\ldots, n$ and $h = 2/n$ produce a cubic spline fit for $n = 4$ and $n = 20$.

6. A glass envelope is required for the structure of a high-power electric light bulb. The profile of the bulb has been specified by the following five points

x	0	150	250	350	365
y	60	92	138	62	0

The dimensions are in millimetres and x is the coordinate along the centreline of the bulb, starting at the cap; y is the magnitude of the radius at each of these positions. In order that a forming tool can be made, a smooth curve must be fitted through these points. Having established the equation for this curve it will be assumed that five intermediate points between each pair are adequate to plot the general profile for a visual inspection.
Calculate the coordinates of these points and draw a graph of the resulting profile.

Chapter Eight

Partial Differential Equations

8.1 INTRODUCTION

When our mathematical models were ordinary differential equations we had to restrict our attention to **lumped parameter** models, where we considered such parameters as mass, stiffness, inductance and capacitance as being concentrated in isolated lumps. For example, the motion of a heavy mass rotating at the end of a light shaft could be modelled by an ordinary differential equation. However, a system where the shaft could not be dismissed as lightweight in comparison with the suspended mass would demand a continuously-distributed approach. Also, more than one independent variable in a problem requires partial derivatives. A partial differential equation is basically an equation involving partial derivatives. The advantage of partial differential equations is that they allow us to study problems in which the model parameters are **continuous.**

Many engineering systems can be modelled by means of partial differential equations. In this chapter we shall concentrate on three basic kinds of linear partial differential equations which can each be used as the model for a wide class of physical phenomena. We shall examine, in the main, numerical and analytical solutions, but in some cases the use of analogues is helpful and we shall mention these for completeness. It will be impossible in a short space to describe all the major applications of our three types of equation, but we hope that through the text and the problems we have given you a reasonably wide compass. We shall sometimes use the abbreviation p.d.e.

8.2 CASE STUDY : STEADY STATE TEMPERATURE DISTRIBUTION IN A PLATE

We wish to find the temperature distribution in a rectangular metal plate under certain conditions. Refer to Figure 8.1(a). The plate is covered on its top and bottom faces by layers of thermally insulating material so that heat is constrained to flow mainly in the x - and y - directions. Along the edges of the plate various conditions are applied – a particular case is shown in Figure 8.1(b). These conditions, being applied at the boundaries of the region of interest, are known as **boundary conditions.**

We now make assumptions about the physical situation in order to formulate a reasonably simple mathematical model.

Assumptions

(i) The metal is *uniform* in the sense that its thermal conductivity is the same at all points of the plate.

(ii) The plate is sufficiently *thin* so that we may neglect any heat flow in the directions perpendicular to its faces.

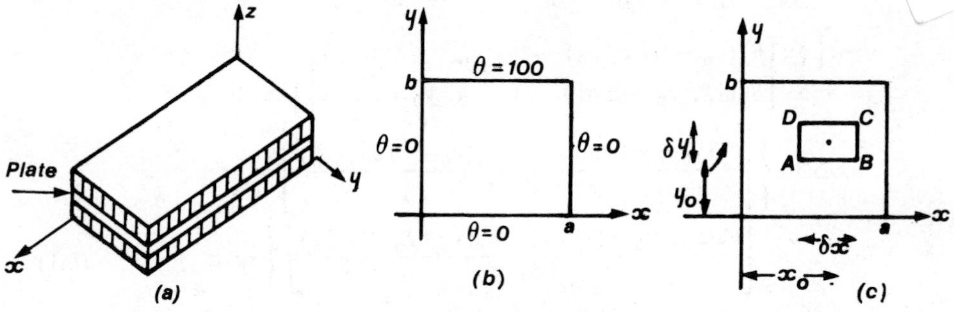

Figure 8.1

(iii) The temperature distribution is in the *steady state*, i.e. the temperature at any point in the plate does not depend on time.

Model

We are now entitled to assume that the model plate is infinitesimally thin and that the temperature function θ depends only on x and y, i.e. $\theta = \theta(x, y)$. We have chosen the origin of coordinates at one corner for convenience. Let the dimensions of the plate be as shown in Figure 8.1(c), so that the plate may be defined as $0 \leqslant x \leqslant a$, $0 \leqslant y \leqslant b$. We now analyse the situation in a small rectangular region ABCD of the plate, whose position is as shown; its centre is the point (x_0, y_0).

Consider what happens at the edge DA of this sub-region. The temperature will change along DA, since the y - coordinate varies, but it will change only with y. We shall ultimately take limits as δx and δy tend to zero, so we may approximate the derivative $\dfrac{\partial\theta}{\partial x} \equiv \theta_x$ at all points of the edge DA by its value at the mid-point of that edge, viz. $(x_0 - \tfrac{1}{2}\delta x,\ y_0)$. Then the heat flowing across DA to the right is approximately given by

$$H_{DA} \simeq -k\delta y\theta_x(x_0 - \tfrac{1}{2}\delta x,\ y_0)$$

This is a mathematical statement of **Fourier's Law** which states that the flow of heat is proportional to the local temperature gradient. The constant of proportionality k is related to the thermal conductivity of the plate material. Now the flow of heat across CB is

$$H_{CB} \simeq -k\delta y\theta_x(x_0 + \tfrac{1}{2}\delta x,\ y_0)$$

Notice that if θ decreases as x increases then $\partial\theta/\partial x$ will be negative and we shall have a positive flow of heat to the right.

Precisely analogous arguments show that the flows of heat across AB and DC are given, respectively, by

$$H_{AB} \simeq -k\delta x\theta_y(x_0,\ y_0 - \tfrac{1}{2}\delta y) \text{ and } H_{DC} \simeq -k\delta x\theta_y(x_0,\ y_0 + \tfrac{1}{2}\delta y)$$

Since we are considering the steady-state, there is no accumulation of heat in the region ABCD and the input of heat flow must equal the output. We shall not lose any generality if we assume that heat flows in through AB and DA and out through CB and DC.

Then $H_{DA} - H_{CB} + H_{AB} - H_{DC} = 0$ i.e.

$$-k\left\{\delta y\left[\theta_x(x_0 - \tfrac{1}{2}\delta x,\ y_0) - \theta_x(x_0 + \tfrac{1}{2}\delta x,\ y_0)\right] + \delta x\left[\theta_y(x_0,\ y_0 - \tfrac{1}{2}\delta y) - \theta_y(x_0,\ y_0 + \tfrac{1}{2}\delta y)\right]\right\} \simeq 0$$

i.e.
$$+k\delta x\delta y\left\{\left[\frac{\theta_x(x_0 + \tfrac{1}{2}\delta x,\ y_0) - \theta_x(x_0 - \tfrac{1}{2}\delta x,\ y_0)}{\delta x}\right] + \left[\frac{\theta_y(x_0,\ y_0 + \tfrac{1}{2}\delta y) - \theta_y(x_0,\ y_0 - \tfrac{1}{2}\delta y)}{\delta y}\right]\right\} \simeq 0 \qquad (8.1)$$

We shall now let the area of the rectangle **ABCD** tend to zero by letting δx and δy tend to zero simultaneously. This will have the following effects

(i) $\left[\dfrac{\theta_x(x_0 + \tfrac{1}{2}\delta x,\ y_0) - \theta_x(x_0 - \tfrac{1}{2}\delta x,\ y_0)}{\delta x}\right]$ will tend to the value

$$\left.\frac{\partial\theta_x}{\partial x}\right|_{(x_0,\ y_0)} \quad \text{i.e.} \quad \theta_{xx}(x_0,\ y_0)$$

(ii) The other term in square brackets will tend to the value $\theta_{yy}(x_0,\ y_0)$

(iii) The approximation becomes an equality.

Of course, as equation (8.1) stands we shall get the result $0 = 0$; therefore we first divide both sides by $k\delta x\delta y$ *before* taking limits. As a result of letting δx and δy tend simultaneously to zero we obtain **Laplace's Equation** in two dimensions

$$\frac{\partial^2\theta}{\partial x^2} + \frac{\partial^2\theta}{\partial y^2} = 0 \qquad (8.2)$$

Boundary Conditions

Equation (8.2) possesses infinitely many solutions. We need to specify extra conditions to cut the possibilities down to a *unique* solution. Too many conditions may not permit *any* solution and too few will not fix a unique solution. These conditions must be of the right kind. We shall study these **boundary conditions** more fully in the next section, but for the moment we shall specify simple conditions as shown in Figure 8.1(b). Three edges are maintained at zero temperature, the fourth is maintained at a temperature of 100°C. Roughly speaking we need two conditions on the temperature for given values of x and two for given values of y. This follows because, in order to solve (8.2) analytically we need two integrations with respect to x and two integrations with respect to y. Each of these will introduce an arbitrary function with respect to the other variable and we need effectively to supply limits of integration.

The conditions can be stated mathematically as

$$\theta = 0 \quad \text{when} \quad x = 0, \quad \text{for} \quad 0 \leqslant y < b \qquad (8.3a)$$

$$\theta = 0 \quad \text{when} \quad y = 0, \quad \text{for} \quad 0 \leqslant x \leqslant a \qquad (8.3b)$$

$$\theta = 0 \quad \text{when} \quad x = a, \quad \text{for} \quad 0 \leqslant y < b \qquad (8.3c)$$

$$\theta = 100 \quad \text{when} \quad y = b, \quad \text{for} \quad 0 \leqslant x \leqslant a \qquad (8.3d)$$

We may write these conditions in a different way; for example (8.3a) becomes $\theta(0, y) = 0$ for $0 \leq y < b$.

Notice that we have a discontinuity in the temperature function at the points $(0, b)$ and (a, b): this is the price we pay for taking a very simple condition along the edge DC.

The equation (8.2) together with the boundary conditions (8.3) forms the mathematical model of the problem. We now examine three methods of solution.

Analogue method

Laplace's equation is extremely widespread in its appearance in mathematical models. It applies to problems which can be expressed in terms of a **potential function** ; for example, electrostatics, magnetostatics, temperature problems and problems in stress analysis, seepage and irrotational motion of fluids. Because of the same equation governing phenomena in electrostatics, we can produce an analogous model using voltages by specifying the appropriate geometry and boundary conditions. Reading the voltages at specified points gives us a picture of the overall distribution of voltage which will be the distribution of temperature in our model. Other analogy methods are considered in Section 8.8.

With regard to our problem we may proceed as follows.

We take a sheet of conducting paper cut to the appropriate rectangular shape. We put a conducting wire along the top edge and cover it with conducting paint. We put a second wire along the remaining three edges and cover it with paint. Then the wires are connected to a variable voltage supply so that we may regard the top edge as being at 100λ volts and the other three edges at 0 volts; λ is a scaling factor. We now select a suitable temperature, say 80°C and set the variable scale of voltage to 80λ. We then take a conducting pencil connected to the output via a galvanometer. If we press hard on the conducting pencil at a point where the potential is 80λ we shall then find no deflection of the galvanometer needle. If we make several sets of exploratory moves, we can find a set of points at which the potential is the same: if we join these points we have a curve of equal potential which corresponds to an isotherm for 80°C. Proceeding in a similar way we may construct further isotherms. In Figure 8.2 we represent schematically the apparatus, while in Figure 8.3 we show a typical set of results; in this latter figure we have constructed a **flow net** for the left-hand part by adding the orthogonal lines of heat flow. Because of the temperature discontinuities at the top corners we have all the isotherms entering and leaving these corners.

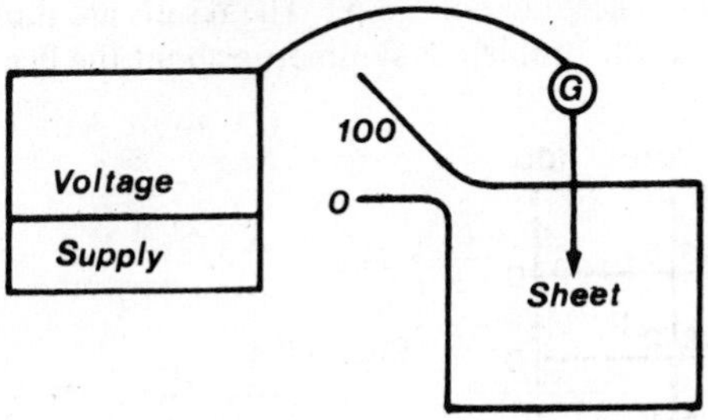

Figure 8.2

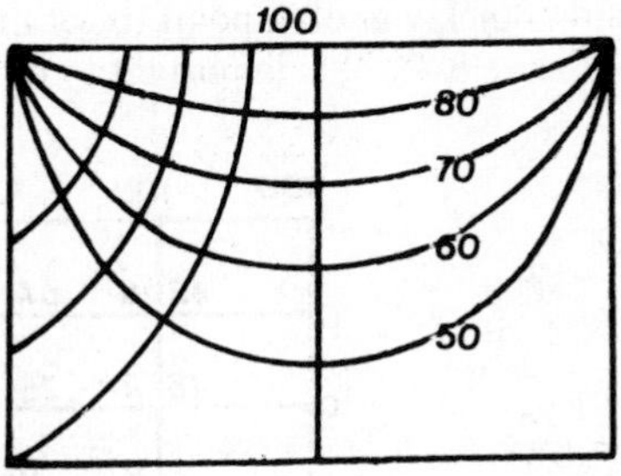

Figure 8.3

Analytical solution – Separation of Variables method

If we can use analytical techniques to solve our problem, we shall hope to arrive at a formula for θ which will be our solution. How useful this formula is remains to be seen. The technique most usefully employed in such problems is the method of *separation of variables.* This is dealt with more fully in Section 8.4, but we outline its main principles here.

The approach we use is a variant of a main line of attack in mathematics: reduce the problem at hand to one (or more) which you *can* solve. In this instance, whilst we know how to solve *ordinary* differential equations (particularly if they are linear with constant coefficients) we have no technique as yet for solving *partial* differential equations. What we aim to do is to separate the independent variables x and y so that we obtain *two* ordinary differential equations. We accomplish this aim by assuming that we can express $\theta(x, y)$ as the product of a function of x only and a function of y only. For example, $\theta(x, y) = 3 \cos 2x\, e^{-2y}$ is of the form we seek, whereas $\theta(x, y) = 4 \log(x + 2y)$ is not; the form $\theta(x, y) = 2 \sin (x + 2y)$ *is* a possibility since it can be expanded into $\theta(x, y) = 2 \sin x \cos 2y + 2 \cos x \sin 2y$ and is a *linear combination* of acceptable forms . Of course, we have not shown that *any* of the three forms given actually satisfies Laplace's equation. (In fact, only the first form does.) What we do is to obtain the most general solution of Laplace's equation which stands a chance of satisfying the boundary conditions and then apply the boundary conditions one at a time, whittling down the possibilities until, if the problem is properly posed, we end up with a unique solution to the problem. As we shall see in Section 8.4, any function of the form $\theta(x,y) = (A \cos kx + B \sin kx)(C \cosh ky + D \sinh ky)$ will satisfy Laplace's equation and also lead to satisfaction of the boundary conditions. There are four independent constants to be determined and we have four conditions to apply. It turns out, having allowed for the linearity of the equation, that the unique solution to the problem is

$$\theta(x,y) = \frac{400}{\pi} \sum_{n=1,3,5\,...}^{\infty} \frac{\sin(n\pi x/a)\sinh(n\pi y/a)}{n \sinh(n\pi b/a)} \tag{8.4}$$

Now this is all very well, but we certainly cannot immediately draw a graphical representation of $\theta(x, y)$ nor is it so easy to calculate the values of θ at given points (x, y). We programmed a digital computer to calculate the value of θ at 9 internal points of the slab, equally spaced, so that we could get a reasonable idea of the temperature distribution. We took $a = b = 1$ for simplicity. The series converged fairly quickly at each point to get the required accuracy of 1 d.p. The results are shown in Figure 8.4. It is important to remark that the problem is symmetric about the line

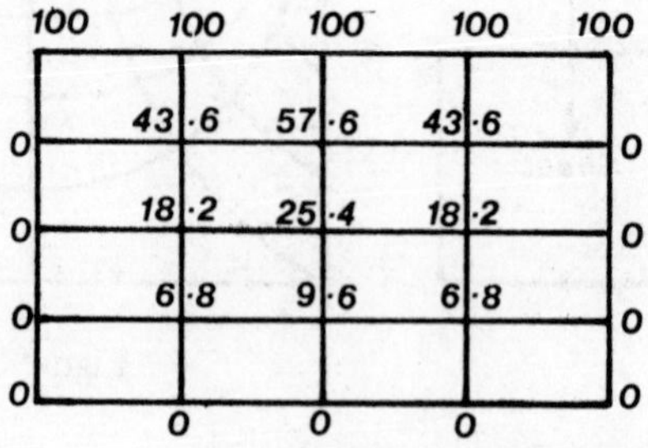

Figure 8.4

$x = a/2$ and therefore we need only calculate the temperature at 6 internal points. We then programmed the computer to deal with 121 equally spaced internal points and the results of this allowed us to sketch the isotherms as in Figure 8.5(a)

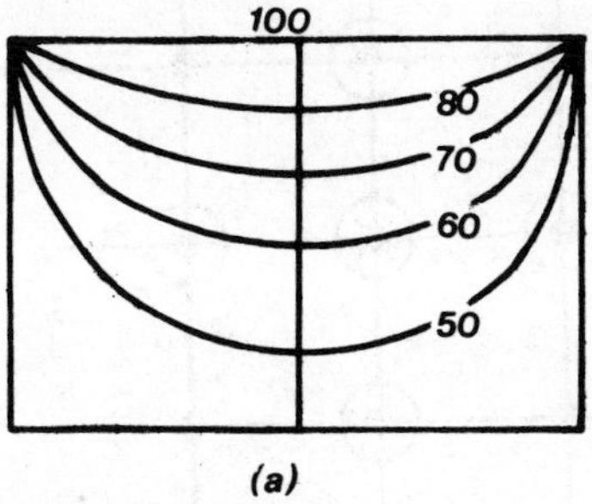

Figure 8.5

It is worth remarking that the symmetry of the problem means that the slopes of the isotherms are parallel to the x - axis at $x = a/2$ and this in turn means that the slab has two distinct regions of heat flow. If we make the line $x = a/2$ an insulating line then we can deal with the problem where no heat flows across a boundary. The two possible cases are shown in Figure 8.5(b) and (c).

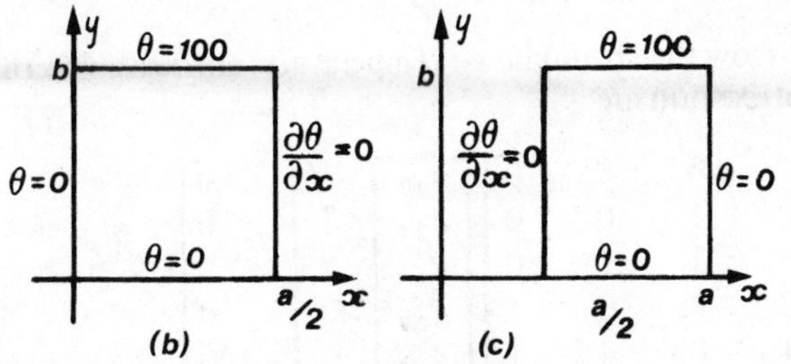

Figure 8.5

We can see that the analytical method can be made to yield both qualitative and quantitative information. We have taken a geometrically simple region and straightforward boundary conditions and the resulting formula for θ was even so an unpleasant entity. If, for example, the geometry of the region is more awkward then the analytical method may become virtually impossible.

Numerical approach

In such a case we can resort to a numerical method of attack. We divide the region of the slab by a mesh and content ourselves with calculating the temperatures at the nodes of the mesh. This was what we effectively did in Figure 8.4 In the numerical approach, however, we replace the differential equation by a finite difference approximation. We shall find in Section 8.8 that an approximation which is $O(h^2)$, where h is the spacing of the mesh nodes in either the x - or the y - direction, is given by

$$\theta(x_r + h,\ y_s) + \theta(x_r - h,\ y_s) + \theta(x_r,\ y_s + h) + \theta(x_r,\ y_s - h) - 4\,\theta(x_r, y_s) = 0 \tag{8.5}$$

In Figure 8.6(a) we show a selection of mesh points around the point $(x_r,\ y_s)$ and we note that (8.5) can be interpreted as stating that the average of the temperatures at the four mesh points nearest to $(x_r,\ y_s)$ is equal to the value of the temperature at $(x_r,\ y_s)$.

Let $\theta_{r,s}$ represent the temperature at the point $(x_r,\ y_s)$. Then the application of equation (8.5) to the 9 internal points of Figure 8.4 will provide 9 equations in 9

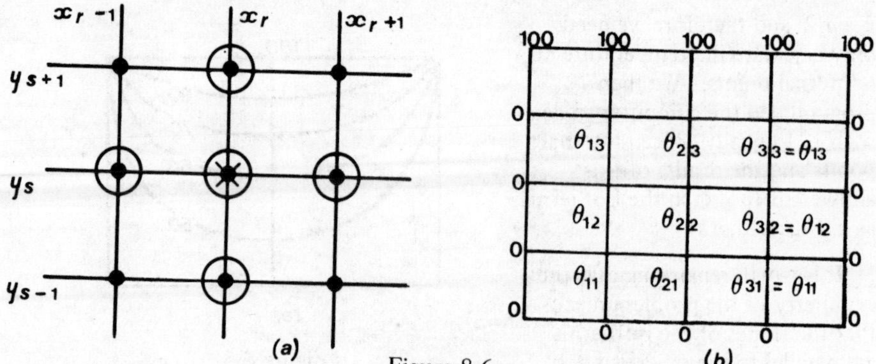

Figure 8.6

unknowns $\theta_{r,s}$. However, we appeal to symmetry to reduce this number to 6. In general, equation (8.5) can be simplified near boundaries. For example, if $r = 1$, then $\theta_{r-1,s} = 0$ for $s = 1, 2, 3$. The case of most interest occurs when $r = 2$ since

$$\theta_{r+1,s} = \theta_{r-1,s} \quad \text{for} \quad s = 1,\ 2,\ 3. \tag{8.6}$$

In Figure 8.6(b) we show the six unknown temperatures and leave you to show that the six equations in matrix form are

$$\begin{bmatrix} -4 & 1 & 1 & 0 & 0 & 0 \\ 2 & -4 & 0 & 1 & 0 & 0 \\ 1 & 0 & -4 & 1 & 1 & 0 \\ 0 & 1 & 2 & -4 & 0 & 1 \\ 0 & 0 & 1 & 0 & -4 & 1 \\ 0 & 0 & 0 & 1 & 2 & -4 \end{bmatrix} \begin{bmatrix} \theta_{11} \\ \theta_{21} \\ \theta_{12} \\ \theta_{22} \\ \theta_{13} \\ \theta_{23} \end{bmatrix} = \begin{bmatrix} 0 \\ 0 \\ 0 \\ 0 \\ -100 \\ -100 \end{bmatrix} \tag{8.7}$$

We can solve these equations by Gauss elimination since the **sparseness** (i.e. presence of many zeros) is not destroyed by the method. This is so because the coefficient matrix is banded. The solution we obtained was, to 1 d.p.,

$$\theta_{11} = 7.1;\ \ \theta_{21} = 9.8;\ \ \theta_{12} = 18.7;\ \ \theta_{22} = 25.0;\ \ \theta_{13} = 42.9;\ \ \theta_{23} = 52.2$$

This agrees with the analytical results only tolerably. We need a more sophisticated numerical approach using a finer mesh.

Problems

Apply the methods used on the case study to the following physical situations. The analytical solution is provided for you.

1. A thin square plate $0 \leqslant x \leqslant a,\ 0 \leqslant y \leqslant a$ has its faces insulated. The edges $x = 0,\ y = a$ are maintained at zero temperature. The edge $y = 0$ is insulated and the edge $x = a$ is kept at a constant temperature 50°C. (An insulated edge implies that the normal derivative of temperature is zero there.) The steady-state temperature is given by

$$\theta(x, y) = \frac{200}{\pi} \sum_{n=0}^{\infty} \frac{(-1)^n}{2n+1} \frac{\sinh \dfrac{(2n+1)\pi x}{2a}}{\sinh \dfrac{(2n+1)\pi}{2}} \cos \frac{(2n+1)\pi y}{2a}$$

2. A rectangular plate $0 \leqslant x \leqslant 20$, $0 \leqslant y \leqslant 10$ has the edges $x = 0$, $x = 20$, $y = 0$ maintained at zero temperature whilst the temperature of the edge $y = 10$ is given by $\theta = 20x - x^2$. The steady-state temperature is given by

$$\theta(x, y) = \frac{3200}{\pi^3} \sum_{n=1}^{\infty} \frac{\sin \dfrac{(2n-1)\pi x}{20} \sinh \dfrac{(2n-1)\pi y}{20}}{(2n-1)^3 \sinh \frac{1}{2}(2n-1)\pi}$$

8.3 SOME BASIC IDEAS

The **order** of a partial differential equation is the order of the highest partial derivative it contains. A partial differential equation is said to be **linear** if all partial derivatives and the dependent variable occur as first degree terms. By an **analytical solution** to a partial differential equation, we mean a function of the independent variables which satisfies identically the equation at every point in a domain of the independent variables. For example, $\theta = A \cos kx \sinh ky$ is a solution of equation (8.2) and so is $\theta = B \sin kx \cosh ky$ for all (x, y).

In general, a partial differential equation of order n has a solution which contains at most n arbitrary functions. It may be, however, that we need to express the **general** solution as a sum of such solutions. This follows from the linearity of the equation: if θ_1 and θ_2 are solutions, so is $a\theta_1 + b\theta_2$ where a and b are constants.

We use the ideas of complementary function and particular integral. If θ_1 is the complementary function of the associated homogeneous equation and θ_2 is a particular integral of the linear p.d.e. then $\theta = \theta_1 + \theta_2$ is a general solution of the full equation.

The general solution can be particularised to a *unique* solution if appropriate extra conditions are provided. These are generally classed as **boundary conditions**; where time, t, is one of the independent variables and we specify a configuration at $t = 0$, we refer to it as an **initial condition.**

The kind of boundary condition we need to specify depends on the nature of the equation. We shall give specific examples in later sections, but we can make some general remarks here.

First, we classify the problems modelled by partial differential equations. We have already remarked that Laplace's equation arises in many different engineering contexts. Although the symbols may be different and the nomenclature may vary, the underlying *structure* is the same.

Equilibrium problems relate to steady-state conditions. The problems which fall into this category include steady-state temperature distributions, steady flows of electric current, equilibrium stress situations and steady ideal fluid flows. We seek configurations of the system studied, for example, displacements, temperatures and velocities.

These problems are **boundary-value problems.** Correspondingly, we need to specify conditions which exist along the entire boundary. We may specify the value of the problem variable, for example temperature, at each point on the boundary. Alternatively we may specify the normal derivative of the variable at some points on the boundary and the problem variable at the others. See Figure 8.7(a).

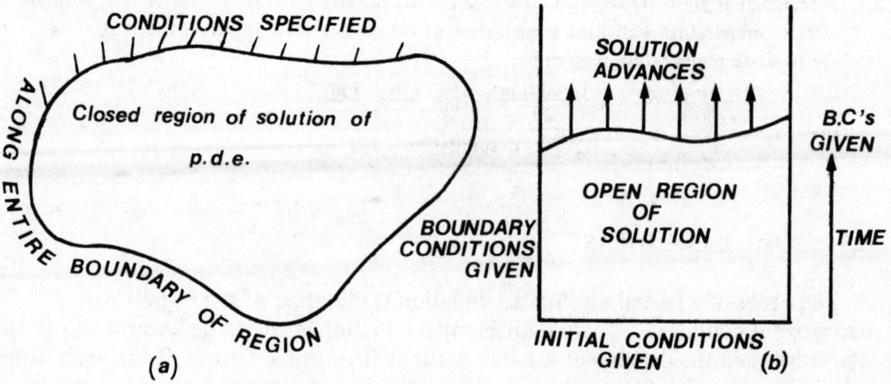

Figure 8.7

Notice that the boundary conditions have a tremendous influence on the solution. They have been likened to a jury demanding that the solution satisfies the equation at all points in the closed region and, simultaneously, satisfies the conditions prescribed at all points on the boundary.

The two equations of this class which we shall study in depth are **Laplace's equation** which in two-dimensional cartesian coordinates is

$$\frac{\partial^2 \theta}{\partial x^2} + \frac{\partial^2 \theta}{\partial y^2} = 0 \tag{8.2}$$

and **Poisson's equation**, an example of which is

$$\frac{\partial^2 \phi}{\partial x^2} + \frac{\partial^2 \phi}{\partial y^2} = f(x, y) \tag{8.8}$$

The function $\theta(x,y)$ in (8.2) could be gravitational potential, electrostatic potential, magnetic potential, velocity potential in irrotational fluid flow as well as steady-state temperature in a uniform solid. The function $\phi\,(x, y)$ in (8.8) could be gravitational potential in a region where $f(x, y)$ is proportional to density of the material, or electrostatic potential in a region where $f(x, y)$ is proportional to charge distribution, or a measure of shear stress in a long bar where $f(x, y)$ is constant.

Laplace's equation is a special case of Poisson's equation.

We may also mention the **biharmonic equation**

$$\frac{\partial^4 \phi}{\partial x^4} + 2\,\frac{\partial^4 \phi}{\partial x^2\,\partial y^2} + \frac{\partial^4 \phi}{\partial y^4} = 0 \tag{8.9}$$

The numerical approach is to use finite difference techniques to transform the differential system into a set of linear simultaneous algebraic equations; the solution will be found at suitable points.

Propagation problems are **initial-value problems**. Here we need to specify conditions for initial time as well as on the space boundaries. Such problems may be concerned with unsteady-state and transient phenomena. Knowledge of a system in an initial state is used to predict its behaviour at later times. The kinds of problem studied include propagation of heat, of displacements and stresses in elastic structures, of pressure waves in air and studies of evaporation. Figure 8.7(b) shows schematically the solution marching out into the future, guided along by the space boundary conditions. The region of solution is therefore an open one.

An equation we shall study is the one-dimensional **diffusion equation**

$$k^2 \frac{\partial^2 \phi}{\partial x^2} = \frac{\partial \phi}{\partial t} \tag{8.10}$$

which has the two-dimensional extension

$$k^2 \left(\frac{\partial^2 \phi}{\partial x^2} + \frac{\partial^2 \phi}{\partial y^2} \right) = \frac{\partial \phi}{\partial t} \tag{8.11}$$

The diffusion can be of heat or mass and (8.10) could model the voltage (or current) in an electrical transmission line where both inductance and leakage are negligible.

In both equations k is a constant.

We shall also study the one-dimensional **wave equation**

$$c^2 \frac{\partial^2 \phi}{\partial x^2} = \frac{\partial^2 \phi}{\partial t^2} \tag{8.12}$$

which has a two-dimensional extension

$$c^2 \left[\frac{\partial^2 \phi}{\partial x^2} + \frac{\partial^2 \phi}{\partial y^2} \right] = \frac{\partial^2 \phi}{\partial t^2} \tag{8.13}$$

The function ϕ could be a component of displacement in a vibrating system, the velocity potential of a gas in acoustic theory or each component of the electric or magnetic vector in the electromagnetic theory of light.

Eigenvalue problems are extensions of equilibrium problems with a leaning towards initial-value problems. In addition to the steady-state configuration or **mode**, it is required to find critical values (eigenvalues) of a scalar problem parameter. The kinds of problem studied include vibrations at natural frequencies, resonance in electrical circuits and acoustics, and the buckling of structures. It is often only the relative displacements or relative amplitudes in a particular mode which can be found.

Some further remarks are in order. Since eigenvalue problems can be regarded as almost equilibrium problems or almost propagation problems, equations under either of these latter categories *may* be models for the former class.

The nature of the problem will allow us to decide if it can be classed as an eigenvalue problem.

Alternative classification

For development of solutions we shall find it convenient to classify the model equations by an alternative scheme.

The most general linear partial differential equation of the second order with two independent variables is

$$A\frac{\partial^2\phi}{\partial x^2} + B\frac{\partial^2\phi}{\partial x\partial y} + C\frac{\partial^2\phi}{\partial y^2} + D\frac{\partial\phi}{\partial x} + E\frac{\partial\phi}{\partial y} + F\phi + G = 0 \tag{8.14}$$

where A, B, C, D, E, F and G are functions of x and y including constants.

By analogy with the general equation for a conic: $ax^2 + bxy + cy^2 + dx + ey + f = 0$ we have the following classification

(i) If $B^2 < 4AC$, the equation is **elliptic**.

(ii) If $B^2 > 4AC$, the equation is **hyperbolic**.

(iii) If $B^2 = 4AC$, the equation is **parabolic**.

Examples

(i) **Laplace's equation** (8.2) **is a special case of** (8.14) with $A = 1$, $C = 1$ and all other constants zero. $B^2 < 4AC$ in this case, hence the equation is elliptic.

(ii) The one-dimensional wave equation (8.12) has $A = c^2$, $C = -1$, $B = 0$ and hence $B^2 > 4AC$ to show that the equation is hyperbolic.

(iii) The diffusion equation (8.10) has $A = k^2$, $E = -1$ and all other constants zero. Hence $B^2 - 4AC = 0$ and the equation is parabolic.

Notice that the extensions to more variables (8.13) and (8.11) are hyperbolic and parabolic respectively. Notice too how often in these extensions the expression $\frac{\partial^2\phi}{\partial x^2} + \frac{\partial^2\phi}{\partial y^2}$ occurs. We may recognise $\left(\frac{\partial^2}{\partial x^2} + \frac{\partial^2}{\partial y^2}\right)$ as the two-dimensional case of the **Laplace operator, ∇^2**, see Section 8.4

Problems

1. Classify each of the following equations as elliptic, hyperbolic or parabolic.

(a) $\frac{\partial^2 u}{\partial x^2} - \frac{\partial^2 u}{\partial y^2} = 0$ (b) $\frac{\partial u}{\partial x} + \frac{\partial^2 u}{\partial x\partial y} = 8$ (c) $\frac{\partial^2 u}{\partial x^2} - 2\frac{\partial^2 u}{\partial x\partial y} + 2\frac{\partial^2 u}{\partial y^2} = x + 3y$

(d) $\frac{\partial^2 u}{\partial x^2} + 3\frac{\partial^2 u}{\partial x\partial y} + 4\frac{\partial^2 u}{\partial y^2} + 5\frac{\partial u}{\partial x} - 2\frac{\partial u}{\partial y} + 4u = 2x - 6y$

(e) $\frac{\partial^2 u}{\partial x^2} - 7\frac{\partial^2 u}{\partial x\partial y} + \frac{\partial^2 u}{\partial y^2} = 0$ (f) $\frac{\partial^2 u}{\partial x^2} + \frac{\partial^2 u}{\partial x\partial y} - 6\frac{\partial^2 u}{\partial y^2} = 0$

(g) $\frac{\partial^2 u}{\partial x^2} + 6\frac{\partial^2 u}{\partial x\partial y} + 9\frac{\partial^2 u}{\partial y^2} = 0$

8.4 SEPARATION OF VARIABLES METHOD

We now develop an analytical method of solution which will cope with the linear partial differential equations we have met so far. Essentially, the aim is to reduce the one *partial* differential equation to two or more *ordinary* differential equations, each one involving only one of the independent problem variables. This will be accomplished by separating these variables from the very beginning. It is easiest to demonstrate this method by applying it to an example. We shall choose for this purpose Laplace's equation representing the steady-state temperature distribution of Section 8.2. For mental refreshment we present the equation and boundary conditions.

$$\text{The equation is} \quad \frac{\partial^2 \theta}{\partial x^2} + \frac{\partial^2 \theta}{\partial y^2} = 0 \tag{8.2}$$

and the boundary conditions are

$$\theta = 0 \text{ when } x = 0;\ \theta = 0 \text{ when } x = a;\ \theta = 0 \text{ when } y = 0;\ \theta = 100 \text{ when } y = b \tag{8.3}$$

The underlying vital assumption is that we *can* find a form for the solution $\theta(x, y)$ as the product of a function of x and a function of y. Therefore, we assume that $\theta = X(x)\ .\ Y(y)$. What we shall do is to cheat a little by producing for illustration a form for θ which does satisfy equation (8.2); we shall develop the first stage of the method for this particular θ in parallel with the development for the general expression for θ.

$$\theta = X(x)\,Y(y) \qquad\qquad \theta = \cos 2x \sinh 2y$$

$$\frac{\partial \theta}{\partial x} = X'(x)\,Y(y) \qquad\qquad \frac{\partial \theta}{\partial x} = -2 \sin 2x \sinh 2y$$

$$\frac{\partial^2 \theta}{\partial x^2} = X''(x)\,Y(y) \qquad\qquad \frac{\partial^2 \theta}{\partial x^2} = -4 \cos 2x \sinh 2y$$

$$\frac{\partial^2 \theta}{\partial y^2} = X(x)\,Y''(y) \qquad\qquad \frac{\partial^2 \theta}{\partial y^2} = \cos 2x\ .\ 4 \sinh 2y$$

$$\frac{\partial^2 \theta}{\partial x^2} + \frac{\partial^2 \theta}{\partial y^2} = X''Y + XY'' = 0 \qquad\qquad \frac{\partial^2 \theta}{\partial x^2} + \frac{\partial^2 \theta}{\partial y^2} = (-4 \cos 2x)(\sinh 2y) + (\cos 2x)(4 \sinh 2y) = 0$$

$$\text{Hence } X''Y = -XY'' \qquad\qquad -4 \cos 2x \sinh 2y = \cos 2x\,(-4 \sinh 2y)$$

$$\text{and } \frac{X''}{X} = \frac{-Y''}{Y} \qquad\qquad \frac{-4 \cos 2x}{\cos 2x} = \frac{-4 \sinh 2y}{\sinh 2y}$$

(We are justified in dividing both sides of the equation by XY unless either $X \equiv 0$ or $Y \equiv 0$; if either of these cases occurs then $\theta \equiv 0$ and this, whilst being a solution of (8.2) does not satisfy the fourth boundary condition.)

You will notice that with the particular form of θ we chose, both sides of the last equation reduce to a constant. In fact, looking at the general case, we see that this must always be so. For the left-hand side (X''/X) being the ratio of two functions of x, is, at worst, also a function of x. Similarly the right-hand side (Y''/Y) must be independent of x, containing at most y. However, the right-hand side and the left-hand side *must* always be in balance and if the right-hand side does not respond to changes in x then nor does the left-hand side. Since this latter cannot respond to changes in y, it must be a constant, as must the right-hand side. Therefore we may conclude that $\dfrac{X''}{X} = \dfrac{-Y''}{Y} = \text{constant}$.

The question that now arises is whether to take the constant to be positive or negative (why not zero?). Consider the two equations

$$Z'' = k^2 Z \qquad\qquad Z'' = -k^2 Z$$

$$\text{or} \quad \frac{d^2 Z}{dz^2} = k^2 Z \qquad\qquad \frac{d^2 Z}{dz^2} = -k^2 Z$$

The general solutions are

$$Z = A' \cosh kz + B' \sinh kz \qquad\qquad Z = C \cos kz + D \sin kz$$

or $Z = Ae^{kz} + Be^{-kz}$

where A, B, A', B', C and D are constants.

It would seem, therefore, that the choice of a positive constant, k^2 will give rise to $X = Ae^{kx} + Be^{-kx}$ and $Y = C \cos ky + D \sin ky$; ($k^2$ is chosen since it is always positive for k real); the choice of a negative constant, $-k^2$ will produce

$$X = C \cos kx + D \sin kx \quad \text{and} \quad Y = Ae^{ky} + Be^{-ky}$$

Which sign we choose for the constant will be governed by the need to satisfy the boundary conditions, which determine the form of the solution. In Figure 8.8 we graph the six basic functions under consideration.

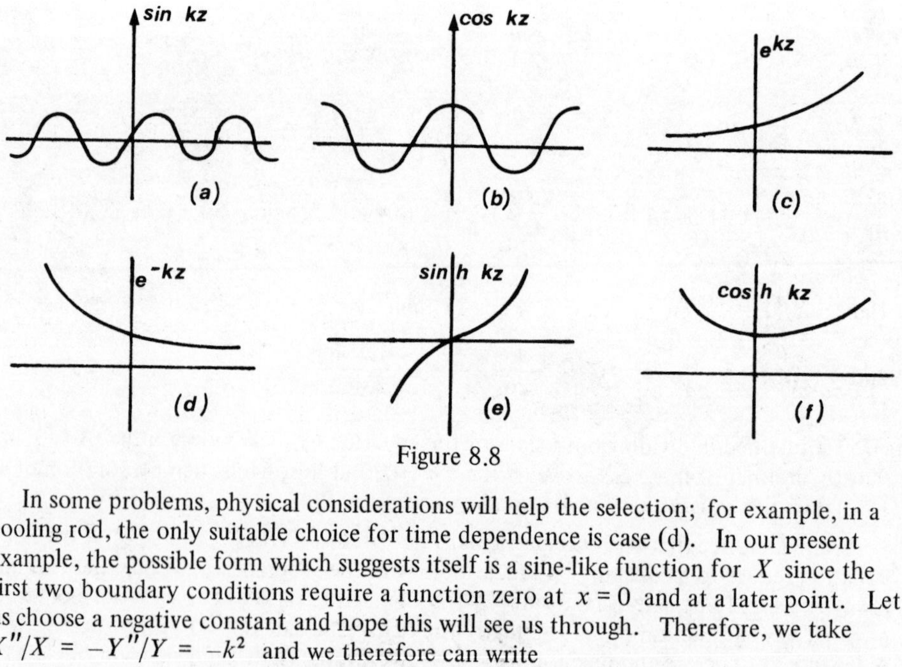

Figure 8.8

In some problems, physical considerations will help the selection; for example, in a cooling rod, the only suitable choice for time dependence is case (d). In our present example, the possible form which suggests itself is a sine-like function for X since the first two boundary conditions require a function zero at $x = 0$ and at a later point. Let us choose a negative constant and hope this will see us through. Therefore, we take $X''/X = -Y''/Y = -k^2$ and we therefore can write

$$\theta(x, y) \equiv X(x) \,.\, Y(y) = (C \cos kx + D \sin kx)(Ae^{ky} + Be^{-ky}) \tag{8.15}$$

We should be quite clear what we are saying here. All expressions of the form (8.15) for θ will satisfy Laplace's equation (8.2); different combinations of values A, B, C and D give different particular solutions. We may need one of these particular solutions or a linear combination of them.

Now apply the first boundary condition; $\theta = 0$ when $x = 0$, for $0 \leqslant y < b$. Then $0 = (C \,.\, 1 + D \,.\, 0)(Ae^{ky} + Be^{-ky})$.

The second bracket cannot be zero for all y in the interval $(0, b)$ unless A and B are both zero; this would give $\theta \equiv 0$ and we therefore reject the possibility. We must have $C = 0$ so that $\theta = D \sin kx\,(Ae^{ky} + Be^{-ky})$

However, in practice we would not distinguish between $\theta = 2 \sin 3x\,(4e^{3y} + 5e^{-3y})$, $\theta = 5 \sin 3x\,(\frac{8}{5}e^{3y} + 2e^{-3y})$ and $\theta = \sin 3x\,(8e^{3y} + 10e^{-3y})$. We shall use this last form. It makes sense to absorb D into the constants A and B, so that from now on A replaces AD and B replaces BD . We then have $\theta = \sin kx\,(Ae^{ky} + Be^{-ky})$.

The third boundary condition requires that $\theta = 0$ when $y = 0$ for all x in $(0, a)$, this implies that $0 = \sin kx\,(A.1 + B.1)$. Since $\sin kx \equiv 0$ in $(0, b)$ would give $\theta \equiv 0$ we require that $A + B = 0$ i.e. $B = -A$ so that $\theta = A \sin kx\,(e^{ky} - e^{-ky})$ or $\theta = A \sin kx \sinh ky$.

Next, we apply the second boundary condition, that $\theta = 0$ when $x = a$ for all y in $(0, b)$ i.e. $0 = A \sin ka \sinh ky$.

You should be able to reason that if either $A = 0$ or $\sinh ky \equiv 0$ then $\theta \equiv 0$, therefore $\sin ka = 0$, i.e. we are restricted in the values of k that we can use to those which make $\sin ka = 0$. But $\sin z = 0$ at $z = n\pi$, n integer. Hence $k = n\pi/a$, n integer.

We are not quite there, however, since we can show that n must be positive. If n were zero then k would be zero and the corresponding value of θ would be zero. Now the effect of choosing k to be negative is to give us no new information. For example, $A \sin \frac{3\pi}{a}x \sinh \frac{3\pi}{a}y$ and $A \sin\left(\frac{-3\pi}{a}\right)x \sinh\left(\frac{-3\pi}{a}\right)y$ are in reality the same expression. Therefore, we restrict k to the values $n\pi/a$, $n = 1, 2, 3, \ldots..$

Each of the functions, $A_1 \sin \frac{\pi x}{a} \sinh \frac{\pi y}{a}$, $A_2 \sin \frac{2\pi x}{a} \sinh \frac{2\pi y}{a}$, $A_3 \sin \frac{3\pi x}{a} \sinh \frac{3\pi y}{a}$ etc. satisfies equation (8.2) and the first three boundary conditions.

If we take stock of the situation we can say that the most general form for θ before we fit the fourth boundary condition is

$$\theta = \sum_{n=1}^{\infty} A_n \sin \frac{n\pi x}{a} \sinh \frac{n\pi y}{a} \tag{8.16}$$

We have chosen different values for the constants $A_1, A_2, A_3, \ldots..$ to be as general as possible.

If we try to fit the fourth boundary condition we arrive at the result

$$100 = \sum_{n=1}^{\infty} A_n \sin \frac{n\pi x}{a} \sinh \frac{n\pi b}{a} = \sum_{n=1}^{\infty} F_n \sin \frac{n\pi x}{a}$$

where we have to determine the coefficients F_n. We have the Fourier series problem of trying to represent the function $f(x) = 100$, $0 \leqslant x \leqslant a$ by a series of sine terms. From the ideas in Section 7.6 we know that the coefficients are given by the formula

$$F_n = \frac{2}{a}\int_0^a 100 \sin\frac{n\pi x}{a}\,dx = \frac{200}{a}\left[-\frac{a}{n\pi}\cos\frac{n\pi x}{a}\right]_0^a$$

$$= \frac{200}{n\pi}\left[1 - \cos n\pi\right]$$

Hence $F_n = 0$ for even n and $F_n = \dfrac{400}{n\pi}$ for odd n. Therefore $A_n = 0$ for even n and $A_n = \dfrac{400}{n\pi} \cdot \dfrac{1}{\sinh(n\pi b/a)}$ for odd n. Substituting for A_n in (8.16), we get

$$\text{finally } \theta = \frac{400}{\pi}\sum_{n=1,3,5,\ldots}^{\infty}\frac{1}{n}\,\frac{\sin(n\pi x/a)\sinh(n\pi y/a)}{\sinh(n\pi b/a)} \tag{8.4}$$

which is the result we quoted in Section 8.2.

Problems

1. Obtain the analytical formula given in Problem 1 of Section 8.2.

2. Repeat for Problem 2 of Section 8.2.

3. Consider the rectangular plate $0 \leqslant x \leqslant a$, $0 \leqslant y \leqslant b$. The temperature distributions along the edges are given by $\theta(0,y) = f_1(y)$; $\theta(a,y) = f_2(y)$; $\theta(x,0) = f_3(x)$; $\theta(x,b) = f_4(x)$. Consider four solutions of Laplace's equation as θ_1, θ_2, θ_3, θ_4 which satisfy the conditions respectively

   $\theta_1(0,y) = f_1(y)$, $\theta_1 = 0$ along the three other edges
   $\theta_2(a,y) = f_2(y)$, $\theta_2 = 0$ along the three other edges
   $\theta_3(x,0) = f_3(x)$, $\theta_3 = 0$ along the three other edges
   $\theta_4(x,b) = f_4(x)$, $\theta_4 = 0$ along the three other edges

   Show that the steady-state solution to the original problem can be written as $\theta(x,y) = \theta_1 + \theta_2 + \theta_3 + \theta_4$. Interpret this result.

4. Find a solution of Laplace's equation which satisfies the conditions

   (i) $\theta \to 0$ as $y \to 0$; (ii) $\theta = 0$ when $x = 0$ and $x = 1$;
   (iii) $\theta = \sin \pi x + \sin 2\pi x$ when $y = 0$.

5. Obtain a solution of the equation $\dfrac{\partial^2 V}{\partial x^2} + 9\dfrac{\partial^2 V}{\partial t^2} = 0$ which satisfies the conditions

   (i) V is periodic in t; (ii) $V = 0$ when $t = 0$; (iii) $\dfrac{\partial V}{\partial x} = 0$ when $x = 0$;
   (iv) $V = 4\cosh 15x$ when $t = \pi/2$.

6. Show that the solution of Laplace's equation which satisfies

   (i) $\theta \to 0$ as $y \to 0$ for $-a < x < a$; (ii) $\partial\theta/\partial x = 0$ when $x = \pm a$;
   (iii) $\theta = -T_0$ when $y = 0$, $-a < x < 0$; (iv) $\theta = T_0$ when $y = 0$, $0 < x < a$

is given by the formula

$$\theta = \frac{T_0}{\pi} \sum_{n=0}^{\infty} \frac{4}{(2n+1)} \exp\left\{-(n+\tfrac{1}{2})\pi y/a\right\} \sin\left\{(n+\tfrac{1}{2})\pi x/a\right\} \qquad \text{(L.U.)}$$

8.5 ORIGIN OF SOME PARTIAL DIFFERENTIAL EQUATIONS

In this section we develop the models for some physical problems and derive their solutions by separating the variables. We believe that the construction of these models is an important part of the process and should be emphasised.

Example 1 Torsion of uniform prism with rectangular section

We wish to calculate the amount of twist produced when a structural member sustains torsional loads; also we wish to calculate the shear stresses set up. We take the simple problem of a solid bar of uniform section twisted by a torque applied to one end, whilst the other end is fixed to prevent rotation. For simplicity we shall consider a rectangular cross-section, but we will impose this particular condition at a later stage. In Figure 8.9, we see the schematic representation of the process.

The free end rotates about the z - axis in such a way that a cross-section distant z from the fixed end rotates through an angle $\theta = \alpha z$, where α is a twist per unit length, assumed constant. Any point in a given section will be displaced in the x - and y - directions by amounts $d_x = -\alpha yz$ $d_y = \alpha xz$ respectively; these results follow from considering the displacement, d_θ in a tangential direction and resolving this in the x - and y - directions.

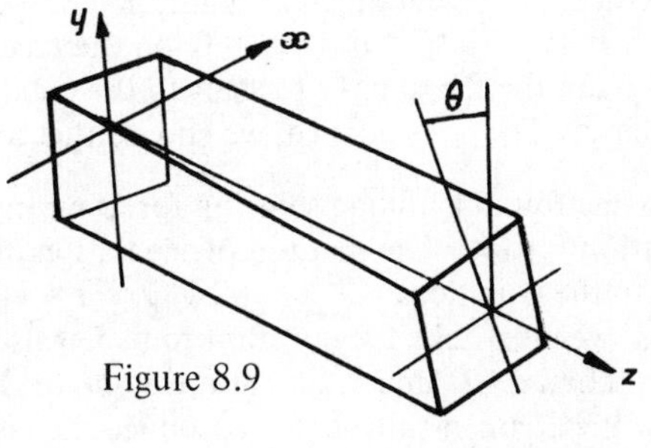

Figure 8.9

To complete the set-up we *assume* that the displacement in the z - direction is a function of x and y only. If we write $d_z = \psi(x, y)$ then $\psi(x, y)$ is called the **warping function.** A section which was originally plane will take up a shape defined by $\psi(x, y)$. The *assumption* implicitly made is that the ends of the bar are free to warp.

We may obtain the strains by differentiating the displacement. For example, $e_x = \frac{\partial}{\partial x}(d_x) = 0$; similarly, $e_y = 0 = e_z$. Since all the direct strains are zero, it follows that throughout the bar, direct stresses are zero. The shear displacements can be found to be $e_{xy} = \frac{1}{2}\frac{\partial}{\partial x}(d_x) + \frac{\partial}{\partial y}(d_y) = 0$, $e_{zx} = \frac{\alpha}{2}(\psi_x - y)$, $e_{zy} = \frac{\alpha}{2}(\psi_y + x)$. The non-vanishing shear stresses are $\sigma_{zx} = G\alpha(\psi_x - y)$ and $\sigma_{zy} = G\alpha(\psi_y + x)$, where G is the shear modulus for the bar. If we consider the three stress equilibrium equations, the one which is not obviously satisfied is $\frac{\partial}{\partial z}\sigma_{zz} + \frac{\partial}{\partial x}\sigma_{zx} + \frac{\partial}{\partial y}\sigma_{zy} = 0$. However, we can ensure that this equation always holds by introducing a **stress function** ϕ which

gives rise to the shear stresses by the equations $\sigma_{zx} = G\alpha \dfrac{\partial \phi}{\partial y}$, $\sigma_{zy} = -G\alpha \dfrac{\partial \phi}{\partial x}$.
Therefore from the two expressions for σ_{zx} we obtain

$$\frac{\partial \phi}{\partial y} = \frac{\partial \psi}{\partial x} - y \tag{8.17a}$$

Similarly

$$\frac{\partial \phi}{\partial x} = -\frac{\partial \psi}{\partial y} - x \tag{8.17b}$$

Now

$$\frac{\partial^2 \phi}{\partial y^2} = \frac{\partial^2 \psi}{\partial y\, \partial x} - 1 \quad \text{and} \quad \frac{\partial^2 \phi}{\partial x^2} = -\frac{\partial^2 \psi}{\partial x\, \partial y} - 1$$

On addition we obtain, for the stress function $\phi(x, y)$ the equation

$$\frac{\partial^2 \phi}{\partial x^2} + \frac{\partial^2 \phi}{\partial y^2} = -2 \tag{8.18}$$

This is a form of Poisson's Equation.
We may also obtain the following equation in $\psi(x,y)$

$$\frac{\partial^2 \psi}{\partial x^2} + \frac{\partial^2 \psi}{\partial y^2} = 0 \tag{8.2}$$

which is Laplace's equation.

We need to find boundary conditions for ϕ. Let the rectangular cross-section be $-a \leqslant x \leqslant a$, $-b \leqslant y \leqslant b$. $\sigma_{zy} = 0$ on the faces $x = \pm a$ and that the shear stress $\sigma_{zx} = 0$ on the faces $y = \pm b$ leads to the condition that ϕ be constant along the entire boundary. For convenience, we choose this constant value to be zero.

Having found a unique solution for ϕ we make use of (8.17a) and (8.17b) to obtain a solution for ψ. The method of separation of variables gives $\phi = X(x).\ Y(y)$ and leads to the equation $X''/X = -2 - Y''/Y =$ constant. To satisfy the conditions at $x = \pm a$ we must take the constant to be negative, i.e. $-k^2$. This gives $X = C\cos kx + D\sin kx$ and also we obtain $Y'' = (k^2 - 2)\,Y$. See whether you can follow the argument through to produce the result

$$\phi = b^2 - y^2 - \frac{32b^2}{\pi^3} \sum_{n=1,3,5,..}^{\infty} (-1)^{(n-1)/2} . \frac{1}{n^3} \operatorname{sech} \frac{n\pi a}{2b} \cosh \frac{n\pi x}{2b} \cos \frac{n\pi y}{2b} \tag{8.19}$$

Note that the terms of this series alternate in sign.

Example 2 Heat flow in one dimension

Suppose that we have a long thin bar of length l which is aligned along the x - axis. We wish to determine the temperature distribution $\theta(x, t)$ in the bar. We make the assumptions that the bar is insulated along its sides and that heat flows in the x - direction only. In Figure 8.10 we examine a section of the bar at a distance x_0 from

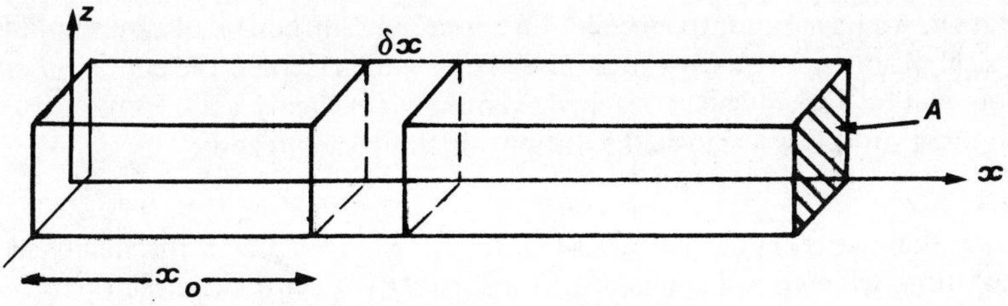

Figure 8.10

one end.

We make use of the following **empirical** laws of heat flow:

(i) The amount of heat in a body is proportional to its mass and to its temperature.

(ii) Heat flows from a point at a higher temperature to one at a lower temperature.

(iii) The rate of flow of heat through a plane surface is proportional to the area of the surface and to the rate of change of temperature with respect to distance in a direction perpendicular to the plane.

The rate of inflow of heat to the section is $-KA\theta_x(x_0, t)$ where K is the thermal conductivity of the material of the bar and A is the (constant) cross-section area of the bar. (Why the negative sign?) Similarly, the rate of outflow of heat is $-KA\theta_x(x_0 + \delta x, t)$. From law (i) we obtain the rate of build-up of heat in the section as $c\rho A\delta x\theta_t(x_0, t)$ where ρ is the density of the material, and c is its specific heat. Let $k = K/c\rho$ be the **diffusivity** of the material. We may equate the rate of build-up of heat to the rate of inflow to the section less the rate of outflow and, by taking the limiting case as $\delta x \to 0$, we eventually obtain the diffusion equation

$$\frac{\partial\theta}{\partial t} = k\frac{\partial^2\theta}{\partial x^2} \tag{8.20}$$

Suitable boundary conditions might be

(i) $\theta = 100$ at $x = 0$ for all times $t > 0$

(ii) $\theta = 50$ at $x = l$ for all times $t > 0$

(iii) The initial temperature distribution is specified, e.g. $\theta = 100$ at $t = 0$ for all points in the bar.

(In this case we may imagine that one end of the bar is suddenly cooled to 50^0C.)

Now the problem is exactly parallel to one where $\theta = 50$ at $x = 0$; $\theta = 0$ at $x = l$ and $\theta = 50$ at $t = 0$. This equation may be slightly easier to solve with these boundary conditions and we may then add 50 to recover the solution of the original problem.

Let $\theta = X(x)T(t)$. Then $\partial\theta/\partial t = XT'$ and $\partial^2\theta/\partial x^2 = X''T$. Substituting into (8.20) we obtain $XT' = kX''T$ or $\dfrac{X''}{X} = \dfrac{1}{k}\dfrac{T'}{T}$ = constant.

Since the temperature of a point on the bar may be expected to decrease with the passage of time we must choose a negative value for the constant, $-\lambda^2$ say, so that we obtain $T' = -\lambda^2 kT$ and hence $T = e^{-\lambda^2 kt}$. Then it follows that $X = C\cos\lambda x + D\sin\lambda x$ and that $\theta = (C\cos\lambda x + D\sin\lambda x)e^{-\lambda^2 kt}$.

Applying the condition that $\theta = 50$ at $x = 0$ we find that $50 = (C.1 + D.0)e^{-\lambda^2 kt}$. Immediately, we have run into trouble. We overcome this difficulty by thinking about the physical problem. If we maintain the end $x = 0$ at 50 and the end $x = l$ at 0 then there will be a steady stage reached eventually (in theory as $t \to \infty$). Superimposed on this is a **transient** solution which decays to zero.

We may then write $\theta(x, t) = \theta_s(x) + \theta_T(x, t)$, where $\theta_s(x)$ is the steady-state part of the solution, which is independent of time, and $\theta_T(x, t)$ the transient part.

Since $\theta_s(x)$ is independent of time $\dfrac{\partial \theta_s}{\partial t} = 0$ and from (8.20) it must satisfy $k \dfrac{d^2\theta_s}{dx^2} = 0$, i.e. $\theta_s(x) = \alpha x + \beta$. Now $\theta_s(0) = 50$ and $\theta_s(l) = 0$, therefore, $\theta_s(x) = 50 - 50x/l$.

Consequently, $\theta_T(0, t) = \theta(0, t) - \theta_s(0) = 0$, $\theta_T(l, t) = \theta(l, t) - \theta_s(l) = 0$ and $\theta_T(x, 0) = \theta(x, 0) - \theta_s(x) = \;\; 50 - 50 \;\; + 50x/l = 50x/l$.

At this stage we note that $\dfrac{\partial \theta}{\partial t} = \dfrac{\partial \theta_s}{\partial t} + \dfrac{\partial \theta_T}{\partial t} = \dfrac{\partial \theta_T}{\partial t}$ and $\partial^2\theta/\partial x^2 = \partial^2\theta_s/\partial x^2 + \partial^2\theta_T/\partial x^2$. Hence both θ_s and θ_T satisfy (8.20) separately.

The most general solution possible which satisfies the first two boundary conditions is

$$\theta_T = \sum_{n=1}^{\infty} D_n \sin \frac{n\pi x}{l} \, e^{-n^2\pi^2 kt/l^2} \qquad \text{(We ignore } n \leqslant 0. \text{ Why?)}$$

The last boundary condition is $\theta_T = 50x/l$ at $t = 0$; therefore

$$\frac{50x}{l} = \sum_{n=1}^{\infty} D_n \sin \frac{n\pi x}{l}$$

Using the ideas applied to the steady-state temperature distribution problem we produce the formula

$$D_n = \frac{2}{l} \int_0^l \frac{50x}{l} \sin \frac{n\pi x}{l} \, dx = \frac{-100}{n\pi} (-1)^n$$

We now add on $\theta_s(x)$ and the 50°C we removed earlier on.

Finally we have the solution for $\theta(x, t)$ given by

$$\theta = 100 - \frac{50x}{l} - \frac{100}{\pi} \sum_{n=1}^{\infty} \frac{(-1)^n}{n} \sin \frac{n\pi x}{l} \, . \, e^{-n^2\pi^2 kt/l^2} \tag{8.21}$$

It is worth noting the presence of $-n^2$ in the index of the exponential; this should indicate that the terms will decay to zero reasonably rapidly and hence only a few terms may be needed to obtain a good approximation to the value of $\theta(x, t)$ being calculated. We programmed a computer to calculate $\theta(x, t)$ for suitable values of x

and t and in Figure 8.11 we show the temperature profile $\theta(x)$ for selected values of t; k and l were taken to be 1. (What do you think will be the effects of changing k and/or l?)

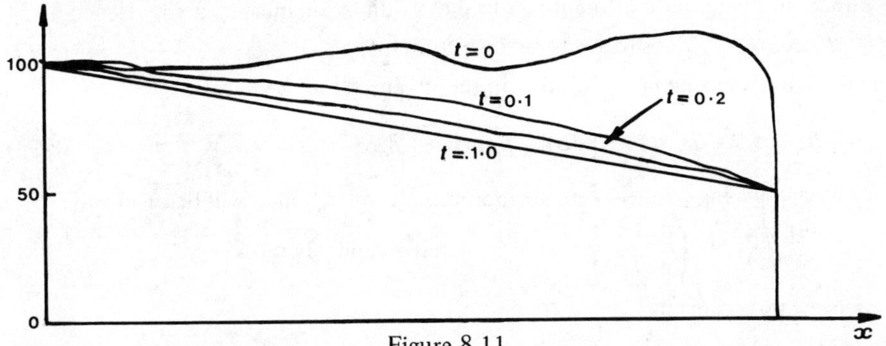

Figure 8.11

Example 3 Vibration of a string

We consider the small transverse vibrations of a stretched string. We know that this is an artificial example but we decided to discuss it not because of its historical importance but because it is easy to visualise the solution in terms of the physical problem.

In Figure 8.12(a) we see the string, initially of length l, fixed at both ends, subject to a constant tension T and in a state of vibration.

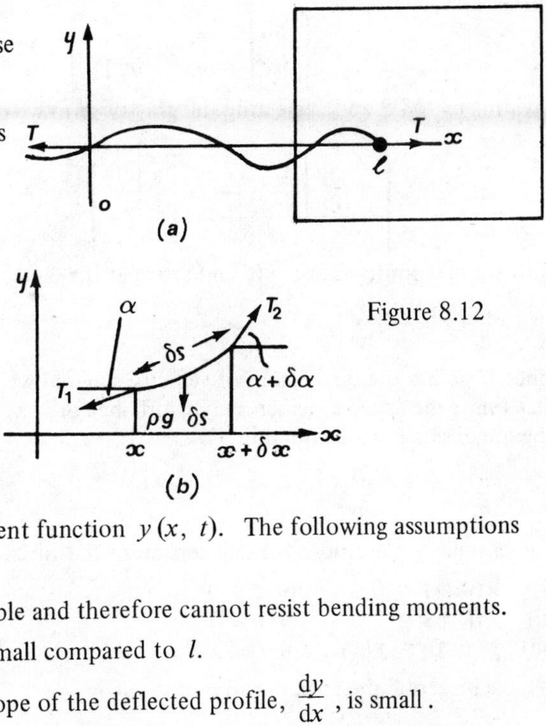

Figure 8.12

We wish to find the displacement function $y(x, t)$. The following assumptions are made.

(i) The string is perfectly flexible and therefore cannot resist bending moments.

(ii) The displacements y are small compared to l.

(iii) At any point of time the slope of the deflected profile, $\frac{dy}{dx}$, is small.

(iv) The tension T remains constant at all times and for all points on the string.

(v) The mass of the string can be neglected.

(vi) The horizontal displacement of a point on the string can be neglected in comparison to the vertical displacement y, i.e. the motion is **transverse**.

(vii) The motion takes place in the x - y plane.

Note that these assumptions are consistent. Inconsistency would arise, if, for example, we required horizontal displacements to be negligible *and* did not require assumption (ii) to hold.

In Figure 8.12(b) we analyse the forces acting on a small part of the string. We let the section be of length δs and it can be considered to be a straight line; ρ is the density of the string per unit length. Since we assume no displacement in the x - direction, the resolute of the forces in the x - direction must be zero. Hence

$$T_1 \cos\alpha = T_2 \cos(\alpha + \delta\alpha) = T, \text{ a constant} \tag{8.22}$$

Using Newton's second law of motion in the y - direction, we have

$$\rho\,\delta s\,\frac{\partial^2 y}{\partial t^2} = T_2 \sin(\alpha + \delta\alpha) - T_1 \sin\alpha \;\; - \rho g \delta s \tag{8.23}$$

Since we are dealing only with small vibrations, $\mathrm{d}y/\mathrm{d}x = \tan\alpha$ will be small and

$$\sin\alpha = \frac{\mathrm{d}y}{\mathrm{d}x}\bigg/\left\{1 + \left(\frac{\mathrm{d}y}{\mathrm{d}x}\right)^2\right\}^{\frac{1}{2}} \simeq \frac{\mathrm{d}y}{\mathrm{d}x} = \tan\alpha \quad \text{and} \quad \delta s \simeq \delta x.$$

Hence, approximately,

$$\rho\,\delta x\,.\,\frac{\partial^2 y}{\partial t^2} = T\tan(\alpha + \delta\alpha) - T\tan\alpha - \rho g \delta x$$

$$= T\left[\frac{\partial y}{\partial x}\bigg|_{x+\delta x} - \frac{\partial y}{\partial x}\bigg|_{x}\right] - \rho g \delta x$$

Dividing by $\rho\delta x$ and neglecting the gravity term we obtain

$$\frac{\partial^2 y}{\partial t^2} = \frac{T}{\rho}\left[\frac{\left.\frac{\partial y}{\partial x}\right|_{x+\delta x} - \left.\frac{\partial y}{\partial x}\right|_{x}}{\delta x}\right]$$

Now we take limits as $\delta x \to 0$ and arrive at the equation

$$\frac{\partial^2 y}{\partial t^2} = \frac{T}{\rho}\,\frac{\partial^2 y}{\partial x^2}$$

Since T/ρ has the dimensions of velocity squared we write it as c^2: it can be shown that this is the speed at which the disturbance or wave travels. We finally produce the one-dimensional wave equation

$$\frac{\partial^2 y}{\partial t^2} = c^2\,\frac{\partial^2 y}{\partial x^2} \tag{8.24}$$

The boundary conditions for one version of the vibrating string problem are:

(i) $y(0, t) = 0$ for $t \geqslant 0$
(ii) $y(l, t) = 0$ for $t \geqslant 0$
(iii) $y(x, 0) = f(x)$ for $0 \leqslant x \leqslant l$

i.e. we prescribe the initial profile of the string

(iv) $\dfrac{\partial y}{\partial t} = 0$ at $t = 0$ for $0 \leqslant x \leqslant l$

i.e. the string is released from rest.

We seek a solution in the form $y = X(x)\,.\,T(t)$.

Separating the variables leads to the result $T''/T = c^2 X''/X =$ constant. Since we are dealing with a vibration, observation suggests that we expect a periodic behaviour in the time variable, i.e. we expect $T = C\cos pt + D\sin pt$.

This implies that we might choose the separation constant to be $-k^2c^2$ and therefore we obtain the equation $X'' = -k^2 X$ so that $X = A\cos kx + B\sin kx$ and $y(x, t) = (A\cos kx + B\sin kx)(C\cos kct + D\sin kct)$.

The solution satisfying (i) and (ii) is of the form

$$y_n(x, t) = \sin\frac{n\pi x}{l}\left(C_n\cos\frac{n\pi ct}{l} + D_n\sin\frac{n\pi ct}{l}\right) \tag{8.25}$$

where we have attached suffices to the constants C and D for ease of calculation. We call such functions as (8.25) **eigenfunctions** and the values $\lambda_n = \frac{n\pi c}{l}$ the **eigenvalues** of the string. Each solution $y_n(x, t)$ represents a periodic motion with a frequency $(nc/2l)$ called the nth **normal mode.** The case $n = 1$ is the **fundamental** mode of vibration. In Figure 8.13 we graph the first three modes of vibration. The **nodes** for each mode are those points which do not move during the vibration. Together with $x = 0$ and $x = l$ they are equally spaced at $x = l/n,\ 2l/n,\ \ldots\ldots,\ \left\{\frac{n-1}{n}\right\} l$. In Figure 8.14 we show the third normal mode at different times.

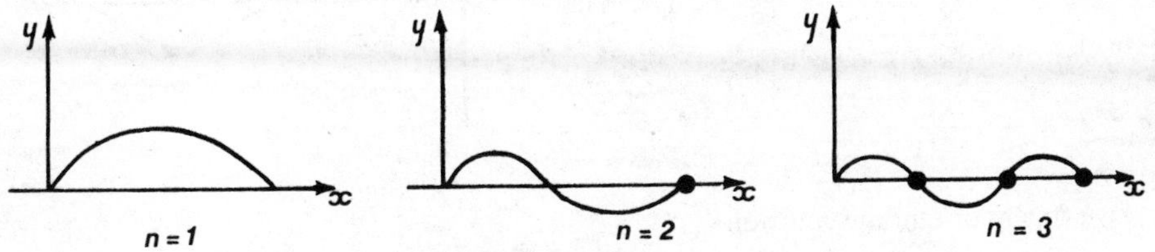

Figure 8.13

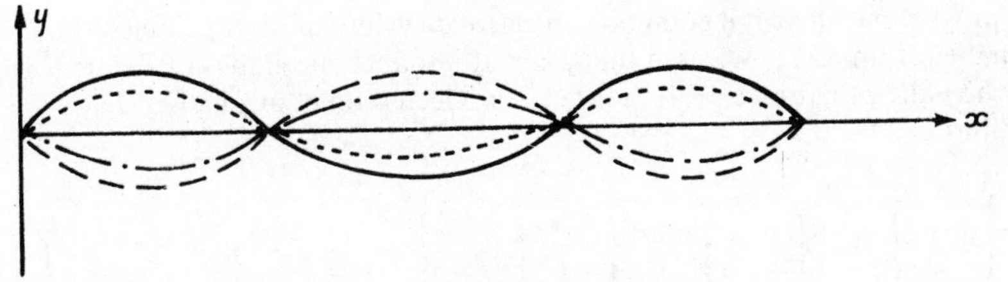

Figure 8.14

Returning to (8.25) we need to produce the most general solution which satisfies the first two boundary conditions. We do this by summing the basic solutions and obtain

$$y(x, t) = \sum_{n=1}^{\infty} \sin\frac{n\pi x}{l}\left(C_n\cos\frac{n\pi ct}{l} + D_n\sin\frac{n\pi ct}{l}\right)$$

Now $\partial y/\partial t = 0$ when $t = 0$; therefore we first find $\partial y/\partial t$;

$$\frac{\partial y}{\partial t} = \sum_{n=1}^{\infty} \sin\frac{n\pi x}{l}\left(-\frac{n\pi c}{l}C_n\sin\frac{n\pi ct}{l} + \frac{n\pi c}{l}D_n\cos\frac{n\pi ct}{l}\right)$$

Applying this boundary condition we have

$$0 = \sum_{n=1}^{\infty} \sin \frac{n\pi x}{l} \frac{n\pi c}{l} \cdot D_n$$

We need to take $D_n = 0$, for all n, and we are now able to state that

$$y(x, t) = \sum_{n=1}^{\infty} C_n \sin \frac{n\pi x}{l} \cos \frac{n\pi ct}{l} \tag{8.26}$$

The last boundary condition prescribes a form $f(x)$ for $y(x, t)$ at $t = 0$. Then (8.26) becomes

$$y(x, 0) = f(x) = \sum_{n=1}^{\infty} C_n \sin \frac{n\pi x}{l}$$

This is a Fourier series problem and we manufacture the function shown in Figure 8.15.

Hence $C_{2n} = 0$ for even n,

$$C_{4n+1} = \frac{8}{(4n+1)^2\pi^2},$$

$$C_{4n+3} = \frac{-8}{(4n+3)^2\pi^2}.$$

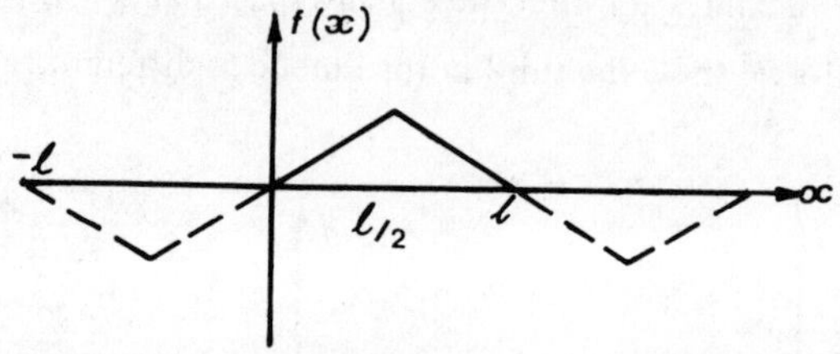

Figure 8.15

We finally obtain the solution

$$y(x, t) = \frac{8}{\pi^2}\left[\frac{1}{1^2} \sin \frac{\pi x}{l} \cos \frac{\pi ct}{l} - \frac{1}{3^2} \sin \frac{3\pi x}{l} \cos \frac{3\pi ct}{l} + \frac{1}{5^2} \sin \frac{5\pi x}{l} \cos \frac{5\pi ct}{l} - \ldots\right] \tag{8.27}$$

We programmed a digital computer to calculate values of y for suitable values of x at a number of times t. We used the results to produce the graphs of Figure 8.16, which show the progressive profiles $y(x)$ for selected times in a half-cycle of oscillations.

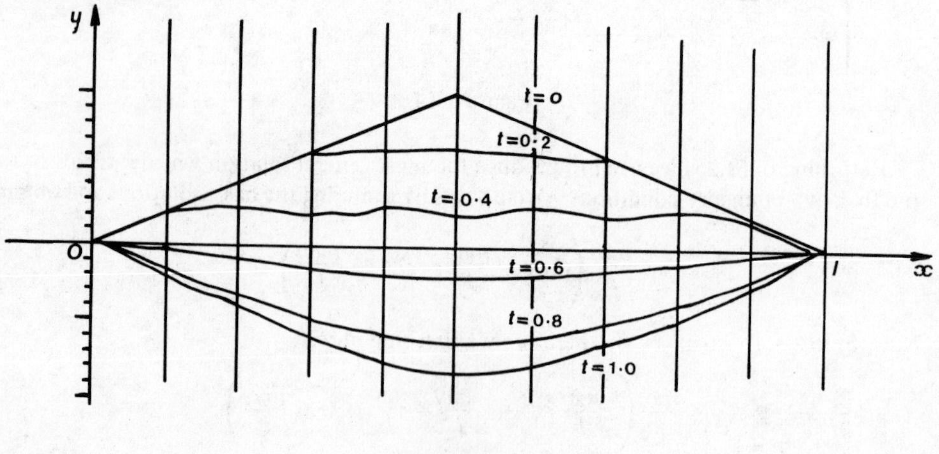

Figure 8.16

Example 4 Vibrating membrane

In Figure 8.17 we depict a rectangular membrane that is tightly stretched and held firmly by a rigid rectangular frame.

When the membrane is disturbed, we seek the deflection $u(x, y, t)$ normal to the x - y plane.

The following assumptions are made.

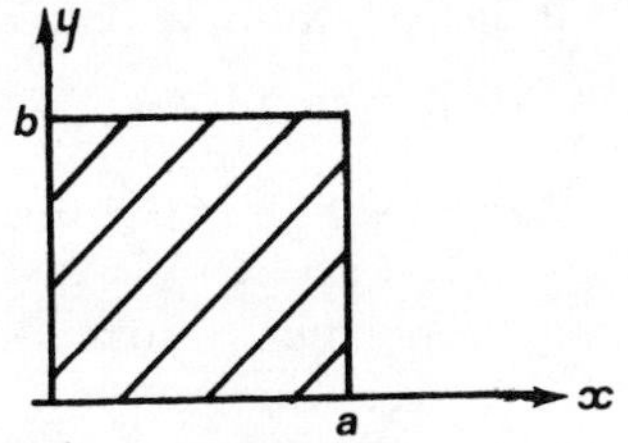

Figure 8.17

(i) The force on the boundary is normal to the boundary and its value per unit length is constant. The total boundary force is large in comparison with the weight of the membrane.

(ii) The membrane is thin and cannot resist bending moments.

(iii) The deflections are small compared to the dimensions a and b.

(iv) The displacements in the x - and y - directions are small compared with the deflection u.

(v) The slopes on any particular deflected profile are small.

The governing partial differential equation is

$$\frac{\partial^2 u}{\partial t^2} = c^2 \left[\frac{\partial^2 u}{\partial x^2} + \frac{\partial^2 u}{\partial y^2} \right] \tag{8.28}$$

where c is a parameter akin to that for a vibrating string.

Suitable boundary conditions are

(i) $u(0, y, t) = 0; \quad 0 \leqslant y \leqslant b, \quad t > 0$
(ii) $u(a, y, t) = 0; \quad 0 \leqslant y \leqslant b, \quad t > 0$
(iii) $u(x, 0, t) = 0; \quad 0 \leqslant x \leqslant a, \quad t > 0$
(iv) $u(x, b, t) = 0; \quad 0 \leqslant x \leqslant a, \quad t > 0$

If we displace the membrane initially to a profile $u = f(x, y)$ and release it from rest, then we have two initial conditions

(v) $u(x, y, 0) = f(x, y); \quad 0 \leqslant x \leqslant a, \quad 0 \leqslant y \leqslant b$
(vi) $u_t(x, y, 0) = 0; \quad 0 \leqslant x \leqslant a, \quad 0 \leqslant y \leqslant b$

If we assume a solution in the form $u = X(x) \, . \, Y(y) \, . \, T(t)$ then we may separate the variables as $\dfrac{X''}{X} = -\dfrac{Y''}{Y} + \dfrac{T''}{c^2 T}$ = constant. Since we require u to vanish both at $x = 0$ and at $x = a$ we expect to fit a sine function, or collection of sine functions in x. This would necessitate choosing the constant to be negative, say $-\alpha^2$, and we then obtain $X'' = -\alpha^2 X$, together with $-Y''/Y + T''/(c^2 T) = -\alpha^2$. Separating the variables in this last equation we obtain $Y''/Y = \alpha^2 + T''/(c^2 T)$ = constant. The boundary conditions on y require the constant to be negative, say $-\beta^2$. Then we obtain $Y'' = -\beta^2 Y$ and $T''/(c^2 T) = -(\alpha^2 + \beta^2) = -\gamma^2/c^2$, say.

We then have cosine and sine solutions for X, Y and T and we may write $u = (A \cos \alpha x + B \sin \alpha x)(C \cos \beta y + D \sin \beta y)(E \cos \gamma t + F \sin \gamma t)$ We apparently have 9 constants to determine and only 6 boundary/initial conditions available.

Now $u = 0$ when $x = 0$, therefore $A = 0$ (check for yourself). Further, $u = 0$ when $x = a$, therefore $\sin \alpha a = 0$ and hence $\alpha = n\pi/a$, n integer.

In an exactly similar manner, we apply the conditions $u = 0$ at $y = 0$ and $y = b$ to obtain $C = 0$ and $\beta = m\pi/b$, m integer.

Then we have that

$$\gamma^2 = c^2(\alpha^2 + \beta^2) = c^2\pi^2\left[\frac{n^2}{a^2} + \frac{m^2}{b^2}\right] \tag{8.29}$$

We shall keep the notation γ to shorten the writing.

We may now state that $u = B \sin \dfrac{n\pi x}{a} \,.\, D \sin \dfrac{m\pi y}{b} \,.(E \cos \gamma t + F \sin \gamma t)$ but we may absorb B and D into the constants E and F, so that

$$u = \sin \frac{n\pi x}{a} \,.\, \sin \frac{m\pi y}{b} \,.(E \cos \gamma t + F \sin \gamma t).$$

We have two constants, E and F, to determine and two initial conditions left. Now m and n can be any positive integers and so we get a whole array of functions satisfying the four boundary conditions. We relabel a particular solution as

$$u_{nm}(x, y, t) = \sin \frac{n\pi x}{a} \,.\, \sin \frac{m\pi y}{b} \,.\, (E_{nm} \cos \gamma_{nm} t + F_{nm} \sin \gamma_{nm} t) \tag{8.30}$$

These particular solutions are the **eigenfunctions** for the problem and the constants γ_{nm} are the **eigenvalues** of the system. The frequencies of these special modes of vibration are $\gamma_{nm}/2\pi = \dfrac{c}{2}\left[\dfrac{n^2}{a^2} + \dfrac{m^2}{b^2}\right]^{1/2}$.

(Note that it is possible to find more than one eigenfunction associated with the same eigenvalue. For example if $a = b$, then certainly $\gamma_{nm} = \gamma_{mn}$ and therefore there are at least two different modes of vibration to each frequency. (We may distinguish between them by examining the nodal curves, i.e. the sets of points on the membrane which do not move during the vibration.)

The general solution to date is $u(x, t) = \displaystyle\sum_{n=1}^{\infty} \sum_{m=1}^{\infty} u_{nm}(x, y, t)$

$$\text{i.e. } u = \sum_{n=1}^{\infty} \sum_{m=1}^{\infty} \sin \frac{n\pi x}{a} \sin \frac{m\pi y}{b} (E_{nm} \cos \gamma_{nm} t + F_{nm} \sin \gamma_{nm} t)$$

Applying the initial conditions and assuming that we can deal with a **double Fourier series** in a way which suggests itself naturally, the solution may be written

$$u = \frac{64a^2b^2}{\pi^6} \sum_{n=1,3,5..}^{\infty} \sum_{m=1,3,5..}^{\infty} \frac{1}{n^3 m^3} \sin \frac{n\pi x}{a} \sin \frac{m\pi y}{b} \cos c\left[\frac{n^2\pi^2}{a^2} + \frac{m^2\pi^2}{b^2}\right]^{1/2} t \tag{8.31}$$

Example 5 Flow of electricity in a cable

We consider finally an example from electrical engineering. Suppose we have a transmission line in which we assume that the resistance, inductance and capacitance

vary linearly with x, the distance from the source of electricity, measured along the line. We shall need the following quantities:
$e(x, t)$ = the potential at a point on the cable; $i(x, t)$ = the current there;
R = the resistance of the cable per unit length; L = inductance per unit length;
G = conductance to ground per unit length; C = capacitance to ground per unit length.

In Figure 8.18(a) we show schematically the system and in Figure 8.18(b) we show the equivalent circuit for the portion PQ.

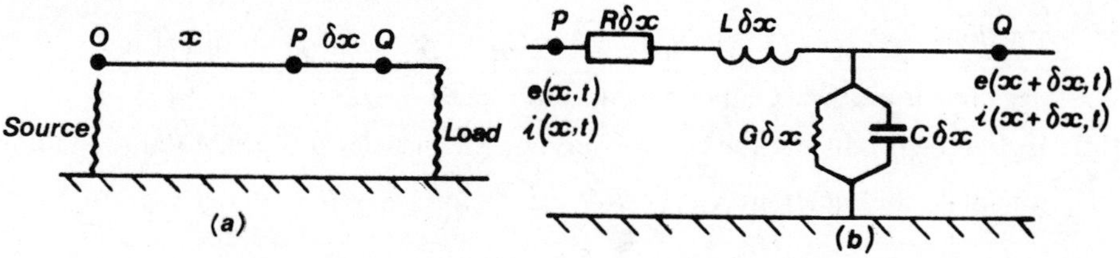

Figure 8.18

We are allowing the cable to leak to earth.

Applying Kirchhoff's second law to PQ, the drop in potential along the section PQ is given by

$$e(x, t) - e(x + \delta x, t) = iR\delta x + \frac{\partial i}{\partial t} L\,\delta x$$

Dividing by δx and letting $\delta x \to 0$ we obtain

$$\frac{\partial e}{\partial x} = -Ri - L\frac{\partial i}{\partial t} \tag{8.32}$$

Applying Kirchhoff's first law to the equivalent circuit

$$i(x, t) = i(x + \delta x, t) + C\,\delta x\frac{\partial e}{\partial t} + G\,\delta x\, e(x, t)$$

Dividing by δx and letting $\delta x \to 0$ we obtain

$$\frac{\partial i}{\partial x} = -Ge - C\frac{\partial e}{\partial t} \tag{8.33}$$

We differentiate (8.32) partially with respect to x to obtain

$$\frac{\partial^2 e}{\partial x^2} = -R\frac{\partial i}{\partial x} - L\frac{\partial^2 i}{\partial x \partial t} \tag{8.34}$$

We differentiate (8.33) partially with respect to t to obtain

$$\frac{\partial^2 i}{\partial x \partial t} = -G\frac{\partial e}{\partial t} - C\frac{\partial^2 e}{\partial t^2} \tag{8.35}$$

We now substitute (8.33) and (8.35) into (8.34) and produce the following equation for e

$$\frac{\partial^2 e}{\partial x^2} = LC\frac{\partial^2 e}{\partial t^2} + (RC + LG)\frac{\partial e}{\partial t} + RG\,e \tag{8.36}$$

Similarly,

$$\frac{\partial^2 i}{\partial x^2} = LC\frac{\partial^2 i}{\partial t^2} + (RC + LG)\frac{\partial i}{\partial t} + RG\,i \tag{8.37}$$

These last two equations are called the **telephone equations.**

We may consider three special cases.

(i) If $G = L = 0$, i.e. we may neglect leakage and inductance, which is a reasonable assumption for a submarine cable, then we have the simplified **telegraph equations**

$$\frac{\partial^2 e}{\partial x^2} = RC\frac{\partial e}{\partial t} \quad \text{and} \quad \frac{\partial^2 i}{\partial x^2} = RC\frac{\partial i}{\partial t} \tag{8.38}$$

(ii) If $R = G = 0$ as is reasonable to assume for high frequencies, we obtain the equations $\dfrac{\partial^2 e}{\partial x^2} = LC\dfrac{\partial^2 e}{\partial t^2}$ and $\dfrac{\partial^2 i}{\partial x^2} = LC\dfrac{\partial^2 i}{\partial t^2}$ which are forms of the one-dimensional wave equation with wave speed $1/\sqrt{LC}$.

(iii) If $L = C = 0$ which is the case for very low frequencies and steady state conditions, we obtain the equations $\dfrac{\partial^2 e}{\partial x^2} = RGe$ and $\dfrac{\partial^2 i}{\partial x^2} = RGi.$

Transformation of equations to non-dimensional form

Consider the equation $\dfrac{\partial \phi}{\partial t} = k\dfrac{\partial^2 \phi}{\partial x^2}$ where k is constant.

Suppose we have a reference value ϕ_0 at time $t = 0$; this may be a typical value, a maximum or minimum value or a constant initial temperature. If l is a suitable reference length in the problem then the transformations $X = x/l$, $\Phi = \phi/\phi_0$ will reduce the equation to the form $\partial\Phi/\partial t = \partial^2\Phi/\partial X^2$ where Φ and X are non-dimensional variables.

The need for a numerical approach

Sometimes the problem at hand cannot be solved by the separation of variables method, especially if the boundary is an awkward one geometrically. Even when such a solution is possible, a glance at a formula such as (8.4) makes one doubt whether there is much to commend it. The formula is an extremely tangled one to use to calculate values of θ and we would be well advised to attempt a numerical solution. In the next few sections we examine some numerical methods of solution of the well-known equations. We should remark perhaps that the equations we have solved by separation of variables are of importance in models of engineering systems.

Problems

1. The ends A and B of a rod 50 cm. long have their temperatures kept at 0° and 100°C, respectively until the temperatures are indistinguishable from those for the steady state. At some time after this, there is a sudden change in the end temperatures. Find the temperature of any point in the rod, $\theta(x, t)$ as a function of x (cm.), the distance from one end A, and t (sec) the time elapsed after the sudden change for each of the following cases.

   (a) The temperature of B is suddenly reduced to 0°C., and kept so, while that of A is kept at 0°C.;

   (b) The temperature of B is kept at 100°C., while that of A is suddenly increased to 100°C., and maintained at this value;

(c) The temperature of B is suddenly reduced to 50°C., and kept at this value while that of A is kept at 0°C.;

(d) The temperature of A is suddenly raised to 25°C., while that of B is suddenly reduced to 75°C., and then these temperatures are maintained;

(e) The temperature of A is suddenly raised to 50°C., while that of B is suddenly raised to 150°C., and then these temperatures are maintained.

2. Find the solution of the one-dimensional diffusion equation $\frac{\partial^2 u}{\partial x^2} = \frac{1}{k}\frac{\partial u}{\partial t}$, where k is a constant, satisfying the boundary conditions $u(0, t) = u(\pi, t) = 0,\ t \geqslant 0$, $u(x, 0) = f(x),\ 0 \leqslant x \leqslant \pi$ where

$f(x) = x$ for $0 \leqslant x \leqslant \frac{\pi}{2}$ and $f(x) = \pi - x$ for $\frac{\pi}{2} \leqslant x \leqslant \pi$. (C.E.I.)

3. In the equation $\frac{\partial u}{\partial t} = h^2 \frac{\partial^2 u}{\partial x^2} - k(u - u_0)$ the change of variable $u = u_0 + e^{-kt} v$ is introduced, where h, k and u_0 are constants: show that the equation in v becomes $\frac{\partial v}{\partial t} = h^2 \frac{\partial^2 v}{\partial x^2}$. Hence given boundary conditions $u(0, t) = u_0 = u(1, t)$ and assuming a solution of the form $u(x, t) = u_0 + X(x)\, e^{-(k+m^2)t}$, m real, obtain a general solution of the given equation.

If the initial condition is $u(x, 0) = u_0 + x(1 - x)$, determine the solution completely. (C.E.I.)

4. The function $\theta(x, t)$ satisfies the differential equation $\frac{\partial^2 \theta}{\partial x^2} = \frac{1}{c^2}\frac{\partial \theta}{\partial t}$ subject to the conditions

(i) $\theta(0, t) = 0$
(ii) $\theta(a, t) = 0$ } for all $t > 0$
(iii) $\theta(x, 0) = \theta_0$ for $0 < x < a$, where θ_0 is a constant.

Use the method of separation of variables to obtain the solution in the form

$$\theta(x, t) = \frac{4\theta_0}{\pi} \sum_{m=1}^{\infty} \frac{1}{(2m-1)} \sin (2m-1)\frac{\pi x}{a}\, e^{-(2m-1)^2 \pi^2 c^2 t/a^2} \qquad \text{(C.E.I.)}$$

5. The temperature $\theta(x, t)$ in a bar of length a is given by $\frac{\partial \theta}{\partial t} = k \frac{\partial^2 \theta}{\partial x^2}$ where k is the constant of diffusivity.

If the ends $x = 0$ and $x = a$ of the bar are maintained at zero temperature and if the initial temperature distribution is: $\theta = \theta_0 x(a - x)$ where θ_0 is constant, show that

$$\theta(x, t) = \frac{8\theta_0 a^2}{\pi^3} \sum_{n}^{\infty} \frac{1}{n^3} \sin \frac{n\pi x}{a} \exp\left(-\frac{kn^2\pi^2 t}{a^2}\right)$$

the summation extending over $n = 1, 3, 5, 7, \ldots$ only. (C.E.I.)

6. In Terzaghi's theory of one-dimensional consolidation, the equation for the excess pore water pressure, u, is $\frac{\partial u}{\partial t} = c_v \frac{\partial^2 u}{\partial z^2}$ where c_v is the (constant) coefficient of consolidation.

A soil sample of thickness $2H$ is tested and allowed to drain at both the upper and lower surfaces: the initial distribution of u is a constant u_i. Write down the three boundary conditions for this case and solve the equations. Defining the degree of consolidation, U, as $\dfrac{u_i - u}{u_i}$ find $\bar{U}$, the average degree of consolidation over the sample layer, and sketch the graph of $\bar{U}$ against t. Explain where the degree of consolidation is a minimum for all time and verify analytically that this is so.

7. A thin rectangular slab of cross-section area A, of thickness t is painted on the edges so that the liquid content of the slab can evaporate only through its faces. Assume that liquid evaporates *immediately* on reaching a face. The constant concentration of liquid at each face is called the **equilibrium concentration.** Assume that the initial concentration is uniform throughout the interior of the slab but falls immediately to the equilibrium value at the surface. What further assumptions are necessary to show that the excess of concentration over the equilibrium value satisfies the one-dimensional diffusion equation? Suggest suitable boundary conditions and hence find the concentration as a function of position and time.

8. The current i in a cable satisfies the equation $\dfrac{\partial^2 i}{\partial x^2} = \dfrac{2}{k}\dfrac{\partial i}{\partial t} + i$, ($k$ constant.)

By assuming a solution of the type $i = XT$, where X is a function of x alone and T is a function of t alone, show that if $i = 0$ when $x = l$ and $\partial i/\partial x = -ae^{-kt}$ when $x = 0$, the current is given by $i = ae^{-kt}\dfrac{\sin(l - x)}{\cos l}$. (L.U.)

9. In a telegraph wire the resistance R and the leakage conductance G are very large in comparison with the capacitance and inductance. The wire is of length l and the voltage at the sending end is V_0 (constant). If the other end of the wire is earthed, show that the voltage V and the current i at distance x from the sending end are given approximately by

$$V = V_0 \frac{\sinh a(l - x)}{\sinh al}, \qquad i = V_0\left(\frac{G}{R}\right)^{1/2}\frac{\cosh a(l - x)}{\sinh al}$$

where $a = (RG)^{1/2}$. (L.U.)

10. A distortionless transmission line ($RC = LG$) of length l is initially charged to unit potential and the end $x = l$ is insulated. If, at time $t = 0$, the end $x = 0$ is earthed, show that the potential at time t and distance x from the earthed end is given by

$$\frac{4}{\pi} e^{-kt} \sum_{r=0}^{\infty} \frac{1}{2r + 1} \sin\frac{(2r + 1)\pi x}{2l} \cos\frac{(2r + 1)\pi ut}{2l},$$

where $u = (LC)^{-1/2}$ and $k = R/L$. (L.U.)

11. A light elastic string of length a has its ends $x = 0$ and $x = a$ fixed. The point where $x = a/3$ is drawn aside a small distance h and released from rest at time $t = 0$. Assume that the displacement $y(x, t)$ of the string satisfies the one-dimensional wave equation $\dfrac{\partial^2 y}{\partial x^2} = \dfrac{1}{c^2}\dfrac{\partial^2 y}{\partial t^2}$ where c is a constant. By separating the variables show that

$$y(x, t) = \frac{9h}{\pi^2} \sum_{n=1}^{\infty} \frac{1}{n^2} \sin\frac{n\pi}{3} \sin\frac{n\pi x}{a} \sin\frac{n\pi ct}{a}$$ (C.E.I.)

12. Show that the Fourier sine series for the function $f(x) = kx$ for $0 \leqslant x < 1$ and $f(x) = -kx + 2k$ for $1 < x \leqslant 2$ is

$$\frac{8k}{\pi^2} \sum_{n=1}^{\infty} \frac{1}{n^2} \sin \frac{n\pi}{2} \sin \frac{n\pi x}{2} .$$

Use the method of separation of variables to solve the differential equation

$\frac{\partial^2 y}{\partial t^2} = a^2 \frac{\partial^2 y}{\partial x^2}$ $(0 < x < 2,\ t > 0)$ satisfying the boundary conditions

(i) $y(0, t) = 0,\ y(2, t) = 0$ for all t

(ii) $\frac{\partial y}{\partial t} = 0$ at $t = 0$ for all x

(iii) $y(x, 0) = f(x)$ as defined earlier.

Obtain the solution in the form

$$y(x, t) = \frac{8k}{\pi^2} \sum_{m=1}^{\infty} \frac{(-1)^{m+1}}{(2m-1)^2} \sin \frac{(2m-1)\pi x}{2} \cos \frac{(2m-1)\pi a t}{2} .$$ (C.E.I.)

13. (i) Show that the general form $y(x, t) = F(x - ct) + G(x + ct)$ satisfies the one-dimensional wave equation $\frac{\partial^2 y}{\partial x^2} = \frac{1}{c^2} \frac{\partial^2 y}{\partial t^2}$.

If, at $t = 0$, $y(x, 0)$ is given by $\alpha(x)$ and $\frac{\partial y(x, 0)}{\partial t}$ by $\beta(x)$, for all values of x, show that

$$y(x, t) = \tfrac{1}{2}\left[\alpha(x - ct) + \alpha(x + ct)\right] + \frac{1}{2c} \int_{x-ct}^{x+ct} \beta(u)du.$$

(ii) Find the solution of the differential equation $\frac{\partial^2 F}{\partial x^2} + \frac{\partial^2 F}{\partial y^2} + 5F = 0$ in the form $F(x, y) = f(x) g(y)$ satisfying the following conditions:

(a) F is periodic in x;
(b) $F = 0$ when $x = 0$ for all values of y;
(c) $F = 3e^{-2y}$ when $x = \pi/6$ for all values of y. (L.U.)

14. Solve the one-dimensional wave equation for a string of length 1 with zero initial velocity and a profile given by $f(x)$ where

(i) $f(x) = A \sin 2\pi x$; (ii) $f(x) = Ax(1 - x^2)$

(iii) $$f(x) = \begin{cases} x, & 0 < x < \tfrac{1}{4} \\ \tfrac{1}{2} - x, & \tfrac{1}{4} < x \leqslant \tfrac{1}{2} \\ 0, & \tfrac{1}{2} \leqslant x \leqslant 1 \end{cases}$$

(iv) $f(x) = Ax^2(1 - x^2)$.

15. The small vertical vibrations of a uniform beam of length l are given by the equation $EI \frac{\partial^4 y}{\partial x^4} + \rho A \frac{\partial^2 y}{\partial t^2} = 0$ where EI is the flexural rigidity of the beam of cross-sectional area A; the uniform density of the beam is ρ. Find the general solution of this equation.

(This beam is referred to in Problems 16, 17 and 18).

16. If the beam is simply supported at both ends find the fundamental mode of vibration.

17. If the beam is cantilevered at $x = 0$ show that the period of oscillation is $2\pi/\omega$ where $\rho A\omega^2 = EIa^4$ and $\cosh al \cos al = -1$.

18. If the beam is clamped horizontally at both ends show that the period of oscillation is $2\pi/\omega$ where $\rho A\omega^2 = EIa^4$ and $\cosh al \cos al = 1$. Find graphically the smallest root of this equation and hence the fundamental period. Show that for vibrations of higher frequency ω is given by $\omega = \left[\frac{EI}{\rho A}\right]^{1/2}\left[\frac{(2n+1)\pi}{2l}\right]^2 \quad n = 2, 3, 4, \ldots\ldots$

19. A long rectangular wave-guide has its axis parallel to the z - axis, its walls being formed by the four planes $x = 0$, $x = a$, $y = 0$, $y = b$. Associated with a transverse magnetic wave travelling in the guide is a function ϕ which vanishes on the four walls of the guide, and satisfies the equation $\frac{\partial^2\phi}{\partial x^2} + \frac{\partial^2\phi}{\partial y^2} + \frac{\partial^2\phi}{\partial z^2} = \frac{1}{c^2}\frac{\partial^2\phi}{\partial t^2}$ inside the guide. In order that a wave of frequency f should be transmitted, ϕ must be of the form $U(x,y)\cos(\beta z - 2\pi ft)$, where β is real. Show that if $U(x,y)$ has the form $X(x)\,Y(y)$, then the only possible solutions of the problem are $\phi = A\sin(m\pi x/a)\sin(n\pi y/b)\cos(\beta z - 2\pi ft)$, where A is a constant, and m, n are integers such that $m^2/a^2 + n^2/b^2 = 4f^2/c^2 - \beta^2/\pi^2$. Deduce that if $a = b$, waves given by a solution of the above form must have a frequency not less than $c/(a\sqrt{2})$. (L.U.)

20. Show that the transmission line equation $\frac{\partial^2 e}{\partial x^2} = LC\frac{\partial^2 e}{\partial t^2} + (RC + GL)\frac{\partial e}{\partial t} + RGe$ with boundary conditions $e(0,t) = E_0\cos\omega t$ (E_0, ω constant) and $e(x,t)$ bounded as $x \to \infty$ or $t \to \infty$, where $e(x,t)$ is the voltage at distance x and time t and R, L, G, C are constants of the line, does not have a separable solution of the form $e(x,t) = X(x).T(t)$.

Further, show that there exists a solution of form $e(x,t) = E_0 e^{-ax}\cos(\omega t + bx)$ provided that $a^2 - b^2 = RG - LC\omega^2$ and $2ab = -(RC + GL)\omega$.

Hence obtain the parameters a and b. (C.E.I.)

21. Show that the equation $\frac{\partial^2 V}{\partial r^2} + \frac{1}{r}\frac{\partial V}{\partial r} + \frac{1}{r^2}\frac{\partial^2 V}{\partial \theta^2} = 0$ has a solution of the form

$V = \left[A^n + \frac{B}{r^n}\right] f(\theta)$ where $f(\theta)$ is a trigonometric function of θ only with period $2\pi/n$.

Find $f(\theta)$.

Find V for the case $n = 2$, given that

$V \to 0$ as $r \to \infty$, $\quad V = 0$ when $\theta = 0$, $\quad V = 1$ when $r = 1$ and $\theta = \pi/4$. (L.U.)

22. (i) The temperature V at any point of a circular metal plate with radius a satisfies, when the steady state has been reached, the partial differential equation

$r^2\frac{\partial^2 V}{\partial r^2} + r\frac{\partial V}{\partial r} + \frac{\partial^2 V}{\partial \theta^2} = 0$, where r, θ are polar coordinates and the centre of the plate is the pole.

Find a solution of the equation which is bounded at the centre of the plate and is such that $V = a^2(1 - \cos 2\theta)$ whenever $r = a$.

(ii) Obtain, by the method of separation of variables, a solution of the equation

$x\frac{\partial f}{\partial x} + x^2\frac{\partial f}{\partial t} = 2f$. Determine the function $f(x,t)$ if $(1,t) = e^{-t}$. (L.U.)

8.6 PARABOLIC EQUATIONS: FINITE DIFFERENCE METHODS

We shall take the one-dimensional heat conduction equation which is an example of a parabolic partial differential equation (see Section 8.3). To avoid cluttering up the arithmetic, we take the equation in the form $\frac{\partial \theta}{\partial t} = \frac{\partial^2 \theta}{\partial x^2}$, i.e. we assume unit diffusivity.

We take the boundary conditions: $\theta = 0$ when $x = 0$ and when $x = 1$ for $t \geqslant 0$ and $\theta = 100$ at $t = 0$ for $0 < x < 1$.

This can be interpreted as the rod $0 \leqslant x \leqslant 1$ being heated to a uniform temperature of 100° and then both ends being simultaneously held in melting ice. Not a very realistic problem, maybe, but one which allows the features of the numerical process to come to the fore.

A simple explicit method

If we are going to use finite differences as approximations to derivatives we can hope to use central differences in space (they are the most accurate) but since we are propagating forwards in time we use a forward time difference. (We will certainly need one at the first step forward, but it might be argued that thereafter we could use a central time difference; at the moment we are only considering a simple method.) We therefore take as an approximation to the differential equation the difference equation

$$\frac{\theta_{i,j+1} - \theta_{i,j}}{k} = \frac{\theta_{i+1,j} - 2\theta_{i,j} + \theta_{i-1,j}}{h^2} \tag{8.39}$$

Reference to Figure 8.19(a) shows the relative positions of the four mesh values of θ which are involved.

The fact that there is only one value at the further forward **time level** $(j+1)$ indicates that the scheme (8.39) is an **explicit** one. Note that the scheme is valid for mesh points in the interior of the domain of the solution. The values on the boundary are specified.

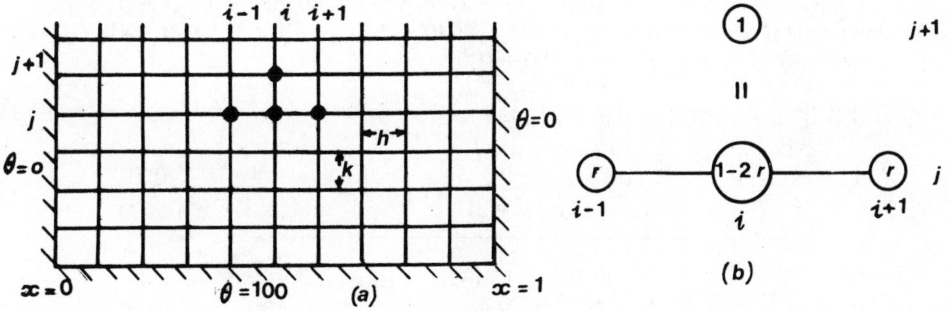

Figure 8.19

We may rearrange the scheme to give explicitly a formula for $\theta_{i,j+1}$. It is

$$\theta_{i,\,j+1} = r\,\theta_{i-1,j} + (1-2r)\theta_{i,j} + r\,\theta_{i+1,j} \tag{8.40}$$

$$\text{where } r = k/h^2 \tag{8.41}$$

In Figure 8.19(b), we show the **computational molecule** which summarises the scheme. It depicts the relative positions of the component mesh values and their weightings in the difference scheme.

At this stage, we ought to remark that we have chosen the simplest forward difference approximation to $\partial\theta/\partial t$ and the simplest central difference approximation to $\partial^2\theta/\partial x^2$. The truncation errors in these approximations are $O(k)$ and $O(h^2)$ respectively.

Before we embark on our solution we need to specify the parameter r. From (8.41) we see that the same value of r can arise from many combinations of k and h. Suppose we decide to divide the interval $0 \leqslant x \leqslant 1$ into 10 equal parts making $h = 0.1$. We should remember that the problem, and hence the solution, is symmetric about $x = \frac{1}{2}$. We shall therefore only need to compute θ for $x = 0.1(0.1)0.5$. What value do we choose for k? Bearing in mind the error magnitudes for the finite difference approximations, it would seem that the smaller we choose k the better. However, the smaller we choose k, the more steps are needed to reach a given time. We need to strike a compromise. Therefore, for simplicity, let us choose $r = 1$, i.e. $k = 0.01$. This does seem a very small time step, but we will progress with it. Then the formula for $\theta_{i,j+1}$ becomes

$$\theta_{i,j+1} = \theta_{i-1,j} - \theta_{i,j} + \theta_{i+1,j} \tag{8.42}$$

We now obtain the solution for the first few time steps and show the results of selected time steps in Table 8.1

For comparison, we show in Table 8.2 the corresponding results from the analytical solution

$$\theta = \frac{400}{\pi} \sum_{n=1,3,5..}^{\infty} \frac{1}{n} e^{-n^2\pi^2 t} \sin n\pi x \tag{8.43}$$

which you can derive by separation of variables.

To demonstrate briefly the method we work through a few calculations. We know the solution at $j = 0$, $i = 0, 1, \ldots, 10$. We also know that $\theta_{0,1} = 0$ from the boundary condition. Then $\theta_{1,1} = \theta_{0,0} - \theta_{1,0} + \theta_{2,0} = 0 - 100 + 100 = 0$, $\theta_{2,1} = \theta_{1,0} - \theta_{2,0} + \theta_{3,0} = 100 - 100 + 100 = 100$. Similarly, $\theta_{3,1} = 100$ etc. Moving to the next time level, $\theta_{1,2} = \theta_{0,1} - \theta_{1,1} + \theta_{2,1} = 0 - 0 + 100 = 100$, $\theta_{2,2} = \theta_{1,1} - \theta_{2,1} + \theta_{3,1} = 0 - 100 + 100$.

Table 8.1 Finite difference solution, $r = 1$

t \ x	0 ($i = 0$)	0.1 ($i = 1$)	0.2 ($i = 2$)	0.3 ($i = 3$)	0.4 ($i = 4$)	0.5 ($i = 5$)
0	0	100	100	100	100	100
0.01	0	0	100	100	100	100
0.02	0	100	0	100	100	100
0.03	0	−100	200	0	100	100
0.04	0	300	−300	300	0	100
0.05	0	−600	900	−600	400	−100

Table 8.2 Analytical solution (1 d.p.)

t \ x	0	0.1	0.2	0.3	0.4	0.5
0	0	100	100	100	100	100
0.001	0	97.5	100	100	100	100
0.002	0	88.6	99.8	100	100	100
0.005	0	68.3	97.1	99.7	100	100
0.01	0	52.0	84.4	98.7	99.7	99.9
0.015	0	43.6	75.2	92.3	99.6	99.9
0.02	0	38.3	68.3	86.8	99.4	99.9
0.025	0	34.5	62.9	81.9	94.6	99.5
0.03	0	31.7	58.5	77.5	90.1	94.7
0.04	0	27.3	51.6	69.8	81.6	85.8
0.05	0	24.4	46.2	63.0	73.9	77.7
0.10	0	14.7	27.9	38.4	45.1	47.5

In Figure 8.20 we display temperature profiles for $t = 0.01, 0.05,$ and 0.1.

Comparison with Table 8.2 is hardly necessary; the results of Table 8.1 are rubbish. They may satisfy (8.42) and the boundary conditions, but they cannot be said to satisfy the differential equation. It would seem as though the time step is too coarse, too insensitive, to adjust to the changes occurring in the temperatures.

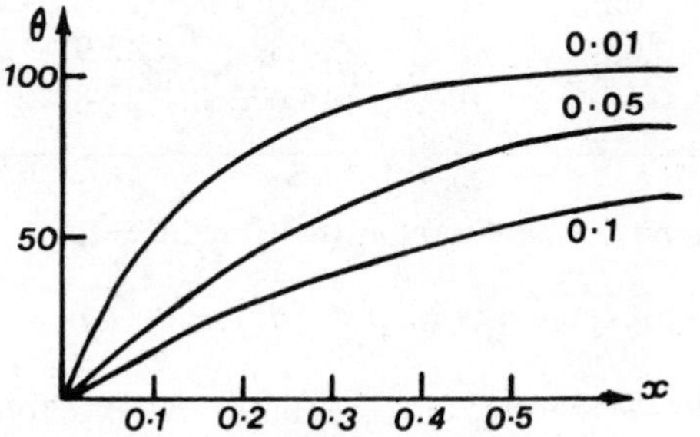

Figure 8.20

Our first reaction might well be to reduce the time step. Let us take $k = 0.005$ i.e. $r = \frac{1}{2}$. Then (8.40) reduces to

$$\theta_{i,j+1} = \tfrac{1}{2}\theta_{i-1,j} + \tfrac{1}{2}\theta_{i+1,j} \tag{8.44}$$

In Table 8.3 we show the results of the first few time steps.

This is a better state of affairs, but there are still some blemishes, particularly in the early stages. Notice that at $t = 0.05$ the values for $x = 0.1$, 0.3 and 0.5 are relatively closer to analytical values than those for $x = 0.2$ and 0.4. If we had taken the results for $t = 0.055$, we should have found the reverse was the case.

To see whether the situation can be improved markedly we decrease k to 0.001 and hence r to 0.1. In Table 8.4, we show the finite difference solution, where we have worked to 3 s.f.

Table 8.3 Finite Difference solution $r = ½$

t \ x	0	0.1	0.2	0.3	0.4	0.5
0	0	100	100	100	100	100
0.005	0	50	100	100	100	100
0.01	0	50	75	100	100	100
0.015	0	37.5	75	87.5	100	100
0.02	0	37.5	62.5	87.5	93.8	100
0.025	0	31.3	62.5	78.1	93.8	93.8
0.05	0	24.4	44.0	63.5	70.3	78.1

Table 8.4 Finite Difference solution ($r = 0.1$)

t \ x	0	0.1	0.2	0.3	0.4	0.5
0	0	100	100	100	100	100
0.001	0	90	100	100	100	100
0.002	0	82	99	100	100	100
0.005	0	65.7	93.4	99.3	100	100
0.01	0	51.1	82.8	95.7	99.2	99.8
0.02	0	38.0	67.6	85.9	94.5	96.9
0.05	0	24.3	45.9	62.6	73.1	76.7
0.10	0	14.6	27.8	38.2	44.9	47.2

With $r = 0.1$, the formula (8.40) reduces to

$$\theta_{i,j+1} = \frac{1}{10}(\theta_{i-1,j} + 8\theta_{i,j} + \theta_{i+1,j}) \tag{8.45}$$

We see that this time the finite difference solution is good. However, we have the drawback that we need ten times as many time steps to reach, say, $x = 0.10$. What we do not yet know is which values of r will give reasonable or good accuracy or which values of r cause problems such as those in Table 8.1. Is it in fact, simply the value of r which determines the nature of the results? We notice that the initial temperature distribution had discontinuities at $x = 0$ and $x = 1$. For small values of time, the finite difference solution is poor near $x = 0$. As the value of t increases so the accuracy improves. This suggests that a method which collected information from other than the first row and the first column near $x = 0$, $t = 0$ would be better; such an approach is to use an **implicit method** such as the Crank-Nicolson method discussed later. However, it must be pointed out that the discontinuity will always cause trouble, no matter which finite difference method is used.

Convergence, Stability and Compatibility

In order that a finite difference scheme should give solutions which are reasonably accurate approximations to the true solution of the appropriate partial differential equation we require three conditions to be met: convergence, stability and compatibility. We briefly examine the ideas behind each of these in turn. Unless stated otherwise, these remarks apply to partial differential equations quite generally.

Convergence

When we solve the finite difference equations, we should remember that even if we were able to find an exact solution, we will not necessarily have solved the partial differential equation exactly.

Let $u(x, t)$ represent the **exact** solution to a partial differential equation and $u^*(x, t)$ the **exact** solution to the corresponding finite difference equation. We define the **convergence** of the finite-difference solution as occurring when $u^*(x_0, t_0) \to u(x_0, t_0)$ as $\delta x = h$ and $\delta t = k$ both tend to zero, for all x_0 and t_0 in the solution domain,

Some authors call the difference $(u - u^*)$ the **discretization error**, some refer to it as the **truncation error**. At any mesh point, the size of the discretization error depends partly on the grid spacing and partly on the accuracy of the finite difference approximations to the derivatives. If we attempt to reduce the discretization error by reducing h and k we shall increase the number of equations to be solved, and we then encounter problems of computer storage space and computing time required. It is, in general, very difficult to obtain criteria for convergence.

However, for the parabolic equation we are considering, it can be shown that the finite-difference solution converges to the true solution of the p.d.e. as $h \to 0$ *provided* $r \leqslant \frac{1}{2}$.

Stability

When we solve the finite difference equations, we cannot generally work with exact arithmetic. Therefore, we will introduce **round-off error** at each stage. As we proceed along a given time-level, and then from one time-level to the next, we shall accumulate this error. We define the finite difference equations to be **stable** when the accumulating effect of round-off error can be neglected, i.e. when this error tends to damp out.

We remark that the trouble with the case $r = 1$ was that the solution was not convergent; stability did not enter into the story, since we worked with exact arithmetic.

There are two well-known methods for determining a stability criterion. The Fourier series method expresses an initial row of errors as a finite Fourier series, with as many terms in the series as there are mesh points in each row. The series is usually formulated in exponential form and the effect is examined for a single term propagating from one time level to the next. The overall effect is determined by superimposition. However, the Fourier series method neglects boundary conditions.

A second method involves the use of matrices and eigenvalues; it is more difficult to apply, but it has the advantage of taking into account the boundary conditions. The discussion of both these methods is beyond the scope of this book.

It can be shown that either method produces the criterion for the heat conduction equation

$$r \leqslant \tfrac{1}{2} \tag{8.46}$$

In this instance, the stability criterion is the same as the convergence criterion.

Compatibility

In an attempt to produce a finite-difference scheme that is stable and effective, it is possible to end up with one which has a solution that, under certain combinations of k and h, will not converge to the solution of the original differential equation as $h \to 0$, but rather converges to the solution of a different differential equation. The scheme is then said to be **incompatible**. The explict scheme (8.39) is compatible with the equation $\partial\theta/\partial t = \partial^2\theta/\partial x^2$.

Crank-Nicolson implicit scheme

We would like to replace the explicit scheme by a method which is convergent and stable for all finite values of r and yet which does not involve an inordinate amount of calculation. The **Crank-Nicolson method** replaces $\partial^2\theta/\partial x^2$ by the average of its finite-difference approximations on the current time level and the previous time level. Then the finite-difference scheme becomes

$$\frac{\theta_{i,j+1} - \theta_{i,j}}{k} = \frac{1}{2}\left\{\frac{\theta_{i-1,j+1} - 2\theta_{i,j+1} + \theta_{i+1,j+1}}{h^2} + \frac{\theta_{i-1,j} - 2\theta_{i,j} + \theta_{i+1,j}}{h^2}\right\}$$

This can be arranged to give

$$-r\theta_{i-1,j+1} + (2+2r)\theta_{i,j+1} - r\theta_{i+1,j+1} = r\theta_{i-1,j} + (2-2r)\theta_{i,j} + r\theta_{i+1,j} \qquad (8.47)$$

where $r = k/h^2$, as before.

If we have mesh points x_0 to x_n for a given time level, then there will be $(n-1)$ simultaneous equations for the $(n-1)$ unknowns.

In Figure 8.21(a) we show the computational molecule. In Figure 8.21(b) we represent the mesh near the point of interest.

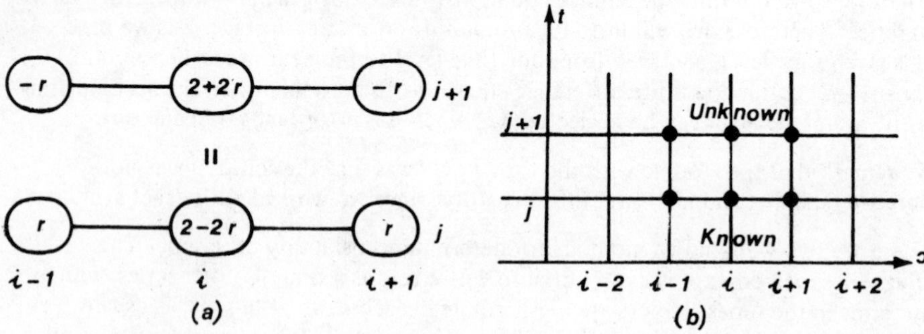

Figure 8.21

Example

We shall now apply the Crank-Nicolson method to the problem we have been studying, namely $\partial\theta/\partial t = \partial^2\theta/\partial x^2$, where $\theta = 0$ at $x = 0$ and $x = 1$ for all $t \geqslant 0$ and where $\theta = 100$ at $t = 0$ for $0 < x < 1$.

Let us choose $h = 0.1$ as before. If we choose $r = 1$, it has the advantage that we shall have simple coefficients to handle and, in particular, the middle term on the right-hand-side of (8.47) disappears. We have

$$-\theta_{i-1,j+1} + 4\theta_{i,j+1} - \theta_{i+1,j+1} = \theta_{i-1,j} + \theta_{i+1,j} \tag{8.48}$$

In our problem, we have symmetry about $x = 0.5$; hence $\theta_{4,j} = \theta_{6,j}$ etc. For simplicity we shall write $\theta_{i,1}$ as θ_i.

In matrix form, we may write the five equations for the five unknowns θ_1, θ_2, θ_3, θ_4 and θ_5 at $j = 1$ as follows.

$$\begin{bmatrix} 4 & -1 & 0 & 0 & 0 \\ -1 & 4 & -1 & 0 & 0 \\ 0 & -1 & 4 & -1 & 0 \\ 0 & 0 & -1 & 4 & -1 \\ 0 & 0 & 0 & -2 & 4 \end{bmatrix} \begin{bmatrix} \theta_1 \\ \theta_2 \\ \theta_3 \\ \theta_4 \\ \theta_5 \end{bmatrix} = \begin{bmatrix} 0 + 100 \\ 100 + 100 \\ 100 + 100 \\ 100 + 100 \\ 100 + 100 \end{bmatrix} \tag{8.49}$$

The basic pattern of coefficients is $-1 \quad 4 \quad -1$, but there are two exceptions. For $j = 1$, $\theta_{i-1,j} = \theta_{0,j} = 0$ by the boundary conditions. For $j = 5$, $\theta_{i+1,j} = \theta_{6,j} = \theta_{4,j} = \theta_{i-1,j}$ by symmetry and hence we get the pattern $-2 \quad 4$.

We saw in Section 1.6 how to solve such a tridiagonal matrix set of equations. The solution is, using the label $\theta_{i,1}$ for θ_i, $\theta_{1,1} = 46.4$, $\theta_{2,1} = 85.6$, $\theta_{3,1} = 96.1$, $\theta_{4,1} = 98.9$, $\theta_{5,1} = 99.4$ Then we may compute the next time step. Putting now $\theta_{i,2} = \theta_i$ we have the equations

$$\begin{bmatrix} 4 & -1 & 0 & 0 & 0 \\ -1 & 4 & -1 & 0 & 0 \\ 0 & -1 & 4 & -1 & 0 \\ 0 & 0 & -1 & 4 & -1 \\ 0 & 0 & 0 & -2 & 4 \end{bmatrix} \begin{bmatrix} \theta_1 \\ \theta_2 \\ \theta_3 \\ \theta_4 \\ \theta_5 \end{bmatrix} = \begin{bmatrix} 0 + 85.6 \\ 46.4 + 96.1 \\ 85.6 + 98.9 \\ 96.1 + 99.4 \\ 98.9 + 98.9 \end{bmatrix}$$

The solution for the first few time steps is shown in Table 8.5. Compare with Table 8.2.

Table 8.5

t \ x	0	0.1	0.2	0.3	0.4	0.5
0	0	100	100	100	100	100
0.01	0	46.4	85.6	96.1	98.9	99.4
0.02	0	38.1	66.8	86.5	94.7	96.8
0.03	0	31.2	58.2	76.9	87.9	91.3
0.04	0	27.4	51.2	69.4	80.4	84.2
0.05	0	24.3	46.0	62.8	73.3	76.9
0.10	0	14.7	27.9	38.4	45.1	47.4

In Table 8.6 we compare the Crank-Nicolson estimated values with the analytical solution for $x = 0.5$

You can see that the Crank-Nicolson solution is more accurate than the explicit method.

We emphasise that the Crank-Nicolson method is stable for all values of r.

Table 8.6

t	Analytical solution (3 d.p.)	Crank-Nicolson solution (8.48)	Percentage error	Explicit method $r = 0.1$	Percentage error
0.01	98.7	96.1	−2.6	95.7	−3.0
0.02	86.8	86.5	−0.03	85.9	−1.0
0.05	63.0	62.8	−0.3	62.6	−0.6
0.10	38.4	38.4	0	38.2	−0.5

Relative computing effort

We briefly compare the amounts of computing effort required by the explicit and implicit methods. As is customary, we consider only multiplications and divisions. Let us suppose that there are $(N-1)$ internal mesh points for each time level and further suppose that there is no symmetry to employ. The explicit method requires $2(N-1)$ multiplications. The Crank-Nicolson method, solved via a Gauss elimination adapted to a tridiagonal matrix system requires $3(N-1)$ multiplications, a relatively minor increase, bearing in mind the lack of restriction on r. If we solve the Crank-Nicolson scheme by using the Gauss-Seidel technique, each **iteration step** takes $2(N-1)$ multiplications and this may be too time-consuming to be preferred to the Gauss elimination approach.

Problems

1. Use the explicit method to solve the equation $u_{xx} = u_t$ where $u(0, t) = u(20, t) = 200$,

   $$u(x, 0) = \begin{cases} 100 - 10x, & 0 \leqslant x \leqslant 10 \\ -200 + 20x, & 10 \leqslant x \leqslant 20 \end{cases}.$$

   Use $h = \dfrac{1}{20}$ and choose k so that $r = 0.2$. Continue the solution for 4 time steps. Then write a computer program to carry out 100 time steps.

2. Repeat Problem 1 for $r = 0.1, 0.4, 0.8, 1.0$.

3. Repeat Problem 1 for the new boundary condition $u(12, 0) = 200$. Comment.

4. Repeat Problem 1 with $u(0, t) = 0$ and initial condition $u(x, 0) = \begin{cases} 0, & 0 \leqslant x \leqslant 10 \\ 20x - 200, & 10 \leqslant x \leqslant 20 \end{cases}$

5. Repeat Problem 1 for conditions $u(0, t) = 0$, $u(20, t) = 0$,

   $$u(x, 0) = \begin{cases} 50x, & 0 \leqslant x \leqslant 4 \\ 400 - 50x, & 4 \leqslant x \leqslant 8 \\ 0, & x \geqslant 8 \end{cases}$$

6. Repeat Problems 1 to 5 using the Crank-Nicolson method with $r = 0.8, 1.0, 2.0, 5.0$. Use a computer program to help you in your calculations.

8.7 CASE STUDY: FORCED CONVECTION

So far the problems in heat conduction we have considered have been ones where the temperature at the boundaries is specified; this is not the usual practical situation. Heat loss from the surface of a solid can be by natural convection, forced convection or radiation. With radiative heat loss, the rate of loss is proportional to the fourth power of the difference between the temperature of the solid surface and the surrounding surfaces. In natural convection, the motion of a surrounding fluid is governed by the thermal gradients built up. In forced convection, the surrounding fluid is made to flow by some means, e.g. a fan, so that it carries away heat from the solid surface.

In this last case, the one which we shall study, we may express the rate of loss of heat from the solid surface by $hA(\theta - \theta_s)$, where A is the area of the surface, θ its temperature, θ_s the temperature of the surrounding medium and h is a heat transfer coefficient.

Let us consider a slab of material as shown in Figure 8.22(a) which has one face free to lose heat to the surroundings, the other faces being insulated. Then we may consider the flow as taking place in the x - direction. We therefore have the differential equation $\dfrac{\partial\theta}{\partial t} = \dfrac{k}{c\rho}\dfrac{\partial^2\theta}{\partial x^2}$ with initial condition $\theta = \theta_0$ at $t = 0$. Now the rate at which heat flows to the surface must be matched by the rate at which heat escapes from the surface. Therefore at the face $x = 0$ we have $-kA\,\partial\theta/\partial x = -hA(\theta - \theta_s)$. At the face $x = a$ we have perfect insulation so that $\partial\theta/\partial x = 0$ there.

The Crank-Nicolson scheme for interior points, taking $r = 1$, is

$$\theta_{i-1,j+1} - 4\theta_{i,j+1} + \theta_{i+1,j+1} = -\theta_{i-1,j} - \theta_{i+1,j} \tag{8.50}$$

At $x = 0$,

$$-k\partial\theta/\partial x \simeq -k(\theta_{2,j} - \theta_{0,j})/2\delta x = h(\theta_{1,j} - \theta_s) \qquad \text{i.e.}$$

$$\theta_{0,j} = \theta_{2,j} - B\theta_{1,j} + B\theta_s \qquad \text{where} \qquad B = 2h\delta x/k$$

What we have done is to introduce a fictitious point $\theta_{0,j}$ which lies outside the solution domain, and which only arises because (8.50) needs to operate at $i = 1$ (the face $x = 0$). See Figure 8.22(b). We may substitute for the fictitious temperatures $\theta_{0,j}$ and $\theta_{0,j+1}$ to obtain the following form of (8.51) which holds at $x = 0$

$$-(4 + B)\theta_{1,j+1} + 2\theta_{2,j+1} = -2\theta_{2,j} + B\theta_{1,j} - 2B\theta_s \tag{8.51}$$

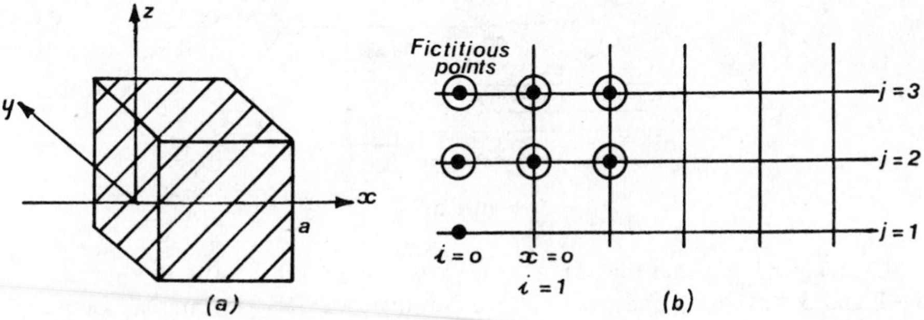

Figure 8.22

Similarly, at $x = a$ (or $i = n$) we use a fictitious point $\theta_{n+1,j}$ so that $-k\partial\theta/\partial x \simeq -k(\theta_{n+1,j} - \theta_{n-1,j})/2\delta x = 0$. This yields $\theta_{n+1,j} = \theta_{n-1,j}$ and $\theta_{n+1,j+1} = \theta_{n-1,j+1}$. Then the appropriate form of (8.50) is

$$2\theta_{n-1,j+1} - 4\theta_{n,j+1} = -2\theta_{n-1,j} \tag{8.52}$$

The program we shall use performs the tasks shown in the flow chart, Figure 8.23

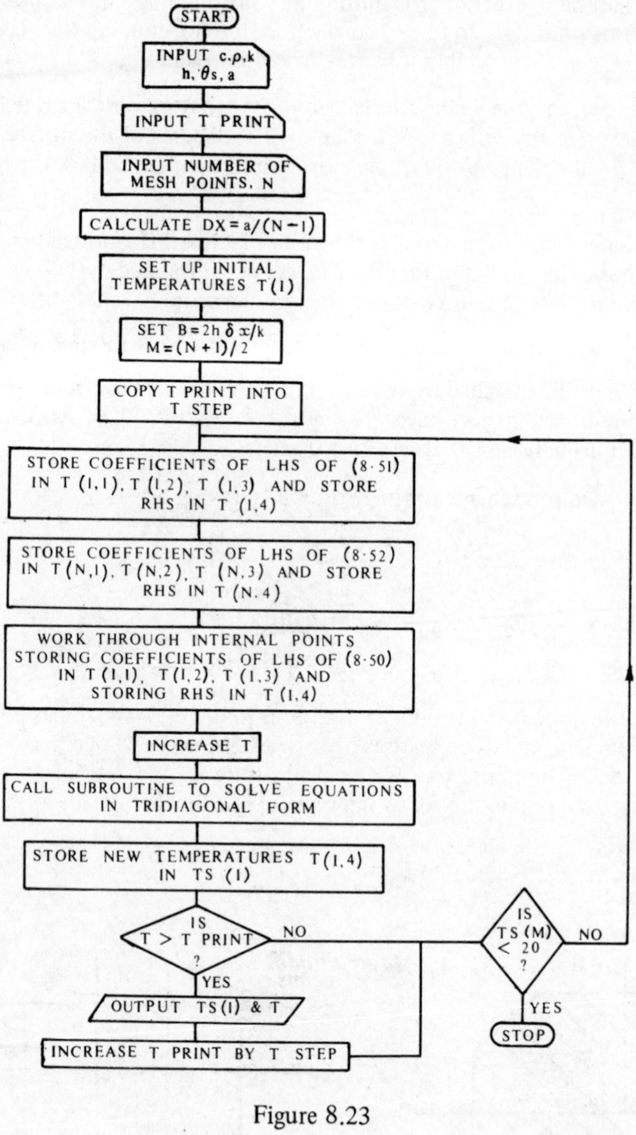

Figure 8.23

Choosing $r = 1$ means that $\delta t = c\rho(\delta x)^2/k$. Note that $x = 0$ corresponds to $I = 1$ and $x = a$ corresponds to $I = N$, so that there are $(N - 1)$ internal points. We have decided to continue the calculations so long as the temperature of the mid-point corresponding to $I = M = (N + 1)/2$ is greater than $20^{\circ}C$. Note that N has to be odd.

The print-out was arranged for every 100 seconds. Had this proved to give either too much or too little information, we could adjust the parameter TPRINT. After each print-out, we increase TPRINT by its starting value.

The coefficients have to be restored after each call of the subroutine since it modifies these coefficients in its operations. As well as storing coefficients, the array T stores the new temperatures calculated by the subroutine.

Results for the program with input values $c = 0.0928$, $\rho = 8.89$, $k = 0.918$, $h = 0.04$, $\theta_s = 80.0$, $\theta_0 = 100$, $a = 100.0$, TPRINT = 100.0, N = 21 are shown in Table 8.7. Times are approximate and temperatures are quoted to 3 s.f.

Table 8.7

t \ x	0	0.25a	0.5a	0.75a	a
0	100	100	100	100	100
100	86.6	92.8	96.8	98.8	99.4
200	84.9	89.9	93.7	96.0	96.8
300	84.0	88.1	91.4	93.5	94.2
400	83.3	86.7	89.5	91.2	91.8
500	82.7	85.6	87.9	89.3	89.8
1000	81.1	82.2	83.1	83.7	83.9

Draw the temperature profiles for these results. Comment.

8.8 ELLIPTIC EQUATIONS

In this section we consider further finite difference approaches and exploit further the idea of analogies.

First we make a few general remarks on improving the accuracy of finite difference solutions. The most obvious way of proceeding is to make the mesh finer by reducing the mesh size. However, for elliptic problems the number of equations to be solved is inversely proportional to the square of the mesh length; an alternative method is therefore desirable. A second possibility is to use more accurate finite-difference formulae for the second partial derivatives $\partial^2\theta/\partial x^2$ and $\partial^2\theta/\partial y^2$.

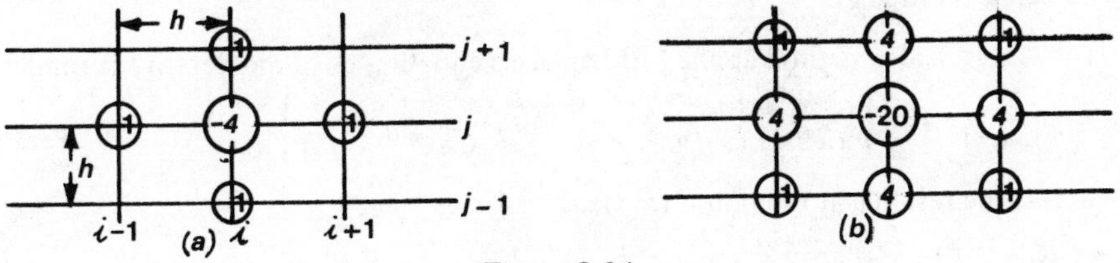

Figure 8.24

In Figure 8.24(a) we show the five points employed by the formulae that we used. The approximations were obtained by using Taylor's series.

$$\theta_{i,j+1} = \theta_{i,j} + h\theta_x\Big|_{(i,j)} + \frac{h^2}{2!}\theta_{xx}\Big|_{(i,j)} + \frac{h^3}{3!}\theta_{xxx}\Big|_{(i,j)} + 0(h^4) \tag{8.53a}$$

$$\theta_{i,j-1} = \theta_{i,j} - h\theta_x\Big|_{(i,j)} + \frac{h^2}{2!}\theta_{xx}\Big|_{(i,j)} - \frac{h^3}{3!}\theta_{xxx}\Big|_{(i,j)} + 0(h^4) \tag{8.53b}$$

Adding these two equations gives

$$\theta_{i,j+1} + \theta_{i,j-1} - 2\theta_{i,j} = 2\frac{h^2}{2!}\theta_{xx}\Big|_{(i,j)} + 0(h^4)$$

Dividing by h^2 produces

$$\theta_{xx}\Big|_{(i,j)} = \frac{\theta_{i,j-1} - 2\theta_{i,j} + \theta_{i,j+1}}{h^2} + 0(h^2)$$

Similarly we may obtain

$$\theta_{yy}\Big|_{(i,j)} = \frac{\theta_{i-1,j} - 2\theta_{i,j} + \theta_{i+1,j}}{h^2} + 0(h^2)$$

Therefore the approximation to Laplace's equation we used, viz.

$$\theta_{i-1,j} + \theta_{i+1,j} + \theta_{i,j-1} + \theta_{i,j+1} = 4\theta_{i,j}$$

has error $0(h^2)$. The weights are shown in Figure 8.24(a).

Suppose we now take all nine points shown in Figure 8.24(b) into account. We have, for example, to consider

$$\theta_{i+1,j+1} = \theta_{i,j} + h\theta_x\Big|_{(i,j)} + h\theta_y\Big|_{(i,j)} + \frac{h^2}{2}\theta_{xx}\Big|_{(i,j)} + h^2\theta_{xy}\Big|_{(i,j)} + \frac{h^2}{2}\theta_{yy}\Big|_{(i,j)}$$
$$+ \frac{h^3}{3!}\left\{\theta_{xxx}\Big|_{(i,j)} + 3\theta_{xxy}\Big|_{(i,j)} + 3\theta_{xyy}\Big|_{(i,j)} + \theta_{yyy}\Big|_{(i,j)}\right\} + 0(h^4)$$

In fact we take the expansions as far as terms $0(h^5)$ and we may derive the nine-point approximation to Poisson's equation $\nabla^2\theta = K$, K constant, which has a truncation error $0(h^6)$. This approximation is

$$\theta_{i-1,j-1} + \theta_{i+1,j-1} + \theta_{i-1,j+1} + \theta_{i+1,j+1} + 4\left[\theta_{i-1,j} + \theta_{i+1,j} + \theta_{i,j-1} + \theta_{i,j+1}\right]$$
$$= 20\theta_{i,j} \tag{8.54}$$

The weights are shown in Figure 8.24(b).

Gauss-Seidel method

The Gauss-Seidel method applied to Laplace's equation gives the iterative formula

$$\theta_{i,j,n+1} = \frac{1}{4}\left\{\theta_{i-1,j,n+1} + \theta_{i,j-1,n+1} + \theta_{i+1,j,n} + \theta_{i,j+1,n}\right\} \tag{8.55}$$

We define the residual at the point (i, j) as

$$R_{i,j} = \theta_{i-1,j,n+1} + \theta_{i,j-1,n+1} + \theta_{i+1,j,n} + \theta_{i,j+1,n} - 4\theta_{i,j,n+1}$$

The Gauss-Seidel method is known as the (unextrapolated) **Liebmann method.**

We choose the starting values by first estimating $\theta_{2,2,0}$ from the nine-point formula; the other values may be guessed by linear interpolation. Since the method converges only slowly, it is best suited to a digital computer.

The method of relaxation

This is a method mainly of historical interest, since the widespread use of digital computers has effectively made it redundant. However, we briefly examine it so that we may set the scene for the next subsection.

Essentially, the method calculates the residual at each mesh point and then selects the largest of these. Since this occurs at the point where the initial approximation is worst we take steps to improve the situation here before any other points. Altering the function value at this point has the effect of changing the residuals at neighbouring points as well at the point under consideration. If we use the five-point formula then, if the largest residual at a point is -40, adding 10 to that value of θ will reduce the residual to zero and add 10 to the four nearest residuals. Then one looks for the next largest residual and so on. However, relaxation has been developed to an art and residuals are not always **relaxed** exactly to zero. Sometimes the residual is **under-relaxed**, sometimes **over-relaxed.** An experienced operator would get a feel for how to proceed at each step. The method is therefore not suitable to programming for a computer. A common device is to multiply all mesh values by 10 when residuals get small, in order to keep to integer arithmetic.

Successive Over-Relaxation

This method, also known as **S.O.R.** or the **Extrapolated Liebmann method** seeks to accelerate the convergence of Liebmann's method and arrive at a technique which combines rapid convergence with the systematic approach needed for digital computation.

For the Gauss-Seidel method we can write the iteration formula as

$$\theta_{i,j,n+1} = \theta_{i,j,n} + \frac{1}{4} R_{i,j} \tag{8.56}$$

In the S.O.R. method we change $\theta_{i,j}$ by a larger amount than $\frac{1}{4} R_{i,j}$ – hence the term **over-relaxation.** The iterative formula we use is

$$\theta_{i,j,n+1} = \theta_{i,j,n} + \frac{\omega}{4} R_{i,j} \tag{8.57}$$

where ω is a positive constant between 1 and 2.

But what value should we choose for ω?

One case which has managed to yield a simple rule for determining ω is a rectangular region with m and n mesh divisions on adjacent sides. It has been proved that the value of ω can be found by solving the equation

$$\left[\cos\frac{\pi}{m} + \cos\frac{\pi}{n}\right]^2 \omega^2 - 16\omega + 16 = 0 \tag{8.58}$$

We shall study the method in operation in the next section when we apply it to **Poisson's equation.**

General remarks on elliptic equations

Boundary conditions specify at each point of the boundary curve *C either* the value of the independent variable *or* its derivative normal to the boundary. They may not specify both at the *same* point.

Further, a result which any finite difference scheme should model is that the solution of Laplace's equation has neither its maximum nor its minimum value on points *inside* the boundary curve.

Experimental Analogies

In Section 8.2 we saw that since electrical potential satisfies Laplace's equation we were able to set up an **electrical analogue** to the steady-state temperature distribution problem. Of course, the method can apply to the solution of Laplace's equation in any context, placing a known potential at some points of the boundary and insulating other points (so that the derivative there normal to the boundary is zero).

A second analogy that is used is the **membrane analogy**. A thin rubber sheet or a soap film can be stretched over a frame which is shaped in such a way that the three-dimensional surface formed by the membrane has contours which in plan view correspond to streamlines of a flow pattern. Should we require the pattern of flow round an immersed body then the membrane can be attached to the body so that it meets it at a constant height. The contours may be traced by a movable point gauge in the case of a rubber membrane or by reflection of light rays in the case of the soap film.

A third analogy, which may be used in the study of irrotational flows, is the **viscous flow analogy**. Viscous fluid is made to flow between horizontal parallel plates separated by a small distance. Suitable obstacles may be placed between the plates. Dye is fed continuously at regular intervals across the boundary at which fluid enters the plates (which are usually sealed along two edges to form a cell) and it traces out the corresponding flow pattern. Near the boundaries of the flow, the analogy becomes somewhat inaccurate, but reducing the spacing of the plates will alleviate this to an extent.

Problems

1. A square plate is heated along its edges to temperatures which vary in a way indicated in the diagram. The mesh is square. We require to find the temperatures at points A, B, C and D. First guess the temperatures T_A, T_B, T_C and T_D by inspection. Use the four-point formula for (8.40) for Laplace's equation, to find a new estimate for T_A; then find new values for T_B, T_C and T_D in that order. Repeat the cycle until the temperatures are obtained to the nearest degree. Write and

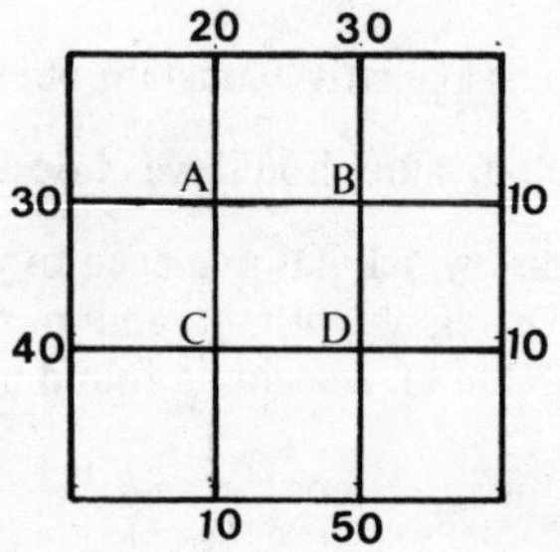

run a computer program to continue the iterations until the temperatures are obtained to a prescribed accuracy. Use the computer program of Problem 1 or a suitable modification to solve Problems 2 to 6 which all relate to Laplace's equation with suitable boundary conditions.

2.

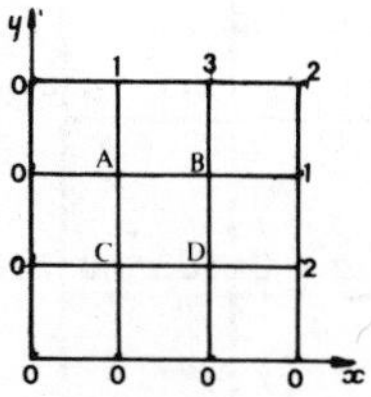

3.

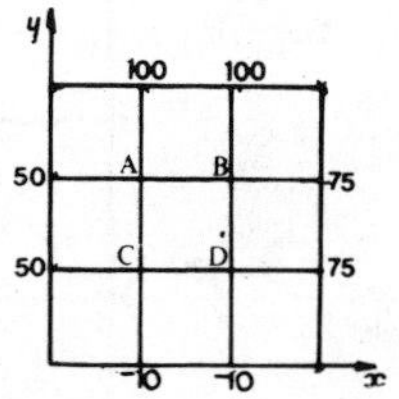

4.

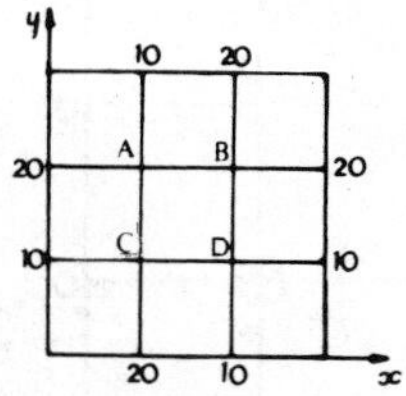

5. A rectangular area is bounded by the lines $x = 0,\ x = 4,\ y = 0,\ y = 5$. A function V satisfies the equation $\dfrac{\partial^2 V}{\partial x^2} + \dfrac{\partial^2 V}{\partial y^2} = 0$ within the area, and $V = 5(2 - x)^2 + y^2$ on the boundary.

Using a mesh of unit squares, find the values of V at the mesh points. (L.U.)

6. A square plate is bounded by the lines $x = \pm 2;\ y = \pm 2$. The temperature θ of the plate obeys Laplace's equation $\dfrac{\partial^2 \theta}{\partial x^2} + \dfrac{\partial^2 \theta}{\partial y^2} = 0$. The boundary temperatures are

on the line $y = 2,\ \theta = -80$; on the lines $x = \pm 2,\ \theta = -40y$;
on the line $y = -2,\ \theta = 240 + 80x$ for $-2 \leqslant x \leqslant 0$.
$\theta = 240 - 80x$ for $0 \leqslant x \leqslant 2$

Find the temperatures at the nine points $(x_i,\ y_j)$ where $i = -1,\ 0,\ 1,\ j = -1,\ 0,\ 1$. (L.U.)

7. A long square pipe with a square hole carries a hot fluid. The bottom half of the outside of the pipe is at a temperature of 0°C. The top surface of the pipe is at 100°C. Assume that the temperature varies linearly along the top half of each vertical side. The temperature of the fluid inside the pipe is 200°C. The inside dimension of the pipe is 4 cm and the outside dimension is 10 cm. Choose a suitable mesh size and find the temperature at the mesh points.

8. Repeat Problems 2, 3 and 4 using the method of successive over-relaxation.

8.9 CASE STUDY : TORSION OF A RECTANGULAR PRISM

We have already derived the formula (8.19) for the stress function ϕ which satisfies Poisson's equation $\partial^2\phi/\partial x^2 + \partial^2\phi/\partial y^2 = -2$ in the rectangular region $-a \leqslant x \leqslant a$, $-b \leqslant y \leqslant b$ with the condition that $\phi = 0$ on the boundaries. Let us consider a bar with square cross-section $-1 \leqslant x \leqslant 1,\ -1 \leqslant y \leqslant 1$. Then the formula reduces to

$$\phi = 1 - y^2 - \frac{32}{\pi^3} \sum_{n=1,3,5\ldots}^{\infty} \frac{(-1)^{(n-1)/2}}{n^3} \operatorname{sech} \frac{n\pi}{2} \cosh \frac{n\pi x}{2} \cos \frac{n\pi y}{2} \qquad (8.59)$$

We shall divide the region into 36 equal squares so that there are 25 internal points. We shall use S.O.R. to solve this problem and we shall make no appeal to symmetry, using a general program to achieve the solution. We compare the results using an initial approximation that $\phi = 0$ at all internal points and an initial approximation obtained as follows. If we divide the region into 9 equal squares then there are 4 internal points. You can see from Figure 8.25(a) that by symmetry, the value at these points should be the same. Using a nine-point formula we find that this value is

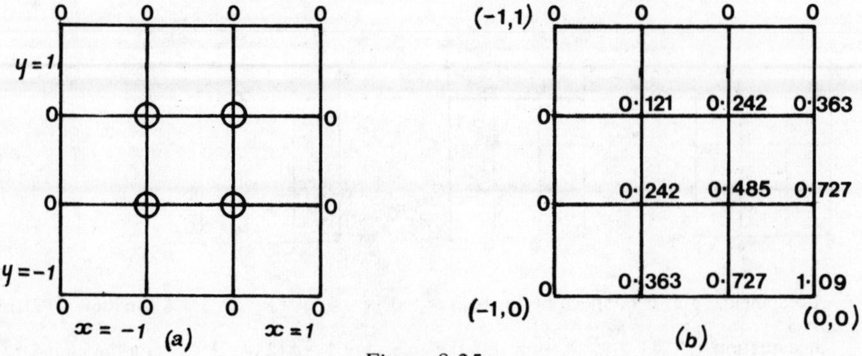

Figure 8.25

$0+\phi+0+0+0+4\phi+4\phi+0-20\phi=6(2/3)^2.(-2)\Rightarrow\phi=16/33=0.485$ (3 d.p.)
We have used the result that the nine-point approximation to $\nabla^2\phi=-2$ is

$$\phi_{i-1,j-1}+\phi_{i+1,j-1}+\phi_{i+1,j-1}+\phi_{i+1,j+1}+4\phi_{i-1,j}+4\phi_{i,j-1}+4\phi_{i+1,j}$$
$$+4\phi_{i,j+1}-20\phi_{i,j}=6h^2(-2) \qquad (8.60)$$

where $h=2/3$.

In Figure 8.25(b) we have filled the other values in by linear interpolation; remaining values can be found by symmetry. These values can be read into the computer program.

What value for the relaxation factor ω should be taken?

Equation (8.58) tells us to solve:

$$\left[\cos\frac{\pi}{6}+\cos\frac{\pi}{6}\right]^2\omega^2-16\omega+16=0 \qquad \text{i.e.}$$

$$3\omega^2-16\omega+16=0$$

This last equation has roots $\omega=4$ or $4/3$. The optimum relaxation factor is found to be the latter value.

In Table 8.8 we show the results with $\omega=4/3$ and initial values $\phi=0$; 8 iterations were used. Similar results were obtained after 6 iterations via the initial values estimated as in Figure 8.25(b). In Table 8.9 we had $\phi=0$ at all points with $\omega=1$ and used 11 iterations.

Table 8.8

0.000	0.000	0.000	0.000	0.000	0.000	0.000
0.000	0.209	0.310	0.340	0.311	0.211	0.000
0.000	0.310	0.470	0.519	0.471	0.312	0.000
0.000	0.340	0.519	0.576	0.521	0.342	0.000
0.000	0.311	0.471	0.521	0.472	0.312	0.000
0.000	0.211	0.312	0.342	0.312	0.212	0.000
0.000	0.000	0.000	0.000	0.000	0.000	0.000

Table 8.9

0.000	0.000	0.000	0.000	0.000	0.000	0.000
0.000	0.198	0.292	0.322	0.297	0.204	0.000
0.000	0.292	0.442	0.491	0.450	0.301	0.000
0.000	0.322	0.491	0.547	0.499	0.331	0.000
0.000	0.297	0.449	0.499	0.455	0.304	0.000
0.000	0.204	0.301	0.331	0.304	0.207	0.000
0.000	0.000	0.000	0.000	0.000	0.000	0.000

The flow chart for the program is shown in Figure 8.26. It is designed for $\nabla^2 \phi = C$ with ϕ specified at the mesh points. The maximum number of iterations permitted is ITMAX and the accuracy required TOL.

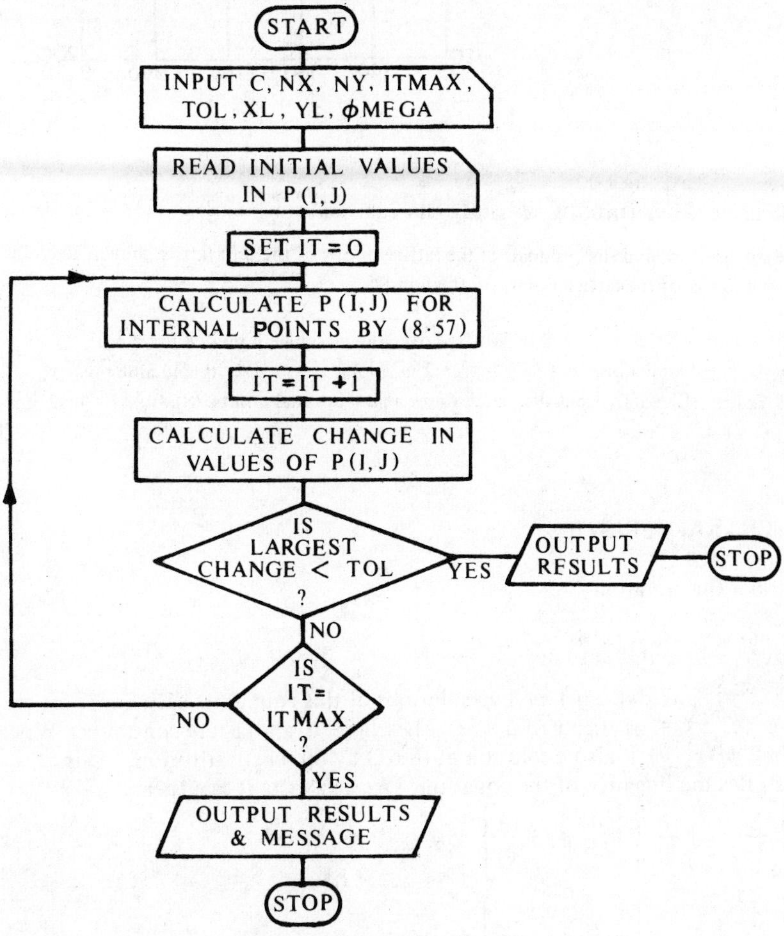

Figure 8.26

Problems

1. The Poisson equation $\dfrac{\partial^2 \phi}{\partial x^2} + \dfrac{\partial^2 \phi}{\partial y^2} + 2 = 0$ is to be solved for the hollow bar shown.

The area is divided into unit squares and the value of ϕ on all boundaries is zero.

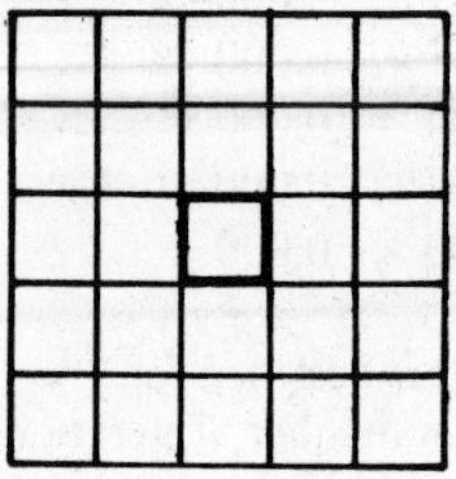

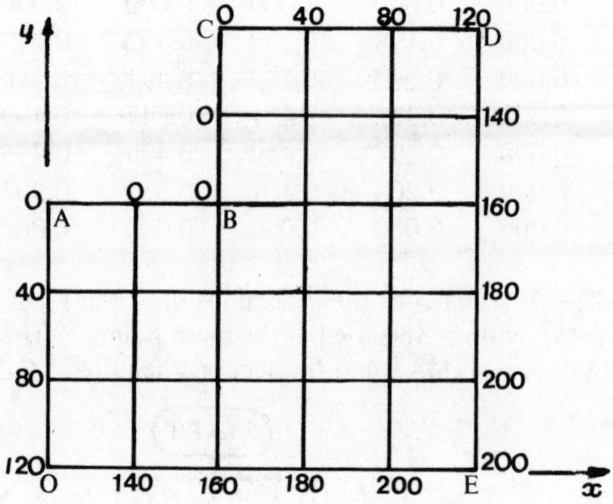

2. Within the region OABCDE, ϕ satisfies the equation $\dfrac{\partial^2\phi}{\partial x^2} + \dfrac{\partial^2\phi}{\partial y^2} = 4x(5-y)$. Determine the values of ϕ and the residuals at the lattice points of the unit lattice shown, given the values shown for ϕ at the lattice points on the boundary. (L.U.)

3. The equation $\nabla^2\phi + 2 = 0$ is satisfied over the rectangle bounded by $x = 0,\ 5,\ y = 0,\ 4$. Along $x = 0$ and along $x = 5$, $\phi = y^2 + 2y$. Along $y = 0$, $\phi = 0$ and along $y = 4$, $\phi = 4(3-x)(2-x)$. Find the values of ϕ at the internal points (x_i, y_j), $i = 1, 2$ and $j = 1, 2, 3$. (L.U.)

8.10 CHARACTERISTICS

Consider the equation

$$\frac{\partial^2\phi}{\partial x^2} + 4\frac{\partial^2\phi}{\partial x\partial y} + 3\frac{\partial^2\phi}{\partial y^2} = 0 \tag{8.61}$$

Let $\phi_1(x, y)$ and $\phi_2(x, y)$ be two solutions of this equation. Then suppose we define $\phi^*(x, y) = \alpha\phi_1(x, y) + \beta\phi_2(x, y)$, where α and β are constants. We can show that $\phi^*(x, y)$ is also a solution of (8.61) by direct substitution. This demonstrates the linearity of the equation. We can write it as either

$$\left[\frac{\partial}{\partial x} + \frac{\partial}{\partial y}\right]\left[\frac{\partial}{\partial x} + 3\frac{\partial}{\partial y}\right]\phi = 0 \tag{8.62a}$$

or

$$\left[\frac{\partial}{\partial x} + 3\frac{\partial}{\partial y}\right]\left[\frac{\partial}{\partial x} + \frac{\partial}{\partial y}\right]\phi = 0 \tag{8.62b}$$

Now it can be shown that the general solution of the equation

$$\left[\frac{\partial}{\partial x} + 3\frac{\partial}{\partial y}\right]\phi \equiv \frac{\partial\phi}{\partial x} + 3\frac{\partial\phi}{\partial y} = 0$$

is $\phi_1 = f(y - 3x)$ for a suitably differentiable function f. It follows from (8.62a) that this form of ϕ is a solution of (8.61). Likewise, starting from the equation $\left[\frac{\partial}{\partial x} + \frac{\partial}{\partial y}\right]\phi = 0$ whose general solution is $\phi_2 = g(y - x)$ for a suitably differentiable g, we find that ϕ_2 is another solution of (8.61). Since arbitrary constants may be absorbed into f and g, it follows that the general solution of (8.61) can be written as

$$\phi = f(y - 3x) + g(y - x) \tag{8.63}$$

Change of variable

We can transform the independent variables x and y into u and v where $u = x - y$, $v = 3x - y$. Applying the chain rule for partial derivatives eventually leads to the result

$$4\frac{\partial^2\phi}{\partial u \partial v} = 0 \tag{8.64}$$

Integrating with respect to v gives $\frac{\partial\phi}{\partial u} = h(v)$ and integrating this last equation with respect to u provides $\phi = f(v) + g(u)$ where $f'(v) \equiv h(v)$. **Hence we recover** solution (8.63).

The curves $3x - y =$ constant and $x - y =$ constant are called **characteristics** of the equation. Along them the solution takes on a particularly simple form. The form (8.64) is called the **canonical form** of the equation (8.61). It can be shown that along the curves $3x - y =$ constant, the value of $\left[\frac{\partial\phi}{\partial x} + \frac{\partial\phi}{\partial y}\right]$ is constant and along the curves $x - y =$ constant, the value of $\left[\frac{\partial\phi}{\partial x} + 3\frac{\partial\phi}{\partial y}\right]$ is constant.

Classification via characteristics

The general second-order quasi-linear partial differential equation with two independent variables, whose coefficients may be functions of x and/or y, is

$$A\frac{\partial^2\phi}{\partial x^2} + B\frac{\partial^2\phi}{\partial x \partial y} + C\frac{\partial^2\phi}{\partial y^2} + D\frac{\partial\phi}{\partial x} + E\frac{\partial\phi}{\partial y} + F\phi + G = 0 \tag{8.65}$$

We shall confine our attention to the case $D = E = F = 0$, although our remarks will generalise. If we employ the notation which is customary, viz. $p = \partial\phi/\partial x$, $q = \partial\phi/\partial y$, $r = \partial^2\phi/\partial x^2$, $s = \partial^2\phi/\partial x\partial y$, $t = \partial^2\phi/\partial y^2$ then the simplified form of (8.65) may be written as

$$Ar + Bs + Ct = -G \tag{8.66}$$

Now

$$dp = \frac{\partial p}{\partial x}dx + \frac{\partial p}{\partial y}dy = rdx + sdy = rdx + sdy + t.0 \tag{8.67a}$$

and

$$dq = sdx + tdy = r.0 + sdx + tdy \tag{8.67b}$$

Suppose that the values of $p = \partial\phi/\partial x$ and $q = \partial\phi/\partial y$ are given on some curve in the x - y plane. From equations (8.66) and (8.67) we can deduce the values of r, s and t on the curve provided that

$$\Delta = \begin{vmatrix} dx & dy & 0 \\ 0 & dx & dy \\ A & B & C \end{vmatrix} \neq 0 \tag{8.68}$$

The requirement $\Delta = 0$ leads to the equation

$$A\left[\frac{dy}{dx}\right]^2 - B\left[\frac{dy}{dx}\right] + C = 0 \tag{8.69}$$

This quadratic equation will have two roots if $B^2 - 4AC > 0$; there are two real characteristics and the original equation is said to be **hyperbolic.** If $B^2 = 4AC$ then there is only one characteristic direction and the equation is **parabolic.** Finally, if $B^2 < 4AC$ the equation is **elliptic**; there are no real characteristic curves.

Hyperbolic equations may be transformed to $\dfrac{\partial^2\phi}{\partial u \partial v} = F\left(\phi, u, v, \dfrac{\partial\phi}{\partial u}, \dfrac{\partial\phi}{\partial v}\right)$.
The one-dimensional wave equation $c^2\partial^2\phi/\partial x^2 - \partial^2\phi/\partial t^2 = 0$ is hyperbolic. It can be reduced to $\partial^2\phi/\partial u\partial v = 0$ where $u = x - ct$, $v = x + ct$ with general solution $\phi = f(x - ct) + g(x + ct)$.

Note that a p.d.e. may be elliptic in one region of the plane, parabolic in another, and hyperbolic in yet a third region; see Problem 1.

The method of characteristics

For reasons which will become clear, this method works only for hyperbolic equations.

We work through one example to show the ideas. Consider the equation

$$\frac{\partial^2\phi}{\partial x^2} - 2\frac{\partial^2\phi}{\partial x \partial y} - 3\frac{\partial^2\phi}{\partial y^2} = -1 \tag{8.70}$$

Suppose we are given that at $y = 0, \phi = x = \dfrac{\partial\phi}{\partial y}$ for $0 < x < 1$.

Then if we refer to the more general case shown in Figure 8.27 we see that from each point two characteristic curves may be drawn. A characteristic of one class from P will intersect one of the other class from Q in a point S. P, Q, R are points on the first level and S and T are points on the second level. It is true in general that as we proceed from one level to the next, the number of points available will become smaller. Returning to our example, the characteristics are given by

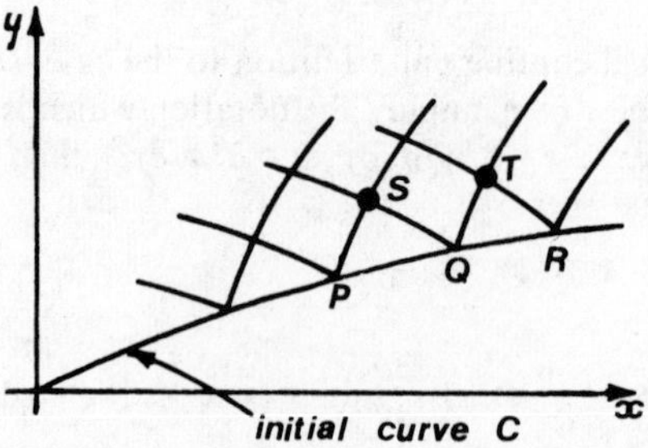

Figure 8.27

$$\left[\frac{dy}{dx}\right]^2 - 2\frac{dy}{dx} - 3 = 0 \text{ and are straight lines } \frac{dy}{dx} = 3, \ \frac{dy}{dx} = -1.$$

Families of these lines emanating from the initial curve $y = 0$ at $x = 0(0.1)1.0$ are shown in Figure 8.28. In general, the pattern of characteristics will not be so simple.

Now we start from the knowledge at $y = 0$, $x = 0(0.1)1.0$ – 11 points – and use this to advance to the 10 points on level 2 then to the 9 points on level 3. Consider $P = (0.4, 0)$ and $Q = (0.5, 0)$ and let characteristics from those points intersect at R. Then we find that the coordinates of R are (0.425, 0.075)

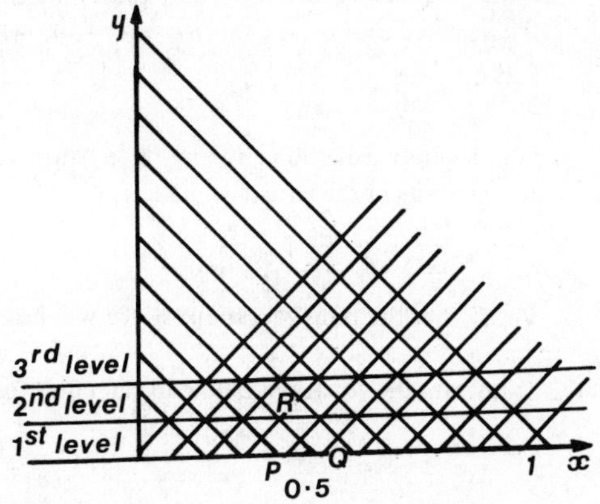

Figure 8.28

Note that we know that at $y = \nu$, $\partial\phi/\partial x = 1$ so that on the initial curve, ϕ, $\partial\phi/\partial x$ and $\partial\phi/\partial y$ are known.

It can be shown, although the proof is beyond the scope of this book, that we can find the values of p and q at R and the knowledge of these allows us to find the values of ϕ at R. We can therefore find all values at the points on the second level and from these proceed to the third level.

Note that if the initial curve is itself a characteristic we will be in trouble. There will be a unique solution only along characteristic curves of the same family.

In physical problems modelled by hyperbolic equations it may be that there are discontinuities in the initial conditions. These discontinuities are propagated along the characteristics as discontinuities in the solution. An example is that of shock waves.

Problems

1. For what regions are the equations below elliptic, parabolic, hyperbolic?

(a) $x\dfrac{\partial^2 u}{\partial x^2} + y\dfrac{\partial^2 u}{\partial y^2} + 3y^2\dfrac{\partial u}{\partial x} = 0$

(b) $x^2\dfrac{\partial^2 u}{\partial x^2} + 2xy\dfrac{\partial^2 u}{\partial x\,\partial y} + y^2\dfrac{\partial^2 u}{\partial y^2} = 0$

(c) $(x^2 - 1)\dfrac{\partial^2 u}{\partial x^2} + 2xy\dfrac{\partial^2 u}{\partial x\,\partial y} + (y^2 - 1)\dfrac{\partial^2 u}{\partial y^2} = x\dfrac{\partial u}{\partial x} + y\dfrac{\partial u}{\partial y}$

(d) $\dfrac{\partial^2 u}{\partial x^2} + x\dfrac{\partial^2 u}{\partial x\,\partial y} + y\dfrac{\partial^2 u}{\partial y^2} - xy\dfrac{\partial u}{\partial x} = 0$

2. The canonical form of a hyperbolic equation given by (8.65) is $\dfrac{\partial^2 u}{\partial\xi\,\partial\eta} = d\dfrac{\partial u}{\partial\xi} + e\dfrac{\partial u}{\partial\eta} + fu$ where d, e and f are real constants and $\xi = \lambda_1 x + y$, $\eta = \lambda_2 x + y$; λ_1 and λ_2 are the roots of the equation $A\lambda^2 + B\lambda + C = 0$.

For a parabolic equation given by (8.65) the transformation $\xi = \lambda x + y$, $\eta = y$, where λ is the repeated root of $A\lambda^2 + B\lambda + C = 0$, will produce the canonical form

$$\frac{\partial^2 u}{\partial \eta^2} = d\frac{\partial u}{\partial \xi} + e\frac{\partial u}{\partial \eta} + fu.$$

For an elliptic equation given by (8.65), the transformation $\xi = ax + y$, $\eta = bx$ where $a \pm bi$ are the roots of the equation $A\lambda^2 + B\lambda + C = 0$, will produce the canonical form

$$\frac{\partial^2 u}{\partial \xi^2} + \frac{\partial^2 u}{\partial \eta^2} = d\frac{\partial u}{\partial \xi} + e\frac{\partial u}{\partial \eta} + fu.$$

Verify that the transformations stated will produce the given canonical forms.

3. Transform the equations below to canonical form

(i) $\dfrac{\partial^2 u}{\partial x^2} - 5\dfrac{\partial^2 u}{\partial x \partial y} + 6\dfrac{\partial^2 u}{\partial y^2} = 0$ (ii) $\dfrac{\partial^2 u}{\partial x^2} - 2\dfrac{\partial^2 u}{\partial x \partial y} - 8\dfrac{\partial^2 u}{\partial y^2} + 9\dfrac{\partial u}{\partial x} = 0$

(iii) $\dfrac{\partial^2 u}{\partial x^2} - 4\dfrac{\partial^2 u}{\partial x \partial y} + 4\dfrac{\partial^2 u}{\partial y^2} = 0$ (iv) $\dfrac{\partial^2 u}{\partial x^2} + 2\dfrac{\partial^2 u}{\partial x \partial y} + \dfrac{\partial^2 u}{\partial y^2} + 3\dfrac{\partial u}{\partial x} + 9u = 0$

4. Find suitable constants a and b so that, under the transformation of co-ordinates $u = x + ay$, $v = x + by$, the differential expression $2\dfrac{\partial^2 z}{\partial x^2} - 3\dfrac{\partial^2 z}{\partial x \partial y} + \dfrac{\partial^2 z}{\partial y^2}$ becomes $\dfrac{\partial^2 z}{\partial u \partial v}$.

Hence derive the general solution of the equation $2\dfrac{\partial^2 z}{\partial x^2} - 3\dfrac{\partial^2 z}{\partial x \partial y} + \dfrac{\partial^2 z}{\partial y^2} = 0.$ (C.E.I.)

5. If in the equation $\dfrac{\partial u}{\partial t} = h^2\dfrac{\partial^2 u}{\partial x^2} - k(u - u_0)$ the change of variable $u = u_0 + e^{-kt}v$ is introduced, where h, k and u_0 are constants, show that the equation in v becomes $\dfrac{\partial v}{\partial t} = h^2\dfrac{\partial^2 v}{\partial x^2}$. Hence, given boundary conditions $u(0, t) = u_0 = u(1, t)$ and assuming a solution of the form $u(x, t) = u_0 + X(x) \, . \, e^{-(k+m^2)t}$, m real, obtain a general solution of the given equation.

If the initial condition is $u(x, 0) = u_0 + x(1 - x)$, determine the solution completely. (C.E.I.)

6. A function $y(x, t)$ satisfies the differential equation $\dfrac{\partial^2 y}{\partial x^2} = \dfrac{\partial^2 y}{\partial t^2}$ subject to the conditions

(a) $y = 0$, which implies $\dfrac{\partial y}{\partial t} = 0$ when $x = 0$ and $x = 10$ for all $t \geqslant 0$;

(b) $\dfrac{\partial y}{\partial t} = 0$ when $t = 0$ for $0 \leqslant x \leqslant 10$;

(c) when $t = 0$, $\dfrac{\partial y}{\partial x}$ has values given by the table

x	0	1	2	3	4	5	6	7	8	9	10
$\dfrac{\partial y}{\partial x}$	2.00	1.90	1.62	1.18	0.62	0.00	−0.62	−1.18	−1.62	−1.90	−2.00

Using the method of characteristics determine the values of $\dfrac{\partial y}{\partial t}$ when $t = 4$ for $x = 1, 5, 8$ explaining clearly the steps in the solution. (L.U.)

Chapter Nine

Transform Methods

9.1 INTRODUCTION

In this chapter we study two examples of integral transforms. An **integral transform** of a function $f(x)$, which is defined for all x in the interval $a \leqslant x \leqslant b$, is a function

$$F(s) = \int_a^b f(x)\, k(x, s)\, dx \tag{9.1}$$

where $k(x, s)$ is a function known as the **kernel** of the transform.

We can regard the process of obtaining the transform as performing an **operation** on f. Sometimes $F(s)$ is said to be defined on the image - space or s - space. (Note that some authors use p instead of s.)

Much of the application of integral transforms is for the purpose of solving differential equations. The method is depicted in Figure 9.1.

(The method can be applied to a model involving simultaneous differential equations.) The idea is that the equation for $F(s)$ is easier to solve than the original model equation; in the case of the original being an ordinary differential equation, we aim to produce an algebraic equation for $F(s)$, whereas a partial differential equation can be converted to an ordinary differential equation.

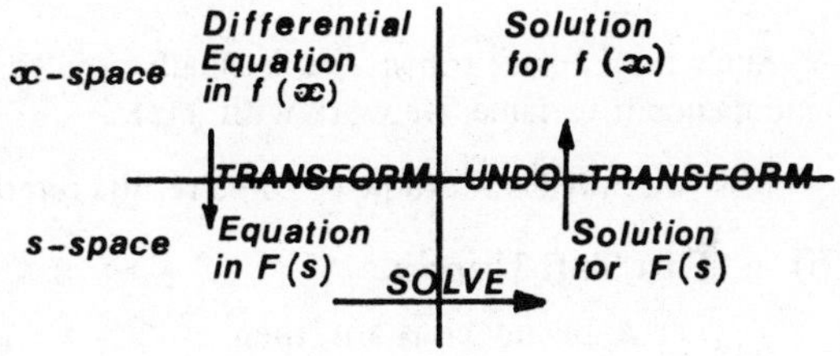

Figure 9.1

For the approach to be useful, we require that the process of taking a transform is a one-to-one operation, i.e. for each function $f(x)$ there is a unique transform $F(s)$ and, conversely, for each transform $F(s)$ there is a unique function $f(x)$ upon inverting the process. We may note in passing that the transform operation is a linear one as a consequence of the definition (9.1).

In Chapter 9 of *Engineering Mathematics* we introduced the Laplace Transform and applied it to some simple ordinary differential equations. We begin this chapter by considering further applications where the forcing term is discontinuous or periodic.

In Chapter 7 of this book, we introduced the Fourier series representation of a function and applied the technique to the solution of partial differential equations in Chapter 8. We had to assume that the function represented was defined on a finite interval and then we were able to extend artificially its domain to make it periodic. In the latter part of this chapter we introduce the idea of a **Finite Fourier Transform** which for the moment can be regarded as a way of transforming the partial differential equation in anticipation of the Fourier series representation. Then we develop the limiting case as the domain of definition becomes infinite (or semi-infinite) and produce the **Infinite Fourier Transforms.**

It is perhaps more an art than a science to decide which transform method to employ on a partial differential equation. Much depends on the boundary conditions and the expected nature of the solution. We shall study the heat conduction problem with differing boundary conditions to help illustrate the contrast between the effectiveness of various transforms.

We can analyse waveforms by Fourier Transforms in an analogous method to Fourier series analysis. We end the chapter by looking at some results in this area.

9.2 BASIC RESULTS ON THE LAPLACE TRANSFORM

Some important results on Laplace transforms were derived in *Engineering Mathematics.* In this section we produce some more. In the following we assume that the Laplace transform of $f(t)$ exists and is defined by

$$L\{f(t)\} \equiv F(s) = \int_0^\infty e^{-st} f(t)\,dt \qquad (9.2)$$

Since the Laplace transform is usually applied to problems where time is the independent variable, we work with $f(t)$.

First we provide a resume of some results referred to above.

(i) **First Shift Theorem**

Let a be any constant, then

$$L\{e^{at} f(t)\} = F(s-a) \qquad (9.3)$$

This represents a shift in s - space.

(ii) **Multiplication by t^n**

If n is any positive integer,

$$L\{t^n f(t)\} = (-1)^n \frac{d^n F(s)}{ds^n} \qquad (9.4)$$

Note the special case where $n = 1$.

(iii) **Division by t**

$$L\left\{\frac{f(t)}{t}\right\} = \int_s^\infty F(s)\,ds \qquad (9.5)$$

(iv) **Transform of Integral of $f(t)$**

$$L\left\{\int_0^t f(u)\,du\right\} = \frac{1}{s}F(s) \qquad (9.6)$$

(v) **Transform of derivatives**

If $f(t)$ is continuous and $\lim_{t\to 0} f(t) = f(0)$ then

$$L\{f'(t)\} = sF(s) - f(0) \qquad (9.7a)$$

If, further, $f'(t)$ is continuous and $\lim_{t\to 0} f'(t) = f'(0)$ then

$$L\{f''(t)\} = s^2F(s) - sf(0) - f'(0) \qquad (9.7b)$$

These results generalise to the result that if $f(t)$ and all its derivatives up to and including order $(n-1)$ are continuous then

$$L\{f^{(n)}(t)\} = s^nF(s) - s^{n-1}f(0) - s^{n-2}f'(0) - \dots - f^{(n-1)}(0) \qquad (9.7c)$$

(vi) **Change of scale**

If a is any positive constant then

$$L\{f(at)\} = \frac{1}{a}F\left[\frac{s}{a}\right] \qquad (9.8)$$

(vii) **Parametric differentiation**

If a is some parameter involved in the basic function then

$$L\left\{\frac{\partial}{\partial a}f(t,a)\right\} = \frac{\partial}{\partial a}\left[F(s,a)\right] \qquad (9.9)$$

The above results allow us to extend the table of standard transforms. A further result we may quote is that $\lim_{s\to\infty} F(s) = 0$.

Initial - and Final - value Theorems

Sometimes we want to know, not the solution of a differential equation but either the initial or, more usually, the final value of the dependent variable. We have two useful results.

Provided the limits exist then

(i) $\lim_{s\to\infty} sF(s) = \lim_{t\to 0} f(t)$ (9.10)

(ii) $\lim_{s\to 0} sF(s) = \lim_{t\to\infty} f(t)$ (9.11)

Example

A mass m falls through a viscous liquid. The equation of motion of the mass in terms of its velocity $v(t)$ is $m\dfrac{dv}{dt} = -kv + mg$ where k is the viscous damping coefficient of the liquid. Taking Laplace transforms, we obtain

$$m[sV(s) - v(0)] = -kV(s) + \frac{mg}{s} \quad \text{so that } V(s) = \frac{mg}{s(ms+k)} + \frac{mv(0)}{(ms+k)}.$$

Then the terminal velocity of the mass is given by

$$\lim_{t\to\infty} v(t) = \lim_{s\to 0} sV(s) = \frac{mg}{k}$$

Problems

1. Find the Laplace transform of each of the following:
$e^{-t}\sin^2 t$; $t^2\cos t$; $\sin kt \cos kt$; $\sin \omega t . te^{-2t}$; $e^{-4t}\cosh 2t$; $e^{2t}(t^2 - 5t + 6)$; $\frac{1}{t}\sinh t$.

2. The concentrations N_1 and N_2 of Iodine and Xenon in a nuclear reactor are given by the differential equations

$$\frac{dN_1}{dt} = a - \lambda_1 N_1, \quad \frac{dN_2}{dt} = \lambda_1 N_1 - \lambda_2 N_2 - bN_2$$

where a, b, λ_1 and λ_2 are constants. Use Laplace transforms to solve these equations subject to the initial conditions $N_1(0) = 0 = N_2(0)$, and obtain the steady-state values A_1 and A_2, say, of N_1 and N_2. If the reactor is then shut down the system becomes

$$\frac{dN_1}{dt} = -\lambda_1 N_1, \quad \frac{dN_2}{dt} = \lambda_1 N_1 - \lambda_2 N_2$$

subject now to the initial conditions $N_1(0) = A_1$, $N_2(0) = A_2$, where $t = 0$ now denotes the time of shut-down.
Solve these equations and show that the Xenon concentration builds up from its equilibrium value A_2 to a maximum concentration before decaying with time. (C.E.I.)

3. The primary and secondary currents, I_1 and I_2 respectively, in a pair of inductively coupled electrical circuits are given by the differential equations

$$L_1\frac{dI_1}{dt} + R_1 I_1 + M\frac{dI_2}{dt} = E, \qquad L_2\frac{dI_2}{dt} + R_2 I_2 + M\frac{dI_1}{dt} = 0$$

in which L_1, R_1, L_2, R_2, M and E are all constant. If initially both I_1 and I_2 are zero use Laplace Transforms to show that the secondary current at time t is given by

$$I_2 = \frac{EM}{L_1 L_2 - M^2}\,\frac{e^{a_1 t} - e^{a_2 t}}{a_2 - a_1}$$

where a_1 and a_2 are the roots of the equation

$$(L_1 L_2 - M^2)a^2 + (L_1 R_2 + L_2 R_1)a + R_1 R_2 = 0$$

Give a rough sketch graph to show the general behaviour of $I_2(t)$, demonstrating in particular that I_2 varies linearly with t for small t. (You may assume that $L_1 L_2 > M$.) (C.E.I.)

9.3 FURTHER FORCING TERMS

In this section we consider some more forcing terms and examine how to build up our stock of techniques.

(a) **Unit Step Function**

Figure 9.2(a) depicts the unit step function $U(t)$ defined by

$$U(t) = \begin{cases} 0 & t < 0 \\ 1 & t \geqslant 0 \end{cases} \qquad (9.12)$$

A generalisation of this is shown in Figure 9.2(b) where the function is denoted by $U(t-a)$ and defined by

$$U(t-a) = \begin{cases} 0 & t < a \\ 1 & t \geqslant a \end{cases} \qquad (9.13)$$

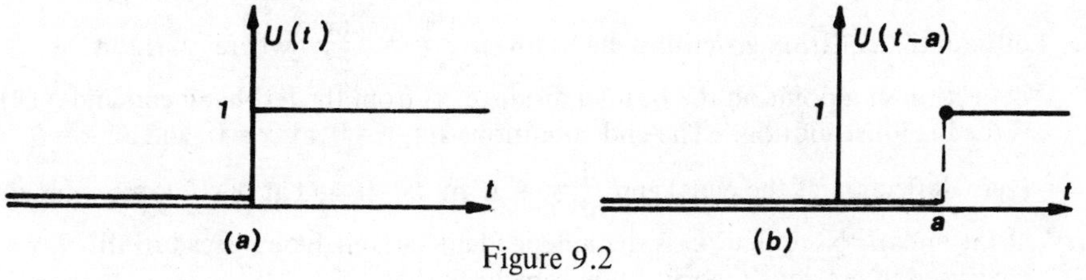

Figure 9.2

The Laplace Transform of $U(t-a)$ is found directly from the definition to be

$$\int_0^\infty e^{-st}\, U(t-a)\, dt = \int_a^\infty e^{-st}\, dt = \frac{1}{s} e^{-as} \qquad (9.14)$$

Obviously the transform of $U(t)$ is $\frac{1}{s}$.

The response of a system to excitation by a unit step function is called its **indicial response** or **step response.**

Other functions can be expressed directly in terms of step functions.

For example, the pulse function in Figure 9.3(a) is defined by $f(t) = k\left\{U(t-a) - U(t-b)\right\}$; it would help you to sketch $-U(t-b)$.

The square curve shown in Figure 9.3(b) is defined by $g(t) = U(t) - 2U(t-a) + 2U(t-2a) - 2U(t-3a) + \ldots..$

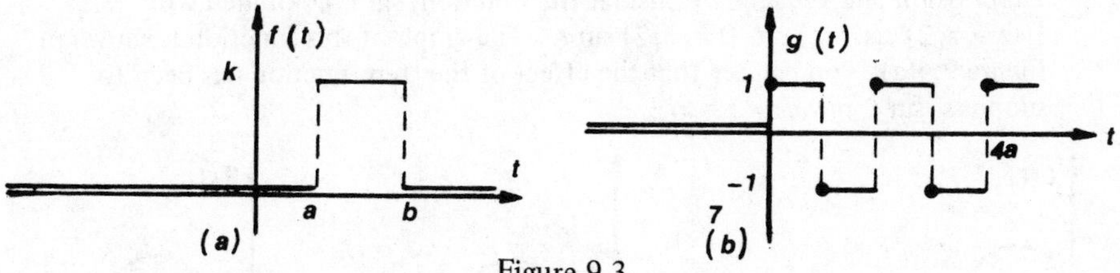

Figure 9.3

As you can verify, the transform of $f(t)$ is $\frac{k}{s}(e^{-as} - e^{-bs})$ and the transform of $g(t)$ is

$$\frac{1}{s} - \frac{2e^{-as}}{s} + \frac{2e^{-2as}}{s} - \frac{2e^{-3as}}{s} + \ldots. = \frac{1}{s} - \frac{2e^{-as}}{s}\left[1 - e^{-as} + e^{-2as} - \ldots..\right]$$

$$= \frac{1}{s} - \frac{2e^{-as}}{s} \cdot \frac{1}{1+e^{-as}} = \frac{1}{s}\frac{(1-e^{-as})}{(1+e^{-as})}$$

Example

A light beam is freely supported at its ends by point supports at the same horizontal level. It carries a uniform load over the middle third of its length. Find the deflected profile of the beam.

If the length of the beam is l and the uniform load is w/unit length then the differential equation governing the beam is $\frac{d^4y}{dx^4} = \frac{f(x)}{EI}$ where y is the deflection of a point on the beam a distance x from the left-hand end and $f(x)$ is the load distribution. The end conditions are $y = 0$ at $x = 0$ and at $x = l$ (zero deflection at the ends) and $\frac{d^2y}{dx^2} = 0$ at $x = 0$ and at $x = l$ (zero moment at the ends). Note that x is the independent variable here instead of the t we have used in previous theory. You can check that

$$f(x) = w[U(x - l/3) - U(x - 2l/3)]$$

and that the transformed equation is

$$EI[s^4Y(s) - s^2y'(0) - y'''(0)] = w\left[\frac{e^{-sl/3} - e^{-2sl/3}}{s}\right]$$

where $y'(0)$ and $y'''(0)$ are unknowns.

It follows then that

$$Y(s) = \frac{w}{EI}\left[\frac{e^{-sl/3} - e^{-2sl/3}}{s^5}\right] + \frac{y'(0)}{s^2} + \frac{y'''(0)}{s^4} \qquad (9.15)$$

We are unable as yet to finish this example; we need to study first the next subsection.

(b) **Combination of step function with other functions**

There are two important effects of combining step functions with other functions: *suppression* and *shifting*. Consider the function $\sin t$ combined with $U(t - \pi/2)$ as $f(t) = U(t - \pi/2)\sin t$. The graph of this function is shown in Figure 9.4(a); you can see that the effect of the step function has been to suppress $\sin t$ prior to $t = \pi/2$.

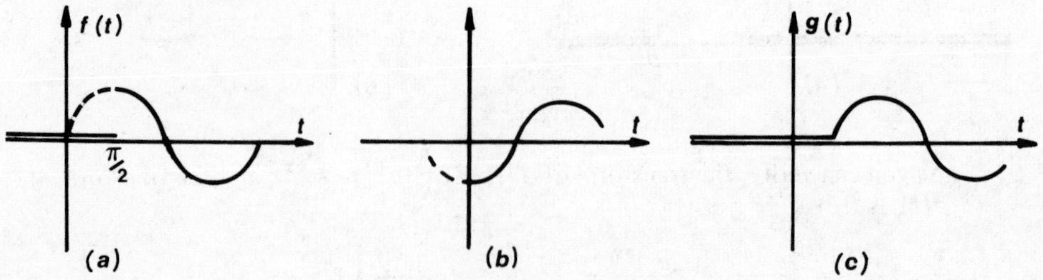

Figure 9.4

However, if we consider $g(t) = U(t - \pi/2)\sin(t - \pi/2)$, a different effect is apparent. The function $\sin(t - \pi/2)$ is shown in Figure 9.4(b): it is merely the sine function shifted along through $\pi/2$. The graph of $g(t)$ is then as displayed in Figure 9.4(c) and you can see that we have in essence shifted the graph of $\sin t$ (for $t > 0$) through $\pi/2$. The transform is easily found by application of the following theorem.

Second shift theorem

If $L\{f(t)\} = F(s)$ then $L\{U(t-a)f(t-a)\} = e^{-as}F(s)$ (9.16)

For the proof of this result we start with the definition

$$L\{U(t-a)f(t-a)\} = \int_0^\infty e^{-st}U(t-a)f(t-a)\,dt = \int_a^\infty e^{-st}f(t-a)\,dt$$

We now change the variable of integration by $u = t - a$ so that

$$L\{U(t-a)f(t-a)\} = \int_0^\infty e^{-s(u+a)}f(u)\,du = e^{-as}\int_0^\infty e^{-su}f(u)\,du$$

$$= e^{-as}F(s)$$

This is the **second shift theorem**, representing a shift in the t - direction.

Example

We are now in a position to complete the previous example. Since we know $L\{x^4\} = 4!/s^5$ it follows that $\dfrac{e^{-sl/3}}{s^5}$ must be the transform of a function similar to $U(x - l/3)(x - l/3)^4$. In fact it is the transform of $\dfrac{1}{4!}U(x - l/3)(x - l/3)^4$ as you can check.

Hence we may invert (9.15) to obtain

$$y(x) = \frac{w}{24EI}\left[(x - l/3)^4U(x - l/3) - (x - 2l/3)^4U(x - 2l/3)\right] + y'(0).x + y'''(0)\frac{x^3}{6}$$

Now $y(l) = 0$ so that

$$0 = \frac{w}{24EI}\left(\left[\frac{2l}{3}\right]^4 - \left[\frac{l}{3}\right]^4\right) + y'(0)l + y'''(0)\frac{l^3}{6} \qquad (9.17a)$$

Furthermore $y''(l) = 0$ so that

$$0 = \frac{w}{24EI}\left(12\left[\frac{2l}{3}\right]^2 - 12\left[\frac{l}{3}\right]^2\right) + y'''(0)l \qquad (9.17b)$$

From (9.17a) and (9.17b) we obtain

$$y'''(0) = \frac{wl}{6EI} \quad \text{and} \quad y'(0) = \frac{13}{648}\frac{wl^3}{EI}$$

We can alternatively express the solution via three formulae:

(i) for $0 \leqslant x \leqslant l/3$, $y(x) = \dfrac{13}{648}\dfrac{wl^3}{EI}x - \dfrac{wlx^3}{36EI}$

(ii) for $l/3 \leqslant x \leqslant 2l/3$, $y(x) = \dfrac{w}{24EI}(x - l/3)^4 + \dfrac{13wl^3}{648EI}x - \dfrac{wlx^3}{36EI}$

(iii) for $2l/3 \leqslant x \leqslant l$, $y(x) = \dfrac{w}{24EI}\left(\left[x - \dfrac{l}{3}\right]^4 - \left[x - \dfrac{2l}{3}\right]^4\right) + \dfrac{13wl^3x}{648EI} - \dfrac{wlx^3}{36EI}$

(c) **Unit Impulse Function**

When a large force acts for a very short time as for example in a shock test, or when a shock voltage passes through an electrical system we can idealise the situation by assuming all the effect to occur at a point in time. We can also idealise in space as with point loads on a structure. In Figure 9.5(a) we see a rectangular pulse of finite duration and of unit strength (measured by the area under the graph). If we keep the strength constant and reduce the duration of the pulse we obtain a function similar to the one depicted in Figure 9.5(b). In the limit as $\epsilon \to 0$ we obtain the idealised **Dirac delta function** $\delta(t - a)$,

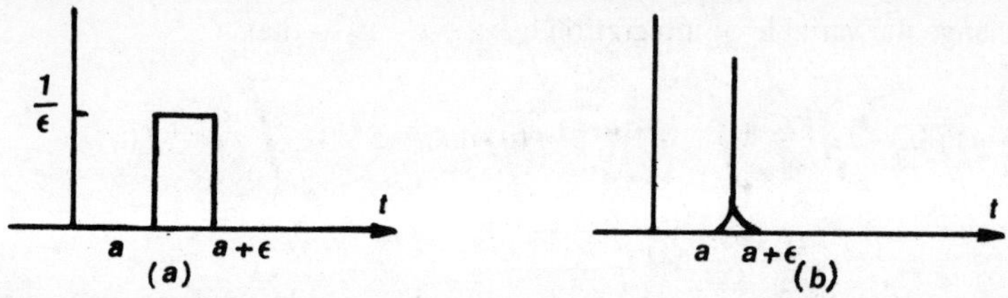

Figure 9.5

defined by

$$\delta(t - a) = \lim_{\epsilon \to 0}\left[\frac{1}{\epsilon}\Big(U(t) - U(t - a)\Big)\right] \qquad (9.18)$$

Note that $\delta(t - a) = 0$ for all $t \neq a$ and that $\displaystyle\int_{-\infty}^{\infty} \delta(t - a)\,dt = 1.$

(Note also the special case when $a = 0$.)

For example, if an impulse voltage of strength E is applied to an electrical circuit at $t = a$ then we represent it as $E\delta(t - a)$. We have the result that if $b < a < c$ and if $f(t)$ is integrable over the interval then

$$\int_b^c f(t)\,\delta(t - a)\,dt = f(a) \qquad (9.19)$$

This result follows from the definition of a definite integral.

The Laplace transform of the delta function is given by

$$L\{\delta(t - a)\} = \int_0^{\infty} e^{-st}\,\delta(t - a)\,dt = e^{-sa} \qquad (9.20a)$$

from result (9.19). Note the special case

$$L\{\delta(t)\} = 1 \qquad (9.20b)$$

An important result is provided by the following equation which you can verify for yourself.

$$L\{f(t)\delta(t-a)\} = \int_0^\infty e^{-st} f(t)\delta(t-a)\,\mathrm{d}t = f(a)\,e^{-sa}$$

Example 1

(i) $L\{t^2\delta(t-3)\} = 3^2 e^{-3s} = 9e^{-3s}$

(ii) $L\{\sin t\,\delta(t-\pi/2)\} = \sin\frac{\pi}{2}\,e^{-\pi s/2} = e^{-\pi s/2}$

We defined the step function $U(t-a)$ to be of value 1 when $t = a$. If we modify the definition slightly so that $U(t-a) = 0$ when $t = a$ then we will not markedly alter any results so far obtained. However, if we denote this new function $U^*(t-a)$ then it can be shown that

$$\delta(t-a) = \frac{\mathrm{d}}{\mathrm{d}t}U^*(t-a) \tag{9.21}$$

Some authors work with $U^*(t-a)$ in preference to $U(t-a)$.

Example 2

The system shown at rest in Figure 9.6 is given an impulsive blow $F\delta(t)$. Describe its response.

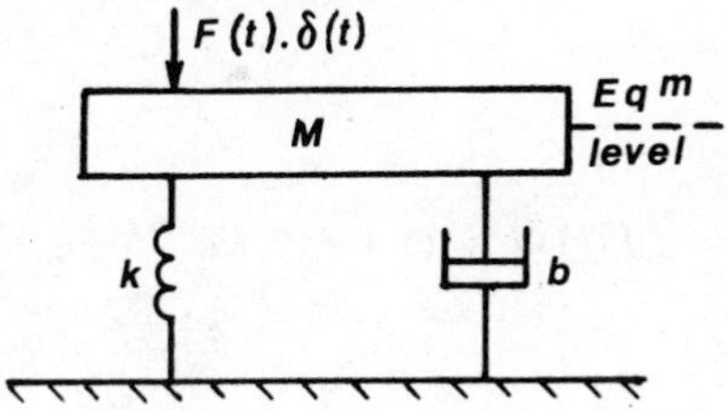

Figure 9.6

The equation of motion for the mass is $M\ddot{x} = F\delta(t) - kx - b\dot{x}$ and the initial conditions are $x = \dot{x} = 0$ at $t = 0$.

The transformed equation is

$$M[s^2 X(s)] = F - kX(s) - b[sX(s)] \Rightarrow X(s) = \frac{F}{(Ms^2 + bs + k)}$$

As you can verify, this is precisely the same result as if we had removed the impulsive blow and given the system an initial velocity $\dot{x}(0) = F/m$, as we might have expected.

(d) **Periodic functions**

Sometimes the input to a system is of a periodic nature or more often the output is periodic. For example, the output of a half-wave rectifier for an input $\sin wt$ is shown in Figure 9.7.

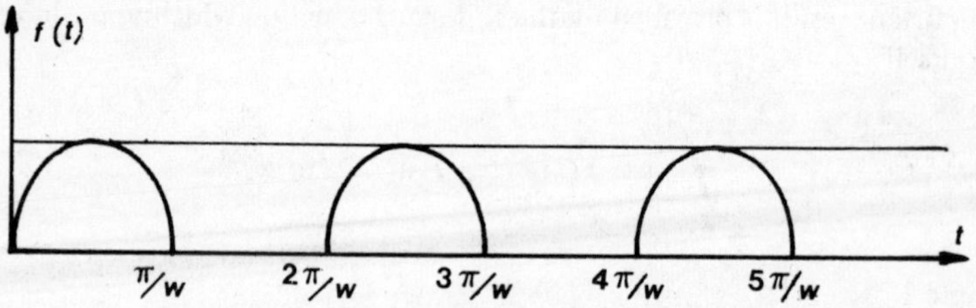

Figure 9.7

The period of $f(t)$ is $2\pi/w$ and is defined in $[0,\ 2\pi/w]$ by

$$f(t) = \begin{cases} \sin wt & 0 \leqslant t \leqslant \pi/w \\ 0 & \pi/w \leqslant t \leqslant 2\pi/w \end{cases}$$

In general, we may consider a function of period T defined on the interval $[0,\ T]$ by $f(t)$. Then $f(t+T) = f(t)$ and

$$F(s) = \int_0^\infty e^{-st} f(t)\,\mathrm{d}t = \int_0^T e^{-st} f(t)\,\mathrm{d}t + \int_T^{2T} e^{-st} f(t)\,\mathrm{d}t + \ldots..$$

$$= \sum_{r=0}^{\infty} \int_{rT}^{(r+1)T} e^{-st}\ f(t)\,\mathrm{d}t$$

$$= \sum_{r=0}^{\infty} \int_0^T e^{-s(t+rT)} f(t+rT)\,\mathrm{d}t \quad \text{(by replacing } t \text{ by } t+rT\text{)}$$

$$= \sum_{r=0}^{\infty} e^{-srT} \int_0^T e^{-st} f(t)\,\mathrm{d}t = (1 + e^{-sT} + e^{-2sT} + \ldots) \int_0^T e^{-st} f(t)\,\mathrm{d}t$$

$$= \frac{1}{(1-e^{-sT})} \int_0^T e^{-st}\ f(t)\,\mathrm{d}t \tag{9.22}$$

Example 1

If we return to the half-rectified sine wave of Figure 9.7 we have

$$F(s) = \frac{1}{(1-e^{-2s\pi/w})} \left[\int_0^{\pi/w} e^{-st} \sin wt\,\mathrm{d}t + \int_{\pi/w}^{2\pi/w} e^{-st}\,.\,0\,\mathrm{d}t \right]$$

Let

$$I = \int_0^{\pi/w} e^{-st} \sin wt\,\mathrm{d}t = \mathcal{I}\left\{ \int_0^{\pi/w} e^{-st}\,.\,e^{iwt}\,\mathrm{d}t \right\}$$

$$= \mathcal{J}\left\{\int_0^{\pi/w} e^{-(s-iw)t}\,dt\right\} = \mathcal{J}\left\{\left[\frac{e^{-(s-iw)t}}{-(s-iw)}\right]_0^{\pi/w}\right\}$$

$$= \mathcal{J}\left\{\frac{1}{(s-iw)}\left[1 - e^{-(s\pi/w - i\pi)}\right]\right\} = \mathcal{J}\left\{\left[\frac{s+iw}{s^2+w^2}\right]\left[1 + e^{-s\pi/w}\right]\right\}$$

$$\therefore F(s) = \frac{w}{s^2+w^2}\,\frac{(1+e^{-s\pi/w})}{(1-e^{-2s\pi/w})} = \frac{w}{s^2+w^2}\cdot\frac{1}{(1-e^{-s\pi/w})} \tag{9.23}$$

(We have used the result that $\dfrac{a+b}{a^2-b^2} = \dfrac{1}{a-b}$, provided $a \neq b$)

Example 2

The function of Figure 9.3(b) is a particular square wave pulse. A domestic example of this general waveform is the pressure wave generated near the valve when a tap is turned off suddenly: the phenomenon of water hammer. Let us now derive its Laplace transform as a periodic function.

From (9.22) this is

$$\frac{1}{(1-e^{-2sa})}\left\{\int_0^a e^{-st}\,.\,1\,dt + \int_a^{2a} e^{-st}(-1)\,dt\right\}$$

$$= \frac{1}{(1-e^{-2sa})}\left\{\left[\frac{e^{-st}}{-s}\right]_0^a + \left[\frac{e^{-st}}{s}\right]_a^{2a}\right\}$$

$$= \frac{1}{(1-e^{-2sa})}\left\{\frac{1}{s}(1-e^{-sa}) + \frac{1}{s}(e^{-2sa} - e^{-sa})\right\}$$

$$= \frac{1}{(1-e^{-2sa})}\cdot\frac{1}{s}(1-e^{-sa})^2 = \frac{1}{s}\cdot\frac{(1-e^{-sa})}{(1+e^{-sa})} \tag{9.24}$$

as we obtained before on page 296.

Problems

1. Prove that

(i) $L\left\{\frac{1}{4}(1-\cos 2t)\right\} = \dfrac{1}{s(s^2+4)}$

(ii) $L\left\{U(t-a)\right\} = \dfrac{e^{-sa}}{s}$, a any positive constant

(iii) $L\left\{U(t-a)f(t-a)\right\} = e^{-sa}F(s)$

Use the above results to solve the equation

$$\frac{d^2y}{dt^2} + 4y = 3U(t-2)$$

given that $y = 1$, $\dfrac{dy}{dt} = 0$ when $t = 0$. (C.E.I.)

2. Evaluate

(i) $f(t) = L^{-1}\left\{\dfrac{1}{p(1-e^{-p\pi})}\right\}$ (ii) $f(t) = L^{-1}\left\{\dfrac{p}{(1+p^2)(1-e^{-p\pi})}\right\}$

Sketch a graph of each function. Solve the differential equation

$$\frac{d^2y}{dt^2} + y = f(t),$$

where $f(t) = n + 1$ for $n\pi < t < (n+1)\pi$ and given that $y = 0 = \dfrac{dy}{dt}$ when $t = 0$. (L.U.)

3. (i) Show that if $L\{f(t)\} = \bar{f}(p)$, and $g(t) \begin{cases} = f(t-a), & t > a, \\ = 0, & t < a, \end{cases}$

then $L\{g(t)\} = e^{-ap}\bar{f}(p)$.

(ii) A function $f(t)$ is periodic, of period $\dfrac{2\pi}{b}$, and is defined as

$$\begin{aligned} f(t) &= 0, & 0 < t < c. \\ &= \sin b(t-c), & c < t < c + \frac{\pi}{b}, \\ &= 0, & c + \frac{\pi}{b} < t < c + \frac{2\pi}{b}. \end{aligned}$$

Obtain the Laplace transform of $f(t)$, (L.U.)

4. If the periodic function $g(t)$ of period T is such that

$g(t) = c \qquad 0 < t < \tfrac{1}{2}T$

$g(t) = 0 \qquad \tfrac{1}{2}T < t < T,$

show that $L\{g(t)\} = \dfrac{c}{p(1+e^{-pT/2})}$.

The variable θ satisfies the differential equation $\dfrac{d^2\theta}{dt^2} + \omega^2\theta = g(t)$ where $g(t)$ is defined above and at time $t = 0$, $\theta = 0$, $d\theta/dt = 0$. Show that at time $t = T\ (>0)$, the value of θ is $\dfrac{c}{\omega^2}(\cos \tfrac{1}{2}\omega T - \cos \omega T)$. (L.U.)

5. (i) If $f(t)$ takes the form

$f(t) = 1 \qquad 0 < t < T$
$f(t) = -1 \qquad T < t < 2T$

and is of period $2T$ show that $\bar{f}(p) = \dfrac{1}{p}\tanh \tfrac{1}{2}pT$.

(ii) If $f(t)$ takes the form

$f(t) = \sin t \qquad 0 < t < \pi$
$f(t) = 0 \qquad \pi < t < 2\pi$

and is of period 2π, show that $\bar{f}(p) = \dfrac{1}{(p^2+1)(1-e^{-\pi p})}$. (L.U.)

6. The input θ_1 and output θ_0 of a servomechanism are related by the equation

$$\frac{d^2\theta_0}{dt^2} + 8\frac{d\theta_0}{dt} + 16\theta_0 = \theta_i,\ t > 0 \text{ and initially } \theta_0 = 0 = \frac{d\theta_0}{dt}.$$

If $\theta_i = f(t)$, where $f(t) = 1 - t, \quad 0 < t < 1$

$f(t) = 0, \quad t > 1$

show that $L\{f(t)\} = \frac{p-1}{p^2} + \frac{1}{p^2} \cdot e^{-p}$

and hence obtain an expression for θ_0 in terms of t. (L.U.)

7. (i) Find $L\{U(t-b)\}$

(ii) Find $L\{(t-b)\,U(t-b)\}$ where $b > 0$

(iii) Show that $L\{(t-b)^3 U(t-b)\} = \frac{6e^{-pb}}{p^4}$

If $\frac{d^2y}{dx^2} = 1 + \frac{(x - \frac{1}{2}a)}{a} U(x - \frac{1}{2}a)$ for $0 < x < a$ and $y = 0$, $dy/dx = 0$ at $x = 0$, find y when $x = a$. (L.U.)

8. A uniform light bar of length l has a load wx/unit length for $0 < x < \frac{1}{2}l$ and $w(l - x)$/unit length for $\frac{1}{2}l < x < l$. The bar is freely hinged at $x = 0$ and clamped horizontally at $x = l$. Find the deflection of the mid-point.

9. An electric circuit comprises an inductance of L henrys, in series with a capacitance C farads. At $t = 0$ an e.m.f. given by

$$E(t) = \begin{cases} E_0 t/T, & 0 < t < T \\ 0 & t > T \end{cases}$$

is applied to the circuit. The initial current and the charge on the capacitor are zero. Find the charge for any time $t > 0$.

9.4 FURTHER RESULTS ON LAPLACE TRANSFORMS

In this subsection we collect together two more aspects of Laplace Transforms which have a wide variety of applications.

Convolution Theorem

We shall examine more closely the concept of convolution in Section 9.8. For the moment we shall regard it as a means of finding the inverse transform when the transform $F(s)$ is a product of the transforms of known functions. For example, if $F(s) = \frac{2}{s^3} \cdot \frac{a}{(s^2 + a^2)}$ then we recognise $\frac{2}{s^3}$ as the transform of t^2 and $a/(s^2 + a^2)$ as the transform of $\sin at$.

We make use of the following **convolution theorem**.

If $H(s)$ is the Laplace transform of $h(t)$ and $G(s)$ is the transform of $g(t)$ then $H(s) \,.\, G(s)$ is the Laplace transform of

$$\int_0^t h(u)\,g(t-u)\,du = \int_0^t h(t-u)\,g(u)\,du \tag{9.25}$$

Either of these integrals is called the **convolution** of $h(t)$ and $g(t)$, and written $h * g$.

With reference to our example, $h(t) = t^2$ and $g(t) = \sin at$ so that $F(s)$ is the transform of $I = \int_0^t u^2 \sin a(t-u)\,du$. We evaluate this integral by parts twice to obtain $I = \frac{t^2}{a} - \frac{2}{a^3}(1 - \cos at)$.

You can verify directly that this is the function whose transform is $\frac{2}{s^3}\,\frac{a}{(s^2+a^2)}$. We now look at a second case.

Example

Find the function whose transform is $s^2/(s^2+4)^2$.

We see that the given expression is the square of $s/(s^2+4)$ which is the transform of $\cos 2t$. Consequently both $h(t)$ and $g(t)$ are $\cos 2t$. By the convolution theorem the function we seek is given by

$$f(t) = \int_0^t \cos 2u \,.\, \cos 2(t-u)\,du = \tfrac{1}{2}\int_0^t [\cos 2t + \cos(4u - 2t)]\,du$$

$$= \tfrac{1}{2}[u\cos 2t + \tfrac{1}{4}\sin(4u - 2t)]_0^t$$

$$= \tfrac{1}{2}(t\cos 2t + \tfrac{1}{4}\sin 2t + \tfrac{1}{4}\sin 2t) = \tfrac{1}{2}t\cos 2t + \tfrac{1}{4}\sin 2t$$

For the moment we may note that if the impulsive response $h(t)$ of a system is known, then the response to any arbitrary excitation $g(t)$ is given by the convolution $h * g$.

Transfer Functions

If a linear system has all initial conditions zero then the **transfer function** of the system is defined to be the ratio of the Laplace transform of the output variable to the Laplace transform of the input variable. For example, the transformed equation for a damped spring-mass system with zero initial conditions may be written

$$Ms^2 Y(s) + ksY(s) + n^2 Y(s) = F(s) \tag{9.26}$$

The transfer function of the system is

$$G(s) = \frac{Y(s)}{F(s)} = \frac{1}{Ms^2 + ks + n^2} \tag{9.27}$$

In general, the transfer function tells us how much of the input variable is transferred to the output variable. Often the transfer function is a ratio of polynomials in s. The values of s which make the denominator zero are called the **poles** of the function. Since the denominator is evolved from the input variable, the poles give information as to the natural motions that the system may possess.
The poles may be complex numbers; in addition we might need to resort to numerical methods for their location.

Inversion of the Transform

A vital stage in the process of solving a differential equation by Laplace transforms is that of inverting the transform to obtain the solution. So far we have not considered this stage other than to refer to standard results or show ways in which the transform can be related to one or more standard transforms. In practice, it might not be always that straightforward. The rigorous definition of the Laplace transform allows the parameter s to be a complex number with positive real part. The appropriate inversion can, in general be achieved via integration in the complex plane; this matter is taken up in Section 12.10.

Problems

1. State the convolution theorem for Laplace transforms. Hence show that the solution of the system of differential equations

$$\dot{x} - 2\dot{y} = e^{-t^2} \qquad \ddot{x} - \ddot{y} + y = 0$$

subject to the conditions $x(0) = \dot{x}(0) = y(0) = \dot{y}(0) = 0$ with the usual notation, may be written as

$$x(t) = A - 2C\cos t - 2S\sin t \qquad y(t) = -C\cos t - S\sin t$$

where

$$A = \int_0^t e^{-u^2}\,du, \quad C = \int_0^t e^{-u^2}\cos u\,du \quad \text{and} \quad S = \int_0^t e^{-u^2}\sin u\,du$$

Note: In the solution of this problem, you will not need to evaluate the Laplace transform of e^{-t^2}. (C.E.I.)

2. The equation representing the forced oscillations of a damped harmonic oscillator is $\dfrac{d^2x}{dt^2} + 2k\dfrac{dx}{dt} + n^2x = f(t)$ subject to $x = x_0$ and $\dfrac{dx}{dt} = v_0$ at $t = 0$. (x_0, v_0, n and k are constants with $k < n$).

Show that $x(t) = x_1(t) + x_2(t)$ where

$$x_1(t) = x_0e^{-kt}\cos\omega t + \frac{1}{\omega}(kx_0 + v_0)e^{-kt}\sin\omega t,$$

$$x_2(t) = \frac{1}{\omega}\int_0^t f(y)\,e^{-k(t-y)}\sin\omega(t-y)\,dy \quad \text{and} \quad \omega^2 = n^2 - k^2. \qquad \text{(C.E.I.)}$$

3. Using the Convolution theorem, evaluate

(a) $L^{-1}\left\{\dfrac{1}{(s+1)(s^2+1)}\right\}$ (b) $L^{-1}\left\{\dfrac{1}{(s-2)(s+1)^2}\right\}$ (C.E.I.)

9.5 FINITE FOURIER TRANSFORMS

Let us suppose that a function $f(x)$ is defined over the interval $0 \leqslant x \leqslant l$ and that it satisfies the Dirichlet conditions (see page 209). We can convert it into an odd function by defining $f(-x) = -f(x)$ and then we can extend its domain by periodicity. Then we can represent $f(x)$ by a half-range sine series as follows:

$$f(x) = \sum_{s=1}^{\infty} b_s \sin\left[\frac{s\pi x}{l}\right] \tag{9.28a}$$

where the coefficients are given by

$$b_s = \int_0^l f(x)\left(\frac{2}{l}\sin\left[\frac{s\pi x}{l}\right]\right) dx \tag{9.28b}$$

Expressed in this form you can see that b_s is the effect of applying an integral transform to $f(x)$. The kernel of the transform is $\frac{2}{l}\sin\left[\frac{s\pi x}{l}\right]$. Formula (9.28b) defines this **finite Fourier sine transform** and, in a sense, formula (9.28a) acts as an **inversion formula.**

Equally well, we can convert the original function $f(x)$ into an even function and then define the **finite Fourier cosine transform** and its inversion by the following formulae

$$f(x) = \tfrac{1}{2}a_0 + \sum_{s=1}^{\infty} a_s \cos\left[\frac{s\pi x}{l}\right] \tag{9.29a}$$

where the coefficients are given by

$$a_s = \int_0^l f(x) \cdot \left(\frac{2}{l}\cos\left[\frac{s\pi x}{l}\right]\right) dx \tag{9.29b}$$

If $f(x)$ is neither even nor odd and defined on the interval $[-l, l]$ then we may use the complex form of Fourier series representation and define the **finite Fourier transform** by

$$f(x) = \sum_{s=-\infty}^{\infty} c_s \exp(is\pi x/l) \tag{9.30a}$$

where

$$c_s = \int_{-l}^{l} f(x)\left(\frac{1}{2l}\exp(-is\pi x/l)\right) dx \tag{9.30b}$$

We shall shortly give two examples of how to use Fourier transforms to solve partial differential equations which we would have solved by separation of variables. First we need the transform of the second derivative of a function, since this is the derivative occurring most frequently in such equations. Note that we use x as the variable in our definition of the transforms.

Let us denote the finite Fourier sine transform of $f(x)$ by $\bar{f}_s(s)$ or $\bar{f}_s$. Then the finite Fourier sine transform of $\partial^2 f/\partial x^2$ is

$$\int_0^l \frac{\partial^2 f}{\partial x^2}\frac{2}{l}\sin\left[\frac{s\pi x}{l}\right] dx = \int_0^l \frac{2}{l}\sin\left[\frac{s\pi x}{l}\right] \cdot \frac{\partial^2 f}{\partial x^2}\, dx$$

Integrating by parts, we obtain

$$\left[\frac{2}{l}\sin\left[\frac{s\pi x}{l}\right]\cdot\frac{\partial f}{\partial x}\right]_0^l - \int_0^l \frac{2s\pi}{l^2}\cos\left[\frac{s\pi x}{l}\right]\frac{\partial f}{\partial x}\,dx$$

$$= 0 - \frac{2s\pi}{l^2}\int_0^l \cos\left[\frac{s\pi x}{l}\right]\frac{\partial f}{\partial x}\,dx$$

$$= \frac{-2s\pi}{l^2}\left\{\left[\cos\left[\frac{s\pi x}{l}\right]f(x)\right]_0^l + \frac{s\pi}{l}\int_0^l \sin\left[\frac{s\pi x}{l}\right]f(x)\,dx\right\}$$

$$= \frac{-2s\pi}{l^2}\left[(-1)^s f(l) - f(0)\right] - \frac{s^2\pi^2}{l^2}\bar{f}_s \tag{9.31}$$

If we know $f(l)$ and $f(0)$ then we can specify this transform completely.

Similarly, if the finite Fourier cosine transform of $f(x)$ be $\bar{f}_c$ then the same transform of $\partial^2 f/\partial x^2$ is

$$\frac{2}{l}\left[(-1)^s\left(\frac{\partial f}{\partial x}\right)_{x=l} - \left(\frac{\partial f}{\partial x}\right)_{x=0}\right] - \frac{s^2\pi^2}{l^2}\bar{f}_c \tag{9.32}$$

If the value of $\dfrac{\partial f}{\partial x}$ is specified at $x = 0$ and $x = l$ then this transform is determined completely.

Before we apply the transforms to differential equations we should note that the factor $(2/l)$ is a slight nuisance and we could re-define our transforms to put it in the inversion formulae. Also, the boundary conditions on the problem will select the transform to be used.

Example 1

The equation of heat conduction for a bar whose sides are insulated and which is of length l is

$$\frac{\partial\theta}{\partial t} = K\frac{\partial^2\theta}{\partial x^2} \tag{9.33}$$

If the bar is insulated across the end faces $x = 0$ and $x = l$ we have the boundary conditions

$$\partial\theta/\partial x = 0 \quad \text{at} \quad x = 0 \quad \text{and} \quad x = l$$

Finally, let

$$\theta = 100x(l - x) \quad \text{at} \quad t = 0 \quad \text{for} \quad 0 < x < l \tag{9.34}$$

Since $\partial\theta/\partial x$ is specified on the boundaries, a cosine transform is required to eliminate the x - dependence. We multiply both sides of (9.33) by

$\cos\left[\dfrac{s\pi x}{l}\right]$ and integrate over the interval $[0, l]$ to obtain

$$\int_0^l \cos\left[\frac{s\pi x}{l}\right] \frac{\partial\theta}{\partial t}\,\mathrm{d}x = K \int_0^l \cos\left[\frac{s\pi x}{l}\right] \frac{\partial^2\theta}{\partial x^2}\,\mathrm{d}x \tag{9.35}$$

If we let

$$\bar{\theta}_c = \int_0^l \theta(x,\ t) \cos\left[\frac{s\pi x}{l}\right] \mathrm{d}x$$

then the left-hand side of (9.35) is simply $\dfrac{\partial\bar{\theta}_c}{\partial t}$. The right-hand-side becomes

$$K\left[\cos\left[\frac{s\pi x}{l}\right] \frac{\partial\theta}{\partial x}\right]_0^l + \frac{Ks\pi}{l} \int_0^l \sin\left[\frac{s\pi x}{l}\right] \cdot \frac{\partial\theta}{\partial x}\,\mathrm{d}x$$

$$= K[0-0] + \frac{Ks\pi}{l} \int_0^l \sin\left[\frac{s\pi x}{l}\right] \frac{\partial\theta}{\partial x}\,\mathrm{d}x$$

$$= \frac{Ks\pi}{l}\left[\sin\left[\frac{s\pi x}{l}\right] \cdot \theta(x,\ t)\right]_0^l - \frac{K\ s^2\pi^2}{l^2} \int_0^l \cos\left[\frac{s\pi x}{l}\right] \theta(x,\ t)\,\mathrm{d}x$$

$$= -\frac{K\ s^2\pi^2}{l^2}\,\bar{\theta}_c$$

(Note the comparison with (9.32).)

The partial differential equation (9.33) has been reduced to an ordinary differential equation

$$\frac{\partial\bar{\theta}_c}{\partial t} = -\frac{Ks^2\pi^2}{l^2}\,\bar{\theta}_c$$

although we retain the partial derivative sign since, strictly, $\bar{\theta}_c = \bar{\theta}_c(s,\ t)$. This has the general solution

$$\bar{\theta}_c = A e^{-Ks^2\pi^2 t/l^2}$$

where the constant A is the value of $\bar{\theta}_c$ at $t = 0$ i.e.

$$\int_0^l \theta(x,\ 0) \cos\left[\frac{s\pi x}{l}\right] \mathrm{d}x = \int_0^l 100x(l-x) \cos\left[\frac{s\pi x}{l}\right] \mathrm{d}x$$

i.e.

$$A = \frac{-100l^3}{s^2\pi^2}(1+\cos s\pi) = \begin{cases} 0 & \text{if } s \text{ is odd} \\ \dfrac{-200l^3}{s^2\pi^2} & \text{if } s \text{ is even} \end{cases}$$

Now the inversion formula would give us

$$\theta(x, t) = \frac{1}{2} \cdot \frac{2}{l} a_0 + \frac{2}{l} \sum_{s=1}^{\infty} a_s \cos\left[\frac{s\pi x}{l}\right]$$

where

$$a_s = \int_0^l \theta(x, t) \cos\left[\frac{s\pi x}{l}\right] dx = \bar{\theta}_c(s, t)$$

Note that the factor $\frac{2}{l}$ now appears in the inversion formula and that

$$a_0 = \int_0^l \theta(x, 0)\, dx = 100 \int_0^l x(l-x)\, dx = 100l^3/6 = \bar{\theta}_c(0, t)$$

Therefore

$$\theta(x, t) = \frac{1}{l} \bar{\theta}_c(0, t) + \frac{2}{l} \sum_{s=1}^{\infty} \bar{\theta}_c(s, t) \cos\left[\frac{s\pi x}{l}\right]$$

$$= 100l^2 \left\{ \frac{1}{6} - \sum_{s=2,4,6\ldots}^{\infty} \frac{4}{s^2\pi^2} \cos\left[\frac{s\pi x}{l}\right] e^{-Ks^2\pi^2 t/l^2} \right\} \tag{9.36}$$

Example 2

Consider the one-dimensional wave equation for a vibrating string of length l, fixed at its ends and released from rest with an initial profile $y = f(x)$.
The governing equation is

$$\frac{1}{c^2} \frac{\partial^2 y}{\partial t^2} = \frac{\partial^2 y}{\partial x^2} \tag{9.37}$$

with boundary conditions

$$y(0, t) = y(l, t) = 0$$

and initial conditions

$$y = f(x) \quad \text{and} \quad \frac{\partial y}{\partial t} = 0 \quad \text{at} \quad t = 0 \tag{9.38}$$

We choose the sine transform to eliminate x since y itself is specified at the boundaries.

We multiply (9.37) by $\sin\left[\frac{s\pi x}{l}\right]$ and integrate over the interval $[0, l]$ to obtain

$$\frac{1}{c^2} \int_0^l \sin\left[\frac{s\pi x}{l}\right] \frac{\partial^2 y}{\partial t^2} dx = \int_0^l \sin\left[\frac{s\pi x}{l}\right] \frac{\partial^2 y}{\partial x^2} dx \tag{9.39}$$

Let $\bar{y}_s(s, t) = \int_0^l \sin\left[\frac{s\pi x}{l}\right] y(x, t)\, dx$

Then the left-hand side of (9.39) is $\frac{1}{c^2}\frac{\partial^2 \bar{y}_s}{\partial t^2}$

The right-hand side becomes, after integrating by parts, $-\frac{s^2\pi^2}{l^2}\bar{y}_s$

Hence we have

$$\frac{1}{c^2}\frac{\partial^2 \bar{y}_s}{\partial t^2} = -\frac{s^2\pi^2}{l^2}\bar{y}_s$$

and this has the general solution

$$\bar{y}_s(s,\ t) = A_s \cos\left[\frac{s\pi ct}{l}\right] + B_s \sin\left[\frac{s\pi ct}{l}\right]$$

where A_s and B_s can be determined from conditions (9.38).

Hence

$$\bar{y}_s(x,\ 0) = \int_0^l f(x)\sin\left[\frac{s\pi x}{l}\right] dx = A_s \quad \text{and} \quad \frac{\partial \bar{y}_s}{\partial t} = 0 = B_s$$

Therefore,

$$\bar{y}_s(s,\ t) = \cos\left[\frac{s\pi ct}{l}\right]\int_0^l f(x)\sin\left[\frac{s\pi x}{l}\right] dx$$

Inverting the transform we obtain

$$y(x,\ t) = \frac{2}{l}\sum_{s=1}^{\infty}\left\{\sin\left[\frac{s\pi x}{l}\right]\cos\left[\frac{s\pi ct}{l}\right]\right\}\int_0^l f(u)\sin\left[\frac{s\pi u}{l}\right] du \qquad (9.40)$$

where we have changed the variable of integration from x to u.

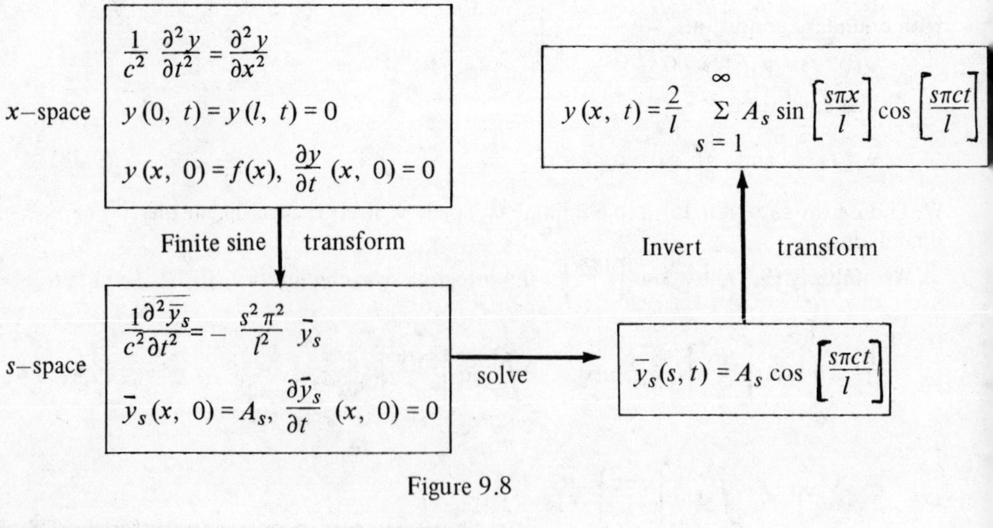

Figure 9.8

Notes on the transforms

We cannot exclude odd order derivatives by the Fourier transforms and therefore in the heat conduction problem we would naturally eliminate x. In the case of the wave equation the boundary conditions on x were the more suited to transformations. The Fourier transformation is less flexible than the Laplace transform but it does produce a quicker result when it is applicable.

For an equation involving three independent variables we can apply transforms repeatedly, but we shall not take the matter further here.

It may be of help to summarise the last problem in diagram form as shown in Figure 9.8.

Problems

1. The temperature distribution $\theta(x, t)$ in a uniform bar of length l is given by $k\dfrac{\partial^2\theta}{\partial x^2} = \dfrac{\partial\theta}{\partial t}$.

   The end $x = 0$ is given a temperature $\theta = f(t)$ and the other end $x = l$ is kept at $\theta = 0$. If the initial temperature is zero throughout the bar, use a finite Fourier sine transform to show that θ at any point x at any time t is

$$\theta(x, t) = \frac{2\pi k}{l^2} \sum_{n=1}^{\infty} n \sin\left[\frac{n\pi x}{l}\right] \int_0^t f(t-u)\, e^{-kn^2\pi^2 u/l^2}\, du$$

2. The concentration $V(x, t)$ at position x and time t of a liquid diffusing along a channel $0 < x < \alpha$ is given by $\dfrac{\partial V}{\partial t} = k\dfrac{\partial^2 V}{\partial x^2}$ where k is the constant diffusion coefficient.

   If $\partial V/\partial x = 0$ at $x = 0$ and $x = \alpha$ and also $V = V_0 x/\alpha$ initially where V_0 is a constant, use a finite Fourier cosine transform to show that

$$V(x, t) = V_0/2 - (4V_0/\pi^2) \sum_n (1/n^2) \cos(n\pi x/\alpha) \exp(-kn^2\pi^2 t/\alpha^2)$$

   where the summation extends over $n = 1, 3, 5, \ldots\ldots$

3. A uniform metal sphere of radius a is heated to a constant temperature θ_0 throughout and is then immersed in a coolant liquid so that its surface is thereafter maintained at zero temperature. Assume that the temperature distribution $\theta(r, t)$ at radial distance r after a time t is given by the heat conduction equation $\dfrac{\partial^2\theta}{\partial r^2} + \dfrac{2}{r}\dfrac{\partial\theta}{\partial r} = \dfrac{1}{k}\dfrac{\partial\theta}{\partial t}$.

   Substitute $\theta(r, t) = r^{-1} f(r, t)$, then use a finite Fourier transform to solve the resulting equation for $f(r, t)$ and show that

$$\theta(r, t) = \frac{2\theta_0 a}{r} \sum_{n=1}^{\infty} \frac{(-1)^{n+1}}{n} \sin\left[\frac{n\pi r}{a}\right] e^{-kn^2\pi^2 t/a^2}$$

9.6 INFINITE FOURIER TRANSFORMS

In some physical problems we consider one or more space dimensions to be infinite or semi-infinite. We may have a bar which is very long in comparison to its width and depth or, in order to neglect edge effects we may wish to choose as a model a plate which is infinite in extent. In such cases we want to consider the domain of the transform to extend to infinity in both directions. Furthermore, the finite transform, or equivalently the Fourier series required a waveform which was effectively periodic or which was defined on a finite interval. If, for example, we wish to study transient waveforms then the finite transforms will not suffice. In this section we examine infinite transforms as applied to partial differential equations. In Section 9.8 we shall study the interpretation of transforms of waveforms.

Fourier's Integral Formula and the Fourier Transforms

We shall merely quote this result, which can be regarded as the starting point for the transforms. It states that for a function $f(x)$ which satisfies Dirichlet's conditions (page 209) we may write

$$f(x) = \frac{1}{\pi} \int_0^\infty \left\{ \int_{-\infty}^{\infty} f(u) \cos\left[s(x-u)\right] du \right\} ds \tag{9.41}$$

Here u is a dummy variable of integration.

The proof follows by considering the limiting case of a Fourier series representation. An alternative formulation is

$$f(x) = \frac{1}{2\pi} \int_{-\infty}^{\infty} e^{isx} \left\{ \int_{-\infty}^{\infty} f(u)\, e^{-isu}\, du \right\} ds$$

$$= \frac{1}{\sqrt{2\pi}} \int_{-\infty}^{\infty} e^{isx} \left\{ \frac{1}{\sqrt{2\pi}} \int_{-\infty}^{\infty} f(u)\, e^{-isu}\, du \right\} ds \tag{9.42}$$

Equate the real part of the right-hand side to $f(x)$.

This last symmetrical form suggests that we define the **ordinary Fourier Transform** by

$$F(s) = \frac{1}{\sqrt{2\pi}} \int_{-\infty}^{\infty} e^{-isu}\, f(u)\, du \tag{9.43a}$$

with inversion formula

$$f(x) = \frac{1}{\sqrt{2\pi}} \int_{-\infty}^{\infty} e^{isx}\, F(s)\, ds \tag{9.43b}$$

If $f(x)$ is defined on the interval $0 \leqslant x$ we may extend the domain by creating it an even function. Then, returning to (9.41) we find that

$$f(x) = \frac{1}{\pi} \int_0^\infty \left\{ \int_{-\infty}^{0} f(u) \cos\left[s(x-u)\right] du + \int_0^\infty f(u) \cos\left[s(x-u)\right] du \right\} ds$$

$$= \frac{1}{\pi} \int_0^\infty \left\{ \int_0^\infty f(u) \cos\left[s(x+u)\right] du + \int_0^\infty f(u) \cos\left[s(x-u)\right] du \right\} ds^{\dagger}$$

$$= \frac{1}{\pi} \int_0^\infty \left\{ \int_0^\infty 2f(u) \cos sx \cos su \, du \right\} ds$$

Hence we may define the **Fourier cosine transform** by

$$F_c(s) = \sqrt{\frac{2}{\pi}} \int_0^\infty f(u) \cos su \, du \tag{9.44a}$$

with the inversion formula

$$f(x) = \sqrt{\frac{2}{\pi}} \int_0^\infty F_c(s) \cos sx \, ds \tag{9.44b}$$

In a similar way we define the **Fourier sine transform** by

$$F_s(s) = \sqrt{\frac{2}{\pi}} \int_0^\infty f(u) \sin su \, du \tag{9.45a}$$

with inversion formula

$$f(x) = \sqrt{\frac{2}{\pi}} \int_0^\infty F_s(s) \sin sx \, ds \tag{9.45b}$$

Example 1

(i) We find the ordinary Fourier transform of the rectangular pulse shown in Figure 9.9(a) whose definition is

$$f(t) = \begin{cases} 1/2T & -T < t < T \\ 0 & \text{otherwise} \end{cases}$$

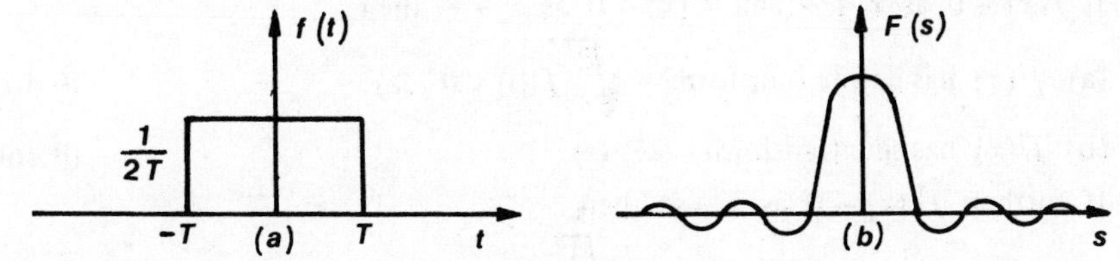

Figure 9.9

† Put $u = -u$ and $f(-u) = f(u)$.

The transform is, from (9.43a)

$$F(s) = \frac{1}{\sqrt{2\pi}} \int_{-T}^{T} \frac{1}{2T} e^{-isu}\, du = \frac{1}{\sqrt{2\pi}} \cdot \frac{1}{2T} \left[\frac{e^{-isu}}{-is} \right]_{-T}^{T}$$

$$= \frac{1}{\sqrt{2\pi}} \cdot \frac{1}{Ts} \frac{1}{2i} (e^{isT} - e^{-isT}) = \frac{1}{\sqrt{2\pi}} \cdot \frac{\sin sT}{sT}$$

Note that as $s \to 0$, $F(s) \to 1/\sqrt{2\pi}$; the graph of $F(s)$ is presented in Figure 9.9(b). If $f(t)$ represents a rectangular pulse then $F(s)$ represents a frequency spectrum. Conversely, if $F(s)$ represents a frequency distribution then $f(t)$ represents the signal which produced it.

In some applications, ω is used as the variable of the transform instead of s.

(ii) Since $f(t)$ is symmetrical we could have found its Fourier cosine transform. It is

$$\sqrt{\frac{2}{\pi}} \int_{0}^{T} \frac{1}{2T} \cos su\, du = \sqrt{\frac{2}{\pi}} \frac{1}{2T} \left[\frac{\sin su}{s} \right]_{0}^{T} = \frac{1}{\sqrt{2\pi}\, T} \frac{\sin sT}{s}, \text{ as before.}$$

(iii) To complete the picture, we shall now find the Fourier sine transform of this function. It is

$$\sqrt{\frac{2}{\pi}} \int_{0}^{T} \frac{1}{2T} \sin su\, du = \frac{1}{\sqrt{2\pi}} \cdot \frac{1}{T} \left[\frac{\cos su}{-s} \right]_{0}^{T} = \frac{1}{\sqrt{2\pi}} \cdot \frac{1}{sT} (1 - \cos sT)$$

Example 2 Unit impulse

This can be regarded as the limit of the rectangular pulse as $T \to 0$. We can see that the spectrum is given by $F(s) = 1$.

Miscellaneous Results

(It is assumed that $f(x)$ has transforms $F(s)$, $F_s(s)$, $F_c(s)$.)

(i) **Transforms of derivatives**

If $f(x) \to 0$ as $x \to \infty$ and $f'(x) \to 0$ as $x \to \infty$ then

(a) $f'(x)$ has cosine transform $-\sqrt{\frac{2}{\pi}} f(0) + sF_s(s)$ (9.46a)

(b) $f'(x)$ has sine transform $-sF_c(s)$ (9.46b)

If, further, $f''(x) \to 0$ as $x \to \infty$ then

(c) $f''(x)$ has cosine transform $-\sqrt{\frac{2}{\pi}} f'(0) - s^2 F_c(s)$ (9.46c)

(d) $f''(x)$ has sine transform $\sqrt{\frac{2}{\pi}} sf(0) - s^2 F_s(s)$ (9.46d)

If $f(x) \to 0$ as $x \to \pm\infty$ and $f'(x) \to 0$ as $x \to \pm\infty$ then

(e) $f'(x)$ has Fourier transform $-isF(s)$ (9.46e)

If, further, $f''(x) \to 0$ as $x \to \pm\infty$ then

(g) $f''(x)$ has Fourier transform $(is)^2 F(s)$ (9.46f)

(ii) **Scaling theorem**

For $\lambda > 0$, $f(\lambda x)$ has a Fourier transform equal to $\frac{1}{\lambda} F(s/\lambda)$ with similar results holding for the other transforms.

(iii) **Shift theorem**

$e^{-\lambda x} f(x)$ has Fourier transform $F(s - \lambda)$ and $f(x - \lambda)$ has transform $e^{i\lambda s} F(s)$

(iv) **Derivatives of transforms**

Assuming that differentiation is possible, $\frac{d}{ds} F(s)$ is the Fourier transform of $isf(x)$, $\frac{dF_c(s)}{ds}$ is the cosine transform of $xf(x)$ and $\frac{dF_s(s)}{ds}$ is the sine transform of $-xf(x)$.

(v) **Convolutions**

The convolution of the functions g and h is defined as

$$f(x) = \int_{-\infty}^{\infty} g(u)\, h(x-u)\, du = \int_{-\infty}^{\infty} g(x-u)\, h(u)\, du$$

The Fourier transform of the convolution is the product of the transforms of $g(x)$ and $h(x)$.

The use of convolutions in solving problems with time-dependent boundary conditions allows reduction to problems with constant boundary conditions. We discuss convolutions more fully in Section 9.8.

Comparison of Fourier and Laplace transforms

Let us first investigate the direct transforms.

The Fourier transform, being defined by an integral with limits from $t = -\infty$ to $t = +\infty$, has the advantage of being applicable to functions involving negative values of the variable t. This may not seem to be a significant advantage if t represents time; very seldom do we use time functions known or existing at $t \to -\infty$. However, if t does not represent time but some other variable, distance for instance, it is quite conceivable for this variable to assume negative values and for a function to be non-zero for such negative values of its variable. Obviously the direct Laplace transform, where the integration is carried from $t = 0$ to $t = \infty$, of such a function does not exist. (Actually, an implicit assumption is made that $f(t) = 0$ for $t < 0$.) A fundamental difference therefore between the Laplace and Fourier transforms lies in the fact that the former is applicable only to functions whose variable is always positive and the latter to functions whose variable may include negative values.

The Laplace transform, however, compensates for this diasdvantage by being applicable to a greater variety of functions (all functions of exponential order). Consider the function $f(t) = e^{at}$ where $a > 0$. Its Fourier transform does not exist since $f(t) \to \infty$ as $t \to \infty$; the function to be integrated is $e^{at}\, e^{-i\omega t}$ $= e^{at}$ (cos ωt – i sin ωt) and the area under the curve e^{at} cos ωt or e^{at} sin ωt is not finite as $t \to \infty$. However, in the case of the Laplace transform ($s = \sigma + i\omega$), the function to be integrated is $e^{at}\left[e^{-st} = e^{-\sigma t}(\cos \omega t - i \sin \omega t)\right]$ and the area under the curve $e^{(a-\sigma)t}$ cos ωt or $e^{(a-\sigma)t}$ sin ωt is finite if $\sigma > a$. Hence the function e^{at}, $(a > 0)$, has a Laplace transform but not a Fourier transform. We see that the Laplace transform has a convergence factor $e^{-\sigma t}$, which gives the advantage over the Fourier transform. It is evident why the Laplace transform cannot involve negative values of t: the factor e^{-at} contributes to convergence only for $t > 0$ and causes divergence for $t < 0$. We may conclude then that the Laplace transform, because of its better convergence properties, is applicable to a greater variety of functions provided these functions are zero for negative values of the variable.

Let us now examine the inverse transforms.

The inverse Fourier transform has the advantage of involving an integration with a real variable ω, whereas the inverse Laplace transform necessitates a contour integration in the complex s – plane. From this point of view it may sometimes be easier to find the inverse Fourier transform than the inverse Laplace transform.

Where $f(t)$ is a time function, the Fourier transform $F(i\omega)$ represents the frequency spectrum of $f(t)$.

Because of the greater variety of functions to which the Laplace transform is applicable, it is used more extensively in engineering than the Fourier transform. Its primary purpose is to facilitate the solution and to give a better insight into differential equations involved in engineering. In circuit theory it gives rise to the concept of operational impedances and admittances.

Since $L[\delta(t)] = 1$, the inverse Laplace transform of operational impedances and admittances as well as of transfer functions has the physical meaning of being the time response to a unit impulse excitation. If $G(s)$ is the transfer function of a linear system (ratio of the Laplace transforms of, let us say, the output to input voltages), then $L^{-1}[G(s)]$ is the output for a unit impulse input.

Problems

1. Obtain

   (i) the Fourier transform of $f(x) = \begin{cases} \dfrac{1}{2a} & |x| \leqslant b, \\ 0 & |x| > b. \end{cases}$

   (ii) The Fourier sine transform of $g(x) = e^{-\pi x}$, $x \geqslant 0$.

   State the reciprocal relationship existingbetween the function $g(x)$ and its Fourier sine transform and use this relationship to evaluate

$$\int_0^\infty \frac{x \sin mx}{x^2 + \pi^2}\, dx$$

(L.U.)

2. (i) Obtain the Fourier sine transform $F_s(p)$ of the function $g(x)$ given by $g(x) = xe^{-|x|}$.

(ii) An even function $f(x)$ is given by $f(x) = \begin{cases} x^2 & \text{for } 0 \leqslant x \leqslant 2 \\ 0 & \text{for } x > 2. \end{cases}$

Find the Fourier cosine transform of $f(x)$ and assuming the reciprocal relation between a function and its Fourier cosine transform show that

$$\int_0^\infty [(2p^2 - 1)\sin 2p + 2p\cos 2p] \frac{\cos p}{p^3} dp = \frac{\pi}{4}.$$

(L.U.)

3. Evaluate $F_c(s)$ when

(i) $f(x) = \begin{cases} \cos x, & 0 < x < \pi \\ 0, & x > \pi \end{cases}$

(ii) $f(x) = e^{-x}$

Assuming the reciprocal relation between a function and its Fourier cosine transform, use the second result to thow that

$$\int_0^\infty \frac{\cos kx}{1 + x^2} dx = \frac{\pi}{2} e^{-k}.$$

(L.U.)

4. Find the cosine transform of $g(x)$ if $g(x) = 1$ for $0 < x < a$, $g(x) = 0$ for $x > a$.

Find the function whose sine transform is $\dfrac{\sin pa - pa\cos pa}{p^2}$ (L.U.)

5. Verify that the sine transform of $f''(x)$ is $pf(0) - p^2F(p)$.

State the inversion formula for $f(x)$ in terms of $F(p)$. (L.U.)

9.7 SOLUTION OF HEAT CONDUCTION EQUATION

We now use various Fourier transforms to solve the heat conduction equation for a semi–infinite bar $x \geqslant 0$ with differing boundary conditions. Basically if the range of definition of the problem is $-\infty < x < \infty$ then the ordinary transform is called for, whereas if the range is $0 < x < \infty$ then either sine or cosine transforms are used to eliminate the **even order derivative**.

The heat conduction equation is

$$\frac{\partial \theta}{\partial t} = K \frac{\partial^2 \theta}{\partial x^2} \tag{9.33}$$

Case I

The bar is lagged at $x = 0$ so that $\dfrac{\partial \theta}{\partial x} = 0$ there, and $\theta = \theta_0(x)$ at $t = 0$.

From result (9.46c) we see that knowledge of $\partial\theta/\partial x$ at $x = 0$ suggests the cosine transform to eliminate x from the problem. We therefore apply the transform to both sides of (9.33) to obtain

$$\sqrt{\frac{2}{\pi}} \int_0^\infty \frac{\partial \theta}{\partial t} \cos sx \, dx = K\sqrt{\frac{2}{\pi}} \int_0^\infty \frac{\partial^2 \theta}{\partial x^2} \cos sx \, dx$$

i.e.

$$\frac{\partial \Theta_c}{\partial t} = K \left\{ -\sqrt{\frac{2}{\pi}}\, \theta'(0) - s^2 \Theta_c \right\}$$

where $\Theta_c \equiv \Theta_c(s, t)$ is the cosine transform of $\theta(x, t)$.

We therefore have the equation

$$\frac{\partial \Theta_c}{\partial t} = -Ks^2 \Theta_c \tag{9.47}$$

which has solution

$$\Theta_c = Ae^{-Ks^2 t} \tag{9.48}$$

where A is a function of s which is effectively the value of Θ_c at $t = 0$ i.e. $\Theta_c(s, 0)$. Now

$$\Theta_c(s, 0) = \sqrt{\frac{2}{\pi}} \int_0^\infty \theta(x, 0) \cos sx \, dx$$

$$= \sqrt{\frac{2}{\pi}} \int_0^\infty \theta_0(x) \cos sx \, dx = F(s), \text{ say.}$$

Note that for $F(s)$ to exist, $\theta_0(x)$ must $\to 0$ as $x \to \infty$.

Hence

$$\Theta_c = F(s)\, e^{-Ks^2 t} \tag{9.49}$$

We invert both sides by (9.44b) and obtain

$$\theta(x, t) = \sqrt{\frac{2}{\pi}} \int_0^\infty F(s)\, e^{-Ks^2 t} \cos sx \, ds$$

To be specific, we let $\theta_0(x)$ be $100e^{-x}$ so that

$$F(s) = \sqrt{\frac{2}{\pi}} \int_0^\infty 100e^{-x} \cos sx \, dx = \frac{100}{(s^2+1)} \sqrt{\frac{2}{\pi}} \tag{9.50}$$

Hence

$$\theta(x, t) = \frac{200}{\pi} \int_0^\infty \frac{1}{(s^2+1)} \cdot e^{-Ks^2 t} \cos sx \, ds$$

It is not always easy to evaluate such integrals analytically.

Case II

The initial temperature distribution in the bar is zero and the end $x = 0$ is maintained at a constant temperature θ_0. On this occasion, the boundary conditions suggest the use of a sine transform.

Applying this transform to both sides of (9.33) we obtain

$$\sqrt{\frac{2}{\pi}}\int_0^\infty \frac{\partial\theta}{\partial t}\sin sx\,dx = K\sqrt{\frac{2}{\pi}}\int_0^\infty \frac{\partial^2\theta}{\partial x^2}\sin sx\,dx$$

i.e. $$\frac{\partial\Theta_s}{\partial t} = K\left\{\sqrt{\frac{2}{\pi}}\,s\theta(0) - s^2\Theta_s\right\}$$

where $\Theta_s = \Theta_s(s, t)$ is the sine transform of $\theta(x, t)$.

Hence we have the equation

$$\frac{\partial\Theta_s}{\partial t} + Ks^2\Theta_s = K\sqrt{\frac{2}{\pi}}\,s\,\theta_0$$

which has solution

$$\Theta_s = Ae^{-Ks^2t} + \sqrt{\frac{2}{\pi}}\frac{\theta_0}{s}$$

where A is a yet undetermined function of s.

When $t = 0$, $\theta = 0$ and hence $\Theta_s(s, 0) = 0$. It follows that $A = -\sqrt{\frac{2}{\pi}}\frac{\theta_0}{s}$ and therefore

$$\Theta_s = \sqrt{\frac{2}{\pi}}\frac{\theta_0}{s}(1 - e^{-Ks^2t}) \tag{9.51}$$

We invert the transform via (9.45b) to obtain

$$\theta(x, t) = \frac{2}{\pi}\theta_0\int_0^\infty \frac{1}{s}(1 - e^{-Ks^2t})\sin xs\,ds \tag{9.52}$$

$$= \frac{2\theta_0}{\pi}\int_0^\infty \frac{1}{s}\sin xs\,ds - \frac{2\theta_0}{\pi}\int_0^\infty \frac{1}{s}e^{-Ks^2t}\sin xs\,ds$$

Now, by result (A3) in Appendix A the first integral is equal to $\frac{\pi}{2}$. The second integral can be reduced by putting $s = \frac{u}{\sqrt{Kt}}$; it then becomes

$$\int_0^\infty \frac{e^{-u^2}\sin xu/\sqrt{(Kt)}}{u}\,du$$

as you can check. Result (A6) tells us that this integral is equal to $\frac{\pi}{2}\,\mathrm{erf}\left(\frac{x}{2\sqrt{Kt}}\right)$

Hence

$$\theta(x,\ t) = \theta_0\left(1 - \text{erf}\left[\frac{x}{2\sqrt{Kt}}\right]\right) \tag{9.53}$$

The solution is sketched in Figure 9.10.

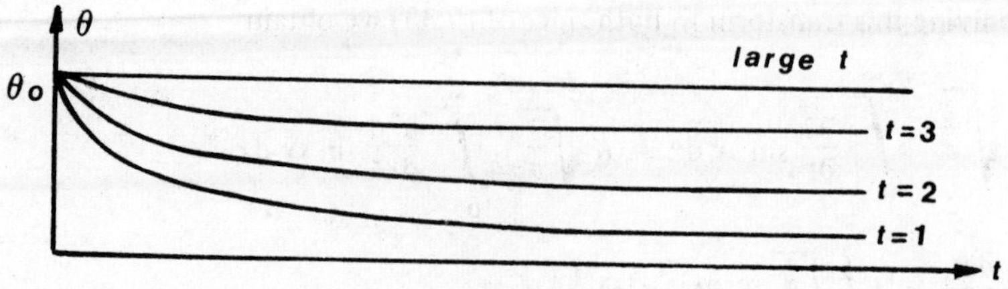

Figure 9.10

Case III

We now repeat Case II using Laplace transforms to eliminate the time dependence (it transpires that this is easier than eliminating x–dependence).

If we let $\Theta(x,\ s)$ be the transform of $\theta(x,\ t)$ then if we multiply both sides of (9.33) by e^{-st} and integrate from $t = 0$ to $t = \infty$ we shall obtain

$$\int_0^\infty \frac{\partial\theta}{\partial t}\ e^{-st}\ dt = K \int_0^\infty \frac{\partial^2\theta}{\partial x^2}\ e^{-st}\ dt$$

i.e. $$s\Theta(x,\ s) - \theta(x,\ 0) = K\,\frac{\partial^2\Theta(x,\ s)}{\partial x^2}$$

i.e. $$\frac{\partial^2\Theta}{\partial x^2} = \frac{s}{K}\,\Theta \tag{9.54}$$

This equation has general solution

$$\Theta(x,\ s) = A\,e^{\sqrt{\frac{s}{K}}\,x} + B\,e^{-\sqrt{\frac{s}{K}}\,x}$$

where A and B are constants to be determined.

The boundary conditions for (9.54) can be found from the boundary conditions of (9.33) which have not yet been employed.

Now $\theta(x,\ t) \to 0$ as $x \to \infty$ and therefore $\Theta(x,\ s) \to 0$ as $x \to \infty$. Hence A must be zero and

$$\Theta(x,\ s) = B\,e^{-\sqrt{\frac{s}{K}}\,x}$$

Now

$$\Theta(0,\ s) = \int_0^\infty \theta_0\ e^{-st}\ dt = \frac{\theta_0}{s} = B$$

and so

$$\Theta(x, s) = \theta_0 \frac{1}{s} e^{-\sqrt{\frac{s}{K}}x} = \theta_0 \frac{1}{s} e^{-\frac{x\sqrt{s}}{\sqrt{K}}} \tag{9.55}$$

Normally, to invert the transform would require integration in the complex plane (see Section 12.9), but on this occasion we can proceed by ingenious manipulation. This involves appealing to results already established, but we shall merely quote the result.

$$\Theta(x, s) = L\left\{\theta_0 \left(1 - \operatorname{erf}\left[\frac{x}{2\sqrt{Kt}}\right]\right)\right\} \tag{9.56}$$

Hence we obtain the solution (9.53) again for $\theta(x, t)$.

We were lucky to be able to invert the transform by appealing to results already established. Furthermore, equation (9.54) was reasonably straightforward; had it been less so we might have needed transforms even to solve it.

Choice of Transform

It would seem that the choice of which transform to use is partly dictated by boundary conditions. Sometimes, however, the choice is not always clear-cut and only by trying two or more approaches can we decide which suits a particular problem best.

9.8 INTERPRETATION OF FOURIER TRANSFORMS

We can place physical interpretation on Fourier transforms, and for those workers whose concern with the transforms lies in studying waveforms, such physical interpretations allow a deeper insight into their subject. For readable accounts on this area, you are referred to the bibliography. Here, we shall merely look at one or two ideas.

The Fourier transform of a waveform is its spectrum and vice-versa; hence we may regard each as equally measurable. An electrical waveform can be visualised on an oscilloscope whilst its spectrum can be analysed on a spectroscope. The waveform is a function of time and the spectrum is a function of frequency. The approach of the transform to a rectangular pulse, say, is to regard it as an isolated event whereas the Fourier series approach requires it to be repeated in both directions; therefore, the transform approach is the more general.

Where system properties are linear, Fourier transforms may be directly applied when it is required to represent a variable of time as a variable of frequency or vice-versa.

Convolutions

One concept which is often treated only as an integral definition is that of **convolution.** Many problems in electricity or optics can be tackled by applying a Fourier transform to a function, solving a governing equation (which may represent the action of some circuit element) and then inverting the solution. We therefore have two transforming processes: one into the intermediate system (circuit element) and one out again. Perhaps this could be accomplished at a stroke by one transformation. In this way the intermediate system is transformed and then applied to the original function. The resultant output is the integral of the product of each point of the original function with the transformed response.

The convolution is the result obtained by scanning one function with each point of the other and summing the products at each point. Note that the scanner may, in fact spread out the function.

System response and the convolution integral

First we note that a function $x(t)$ can be represented as a continuum of impulses

$$x(t) = \int_{-\infty}^{\infty} x(\lambda)\,\delta(\lambda - t)\,d\lambda \tag{9.57}$$

The input $x(t)$ to a linear system produces an output $y(t)$ which we wish to find. We first find the response to elementary input signals and then use superposition to obtain the total response.

Suppose the response to $\delta(t)$ is known to be $h(t)$; then the response to $x(\lambda)\,\delta(t - \lambda)$ is $x(\lambda)\,h(t - \lambda)$ and the total response of the system is given by

$$y(t) = \int_{-\infty}^{\infty} x(\lambda)\,h(t - \lambda)\,d\lambda = x(t) * g(t) \tag{9.58}$$

This is called the **convolution integral.**

Example

The low pass filter shown in Figure 9.11 has a unit impulse response $h(t) = e^{-t}$. Sketch the time response of the filter to a unit step function of voltage.

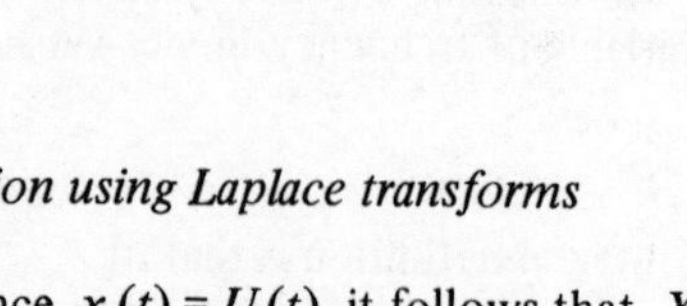

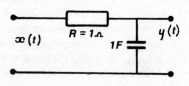

Figure 9.11

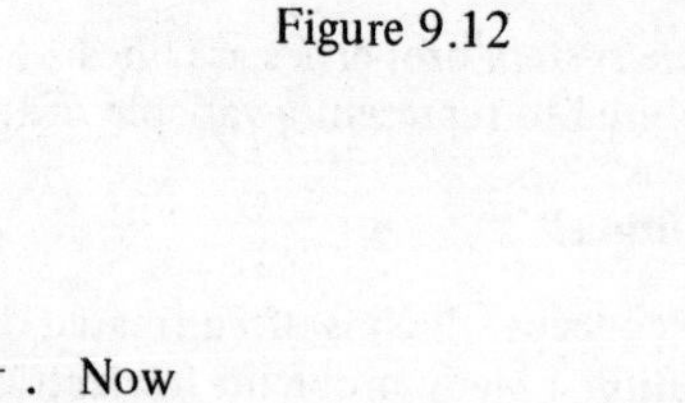

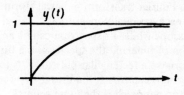

Figure 9.12

Solution using Laplace transforms

Since $x(t) = U(t)$ it follows that $X(s) = \dfrac{1}{s}$. Now

$$Y(s) = H(s)\,X(s) = \frac{1}{(s+1)} \cdot \frac{1}{s} = \frac{1}{s} - \frac{1}{(s+1)}$$

Therefore

$$y(t) = U(t) - U(t)\,e^{-t}$$

We obtain the sketch of $y(t)$ shown in Figure 9.12.

Time domain solution

$$y(t) = \int_{-\infty}^{\infty} x(\lambda)\, h(t-\lambda)\, d\lambda = \int_{-\infty}^{\infty} U(\lambda)\, e^{-(t-\lambda)}\, d\lambda$$

Now convolution is the result of scanning one function with each point of the other and summing the products at each point. This is illustrated in Figure 9.13.

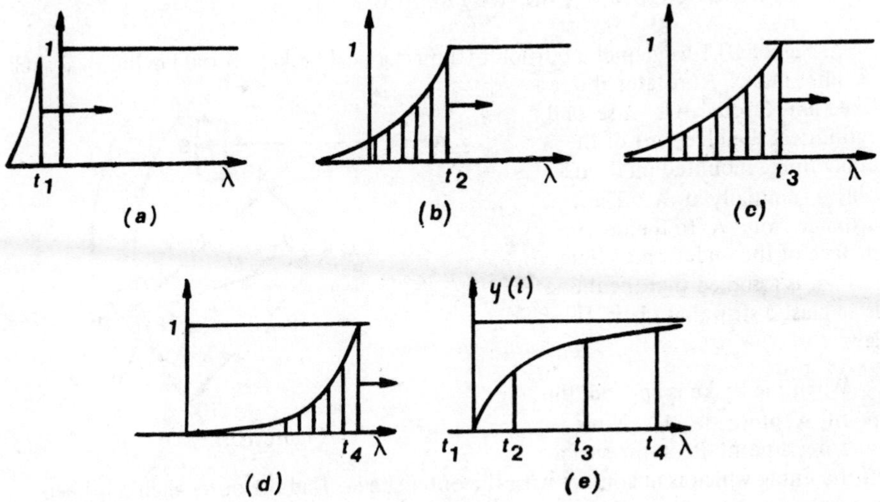

Figure 9.13

Figures (a) to (d) show how the value of the convolution integral is obtained by sliding one function over the other. The shaded area in each case represents the product for the particular t. Figure (e) shows the accumulated results.

Chapter Ten

Integration

10.1 INTRODUCTION A BRAKE PROBLEM

In Figure 10.1 we depict a portion of a mechanical brake. A rigid cylindrical shell of inner radius R rotates about a fixed axis through O. A second cylindrical shell has part of the brake lining mounted on it; its centre is initially at A. The distance from A to the outer surface of the undeformed lining is r; it is assumed that the lining is an elastic strip that obeys Hooke's law.

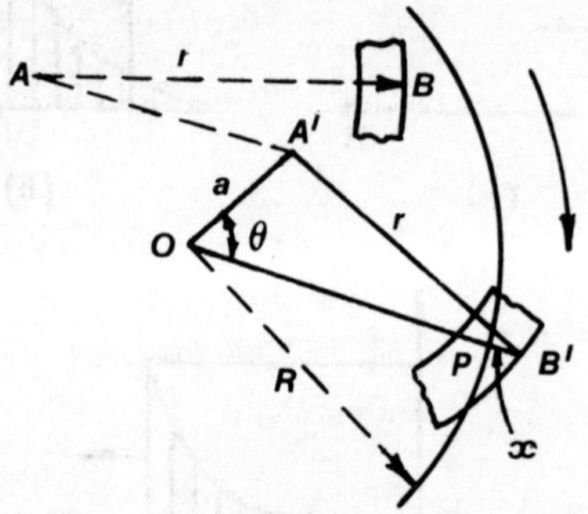

Figure 10.1

When the brake is applied, the point A moves to A′. Now consider a point P on the surface of the lining which is in contact with the outer shell. Had the outer shell not been present, the point P would have been at B′ where $A'B' = r$. Hence the deformation of the elastic lining is $PB' = x$, say. Let the angle POA′ be θ radians and let the distance OA′ be $a\,(a < r)$. Then $OB' = a\cos\theta + (r^2 - a^2\sin^2\theta)^{1/2}$. (It may help to drop a perpendicular from A′ to OP.) Further, $OP = R$ since P is on the outer circular shell. Therefore,

$$PB' = x = a\cos\theta + (r^2 - a^2\sin^2\theta)^{1/2} - R \tag{10.1}$$

Now the radial force acting on an element of arc $R\Delta\theta$ of the outer shell is $\lambda x \Delta\theta$ where λ is a constant of proportionality in Hooke's law. Let μ be the coefficient of friction between the brake lining and the outer shell; then the friction force on the brake element is $\mu\lambda x\Delta\theta$ and the torque exerted by the braking action is $R\mu\lambda x\Delta\theta$. Finally, let α and β be those values of θ corresponding to the extreme positions of the points of contact of the brake with the outer lining. Then the resultant torque for the whole brake is given by

$$T = R\mu\lambda \int_\alpha^\beta (a\cos\theta + (r^2 - a^2\sin^2\theta)^{1/2} - R)\,d\theta$$

i.e. $$T = R\mu\lambda[a(\sin\beta - \sin\alpha) - R(\beta - \alpha)] + R\mu\lambda r\int_\alpha^\beta (1 - (a^2/r^2)\sin^2\theta)^{1/2}\,d\theta \tag{10.2}$$

Since we cannot in general carry out the integration in (10.2) analytically (we require a knowledge of *elliptic functions* which is beyond our present scope) we shall consider a numerical approach.

We shall assume familiarity with the trapezoidal and Simpson's rules; these were discussed in Chapter 8 of *Engineering Mathematics.* In this chapter we shall first study further numerical methods of integration. In Sections 10.5 to 10.8 we shall consider the problem of integration where two or more variables are involved. All the numerical methods can be extended to such cases.

To make the brake problem more specific we take $\alpha = \pi/6$, $\beta = \pi/3$, $r = 2a$, $R = 5a/2$ and $\mu = 0.4$, also, we shall take $\lambda = 1$ for simplicity so that

$$T = a^2(\tfrac{1}{2}(\sqrt{3} - 1) - 5\pi/12) + 2a^2 \int_{\pi/6}^{\pi/3} (1 - \tfrac{1}{4}\sin^2\theta)^{\frac{1}{2}}\, d\theta \tag{10.3}$$

10.2 NEWTON-COTES FORMULAE

The trapezoidal and Simpson's rules are special cases of a class of closed integration rules known as **Newton-Cotes** formulae. The use of the term *closed* indicates that the end-points of the range of integration are included as nodes, or values used, in the formulae. The essential idea is that of passing a polynomial $p_n(x)$ of degree n through $(n + 1)$ specified equally-spaced nodes $(x_i, f(x_i))$. We denote the end-points $x_0 = a$ and $x_{n+1} = b$. Then we approximate

$$\int_a^b f(x)\, dx \quad \text{by} \quad \int_a^b p_n(x)\, dx$$

In the case of the trapezoidal rule the only nodes are the end-points, whereas for Simpson's rule we require a third node at $\{\tfrac{1}{2}[a + b], f(\tfrac{1}{2}[a + b])\}$.

Note that in a general definite integration problem we divide the range of integration into a number of strips which is a multiple of the number required for application of the basic rule. For Simpson's rule the basic requirement is two strips and therefore in any application the range of integration must be divided into an even number of strips.

We quote the basic trapezoidal rule as

$$\int_a^b f(x)\, dx \simeq \frac{h}{2}[f_0 + f_1] \tag{10.4}$$

where $x_0 = a$, $x_1 = b$ and Simpson's rule as

$$\int_a^b f(x)\, dx \simeq \frac{h}{3}[f_0 + 4f_1 + f_2] \tag{10.5}$$

where $x_0 = a$, $x_1 = \dfrac{a + b}{2}$, $x_2 = b$.

In both cases, h is the strip width.

In general then, fitting a polynomial of degree n will lead to a formula

$$\int_a^b f(x)\,dx \simeq \sum_{i=0}^{n} a_i f_i \tag{10.6}$$

where the a_i are **weights** to be determined so that the formula will give results which are exact for any polynomial of degree n or less. We have $(n+1)$ coefficients a_i to find from $(n+1)$ nodes. We shall find these values by evaluating the left-hand side of (10.6) analytically for $f(x) = 1, x, x^2, \ldots\ldots, x^n$ in turn.

Example

Let us take the case $n = 3$. Then we may, without loss of generality put $a = 0$, $b = 3h$ and obtain successively

$$\int_0^{3h} 1\,dx = 3h = a_0 + a_1 + a_2 + a_3$$

$$\int_0^{3h} x\,dx = 9h^2/2 = ha_1 + 2ha_2 + 3ha_3$$

$$\int_0^{3h} x^2\,dx = 9h^3 = h^2a_1 + 4h^2a_2 + 9h^2a_3$$

$$\int_0^{3h} x^3\,dx = 81h^4/4 = h^3a_1 + 8h^3a_2 + 27h^3a_3$$

Solving for a_i, we obtain the **three-eighth's rule**

$$\int_a^b f(x)\,dx \simeq \frac{3h}{8}\,[f_0 + 3f_1 + 3f_2 + f_3] \tag{10.7}$$

where $x_0 = a,\ x_1 = \dfrac{2a+b}{3},\ x_2 = \dfrac{a+2b}{3},\ x_3 = b$

We have said nothing yet about errors involved. The truncation error was calculated by Ralston for the general case. We may modify (10.6) to

$$\int_a^b f(x)\,dx = Ah \sum_{i=0}^{n} w_i f_i + Bh^{k+1} f^{(k)}(\xi) \tag{10.8}$$

where $a \leqslant \xi \leqslant b$.

Note that the coefficients a_i (and hence w_i) are symmetrical. The formulae which involve an even number of strips n also give zero error for polynomials of degree up to $(n+1)$. In such cases k in (10.8) is $(n+1)$ otherwise it has the value n. The use of some formulae of high order may mean that the error term involves high derivatives of the function and this could cause large truncation error. Also round-off error may be a problem if many strips are used.

Table 10.1 shows the coefficients for some of these rules.

Table 10.1

n	A	w_0	w_1	w_2	w_3	w_4	w_5	B
1	1/2	1	1					−1/12
2	1/3	1	4	1				−1/90
3	3/8	1	3	3	1			−3/80
4	2/45	7	32	12	32	7		−8/945
5	5/288	19	75	50	50	75	19	−275/12096

We estimated the value of the integral in (10.3) using the trapezoidal rule, Simpson's rule and the three-eighth's rule each with 6 strips. The results agreed to 5 d.p.

Problems

1. Use the trapezoidal rule, Simpson's rule and the three-eighth's rule on the following integrals comparing your answers with the analytical results. Choose the number of strips yourself.

(a) $\displaystyle\int_0^1 \sqrt{x}\,dx$ (b) $\displaystyle\int_0^1 |x - ½|dx$ (c) $\displaystyle\int_0^1 \frac{\sin x}{x}\,dx$ (you may take the value of the integrand to be 1 at $x = 0$.)

(d) $\displaystyle\int_{1.6}^{3.2} e^x\,dx$ (e) $\displaystyle\int_0^1 \sin 101\pi x\,dx$

2. A particle moves along a straight line so that at time t its distance s from a fixed point on the line is given by $\dfrac{ds}{dt} = t\sqrt{8 - t^3}$.

Use Simpson's rule with 8 strips to calculate the approximate distance travelled by the particle from $t = 0$ to $t = 2$. (L.U.)

3. Repeat Problem 2 using the three-eighth's rule and the rules with 4 and 5 ordinates.

4. Evaluate as accurately as you can $\displaystyle\int_0^4 f(x)\,dx$ from the following table.

x	0	1	2	3	4
$f(x)$	0	1	0	−2	0

5. Use any of the integration formulae to verify the following results correct to 4 d.p.

(a) $\displaystyle\int_0^1 \frac{dx}{1+x^2} = 0.7854$ (b) $\displaystyle\int_0^{\pi/2} \sqrt{1 - \tfrac{1}{4}\sin^2 t}\, dt = 1.4675$

(c) $\displaystyle\int_0^{\pi/2} \frac{dx}{\sin^2 x + \tfrac{1}{4}\cos^2 x} = 3.1416$

10.3 ROMBERG INTEGRATION

It can be shown that the truncation error in applying the trapezoidal rule is bounded by the result

$$|E_t| \leqslant (b-a)\frac{h^2}{12} M \tag{10.9}$$

where M is the maximum value taken by $|f''(x)|$ in $[a, b]$.

Since $|E_t| \propto h^2$ it follows that, round-off error apart, halving the step size h will divide the error in the approximation by 4.

Let the exact value of an integral be I, let the trapezoidal approximation with n strips be T_n and that with $2n$ strips be T_{2n}; for simplicity's sake we take n to be a power of 2. Let $E_n = T_n - I$ and $E_{2n} = T_{2n} - I$. But $E_{2n} \simeq \tfrac{1}{4} E_n$ so that it follows that

$$I \simeq (4T_{2n} - T_n)/3 = T_{2n}^{(1)} \tag{10.10}$$

We use the notation $T_{2n}^{(1)}$ to indicate that we have used the estimates T_{2n} and T_n to produce a revised estimate for I. To see that we are improving estimates let us evaluate $I = \displaystyle\int_0^1 (5 - 5x^4)\, dx = 4$ by the Trapezoidal rule alone.

In Table 10.2 we show the results of our computations. Note that the error decreases until the number of strips causes round-off error to grow sufficiently to spoil matters. Of course there is the very real practical objection that the use of 4000 strips requires a lot of computation and would suggest a more accurate formula be used. Suppose we take $T_2 = 3.59375$ and $T_1 = 2.5$; the improved estimate is $T_2^{(1)} = (4T_2 - T_1)/3 = (4 \times 3.59375 - 2.5)/3 = 3.95833$ which is closer to the true value I. We can repeat the process and obtain $T_4^{(1)} = (4T_4 - T_2)/3 = (4 \times 3.89648 - 3.59375)/3 = 3.99739$ which is even better.

We continue these computations in Table 10.3. Having got a new column of estimates we can use the further result, which we do not prove, that a better estimate is

$$T_{2n}^{(2)} = (16T_{2n}^{(1)} - T_n^{(1)})/15 \tag{10.11}$$

Table 10.2

Number of Strips	Trapezoidal Approximation	Error
1	2.5	1.5
2	3.59375	0.40625
4	3.89648	0.103516
8	3.974	0.026001
16	3.99349	6.50787E–03
32	3.99837	1.62745E–03
64	3.99959	4.06742E–04
128	3.9999	1.00136E–04
256	3.99998	2.43187E–05
512	3.99999	1.04904E–05
1024	4	4.29153E–06
2048	3.99999	1.04904E–05
4096	4.00001	–6.67572E–06

For example we calculate

$$T_4^{(2)} = (16T_4^{(1)} - T_2^{(1)})/15 = (16 \times 3.99739 - 3.95833)/15 = 3.99999$$

As you can see from Table 10.3, we have really got so close that it is not worth continuing the process. Were it required, we could continue the tabulation by writing each succeeding column from the diagonal element downwards.

Table 10.3

2.5		
3.59375	3.95833	
3.89648	3.99739	3.99999
3.974	3.99984	4.00000
3.99349	3.99999	4.00000
3.99837	4.00000	4.00000

This process of producing improved estimates is called **Romberg integration.** Our big worry is that truncation error will not become sufficiently low before round-off error takes over.

The generalisation of (10.11) is

$$T_{2n}^{(j)} = (4^j \, T_{2n}^{(j-1)} - T_n^{(j-1)})/(4^j - 1) \qquad (10.12)$$

In Table 10.3 we have indicated a sensible order of computation so that we can stop as soon as we get agreement to the required accuracy.

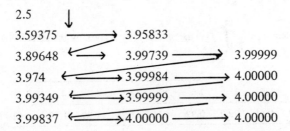

Problems

1. Draw a flow chart for the Romberg process and write a computer program from it. Test your program on the text example.

2. Carry out the integrals in Problem 1 of the last section using Romberg integration.

3. Repeat Problem 4 of the last section using Romberg integration.

4. Apply the integration methods so far introduced to the problem of approximating $\int_1^2 f(x)\,dx$ where $f(x)$ is given by the table below. How much confidence do you place in your results?

x	1.0	1.2	1.4	1.6	1.8	2.0
$f(x)$	1.0000	0.8333	0.7143	0.6250	0.5556	0.5000

With the knowledge that $f(x) = 1/x$ what comments can you make?

5. Repeat Problem 4 for $\int_0^2 f(x)\,dx$ where $f(x)$ is given by the table below.

x	0	0.25	0.50	0.75	1.00	1.25	1.50	1.75	2.00
$f(x)$	1.000	1.284	1.649	2.117	2.718	3.490	4.482	5.755	7.389

$f(x)$ is, in fact, e^x.

6. Repeat Problem 4 for $\int_1^5 f(x)\,dx$ where $f(x)$ is given by the table below.

x	1	1.5	2	2.5	3	3.5	4	4.5	5
$f(x)$	0	0.41	0.69	0.92	1.10	1.25	1.39	1.50	1.61

$f(x)$ is, in fact, $\log_e x$.

7. Use the computer program of Problem 1 to evaluate

(a) $\int_0^{\pi/2} \frac{dx}{\sqrt{(1 - \sin^2 x)}}$ (7 d.p.) (b) $\int_0^{\pi/2} \sqrt{1 - \frac{1}{2}\sin^2 x}\,dx$ (7 d.p.)

8. Apply Romberg integration to $\int_0^{\pi/2} \sin x\,dx$ and note how slow the convergence is.

10.4 GAUSSIAN INTEGRATION

The rules we have so far studied have calculated function values at equally spaced values of x, including the end points of the range of integration. Suppose we relax the condition that the spacing must be uniform and aim to place the values of x so that for a fixed number of x values the approximation is as good as possible. In Figure 10.2(a) we see the nodes of the trapezoidal rule and in Figure 10.2(b) we see the principle behind Gaussian integration; what we now need to do is to locate the nodes.

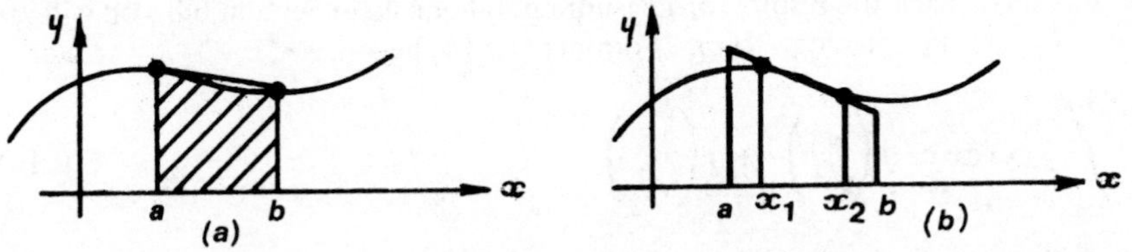

Figure 10.2

We shall study only **Gauss-Legendre** integration. The Legendre polynomials form an orthogonal family and other orthogonal polynomials can be used to handle awkward integrals. The Gauss-Legendre approach requires the integration to be carried out over the range $[-1,\ 1]$. We have to scale the variable x by means of the formula

$$u = (2x - [b+a])/(b-a) \tag{10.13a}$$

so that

$$x = ½\,[b-a]u + ½\,[b+a] \tag{10.13b}$$

Then

$$\int_a^b f(x)\,\mathrm{d}x = \int_{-1}^{1} \phi(u)\,\mathrm{d}u$$

where

$$\phi(u) = ½\,(b-a)\,f(½\,[b-a]u + ½\,[b+a])$$

Two-ordinate case

We derive the formula for the case shown in Figure 10.2(b). We seek an approximation of the form

$$\int_{-1}^{1} f(x)\,\mathrm{d}x \simeq w_1 f(x_1) + w_2 f(x_2) \tag{10.14}$$

You can see that we have four unknowns: the locations x_1 and x_2 and the two weights w_1 and w_2. In effect this means that we can aim to give exact results for $\phi(x)$ a polynomial of degree up to 3 (the trapezoidal rule could only manage degree 1). We therefore seek exact results for $\phi(x) = 1,\ x,\ x^2,\ x^3$. Taking these in turn we have

$$\int_{-1}^{1} 1\,\mathrm{d}x = 2 = w_1 + w_2 \qquad \int_{-1}^{1} x\,\mathrm{d}x = 0 = w_1 x_1 + w_2 x_2$$

$$\int_{-1}^{1} x^2 \, dx = \frac{2}{3} = w_1 x_1^2 + w_2 x_2^2 \qquad \int_{-1}^{1} x^3 \, dx = 0 = w_1 x_1^3 + w_2 x_2^3 \tag{10.15}$$

You can check that the solution of these equations is

$$w_1 = w_2 = 1, \quad x_1 = -1/\sqrt{3}, \quad x_2 = 1/\sqrt{3}$$

(Note that we effectively made $x_2 > x_1$ in Figure 10.2(b))

If you check back the results for Legendre polynomials in Section 6.8 you will see that x_1 and x_2 are the zeros of $P_2(x)$. Formula (10.14) becomes

$$\int_{-1}^{1} f(x) \, dx \simeq f\left(\frac{-1}{\sqrt{3}}\right) + f\left(\frac{1}{\sqrt{3}}\right) \tag{10.16}$$

Apart from the necessity to evaluate $f(x)$ at irrational values of x the formula is straightforward.

Example

We consider

$$I = \int_{-1}^{1} (5 - 5x^4) \, dx = 8$$

Now $f\left(\frac{-1}{\sqrt{3}}\right) = 5 - 5 \cdot \frac{1}{9} = \frac{40}{9} = f\left(\frac{1}{\sqrt{3}}\right)$

Hence $I \simeq \frac{40}{9} + \frac{40}{9} = \frac{80}{9}$. By symmetry, we estimate $\int_0^1 (5 - 5x^4) \, dx$ to be 40/9. However, we may evaluate this latter integral directly by changing the scale via

$$u = (2x - [1 + 0])/(1 - 0), = 2x - 1 \quad \text{or} \quad x = \tfrac{1}{2}(1 + u)$$

We then have

$$\int_0^1 (5 - 5x^4) \, dx = \int_{-1}^{1} \left(5 - 5\left[\frac{1 + u}{2}\right]^4\right) \frac{du}{2}$$

i.e. $$\int_{-1}^{1} \left(\frac{5}{2} - \frac{5}{32} [1 + u]^4\right) du$$

Now

$$f\left(\frac{1}{\sqrt{3}}\right) = \frac{5}{2} - \frac{5}{32}\left[1 + \frac{1}{\sqrt{3}}\right]^4 = 1.532764 \quad (6 \text{ d.p.})$$

Similarly,

$$f\left(\frac{-1}{\sqrt{3}}\right) = 2.495014 \quad (6 \text{ d.p.})$$

The Gauss approximation is therefore 4.027778 (6 d.p.) from (10.16) and the error is 0.027778 (6 d.p.).

General case with n ordinates

The nodes are located with the x-values at the zeros of $P_n(x)$. Note that these values split into pairs symmetrically placed about the origin with the addition of $x = 0$ if n is odd. The weights and locations for the first few cases are shown in Table 10.4.

Table 10.4

Number of ordinates	Locations	Weights
2	±0.5773502692	1
3	±0.7745966692	5/9
	0	8/9
4	±0.8611363116	0.3478548451
	±0.3399810436	0.6521451549
5	±0.9061798459	0.2369268851
	±0.5384693101	0.4786286705
	0	0.5688888889
6	±0.9324695142	0.1713244924
	±0.6612093865	0.3667615730
	±0.2386191861	0.4679139346

Example

For $I = \displaystyle\int_0^1 (5 - 5x^4)\,dx$ we shall approximate with 4 ordinates after transforming as in the last subsection. The approximation is

$$0.3478548451 \left\{\left(\frac{5}{2} - \frac{5}{32}[1 - 0.8611363116]^4\right) + \left(\frac{5}{2} - \frac{5}{32}[1 + 0.8611363116]^4\right)\right\}$$

$$+ 0.6521451549\left\{\left(\frac{5}{2} - \frac{5}{32}[1 - 0.3399810436]^4\right)\right.$$

$$\left. + \left(\frac{5}{2} - \frac{5}{32}[1 + 0.3399810436]^4\right)\right\} = 4.00000000 \text{ (8 d.p.)}$$

This is far more accurate than Romberg integration, even employing a trapezoidal approximation with 8 strips.

Truncation error

For completeness we quote the truncation error for the general case. Assuming that

$\int_{-1}^{1} f(x)\,dx$ is approximated by $\sum_{i=1}^{n} w_i f(x_i)$ then Lanczos' approximation to the truncation error is given by

$$\epsilon_t \simeq \frac{1}{2n+1}\left[f(1) + f(-1) - I_G - \sum_{i=1}^{n} w_i x_i f'(x_i) \right] \qquad (10.17)$$

where I_G is the Gaussian estimate of the integral I.

This does involve applying the Gauss formula to $x f'(x)$ as well as to $f(x)$ but it avoids the need to calculate $f^{(2n)}(x)$ as required by the standard formula for truncation error. However, it should be pointed out that the approximation (10.17) works well only if the integrand is reasonably smooth.

General Remarks

For roughly the same accuracy, Gaussian integration requires about half as many ordinates as Simpson's rule. However, it does require awkward values of x and this may be difficult if a good calculating aid is not at hand. Further, if we double the number of strips, the Gauss method does require completely new values of x. If experimental data is the only information about the function to be integrated then Gaussian integration is unlikely to be useful because of the x - values not being properly spaced. Indeed, if the x - values are haphazardly chosen we have to resort to the Trapezoidal rule. Sometimes, the range of integration is divided into sub-intervals and Gaussian integration can be used over each of these. Usually only the two-point or three-point formulae are then used in each interval.

One final point: the transformations sometimes necessary for Gaussian integration can be tedious.

Comparison of methods

We now apply Simpson's rule, Romberg integration and Gauss integration to a more taxing problem viz.

$$I = \int_{-4}^{4} e^{-x^2}\,dx \qquad (10.18)$$

We apply the trapezoidal rule with $h = 2$ and $h = 1$.

With $h = 2$ the trapezoidal approximation is

$$T_4 = \frac{2}{2}\left(e^{-16} + 2e^{-4} + 2e^{0} + 2e^{-4} + e^{-16} \right) = 2.073261 \quad \text{(5 d.p.)}$$

With $h = 1$,

$$T_8 = \tfrac{1}{2}\left[e^{-16} + 2e^{-9} + 2e^{-4} + 2e^{-1} + 2e^{0} + 2e^{-1} + 2e^{-4} + 2e^{-9} + e^{-16}\right] = 1.77264 \quad \text{(5 d.p.)}$$

The Romberg process requires an approximation

$$T_8^{(1)} = (4T_8 - T_4)/3 = 1.67243 \quad \text{(5 d.p.)}$$

Simpson's Rule with four strips gives the approximation

$$I_5 = \frac{2}{3}\left[e^{-16} + 4e^{-4} + 2e^{0} + 4e^{-4} + e^{-16} \right] = 1.43102 \quad (5 \text{ d.p.})$$

Gauss' method with four ordinates will need the transformation $x = ½ \,.\, 8u + ½(4-4) = 4u$ so that the integral becomes

$$\int_{-1}^{1} 4e^{-16u^2}\, du$$

and the approximation is

$$0.3478548451\left\{4[e^{-16 \times (0.8611363116)^2} + e^{-16 \times (-0.8611363116)^2}]\right\}$$

$$+ 0.6521451549\left\{4[e^{-16 \times (0.3399810436)^2} + e^{-16 \times (-0.3399810436)^2}]\right\}$$

$$= 0.82085 \quad (5 \text{ d.p.})$$

We continued the calculations and the results are presented in Table 10.5.

Table 10.5

Romberg integration

Number of strips					
4	2.07326				
8	1.77264	1.67243			
16	1.77245	1.77239	1.77906		
32	1.77245	1.77245	1.77246	1.77235	
64	1.77245	1.77245	1.77245	1.77245	1.77245

Simpson's Rule

Number of strips	4	8	16	32	64
Estimate	1.43102	1.67243	1.77239	1.77245	1.77245

Gaussian Integration

Number of points	2	3	4	5	6
Estimate	0.038624	3.55586	0.82085	2.31257	1.57087

The function decays too rapidly for a Gauss formula with a small number of ordinates to get a good accuracy. We tried again with a Gauss 15 - point formula and obtained the result 1.77247 (5 d.p.)

Problems

1. Write a computer program to store the coefficients of Table 10.4 and to allow selection of the number of ordinates up to 6. Apply the program to the example of Section 10.1

2. Repeat Problem 1 of Section 10.2 using Gauss integration.

3. Repeat Problem 4 of Section 10.2 using Gauss integration.

4. Repeat Problems 4 to 6 of the last Section using Gaussian integration.

5. Repeat Problem 7 of the last Section using Gaussian integration.

6. Use the computer program of Problem 1 to evaluate the integrals in Problem 8 of the last section.

7. Try various methods of integration on $\int_{-1}^{1} \frac{x^2\, dx}{\sqrt{1-x^2}}$. What difficulties do you meet and how could you overcome them?

8. Apply the two-point formula to $\int_{0}^{\pi/2} \sin t\, dt$. Repeat with the four-point formula.

9. Apply the two-point formula to $\int_{-1}^{1} \frac{1}{1+t^2}\, dt$ and compute the actual error and the error estimated by (10.17). Repeat for the four-point and six-point formulae.

10. Repeat Problem 9 for $\int_{0}^{\pi/2} \frac{\cos x}{1+x}\, dx$; the analytical result is 0.6736 (4 d.p.).

10.5 DOUBLE INTEGRATION

Consider the following examples of the need to sum a physical quantity over an area.

Example 1

A bar of length L with a cross-section shown in Figure 10.3 is subjected to a torque T applied at one end while the other end is prevented from rotating, The bar will attempt to rotate about the z - axis which passes through the centroid of each cross-section. If the angle of twist is θ/unit length of bar, then the torsional rigidity is given by $C = \frac{2}{\theta} \iint_A \psi(x, y)\, dA$ where A is the area of cross-section, dA an element of that cross-section, $\psi(x, y)$ is a *stress function*

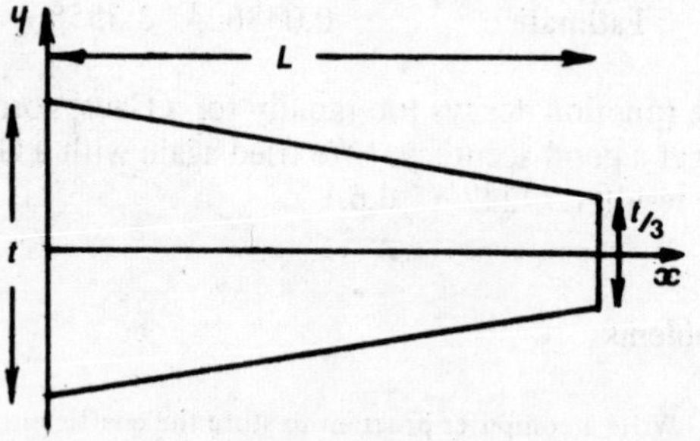

Figure 10.3

which satisfies $\frac{\partial^2 \psi}{\partial x^2} + \frac{\partial^2 \psi}{\partial y^2} = 2G\theta$ inside A and $\psi = 0$ on the perimeter of A;

and G is the shear modulus of the material comprising the bar. The symbolism $\iint_A$ means that the contributions $\psi(x, y)\,\mathrm{d}A$ are summed in a sense to be developed later; for the moment we merely remark that in this summation we cover A.

Example 2

The total strain energy in a plate is given by

$$U = \frac{1}{2}\iint_A \left[\sigma_{xx}\frac{\partial u_x}{\partial x} + \sigma_{yy}\frac{\partial u_y}{\partial y} + \sigma_{xy}\left(\frac{\partial u_y}{\partial x} + \frac{\partial u_x}{\partial y}\right)\right]\mathrm{d}A$$

where σ_{xx}, σ_{xy} and σ_{yy} are direct and shear stresses and u_x, u_y are displacements in the x- and y- directions.

Example 3

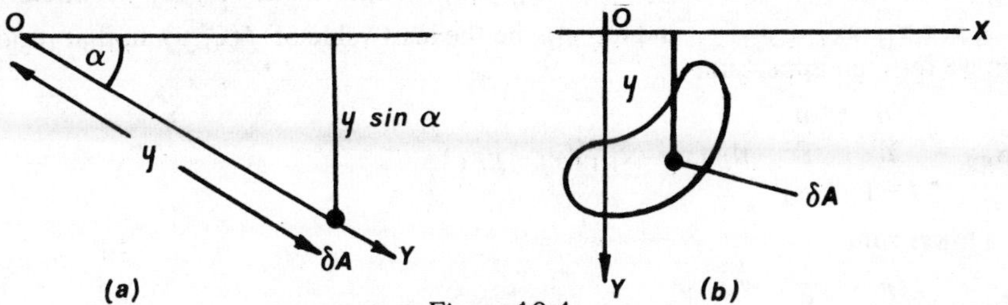

Figure 10.4

A plane area R is submerged in water. See Figure 10.4. If the weight of unit volume of water be w at any point in the fluid then the total thrust on the plane area is $\iint_R wy \sin\alpha\,\mathrm{d}A$. We may require to find the coordinates of the centre of pressure.

In each of these summations we need a double summation to cover the area. We shall now develop some techniques of double integration.

First consider a rectangular region R shown in Figure 10.5(a).

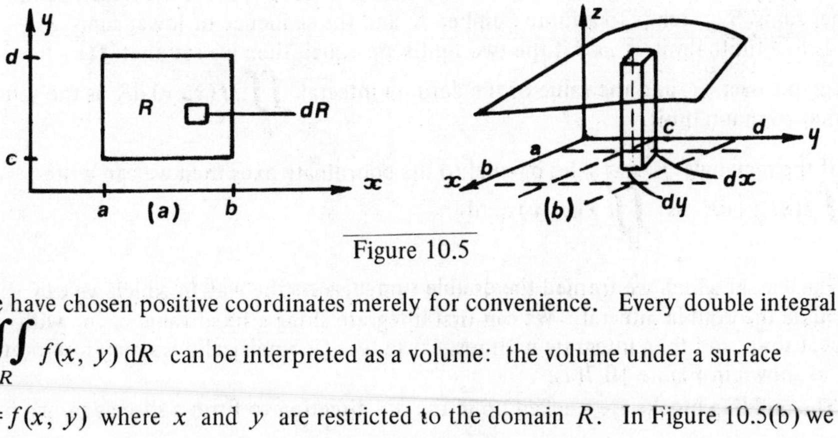

Figure 10.5

We have chosen positive coordinates merely for convenience. Every double integral $\iint_R f(x, y)\,\mathrm{d}R$ can be interpreted as a volume: the volume under a surface $z = f(x, y)$ where x and y are restricted to the domain R. In Figure 10.5(b) we show the building block from which we construct that volume. Its volume is $f(\xi, \eta) . dx . dy$, where (ξ, η) is a point somewhere in the base. Just as for an integral of a function of one variable, so here we can form an **upper sum** and a **lower sum.** Suppose we divide R into small rectangles of equal area as shown in Figure 10.6(a).

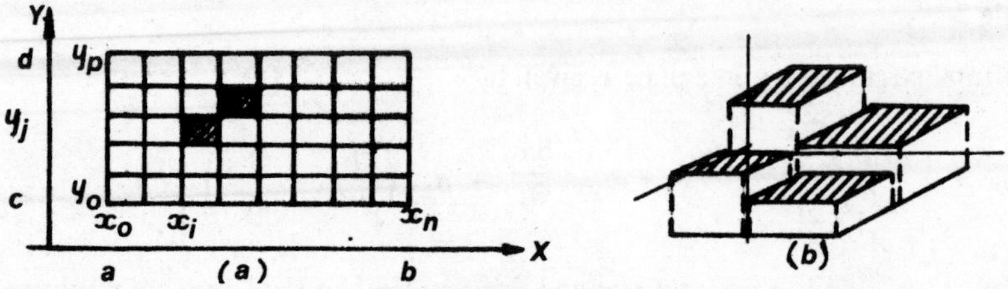

Figure 10.6

The rectangle whose right-hand boundary is $x = x_i$ and whose top boundary is $y = y_j$ may be called the i, jth rectangle. Let M_{ij} be the largest value of $f(x, y)$ where $x_{i-1} < x < x_i$, $y_{j-1} < y < y_j$ and let m_{ij} be the least value of $f(x, y)$ in that region. Then we form an upper sum

$$\overline{S}_{np} = \sum_{i=1}^{n} \sum_{j=1}^{p} M_{ij} (x_i - x_{i-1})(y_j - y_{j-1})$$

and a lower sum

$$\underline{S}_{np} = \sum_{i=1}^{n} \sum_{j=1}^{p} m_{ij} (x_i - x_{i-1})(y_j - y_{j-1})$$

Note that we could write the double sums as $\sum_{j=1}^{p} \sum_{i=1}^{n}$ just as easily. This would imply that we carried out the summation first along a column, then from row to row, instead of vice-versa. If Figure 10.6(b) we show schematically a few adjacent rectangles and corresponding upper sums for each rectangle. Suppose we divide the rectangular region R into smaller rectangles, thereby increasing their number. We carry out this process by saying that if the maximum diagonal of all rectangles in a particular subdivision of R be d, then we arrange that $d \searrow 0$. If the sequence of upper sums $\overline{S}_{np}$ tends to a finite number $\overline{S}$ and the sequence of lower sums $\underline{S}_{np}$ tends to a finite limit $\underline{S}$ *and* if the two limits are equal, then we say that $f(x, y)$ is integrable over R and the value of the definite integral $\iint_R f(x, y)\, dR$ is the value of that common limit.

If the rectangle R has sides parallel to the coordinate axes then we can write $\iint_R f(x, y)\, dR$ as $\iint_R f(x, y)\, dx\, dy$.

The way in which we formed the double sum suggests the way in which we can calculate the double integral. We can first integrate along a fixed value of y, with respect to x and then integrate with respect to y. Geometrically we may interpret this as shown in Figure 10.7(a).

The building blocks are stacked up in the x - direction to form a slice as shown in

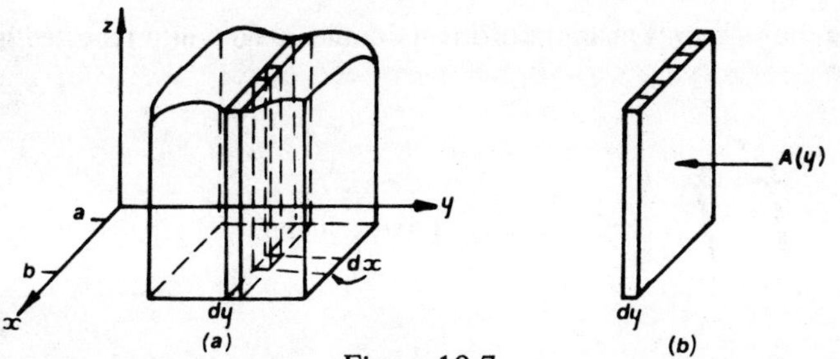

Figure 10.7

Figure 10.7(b); this will have cross-section area $A(y)$, to emphasise that this area depends on the position of the slice, and thickness dy. We then stack together these slices in the y - direction to produce the volume required. We symbolise this process as

$$\int_c^d \int_a^b f(x, y)\, dx\, dy$$

where the interpretation is that the inner integration $\int_a^b f(x, y)\, dx$ is carried out first to produce a function of y and then the resulting function of y, $A(y)$, integrated as $\int_c^d A(y)\, dy$. Some authors use the notation $\int_c^d dy \int_a^b f(x,y)\, dx$. Notice that, since R is rectangular, the limits of integration on x are the same for any value of y.

We now work through three simple examples to fix ideas.

Example 1

Find $\iint_R (3x^2 + y^2)\, dR$ where R is the region $2 \leqslant x \leqslant 3$, $1 \leqslant y \leqslant 2$. We first formulate the problem as $\int_1^2 \int_2^3 (3x^2 + y^2)\, dx\, dy$. The first step is to calculate $A(y) = \int_2^3 (3x^2 + y^2)\, dx$. Now, in this integration, y is regarded as a constant, hence

$$A(y) = [x^3 + xy^2]_2^3 = (27 + 3y^2) - (8 + 2y^2) = 19 + y^2$$

This gives the cross-section area of the volume we seek for any value of y; hence the cross-section increases from its value of 20 at the boundary $y = 1$ to 23 at the boundary $y = 2$.

We now calculate

$$\int_1^2 (19 + y^2)\, dy = \left[19y + \frac{y^3}{3}\right]_1^2 = \left(38 + \frac{8}{3}\right) - \left(19 + \frac{1}{3}\right) = 21\tfrac{1}{3}$$

This example shows how we can transform a double integral into **repeated integration.**

Example 2

Find $I = \int_0^2 \int_1^4 (x + 2y)\, dx\, dy$. First, calculate

$$\int_1^4 (x + 2y)\, dx = \left[\tfrac{1}{2}x^2 + 2yx\right]_1^4 = (8 + 8y) - (\tfrac{1}{2} + 2y) = 7\tfrac{1}{2} + 6y$$

Then calculate

$$\int_0^2 (7\tfrac{1}{2} + 6y)\, dy = \left[7\tfrac{1}{2}y + 3y^2\right]_0^2 = 15 + 12. \quad \text{Hence } I = 27$$

If we refer back to the definition of double integral, we said that the double summation was carried out first in the direction of x and then in the direction of y or vice-versa. Suppose we now reverse the order of integration i.e.

$I = \int_1^4 \int_0^2 (x + 2y)\, dy\, dx$. Refer to Figure 10.8

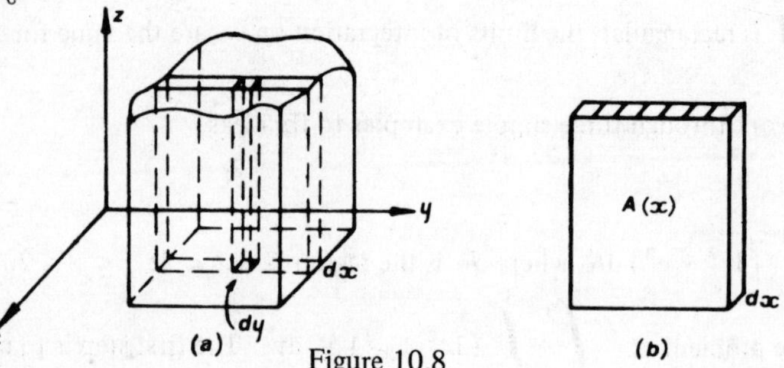

Figure 10.8

First we evaluate $\int_0^2 (x + 2y)\, dy = [xy + y^2]_0^2 = 2x + 4$ and then we evaluate

$$\int_1^4 (2x + 4)\, dx = [x^2 + 4x]_1^4 = (16 + 16) - (1 + 4). \quad \text{Hence } I = 27, \text{ as before.}$$

Example 3

Find $I = \int_0^2 \int_1^4 xy^2\, dx\, dy$

In this example, we can **separate** the function xy^2 into the product of a function of x and a function of y. We can then write

$$I = \left(\int_0^2 y^2\,dy\right)\left(\int_1^4 x\,dx\right) = \left[\frac{y^3}{3}\right]_0^2 \left[\frac{x^2}{2}\right]_1^4 = \frac{8}{3}\cdot\frac{15}{2} = 20$$

You can check for yourself that this is the same result as would have been obtained by the method of the previous examples. We have, in effect, reduced the problem to one of two separate integrations.

We have not rigorously justified the methods used but we shall take the results on trust.

Extension of ideas to non-rectangular regions

Figure 10.9 shows a non-rectangular region D over which a double integral is to be defined. In order that an extension can be carried out we require that the region D have a boundary C which is a simple closed curve. We also assume that D can be enclosed in a circle of sufficiently large, but finite, radius: this avoids improper integrals. We can enclose D by a rectangle R and then, if we wish to find $\iint_D f(x, y)\,dx\,dy$ we can define a function

$$g(x, y) = \begin{cases} f(x, y) & (x, y) \in D \\ 0 & (x, y) \notin D \end{cases}$$

so that $\iint_D f(x, y)\,dx\,dy = \iint_R g(x, y)\,dx\,dy$. See Figure 10.9(a)

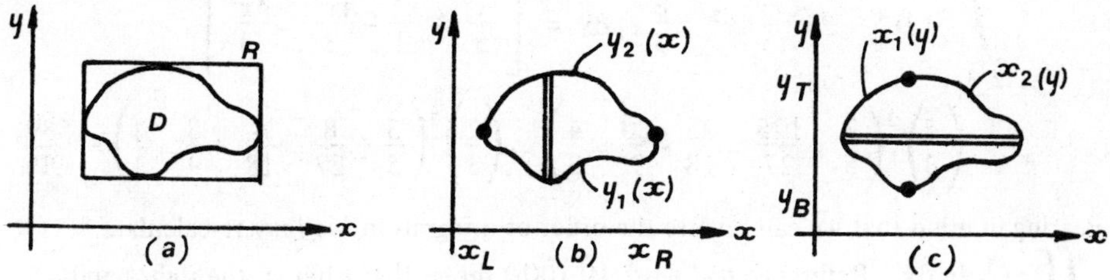

Figure 10.9

When we come to evaluate the double integral we shall first vary y and then vary x, as shown in Figure 10.9(b). Since D is restricted as shown above then there will be a least value of x, x_{min} say, and a greatest value x_{max}. As we traverse the boundary curve C from x_{min} to x_{max} we proceed along the upper part prescribed by a curve $y = y_2(x)$ and along the lower branch by a curve $y = y_1(x)$. A strip such as the one shown has upper and lower limits specified by $y_2(x)$ and $y_1(x)$ respectively. Hence

$$\iint_D f(x, y)\,dy\,dx = \int_{x_{min}}^{x_{max}} \int_{y_1(x)}^{y_2(x)} f(x, y)\,dy dx$$

Equally, since D is closed and bounded we can have a greatest y - value, y_{max} and a least y - value, y_{min} as shown in Figure 10.9(c). We shall then specify the limits on x by $x = x_1(y)$ and $x = x_2(y)$ respectively.

Example 1

Let D be the shaded region shown in Figure 10.10(a).

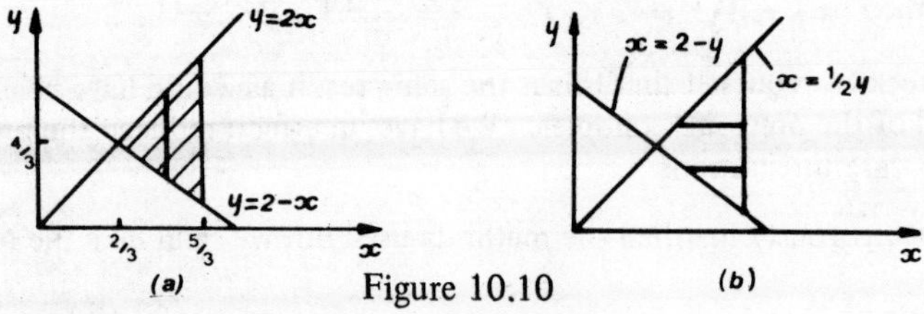

Figure 10.10

If we wish to integrate a function xy^2 over D then

$$I = \iint_D xy^2 \, dy \, dx = \int_{2/3}^{5/3} \int_{2-x}^{2x} xy^2 \, dy \, dx$$

The first step in the evaluation would be

$$\int_{2-x}^{2x} xy^2 \, dy = \left[\frac{xy^3}{3}\right]_{2-x}^{2x} = \frac{x(8x^3)}{3} - \frac{x(2-x)^3}{3}$$

$$= \frac{8x^4 - 8x + 12x^2 - 6x^3 + x^4}{3} = 3x^4 - 2x^3 + 4x^2 - \frac{8}{3}x$$

Then we calculate

$$\int_{2/3}^{5/3} (3x^4 - 2x^3 + 4x^2 - \frac{8}{3}x) \, dx = \left[\frac{3x^5}{5} - \frac{x^4}{2} + \frac{4x^3}{3} - \frac{4x^2}{3}\right]_{2/3}^{5/3}$$

$$= \left(\frac{5}{3}\right)^2\left(\frac{3}{5}\cdot\frac{125}{27} - \frac{25}{18} + \frac{20}{9} - \frac{4}{3}\right) - \left(\frac{2}{3}\right)^2\left(\frac{3}{5}\cdot\frac{8}{27} - \frac{4}{18} + \frac{8}{9} - \frac{4}{3}\right) = \frac{589}{90}$$

Bearing in mind that we can reverse the order of integration we may re-calculate $\iint_D xy^2 \, dy \, dx$. Reference to Figure 10.10(b) shows that whereas the right-hand boundary of D is always prescribed by $x = \frac{5}{3}$ the left-hand boundary is given by $x = \frac{1}{2}y$ for y in the range $\left[\frac{4}{3}, \frac{10}{3}\right]$ and by $x = 2 - y$ for y in the range $\left[\frac{1}{3}, \frac{4}{3}\right]$. Hence we need to express the problem as a sum of double integrals, viz.

$$\int_{1/3}^{4/3} \int_{2-y}^{5/3} xy^2 \, dx \, dy + \int_{4/3}^{10/3} \int_{y/2}^{5/3} xy^2 \, dx \, dy = I_1 + I_2$$

For I_1, we first calculate

$$\int_{2-y}^{5/3} xy^2 \, dx = \left[\frac{x^2y^2}{2}\right]_{2-y}^{5/3} = \frac{y^2}{2}\left(\frac{25}{9} - 4 + 4y - y^2\right)$$

$$= \frac{y^2}{2}\left(\frac{-11}{9} + 4y - y^2\right)$$

and then we evaluate

$$\int_{1/3}^{4/3} \left(-\frac{11}{18}y^2 + 2y^3 - \frac{1}{2}y^4\right) dy = \left[-\frac{11}{54}y^3 + \frac{y^4}{2} - \frac{1}{10}y^5\right]_{1/3}^{4/3}$$

$$= \frac{4941}{10 \times 3^6}$$

For I_2, we first calculate

$$\int_{\frac{1}{2}y}^{5/3} xy^2 \, dx = \left[\frac{x^2y^2}{2}\right]_{\frac{1}{2}y}^{5/3} = \frac{y^2}{2}\left(\frac{25}{9} - \frac{y^2}{4}\right)$$

and then we evaluate

$$\int_{4/3}^{10/3} \left(\frac{25}{18}y^2 - \frac{y^4}{8}\right) dy = \left[\frac{25}{54}y^3 - \frac{y^5}{40}\right]_{4/3}^{10/3} = \frac{42768}{10 \times 3^6}$$

Hence

$$I_1 + I_2 = \frac{47709}{10 \times 3^6} = \frac{589}{90} \quad \text{as before.}$$

Notice that this way round required two double integrals and it would obviously be preferable to choose that order of integration which avoided this difficulty. The problem may be such, however, that one way round the first integral may be hard or even impossible to evaluate. If this should be the case, we have to reverse the order of integration even if it means performing two double integrals. However, it may not always be easy to spot which way round is best.

Example 2

Consider Figure 10.11. We wish to calculate

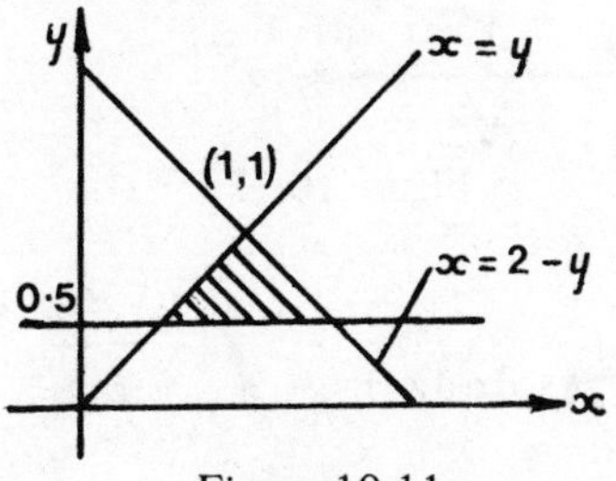

Figure 10.11

$$\iint_D xy^2 \, dx \, dy = \int_{0.5}^{1} \int_{y}^{2-y} xy^2 \, dx \, dy = \int_{y_{min}}^{y_{max}} \int_{x_1(y)}^{x_2(y)} xy^2 \, dx \, dy$$

First we calculate

$$\int_y^{2-y} xy^2\,dx = \left[\frac{x^2y^2}{2}\right]_y^{2-y} = \frac{(2-y)^2}{2}y^2 - \frac{y^4}{2} = 2y^2 - 2y^3$$

Then we evaluate

$$\int_{0.5}^{1} (2y^2 - 2y^3)\,dy = \left[2y^3/3 - y^4/2\right]_{0.5}^{1}$$

$$= \left[\frac{2}{3} - \frac{1}{2}\right] - \left[\frac{1}{12} - \frac{1}{32}\right] = \frac{1}{6} - \frac{5}{96} = \frac{11}{96}$$

Example 3

Evaluate

$$I = \int_0^1 \int_x^{\sqrt{2-x^2}} \frac{x}{\sqrt{x^2+y^2}}\,dy\,dx$$

As the integral stands, the first step would be to calculate

$$\int_x^{\sqrt{2-x^2}} \frac{x}{\sqrt{x^2+y^2}}\,dy$$

Since x is constant we are able to obtain a solution by putting $y = x\cosh\theta$, but we are in for some tedious algebra. Let us reverse the order of integration; and first we shall sketch the region over which the integral is defined; see Figure 10.12. Notice the price we shall have to pay: there are two double integrals to evaluate.

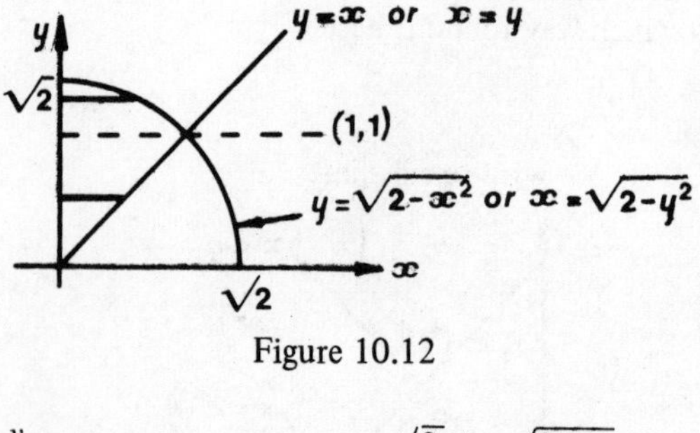

Figure 10.12

$$I = \int_0^1 \int_0^y \frac{x}{\sqrt{x^2+y^2}}\,dx\,dy + \int_1^{\sqrt{2}} \int_0^{\sqrt{2-y^2}} \frac{x}{\sqrt{x^2+y^2}}\,dx\,dy = I_1 + I_2$$

Now the first step for I_1 is to evaluate

$$\int_0^y \frac{x}{\sqrt{x^2+y^2}}\,dx = \left[(x^2+y^2)^{1/2}\right]_0^y = \sqrt{2}\,y - y$$

Then the second step gives

$$I_1 = \int_0^1 (\sqrt{2}\,y - y)\,dy = \left[(\sqrt{2} - 1)\,y^2/2\right]_0^1 = (\sqrt{2} - 1)/2$$

The first step for I_2 is to evaluate

$$\int_0^{\sqrt{2-y^2}} \frac{x}{\sqrt{x^2 + y^2}}\,dx = \left[(x^2 + y^2)^{1/2}\right]_0^{\sqrt{2-y^2}} = \sqrt{2} - y$$

and the second step gives

$$I_2 = \int_1^{\sqrt{2}} (\sqrt{2} - y)\,dy = \left[\sqrt{2}\,y - y^2/2\right]_1^{\sqrt{2}} = 2 - 1 - \sqrt{2} + 1/2 = 3/2 - \sqrt{2}$$

Therefore,

$$I = I_1 + I_2 = (1 - \sqrt{2}/2)$$

Example 4

Find $\displaystyle\int_1^2 \int_{-\sqrt{2-y}}^{\sqrt{2-y}} 2x^2y^2\,dx\,dy$. The region over which the integral is defined is shown in Figure 10.13. Performing the first integration,

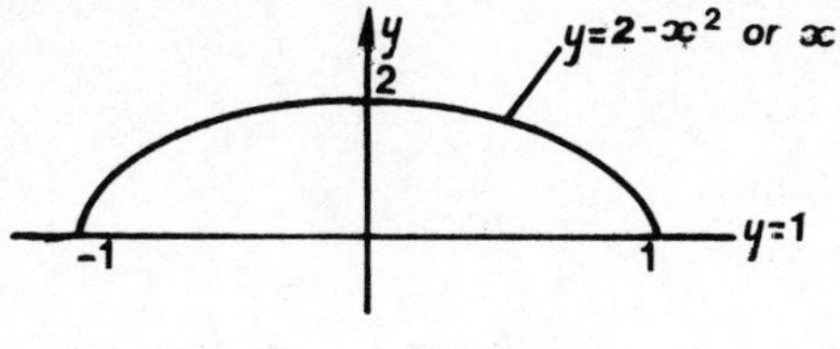

Figure 10.13

$$\int_{-\sqrt{2-y}}^{\sqrt{2-y}} 2x^2y^2\,dx = \left[\frac{2}{3}x^3y^2\right]_{-\sqrt{2-y}}^{\sqrt{2-y}} = \frac{4}{3}y^2\,(2 - y)^{3/2}$$

To carry out the operation $\displaystyle\int_1^2 \frac{4}{3}y^2\,(2 - y)^{3/2}\,dy$ would be tedious, but this is not necessarily obvious at first.

If we reverse the order of integration, we would need to find

$\displaystyle\int_{-1}^{1}\int_{1}^{2-x^2} 2x^2y^2\,dy\,dx$. First we calculate

$$\int_1^{2-x^2} 2x^2y^2\,dy = \left[\frac{2}{3}x^2y^3\right]_1^{2-x^2} = \frac{2}{3}x^2\left[(2-x^2)^3 - 1\right]$$

Then we evaluate

$$\int_{-1}^{1} \frac{2}{3}x^2\,[7 - 12x^2 + 6x^4 - x^6]\,dx$$

$$= 2\int_0^1 \frac{2}{3}x^2\,[7 - 12x^2 + 6x^4 - x^6]\,dx, \text{ by symmetry}$$

$$= \frac{4}{3}\left[\frac{7}{3}x^3 - \frac{12}{5}x^5 + \frac{6}{7}x^7 - \frac{x^9}{9}\right]_0^1 = \frac{4}{3}\left(\frac{7}{3} - \frac{12}{5} + \frac{6}{7} - \frac{1}{9}\right) = \frac{856}{945}$$

Problems

1. Evaluate the following integrals and sketch the region over which each is taken.

(a) $\displaystyle\int_0^3\int_1^3 xy\,dy\,dx$ (b) $\displaystyle\int_0^2\int_{-x}^{x} (x^2 + y^2)\,dy\,dx$

(c) $\displaystyle\int_{-1}^{1}\int_{-1}^{x} (2xy + x^2y^2)\,dy\,dx$ (d) $\displaystyle\int_{-1}^{1}\int_0^{\sqrt{1-y^2}} 2xy^2\,dx\,dy$

(e) $\displaystyle\int_{-1}^{2}\int_{x^2-2}^{x} (x + y)\,dy\,dx$

2. Convert each integral below into one or more integrals in which the order is reversed. Check by evaluating both forms.

(a) $\displaystyle\int_0^1\int_{x^2}^{x} y\,dy\,dx$ (b) $\displaystyle\int_{-2}^{2}\int_0^{4-x^2} dy\,dx$

(c) $\displaystyle\int_0^1\int_{1-x^2}^{\sqrt{1-x^2}} x\,dy\,dx$

3. Evaluate the double integrals below over the given regions. In each case sketch the region R.

(a) $\displaystyle\iint_R (x + y)\,dA;\ 0 \leqslant y \leqslant 2,\ y^2 \leqslant x \leqslant 4$ (b) $\displaystyle\iint_R dA;\ -1 \leqslant x \leqslant 1,\ -\tfrac{1}{2} \leqslant y \leqslant x^2 + 1$

(c) $\iint_R (x^2+y^2)\,dA \, ; \; \frac{y}{2} \leqslant x \leqslant 3-y, \; 0 \leqslant y \leqslant 2$

(d) $\iint_R \frac{x}{x^2+y^2}\,dA \, ; \; R$ bounded by $y=0, \; y=x, \; x=1, \; x=\sqrt{3}$

4. Find in each case below the volume of the solid bounded above by the given surface and below by the given region of the xy - plane.

(a) $z = e^x \cos y; \; 0 \leqslant x \leqslant \ln 3, \; -\frac{\pi}{2} \leqslant y \leqslant \frac{\pi}{2}$

(b) $z = c\sqrt{1 - \frac{x^2}{a^2} - \frac{y^2}{b^2}} \; ; \; \frac{x^2}{a^2} + \frac{y^2}{b^2} \leqslant 1$ (half an ellipsoid)

(c) $z = c\left(1 - \frac{x}{a} - \frac{y}{b}\right) \; ; \; 0 \leqslant x \leqslant a, \; 0 \leqslant y \leqslant b\left(1 - \frac{x}{a}\right)$ (a tetrahedron)

(d) $z = xy$; the region bounded by $y = x^2$ and $y = x(2-x)$

(e) $z = 4y^2 - x^2$; the region bounded by $x = 2y$ and $x = y^2$

5. Find the volume of the solid bounded by the given surfaces in each case below

(a) The bounding surfaces are $z = y, \; z = y+1, \; y = 0, \; y = 1, \; x = 0$ and $x = 1$

(b) The bounding surfaces are $z = 2y+4, \; z = y+2, \; y^2 = 4-x$ and $x = 0$

(c) The bounding surfaces are $x^2+y^2 = 1, \; x = 0, \; y = 0, \; z = y$ and $z = 2$

(d) The bounding surfaces are $x^2 = y+z, \; x = 2, \; y = 0$ and $z = 0$

(e) The bounding surfaces are $z = x^2+y^2, \; z = 0$ and $x^2+y^2 = 4$

10.6 FURTHER FEATURES OF DOUBLE INTEGRALS

We first quote, without proof, some properties of double integrals (we have tacitly assumed some of these). Suitable conditions on continuity can be assumed.

(i) If c is a constant then $\iint_D c\,f(x,y)\,dx\,dy = c\iint_D f(x,y)\,dx\,dy$

(ii) $\iint_D \{f(x,y)+g(x,y)\}\,dx\,dy = \iint_D f(x,y)\,dx\,dy + \iint_D g(x,y)\,dx\,dy$

(iii) If $D_1 \cap D_2 = \phi$ then

$$\iint_{D_1 \cap D_2} f(x,y)\,dx\,dy = \iint_{D_1} f(x,y)\,dx\,dy + \iint_{D_2} f(x,y)\,dx\,dy$$

(iv) If $\iint_D [f(x,y)]^2\,dx\,dy = 0$ and $D \neq \phi$ then $f(x,y) \equiv 0$ throughout D.

(v) If $m < f(x, y) < M$ *for* $(x, y) \in D$ and we denote $\iint_D 1 \, . \, dx \, dy$ by A then

$$mA < \iint_D f(x, y) \, dx \, dy < MA$$

(vi) Mean value theorem: there is a point $(x_n, y_n) \in D$ such that

$$\iint_D f(x, y) \, dx \, dy = f(x_n, y_n) \, . \, A \qquad [A \text{ as defined in (v)}]$$

(vii) $$\left| \iint_D f(x, y) \, dx \, dy \right| \leqslant \iint_D |f(x, y)| \, dx \, dy$$

Applications of double integrals

(i) *Area* If $f(x, y) \equiv 1$ in D then $\iint_D f(x, y) \, dy \, dx$ is the area of D.

Note the relationship with single integration to find an area.

(ii) *Mass* If $\rho(x, y)$ is mass/unit area, i.e. a surface density, then

$M = \iint_D \rho(x, y) \, dx \, dy$ is the mass of a lamina whose profile is D.

(iii) *Centre of mass* The coordinates of the centre of mass of the above lamina are given by $M\bar{x} = \iint_D x\rho(x, y) \, dx \, dy$ and $M\bar{y} = \iint_D y\rho(x, y) \, dx \, dy$

(iv) *Moments of inertia* The moment of inertia of the lamina about Ox is $\iint_D \rho(x, y) \, . \, y^2 \, dx \, dy$, about Oy is $\iint_D \rho(x, y) x^2 \, dx \, dy$ and the product of inertia about both axes is $\iint_D \rho(x, y) \, . \, xy \, dx \, dy$. The moment of inertia about the axis Oz is $\iint_D (x^2 + y^2) \rho(x, y) \, dx \, dy = \iint_D x^2 \rho(x, y) \, dx \, dy + \iint_D y^2 \rho(x, y) \, dx \, dy$.

(This is the **perpendicular axes theorem.**)

Example

Find the moments and product of inertia about the coordinate axes of the triangular lamina shown in Figure 10.14 where the mass distribution is given by $\rho(x, y) = xy$.

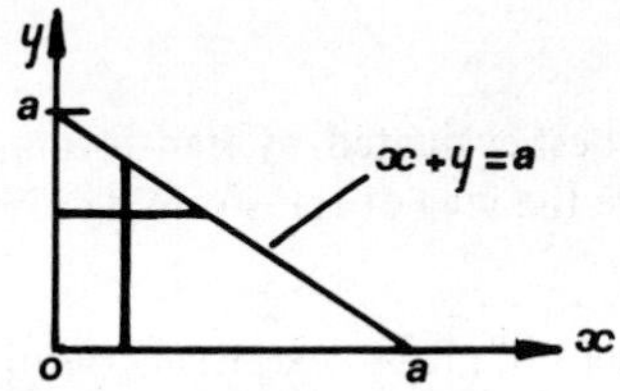

Figure 10.14

The moment of inertia about the axis Ox is $I_{xx} = \int_0^a \int_0^{a-y} xy \,.\, y^2 \, dx \, dy$

(we have taken a horizontal strip whose distance from the x - axis is y). Now

$$\int_0^{a-y} xy^3 \, dx = \left[\frac{x^2}{2} y^3 \right]_0^{a-y} = \frac{(a-y)^2}{2} y^3$$

Hence

$$I_{xx} = \int_0^a \frac{(a-y)^2}{2} y^3 \, dy = \frac{1}{2} \int_0^a (a^2 y^3 - 2ay^4 + y^5) \, dy$$

$$= \frac{1}{2} \left[a^2 \frac{y^4}{4} - \frac{2}{5} ay^5 + \frac{y^6}{6} \right]_0^a = \frac{a^6}{2} \left(\frac{1}{4} - \frac{2}{5} + \frac{1}{6} \right) = \frac{a^6}{120}$$

Note that the moment of inertia about the axis Oy, $I_{yy} = \dfrac{a^6}{120}$ by symmetry.
It is customary to quote moments of inertia in terms of the mass of the lamina,

$$M = \int_0^a \int_0^{a-y} xy \, dx \, dy = \int_0^a \frac{y(a-y)^2}{2} \, dy = \frac{a^4}{24}$$

Hence $I_{xx} = I_{yy} = \dfrac{Ma^2}{5}$

By the perpendicular axes theorem, $I_{zz} = \dfrac{Ma^2}{5} + \dfrac{Ma^2}{5} = \dfrac{2Ma^2}{5}$

The product of inertia

$$I_{xy} = \int_0^a \int_0^{a-y} (xy)(xy) \, dx dy \quad = \int_0^a \frac{(a-y)^3}{3} y^2 \, dy$$

$$= \frac{1}{3} \int_0^a (a^3 y^2 - 3a^2 y^3 + 3ay^4 - y^5) \, dy = \frac{a^6}{180} = \frac{2Ma^2}{15}$$

Polar coordinates

Some double integrals may be best evaluated by transforming to a new system of coordinates. We merely introduce the idea of transforming to polar coordinates via a geometrical approach.

Figure 10.15(a) shows an element of area in cartesian coordinates bounded by lines x = constant and y = constant; Figure 10.15(b) shows a corresponding element of area in polar coordinates, this time bounded by the curves r = constant and θ = constant. For completeness, we show in Figure 10.15(c) an element of area bounded by curves $\phi(x, y)$ = constant, $\psi(x, y)$ = constant; note that ϕ and ψ are orthogonal coordinates, i.e. curves of constant ψ and of constant ϕ intersect at right angles.

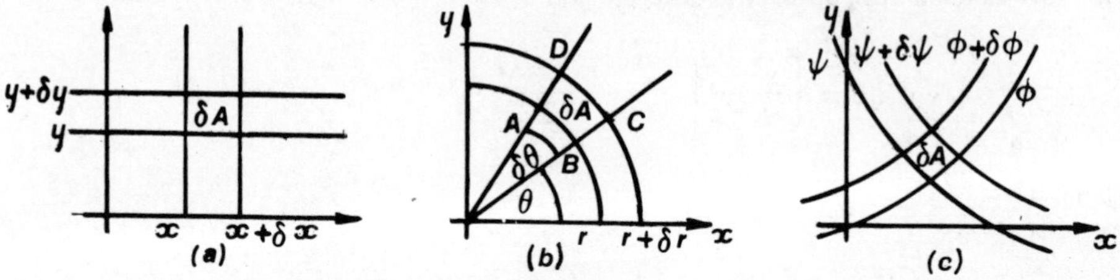

Figure 10.15

In polar coordinates, $\delta A \simeq r\delta\theta\, \delta r$, since we can treat the element of area, δA, as approximately rectangular and AB = $r\delta\theta \simeq$ DC. This will allow us to state that

$$\iint_A f(x, y)\, \mathrm{d}x\, \mathrm{d}y = \iint_A g(r, \theta)\, r\mathrm{d}r\, \mathrm{d}\theta \qquad (10.19)$$

where $g(r, \theta)$ is $f(x, y)$ expressed in polar coordinates via the equations

$$x = r\cos\theta, \quad y = r\sin\theta \qquad (10.20)$$

For example, $f(x, y) = xy^2 \Rightarrow r^3 \cos\theta \sin^2\theta = g(r, \theta)$.

Of course, we shall have to express the boundaries of A in terms of polar coordinates.

Example 1

Find the mass of a circular disc of radius a where the density is proportional to distance from the origin.

Clearly, the area over which the integral is defined can be expressed as $0 < r < a$, $0 < \theta < 2\pi$. The density function is $\rho(r, \theta) = kr$. The mass we require is given by

$$\int_0^{2\pi}\int_0^{a} kr\,.\,r\, \mathrm{d}r\, \mathrm{d}\theta = \int_0^{2\pi} \left[\frac{kr^3}{3}\right]_0^a \mathrm{d}\theta = \frac{ka^3}{3}\int_0^{2\pi} \mathrm{d}\theta = \frac{2\pi k a^3}{3}$$

To emphasise the simplification, you try and evaluate the integral in cartesian coordinates.

Example 2

Find the volume of the solid which is bounded above by the sphere $x^2 + y^2 + z^2 = a^2$ and bounded below by the cone $z^2 = x^2 + y^2$. The solid is shown in Figure 10.16(a).

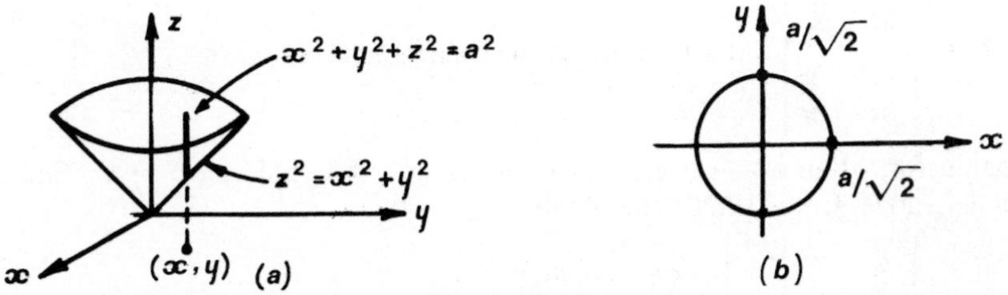

Figure 10.16

An element of volume whose base is centred at (x, y) in the plane $z = 0$ can be regarded as the difference between two volumes, one from the sphere down to $z = 0$ and the other from the cone down to $z = 0$. This element of volume is $\left\{(a^2 - x^2 - y^2)^{1/2} - (x^2 + y^2)^{1/2}\right\} \mathrm{d}x\,\mathrm{d}y$. We now have to define the region over which x and y are allowed to move. This is found by eliminating z between the equations of the two bounding surfaces, viz. $2(x^2 + y^2) = a^2$ i.e. $D \equiv x^2 + y^2 < \dfrac{a^2}{2}$; see Figure 10.16(b). Hence the volume we seek is

$$V = \iint_D \left\{(a^2 - x^2 - y^2)^{1/2} - (x^2 + y^2)^{1/2}\right\} \mathrm{d}x\,\mathrm{d}y$$

Converting to polar coordinates, we have $x^2 + y^2 = r^2$ and therefore

$$V = \int_0^{2\pi} \int_0^{a/\sqrt{2}} \left\{ (a^2 - r^2)^{1/2} - r \right\} r\,\mathrm{d}r\,\mathrm{d}\theta$$

$$= \int_0^{2\pi} \left[\frac{-(a^2 - r^2)^{3/2}}{3} - \frac{r^3}{3} \right]_0^{a/\sqrt{2}} \mathrm{d}\theta$$

$$= \int_0^{2\pi} \left(\frac{-a^3}{3.2\sqrt{2}} - \frac{a^3}{3.2\sqrt{2}} + \frac{a^3}{3} \right) \mathrm{d}\theta$$

$$= \frac{a^3}{6}(2 - \sqrt{2}) \int_0^{2\pi} \mathrm{d}\theta = \frac{\pi a^3}{3}(2 - \sqrt{2})$$

Notice that by exploiting the circular symmetry of the problem we were able to separate the integrals.

Some further remarks will be made on general transformation of coordinate systems.

We have added a scale factor to convert $dr\,d\theta$ into an element of area ($dr\,d\theta$ has dimensions of length). It can be shown in general that if we convert from one coordinate system (u, v) to a second system (s, t), where the connecting equations are $u = u(s, t)$, $v = v(s, t)$ then the scale factor is provided by

$$J = \begin{vmatrix} \dfrac{\partial u}{\partial s} & \dfrac{\partial u}{\partial t} \\ \dfrac{\partial v}{\partial s} & \dfrac{\partial v}{\partial t} \end{vmatrix}, \text{ the **Jacobian** of the transformation.}$$

In the transformation from cartesian coordinates (x, y) to polars (r, θ) we find the partial derivatives from (10.20) and obtain

$$J = \begin{vmatrix} \dfrac{\partial x}{\partial r} & \dfrac{\partial x}{\partial \theta} \\ \dfrac{\partial y}{\partial r} & \dfrac{\partial y}{\partial \theta} \end{vmatrix} = \begin{vmatrix} \cos\theta & -r\sin\theta \\ \sin\theta & r\cos\theta \end{vmatrix} = r\cos^2\theta + r\sin^2\theta = r$$

Area of curved surfaces

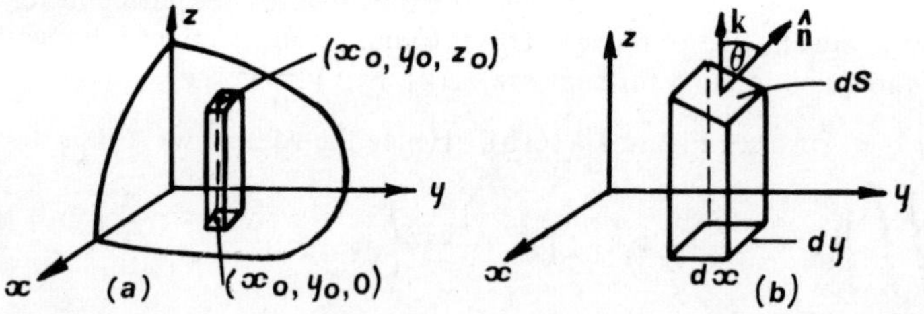

Figure 10.17

In Figure 10.17(a) we show a portion of a curved surface and we require to find its area. We show a small patch of the surface and one point on that patch; that point (x_0, y_0, z_0) is projected down onto the x - y plane. In the same way the small surface patch can be projected down onto the x - y plane. In Figure 10.17(b) we see the process in reverse: the element of area $dx\,dy$ in the x - y plane is projected back onto the curved surface to form an elementary area dS. If this elementary area is sufficiently small then it can be regarded as a plane. We know from earlier work that this plane has slopes given by $f_x(x_0, y_0)$ and $f_y(x_0, y_0)$ where the equation of the surface is expressed as $z = f(x, y)$. The equation of the tangent plane is

$$z = f(x_0, y_0) + (x - x_0)f_x(x_0, y_0) + (y - y_0)f_y(x_0, y_0) \quad \text{or}$$

$$[f(x_0, y_0) - x_0 f_x(x_0, y_0) - y_0 f_y(x_0, y_0)] + x f_x(x_0, y_0) + y f_y(x_0, y_0) - z = 0$$

The unit vector $\hat{\mathbf{n}}$ is given by

$$\left(\frac{f_x(x_0, y_0)}{D}, \frac{f_y(x_0, y_0)}{D}, \frac{-1}{D}\right) \text{ where } D = \sqrt{[f_x(x_0, y_0)]^2 + [f_y(x_0, y_0)]^2 + 1}$$

The area of the surface patch $dS = dx\,dy/(\hat{\mathbf{n}}\,.\,\mathbf{k}) = dx dy/\cos\theta$ where $\mathbf{k}$ is the unit vector in the z - direction. But $\cos\theta = 1/D$ and therefore $dS = dx\,dy\,.\,D$ i.e.

$$dS = dx\,dy\sqrt{[f_x(x_0, y_0)]^2 + [f_y(x_0, y_0)]^2 + 1} \tag{10.21}$$

Example 1

Find the surface area of the hemispherical shell $x^2 + y^2 + z^2 = a^2$, $z > 0$. It should be clear that the projection of the hemispherical shell on the x - y plane is the disk $x^2 + y^2 \leqslant a^2$. Then, from (10.21), the surface area is, by symmetry,

$$4 \int_0^a \int_0^{\sqrt{a^2 - x^2}} \sqrt{[f_x(x_0, y_0)]^2 + [f_y(x_0, y_0)]^2 + 1} \; dy\, dx$$

But for the surface $z = (a^2 - x^2 - y^2)^{1/2} = f(x, y)$,

$$f_x = \frac{-x}{(a^2 - x^2 - y^2)^{1/2}}, \quad f_y = \frac{-y}{(a^2 - x^2 - y^2)^{1/2}}$$

Hence

$$\left([f_x(x_0, y_0)]^2 + [f_y(x_0, y_0)]^2 + 1\right)^{1/2} = \left[\frac{x^2 + y^2 + (a^2 - x^2 - y^2)}{(a^2 - x^2 - y^2)}\right]^{1/2}$$

$$= a/(a^2 - x^2 - y^2)^{1/2}$$

Therefore the area we require,

$$S = 4 \int_0^a \int_0^{\sqrt{(a^2 - x^2)}} \frac{a}{(a^2 - x^2 - y^2)^{1/2}} \, dy\, dx$$

Converting this to polar coordinates, we obtain

$$S = 4 \int_0^{\pi/2} \int_0^a \frac{a}{(a^2 - r^2)^{1/2}} \, r\, dr\, d\theta$$

i.e. $$S = 4a \left[\int_0^{\pi/2} d\theta\right] \left[\int_0^a \frac{r}{(a^2 - r^2)^{1/2}} \, dr\right] = 4a \, . \frac{\pi}{2} \, . \, a = 2\pi a^2$$

Example 2

Find the area of that part of the surface of the cylinder $x^2 + z^2 = a^2$ which lies inside the cylinder $x^2 + y^2 = a^2$. The appropriate surface is symmetrical about all three axes, one-eighth of it is shown in Figure 10.18.

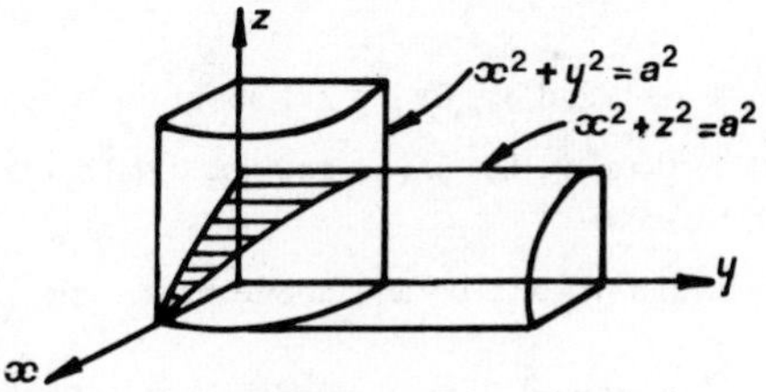

Figure 10.18

Let us find the shaded area A.

The equation of the curved surface is $z = \sqrt{a^2 - x^2}$ (note $+\sqrt{\ }$) and hence $f_x = -x/\sqrt{a^2 - x^2}$, $f_y = 0$. Therefore

$$dS = \sqrt{\frac{x^2 + 0 + (a^2 - x^2)}{a^2 - x^2}}\,dxdy = \frac{a}{\sqrt{a^2 - x^2}}\,dx\,dy$$

The projection of the surface on the x - y plane is the quadrant of the circle $x^2 + y^2 \leqslant a^2$, $x > 0$, $y > 0$. The equation is best expressed in polars but the integrand is not and the area we seek is best given by

$$S = \int_0^a \int_0^{\sqrt{a^2 - x^2}} \frac{a}{(a^2 - x^2)^{1/2}}\,dy\,dx$$

$$= a \int_0^a \frac{1}{(a^2 - x^2)^{1/2}} \int_0^{\sqrt{a^2 - x^2}} dy\,dx = a \int_0^a 1\,.\,dx = a^2$$

The total surface is thus $8a^2$.

Problems

1. Sketch the region R and compute its area by means of a double integral for the following:

   (a) R is bounded by $y = x^3$ and $y = x^{1/2}$

   (b) R is bounded by $y = \sin x$ and $y = \dfrac{2}{\pi}x$

   (c) R is determined by the inequalities $x^2 + (y - 1)^2 \leqslant 1$, $y \geqslant 2 - x^2$

2. Find the mass of the region R corresponding to the given density for the following:

   (a) R is the triangle with vertices (0, 0), (−2, 0) and (−2, 2); $\rho = x^2 + y^2$

   (b) R is the circular disk $x^2 + y^2 \leqslant 25$; $\rho = x^2 + y^2$

   (c) R is bounded by $y = 0$ and $y = \sqrt{4 - x^2}$; $\rho = x^2$

3. Compute the mean value of the given function in the given region for the following:

   (a) $f(x, y) = x + y$; R the region $x^2 + y^2 \leqslant 1$

   (b) $f(x, y) = x$; R the region $y \geqslant 0$, $y \leqslant \sin x$, $0 \leqslant x \leqslant \pi$

4. Find the moment of inertia of the given mass distribution in the xy - plane, about the given axis for the following:

   (a) R bounded by $y = 4 - x^2$ and $y = 0$; $\rho = 2y$; about the y - axis

   (b) R the triangle with vertices (0, 0), $(a, 0)$ and (b, c):$(a, b, c > 0)$; $\rho = k$, a positive constant; about the x - axis

   (c) R bounded by $y = x$ and $y = x^2$; $\rho = xy$; about the z - axis

5. Find the centre of mass of the region R for the following:

   (a) R the triangle with vertices (0, 0), (1, 0) and (1, 1); $\rho = x^2 + y^2$

(b) R the region $0 \leqslant y \leqslant e^x$, $0 \leqslant x \leqslant 1$; $\rho = x$

6. Transform the following integrals into polar coordinates and evaluate them.

(a) $$\int_{-1}^{1} \int_{-\sqrt{1-y^2}}^{\sqrt{1-y^2}} e^{-(x^2+y^2)} \, dx \, dy$$

(b) $$\int_{0}^{\pi/2} \int_{0}^{\sqrt{\frac{1}{4}\pi^2 - x^2}} \frac{1 + x^2 + y^2}{\sqrt{x^2 + y^2}} \, dy \, dx$$

(c) $$\int_{0}^{1} \int_{0}^{\sqrt{1-y^2}} xy \sqrt{x^2 + y^2} \, dx \, dy$$

7. Evaluate the integrals below.

(a) $\displaystyle\iint_R (x^2 + y^2)^{-3/2} \, dA$; $-3 \leqslant x \leqslant 0$, $-\sqrt{9 - x^2} \leqslant y \leqslant 0$

(b) $\displaystyle\iint_R e^{-(x^2+y^2)/2} \, dA$; R the sector of $x^2 + y^2 \leqslant 1$ cut out by $y = |x|$

(c) $\displaystyle\iint_R \sqrt{x^2 + y^2} \, dA$; R bounded by $x^2 + y^2 = 2x$

(d) $\displaystyle\iint_R y \, dA$; R the first quadrant region bounded by $y = \sqrt{3}x$ and $x^2 + y^2 = 4y$

8. Find the area of the given regions
   (a) The region inside the circle $r = 8 \cos \theta$ and outside the circle $r = 4$
   (b) The region bounded by $y^2 = 4x$ and $y = x/2$
   (c) The region inside the cardioid $r = 1 + \cos \theta$ and to the left of the line $4x = 3$
   (d) The region bounded on the left by the parabola $y^2 = 1 - 2x$ and on the right by $x^2 + y^2 = 1$

9. Find the Jacobian of each of the given transformations

   (a) $x = u^2 + v^2$, $y = u^2 - v^2$
   (b) $x = 2uv - v^2$, $y = v^2 - 2u$
   (c) $x = \sin r \cos t$, $y = \cos r \cos t$

10. Show that the area bounded by the isothermal curves $pv = a$, $pv = b$ and the adiabatic curves $pv^\gamma = c$, $pv^\gamma = d$ is $\dfrac{b - a}{\gamma - 1} \log \left[\dfrac{d}{c}\right]$

11. Find the area of the given surface over the given region for the following:

(a) $z = 2x - 2y + 1;\ 0 \leqslant x \leqslant 1,\ 0 \leqslant y \leqslant 2$

(b) $z = x + 2y + 3;\ 1 \leqslant y \leqslant 4,\ y \leqslant x \leqslant y^2$

(c) $z = xy;\ x^2 + y^2 \leqslant 1,\ x \geqslant 0,\ y \geqslant 0$

(d) $x^2 + y^2 + z^2 = a^2$ above one loop of $r = a \cos 3\theta$

(e) Find the surface area of the portion of the paraboloid $y^2 + z^2 = 4x$ intercepted by the planes $y = 0,\ x = 3;\ y \geqslant 0$

(f) A cylindrical hole of radius b is cut through the centre of a sphere of radius $a > b$. Find the surface area removed from the sphere.

12. Show in a diagram the region over which the integral

$$\int_0^1 dx \int_0^{\sqrt{(x - x^2)}} \frac{4xy}{x^2 + y^2}\, e^{-(x^2 + y^2)}\, dy$$

extends. Transform to polar coordinates and hence or otherwise show that the integral has the value e^{-1}. (L.U.)

13. A plate in the form of a quadrant of the ellipse $x^2/a^2 + y^2/b^2 = 1$ is of small but varying thickness which is at any point proportional to the product of its distances from the axes. Show that the coordinates of the centroid are $\left(\frac{8a}{15}, \frac{8b}{15}\right)$. (L.U.)

14. $x^2/a^2 + y^2/b^2 = 1$ gives the contour of the base of a right circular cylinder; the height is c. The cylinder is bevelled down so that the height z of any point (x, y) of the base in the first quadrant is given by $z = c\left(1 - \frac{x}{a}\right)\left(1 - \frac{y}{b}\right)$. Express the volume left in this quadrant as a double integral and show that its value is $\frac{1}{4}abc(\pi - \frac{13}{6})$. (L.U.)

15. The length of the side of a square plate ABCD is $2a$, and O is the mid point of AB. If the surface density at any point P on the plate is $\lambda \mathrm{OP}^2$, λ being a constant, express the mass of the plate as a double integral. Show that the distance of the centre of mass of the plate from AB is $1.4a$ and find the moment of inertia of the plate about AB. (L.U.)

16. A footstep bearing carries a total load W distributed over a circle of radius a; the coefficient of friction between the bearing surfaces is μ. When new the load is uniformly distributed over the circle, but when worn the intensity varies inversely as the distance from the centre. Calculate the frictional torque in each case and show that for a worn bearing it is ¾ of that for a new one. (L.U.)

17. A heavy uniform plate of weight W in the form of the cardioid $r = a(1 + \cos\theta)$ lies on a rough horizontal table (coefficient of friction μ), and is movable about a vertical axis through the pole of coordinates. Show that the couple which must be applied to the plate in order to make it move is $\frac{10}{9}\mu a W$. (L.U.)

18. The electrical attraction at a point situated at a distance r from an infinite plane due to a surface density of electricity γ is given by

$$r\gamma \int_0^{\infty} \int_0^{2\pi} \frac{x\,d\theta\,dx}{(r^2 + x^2)^{3/2}}$$

Find the value of this attraction.

19. Find $\iint (a - x)^2\,dx\,dy$ taken over half the circle $x^2 + y^2 = a^2$. A horizontal boiler has a flat bottom, and its ends are plane and semicircular. If it is just filled with water, show that the depth of the centre of pressure of either end is 0.7 x total depth very nearly. (L.U.)

20. Let R_{uv} be the rectangular region $1 \leqslant u \leqslant 2, -1 \leqslant v \leqslant 1$. Evaluate

$$\iint_{R_{xy}} x\,dA_{xy} \quad \text{when} \quad x = u^2 - v^2, \; y = 2uv.$$ Sketch the region R_{xy}.

10.7 VOLUMES OF INTERPENETRATION

In this section, we extend our techniques for finding volumes of the intersection of two solids. We shall accomplish our aim by means of four examples.

Example 1

Find the volume of that part of the cylinder $x^2 + y^2 = a^2$ which is cut off by the plane $z = 0$ and the paraboloid $x^2 + y^2 = 4a^2 - z$. The volume under consideration is shown in Figure 10.19(a).

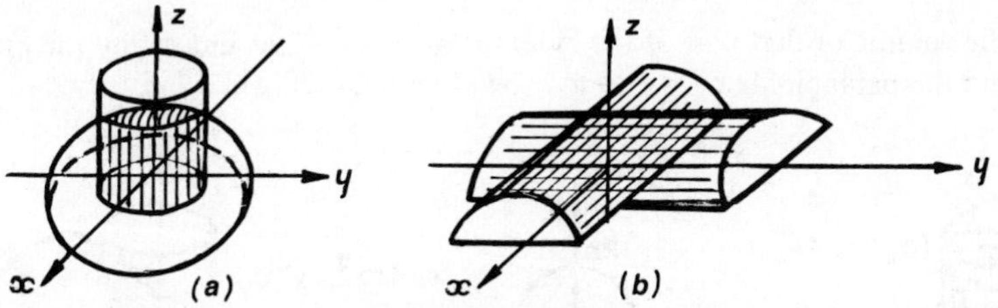

Figure 10.19

The height of an element of volume positioned at (x, y) is $z = 4a^2 - x^2 - y^2$ and the domain of (x, y) is the disk $x^2 + y^2 \leqslant a^2$. Hence the required volume is

$$V = 4 \int_0^a \int_0^{\sqrt{a^2 - x^2}} (4a^2 - x^2 - y^2)\,dy\,dx$$

where we have made an appeal to symmetry.

Converting to polar coordinates we obtain

$$V = 4 \int_0^{\pi/2} \int_0^a (4a^2 - r^2)\,dr\,r\,d\theta = 4 \int_0^{\pi/2} \left(\frac{4a^4}{2} - \frac{a^4}{4}\right) d\theta$$

$$= 4 \int_0^{\pi/2} \frac{7}{4} a^4 \, d\theta = \frac{7\pi a^4}{2}$$

Example 2

Find the volume which is common to the cylinders $x^2 + z^2 = b^2$ and $z^2 + y^2 = b^2$ see Figure 10.19(b).

By symmetry, the required volume is eight times that in the octant $x > 0$, $y > 0$, $z > 0$. We shall find this latter volume V.

The height of the elemental volume positioned at (x, y) is not easy to determine and this time we shall integrate with respect to x and z and state that y is given by $\sqrt{b^2 - z^2}$. The limits on (x, y) are $x^2 + z^2 < b^2$ and therefore

$$V = \int_0^b \int_0^{\sqrt{b^2 - z^2}} (b^2 - z^2)^{1/2} \, dx \, dz$$

$$= \int_0^b (b^2 - z^2)^{1/2} (b^2 - z^2)^{1/2} \, dz = \int_0^b (b^2 - z^2) \, dz = \frac{2b^3}{3}$$

The original volume asked for is $\dfrac{16b^3}{3}$.

Example 3

Find the volume of that part of the cylinder $x^2 + y^2 = 2ay$ cut off by the plane $z = \frac{1}{2} y$ and the paraboloid $x^2 + y^2 = az$. See Figure 10.20(a).

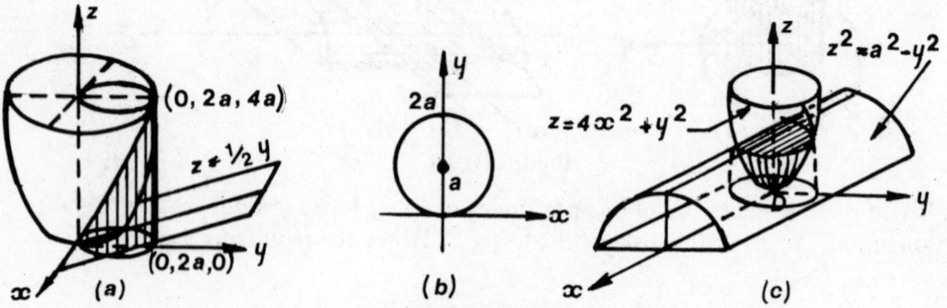

Figure 10.20

The domain D over which (x, y) can move is the disk $x^2 + (y - a)^2 \leqslant a^2$ (or $x^2 + y^2 \leqslant 2ay$) and the values of z at the position (x, y) are $\frac{1}{2}y < z < \frac{1}{a}(x^2 + y^2)$. This would suggest that the volume is given by

$$V = \iint_D \left(\frac{1}{a}(x^2 + y^2) - \frac{1}{2}y \right) dx \, dy.$$

The difficulty is to describe the domain D adequately in cartesian coordinates. To describe it in polar coordinates we first obtain for the boundary $r^2 = 2a\, r \sin\theta$ i.e. $r = 2a \sin\theta$. The region D is therefore described by $0 < r < 2a \sin\theta$, $0 < \theta < \pi$; see Figure 10.20(b). Hence the volume

$$V = \int_0^{\pi} \int_0^{2a\sin\theta} \left(\frac{r^2}{a} - \frac{1}{2} r \sin\theta\right) r dr\, d\theta$$

Now $$\int_0^{2a\sin\theta} \left(\frac{r^3}{a} - \frac{1}{2} r^2 \sin\theta\right) dr = \left[\frac{r^4}{4a} - \frac{1}{6} r^3 \sin\theta\right]_0^{2a\sin\theta}$$

$$= 4a^3 \sin^4\theta - \frac{4}{3} a^3 \sin^4\theta = \frac{8}{3} a^3 \sin^4\theta$$

Therefore

$$V = \frac{8a^3}{3} \int_0^{\pi} \sin^4\theta\, d\theta = \frac{8a^3}{3}\left[\frac{3\theta}{8} - \frac{\sin 2\theta}{4} + \frac{\sin 4\theta}{32}\right]_0^{\pi}$$

$$= \frac{8a^3}{3}\left(\frac{3\pi}{8}\right) = \pi a^3$$

Example 4

Find the volume common to the parabolic cylinder $z = a^2 - y^2$ and the paraboloid $z = 2x^2 + y^2$. See Figure 10.20(c).

The domain D is found by projecting on the xy - plane the intersection of the two solids which is found by eliminating z between the two equations; this intersection is $a^2 - y^2 = 2x^2 + y^2$ i.e. $2x^2 + 2y^2 = a^2$, i.e. a circle. The resulting volume is given by

$$V = \int_{-a}^{0} \int_{-\frac{\sqrt{a^2-2x^2}}{\sqrt{2}}}^{\frac{\sqrt{a^2-2x^2}}{\sqrt{2}}} \left(a^2 - y^2 - (2x^2 + y^2)\right) dy\, dx$$

Employing symmetry about the $x-$ and $y-$ axes and transforming to polar coordinates we obtain

$$V = 4 \int_0^{\frac{\pi}{2}} \int_0^{\frac{a}{\sqrt{2}}} \left\{a^2 - r^2 \sin^2\theta - (2r^2 \cos^2\theta + r^2 \sin^2\theta)\right\} r dr d\theta$$

$$= 4 \int_0^{\frac{\pi}{2}} \int_0^{\frac{a}{\sqrt{2}}} \left\{a^2 - 2r^2\right\} r dr d\theta$$

$$= 4 \int_0^{\frac{\pi}{2}} d\theta \int_0^{\frac{a}{\sqrt{2}}} (a^2 r - 2r^3)\, dr = 4 \cdot \frac{\pi}{2} \cdot \left[a^2 r^2/2 - r^4/2 \right]_0^{\frac{a}{\sqrt{2}}}$$

$$= 4 \cdot \frac{\pi}{2} \cdot \frac{a^4}{2} \cdot \frac{1}{4} = \frac{\pi a^4}{4}$$

Problems

1. Find the volume of the given regions:

   (a) The region bounded above by the plane $z = x$ and below by the paraboloid $z = x^2 + y^2$.

   (b) The region bounded by the cone $z^2 = x^2 + y^2$ and the paraboloid $3z = x^2 + y^2$.

   (c) The region above the xy - plane bounded by the paraboloid $z = x^2 + y^2$ and the cylinder $x^2 + (y + 2)^2 = 4$.

   (d) The region bounded above by the cone $z^2 = x^2 + y^2$ and below by the region of the xy - plane inside the circle $x^2 + y^2 = 6x$.

2. Prove that the volume bounded by the plane $z = 0$, the cylinder $x^2 + y^2 = a^2$ and the paraboloid $x^2 + y^2 = 4b(c - z)$ is $\pi a^2 \left[c - \frac{5a^2}{48b} \right]$.

3. Prove that the volume cut off the cylinder $(x - a)^2 + (y - b)^2 = c^2$ by the plane $z = 0$ and the hyperbolic paraboloid $xy = dz$ is $\frac{\pi abc^2}{d}$.

4. Prove that the volume cut from the surface $z^n = ax^2 + by^2$ by any plane parallel to $z = 0$ is $\frac{1}{(n+1)}$ th of part of the cylinder standing on the same section and terminated by the plane $z = 0$.

5. A circular hole of radius b is made centrally through a sphere of radius a. Find the volume of the sphere remaining.

6. Find the volume of the part of the cylinder $x^2 + y^2 - 2ax = 0$ cut off by the two planes $z = x \tan \alpha$, $z = x \tan \beta$ $(\alpha > \beta > 0)$.

7. A right cone, semivertical angle α, has its vertex on the surface of a sphere and its axis along a diameter of the sphere. Find the volume common to the cone and the sphere. (Use polar coordinates, pole at vertex, initial line along the diameter. Equation of sphere is $r = 2a \cos \theta$).

8. Find the volume included between a right circular cone, semivertical angle 30°, and a sphere of radius a touching it along a circle.

9. Find the volume cut off the sphere $x^2 + y^2 + z^2 = a^2$ by the cone $x^2 + y^2 = z^2$.

10. Find the volume enclosed by the surfaces $x^2 + y^2 = cz$, $x^2 + y^2 = 2ax$ and $z = 0$.

11. Find the volume included between the elliptic paraboloid $2z = \frac{x^2}{p} + \frac{y^2}{q}$, the cylinder $x^2 + y^2 = a^2$ and the plane $z = 0$.

10.8 TRIPLE INTEGRALS

The techniques obtained for double integrals may be extended to triple integrals. We shall restrict our attention here to using triple integration to find volumes. The basic idea will be to take an element of volume $dxdydz$ in cartesian coordinates and perform three integrations in turn with respect to x, y and z.

Example 1

Find the volume of the solid bounded by the surfaces $z = x + 2$, $z = 0$, $y = x^2$ and $y = 2x + 3$.

The relevant solid is shown in Figure 10.21(a). The vertical walls are the parabolic cylinder $y = x^2$ and the plane $y = 2x + 3$. The base of the solid is shown in Figure 10.21(b) and is the projection on the x - y plane of any cross-section of the solid.

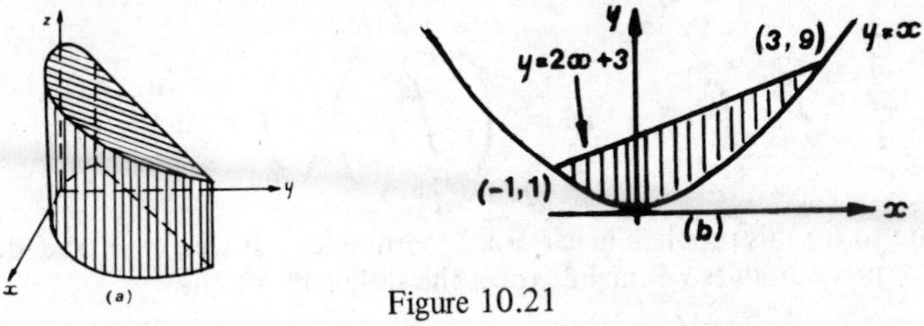

Figure 10.21

The volume V of the solid is found by integrating the element of volume $dxdydz$ over the region specified by $0 < z < x + 2$, $x^2 < y < 2x + 3$, $-1 < x < 3$. We could change the order of integration so as not to integrate with respect to x last, but this would involve a rewriting of the equations of the boundaries of the solid. Therefore

$$V = \int_{-1}^{3} \int_{x^2}^{2x+3} \int_{0}^{x+2} dz\,dy\,dx = \int_{-1}^{3} \int_{x^2}^{2x+3} (x+2)\,dy\,dx$$

$$= \int_{-1}^{3} (x+2)(2x+3-x^2)\,dx$$

$$= \int_{-1}^{3} (-x^3+7x+6)\,dx = \left[-\frac{x^4}{4} + \frac{7x^2}{2} + 6x \right]_{-1}^{3}$$

$$= \left(\frac{-81}{4} + \frac{63}{2} + 18 \right) - \left(\frac{-1}{4} + \frac{7}{2} - 6 \right) = 32$$

We could, of course, have carried out the solution by means of a double integration. This would have meant doing the z integration mentally.

We can usefully employ other coordinate systems. Two of the most common are cylindrical coordinates and spherical coordinates. In Figure 10.22(a) we show the element of volume in **cylindrical coordinates**; it is $r dr d\theta dz$. The transforming equations are $x = r\cos\theta$, $y = r\sin\theta$, $z = z$. (10.22)

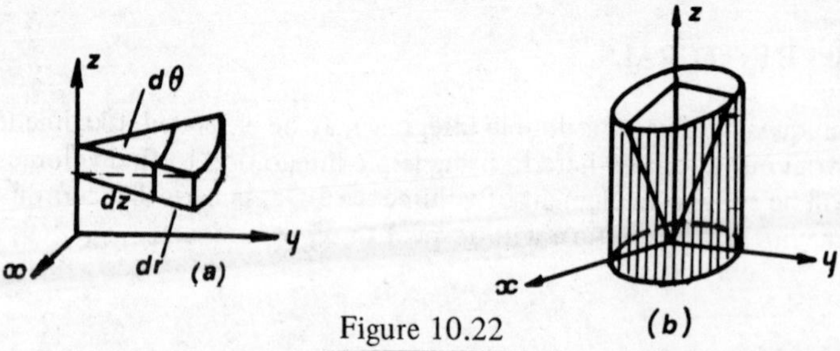

Figure 10.22

Example 2

Find the volume of the solid in the first octant bounded by the cone $z^2 = x^2 + y^2$ and the cylinder $x^2 + y^2 = a^2$; See Figure 10.22(b). In cylindrical polars, the cylinder has equation $r = a$ and the cone has the equation $z = r$. The solid can be defined by the conditions $0 < r < a$, $0 < \theta < \pi/2$, $0 < z < r$. Therefore the volume required,

$$V = \int_0^{\frac{\pi}{2}} \int_0^a \int_0^r r \, dz \, dr \, d\theta = \left(\int_0^{\frac{\pi}{2}} d\theta \right) \left(\int_0^a r^2 . dr \right) = \frac{\pi}{2} \cdot \frac{a^3}{3} = \frac{\pi a^3}{6}$$

You might care to try this problem in cartesian coordinates. If we extend the idea of a Jacobian to three variables we might expect the scale factor to be

$$\begin{vmatrix} \frac{\partial x}{\partial r} & \frac{\partial x}{\partial \theta} & \frac{\partial x}{\partial z} \\ \frac{\partial y}{\partial r} & \frac{\partial y}{\partial \theta} & \frac{\partial y}{\partial z} \\ \frac{\partial z}{\partial r} & \frac{\partial z}{\partial \theta} & \frac{\partial z}{\partial z} \end{vmatrix} \quad \text{i.e.} \quad \begin{vmatrix} \cos\theta & -r\sin\theta & 0 \\ \sin\theta & r\cos\theta & 0 \\ 0 & 0 & 1 \end{vmatrix} = r$$

Spherical coordinates are defined by the following rules. Choose a value of r. The condition $r = a$ fixes a spherical shell centred at the origin of radius a; a point in space must lie on one such shell and we fix its position on that shell by two angles. The angle ϕ measures declination from the vertical and the curve $\phi =$ constant is a cone; see Figure 10.23(a). Such a cone will intersect the spherical shell $r = a$ in a horizontal circle; see Figure 10.23(b). The angle θ is made by a line joining our point in space to the origin with a fixed line through the origin. In a sense, θ is a measure of longitude (though in our system $0 < \theta < 2\pi$) and ϕ is a kind of co-latitude, where $0 < \phi < \pi$.

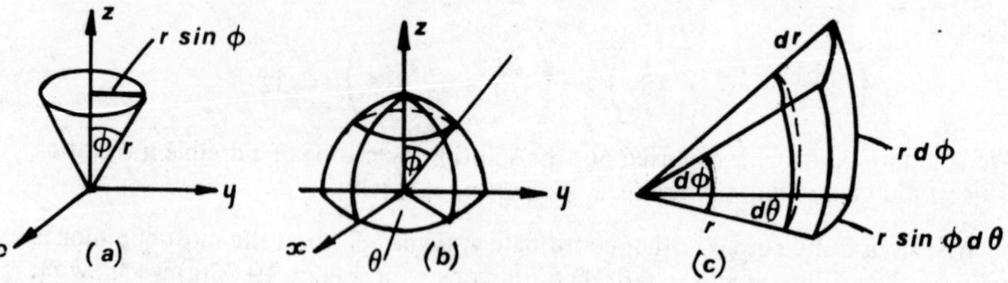

Figure 10.23

The transforming equations are

$$x = r\sin\phi\cos\theta, \quad y = r\sin\phi\sin\theta, \quad z = r\cos\phi \tag{10.23}$$

The element of volume is found by considering Figure 10.23(c) to be $r^2 \sin\phi \, dr \, d\phi \, d\theta$.

Alternatively, we find the scale factor from the Jacobian

$$\begin{vmatrix} \dfrac{\partial x}{\partial r} & \dfrac{\partial x}{\partial \phi} & \dfrac{\partial x}{\partial \theta} \\ \dfrac{\partial y}{\partial r} & \dfrac{\partial y}{\partial \phi} & \dfrac{\partial y}{\partial \theta} \\ \dfrac{\partial z}{\partial r} & \dfrac{\partial z}{\partial \phi} & \dfrac{\partial z}{\partial \theta} \end{vmatrix} = \begin{vmatrix} \sin\phi\cos\theta & r\cos\phi\cos\theta & -r\sin\phi\sin\theta \\ \sin\phi\sin\theta & r\cos\phi\sin\theta & r\sin\phi\cos\theta \\ \cos\phi & -r\sin\phi & 0 \end{vmatrix}$$

$$= \cos\phi \begin{vmatrix} r\cos\phi\cos\theta & -r\sin\phi\sin\theta \\ r\cos\phi\sin\theta & r\sin\phi\cos\theta \end{vmatrix} + r\sin\phi \begin{vmatrix} \sin\phi\cos\theta & -r\sin\phi\sin\theta \\ \sin\phi\sin\theta & r\sin\phi\cos\theta \end{vmatrix}$$

$$= [\cos\phi] r^2 \left(\cos\phi\sin\phi\cos^2\theta + \cos\phi\sin\phi\sin^2\theta\right) + r^2\sin\phi\,[\sin^2\phi\cos^2\theta + \sin^2\phi\sin^2\theta]$$

$$= r^2\cos^2\phi\,\sin\phi + r^2\sin^3\phi = r^2\sin\phi$$

Example 1

Find the volume of the solid bounded above by the sphere, $x^2 + y^2 + (z - a)^2 = a^2$ and below by the cone $z^2 = x^2 + y^2$. See Figure 10.24

The cone has the equation $r^2\cos^2\phi = r^2\sin^2\phi$ i.e. $\tan\phi = 1$ and hence $\phi = \pi/4$. The sphere has the equation $r^2\sin^2\phi + (r\cos\phi - a)^2 = a^2$ i.e. $r^2 - 2ar\cos\phi + a^2 = a^2$ or $r = 2a\cos\phi$.

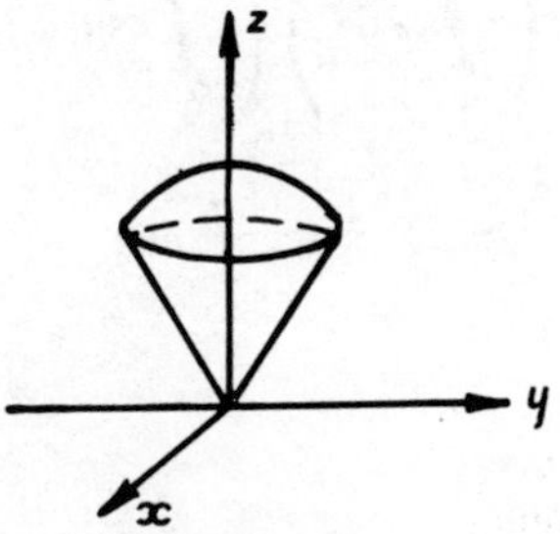

Figure 10.24

The solid is defined by the conditions $0 \leqslant \theta \leqslant 2\pi$, $0 \leqslant \phi \leqslant \pi/4$, $0 < r < 2a\cos\phi$. Using symmetry, we have that the volume

$$V = 4\int_0^{\frac{\pi}{2}}\int_0^{\frac{\pi}{4}}\int_0^{2a\cos\phi} r^2\sin\phi\,d r d\phi d\theta$$

$$= 4\left(\int_0^{\frac{\pi}{2}} d\theta\right)\int_0^{\frac{\pi}{4}} \sin\phi\left[r^3/3\right]_0^{2a\cos\phi} d\phi$$

$$= 4\,\frac{\pi}{2}\,.\,8\int_0^{\frac{\pi}{4}} \frac{a^3}{3}\quad \cos^3\phi\,\sin\phi\,d\phi$$

$$= -\frac{16\pi a^3}{3}\left[\frac{\cos^4\phi}{4}\right]_0^{\frac{\pi}{4}} = \frac{4a^3\pi}{3}\left(1-\frac{1}{4}\right) = \pi a^3$$

Since the volume of the spherical 'hat' is $\frac{2}{3}\pi a^3$ it follows that the cone contributes $\frac{1}{3}\pi a^3$ to the total.

Example 2

Find the volume of the ellipsoid $\frac{x^2}{a^2} + \frac{y^2}{b^2} + \frac{z^2}{c^2} \leqslant 1$.

We shall find it helpful to make a substitution not quite to spherical coordinates but to transfer to coordinates by $x = ar\sin\phi\cos\theta$, $y = br\sin\phi\sin\theta$, $z = cr\cos\phi$. The Jacobian as you can verify is $abcr^2\sin\phi$. The solid is now described by $0 \leqslant \theta \leqslant 2\pi$, $0 \leqslant \phi \leqslant \pi$, $0 \leqslant r \leqslant 1$. Its volume is given by

$$V = \int_0^{2\pi}\int_0^{\pi}\int_0^{1} abc\,r^2\,\sin\phi\,dr\,d\phi\,d\theta$$

and we may separate the integrals to give

$$V = abc\left(\int_0^{2\pi} d\theta\right)\left(\int_0^{\pi}\sin\phi\,d\phi\right)\left(\int_0^{1} r^2\,dr\right)$$

$$= abc\,.\,2\pi\,.\,2\,.\,\frac{1}{3} = \frac{4\pi}{3}\,abc$$

Problems

1. Evaluate the integrals

(a) $\displaystyle\int_0^{2}\int_0^{x}\int_0^{1-x-y} x\,dz\,dy\,dx$ (b) $\displaystyle\int_0^{1}\int_{y^2}^{\sqrt{y}}\int_0^{y+z} y\,dx\,dz\,dy$

2. Evaluate $\displaystyle\iiint_R f(x,y,z)\,dV$ for the given function f and region R.

(a) $f(x,y,z) = xyz$; R the rectangular solid bounded by the coordinate planes and the planes $x = a$, $y = b$, $z = c$; $a, b, c > 0$.

(b) $f(x,y,z) = x^2$; R the region $x^2 + y^2 + z^2 \leqslant 1$.

(c) $f(x,y,z) = x^2 + y^2$; R the region bounded by the cylinder $x^2 + z^2 = 4$, and by the planes $y = 0$, $z = 0$, and $y + z = 2$; $z > 0$.

(d) $f(x, y, z) = 1$; R the region bounded by $x^2 + y^2 = 1$ and $z = 4 - x^2 - y^2$, $z \geqslant 0$.

3. Use cylindrical coordinates to evaluate $\iiint_R f(x, y, z)\, dV$

(a) $f(x, y, z) = z$; R the region above the cone $z^2 = x^2 + y^2$ and below the plane $z = a$, $a > 0$.

(b) $f(x, y, z) = e^{(x^2 + y^2)^{3/2}}$; R the region bounded by the planes $z = 0$ and $y = z$ and by the cylinder $x^2 + y^2 = 1$, $z \geqslant 0$.

4. Use the spherical coordinates to evaluate $\iiint_R f(x, y, z)\, dV$

(a) $f(x, y, z) = x^2 + y^2 + z^2$; R the region above the cone $z^2 = x^2 + y^2$ and below the plane $z = a$, $a > 0$.

(b) $f(x, y, z) = (x^2 + y^2 + z^2)^{-3/2}$; R the region below the plane $z = a$ $(a > 0)$ and above the sphere $x^2 + y^2 + z^2 = 4a^2$.

5. Find the volumes of the given regions.

(a) R the region between the cylinders $x^2 + y^2 = a^2$ and $x^2 + z^2 = a^2$.

(b) R the region above the cone $z^2 = x^2 + y^2$ and below the sphere $x^2 + y^2 + z^2 = a^2$.

(c) R the region bounded above by the cone $z^2 = 25(x^2 + y^2)$ and below by the paraboloid $z = x^2 + y^2 + 4$.

(d) R the region bounded by the plane $z = 0$ and the cylinders $x^2 + y^2 = a^2$ and $az = a^2 - x^2$.

6. (a) Find the moment of inertia about a diameter of a mass distributed with constant density $\rho = k$ through a spherical ball of radius a.

(b) Find the moment of inertia about the z - axis of the ellipsoid $x^2/a^2 + y^2/b^2 + z^2/c^2 \leqslant 1$, given that $\rho = k$, a constant.

7. (a) Find the value of $\iiint x^2\, dx dy\, dz$ through the volume bounded by the ellipsoid $\dfrac{x^2}{a^2} + \dfrac{y^2}{b^2} + \dfrac{z^2}{c^2} = 1$. (Put $x = aX$, $y = bY$, $z = cZ$, and transform to polars.)

(b) Find the value of $\iiint xyz\,(x^2 + y^2 + z^2)\, dx\, dy\, dz$ through the positive spherical octant for which $x^2 + y^2 + z^2 \leqslant a^2$.

8. The water face of a dam is vertical and has the form of a trapezium height h and length $2a$ at the top and a at the bottom. The thickness of the dam increases uniformly from zero at the top to a at the bottom. Show that the C.G. of the dam is $5a/16$ from the vertical face and find its distance below the top of the dam. (L.U.)

Chapter Eleven

Vector Field Theory

11.1 INTRODUCTION. SCALAR AND VECTOR FIELDS

In many areas of applied mathematics the governing equation takes on different forms according to the geometry of the particular problem under consideration. For example, steady state temperature distribution in a long bar is governed by $\frac{d^2\theta}{dx^2} = 0$, in a rectangular plate by $\frac{\partial^2\theta}{\partial x^2} + \frac{\partial^2\theta}{\partial y^2} = 0$ and in a cylindrical conductor with radial flow by $\frac{\partial^2\theta}{\partial r^2} + \frac{1}{r}\frac{\partial\theta}{\partial r} = 0$. Yet the same physical process underlies all three problems. In order to remove the effects of geometrical variation, and hence to produce an equation independent of coordinate systems we resort to **vector analysis.** The governing equation, valid for steady state heat conduction processes where thermal conductivity is constant with one, two or three space dimensions and any shape of conductor is

$$\nabla^2\theta = 0 \tag{11.1}$$

This is familiar to us as Laplace's equation and takes on the form apposite to the appropriate goemetry in each particular problem. The equation can be obtained for the general case using vector methods. The great power of vector analysis lies in its generality of approach.

In measurements of atmospheric pressure, balloons are used to record the pressure p at various points (x, y, z) in space. The totality of possible measurements $p(x, y, z)$ form a **scalar field**; regarded as a function, $p(x, y, z)$ is called a **scalar point function.** We could take a fixed value of the pressure, p_0 say and consider the set of all points in space at which the pressure is p_0; these form a surface, known as a **level surface** of pressure (though 'level' does not necessarily imply 'horizontal'). Through each point in space one and only one level surface will pass. Weather maps show isobars which are two-dimensional projections of these surfaces onto the plane of the map. We could also talk about the density $\rho(x, y, z)$ of the air as a scalar point function, as well as the temperature $\theta(x, y, z)$.

In general, the air will be in motion and the velocity of the air at each point will have a specific direction as well as magnitude, i.e. it will be a vector quantity. We may thus describe the velocity $\mathbf{v}(x, y, z)$ as a **vector point function** which will give rise to a **vector field.**

So far we have ignored time changes in our discussion, i.e. we have assumed **time - independent** or **steady fields.** If we generalise our arguments then we must consider $\rho(x, y, z, t)$, $\mathbf{v}(x, y, z, t)$ etc.

Problems

1. Describe the level surfaces of the scalar point functions

   (i) $\phi = x + y + z$ (ii) $\phi = x^2 + y^2 - z^2$

2. Explain the nature of the vector field $\mathbf{v} = x\mathbf{i} + y\mathbf{j} + z\mathbf{k}$.

3. An electrostatic field $\mathbf{F}$ is given by

   $$\mathbf{F} = \frac{1}{[(x+1)^2 + y^2][(x-1)^2 + y^2]}[2(x^2 - y^2 - 1)\mathbf{i} + 4xy\mathbf{j}]$$

   Find the directions of this field at points

   (i) on the x - axis (ii) on the y - axis (iii) on the lines $y = \pm x$ (iv) at the points $(\pm 1, 0)$.

   Hence sketch the field.

4. Prove that the field $\mathbf{v} = (x^2 - y^2)\mathbf{i} + 2xy\mathbf{j}$ consists of a system of circles, all having their centres on the y- axis and all passing through the origin.

11.2 DIFFERENTIATION OF VECTORS

We are frequently concerned with the rate of change of a vector quantity in time or in space. In this section we examine the ways of calculating these rates of change.

Differentiation with respect to one scalar

If a vector quantity $\mathbf{v}$ depends upon one scalar variable t so that $\mathbf{v} = \mathbf{v}(t)$ then let $\mathbf{v}$ change to $\mathbf{v} + \delta\mathbf{v}$ whilst t changes to $t + \delta t$. If in the limit as $\delta t \to 0$, the ratio $\frac{\delta \mathbf{v}}{\delta t}$ tends to a limiting value, this limiting value, denoted $\frac{d\mathbf{v}}{dt}$, is the derivative of $\mathbf{v}$ with respect to t.

If in cartesian coordinates $\mathbf{v} = v_1\mathbf{i} + v_2\mathbf{j} + v_3\mathbf{k}$ then

$$\frac{d\mathbf{v}}{dt} = \frac{dv_1}{dt}\mathbf{i} + \frac{dv_2}{dt}\mathbf{j} + \frac{dv_3}{dt}\mathbf{k} \tag{11.2}$$

since v_1, v_2, v_3 are functions of t only.

For example, if $\mathbf{v} = t^2\mathbf{i} + \sqrt{t}\,\mathbf{j} - \sin t\,\mathbf{k}$ then $\frac{d\mathbf{v}}{dt} = 2t\mathbf{i} + \frac{1}{2\sqrt{t}}\mathbf{j} - \cos t\,\mathbf{k}$

Formal rules for differentiation

Assume that $\mathbf{a}, \mathbf{b}$ and ϕ are functions of one scalar t and that the derivatives exist as necessary. Then the following rules hold as you can verify.

(i) $$\frac{d\mathbf{a}}{dt} = 0 \text{ if and only if } \mathbf{a} \text{ is a constant vector} \tag{11.3}$$

(ii) $$\frac{d}{dt}(\mathbf{a} + \mathbf{b}) = \frac{d}{dt}\mathbf{a} + \frac{d}{dt}\mathbf{b} \tag{11.4}$$

(iii) $\dfrac{d}{dt}(\phi\ \mathbf{a}) = \dfrac{d\phi}{dt}\mathbf{a} + \phi\dfrac{d\mathbf{a}}{dt}$ (11.5)

(iv) $\dfrac{d}{dt}(\mathbf{a}\,.\,\mathbf{b}) = \dfrac{d}{dt}\mathbf{a}.\ \mathbf{b} + \mathbf{a}\,.\,\dfrac{d\mathbf{b}}{dt}$ (11.6)

(v) $\dfrac{d}{dt}(\mathbf{a}_\wedge \mathbf{b}) = \dfrac{d}{dt}\mathbf{a}_\wedge\ \mathbf{b} + \mathbf{a}_\wedge\dfrac{d\mathbf{b}}{dt}$ (11.7)

Other rules follow as one might expect. For example, if $\mathbf{a} = \mathbf{b}$ in (11.6) then $\dfrac{d}{dt}(\mathbf{a}\,.\,\mathbf{a}) = 2\mathbf{a}\,.\,\dfrac{d\mathbf{a}}{dt}$ and if $|\mathbf{a}|$ is constant then $2\mathbf{a}\,.\,\dfrac{d\mathbf{a}}{dt} = 0$ and $\mathbf{a}$ is perpendicular to $\dfrac{d\mathbf{a}}{dt}$. In the special case where $\mathbf{a}$ is the **position vector**

$$\mathbf{r}(t) = x(t)\mathbf{i} + y(t)\mathbf{j} + z(t)\mathbf{k}$$

then $\dfrac{d\mathbf{r}}{dt}$ is the velocity vector $\mathbf{v}$ and we have the result that if a particle moves so that its distance from the origin, $|\mathbf{r}|$, is constant then its velocity is in a direction perpendicular to the line joining it to the origin. What interpretation can you draw from this?

Example

A particle of mass m and charge q moves with velocity v in a uniform magnetic field $\mathbf{B}$. Let us choose cartesian axes so that the z - axis is parallel to $\mathbf{B}$, i.e. that $\mathbf{B} = (0, 0, B)$ and let $\mathbf{v} = (v_x, v_y, v_z)$. The governing equation of motion is known to be

$$m\frac{d\mathbf{v}}{dt} = q(\mathbf{v}_\wedge \mathbf{B}) \tag{11.8}$$

If we resolve (11.8) into three component equations and introduce the constant $\omega = qB/m$ we find that $v_z = V$, a constant, and that $\ddot{v}_x = -\omega^2 v_x$, $\ddot{v}_y = -\omega^2 v_y$, where the dots above the components indicate time differentiation. If we align the x - and y - axes so that the initial velocity, U, was along the latter and if the particle is at the origin at $t = 0$, then the position vector at any subsequent time is given by

$$\mathbf{r} = \left(\frac{U}{\omega}[1 - \cos\omega t],\ \frac{U}{\omega}\sin\omega t,\ Vt\right) \tag{11.9}$$

If we eliminate t from the equations for x and y we obtain the projection of the path of the particle on the x - y plane. It is

$$(x - U/\omega)^2 + y^2 = U^2/\omega^2 \tag{11.10}$$

This is clearly a circle, centre $(U/\omega, 0)$ radius U/ω; the path is therefore a helix, (Figure 11.1), which is wrapped round the cylinder whose equation is (11.10). The velocity vector is $\mathbf{v} = (U\sin\omega t,\ U\cos\omega t,\ V)$ and we see that the particle rotates uniformly about the axis of the cylinder whilst moving uniformly parallel to this axis. Note that

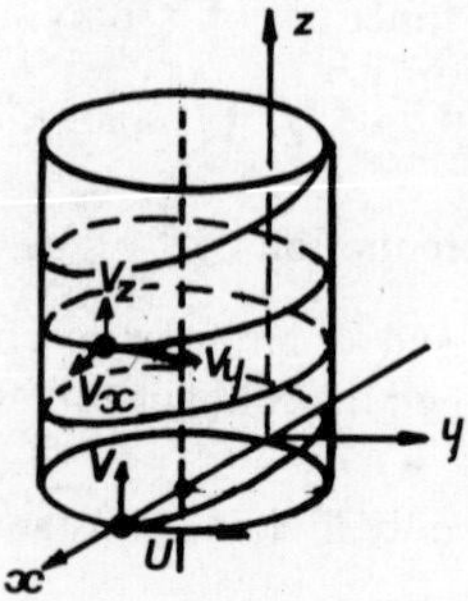

Figure 11.1

from (11.6) $\frac{d}{dt}(\mathbf{v}^2) = 2\mathbf{v} \cdot \frac{d\mathbf{v}}{dt} = 2\frac{q}{m}\mathbf{v} \cdot (\mathbf{v} \wedge \mathbf{B}) = 0$ and hence $\mathbf{v}^2$ and the kinetic energy $\frac{1}{2}m\mathbf{v}^2$ are constant. Further, if the acceleration vector $\frac{d\mathbf{v}}{dt}$ be denoted $\mathbf{f}$, then $\mathbf{f}$ is perpendicular to $\mathbf{v}$ at all stages of the motion and is directed to the axis of the cylinder.

Partial differentiation

Now consider a vector $\mathbf{A}$ which is a function of x, y and z so that $\mathbf{A} = \mathbf{A}(x, y, z)$ and each component is a function of x, y and z. Then partial derivatives of $\mathbf{A}$, provided that they exist, may be calculated in the usual way. For example, if $\mathbf{A} = 2xy^2\mathbf{i} - y^2z\mathbf{j} + xz^2\mathbf{k}$, then $\frac{\partial \mathbf{A}}{\partial x} = 2y^2\mathbf{i} + z^2\mathbf{k}$, $\frac{\partial \mathbf{A}}{\partial y} = 4xy\mathbf{i} - 2yz\mathbf{j}$, $\frac{\partial \mathbf{A}}{\partial z} = -y^2\mathbf{j} + 2xz\mathbf{k}$ with results for higher derivatives following as is obvious. If, in turn, x, y and z are functions of one scalar variable t then the **total derivative**

$$\frac{d\mathbf{A}}{dt} = \frac{\partial \mathbf{A}}{\partial x}\frac{dx}{dt} + \frac{\partial \mathbf{A}}{\partial y}\frac{dy}{dt} + \frac{\partial \mathbf{A}}{\partial z}\frac{dz}{dt} \tag{11.11}$$

The **total differential** of $\mathbf{A}$ is

$$d\mathbf{A} = \frac{\partial \mathbf{A}}{\partial x}dx + \frac{\partial \mathbf{A}}{\partial y}dy + \frac{\partial \mathbf{A}}{\partial z}dz \tag{11.12}$$

The differential $d\mathbf{r} = (dx\mathbf{i} + dy\mathbf{j} + dz\mathbf{k})$ (11.13)

If we use the notation (in cartesian coordinates)

$$\nabla = \mathbf{i}\frac{\partial}{\partial x} + \mathbf{j}\frac{\partial}{\partial y} + \mathbf{k}\frac{\partial}{\partial z} \tag{11.14}$$

which we introduced in Section 4.4 then

$$d\mathbf{A} = (\nabla \cdot d\mathbf{r})\mathbf{A} \tag{11.15}$$

Ordinary integrals of vectors

Just as we can differentiate a position vector to obtain a velocity vector, we can reverse the process; corresponding to the arbitrary constant of scalar integration we have to add a time-independent arbitrary vector. We give one example as an illustration of the general process of integration.

If the velocity vector is $\mathbf{v} = 2\sin t\mathbf{i} + (\cos t - 1)\mathbf{j} + 6t\mathbf{k}$ and $\mathbf{r} = \mathbf{0}$ at $t = 0$ we first find

$$\begin{aligned}\mathbf{r} &= \int (2\sin t\mathbf{i} + (\cos t - 1)\mathbf{j} + 6t\mathbf{k})\,dt \\ &= \int 2\sin t\,dt\;\mathbf{i} + \int (\cos t - 1)\,dt\;\mathbf{j} + \int 6t\,dt\;\mathbf{k} \\ &= -2\cos t\mathbf{i} + (\sin t - t)\mathbf{j} + 3t^2\mathbf{k} + \mathbf{d}\end{aligned}$$

where $\mathbf{d}$ is an arbitrary time-independent vector.

Since $\mathbf{r} = \mathbf{0}$ when $t = 0$, $\mathbf{0} = -2\mathbf{i} + \mathbf{d}$ and hence $\mathbf{d} = 2\mathbf{i}$. Therefore

$$\mathbf{r} = (2 - 2\cos t)\mathbf{i} + (\sin t - t)\mathbf{j} + 3t^2\mathbf{k}$$

Problems

1. Find $\frac{d\mathbf{A}}{dt}$, $\frac{d^2\mathbf{A}}{dt^2}$ for $\mathbf{A} = e^{-t}\mathbf{i} + 4\cos 3t\mathbf{j} + 2\sin t\mathbf{k}$.

2. If $\mathbf{A} = (2x^2y - x^4)\mathbf{i} + y\sin x\mathbf{j} + \cos y\mathbf{k}$, find $\frac{\partial \mathbf{A}}{\partial x}$, $\frac{\partial \mathbf{A}}{\partial y}$.

3. Show that a vector function $\mathbf{v}(t)$ of the variable t will have

   (i) a constant magnitude if and only if $\mathbf{v} \cdot \frac{d\mathbf{v}}{dt} = 0$;

   (ii) will remain parallel to a fixed line if and only if $\mathbf{v} \wedge \frac{d\mathbf{v}}{dt} = 0$. (L.U.)

4. The position vector $\mathbf{r}$ of a point P satisfies the equation $\frac{d\mathbf{r}}{dt} = \omega \wedge \mathbf{r}$, where ω is a constant vector. Show that $|\mathbf{r}|$ is a constant and P lies in a fixed plane. Deduce that P describes a circle with constant speed. (L.U.)

5. The position vector $\mathbf{r}$, measured from an origin O, of a particle satisfies the differential equation

$$\frac{d^2\mathbf{r}}{dt^2} + 2n\frac{d\mathbf{r}}{dt} + 2n^2\mathbf{r} = \mathbf{0}$$

where n is a positive constant and $\mathbf{0}$ is the zero vector. Find the values of the constant m which make $\mathbf{A}e^{mt}$, where $\mathbf{A}$ is a constant vector, a solution of the equation, and deduce the general solution for $\mathbf{r}$. Find the particular solution if at $t = 0$ the particle starts from $\mathbf{O}$ with velocity $\mathbf{V}$. (L.U.)

11.3 THE GRADIENT OF A SCALAR FIELD

Let $\phi(x, y, z)$ constitute a scalar field. Then let us examine how to calculate the rate of change of ϕ with distance in any given direction. In Figure 11.2 we show two level surfaces corresponding to two constant values of ϕ, ϕ_0 and ϕ_1, say. Let $\phi(\text{P}) = \phi_0$ and $\phi(Q_1) = \phi(Q_2) = \phi(Q_3) = \phi_1$. If we consider the values of $\frac{\phi(Q) - \phi(P)}{\text{PQ}}$ where Q represents points like Q_1, Q_2, Q_3 on the surface $\phi = \phi_1$ then these values depend only on the distance PQ; the ratio will have its maximum value when PQ is least, that is when PQ is normal to the level surface at P (why is this?). In general, we write $\frac{\phi(Q) - \phi(P)}{\text{PQ}}$ as $\frac{\delta\phi}{\delta s}$ and if we let $\delta s \to 0$ then if a limit exists we

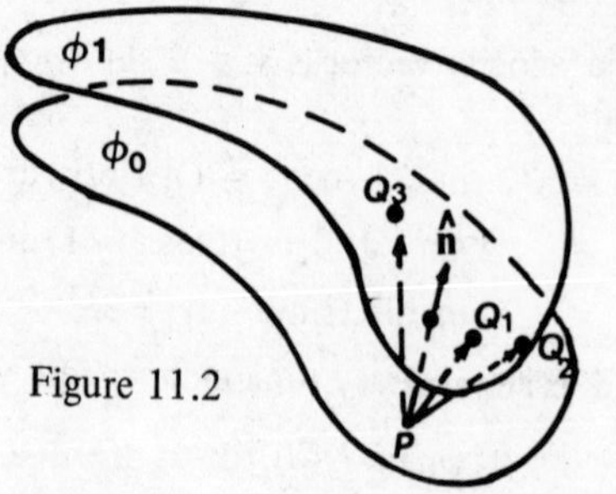

Figure 11.2

denote it $\frac{\partial \phi}{\partial s}$ and call it a **directional derivative** of ϕ. The maximum value is denoted $\frac{\partial \phi}{\partial n}$ in direction **n**, where $\hat{\mathbf{n}}$ is a unit vector normal to the level surface at P. We call this vector the **gradient** of ϕ, written grad ϕ and therefore

$$\text{grad}\,\phi = \frac{\partial \phi}{\partial n}\hat{\mathbf{n}} \tag{11.16}$$

Let PQ_n be the normal at P to the surface on which P lies; Figure 11.3. Then

$$\frac{\partial \phi}{\partial s} = \lim_{\delta s \to 0}\left(\frac{\delta \phi}{\delta s}\right)$$

$$= \lim_{\delta s \to 0}\left(\frac{\delta \phi}{\delta n}\cdot\frac{\delta n}{\delta s}\right)$$

$$= \frac{\partial \phi}{\partial n}\cos\theta$$

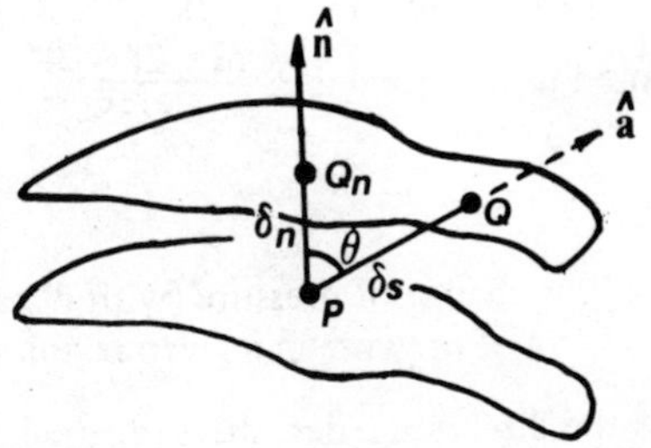

Figure 11.3

If $\hat{\mathbf{a}}$ is a unit vector in the direction of PQ then

$$\cos\theta = \hat{\mathbf{n}}\,.\,\hat{\mathbf{a}} \quad \text{and} \quad \frac{\partial \phi}{\partial s} = \frac{\partial \phi}{\partial n}(\hat{\mathbf{n}}\,.\,\hat{\mathbf{a}}) = \text{grad}\,\phi\,.\,\hat{\mathbf{a}}$$

We see that $\frac{\partial \phi}{\partial s}$ is the resolved component of grad ϕ in the direction **a**. In particular, if we take $\hat{\mathbf{a}}$ to be **i** then $\frac{\partial \phi}{\partial x}$ = grad ϕ . **i**. Similarly $\frac{\partial \phi}{\partial y}$ = grad ϕ . **j** and $\frac{\partial \phi}{\partial z}$ = grad ϕ . **k**. It follows that in cartesian coordinates

$$\text{grad}\,\phi = \frac{\partial \phi}{\partial x}\mathbf{i} + \frac{\partial \phi}{\partial y}\mathbf{j} + \frac{\partial \phi}{\partial z}\mathbf{k} \tag{11.17}$$

Now we may via (11.14) identify grad ϕ with $\nabla\phi$. ∇ is a **vector operator** which associates with a scalar ϕ a unique vector $\nabla\phi$ or grad ϕ.

We should emphasise that result (11.16) is *independent of the coordinate system chosen* and this demonstrates the power of vector field theory. The expressions (11.14) and (11.17) are merely one particular form of (11.16).

Example 1

The temperature of a gas in a room is given by $T = T_0(x^2 - y^2 + xyz) + 273$. We wish to find the maximum rate of change of temperature at the point (1, 1, 1) and the direction in which this occurs.

First we find grad $T = T_0[(2x + yz)\mathbf{i} + (xz - 2y)\mathbf{j} + xy\mathbf{k}]$ and then evaluate it at (1, 1, 1) to obtain $T_0(3\mathbf{i} - \mathbf{j} + \mathbf{k})$. The magnitude of this vector is $T_0\sqrt{11}$ which is the required maximum rate of change and its direction is given by

$$\hat{\mathbf{n}} = \frac{3\mathbf{i} - \mathbf{j} + \mathbf{k}}{\sqrt{11}}.$$

Example 2

We wish to find a unit normal to the surface $x^2y + y^2z + z^2x = 5$ at the point $(2, 1, -1)$.

Let $\phi = x^2y + y^2z + z^2x$; then we are concerned with the level surface $\phi = 5$ which is the only one to pass through the point $(2, 1, -1)$. The value of grad ϕ at this point is easily shown to be $5\mathbf{i} + 2\mathbf{j} - 3\mathbf{k}$. Since grad ϕ is normal to the level surface then the required unit normal is $\pm \dfrac{5\mathbf{i} + 2\mathbf{j} - 3\mathbf{k}}{\sqrt{38}}$.

Example 3

Find the rate of change of pressure (with distance) in the direction $2\mathbf{i} + \mathbf{j} - \mathbf{k}$ at the point $(2, 3, 1)$ if pressure in a given region is $p = p_0xyz^2$.

We require the directional derivative $\hat{\mathbf{a}}$. grad ϕ where $\hat{\mathbf{a}}$ is the unit vector $(2\mathbf{i} + \mathbf{j} - \mathbf{k})/\sqrt{6}$. The value of grad ϕ at $(2, 3, 1)$ is $p_0(3\mathbf{i} + 2\mathbf{j} + 12\mathbf{k})$ and the required result is $p_0(2\mathbf{i} + \mathbf{j} - \mathbf{k}) . (3\mathbf{i} + 2\mathbf{j} + 12\mathbf{k})/\sqrt{6} = -4p_0/\sqrt{6}$.

Further results on gradients

(i) Certain results follow fairly obviously from the definition of grad. For example,

$$\nabla(\phi_1 + \phi_2) = \nabla\phi_1 + \nabla\phi_2$$

(ii) If $\mathbf{r}$ is the position vector then $r = |\mathbf{r}| = (x^2 + y^2 + z^2)^{1/2}$ and it is easily shown that

$$\nabla r = \mathbf{r}/r = \hat{\mathbf{r}} \qquad (11.18)$$

(iii) In general,

$$\nabla\{f(r)\} = \hat{\mathbf{r}}f'(r) \qquad (11.19)$$

(iv) If $\mathbf{A}$ is a constant vector then

$$\nabla(\mathbf{A} . \mathbf{r}) = \mathbf{A} \qquad (11.20)$$

(v) If $\mathbf{B}$ is an arbitrary vector then $(\mathbf{B} . \nabla)\phi = \mathbf{B} . (\nabla\phi)$. In particular,

$$(\mathbf{B} . \nabla)r = \mathbf{B} . \mathbf{r} \qquad (11.21)$$

Example

Consider the transport of heat or mass by a fluid. Let this quantity be specified by $\phi(x, y, z, t) = \phi(\mathbf{r}, t)$. Then

$$\delta\phi = \frac{\partial\phi}{\partial t}\delta t + \frac{\partial\phi}{\partial x}\delta x + \frac{\partial\phi}{\partial y}\delta y + \frac{\partial\phi}{\partial z}\delta z = \frac{\partial\phi}{\partial t}\delta t + (\delta\mathbf{r} . \nabla)\phi$$

The **total rate of change of** ϕ with time experienced by a particle of fluid as it moves (called differentiation following the fluid) is given by $\dfrac{d\phi}{dt} = \dfrac{\partial\phi}{\partial t} + \left[\dfrac{d\mathbf{r}}{dt} \cdot \nabla\right]\phi$. The **local rate of change** $\dfrac{\partial\phi}{\partial t}$ represents the changes taking place at a fixed point in space and the **convective rate of change** $\left[\dfrac{d\mathbf{r}}{dt} \cdot \nabla\right]\phi$ represents the changes that take place independent of time variation as the property ϕ is convected with the fluid at its speed $\dfrac{d\mathbf{r}}{dt}$.

(vi) In cylindrical polar coordinates (r, θ, z)

$$\nabla V = \hat{\mathbf{r}}\frac{\partial V}{\partial r} + \hat{\boldsymbol{\theta}}\frac{1}{r}\frac{\partial V}{\partial \theta} + \mathbf{k}\frac{\partial V}{\partial z} \qquad (11.22)$$

(vii) In spherical polar coordinates (r, θ, ϕ)

$$\nabla V = \hat{\mathbf{r}}\frac{\partial V}{\partial r} + \hat{\boldsymbol{\theta}}\frac{1}{r}\frac{\partial V}{\partial \theta} + \hat{\boldsymbol{\phi}}\frac{1}{r\sin\theta}\frac{\partial V}{\partial \phi} \qquad (11.23)$$

Problems

1. Find grad ϕ and $\dfrac{\partial \phi}{\partial n}$ for the following

   (a) $\phi = x + y + z$ (b) $\phi = x^2 + y^2 - z^2$ (c) $\phi = 2xz^4 - x^2 y$ (d) $\phi = 2z - x^3 y$

2. Find the directional derivative of ϕ at point P in the direction **a**

   (i) $\phi = x^2 + y^2 + z^2$; P = (1, 1, 2); $\mathbf{a} = \mathbf{i} - \mathbf{j}$

   (ii) $\phi = 2x^2 - 4y^2 + z^2$; P = (0, 1, 2); $\mathbf{a} = \mathbf{i} - 2\mathbf{j} + 3\mathbf{k}$

   (iii) $\phi = xyz$; P = (−1, 1, 2); $\mathbf{a} = \mathbf{i} + 2\mathbf{j} + \mathbf{k}$

3. Find ϕ if $\mathbf{v} = \nabla \phi$

   (i) $\mathbf{v} = x\mathbf{i} + y\mathbf{j}$ (ii) $\mathbf{v} = 3x\mathbf{i} - 2y\mathbf{j} + z\mathbf{k}$

4. Find a unit normal vector to the surface at P.

   (i) $x + y + z = 2$; P = (2, 1, −3) (ii) $x^2 + y^2 + 3z^2 = 4$; P = (2, 2, 0).

5. What are the direction cosines of ∇f at the point (1, 2, 3), if $f = (x^2 + 2xyz + 3y^2 - z^2)$?

6. Given that $f = 2x + 3y^2 - yz$, find the directional derivative at (2, −1, 5) along the line whose direction ratios are 1 : 1 : 2.

7. A scalar point function ϕ is given by $\phi = \dfrac{2}{x^2 + y^2 + z^2}$. Describe the level surface which passes through the point P = (6, 2, 3). What is the unit normal to this surface at P and what is the normal rate of change of ϕ across the surface?

8. Show that if r is the distance of a point from the origin, $\nabla r^3 = 3r\mathbf{r}$ and $\nabla r^m = mr^{m-2}\mathbf{r}$.

11.4 DIVERGENCE OF A VECTOR FIELD

In Section 7.3 of *Engineering Mathematics* we considered the two-dimensional flow of a fluid into a cuboidal region. Let us now extend the argument to allow the flow to be three-dimensional. (Although we have taken fluid flow as our practical application the same argument would apply to any vector field.)

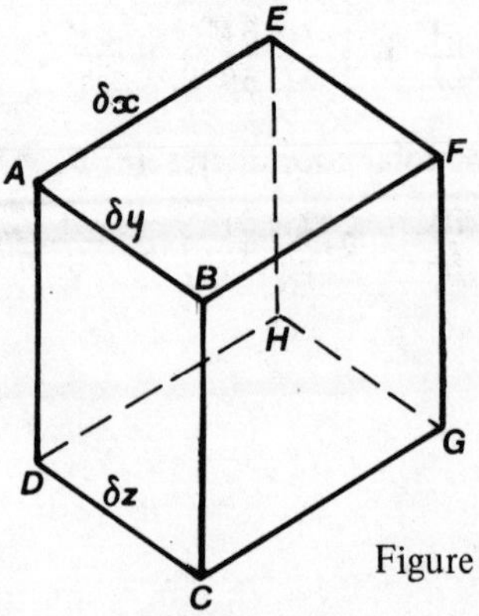

Figure 11.4

Figure 11.4 shows an infinitesimal element of volume with sides parallel to the coordinate axes, lying within the region of fluid flow. Its centre is at the point (x, y, z). Let $\mathbf{v}(x, y, z) = (v_1, v_2, v_3)$ be the time-independent velocity of the fluid and we consider the flux of $\mathbf{v}$ through the volume; (the flux is the volume of fluid crossing the faces of the box in unit time). We shall then let the volume shrink to the point (x, y, z) just as we did with the two-dimensional case and we arrive at the results that the net velocity flux out of the volume element is $\left(\frac{\partial v_1}{\partial x} + \frac{\partial v_2}{\partial y} + \frac{\partial v_3}{\partial z}\right) \delta x\, \delta y\, \delta z$ and that the amount of outward flux per unit volume, called the **divergence** of $\mathbf{v}$ is given by

$$\operatorname{div} \mathbf{v} = \frac{\partial v_1}{\partial x} + \frac{\partial v_2}{\partial y} + \frac{\partial v_3}{\partial z} \tag{11.24}$$

It would seem reasonable to represent $\operatorname{div} \mathbf{v}$ by $\nabla . \mathbf{v}$ although we have only obtained a correspondence in cartesian coordinates; we shall accept the general result.

$$\nabla . \mathbf{v} = \operatorname{div} \mathbf{v} \tag{11.25}$$

Note that certain results follow fairly straightforwardly in cartesian coordinates but hold in any coordinate system; it is usual to verify them in cartesian coordinates. For example

(i) $\nabla . (\mathbf{A} + \mathbf{B}) = \nabla . \mathbf{A} + \nabla . \mathbf{B}$ (11.26)

(ii) $\nabla . (\phi \mathbf{A}) = (\nabla \phi) . \mathbf{A} + \phi(\nabla . \mathbf{A})$ (11.27)

Notice in this last result that the expected rule for differentiation also has the feature that the scalar product is preserved in both terms on the right-hand side.

Example 1

$$\operatorname{div} \mathbf{r} = \nabla . (x\mathbf{i} + y\mathbf{j} + z\mathbf{k}) = \frac{\partial}{\partial x}(x) + \frac{\partial}{\partial y}(y) + \frac{\partial}{\partial z}(z) = 3$$

Example 2

The electrostatic field associated with an isolated charge e is given by $\mathbf{E} = -\operatorname{grad} \phi$ where $\phi = e/r$. $\mathbf{E} = -\nabla\left[\frac{e}{(x^2 + y^2 + z^2)^{1/2}}\right]$ and the symmetry involved will mean that we need only evaluate the first component E_x. In fact

$$E_x = -e \frac{\partial}{\partial x}(x^2 + y^2 + z^2)^{-1/2} = ex(x^2 + y^2 + z^2)^{-3/2} = \frac{ex}{r^3}$$

By using symmetry to obtain the other components we may deduce that

$$\mathbf{E} = \frac{e}{r^3}(x\mathbf{i} + y\mathbf{j} + z\mathbf{k}) = e\mathbf{r}/r^3 = e\hat{\mathbf{r}}/r^2$$

Now

$$\text{div}\,\mathbf{E} = e\left[\frac{\partial}{\partial x}\left(\frac{x}{r^3}\right) + \frac{\partial}{\partial y}\left(\frac{y}{r^3}\right) + \frac{\partial}{\partial z}\left(\frac{z}{r^3}\right)\right]$$

But

$$\frac{\partial}{\partial x}\left(\frac{x}{r^3}\right) = \frac{1}{r^3} + x\frac{\partial}{\partial x}\left([x^2 + y^2 + z^2]^{-3/2}\right) = \frac{1}{r^3} - \frac{3x^2}{r^5}$$

Hence

$$\text{div}\,\mathbf{E} = \frac{3}{r^3} - \frac{3}{r^5}(x^2 + y^2 + z^2) = 0$$

A vector field whose divergence is zero is termed **solenoidal**. We may interpret this by saying that the lines of flow form closed curves.

Remarks on divergence

Returning to our fluid flow example, if the divergence is non-zero at a point in the fluid, it expresses the rate/unit volume at which fluid is flowing away from (if positive) or towards (if negative) the point. When the divergence is positive, then either the fluid is expanding and its density is falling with time or there is a **source** of fluid at the point where fluid is entering the field of flow. For negative divergence the opposite is true; that is either the fluid is contracting and the density rising with time at the point or there is a **sink** of fluid there at which fluid is leaving the field.

In an electric field for example the existence of positive divergence at a point can indicate the presence of an electric pole at that point.

The case of zero divergence indicates that flux entering any element of space is exactly balanced by that leaving it. This implies that fluid is neither 'created' nor 'destroyed' there which implies that there is no source or sink of fluid at that point. If $\text{div}\,\mathbf{v} = 0$ everywhere in a region then no flux can have been generated within it and the lines of flow of the vector $\mathbf{v}$ must either form closed curves (the examples are magnetic field lines) or terminate on a boundary; an example is the electric field lines in a condenser.

Problems

1. Find the divergence of the following

   (i) $3x\mathbf{i} + y\mathbf{j} + 2z\mathbf{k}$ (ii) $x^2\mathbf{i} + y^2\mathbf{j} + z^2\mathbf{k}$ (iii) $xy\mathbf{i} + yz\mathbf{j} + zx\mathbf{k}$

   (iv) $(x\mathbf{i} + y\mathbf{j} + z\mathbf{k})/r^{3/2}$

2. Verify (11.26) and (11.27).

11.5 LINE INTEGRALS

We concern ourselves in this section with a class of line integrals which arise frequently in practice; these are **tangential line integrals**. In Figure 11.5 let C be a curve drawn in a vector field and let δs be an element of arc along it, starting at a point P. Let **v** denote a vector from P making an angle θ with the length element. Then **v** . d**r** $= v \cos\theta\, \delta r$ and if we sum such contributions from A to B along the curve we obtain in the limit as $\delta s \to 0$ the **line integral**

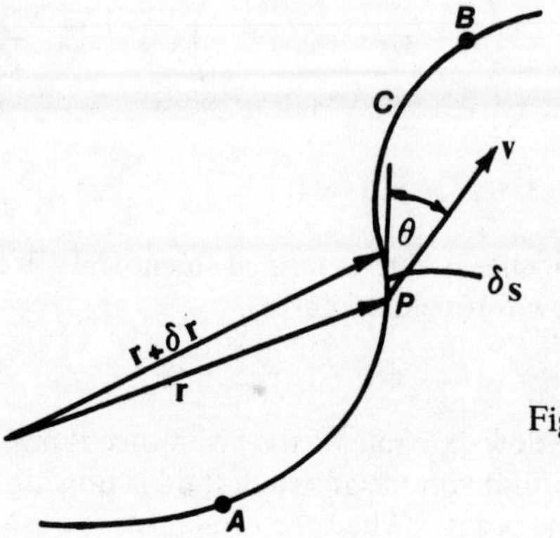

Figure 11.5

$$\int_A^B \mathbf{v} \cdot \mathrm{d}\mathbf{s} = \int_A^B \mathbf{v} \cdot \mathrm{d}\mathbf{r} = \int_A^B v \cos\theta \, \mathrm{d}r \qquad (11.28)$$

where d**r** has been identified with ds. (Why?).

If **v** represents a mechanical force then (11.28) will represent the **work done** in displacing a particle along the curve C from A to B. If **v** is electric field strength then (11.28) represents the **potential difference** between A and B. Methods of evaluation are illustrated in the following examples. Notation relates to the above discussion.

Example 1

Given $\mathbf{v} = (2x + y)\mathbf{i} + (x - 3y)\mathbf{j}$, A = (0, 0), B = (1, 1)

evaluate the integrals of **v** . d**r** from A to B (see Figure 11.6),

(i) along the straight lines AD and DB

(ii) along the straight lines AE and EB

(iii) along the straight line AB

(iv) along the arc AB of the circle, centre D

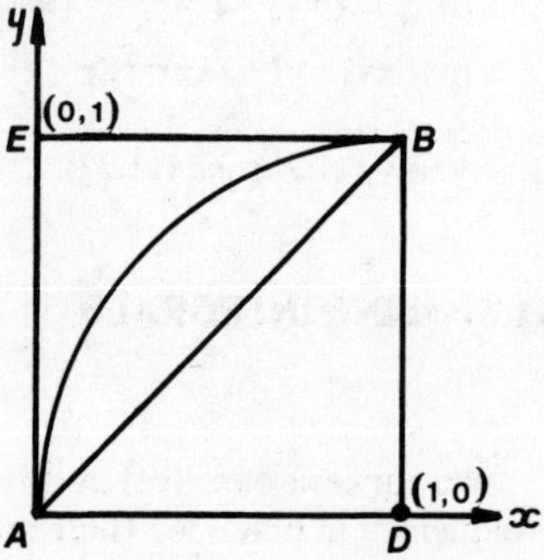

Figure 11.6

(i) $$\int_A^B \mathbf{v}.\mathrm{d}\mathbf{r} = \int_{AD} \mathbf{v} \cdot \mathrm{d}\mathbf{r} + \int_{DB} \mathbf{v} \cdot \mathrm{d}\mathbf{r}$$

Along AD, $y = 0$, d**r** = dx**i** and **v** = 2x**i** + x**j**. Then **v** . d**r** = 2x dx and the required integral becomes

$$\int_0^1 2x\,dx = 1$$

Along DB, $x = 1$, $\mathbf{dr} = dy\mathbf{j}$ and $\mathbf{v} = (2 + y)\mathbf{i} + (1 - 3y)\mathbf{j}$. Then $\mathbf{v} \,.\, \mathbf{dr} = (1 - 3y)\,dy$ and the integral becomes

$$\int_0^1 (1 - 3y)\,dy = 1 - \frac{3}{2} = -\frac{1}{2}$$

On addition we obtain the result ½.

(ii) Along AE, $x = 0$, $\mathbf{dr} = dy\mathbf{j}$, $\mathbf{v} = y\mathbf{i} - 3y\mathbf{j}$ so that $\mathbf{v} \,.\, \mathbf{dr} = -3y\,dy$ and $\int_0^1 -3y\,dy = -3/2$. Along EB, $y = 1$, $\mathbf{dr} = dx\mathbf{i}$, $\mathbf{v} = (2x + 1)\mathbf{i} + (x - 3)\mathbf{j}$ so that $\mathbf{v} \,.\, \mathbf{dr} = (2x + 1)\,dx$ and $\int_0^1 (2x + 1)\,dx = 2$. On addition we obtain the result ½ as before.

(iii) The equation of the line AB is $y = x$ and we may write $x = y = t$ and $dx = dy = dt$, $A \equiv t = 0$ and $B \equiv t = 1$. Then

$$\int_A^B \mathbf{v} \,.\, \mathbf{dr} = \int_A^B [(2x + y)\mathbf{i} + (x - 3y)\mathbf{j}] \,.\, [dx\mathbf{i} + dy\mathbf{j}]$$

$$= \int_A^B [(2x + y)\,dx + (x - 3y)\,dy]$$

$$= \int_0^1 (3t\,dt - 2t\,dt) = \int_0^1 t\,dt = ½.$$

(iv) The circle in question has equation $(x - 1)^2 + y^2 = 1$. It may be represented parametrically by $x = 1 + \cos\theta$, $y = \sin\theta$. Then $A \equiv \theta = \pi$, $B \equiv \theta = \pi/2$, $dx = -\sin\theta\,d\theta$ and $dy = \cos\theta\,d\theta$. Also

$$\int_A^B \mathbf{v} \,.\, \mathbf{dr} = \int_A^B [(2x + y)\,dx + (x - 3y)\,dy]$$

$$\int_\pi^{\pi/2} [-(2 + 2\cos\theta + \sin\theta)\sin\theta\,d\theta + (1 + \cos\theta - 3\sin\theta)\cos\theta\,d\theta]$$

$$\int_{\pi}^{\pi/2} [-2 \sin\theta - 2\cos\theta \sin\theta - \sin^2\theta + \cos\theta + \cos^2\theta - 3\sin\theta\cos\theta]\,d\theta$$

$$= \left[2\cos\theta + \frac{5}{4}\cos 2\theta + ½ \sin 2\theta + \sin\theta\right]_{\pi}^{\pi/2} = ½$$

Example 2

Repeat with paths (i), (ii) and (iii) for $\mathbf{v} = (1 + x^2 y)\mathbf{i} + 2xy\mathbf{j}$. Now

$$\int_A^B \mathbf{v} \,.\, d\mathbf{r} = \int_A^B [(1 + x^2 y)\,dx + 2xy\,dy]$$

(i) Along AD, $\mathbf{v} \,.\, d\mathbf{r} = 1\,dx$ and so $\int_{AD} \mathbf{v} \,.\, d\mathbf{r} = \int_0^1 dx = 1$. Along DB, $\mathbf{v} \,.\, d\mathbf{r} = 2y\,dy$ and so $\int_{DB} \mathbf{v} \,.\, d\mathbf{r} = \int_0^1 2y\,dy = 1$. Adding, we get the result of 2.

(ii) Along AE, $\mathbf{v} \,.\, d\mathbf{r} = 0$ and so $\int_{AE} \mathbf{v} \,.\, d\mathbf{r} = 0$. Along EB, $\mathbf{v} \,.\, d\mathbf{r} = (1 + x^2)\,dx$ and so $\int_{EB} \mathbf{v} \,.\, d\mathbf{r} = \int_0^1 (1 + x^2)\,dx = \frac{4}{3}$. Adding, we get the result of $\frac{4}{3}$.

(iii) Along AB, $\mathbf{v} \,.\, d\mathbf{r} = [(1 + t^3)\mathbf{i} + 2t^2\mathbf{j}] \,.\, [dt\mathbf{i} + dt\mathbf{j}] = (1 + t^3 + 2t^2)\,dt$ so that

$$\int_A^B \mathbf{v} \,.\, d\mathbf{r} = \int_0^1 (1 + t^3 + 2t^2)\,dt = \frac{23}{12}.$$

At once we have come across a conflict. Not only have we found that $\int_A^B \mathbf{v} \,.\, d\mathbf{r}$ is dependent on the path, as well as on the end points A and B but we now *cannot be sure* that in Example 1 any other path would give the result ½. We might have just chosen the paths which give that result: it is unlikely but possible. We need to sort out whether an integral is path independent and if so *then* we can proceed to evaluate it by choosing any suitable path.

Consider Figure 11.7. Let us consider the result of adding $\int_{AEB} \mathbf{v}\,.\,d\mathbf{r}$ to $\int_{BDA} \mathbf{v}\,.\,d\mathbf{r}$. We denote this, $\oint \mathbf{v}\,.\,d\mathbf{r}$ and it is understood that such an integral is called the **circulation**. The anti-clockwise direction is positive; now if $\int_A^B \mathbf{v}\,.\,d\mathbf{r}$ is path independent

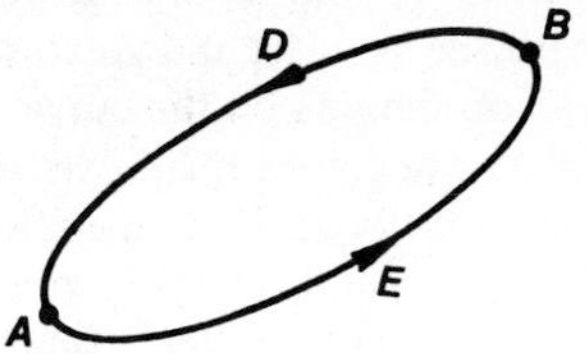

Figure 11.7

then $\int_{BEA} = -\int_{AEB} = -\int_{ADB}$ and therefore $\oint \mathbf{v}\,.\,d\mathbf{r} = 0$. If we consider a particle moving around a closed curve in a force field, no net work would be done and hence energy conserved. In general when $\oint \mathbf{v}\,.\,d\mathbf{r} = 0$ we say the field is **conservative.** In the case where $\int_A^B \mathbf{v}\,.\,d\mathbf{r}$ is path dependent $\oint \mathbf{v}\,.\,d\mathbf{r} \neq 0$ and the field is non-conservative.

Condition for path independence

If $\int_A^B \mathbf{v}\,.\,d\mathbf{r}$ is to be path independent then $\mathbf{v}\,.\,d\mathbf{r}$ must be an exact differential of some function ϕ i.e. $\mathbf{v}.\ d\mathbf{r} = d\phi$ so that $\int_A^B \mathbf{v}\,.\,d\mathbf{r} = \int_A^B d\phi = \phi(B) - \phi(A)$.

But $d\phi = \frac{\partial \phi}{\partial x}dx + \frac{\partial \phi}{\partial y}dy$ so that if $\mathbf{v}\,.\,d\mathbf{r} = P\,dx + Q\,dy$ where P and Q are functions of x and y then $P = \frac{\partial \phi}{\partial x}$ and $Q = \frac{\partial \phi}{\partial y}$ so that if $\frac{\partial^2 \phi}{\partial x \partial y} = \frac{\partial^2 \phi}{\partial y \partial x}$ as is highly likely then

$$\frac{\partial P}{\partial y} = \frac{\partial Q}{\partial x} \tag{11.29}$$

By reversing the argument we may show that (11.29) is a necessary and sufficient condition for path independence.

In the case of Example 1, $P = 2x + y$ and $Q = x - 3y$ so that $\frac{\partial P}{\partial y} = 1 = \frac{\partial Q}{\partial x}$ and any path between A and B would have given the result of ½. However in Example 2, $P = 1 + x^2y$ and $Q = 2xy$. Here $\frac{\partial P}{\partial y} = x^2 \neq 2y = \frac{\partial Q}{\partial x}$ and the integral is path dependent.

Green's theorem in the plane

In Figure 11.8 we have a closed curve which lies in the x - y plane and passes through points A, D, B and E. A and B are the points with the smallest and greatest x - coordinates on the curve. The lower portion of the curve, which passes through D, is specified by $y = y_1(x)$ and the upper portion by $y = y_2(x)$. Green's theorem states that if R is a closed region in the x - y plane bounded by a simple closed curve C and if P and Q are continuous functions of x and y having continuous first derivatives in R then

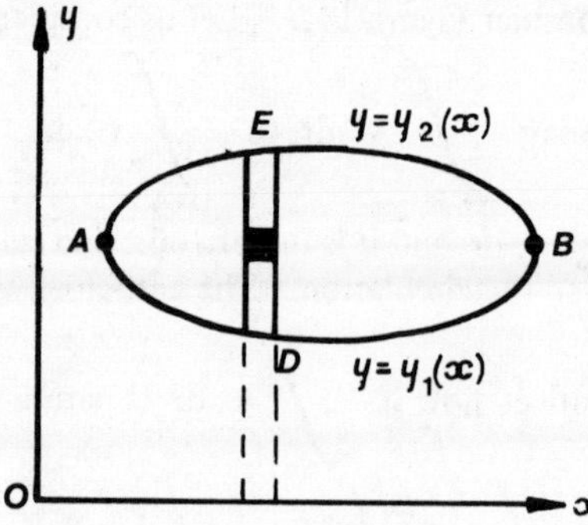

Figure 11.8

$$\oint_C (P\mathrm{d}x + Q\mathrm{d}y) = \iint_R \left(\frac{\partial Q}{\partial x} - \frac{\partial P}{\partial y}\right) \mathrm{d}x\, \mathrm{d}y$$

The proof is outlined as follows.

$$\iint_R \frac{\partial P}{\partial y}\,\mathrm{d}x\,\mathrm{d}y = \int \mathrm{d}x \int \frac{\partial P}{\partial y}\,\mathrm{d}y \quad \text{between appropriate limits}$$

$$= \int \mathrm{d}x \Big[P\Big]_{y_1}^{y_2} = \int [P(x, y_2) - P(x, y_1)]\,\mathrm{d}x$$

$$= \int_{AEB} P\,\mathrm{d}x - \int_{ADB} P\,\mathrm{d}x = -\oint_C P\,\mathrm{d}x$$

Similarly

$$\iint_R \frac{\partial Q}{\partial x}\,\mathrm{d}x\,\mathrm{d}y = \oint_C Q\,\mathrm{d}y$$

and the result follows.

Example 3

Let us evaluate $\oint_C \mathbf{v} \,.\, \mathrm{d}\mathbf{r}$ where $\mathbf{v}$ is the vector of **Example 2** and C is the path ADBA (See Figure 11.6) by applying Green's theorem,

We calculate $I = \int_0^1 \mathrm{d}x \int_0^x (2y - x^2)\,\mathrm{d}y.$

Now $\displaystyle\int_0^x (2y - x^2)\,dy = [y^2 - x^2y]_0^x = x^2 - x^3$ and

$$\int_0^1 (x^2 - x^3)\,dx = \left[\frac{x^3}{3} - \frac{x^4}{4}\right]_0^1 = \frac{1}{12}$$

Note that, from parts (i) and (iii) of Example 2 we find that

$$\int_{ADBA} = \int_{ADB} + \int_{BA} = \int_{ADB} - \int_{AB} = 2 - \frac{23}{12} = \frac{1}{12}.$$

In a sense, $\left(\dfrac{\partial Q}{\partial x} - \dfrac{\partial P}{\partial y}\right)$ represents some residual quantity picked up on going once round the closed path C. If it is zero then the field is conservative.

Extension to three dimensions

If we extend our ideas to three dimensions we can place a conservative field in a new light. The actual calculation of an integral proceeds much as we might expect. Here

$$\int \mathbf{v}\,.\,d\mathbf{r} = \int (v_x\,\mathbf{i} + v_y\,\mathbf{j} + v_z\,\mathbf{k})\,.\,(dx\mathbf{i} + dy\mathbf{j} + dz\mathbf{k})$$

$$= \int (v_x\,dx + v_y\,dy + v_z\,dz)$$

If this integral is to be path independent, then

$$\mathbf{v}\,.\,d\mathbf{r} = v_x\,dx + v_y\,dy + v_z\,dz = d\phi$$

Now

$$d\phi = \frac{\partial \phi}{\partial x}\,dx + \frac{\partial \phi}{\partial y}\,dy + \frac{\partial \phi}{\partial z}\,dz = \nabla\phi.\,d\mathbf{r}$$

Hence $(\mathbf{v} - \nabla\phi)\,.\,d\mathbf{r} = 0$ and we conclude that since $d\mathbf{r}$ is arbitrarily orientated and non-zero

$$\mathbf{v} = \nabla\phi \qquad (11.30)$$

that is, $\mathbf{v}$ is the gradient of some scalar ϕ, called the **scalar potential** of the conservative field. (Note that ϕ is determined apart from a constant and that in some branches of application a minus sign is introduced, thus $\mathbf{E} = -\text{grad}\,\phi$ where $\mathbf{E}$ is the electric field and ϕ the electrostatic potential.) You are probably familiar with gravitational potential where the gravity field may be approximately represented by $\mathbf{F} = -mg\mathbf{k}$ and the potential as mgz; $\nabla(mgz) = mg\mathbf{k}$.

The steps taken to obtain (11.30) are reversible and so a *necessary and sufficient condition* that $\displaystyle\oint_C \mathbf{v}\,.\,d\mathbf{r}$ be zero is that $\mathbf{v} = \nabla\phi$, for some scalar ϕ.

Example

If $\mathbf{v} = (2x + y)\mathbf{i} + (x - 3y)\mathbf{j}$ then a suitable candidate for ϕ is $\phi = x^2 + xy - \frac{3}{2}y^2$ as you can verify by forming $\nabla \phi$.

However, it is a tedious task to try and find ϕ in cases where none exists and we look for a simple test on $\mathbf{v}$ itself to discover whether it is conservative. We effectively want a generalisation of condition (11.29). See the next section.

Problems

1. Evaluate $$\oint_C \frac{(x^3\,dy - y^3\,dx)}{x^2 + y^2}$$

(i) when C is the boundary of the rectangle whose sides lie along the lines $x = \pm a, y = \pm b$,

(ii) when C is the circle $x^2 + y^2 = k^2$,

(iii) the triangle with vertices at $(0, 0)$, (c, c), $(0, c)$. (L.U.)

2. Evaluate the line integral

$$I = \int_C \left\{(x^2 + y^2)\,dx + kxy\,dy\right\}$$

where C is the straight line path, $x + y = a$, starting at A $(a, 0)$ and ending at B $(0, a)$; k is a constant.

Evaluate I also when C is the quadrant of a circle joining A to B given by $x = a\cos\theta$, $y = a\sin\theta$, $0 \leqslant \theta \leqslant \pi/2$. Show that the two integrals are equal in value when $k = 2$. (L.U.)

3. Evaluate $$\oint_C \left\{(y - \cos x)\,dx + \sin x\,dy\right\}$$

where C is the perimeter of the triangle formed by the lines $y = 2x/\pi$, $x = \pi/2$, $y = 0$,
(i) directly, (ii) by using Green's theorem. (L.U.)

4. (a) Show that the line integral

$$\int \frac{yz}{x^2 + y^2}\,ds$$

taken along the arc of the curve in space given by the equations $x = 3at$, $y = 3at^2$, $z = 2a(1 + t^3)$ between the points for which $t = 0$ and $t = 1$, has the value $a[4 - \pi/2 + \log 2]$.

(b) Find the value of the integral

$$\int \left\{(y + 3z)\,dx + (2z + x)\,dy + (3x + 2y)\,dz\right\}$$

taken along the arc of the helix $x = a\cos\theta$, $y = a\sin\theta$, $z = 2a\theta/\pi$ between the points $(a, 0, 0)$ and $(0, a, a)$. (L.U.)

5. Evaluate the work done by the force $\mathbf{F} = 2xy\mathbf{i} - 2\mathbf{j} + z\mathbf{k}$ in moving along the curve specified from point A to point B.

(i) $y = 3x,\ z = 1;\ A = (1, 3, 1),\ B = (3, 9, 1)$

(ii) $y = 3x,\ z = 0;\ A = (3, 9, 0),\ B = (0, 0, 0)$

(iii) $x^2 + y^2 = 16,\ z = 1;\ A = (4, 0, 1),\ B = (0, 4, 1)$

6. Show that $\operatorname{div}\left\{f(r)\hat{\mathbf{r}}\right\} = \dfrac{1}{r^2}\dfrac{d}{dr}\left\{r^2 f(r)\right\}$.

7. Evaluate $\displaystyle\int_C \left\{(20x^4 - 4xy^3)\,dx - 6x^2y^2\,dy\right\}$ where C is that part of the curve $x^4 - 6xy^3 = 4y^2$ from $(0, 0)$ to $(2, 1)$. (Hint: Use Green's theorem.)

8. Verify Green's theorem for $\displaystyle\oint_C \left\{(1.5x^2 - 4y^2)\,dx + (2y - 3xy)\,dy\right\}$ where C is the boundary of the region defined by

(a) $y = x^2,\ x = y^2$ (b) $x = 0,\ y = 0,\ x + y = 2$.

9. Verify that

$\mathbf{F} = (3x^2 - 3yz + 2xz)\mathbf{i} + (3y^2 - 3xz + z^2)\mathbf{j} + (3z^2 - 3xy + x^2 + 2yz)\mathbf{k}$

represents a conservative field of force. Find a scalar ϕ such that $\mathbf{F} = \operatorname{grad}\phi$.

Evalaute $\displaystyle\int_C \mathbf{F}\cdot d\mathbf{r}$ where C is any path joining $(-1, 2, 3)$ to $(3, 2, -1)$.

What would be the value of $\oint \mathbf{F}\cdot d\mathbf{r}$ once round the circle $x^2 + y^2 = 1;\ z = 0$?

11.6 CURL OF A VECTOR FIELD

There are examples in the flow of fluids where although the fluid appears to be moving in straight lines, the individual particles are in fact rotating about their axes. Consider Figure 11.9. An incompressible fluid is in steady one-dimensional flow between two plane walls $y = 0$ and $y = 1$. The velocity vector is $\mathbf{v} = 4Uy(1 - y)\,\mathbf{i}$ where U is the velocity of the fluid along $y = \frac{1}{2}$. If a small paddle wheel is placed in the fluid with its axis in the z - direction then it will spin as it moves

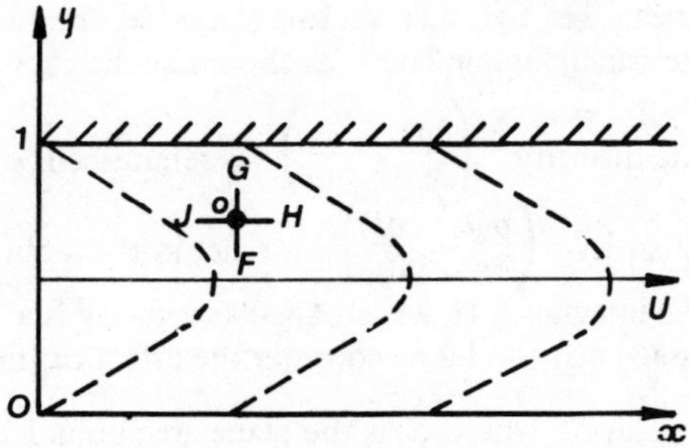

Figure 11.9

downstream, if the velocity of the fluid hitting blade OF is different from that of the fluid hitting blade GO. It is observed that on the line $y = ½$ no spinning occurs; that the spinning is anticlockwise above the line $y = ½$ and clockwise below it; moreover, the angular velocity of the wheel increases the nearer it is placed to either wall.

Let us consider a more general two-dimensional motion in which the velocity components at P are u and v which are both functions of x and y. Refer to Figure 11.10. In a small interval of time let the fluid particle at P move to P′ by when its velocity components have become, to first order $u + (\partial u/\partial y)\,\delta y$ and $v + (\partial v/\partial x)\,\delta x$. The angular velocity of the line PQ is $\dfrac{v + (\partial v/\partial x)\,\delta x - v}{\delta x}$ which becomes $\dfrac{\partial v}{\partial x}$ in the limit as $\delta x \to 0$. Similarly we have a limiting angular velocity of the line PR of $-(\partial u/\partial y)$. Consequently, the rotation of the fluid element at P is the average of these, viz. $\dfrac{1}{2}\left(\dfrac{\partial v}{\partial x} - \dfrac{\partial u}{\partial y}\right) = \omega_z$.

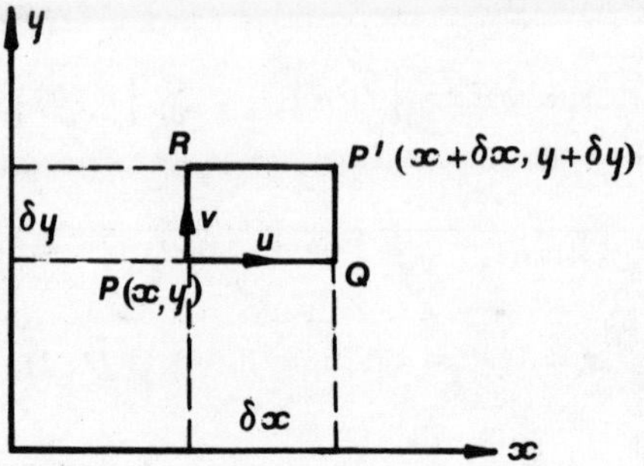

Figure 11.10

Note that in our particular example $u = 4Uy(1 - y)$ and $v = 0$ so that $\omega_z = ½(0 - 4U[1 - 2y]) = -2U(1 - 2y)$. Along $y = ½$ there is no angular velocity; $y < ½$ gives a negative velocity whilst $y > ½$ gives a positive velocity; as $y \to 0$ or $y \to 1$ the *magnitude* of ω_z increases. This bears out the experimental observations.

We have taken a special case of motion in the $x - y$ plane and in a general motion the angular velocity could be about any axis. Let us associate with the angular velocity a vector whose magnitude is that of the angular velocity and whose direction is along the axis of rotation in the sense indicated by Figure 11.11; this, as it can be shown, will mean that two such vectors can in fact be added by the parallelogram law – as should be the case.

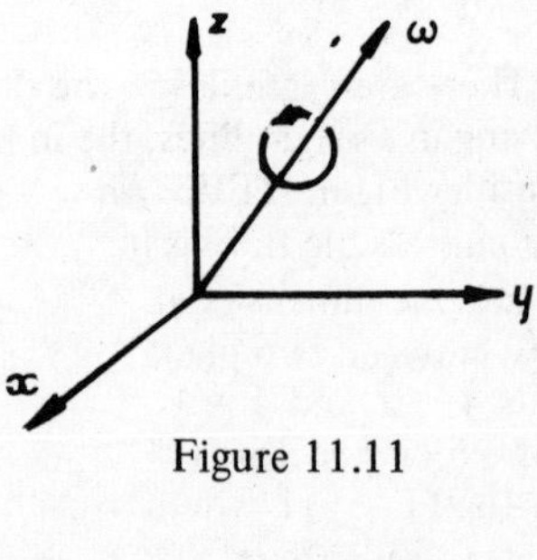

Figure 11.11

The quantity $\left(\dfrac{\partial v}{\partial x} - \dfrac{\partial u}{\partial y}\right)$ is reminiscent of the quantity $\left(\dfrac{\partial Q}{\partial x} - \dfrac{\partial P}{\partial y}\right)$ we met in the section on line integrals. However, angular velocity is a localised entity so let us consider the effect of finding the limit as $S \to 0$ of the ratio $\oint_C \mathbf{v}\,.\,d\mathbf{r}/S$, where S is the plane area bounded by the curve C. Integrating around the element PQP′R of Figure 11.10, and neglecting velocity terms of second order, we obtain

$$\Gamma = \oint \mathbf{v} \,.\, d\mathbf{r} = u\delta x + [v + (\partial v/\partial x)\delta x]\,\delta y - [u + (\partial u/\partial y)\delta y]\,\delta x - v\delta y$$

$$= \left(\frac{\partial v}{\partial x} - \frac{\partial u}{\partial y}\right)\delta x\,\delta y$$

Then

$$\Gamma/S \simeq \left(\frac{\partial v}{\partial x} - \frac{\partial u}{\partial y}\right)$$

and in the limit as $S \to 0$, the approximation, due to the neglected terms, becomes an exact relationship. Notice that this is *twice* the angular velocity.

We define the **curl** of a vector $\mathbf{v}$ by

$$\text{curl } \mathbf{v} = \lim_{S \to 0} \left\{ \oint_C \mathbf{v} \,.\, d\mathbf{r}/S \right\} \tag{11.31}$$

where S and C have the meanings stated earlier.

In fluid dynamics, the curl of the velocity vector is called the **vorticity**. Note that the curl operator gives a vector result.

The current flowing in a wire produces a magnetic field. We generalise this situation to say that a volume distribution of current $\mathbf{J}$ gives rise to a magnetic field strength $\mathbf{H}$ which is given by curl $\mathbf{H} = \mathbf{J}$.

Cartesian components of curl

Definition (11.31) is quite general and it means that curl $\mathbf{v}$ can in general be inclined to all three coordinate axes. By extending the argument earlier it can be shown that if the vector $\mathbf{v} = v_1\mathbf{i} + v_2\mathbf{j} + v_3\mathbf{k}$ then

$$\text{curl } \mathbf{v} = \left(\frac{\partial v_3}{\partial y} - \frac{\partial v_2}{\partial z}\right)\mathbf{i} + \left(\frac{\partial v_1}{\partial z} - \frac{\partial v_3}{\partial x}\right)\mathbf{j} + \left(\frac{\partial v_2}{\partial x} - \frac{\partial v_1}{\partial y}\right)\mathbf{k} \tag{11.32}$$

We were earlier looking at the $\mathbf{k}$ component only. More compactly we may write

$$\text{curl } \mathbf{v} = \begin{vmatrix} \mathbf{i} & \mathbf{j} & \mathbf{k} \\ \dfrac{\partial}{\partial x} & \dfrac{\partial}{\partial y} & \dfrac{\partial}{\partial z} \\ v_1 & v_2 & v_3 \end{vmatrix} = \nabla_\wedge \mathbf{v} \tag{11.33}$$

both these forms following by analogy with the simple vector product. How nice it is to see the operation $\nabla_\wedge \mathbf{v}$ arising; it somehow brings us round full circle.

Note that the form $\nabla_\wedge \mathbf{v}$ holds generally; the form for ∇ will be chosen to suit the particular coordinate system of any particular problem.

Conservative fields

If we recall the condition for a field in the x - y plane to be conservative then we recognise this as being that the z - component of curl $\mathbf{v}$ be zero; since the two other components do not exist this means that a necessary and sufficient condition for the field to be conservative is that its curl should vanish, i.e. curl $\mathbf{v} = \mathbf{0}$ if and only if $\mathbf{v}$ is conservative. *This result can be shown to hold quite generally.*

Example

Find curl $\mathbf{v}$ if (i) $\mathbf{v} = x^2\mathbf{i} + x^2y\mathbf{j} + yz\mathbf{k}$ (ii) $\mathbf{v} = \text{grad}\,\phi$, for some scalar ϕ.

$$\text{(i)} \quad \text{curl}\,\mathbf{v} = \mathbf{i}\left(\frac{\partial}{\partial y}[yz] - \frac{\partial}{\partial z}[x^2y]\right) + \mathbf{j}\left(\frac{\partial}{\partial z}[x^2] - \frac{\partial}{\partial x}[yz]\right)$$

$$+ \mathbf{k}\left(\frac{\partial}{\partial x}[x^2y] - \frac{\partial}{\partial y}[x^2]\right) = z\mathbf{i} + 2xy\mathbf{k}$$

$$\text{(ii)} \quad \text{curl}\,\mathbf{v} = \mathbf{i}\left(\frac{\partial}{\partial y}\left[\frac{\partial \phi}{\partial z}\right] - \frac{\partial}{\partial z}\left[\frac{\partial \phi}{\partial y}\right]\right) + \ldots\ldots$$

$$= \mathbf{i}\left(\frac{\partial^2\phi}{\partial y\partial z} - \frac{\partial^2\phi}{\partial z\partial y}\right) + \mathbf{j}\left(\frac{\partial^2\phi}{\partial z\partial x} - \frac{\partial^2\phi}{\partial x\partial z}\right) + \mathbf{k}\left(\frac{\partial^2\phi}{\partial x\partial y} - \frac{\partial^2\phi}{\partial y\partial x}\right) = 0$$

Hence the curl of any gradient is zero, i.e. curl grad $\phi \equiv \mathbf{0}$. Note also that curl $\mathbf{r} = \mathbf{0}$. A vector whose curl is zero is termed **irrotational**.

Classification of Vector Fields

In practice we meet four kinds of vector fields classified as follows

(a) curl $\mathbf{v} = \mathbf{0}$ and div $\mathbf{v} = 0$. Here since curl $\mathbf{v} = \mathbf{0}$, then $\mathbf{v} = \text{grad}\,\phi$ and the second condition gives div grad $\phi = 0 = \nabla^2\phi$ which is **Laplace's equation**. Typical fields of this kind occur in irrotational motion of incompressible fluids and in electric fields arising from static charges on boundary conductors.

(b) curl $\mathbf{v} = \mathbf{0}$ but div $\mathbf{v} \neq 0$. Again, $\mathbf{v} = \text{grad}\,\phi$ but div grad $\phi = \nabla^2\phi \neq 0$ and we obtain **Poisson's equation** which arises for example in the electric fields of volume distribution of charges (electrons in a thermionic tube) and in the irrotational, compressible flow of a fluid.

(c) curl $\mathbf{v} \neq \mathbf{0}$ but div $\mathbf{v} = 0$. The field is rotational and no scalar potential exists. Such a field is typified by rotational, incompressible fluid flow or the magnetic field within a conductor carrying a current.

(d) curl $\mathbf{v} \neq \mathbf{0}$ and div $\mathbf{v} \neq 0$. This is the most general type of vector field and is found in for example the rotational motion of a compressible fluid. It can be shown that such a field is expressible as the sum of two fields, one of type (b) and one of type (c).

Problems

1. Find the curl of the vectors of Problem 1 of Section 11.4.

2. Find the curl of

   (i) $\mathbf{v} = x\mathbf{i} + 5xy\mathbf{j}$ (ii) $\mathbf{v} = 3x\mathbf{i} + 2y\mathbf{j} - 4z\mathbf{k}$ (iii) $\mathbf{v} = k\hat{\mathbf{r}}/r^2$.

3. Show that the vector $\mathbf{A} = (4xy - z^3)\mathbf{i} + 2x^2\mathbf{j} - 3xz^2\mathbf{k}$ is irrotational and find a function ϕ such that $\mathbf{A} = \text{grad}\,\phi$.

11.7 VECTOR IDENTITIES

We now present some results which are left to you to prove. We assume $\phi(x, y, z)$ and $\psi(x, y, z)$ are scalar point functions and $\mathbf{u}(x, y, z)$, $\mathbf{v}(x, y, z)$ are vector point functions. Note that these results are independent of coordinate systems; you will most easily prove them in Cartesians.

(i) $\text{grad}\,(\phi\ \psi) = \psi\ \text{grad}\ \phi + \phi\ \text{grad}\ \psi$ (11.34)

(ii) $\text{div}\,(\phi\ \mathbf{v}) = (\text{grad}\ \phi)\ .\ \mathbf{v} + \phi\ \text{div}\ \mathbf{v}$ (11.35)

(iii) $\text{curl}\,(\phi\ \mathbf{v}) = (\text{grad}\ \phi) \wedge \mathbf{v} + \phi\ \text{curl}\ \mathbf{v}$ (11.36)

(iv) $\text{div}\,(\mathbf{u} \wedge \mathbf{v}) = \mathbf{v}\ .\ \text{curl}\ \mathbf{u} - \mathbf{u}\ .\ \text{curl}\ \mathbf{v}$ (11.37)

(v) $\text{curl}\,(\mathbf{u} \wedge \mathbf{v}) = \mathbf{u}\ \text{div}\ \mathbf{v} - (\mathbf{u}\ .\ \nabla)\ \mathbf{v} + (\mathbf{v}\ .\ \nabla)\ \mathbf{u} - \mathbf{v}\ \text{div}\ \mathbf{u}$ (11.38)

(vi) $\text{grad}\,(\mathbf{u}\ .\ \mathbf{v}) = (\mathbf{v}\ .\ \nabla)\ \mathbf{u} + (\mathbf{u}\ .\nabla)\ \mathbf{v} + \mathbf{v} \wedge \text{curl}\ \mathbf{u} + \mathbf{u} \wedge \text{curl}\ \mathbf{v}$ (11.39)

Second order operators

We have seen that curl grad $\phi \equiv \mathbf{0}$; similarly, we can show that div curl $\mathbf{v} \equiv 0$.

The other second order operators which exist are grad div $\mathbf{v}$, div grad ϕ and curl curl $\mathbf{v}$ We examine these in turn.

(i) If $\mathbf{v}$ represents electric force, div $\mathbf{v}$ is the space charge density and grad div $\mathbf{v}$ represents the greatest rate of increase of charge (in magnitude and direction) at any point. If $\mathbf{v}$ represents fluid velocity, grad div $\mathbf{v}$ represents the greatest rate of space increase of density.

(ii) div grad ϕ is written $\nabla^2\phi$ and ∇^2 is called **Laplace's operator**. In cartesian coordinates $\nabla^2\phi = \dfrac{\partial^2\phi}{\partial x^2} + \dfrac{\partial^2\phi}{\partial y^2} + \dfrac{\partial^2\phi}{\partial z^2}$. $\nabla^2\phi = 0$ is, of course, **Laplace's equation.** Note that if $\mathbf{A} = A_1\mathbf{i} + A_2\mathbf{j} + A_3\mathbf{k}$ then $\nabla^2\mathbf{A} = \nabla^2 A_1\mathbf{i} + \nabla^2 A_2\mathbf{j} + \nabla^2 A_3\mathbf{k}$.

(iii) It can be shown that

$$\text{curl curl}\ \mathbf{v} = \text{grad div}\ \mathbf{v} - \nabla^2\mathbf{v} \qquad (11.40)$$

In proving some of these vector identities we may note that if each side is a vector it often suffices to show that the $\mathbf{i}$ - component of each side is the same and appeal to symmetry to establish the result.

Example

If $\mathbf{E}$ represents electric field strength, $\mathbf{H}$ magnetic field strength and if we assume unit magnetic permeability and dielectric constant and further a charge-free and current-free region then **Maxwell's equations** for the electromagnetic field reduce to

$$\text{div}\ \mathbf{E} = 0 = \text{div}\ \mathbf{H},\ \text{curl}\ \mathbf{E} = -\frac{\partial\mathbf{H}}{\partial t},\ \text{curl}\ \mathbf{H} = \frac{\partial\mathbf{E}}{\partial t}$$

Now $\text{curl curl}\ \mathbf{H} = \text{curl}\,(\partial\mathbf{E}/\partial t) = \partial\,(\text{curl}\ \mathbf{E})/\partial t$ i.e.

$$\text{grad div}\ \mathbf{H} - \nabla^2\mathbf{H} = \frac{\partial}{\partial t}\left(-\frac{\partial\mathbf{H}}{\partial t}\right) = -\frac{\partial^2\mathbf{H}}{\partial t^2}$$

But div $\mathbf{H} = 0$, therefore $\mathbf{H}$ satisfies the **wave equation**

$$\nabla^2 \mathbf{H} = \partial^2 \mathbf{H}/\partial t^2$$

and a similar equation is satisfied by $\mathbf{E}$.

Problems

1. Prove (11.34) to (11.40).

2. Verify (11.34) to (11.40) for $\mathbf{u} = yz\mathbf{i}$, $\mathbf{v} = x^2y\mathbf{i} + 2xz\mathbf{j} + 6z^2\mathbf{k}$, $\phi = xyz$ and $\psi = x + y + z$.

3. If $\mathbf{r} = x\mathbf{i} + y\mathbf{j} + z\mathbf{k}$ where $\mathbf{i}, \mathbf{j}, \mathbf{k}$ are orthogonal cartesian unit vectors, use (11.34) to (11.40) to show that if n is a constant

   (a) $\text{div}(r^n\mathbf{r}) = (n+3)r^n$, (b) $\text{curl}(r^n\mathbf{r}) = \mathbf{0}$, (c) $\text{curl}(r^n\mathbf{r} \times \mathbf{k}) = nr^{n-2}z\mathbf{r} - (n+2)r^n\mathbf{k}$,

   (d) $\text{div}(r^n\mathbf{r} \times \mathbf{k}) = 0$. [x has the same meaning as $\wedge$] (L.U.)

4. If $\phi = \dfrac{\mathbf{r} \cdot \mathbf{i}}{r^3}$ where $\mathbf{r} = x\mathbf{i} + y\mathbf{j} + z\mathbf{k}$, show that $\text{grad}\,\phi = \dfrac{1}{r^5}\left\{(r^2\mathbf{i} - 3(\mathbf{r} \cdot \mathbf{i})\,\mathbf{r}\right\}$. (L.U.)

5. If $\phi = \dfrac{\mathbf{M} \cdot \mathbf{r}}{r^3}$, $\mathbf{A} = \dfrac{\mathbf{M} \times \mathbf{r}}{r^3}$ where $\mathbf{M}$ is a constant vector, $\mathbf{r} = x\mathbf{i} + y\mathbf{j} + z\mathbf{k}$ and $r = |\mathbf{r}|$, show that curl $\mathbf{A} = -\text{grad}\,\phi$. (L.U.)

6. Show that if $\phi = 1/r$, where $r^2 = x^2 + y^2 + z^2$ and $\mathbf{A} = \text{grad}\,\phi$, then $\text{curl}(\phi\mathbf{A}) = \mathbf{0}$. Find $\text{div}(\phi\mathbf{A})$ in this case. (L.U.)

7. Given $\mathbf{r} = x\mathbf{i} + y\mathbf{j} + z\mathbf{k}$ where $\mathbf{i}, \mathbf{j}, \mathbf{k}$ are cartesian unit vectors, and $\mathbf{p} = a\mathbf{i} + b\mathbf{j} + c\mathbf{k}$ is a constant vector, show that, if $\mathbf{u} = (\mathbf{p} \cdot \mathbf{r})\mathbf{r}$,

   div $\mathbf{u} = 4\mathbf{p} \cdot \mathbf{r}$, curl $\mathbf{u} = \mathbf{p} \times \mathbf{r}$, curl $(\mathbf{p} \times \mathbf{r}) = 2\mathbf{p}$. (L.U.)

8. Show that div $(\mathbf{A} \times \mathbf{r}) = 0$ if curl $\mathbf{A} = \mathbf{0}$, where $\mathbf{r}$ denotes the position vector of any point (x, y, z). Also prove that $\nabla \cdot (r^3\mathbf{r}) = 6r^3$. (C.E.I.)

9. If $\mathbf{A}$ and ϕ are a vector point function and a scalar point function respectively, prove that curl $(\phi\mathbf{A}) = \phi$ curl $\mathbf{A}$ + grad $\phi \wedge \mathbf{A}$.

   Verify the above result when $\mathbf{A} = 2xz^2\mathbf{i} - yz\mathbf{j} + 3xz^3\mathbf{k}$ and $\phi = x^3yz$ at the point (1, 1, 1). (C.E.I.)

11.8 SURFACE INTEGRALS : STOKES' THEOREM

In the course of developing the concept of divergence, flux through a surface is considered. We now look at flux through an arbitrary surface. In Figure 11.12 we consider a point P on a given surface S and an element of area δS surrounding P.

The outward (positive) unit normal vector to the surface at P is denoted $\hat{\mathbf{n}}$. (If the surface is not closed we always define $\hat{\mathbf{n}}$ on the same side of the surface as P moves over the surface.) The vector element of area is defined as $\delta\mathbf{S} = \hat{\mathbf{n}}\,\delta S$. Let $\mathbf{v}$ be a vector field in which the surface lies. Then $\mathbf{v}\,.\,\delta\mathbf{S} = \mathbf{v}\,.\,\hat{\mathbf{n}}\,\delta S = |\mathbf{v}|\cos\theta\delta S$ represents the **flux** of $\mathbf{v}$ through the element of surface δS. The total flux of $\mathbf{v}$ through the surface S is

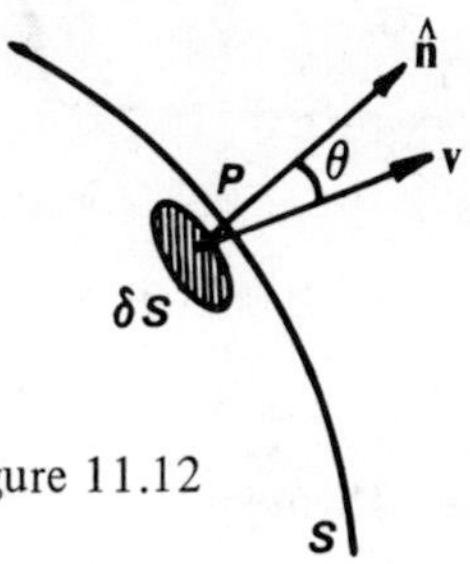

Figure 11.12

$$\iint_S \mathbf{v}\,.\,\mathrm{d}\mathbf{S} \qquad (11.41)$$

where the integral is taken over the surface S.

If $\mathbf{v}$ represents fluid velocity, then (11.41) represents the net amount of fluid crossing the surface in unit time. The same formula can represent current flow, magnetic flux, flow of heat etc.

Example

Evaluate (11.41) where $\mathbf{v} = z\mathbf{i} + x\mathbf{j} - 2y^2z\mathbf{k}$ and S is the curved surface of the cylinder $x^2 + y^2 = a^2$ included in the first octant and lying between $z = 0$ and $z = b$; see Figure 11.13. The steps are straightforward.

(i) Find a unit normal $\hat{\mathbf{n}}$ to the surface.

(ii) Form $\mathbf{v}\,.\,\hat{\mathbf{n}}$.

(iii) Project $\mathrm{d}S$ on to the appropriate plane (in this case the x - z plane will do, as would the y - z plane).

(iv) Perform the double integral which results.

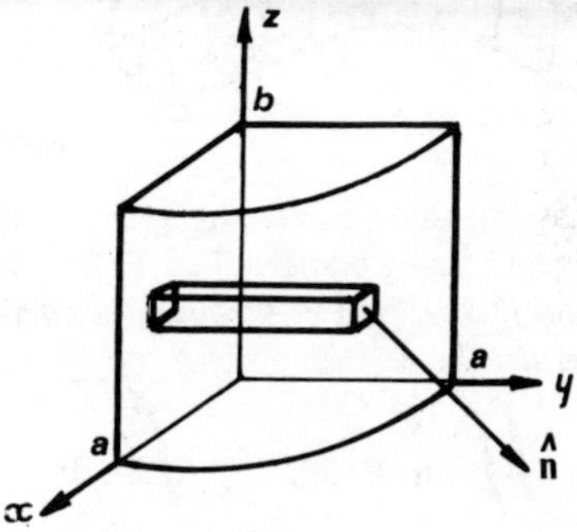

Figure 11.13

In our example, the surface may be written as $\phi(x, y, z) = x^2 + y^2 = a^2$ and a normal vector is $\nabla\phi = 2x\mathbf{i} + 2y\mathbf{j}$.

(i) A suitable unit normal is $\hat{\mathbf{n}} = (2x\mathbf{i} + 2y\mathbf{j})/\sqrt{(2x)^2 + (2y)^2} = (x\mathbf{i} + y\mathbf{j})/a$.

(ii) Then $\mathbf{v}\,.\,\hat{\mathbf{n}} = (zx + xy)/a$.

(iii) The projection of $\mathrm{d}S$ on the x - z plane is $\mathrm{d}x\,\mathrm{d}z/(\hat{\mathbf{n}}\,.\,\mathbf{j})$, since $\mathbf{j}$ is a unit vector normal to the x - z plane, i.e. $\mathrm{d}x\,\mathrm{d}z(a/y)$

(iv) If R is the projection of S on the x - z plane then

$$\iint_S \mathbf{v}\,.\,\mathrm{d}\mathbf{S} = \iint_R \frac{(zx + xy)}{a}\,\frac{\mathrm{d}x\,\mathrm{d}z\,a}{y}$$

$$= \int_0^b \mathrm{d}z \int_0^a \mathrm{d}x \left(\frac{zx}{\sqrt{a^2 - x^2}} + x\right)$$

$$= \int_0^b dz \left[-z(a^2 - x^2)^{1/2} + \tfrac{1}{2}x^2\right]_0^a$$

$$= \int_0^b dz\,(za + \tfrac{1}{2}a^2) = \tfrac{1}{2}ab\,(a+b).$$

Stokes' Theorem

We now present a generalisation of Green's Theorem in the plane, which allows us to convert a line integral into a surface integral. We shall first state the theorem and then outline how it may be proved.

Let an open surface S have a bounding edge C (as in Figure 11.14(a));

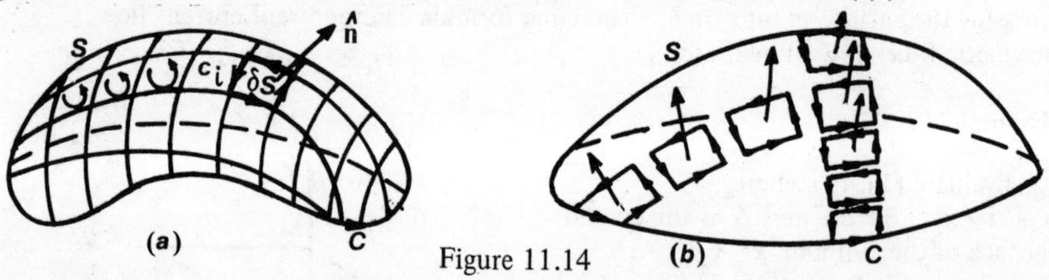

Figure 11.14

you may imagine a basin and its rim – although there is no restriction in the theorem on the edge C being planar. Let **F** be a vector field. Then Stokes' theorem states that *the total flux of* curl **F** *through the surface* S *is equal to the circulation of* **F** *round the circuit* C i.e.

$$\iint_S \text{curl}\,\mathbf{F}\,.\,d\mathbf{S} = \oint_C \mathbf{F}\,.\,d\mathbf{r} \tag{11.42}$$

The proof is effected by dividing S into elementary areas δS_i with boundaries C_i as shown in Figure 11.14(a). From the definition of curl **F**,

$$\text{curl}\,\mathbf{F}\,.\,\hat{\mathbf{n}}\,\delta S_i \simeq \oint_{C_i} \mathbf{F}\,.\,d\mathbf{r} \tag{11.43}$$

If we sum the left-hand side of (11.43) over all elementary areas we obtain the left-hand side of (11.42) after letting $\delta S_i \to 0$. If we sum the right-hand side of (11.43) and note as shown in Figure 11.14(b) that parts of boundaries C_i in common with neighbouring areas will lead to cancellation then the net sum is simply the right-hand side of (11.42), i.e. the integral around the external boundary C. Convince yourself by dividing S into 4 areas, say, and noting the cancellation. Hence the result may be established.

Note that if S lies in the x - y plane, we recover Green's Theorem, since $d\mathbf{S} \equiv dx\,dy\,\mathbf{k}$. Note further that if $\mathbf{F} = \text{grad}\,\phi$ then the left-hand side of (11.42) vanishes identically and we are left with a zero line integral as we should expect from a conservative field.

Problems

1. Evaluate $\iint_S \mathbf{A} \cdot \hat{\mathbf{n}}\, dS$ for

(a) $\mathbf{A} = y\mathbf{i} + 3x\mathbf{j} - z\mathbf{k}$; S is the surface of the plane $x + 2y = 3$ in the first octant cut off by the plane $z = 2$.

(b) $\mathbf{A} = xy\mathbf{i} - 2x\mathbf{j} + 2z\mathbf{k}$; S is the surface of the cylinder $y^2 = 8x$ in the first octant bounded by $y = 4$ and $z = 6$.

(c) $\mathbf{A} = 6z\mathbf{i} + (x + 2y)\mathbf{j} - x\mathbf{k}$; S is the *entire* surface of the region bounded by the cylinder $x^2 + z^2 = 9$ by $x = 0$, $y = 0$, $z = 0$ and $y = 8$.

(d) $\mathbf{A}$ as in (c); S is the surface with $z > 0$ bounded by the cone $z^2 = x^2 + y^2$ and the plane $z = 4$.

2. The electrostatic potential at $(0, 0, a)$ due to a charge of density σ on the hemispherical surface $S = x^2 + y^2 + z^2 = a^2$, $z \geqslant 0$ is given by

$$\phi = \iint_S \frac{\sigma\, dS}{\left\{x^2 + y^2 + (z - a)^2\right\}^{1/2}} .$$

Find ϕ.

3. Using Stokes' theorem, or otherwise, evaluate $\oint_C \mathbf{B} \cdot d\mathbf{s}$, where C is the circle

$x^2 + y^2 = a^2$, $z = 0$, and $\mathbf{B} = y^3\mathbf{i} - x^3\mathbf{j} + xyz\mathbf{k}$. (L.U.)

4. Show, using Stokes' theorem, that

$$\iint_\sigma \phi \text{ curl } \mathbf{H} \cdot d\mathbf{S} = \oint_\gamma \phi \mathbf{H} \cdot d\mathbf{r} - \iint_\sigma (\text{grad } \phi \times \mathbf{H}) \cdot d\mathbf{S}$$

where the open surface σ is bounded by the closed curve γ. (C.E.I.)

5. State Stokes' Theorem.

Use the above to evaluate $\iint_S \text{curl } \mathbf{A} \cdot d\mathbf{S}$, where $\mathbf{A} = (x^2 + y - 4)\mathbf{i} + 3xy\mathbf{j} + (2xz + z^2)\mathbf{k}$

and S is the surface of the paraboloid $z = 4 - (x^2 + y^2)$ above the xy - plane (C.E.I.)

6. By using Stokes' theorem, or otherwise, evaluate $\iint_K \text{curl } \mathbf{Q} \cdot d\mathbf{S}$ where $\mathbf{Q} = y\mathbf{i} - x\mathbf{j}$

and K is that part of the surface of the sphere $x^2 + y^2 + z^2 = b^2$ for which $z \geqslant 0$. (C.E.I.)

7. State Stokes' theorem and also give a brief derivation of the relation $\text{curl}\,\phi\mathbf{B} = \text{grad}\,\phi \times \mathbf{B} + \phi\,\text{curl}\,\mathbf{B}$ where ϕ is a scalar and $\mathbf{B}$ a vector point function.

Hence deduce the result

$$\iint_S \phi\,\text{curl}\,\mathbf{B}\,.\,d\mathbf{S} = \oint_C \phi\mathbf{B}\,.\,d\mathbf{r} - \iint_S (\text{grad}\,\phi \times \mathbf{B})\,.\,d\mathbf{S}$$

where the closed curve C bounds the open surface S.

Further, by putting $\mathbf{B} = \text{grad}\,\psi$ in this result, prove that

$$\iint_S \nabla\phi \times \nabla\psi . d\mathbf{S} = \oint_C \phi\,\nabla\psi\,.\,d\mathbf{r} = -\oint_C \psi\,\nabla\phi\,.d\mathbf{r},$$

giving a short proof of any result which you use. (C.E.I.)

11.9 GAUSS' DIVERGENCE THEOREM

We first produce an alternative definition of divergence based on the ideas in the definition of curl. Let V be the volume enclosed by a surface S then if $\mathbf{A}$ be a vector field which operates throughout the volume V we have the definition

$$\text{div}\,\mathbf{A} = \lim_{V\to 0}\left\{\frac{1}{V}\iint_S \mathbf{A}\,.\,d\mathbf{S}\right\} \tag{11.44}$$

This generalises our previous approach where V was a cuboid.

Gauss' Divergence Theorem

We now state a result which links surface integrals and volume integrals. With the notation above the theorem states that *the total divergence of* $\mathbf{A}$ *through the volume* V *is equal to the net flux of* div $\mathbf{A}$ *through the surface* S, i.e.

$$\iiint_V \text{div}\,\mathbf{A}\,dV = \iint_S \mathbf{A}\,.\,d\mathbf{S} \tag{11.45}$$

If we have an incompressible fluid and draw an imaginary volume in the field of flow, the theorem effectively states that as much fluid leaves through the surface as enters, *unless* there are sources or sinks inside the volume.

The proof of the theorem follows a similar line to that for Stokes' theorem. Refer to Figure 11.15. For the volume element shown we have, using (11.44) that

$$\text{div}\,\mathbf{A}\,\delta V_i \simeq \iint_{S_i} \mathbf{A}\,.\,d\mathbf{S}$$

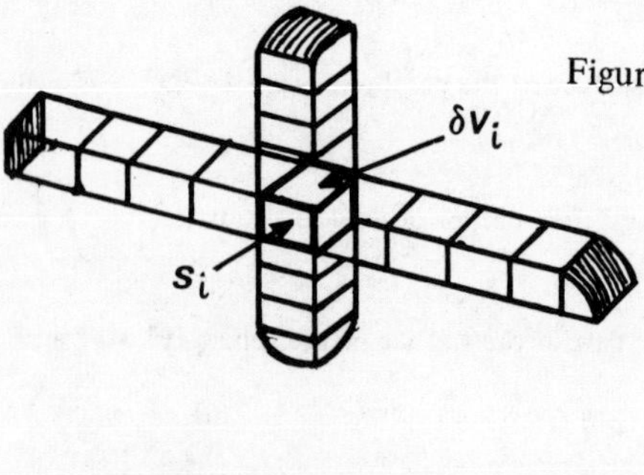

Figure 11.15

Summation of the left-hand side over all blocks produces the left-hand side of (11.45) after letting $\delta V_i \to 0$. Summing the right-hand side and allowing for cancellation at common faces leads to the right-hand side of (11.45); the cancellation is due to the fact that a flow *outwards* from one face of an elementary volume is a flow *inwards* (i.e. a negative outwards flow) to the same face of the neighbouring elementary volume.

Example

Consider a compressible fluid with density $\rho(x, y, z, t)$ flowing with velocity **v**. Consider a small fixed volume V, surrounded by a surface S, which lies in the field of flow. The total mass of fluid inside V at time t is $\iiint_V \rho \, dV$. The rate of increase of mass is $\frac{\partial}{\partial t} \iiint_V \rho \, dV = \iiint_V \frac{\partial \rho}{\partial t} \, dV$. But the net *outward* flow of fluid across the surface S in unit time is $\iint_S \rho \mathbf{v} \cdot d\mathbf{S} = \iiint_V \text{div}(\rho \mathbf{v}) \, dV$ by the divergence theorem. Therefore $\iiint_V \frac{\partial \rho}{\partial t} \, dV = -\iiint_V \text{div}(\rho v) \, dV$ and since V is arbitrary it follows that

$$\frac{\partial \rho}{\partial t} + \text{div}(\rho \mathbf{v}) = 0 \qquad (11.46)$$

This is the **continuity theorem** for fluid flow.

If the fluid in incompressible then ρ is constant and $\text{div}\, \mathbf{v} = 0$.

Problems

1. Verify the Divergence Theorem $\iiint_V \text{div}\, \mathbf{A} \, dV = \iint_S \mathbf{A} \cdot d\mathbf{S}$ where $\mathbf{A} = x\mathbf{i} + y\mathbf{j} + z\mathbf{k}$ and V is the volume of the cube defined by the surfaces $x = 0$, $x = 1$, $y = 0$, $y = 1$, $z = 0$, $z = 1$.

2. Use the divergence theorem to prove that

$$\iint_\Sigma \mathbf{P} \cdot d\mathbf{S} = \frac{12\pi a^5}{5}$$

where Σ is the surface of the sphere $x^2 + y^2 + z^2 = a^2$ and $\mathbf{P} = x^3\mathbf{i} + y^3\mathbf{j} + z^3\mathbf{k}$.

3. (i) If **A** and ϕ are a vector point function and a scalar point function respectively, prove $\text{div}\, \text{curl}\, \mathbf{A} = 0$ and $\text{curl}\, \text{grad}\, \phi = 0$.

(ii) Use the theorem to express $\iint_S \mathbf{r} \cdot d\mathbf{S}$ in terms of the volume V bounded by a closed surface S.

Verify the divergence theorem for $\mathbf{A} = x^2\mathbf{i} + y^2\mathbf{j} + z^2\mathbf{k}$ taken over the surface of the cube: $0 \leqslant x \leqslant 1,\ 0 \leqslant y \leqslant 1,\ 0 \leqslant z \leqslant 1.$ (C.E.I.)

4. Show briefly that $\operatorname{div}(\mathbf{u} \times \mathbf{v}) = \mathbf{v} \cdot \operatorname{curl} \mathbf{u} - \mathbf{u} \cdot \operatorname{curl} \mathbf{v}$ and $\operatorname{curl} \phi\mathbf{v} = (\operatorname{grad} \phi) \times \mathbf{v} + \phi \operatorname{curl} \mathbf{v}$, where $\mathbf{u}$ and $\mathbf{v}$ are vector point functions and ϕ is a scalar point function.

If $\mathbf{A} = \operatorname{curl} \mathbf{B}$ and $\mathbf{B} = \operatorname{curl} \mathbf{C}$ show by applying Gauss' divergence theorem to $\mathbf{C} \times \mathbf{B}$ that

$$\iiint_V \mathbf{B}^2\, dV = \iint_S (\mathbf{C} \times \mathbf{B}) \cdot d\mathbf{S} + \iiint_V \mathbf{C} \cdot \mathbf{A}\, dV,$$

where the volume V is enclosed by surface S. (C.E.I.)

5. Give a brief proof of the relation $\operatorname{div}(\mathbf{a} \times \mathbf{b}) = \mathbf{b} \cdot \operatorname{curl} \mathbf{a} - \mathbf{a} \cdot \operatorname{curl} \mathbf{b}$ where $\mathbf{a}$ and $\mathbf{b}$ are vector point functions.

In a certain fluid flow problem $\mathbf{v}$ is the fluid velocity and the vorticity Ω and vector potential $\mathbf{u}$ are defined by the equations $\Omega = \tfrac{1}{2} \operatorname{curl} \mathbf{v}$, $\mathbf{v} = \operatorname{curl} \mathbf{u}$. Show that

$$\iiint_V v^2\, dV = \iint_S (\mathbf{u} \times \mathbf{v}) \cdot d\mathbf{S} + 2 \iiint_V \mathbf{u} \cdot \Omega\, dV$$

where the closed surface S encloses a volume V of the fluid. (You may assume Gauss' divergence theorem.) (C.E.I.)

6. $\mathbf{A}$ and ϕ are differentiable vector and scalar functions of position, respectively, defined in a region of space V bounded by a closed surface S.

(i) Give an outline of the proof of the divergence theorem, that

$$\iint_S \mathbf{A} \cdot d\mathbf{S} = \iiint_V \operatorname{div} \mathbf{A}\, d\tau$$

and state the sign convention adopted.

(ii) By replacing $\mathbf{A}$ in turn by $\phi\mathbf{t}$ and $\mathbf{A} \wedge \mathbf{t}$ in (i), where $\mathbf{t}$ is an arbitrary constant vector, or otherwise, prove that

$$\iint_S \phi\, d\mathbf{S} = \iiint_V \operatorname{grad} \phi\, d\tau, \qquad \iint_S d\mathbf{S} \wedge \mathbf{A} = \iiint_V \operatorname{curl} \mathbf{A}\, d\tau.$$

(iii) If $\mathbf{r}$ is the position vector of the element dS, prove that,

$$\iint_S d\mathbf{S} \times (\mathbf{t} \wedge \mathbf{r}) = 2\mathbf{t} \iiint_V d\tau.$$

(L.U.)

11.10 MISCELLANEOUS RESULTS

We now collect a few results to end this chapter.

Green's Theorems

(i) If ϕ_1 and ϕ_2 are two finite scalar fields and they possess continuous first and second derivatives in some region V enclosed by a surface S then

$$\iiint_V \left\{(\nabla\phi_1)(\nabla\phi_2) + \phi_1\nabla^2\phi_2\right\} \mathrm{d}V = \iint_S \phi_1(\nabla\phi_2) \,.\, \mathrm{d}\mathbf{S} \qquad (11.47)$$

(ii) By interchanging ϕ_1 and ϕ_2 in (11.47) and subtracting the result from (11.47) we obtain

$$\iiint_V (\phi_1\nabla^2\phi_2 - \phi_2\nabla^2\phi_1)\,\mathrm{d}V = \iint_S (\phi_1 \frac{\partial\phi_2}{\partial n} - \phi_2 \frac{\partial\phi_1}{\partial n})\,\mathrm{d}S \qquad (11.48)$$

where $\partial\phi/\partial n$ is the derivative in the direction of $\hat{\mathbf{n}}$, the outward normal to S.

Note that $\iint_S \phi_1(\nabla\phi_2).\mathrm{d}\mathbf{S}$ has been written $\iint_S \phi_1 \frac{\partial\phi_2}{\partial n}\,\mathrm{d}S$

Note the case when ϕ_1 and ϕ_2 are **harmonic**, i.e. $\nabla^2\phi_1 = 0 = \nabla^2\phi_2$.

These theorems help us, for example, to determine a solution of Laplace's equation under given boundary conditions. Also we can express the scalar potential at a point in some closed region in terms of its distribution inside the region and values on the bounding surface.

Solid angle

In Figure 11.16 we show an element δS of a given surface with $\hat{\mathbf{n}}$ the unit normal to it at P. If each point of δS is joined to O we form a cone of vectors from O; where this cone cuts the unit sphere (radius = 1) an area $\delta\omega$ is produced. Now let θ be the angle between $\delta\omega$ and δS . Then

$$\frac{\delta\omega}{1^2} = \frac{\cos\theta\,\delta S}{r^2}$$

We define the **solid angle** which δS subtends at O to be $\delta\omega$.

The total solid angle subtended by a finite surface S is

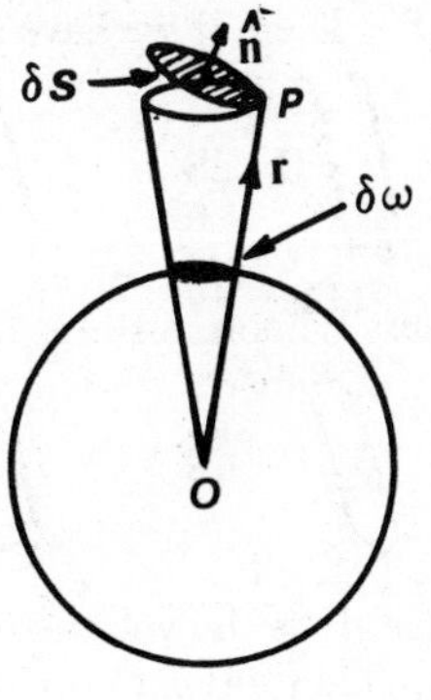

Figure 11.16

$$\omega = \iint_S \frac{\cos\theta}{r^2}\,\mathrm{d}S = \iint_S \frac{\hat{\mathbf{n}}\,.\,\mathbf{r}}{r^3}\,\mathrm{d}S = \iint_S \frac{\mathbf{r}\,.\,\mathrm{d}\mathbf{S}}{r^3} \qquad (11.49)$$

Note that if S is a spherical surface $r = a$ containing O, then

$$\omega = \iint_S \frac{a}{a^3}\, dS = \frac{1}{a^2}\, 4\pi a^2 = 4\pi$$

This result holds for any surface S containing O.

If O lies outside S it can be shown that $\omega = 0$. (What happens if O lies on S?)

Gauss' Law

Many fields in physics are 'inverse square law' fields. That is they possess potentials of the form $\phi = k/r$ so that $\mathbf{F} = \text{grad}\,\phi = -k\hat{\mathbf{r}}/r^2 = -k\mathbf{r}/r^3$ and $\mathbf{F}$ varies inversely as the square of distance from the origin. For example Coulomb's law of electrostatic attraction and the law of gravitational attraction follow this rule and point sources or point sinks in hydromechanics exhibit this law. Note that $\nabla^2\phi = 0$. Such a potential is often termed a **Newtonian potential.**

Consider for example the potential field $\phi = m\gamma/r$ due to a point mass m at O. Then

$$\iint_S \mathbf{F} \,.\, d\mathbf{S} = -\iint_S \frac{m\gamma\mathbf{r}}{r^3} \,.\, d\mathbf{S} = -m\gamma\omega$$

represents the total flow of gravitational force from the point mass. When the surface S encloses O the total flux$= -4\pi\gamma m$, but when O lies outside S then the total flux is zero; these statements comprise Gauss' law. Note that the actual position inside (or outside) does not affect the result.

A similar result would hold if we replaced the point mass m by a point charge e. In this case let $\mathbf{E} = \dfrac{e\hat{\mathbf{r}}}{4\pi\epsilon_0 r^2}$ where ϵ_0 is the permittivity of a vacuum, and define $\mathbf{D} = \epsilon_0\mathbf{E}$. Then, if we have a volume, charge density ρ

$$\iint_S \mathbf{D} \,.\, d\mathbf{S} = \iiint_V \rho\, dV$$

But Gauss' Theorem states that

$$\iint_S \mathbf{D} \,.\, d\mathbf{S} = \iiint_V \text{div}\,\mathbf{D}\, dV$$

and since V is the volume of an arbitrary region, it follows that $\text{div}\,\mathbf{D} = \rho$. Now since $\mathbf{D} = -\epsilon_0\,\text{grad}\,(e/4\pi\epsilon_0 r)$ we deduce that $\epsilon_0\nabla^2\phi = -\rho$ which is **Poisson's equation.** Here, ϕ is the electrostatic potential.

Chapter Twelve

Functions of a Complex Variable

12.1 INTRODUCTION

A large number of problems in engineering can be solved by methods involving complex variables. The theory of analytic functions is important because the results have wide application to problems of heat flow, fluid dynamics, electrostatics and many other areas of engineering.

The real and imaginary parts of an analytic function are each solutions of Laplace's equation and hence two-dimensional potential problems can be dealt with via analytic functions. In particular the technique of conformal mapping allows the solution of a boundary-value problem in potential theory to be effected by transforming a complicated region into a simpler one.

Many integrals that arise in practice cannot be solved by real integration techniques and require methods of complex integration for their evaluation. Indeed, the basis of transform calculus is the integration of functions of a complex variable.

12.2 ANALYTIC FUNCTIONS. THE CAUCHY-RIEMANN EQUATIONS

Let x and y be real variables; then $z = x + \mathrm{i}y$ is called a **complex variable.** Let u and v be two further real variables then $w = u + \mathrm{i}v$ is a second complex variable. Suppose that there is a relationship between these complex variables such that to each value of z in a given region of the z - plane there is associated one and only one value of w; then w is said to be a **function** of z, defined on the given region and we write $w = f(z)$.

Examples

(i) $\underline{w = z^2 - z.}$ This function is defined for all values of z. Note that we may write $w = u + \mathrm{i}v = (x + \mathrm{i}y)^2 - (x + \mathrm{i}y)$ and, separating real and imaginary parts, we obtain $u = x^2 - y^2 - x$, $v = 2xy - y$. If we wish to find $f(3 + 2\mathrm{i})$ we merely calculate $(3 + 2\mathrm{i})^2 - (3 + 2\mathrm{i}) = 9 - 4 + 12\mathrm{i} - 3 - 2\mathrm{i} = 2 + 10\mathrm{i}$; alternatively, we can proceed via the equations for u and v.

(ii) $\underline{w = 3\bar{z} + z.}$ This function is defined for all z. Note that $w = u + \mathrm{i}v = 3(x - \mathrm{i}y) + (x + \mathrm{i}y)$ so that $u = 4x$ and $v = -2y$. As an example $f(4 - \mathrm{i}) = 3(4 + \mathrm{i}) + 4 - \mathrm{i} = 16 + 2\mathrm{i}$.

Representation of a complex function

We are accustomed to drawing a graph of the real-variable function $y = f(x)$ on a plane. However, we are in difficulties when we try to represent a complex function since points on the z - plane will be confused with points on the w - plane. The answer is to keep the planes separate and use *two* Argand diagrams, one for the z - plane and one for the w - plane.

Examples

(i) In Figure 12.1 we show the points $z = 1$, $2 + \mathrm{i}$ and $3 + 2\mathrm{i}$ being mapped by the function $w = 3z + 2$ into the points $w = 5$, $8 + 3\mathrm{i}$, $11 + 6\mathrm{i}$ respectively.

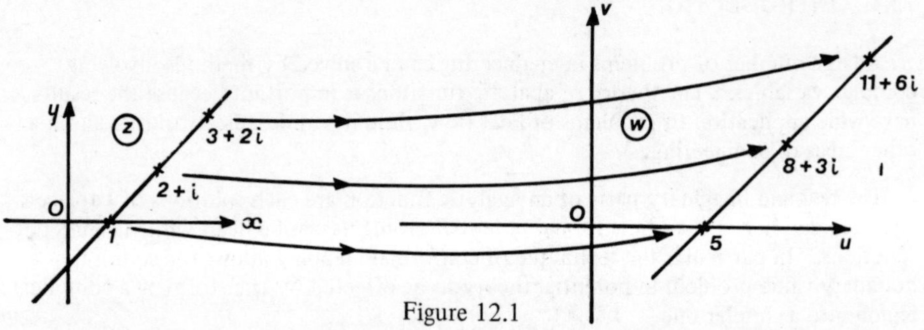

Figure 12.1

Note that the line $y = x - 1$, on which lie all three chosen points in the z - plane can be shown to map into the line $v = u - 5$ in the w - plane and the images of the z - values all lie on this second line. Since $w = u + \mathrm{i}v = 3(x + \mathrm{i}y) + 2$ we find that $u = 3x + 2$ and $v = 3y$ so that $y = x - 1$ becomes $\frac{1}{3}v = \frac{1}{3}(u - 2) - 1$ or $v = u - 5$. The mapping $w = 3z + 2$ is an example of a **translation**. See what happens to points on the circle $x^2 + y^2 = 1$. Also investigate the effect on the region $|z| \leqslant 1$.

(ii) $\underline{w = z^2}$

Here $u + \mathrm{i}v = (x + \mathrm{i}y)^2 = x^2 - y^2 + 2\mathrm{i}xy$ so that $u = x^2 - y^2$, $v = 2xy$. If we take the line $y = x$ in the z - plane we obtain $u = 0$, a straight line, but if we consider $y = x + 1$, then $u = x^2 - (x + 1)^2 = -2x - 1$, $v = 2x(x + 1)$. Hence eliminating x we obtain $u^2 = 2v + 1$. This is a parabola. Any other straight lines of the form $y = x + k\,(k \geqslant 0)$ will transform into parabolas, as can be shown. Let us now use polar co-ordinates, $z = re^{\mathrm{i}\theta}$. Hence $w = z^2 = r^2 e^{\mathrm{i}2\theta}$ i.e. a point (r, θ) in the z - plane, produces a point $(r^2, 2\theta)$ in the w - plane; see Figure 12.2.

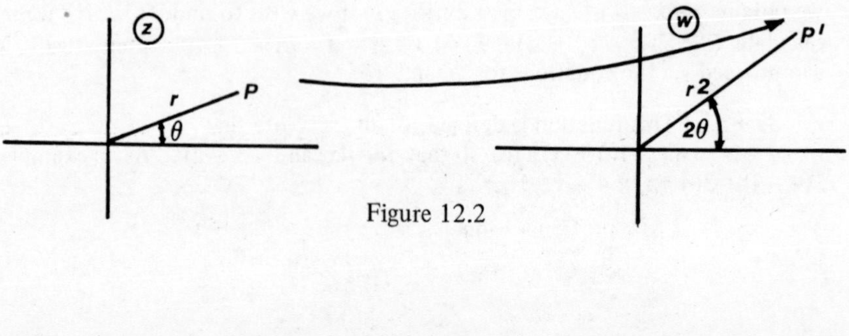

Figure 12.2

Taking $r =$ constant and letting θ vary between 0 and π, then P moves on a semi-circle in the z - plane while in the w - plane, P$'$ will move round a full circle of radius r^2. We take this matter up again in the next section.

(iii) $\underline{w = z^{1/2}}$. In this case there are two values of w for each value of $z \neq 0$. We say that the relationship between w and z is multi-valued. To be consistent with real variable work we should not regard this relationship as a function. However, most authors will regard a function as associating one *or more* values of w with each value of z. It is customary to speak of single-valued and multi-valued functions and regard a multi-valued function as a collection of **branches.** In this example we may refer to the **principal branch** as giving a value of $z^{1/2}$ in the upper half w - plane and the secondary branch as giving a value in the lower half w - plane.

Limit of a function

We define the limit of $w = f(z)$ as $z \to z_0$ as some number l such that $|f(z) - l|$ can be made as small as we like by making $|z - z_0|$ sufficiently small. For example, the limit of $(z^2 - z)$ as $z \to \mathrm{i}$ is, obviously enough, $\mathrm{i}^2 - \mathrm{i} = -1 - \mathrm{i}$. However, we at once encounter a difference from real variables. There, the statement $x \to x_0$ meant a journey along the real axis towards x_0 either from the left or from the right: two directions only. Here, the statement $z \to z_0$ must be interpreted as z approaches the point z_0 *along any path whatsoever.* Since $|z - z_0|$ represents the distance between z and z_0 our requirement is merely that this distance decreases to zero.

Example 1

Find $\lim\limits_{z \to 3+\mathrm{i}} (z^2 - z)$ along the paths shown in Figure 12.3

(a) Along this path $z = x + \mathrm{i}$ and
$z^2 - z = (x + \mathrm{i})^2 - (x + \mathrm{i}) = x^2 - x - 1 + \mathrm{i}(2x - 1)$.
As $z \to 3 + \mathrm{i}$, $x \searrow 3$ and, assuming that we can add the separate limits of the real and imaginary parts,
$z^2 - z \to 3^2 - 3 - 1 + \mathrm{i}(6 - 1) = 5 + 5\mathrm{i}$.

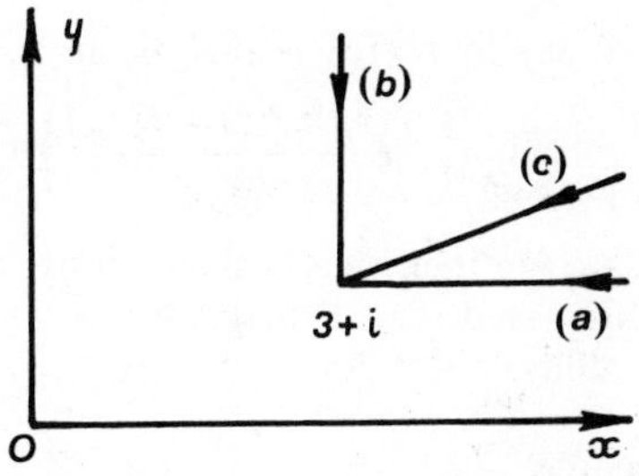

Figure 12.3

(b) Here $z = 3 + \mathrm{i}y$, $z^2 - z = 6 - y^2 + 5y\mathrm{i}$
as $z \to 3 + \mathrm{i}$, $y \searrow 1$ and
$z^2 - z \to 5 + 5\mathrm{i}$.

(c) $z = k(3 + \mathrm{i})$ where k is real;
$z^2 - z = k^2(8 + 6\mathrm{i}) - k(3 + \mathrm{i})$; as
$k \searrow 1$, $z^2 - z \to 5 + 5\mathrm{i}$.

In each case the limit is the same.

Example 2

Find $\lim\limits_{z \to 0} \dfrac{\bar{z}}{z}$ (a) along the x - axis (b) along the y - axis.

(a) $z = x$, $\bar{z} = x$ so that $\bar{z}/z = 1$, which is the limit along the x - axis.

(b) $z = \mathrm{i}y$, $\bar{z} = -\mathrm{i}y$ so that $\bar{z}/z = -1$, which is the limit along the y - axis.

Note that, converting to polars, $\overline{z}/z = e^{-i\theta}/e^{i\theta} = e^{-2i\theta}$ and this will give different values along different straight lines into the origin.

Rules for limits of the sums of functions, the difference between functions, the product of functions and the quotient of functions are analogous to those rules for real variables.

Continuity

As you might expect, $f(z)$ is defined to be **continuous** at z_0 if (i) $f(z_0)$ exists and (ii) $\lim_{z\to z_0} f(z)$ exists and is equal to $f(z_0)$. Again the rules for the sum of two continuous functions, etc. are similar to those for real variables.

Example 1

$$\lim_{z\to 1+i}\left\{\frac{(z^2+1)(z-i)}{z^2+z+1}\right\} = \frac{\lim_{z\to 1+i}\left\{z^2+1\right\}\cdot\lim_{z\to 1+i}\left\{z-i\right\}}{\lim_{z\to 1+i}\left\{z^2+z+1\right\}}$$

$$= [(1+i)^2+1][1+i-i]/[(1+i)^2+(1+i)+1]$$

$$= (8+i)/13 = f(1+i)$$

Example 2

$f(z) = z/(z^2+4)$ is continuous everywhere except at $z = \pm 2i$ when $f(z)$ does not exist.

Derivatives of a function

We say that $f(z)$ is **analytic** at z_0 if

$$\lim_{\Delta z\to 0}\left\{\frac{f(z_0+\Delta z)-f(z_0)}{\Delta z}\right\} \tag{12.1}$$

exists. We then denote the limit by $f'(z_0)$ and call it the **derivative of** $f(z)$ at z_0. Again, note that Δz may tend to zero in *any way whatsoever.* Many simple-looking functions do not possess a derivative at any point; one example of this follows.

Example

$$f(z) = \overline{z} = x - iy$$

Let $R = \dfrac{f(z_0+\Delta z)-f(z_0)}{\Delta z} = \dfrac{(x_0+\Delta x)-i(y_0+\Delta y)-(x_0-iy_0)}{\Delta x+i\Delta y}$

$$= \frac{\Delta x - i\Delta y}{\Delta x + i\Delta y}$$

Consider the path parallel to the x - axis so that $\Delta y = 0$
then $R = \Delta x/\Delta x = 1$, which is the limit as $\Delta x \to 0$.

However, the path parallel to the y - axis is defined by $\Delta x = 0$
then $R = -i\Delta y/i\Delta y = -1$, which is the limit as $\Delta y \to 0$.

Clearly $f(z)$ is not analytic at any point.

Where a function $f(z)$ is analytic, the differentiation proceeds as you might expect. For example, if $f(z) = 3z^2 - z$, $f'(z) = 6z - 1$.

The definition of derivative may seem very restrictive, as it is not enough for the limits along directions parallel to the axes to be equal; we must have limits along all possible paths giving the same value. However, most of the common functions of mathematics are analytic at all but a finite number of points. These points are called **singularities.** We shall see that these points have important practical applications; for example, a fluid source or vortex will be represented by singularities of the function which describes the flow pattern.

Cauchy-Riemann conditions (C - R conditions)

We now seek a set of conditions on the components u and v so that w is analytic. We derive the conditions at an unspecified point z.

If $w = f(z)$ is analytic then

$$f'(z) = \lim_{\Delta z \to 0} \left\{ \frac{f(z + \Delta z) - f(z)}{\Delta z} \right\}$$

$$= \lim_{\Delta z \to 0} \left\{ \frac{[u(x + \Delta x, y + \Delta y) + iv(x + \Delta x, y + \Delta y] - [u(x, y) + iv(x, y)]}{\Delta x + i\Delta y} \right\}$$

The limit exists no matter what path is taken. Consider two particular paths.

(i) We approach z along a path parallel to the x - axis, i.e. $\Delta y = 0$ and we let $\Delta x \to 0$. Then

$$f'(z) = \lim_{\Delta x \to 0} \left\{ \frac{u(x + \Delta x, y) + iv(x + \Delta x, y)}{\Delta x} - \frac{u(x, y) + iv(x, y)}{\Delta x} \right\}$$

$$= \lim_{\Delta x \to 0} \left\{ \frac{u(x + \Delta x, y) - u(x, y)}{\Delta x} + i \frac{v(x + \Delta x, y) - v(x, y)}{\Delta x} \right\}$$

i.e.

$$f'(z) = \frac{\partial u}{\partial x} + i \frac{\partial v}{\partial x} = \frac{\partial f}{\partial x} \tag{12.2}$$

(ii) We now approach z by a path parallel to the y - axis, i.e. $\Delta x = 0$ and we let $\Delta y \to 0$. We leave it to you to show that

$$f'(z) = \frac{1}{i} \frac{\partial u}{\partial y} + \frac{\partial v}{\partial y} = \frac{1}{i} \frac{\partial f}{\partial y} \tag{12.3}$$

By comparing the real and imaginary parts of (12.2) and (12.3) we find that

$$\frac{\partial u}{\partial x} = \frac{\partial v}{\partial y} \quad \text{and} \quad \frac{\partial u}{\partial y} = -\frac{\partial v}{\partial x} \tag{12.4}$$

The conditions obtained in (12.4) are known as the **Cauchy-Riemann equations.** We state without proof the converse result: if $u(x, y)$ and $v(x, y)$ and their partial derivatives with respect to x and y are continuous and satisfy (12.4) in a neighbourhood of z_0 then $f(z)$ is analytic at z_0 and $f'(z_0)$ is given by (12.2) or (12.3).

(By a neighbourhood of z_0 we mean a region $|z - z_0| < \rho$ for some real number ρ.)

Example 1

$$w = f(z) = \bar{z} = x - iy$$

Here $u = x$, $v = -y$ so that $\frac{\partial u}{\partial x} = 1$, $\frac{\partial u}{\partial y} = 0$, $\frac{\partial v}{\partial x} = 0$, $\frac{\partial v}{\partial y} = -1$, and therefore the first equation of (12.4) is not satisfied. Hence $f(z)$ is not analytic anywhere.

Example 2

$$f(z) = z^2 = x^2 - y^2 + 2ixy$$

Here $u = x^2 - y^2$ so that $\frac{\partial u}{\partial x} = 2x$ and $\frac{\partial u}{\partial y} = -2y$. Also $v = 2xy$ so that $\frac{\partial v}{\partial x} = 2y$ and $\frac{\partial v}{\partial y} = 2x$. Then both equations (12.4) are satisfied for all x and y; hence $f(z)$ is analytic everywhere. From (12.2), $f'(z) = 2x + i2y = 2z$.

Example 3

$$f(z) = \frac{1}{z} = \frac{x - iy}{x^2 + y^2}$$

Here $u = \frac{x}{x^2 + y^2}$, $v = \frac{-y}{x^2 + y^2}$. We leave it to you to show that (12.4) are both satisfied, except at $x = y = 0$. You should also obtain $f'(x) = \frac{-1}{z^2}$ using (12.2) or (12.3). Note that this derivation assumes that not both x and y are zero.

Example 4

$$f(z) = z\bar{z} = x^2 + y^2$$

We leave it to you to show that $f'(z)$ exists only at $z = 0$.

Connection with Laplace's equation

Suppose $f(z)$ is analytic in some region, then

$$\frac{\partial^2 u}{\partial x \partial y} = \frac{\partial}{\partial x}\left(\frac{\partial u}{\partial y}\right) = \frac{\partial}{\partial x}\left(-\frac{\partial v}{\partial x}\right) = -\frac{\partial^2 v}{\partial x^2}$$

But

$$\frac{\partial^2 u}{\partial y \partial x} = \frac{\partial}{\partial y}\left(\frac{\partial u}{\partial x}\right) = \frac{\partial}{\partial y}\left(\frac{\partial v}{\partial y}\right) = \frac{\partial^2 v}{\partial y^2}$$

Under the conditions demanded of the partial derivatives, we may equate the mixed partial derivatives of u and obtain

$$\frac{\partial^2 v}{\partial x^2} + \frac{\partial^2 v}{\partial y^2} = 0 \qquad (12.5)$$

In a similar way we find that

$$\frac{\partial^2 u}{\partial x^2} + \frac{\partial^2 u}{\partial y^2} = 0 \qquad (12.6)$$

We call u and v **conjugate harmonic functions.**

Conversely, if we take two functions u and v satisfying equations (12.6) and (12.5) in some region then $f(z) = u + iv$ is analytic in that region. For example, if $u = xy$, then it is easily seen that u satisfies (12.6). Let us find the conjugate harmonic function $v(x, y)$. We know that $\dfrac{\partial v}{\partial y} = \dfrac{\partial u}{\partial x} = y$ and $\dfrac{\partial v}{\partial x} = -\dfrac{\partial u}{\partial y} = -x$; hence $v = \frac{1}{2}(y^2 - x^2) + c$, where c is a constant. Also we may form $f(z) = u + iv = -\frac{1}{2} iz^2 + c$ which is analytic everywhere.

We close this section by noting that it can be shown that if $f(z)$ is analytic in a region then it possesses derivatives of *any order* in that region. This emphasises what a strong condition the existence of a first derivative demands.

Problems

1. Find the real and imaginary parts $u(x, y)$ and $v(x, y)$ of the functions

   (a) z^3 (b) $|z|$ (c) $R(z)$ (d) e^z (e) $1/z$ (f) $z/(z^2 + 1)$ (g) $(2z + 3)/(z + 2)$

2. Plot the z - values given and the corresponding w - values under the function stated. Try to generalise how the z - plane is transformed by the function.

   (a) $z = 0,\ 1 + i,\ -1,\ -3 + 2i,\ -i;\ w = z + 1$

   (b) $z = 0,\ 1,\ 2,\ 3,\ 1 + i,\ 2 + 2i,\ i,\ 2i,\ 3i;\ w = iz$

   (c) $z = i,\ 1,\ -i,\ -1,\ 1 + i,\ -1 - i,\ 2,\ -4i;\ w = i\bar{z}$

3. Consider $w = z^3$. First express the function in the form $w = u + iv$ where u and v are functions of r and θ. Then show that the region $0 < \arg z < \pi/3$ is mapped onto the upper half-plane $0 < \arg w < \pi$. What region in the z - plane will be mapped into the entire w - plane? Suppose every point of the z - plane is mapped onto the w - plane. Show that apart from $w = 0$, every point on the w - plane is the image under the mapping of three distinct points in the z - plane. How are the arguments and moduli of these three points related?

4. Evaluate the following limits

   (a) $\displaystyle\lim_{z\to 1+i}\left\{\frac{z^2 - z + 1 - i}{z^2 - 2z + 2}\right\}$ (b) $\displaystyle\lim_{z\to 2+i}\left\{\frac{1 - z}{1 + z}\right\}$ (c) $\displaystyle\lim_{z\to 2+i}\left\{\frac{z^2 - 2iz}{z^2 + 4}\right\}$

   (d) $\displaystyle\lim_{z\to e^{i\pi/4}}\left\{\frac{z^2}{z^4 + z + 1}\right\}$ (e) $\displaystyle\lim_{z\to 1+i}\left\{\frac{z - 1 - i}{z^2 - 2z + 2}\right\}^2$

5. Evaluate $\displaystyle\lim_{z\to 0}\left\{\frac{2xy}{x^2 + y^2} - \frac{y^2}{x^2}i\right\}$ along the following paths

   (a) the line $y = x$ (b) the line $y = 2x$ (c) the parabola $y = x^2$

   What conclusions do you draw?

6. Show that as $z \to 0$ the limit of each of the following functions does not exist

   (a) $\dfrac{xy}{x^4 + y^4} + 2xi$ (b) $\dfrac{x^3y^2}{x^6 + y^4} - \dfrac{x}{y}i$

7. Let $f(z) = (z^2 + 4)/(z - 2i)$ for $z \neq 2i$. Define $f(2i) = 3 + 4i$. Prove that $\displaystyle\lim_{z\to i} f(z)$ exists and state its value.

8. Where are the following functions discontinuous?

(a) $\dfrac{2z-3}{z^2+2z+2}$; (b) $\dfrac{3z^2+4}{z^4-16}$; (c) $\dfrac{z^2+1}{z^3+9}$ for $|z| \leqslant 2$

9. From the definition, find the derivatives of

(a) z^2+3z (b) z^{-1}

Evaluate these derivatives at $z = -1+i$

10. For the following functions find the singular points. Determine the derivatives at other points.

(a) $z/(z+i)$ (b) $(3z-2)/(z^2+2z+5)$

11. Show that the real and imaginary parts of the following functions satisfy the Cauchy-Riemann equations and state whether the functions are analytic.

(a) $z^2+5iz+3-i$ (b) x^2-iy^2 (c) $\cos y - i\sin y$ (d) $e^x(\cos y + i\sin y)$

12. The polar form of the Cauchy-Riemann equations is

$$r\frac{\partial u}{\partial r} = \frac{\partial v}{\partial \theta}, \quad r\frac{\partial v}{\partial r} = -\frac{\partial u}{\partial \theta} \quad \text{where } f(z) = u(r,\theta) + i\,v(r,\theta)$$

Show that these conditions are satisfied by the functions

(a) $\log_e r + i\theta$ (b) z^n (c) $|z|$ (d) $|z|^2$

13. Show that the real and imaginary components of each of the following functions are harmonic.

(a) z^2+z (b) z^4 (c) $1/z$ (d) $e^x(\cos y + i\sin y)$

(d) $z^4+5iz+3-i$ (f) ze^z (g) $\sin 3z$

14. Show that each of the following functions is harmonic. Then find its conjugate harmonic v and form an analytic function $f(z) = u + iv$

(a) $u = x$ (b) $u = e^x\cos y$ (c) $u = \log_e(x^2+y^2)$ (d) $xe^x\cos y - ye^x\sin y$

(e) $u = x^2-y^2-2xy-2x+3y$ (f) $u = 3x^2y+2x^2-y^3-2y^2$

15. State the Cauchy-Riemann equations for the function $w = f(z) = u(x,y) + jv(x,y)$, where $z = x+jy$. Show that $u = 4xy - x^3 + 3xy^2$ is harmonic and find the conjugate function v. Hence express $u+jv$ as an analytic function of z. (C.E.I.)

16. Determine which of the following functions $u(x,y)$ are harmonic. Find the conjugate harmonic function $v(x,y)$, if it exists, and express $u+iv$ as an analytic function of $z = x+iy$.

(a) $u = 2xy+3xy^2-2y^3$ (b) $u = 2x(1-y)$ (C.E.I.)

12.3 STANDARD FUNCTIONS OF A COMPLEX VARIABLE

In this section we take a brief look at some properties of the more common functions.

(a) Exponential function

We define $$e^z = e^{(x+iy)} = e^x(\cos y + i\sin y) \tag{12.7}$$

Note that

(i) e^z reduces to e^x when $y = 0$

(ii) e^z is single-valued

(iii) $e^{\bar{z}}$ is analytic everywhere and its derivative is e^z

(iv) $|e^z| = e^x$

(v) $\arg e^z = y$

(vi) $e^{(z+2k\pi i)} = e^z$ for any integer k

For example, if $w = u + iv = e^z$ then $u = e^x \cos y$, $v = e^x \sin y$ and it is easy to check that the Cauchy-Riemann conditions are satisfied, which proves the analyticity. Also $dw/dz = \partial u/\partial x + i\,\partial v/\partial x = e^x \cos y + i\,e^x \sin y = e^z$. We leave you to prove properties (i) to (vi).

Example

Find all roots of the equation $e^z = -i$. We have $e^x \cos y = 0$ and $e^x \sin y = -1$ so that $\cos y = 0$ and $y = \pi/2 + k\pi$, k integral. Then $e^x \sin y = \pm e^x$.
Now $e^x \neq -1$ for any x so that $e^x = 1$ and $x = 0$. However, you can easily show (and you should) that k cannot be even, i.e. we cannot have
$\sin\left(\frac{\pi}{2} + 2k\pi\right) = -1$ so that the roots we require are given by

$$z = \left(-\frac{\pi}{2} + 2k\pi\right) i, \ k \text{ integer.}$$

(b) **Trigonometric functions**

We shall define

$$\cos z = \frac{(e^{iz} + e^{-iz})}{2}, \quad \sin z = \frac{(e^{iz} - e^{-iz})}{2i} \tag{12.8}$$

which is consistent with our results when z is the real variable x. We may define other trigonometric functions in ways analogous to results for real variables; thus for example, $\tan z = \sin z/\cos z$. You should be able to show easily that $\sin z$ and $\cos z$ are analytic everywhere and that

$$\frac{d}{dz}(\sin z) = \cos z, \quad \frac{d}{dz}(\cos z) = -\sin z \tag{12.9}$$

All the familiar trigonometric identities hold in the case of complex functions; for example

$$\sin^2 z + \cos^2 z \equiv 1 \tag{12.10}$$

We can obtain representations for $\sin z$ and $\cos z$ in terms of functions of x and y. Thus

$$\begin{aligned}
\sin z &= \sin(x + iy) = \sin x \cos(iy) + \cos x \sin(iy) \\
&= \sin x\left[\frac{e^{-y} + e^{+y}}{2}\right] + \cos x\left[\frac{e^{-y} - e^{+y}}{2i}\right] \\
&= \sin x \cosh y - \cos x \sinh y/i \\
&= \sin x \cosh y + i\cos x \sinh y
\end{aligned} \tag{12.11}$$

Similarly

$$\cos z = \cos x \cosh y - \mathrm{i} \sin x \sinh y \tag{12.12}$$

(c) **Hyperbolic functions**

We shall define

$$\cosh z = \tfrac{1}{2}(e^z + e^{-z}) \quad \text{and} \quad \sinh z = \tfrac{1}{2}(e^z - e^{-z}) \tag{12.13}$$

Then we can easily deduce that

(i) $\cosh z = \cosh x \cos y + \mathrm{i} \sinh x \sin y$ (12.14)
$\sinh z = \sinh x \cos y + \mathrm{i} \cosh x \sin y$ (12.15)

(ii) $\cosh z$ and $\sinh z$ are analytic for all z

(iii) $\dfrac{\mathrm{d}}{\mathrm{d}z}(\sinh z) = \cosh z, \quad \dfrac{\mathrm{d}}{\mathrm{d}z}(\cosh z) = \sinh z$ (12.16)

The formulae involving these and other hyperbolic functions extend without difficulty into complex variables.

(d) **Logarithmic function**

Since the exponential function is one-to-one, it possesses an inverse function which we call $\log z$ (or $\ln z$), defined by $e^w = z\ (z \neq 0)$ so that $w = \log z$. Suppose $w = u + \mathrm{i}v$ and $z = re^{\mathrm{i}\theta}$, then $e^{(u+\mathrm{i}v)} = re^{\mathrm{i}\theta}$, so that $e^u = r$ and $e^{\mathrm{i}v} = e^{\mathrm{i}\theta}$. Then clearly

$$u = \log_e r = \log_e |z| \tag{12.17}$$

and

$$v = \theta = \arg z \tag{12.18}$$

However, although $\arg z$ is usually confined to $-\pi < \theta \leqslant \pi$, it is possible for v to take an infinity of values since $e^{\mathrm{i}\theta} = e^{\mathrm{i}\theta} \cdot e^{2\pi k\mathrm{i}}$ where k is any integer and therefore $v = \theta + 2k\pi\mathrm{i}$

Now we can confine v to the range $-\pi < v \leqslant \pi$, by taking $k = 0$; then the corresponding value of $\log z$ is called the **principal value** of $\log z$ and denoted by $\mathrm{Log}\, z$; it is a function. In general, to each value of $z \neq 0$ there are an infinite number of values of $\log z$ each with the same real part.

We can divide up these values into **branches** of range 2π by selecting $k = 0, \pm1, \pm2, \pm3, \ldots\ldots$ in turn. (This is the customary way of making a **branch cut** for this function.) Notice that each branch is defined on the z – plane without the origin. We shall see later why $z = 0$ is specifically excluded. On each branch $\log z$ is analytic and has derivative $\dfrac{1}{z}$ except along the negative real axis and at the origin; this is represented schematically in Figure 12.4.

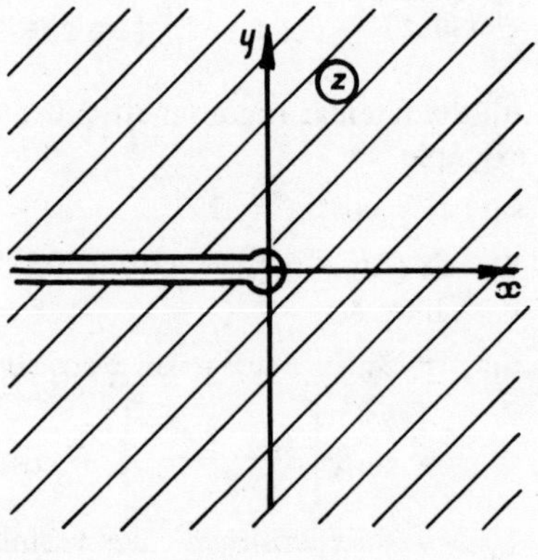

Figure 12.4

The familiar properties of logarithm

extend in complex variables to $\log z$, bearing in mind that with $\operatorname{Log} z$ we may have to adjust the value of argument by a multiple of $2\pi i$ to comply with $-\pi < \arg z \leqslant \pi$.

Examples

(i) $\log(1+i) = \log(\sqrt{2}\, e^{i\pi/4}) = \log\sqrt{2} + i(\pi/4 + 2k\pi)$
$= \tfrac{1}{2}\log 2 + i(\pi/4 + 2k\pi)$

(ii) $\operatorname{Log}(1+i) = \tfrac{1}{2}\log 2 + i\pi/4$

(iii) $\log z = 1 - i\pi \Rightarrow z = e^{1-i\pi} = -e$

(e) **Powers of a complex number, z^m**

If m is a positive integer $w = z^m$ is analytic. Consider though $w = z^2$; as $\arg z$ goes from 0 to π, $\arg w$ goes from 0 to 2π. When $\arg z$ increases from π to 2π, $\arg w$ goes from 2π to 4π.

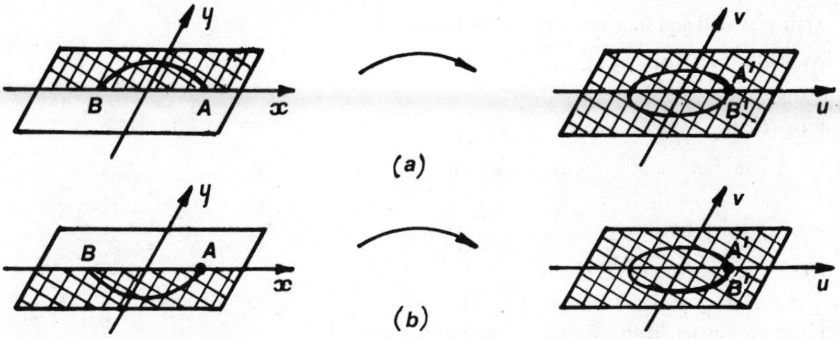

Figure 12.5

In Figure 12.5(a) we consider the effect of $w = z^2$ on the upper half z - plane. The semicircle is mapped into the whole-circle in the w - plane and the upper half z - plane is mapped into all the w - plane. In Figure 12.5(b) $\pi \leqslant \arg z < 2\pi$, yet the lower half z - plane is mapped into the entire w - plane. We have represented the mapping $w = z^2$ associated with the **function** $f(z) = z^2$ on two **sheets** in the w - plane.

If m is a non-integer, z^m is multi-valued and it has m branches, all analytic. For example $z^{1/2}$ has two branches on each of which $z^{1/2}$ is single-valued. If $z = re^{i\theta}$, then one branch is found by taking $0 \leqslant \theta < 2\pi$ and the other by $2\pi < \theta \leqslant 4\pi$. Other ranges of 2π for θ merely repeat these branches. Since the transition from one branch to another is caused by a rotation of z about the origin $z = 0$, we call the origin a **branch point** of $w = z^{1/2}$. This is one kind of **singularity** of a function. (Note that $z = 0$ is a branch point of $\log z$.)

(f) **Reciprocal function**

The function $f(z) = \dfrac{1}{z}$ is defined for all $z \neq 0$ and is analytic for all $z \neq 0$ with derivative $\dfrac{-1}{z^2}$. (We can complete the domain of $f(z)$ to include $z = 0$ if we introduce an ideal '*point at infinity*' such that as $z \to 0$, $\dfrac{1}{z} \to$ the point at infinity.)

Problems

1. Using (12.7), prove properties (i) to (vi) for the exponential function.

2. Prove that

   (i) $e^{z_1/z_2} = e^{z_1 - z_2}$ (ii) $|e^{iz}| = e^{-y}$ (iii) $e^z \neq 0$ for any finite z.

3. Find all values z for which

   (i) $e^{3z} = 1$ (ii) $e^{4z} = \mathrm{i}$ (iii) $e^z = 1 - \mathrm{i}$

4. Find the derivatives of $\sin z$, $\cos z$.

5. Prove that

   (i) $\sin 2z = 2 \sin z \cos z$ (ii) $\cos 2z = \cos^2 z - \sin^2 z$ (iii) $\sin(-z) = -\sin z$

   (iv) $\cos(-z) = \cos z$.

6. Split into real and imaginary parts the following:

   (i) $\cos(-\mathrm{i})$ (ii) $\sin(1 + \mathrm{i})$

7. Prove that

   (i) $1 + \tan^2 z = \sec^2 z$ (ii) $\overline{\sin z} = \sin \bar{z}$

8. Find all the values of

   (i) $\sin^{-1} 2$ (ii) $\cos^{-1} \mathrm{i}$

9. Prove properties (i) to (iii) for hyperbolic functions.

10. Prove that

    (i) $\cos(\mathrm{i}z) = \cosh z$ (ii) $\sin(\mathrm{i}z) = \mathrm{i} \sinh z$

11. Find the value of

    (i) $2 \sinh(\pi \mathrm{i}/3)$ (ii) $\coth(3\pi \mathrm{i}/4)$

12. Find the values of $\cosh^{-1} \mathrm{i}$.

13. Find the logarithm of

    (i) $-\mathrm{i}$ (ii) $3 + 4\mathrm{i}$ (iii) $2 - \mathrm{i}$

14. Find the values of z for which $\log(z + 1) = \pi \mathrm{i}$.

15. Show that $\log(z - 1) = \frac{1}{2} \log\left\{(x - 1)^2 + y^2\right\} + \mathrm{i} \tan^{-1}\left\{y/(x - 1)\right\}$.

16. Prove that

    (i) $\log(z_1 z_2) = \log(z_1 + z_2)$ (ii) $\log e^z = z$

17. Find the values of

    (i) $(1 + \mathrm{i})^{\mathrm{i}}$ (ii) $(2 - \mathrm{i})^{\mathrm{i}+1}$ (iii) $2^{2\mathrm{i}}$.

18. If $w = (z^2 + 1)^{1/2}$ express arg w in terms of arg $(z - \text{i})$ and arg $(z + \text{i})$. Let C be a closed curve around the point i which does not include $-\text{i}$. Then let z go once round C anticlockwise and consider the changes in arg $(z - \text{i})$ and arg $(z + \text{i})$ to show that the change in arg $w \neq 0$ and hence that $z = \text{i}$ is a branch point for $w = f(z)$. Similarly show that $z = -\text{i}$ is also a branch point.

12.4 COMPLEX POTENTIAL AND CONFORMAL MAPPING

At the end of Section 12.2 we stated that the real and imaginary parts of an analytic function each satisfied Laplace's equation. We can go further and show that the curves $u(x, y) =$ constant and $v(x, y) =$ constant intersect orthogonally.

This follows because along $u(x, y) =$ constant, $\text{d}u = \dfrac{\partial u}{\partial x}\,\text{d}x + \dfrac{\partial u}{\partial y}\,\text{d}y = 0$ and so $\dfrac{\text{d}y}{\text{d}x} = \dfrac{-(\partial u/\partial x)}{(\partial u/\partial y)}$ whereas along $v(x, y) =$ constant $\dfrac{\text{d}y}{\text{d}x} = -\dfrac{(\partial v/\partial x)}{(\partial v/\partial y)}$. The product of these gradients is $\dfrac{(\partial u/\partial x)}{(\partial u/\partial y)} \cdot \dfrac{(\partial v/\partial x)}{(\partial v/\partial y)} = \dfrac{(\partial u/\partial x)}{(\partial u/\partial y)} \cdot -\dfrac{(\partial u/\partial y)}{(\partial u/\partial x)} = -1$, by the Cauchy-Riemann conditions. Hence the curves cut at right angles. This property is of great use in potential theory. For example in two-dimensional electrostatic problems if $u =$ constant gives the **equipotential lines** then $v =$ constant are the **electric lines of force**. In Figure 12.6(a) we show some of these two sets of curves for the field due to two oppositely charged particles. In the steady irrotational flow of an ideal fluid in two dimensions if $u =$ constant gives the equipotential lines, $v =$ constant gives the **streamlines** of the flow. In Figure 12.6(b) we show these two sets of curves for flow round a cylinder.

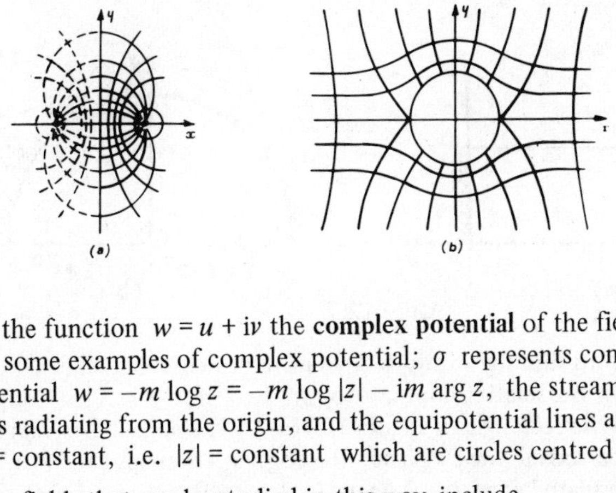

Figure 12.6

We call the function $w = u + \text{i}v$ the **complex potential** of the field. In Table 12.1 we present some examples of complex potential; σ represents conductivity. In the case of the potential $w = -m \log z = -m \log |z| - \text{i}m \arg z$, the streamlines are $\arg z =$ constant, i.e. lines radiating from the origin, and the equipotential lines are given by $\log |z| =$ constant, i.e. $|z| =$ constant which are circles centred at the origin.

Other fields that can be studied in this way include

(i) steady heat flow where $u =$ constant represent isothermals and $v =$ constant represent heat flow lines

(ii) gravitation where $u =$ constant represent equipotential lines and $v =$ constant represent lines of force

Table 12.1

Electrostatics/ Magnetostatics	Current flow	Hydrodynamics
Uniform field E making angle α with real axis $w = E\,e^{-i\alpha}\,z$	Uniform current J making α with real axis $w = (J/\sigma)\,e^{-i\alpha}\,z$	Uniform stream U making α with real axis $w = U\,e^{-i\alpha}\,z$
Line charge or pole e at origin $w = -2e \log z$	Electrode with strength J at origin $w = -(J/\sigma) \log z$	Line source of strength m at origin $w = -m \log z$
Dipole M $w = -2M/z$		Dipole M $w = -M/z$
Intensity of field $\lvert E\rvert = \lvert dw/dz\rvert$	Magnitude of current $\lvert i\rvert = \sigma\,\lvert dw/dz\rvert$	Speed of fluid $\lvert q\rvert = \lvert dw/dz\rvert$

Conformal Mapping

Consider the physical situation represented in Figure 12.7(a); a rectangular plate has one edge maintained at 100° and the opposite edge at 0° whilst the other two edges are insulated. We may assume that flow of heat takes place in the u - v plane until a steady state is reached. It is easy enough to solve Laplace's equation for the steady temperature $T(u,\ v)$ to obtain

$$T = 100v/\pi \qquad (12.19)$$

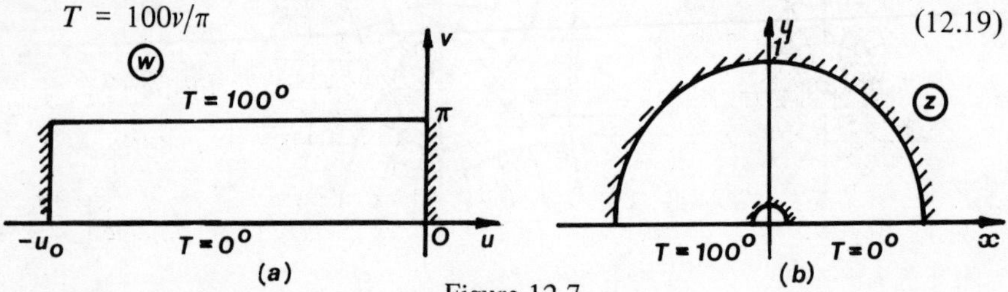

Figure 12.7

Now suppose we wish to find the steady temperature in the semicircular plate of Figure 12.7(b). Consider the relationship $w = \log z$. Now if $z = re^{i\theta}$, $u = \log r$ and $v = \theta$. What happens to the rectangle of Figure 12.7(a)? The edge $v = 0$ maps to $\theta = 0$, i.e. the positive real axis in the z - plane; the edge $v = \pi$ maps to $\theta = \pi$, i.e. the negative real axis; the insulated edge $u = 0$ maps to $\log r = 0$, i.e. $r = 1$ or $|z| = 1$; the insulated edge $u = -u_0$ maps to $\log r = -u_0$, i.e. $r = e^{-u_0}$, which implies that the further the left-hand edge is to the left the larger u_0 and the smaller r. Hence a semi-infinite rectangle maps to a semicircle with a blob of insulation at the origin.

Suppose we now map (12.19). Then

$$T = 100\theta/\pi, \quad 0 \leqslant \theta \leqslant \pi$$

$$= \frac{100}{\pi}\tan^{-1}\left(\frac{y}{x}\right) \qquad (12.20)$$

Does this provide the temperature distribution in the plate? It is easy to show by substitution that (12.20) satisfies Laplace's equation. In general, it is also true that the result of transforming a solution of Laplace's equation by an analytic function leads to a function which also satisfies Laplace's equation. Also, since we use the same function to obtain the transformed temperature as to obtain the transformed region, it follows that temperatures on the transformed boundaries are the same as the temperatures on the corresponding original boundaries. We shall state without proof that the rate of change of T perpendicular to a boundary in the u - v plane is proportional to the rate of change of T perpendicular to the corresponding boundary in the x - y plane; hence insulated surfaces map to insulated surfaces.

Now consider a general transformation $w = f(z)$ where $f(z)$ is analytic at $z = z_0$ and $f'(z_0) \neq 0$. See Figures 12.8(a) and (b).

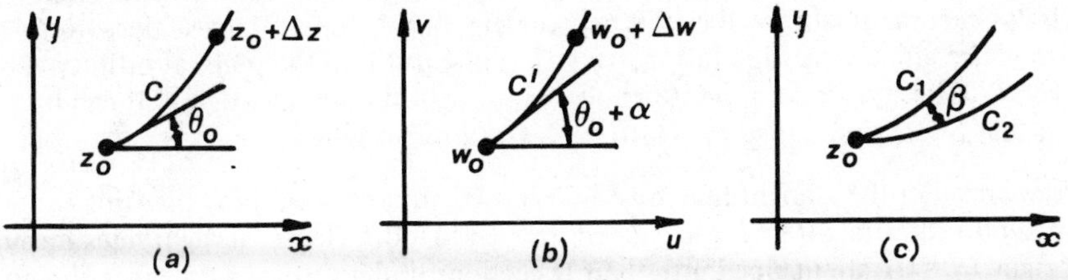

Figure 12.8

As a point moves from z_0 to $z_0 + \Delta z$ along C its image under w moves from w_0 to $w_0 + \Delta w$ along C'. Suppose we characterise the journey along C by a parameter t so that $z = z(t)$; then we may write $w = w(t)$. Now $\frac{dw}{dt} = \frac{dw}{dz} \cdot \frac{dz}{dt}$ and this holds at z_0; dz/dt represents the tangent vector at a point on C and dw/dt represents the tangent vector at the corresponding point on C'. The directions are given by $\theta_0 = \arg(dz/dt)$ and $\theta_0 + \alpha = \arg(dw/dt)$ respectively. Hence $\alpha = \arg(f'(z))$ and has a fixed value at z_0.

Now if two curves C_1 and C_2 pass through z_0 making an angle β with each other there; see Figure 12.8(c), it follows that the angle β is preserved after mapping by an analytic function $f(z)$.

In our particular problem, since isotherms and lines of heat flow are at right angles, they will remain so after the mapping $w = \log z$. Hence we have a complete identity between the two physical situations.

In general to solve a potential problem with a complicated geometry, we merely need to find a transformation from a simpler region where the solution can readily be found via an analytic function. Then we transform the solution. Note that sinks and sources and other **singular points** will map into corresponding singular points.

The property above of preserving the angle, which will be so for any analytic function makes the mapping **conformal**. We may remark that areas of *small* closed regions around z_0 are scaled by an approximate factor $|f'(z_0)|^2$.

We now present some basic mappings.

(a) Translation $\quad w = z + b, \quad b$ constant (12.21a)

(b) Magnification $\quad w = az, \quad a$ real and constant (12.21b)

(c) Rotation $\quad w = iaz, \quad a$ real and constant (12.21c)

(d) Inversion $\qquad w = 1/z \qquad$ (12.21d)

The inversion mapping is worthy of detailed study. We wish to examine the effects on special curves in the z - plane. For this purpose we write the mapping as $z = 1/w$ and it follows that $x = \dfrac{u}{u^2 + v^2}$, $y = \dfrac{-v}{u^2 + v^2}$. Hence the circle $x^2 + y^2 = a^2$ is mapped into $\dfrac{u^2}{(u^2 + v^2)^2} + \dfrac{v^2}{(u^2 + v^2)^2} = a^2$ i.e. $\dfrac{1}{a^2} = u^2 + v^2$ which is a circle in the w - plane.

In polar coordinates let $z = re^{i\theta}$ and $w = Re^{i\Theta}$. Then $R = 1/r$ and $\Theta = -\theta$. Try to describe the effects of the mapping via this approach.

Notice that the radius of this circle is the reciprocal of that of the original circle; big circles become small, small circles become big, but the circle $|z| = 1$ does not change. The circle of zero radius, i.e. $z = 0$ is mapped into the point at infinity and vice-versa. These remarks apply to circles whose centres are the origin. It can be shown that other circles may map into circles or straight lines.

Now consider the straight line $\alpha x + \beta y + \gamma = 0$. It is easy to show that this is mapped into the straight line $\alpha u - \beta v + \gamma(u^2 + v^2) = 0$. Hence, straight lines through the origin ($\gamma = 0$) are mapped into their reflections in the vertical axis, other straight lines are mapped into circles.

Example

Find the image of the circle $|z - 2i| = 2$ under the inversion mapping.

Three points define the circle; we take 0, $4i$, $2 + 2i$ for algebraic simplicity. The corresponding points in the w - plane are the point at infinity, $-\frac{1}{4}i$ and $\frac{1}{4}(1 - i)$. These define a straight line viz. the line parallel to the real axis through the point $-\frac{1}{4}i$. Take other points, equally spaced around the given circle and trace the path in the w - plane as we move once round the circle. Note that the point at infinity can appear at $u = -\infty$ and $u = +\infty$.

Application

In electrical circuit theory the **impedance** due to a resistance R is simply R, due to an inductance L is $j\omega L$ ($j^2 = -1$) and due to a capacitance C is $1/j\omega C$. The **admittance** of a circuit is the reciprocal of its total impedance. The impedance of a circuit comprising a resistance and an inductance in parallel is $\dfrac{1}{Z} = \dfrac{1}{R} - \dfrac{j}{\omega L}$. The **impedance locus** as L varies between 0 and ∞ is shown in Figure 12.9(a).

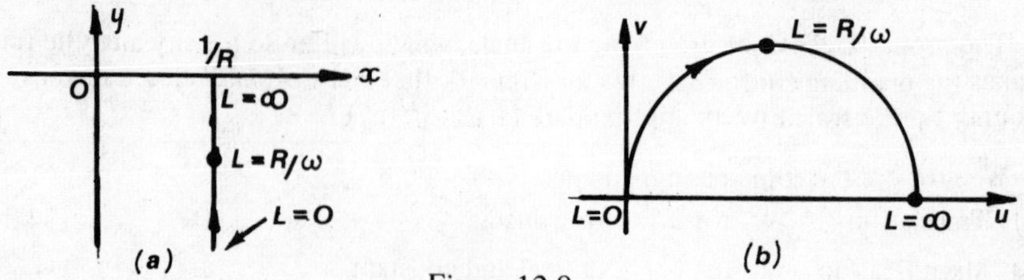

Figure 12.9

The corresponding admittance locus is found by the inversion mapping to be that shown in Figure 12.9(b). Such loci are used frequently in the design of electrical circuits.

Example

In the circuit shown in Figure 12.10(a), C is a variable capacitance, R is a fixed resistance and L is a fixed inductance. An alternating voltage of constant frequency $\omega/2\pi$ is applied. We find the impedance and admittance loci as C varies.

The impedance Z is given by $\dfrac{1}{Z} = \dfrac{1}{R+j\omega L} + j\omega C$. The admittance Y is thus $Y = \dfrac{1}{Z} = \dfrac{R-j\omega L}{R^2+\omega^2L^2} + j\omega C$.

Let $Y = u + iv$, so that $u = \dfrac{R}{R^2+\omega^2L^2}$ and $v = \dfrac{\omega C(R^2+\omega^2L^2) - \omega L}{R^2+\omega^2L^2}$

Since R and L are fixed, the Y- locus as ω and C vary is the straight line $u = \dfrac{R}{R^2+\omega^2L^2}$ between $v = \dfrac{-\omega L}{R^2+\omega^2L^2}$ and ∞.

For the impedance locus we have $\dfrac{1}{Z} = \dfrac{1}{x+iy} = \dfrac{1}{R+j\omega L} + j\omega C$. That is

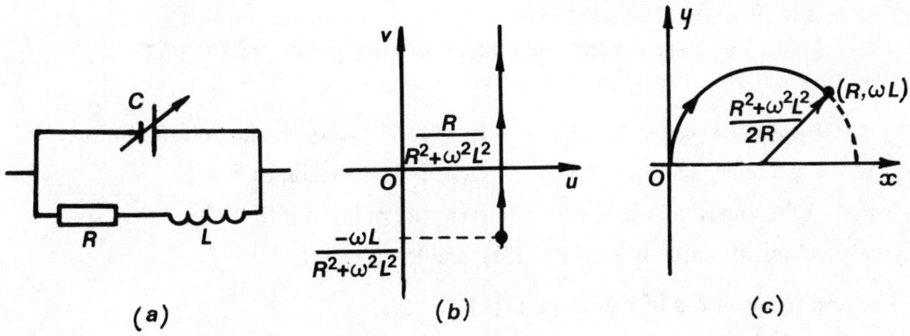

Figure 12.10

$\dfrac{x-iy}{x^2+y^2} = \dfrac{R-j\omega L}{R^2+\omega^2L^2} + j\omega C$. Hence $\dfrac{x}{x^2+y^2} = \dfrac{R}{R^2+\omega^2L^2}$ giving $x^2+y^2 - \left[\dfrac{R^2+\omega^2L^2}{R}\right]x = 0$. This is a circle centre $\left(\dfrac{R^2+\omega^2L^2}{2R}, 0\right)$ and radius $\left[\dfrac{R^2+\omega^2L^2}{2R}\right]$.

The variation in C is from $C=0$ to $C=\infty$ and when $C=0$ we find $\dfrac{1}{Z} = \dfrac{1}{R+j\omega L}$, i.e. $x=R$, $y=\omega L$. When $C=\infty$ we get $\dfrac{1}{Z} = \infty$ or $Z=0$, i.e. $x=0$, $y=0$. The loci are shown in Figure 12.10(b) and (c).

Bilinear mapping

$$w = \frac{az+b}{cz+d} \tag{12.22}$$

where $bc - ad \neq 0$.

You should show that we may rewrite this mapping as

$$w = \frac{(bc - ad)/c^2}{z + d/c} + \frac{a}{c} \qquad (12.23)$$

provided $c \neq 0$. (If $c = 0$, the mapping reduces to a magnification followed by a translation.)

Looking at (12.23) we see that the mapping may be considered as a succession of four simpler mappings:

(i) $z_1 = z + d/c$ (translation)

(ii) $z_2 = 1/z_1$ (inversion)

(iii) $z_3 = (bc - da)\, z_2/c^2$ (magnification)

(iv) $w = z_3 + a/c$ (translation)

Note that if we restrict our attention to straight lines and circles we can find the result of applying (12.22) by its effect on three points which lie on the original curve.

Example

Consider $w = \dfrac{z}{z-1}$. We wish to find its effect on the circle $|z| = 1$, the real axis and the imaginary axis. The eight regions of the z - plane thus formed are shown in Figure 12.11(a). We consider the mapping on the five points marked on that diagram together with the point at infinity.

(i) $z = 0$ becomes $w = 0$ (ii) $z = 1$ becomes the point at infinity

(iii) $z = \mathrm{i}$ becomes $w = \frac{1}{2} - \frac{1}{2}\mathrm{i}$ (iv) $z = -1$ becomes $w = \frac{1}{2}$

(v) $z = -\mathrm{i}$ becomes $w = \frac{1}{2} + \frac{1}{2}\mathrm{i}$ (vi) the point at infinity becomes $w = 1$

(For this last result, write $w = 1/(1 - 1/z)$ and let $z \to \infty$.)

The results are shown in Figure 12.11(b)

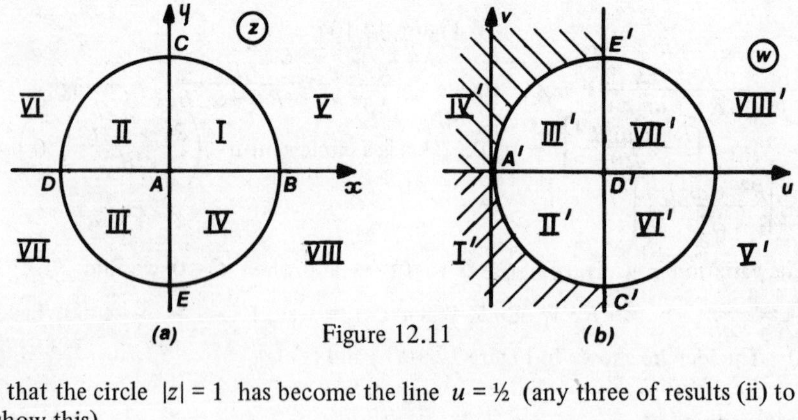

Figure 12.11

Note that the circle $|z| = 1$ has become the line $u = \frac{1}{2}$ (any three of results (ii) to (v) will show this).

Furthermore, the real axis, using (iv), (i) and (ii) becomes the real axis; but note where the points moved by considering the mapped images of $z = 0,\ 1,\ \infty$ and $z = -\infty$ (still the point at infinity), $-1,\ 0$. Take $z = 2$ and $z = \frac{1}{2}$ as an aid.

Finally, the imaginary axis, defined by (v), (i) and (iii), has become the circle $|w - \frac{1}{2}| = \frac{1}{2}$. Examine Figure 12.11 carefully and study the images of the eight regions of the first diagram.

Application

A circuit containing a resistance R and a capacitance C in parallel has impedance Z given by $\dfrac{1}{Z} = \dfrac{1}{R} + j\omega C$ i.e. $Z = R/(1 + j\omega CR)$. The impedance locus as C varies from 0 to ∞ is required. Now the mapping from C to Z is a special case of (12.22) with $a = 0$, $b = R$, $c = j\omega R$, $d = 1$. If we let $C = 0$, $1/(\omega R)$, ∞ in turn we find $Z = R$, $\frac{1}{2}R(1 - j)$, 0 and the locus is the arc of a circle as you can verify.

Problems

1. Sketch the following regions in the z - plane

   (i) $|z + 2 - i| \leqslant 4$ (ii) $1 \leqslant |z - 2i| \leqslant 2$ (iii) $|\arg z| < \pi/2$

   (iv) $|z - i| < 2|z + i|$ (v) $\mathrm{R}(z^2) < 2$ (vi) $|z + 3| + |z - 3| \leqslant 10$

2. Find the family of curves orthogonal to

   (i) $x^3y - xy^3 =$ constant (ii) $e^{-x} \cos y + xy =$ constant (iii) $r^2 \cos 2\theta =$ constant

3. Which of the following functions ϕ could be a possible velocity potential for a fluid flow? For those which are, sketch the streamlines and describe the flow.

   (i) $A(x^2 - y^2)$ (ii) $\sin(x + y)$ (iii) $Ux + Vy$

4. For the following complex potentials plot the lines $u =$ constant and $v =$ constant.

   (i) $w = z^2 + 2z$ (ii) $w = z^4$ (iii) $w = A \sin z$

5. Interpret the complex potentials w_1, w_2 and $w_1 + w_2$ in terms of electrostatics, where $w_1 = C \log(z - x_1)$, $w_2 = -C \log(z - x_2)$ for C a constant.

6. Sketch the equipotential lines of

   (i) $w = z^2$ (ii) $w = 1/z$

7. Interpret the potential $w = -(iK/2\pi)\log z$ in terms of fluid flow. Repeat for $w = (K/2\pi)\log z$. K is a positive real constant.

8. Show that $w = \cosh^{-1} z$ may correspond to a flow through an aperture. Sketch the streamlines for the flow.

9. Find the images of the curves $\arg z = \pi/6$, $|z| = 2$, $\mathcal{R}(z) = -2$, $\mathcal{I}(z) = 1$ under each of the following transformations.

   (i) $w = iz$ (ii) $w = -iz + 2i$ (iii) $w = 2i(z + 1 - i)$

10. Find the image of the parabola $y = x^2$ under $w = -2z + 2i$.

11. Find the image of the curves (i) to (iv) under the mapping $w = z^2$

(i) $y = 1 - x$ (ii) $|z| > 2$ (iii) $1 < R(z) < 2$ (iv) $1 < |z| < 2$ and $|\arg z| < \pi/4$

12. Find the image under $w = 1/z$ of the following

(i) $y = x - 1$ (ii) $x^2 + (y - 1)^2 = 1$ (iii) $x = 0$ (iv) $(x + 1)^2 + (y - 2)^2 = 4$

(v) $|z + 1| = 1$

13. What points are not altered by $w = 1/z$?

14. Verify that the angle of intersection at each of the points where $|z + 1 + 2i| = 2$ and $y = x + 1$ intersect is unaltered by $w = 1/z$.

15. Discuss the mapping $w = 1/(z + 1)$.

16. Find the bilinear mapping which maps $z = 0$, i, -1 onto $w = 12$, $11 + i$ and 11 respectively.

17. Repeat 16 for the z - values and corresponding w - values shown

(i) 0, 1, 2; 4, 10, 16 (ii) 0, 1, ∞; ∞, 1, 0 (iii) 1, i, -1; i, -1, $-$i

(iv) z_1, z_2, z_3; 0, 1, ∞.

18. (i) Show that $w = \dfrac{z - i}{z + i}$ maps the upper half z - plane onto the interior of the unit circle and indicate the other main features of this transformation.

(ii) Show that $\phi = \dfrac{1}{\pi} \tan^{-1}\left(\dfrac{y}{x}\right)$, where $0 \leqslant \phi \leqslant 1$, is harmonic in the upper half-plane and has values 0 and 1 respectively on the positive and negative axes. (C.E.I.)

19. If $|z - c| = m|z + c|$, where $m\,(\neq 1)$, c are positive constants and $z = x + iy$, prove that z lies on a circle of radius $2mc/|1 - m^2|$ and centre $z = c\,(1 + m^2)/(1 - m^2)$.

In the transformation between the z - plane and the t - plane defined by $z = c\,(1 + t)/(1 - t)$, prove that the annulus between the circles $|t| = e^{\alpha}$ and $|t| = e^{-\alpha} (\alpha > 0)$ is mapped on to the region in the z - plane outside the equal circles $|z \pm d| = a$, where $d = c \coth \alpha$ and $a = c \operatorname{cosech} \alpha$.

Illustrate the mappingby a sketch, indicating the points in the z - plane which correspond to the intersections of the circles in the t - plane with the real axis. (L.U.)

20. (a) If $w = (j - z)/(j + z)$, where $w = u + jv$ and $z = x + jy$, show that the interior of the circle $|z| = 1$ transforms into that half of the w - plane for which $u > 0$.

(b) If $w = \cosh(\pi z/a)$, where a is real and positive, find the region in the w - plane which corresponds to the rectangle bounded by the lines $x = 0$, $y = 0$, $x = N > 0$, $y = a$. (L.U.)

21. A bilinear transformation is such that the points $z_1 = 0$, $z_2 = 1$, $z_3 = \infty$ in the z - plane are mapped on the w - plane as $w_1 = 1$, $w_2 = e^{j\pi/4}$, $w_3 = j$. Obtain this transformation and find the curve in the w - plane corresponding to the line $y = 0$. Into what area in the w - plane is the half plane $y > 0$ mapped? (L.U.)

22. If $w = \dfrac{z - 1}{z + 1}$ where $w = u + iv$ and $z = x + iy$, express u and v in terms of x and y.

Show that the straight line $u = k$ (where k is a real constant) in the w - plane transforms into a circle through the point -1 in the z - plane. Sketch the circles givenby $k = 0$ and $k = \tfrac{1}{2}$.

23. Find the image under $w = e^z$ of the following

(i) $x = -2,\ -\pi/2 < y < \pi$ (ii) $x = 1,\ 0 \leqslant y \leqslant \pi$

24. Find the image under $w = \log z$ of the following.

(i) $|z| = 9$ (ii) $\arg z = \pi/4$

25. Find the image under $w = \sin z$ of

(i) $y \geqslant 0,\ -\pi/2 < x < \pi/2$ (ii) $0 < x < \pi/2,\ y \geqslant 0$

26. Show that $w = \cosh z$ maps the semi-infinite strip $x \geqslant 0,\ 0 \leqslant y \leqslant \pi/2$ into the first quadrant of the w - plane.

12.5 FURTHER CONFORMAL MAPPINGS

We open this section with a slightly more complicated problem in mapping. Fluid flows through a small orifice BD in a plane boundary. Refer to Figure 12.12. The stream is observed to contract from width $2a$ to $2c$. The complex velocity is $\dfrac{dw}{dz} = u - iv = qe^{-i\theta}$ where q is the magnitude of the velocity and θ is the direction of flow. The points A, C and E are assumed to be at great distances from the origin. The 'free' streamlines BC and DC have the property that $q = U =$ constant.
On AB, $\theta = 0,\ 0 \leqslant q \leqslant U$;
on BC, $q = U,\ 0 \geqslant \theta \geqslant -\pi/2$;
on ED, $\theta = -\pi,\ 0 \leqslant q \leqslant U$;
on CD, $q = U,\ -\pi/2 \geqslant \theta \geqslant -\pi$

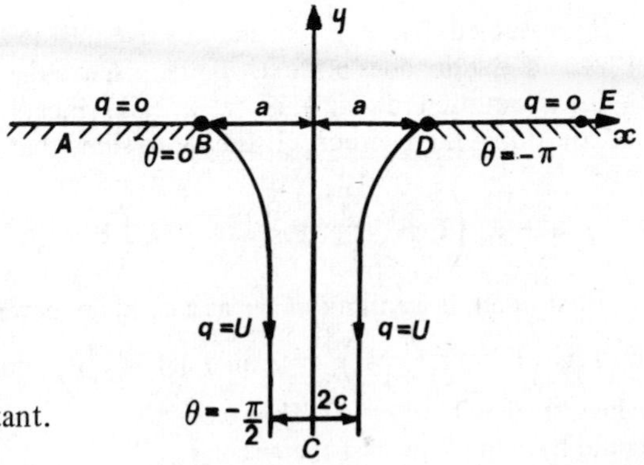

Figure 12.12

Now consider $\xi = \dfrac{1}{2}\left(\dfrac{U}{q}e^{i\theta} + \dfrac{q}{U}e^{-i\theta}\right)$. On AB, $\xi = ½\left(\dfrac{U}{q} + \dfrac{q}{U}\right)$ and takes values from $+\infty$ to $+1$. On BCD, $\xi = \cos\theta$ and takes values from $+1$ at B to -1 at D. On DE, $\xi = -½\left(\dfrac{U}{q} + \dfrac{q}{U}\right)$ and takes values from -1 to $-\infty$. Note that along ABCDE, ξ takes real values only.

In Figure 12.13 we show the equivalent flow in the ξ - plane; it is flow into a sink at $C(\xi = 0)$. The strength of the sink is $m = 2cU/\pi$ found by equating flows in the two planes. The required complex potential is $w' = -m\log\xi = -\dfrac{2cU}{\pi}\log\xi$. Now

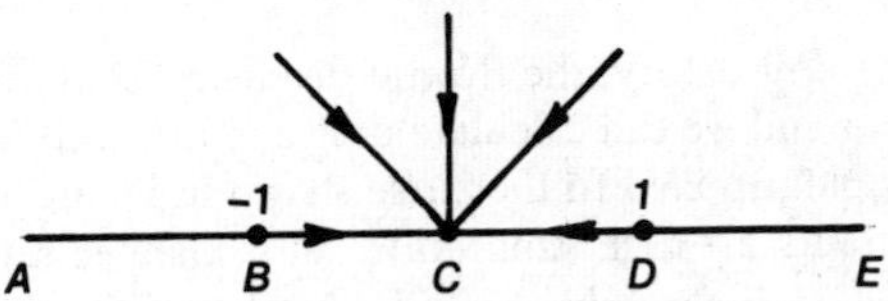

Figure 12.13

along the streamline BC, $\dfrac{dw'}{dz} = Ue^{-i\theta} = \dfrac{-2cU}{\pi\xi} \cdot \dfrac{d\xi}{dz}$ and we know that $\xi = \cos\theta$.

Hence $e^{-i\theta} = \dfrac{-2c}{\pi\cos\theta}(-\sin\theta)\dfrac{d\theta}{dz}$. Therefore

$$\int dz = \int \frac{2c}{\pi} e^{i\theta} \tan\theta \, d\theta \text{ and taking real parts } \int dx = \int \frac{2c}{\pi} \sin\theta \, d\theta$$

At B, $x = -a$ and $\theta = 0$; at C, $x = -c$ and $\theta = -\pi/2$ so that

$$c - a = \int_C^B dx = \int_{-\pi/2}^{0} \frac{2c}{\pi} \sin\theta \, d\theta = \frac{-2c}{\pi}$$

Hence the coefficient of contraction $\dfrac{c}{a} = \dfrac{\pi}{\pi + 2}$

The mapping $w = z + a^2/z$

It is assumed that a is real and positive. Note that at $z = 0$ there is a singularity of w and dw/dz does not exist there. Hence the mapping fails to be conformal at $z = 0$. In addition $dw/dz = 1 - a^2/z^2$ and since this is zero at $z = \pm a$ the mapping is not conformal there either. It is easy to show that

$$u = x\left(1 + \frac{a^2}{x^2 + y^2}\right), \quad v = y\left(1 - \frac{a^2}{x^2 + y^2}\right) \tag{12.24}$$

First of all, if we think of w as a complex potential then the stream function $\psi = y\left(1 - \dfrac{a^2}{x^2 + y^2}\right) = r\sin\theta\left(1 - \dfrac{a^2}{r^2}\right)$ in polar coordinates. Now $\psi = 0$ when $\sin\theta = 0$ or $r = a$. Thus we could have the flow past the circular cylinder $r = a$ as shown in Figure 12.14. The stagnation points of the flow are where the fluid is at rest, i.e. $u = v = 0$ and these occur on the cylinder where $r = a$ and $y = 0$: the points A and B. In practice the viscosity of a liquid would cause frictional effects to alter markedly the downstream pattern of flow.

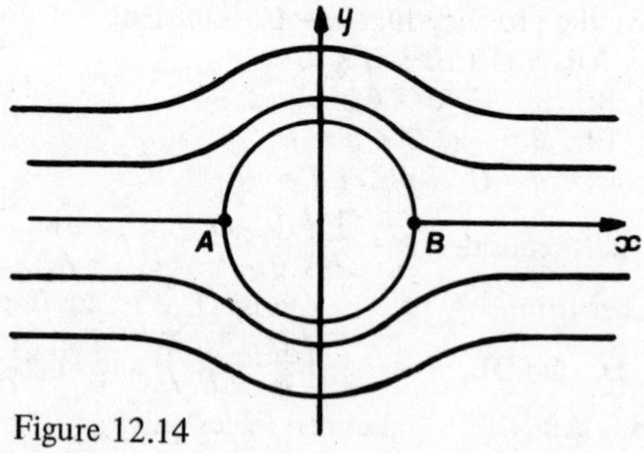

Figure 12.14

In aerofoil theory, the flow at distances from the aerofoil is virtually non-viscous in behaviour and we can calculate drag and lift fairly well via potential theory. Consider our mapping applied to the circle shown in Figure 12.15(a). It passes through $z = -a$ and contains $z = a$ in its interior. It is mapped into a shape depicted in Figure 12.15(b): the so-called **Joukowski aerofoil**. Note that the sharp point or cusp at $w = -2a$ is the image of the region near $z = -a$, which was one of the points of non-conformality. Work has been carried out to modify the transformation so as to 'round off' the cusp.

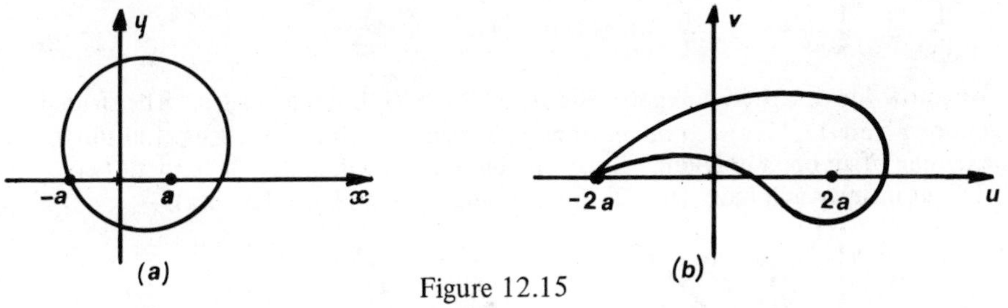

Figure 12.15

The mapping $w = z + e^z$

Consider the strip $-\pi \leqslant y \leqslant \pi$, $-\infty < x < \infty$ depicted in Figure 12.16(a). The components of w are $u = x + e^x \cos y$ and $v = y + e^x \sin y$. The line $y = \pi$ maps into (u, v) where $u = x - e^x$ and $v = \pi$; now as x varies from $-\infty$ to 0, u varies from $-\infty$ to -1. However, as x varies from 0 to ∞, u retraces its steps from -1 to $-\infty$.

If we consider Figure 12.16(a) as representing flow through an open-ended channel then this result implies that a line of flow close to the channel boundary is virtually bent back on itself on reaching the open field. Examine for yourself the images of the lines $y = -\pi$ and $y = 0$.

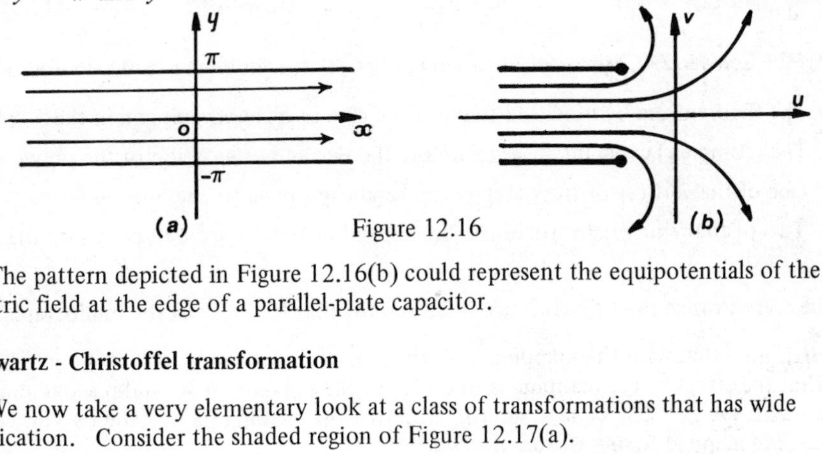

Figure 12.16

The pattern depicted in Figure 12.16(b) could represent the equipotentials of the electric field at the edge of a parallel-plate capacitor.

Schwartz - Christoffel transformation

We now take a very elementary look at a class of transformations that has wide application. Consider the shaded region of Figure 12.17(a).

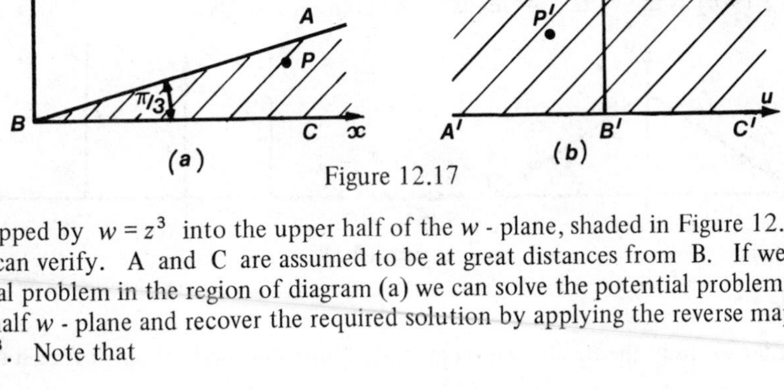

Figure 12.17

It is mapped by $w = z^3$ into the upper half of the w - plane, shaded in Figure 12.17(b) as you can verify. A and C are assumed to be at great distances from B. If we have a potential problem in the region of diagram (a) we can solve the potential problem in the upper half w - plane and recover the required solution by applying the reverse mapping $z = w^{1/3}$. Note that

$$\frac{dz}{dw} = \frac{1}{3} w^{-2/3} = \frac{1}{3} w^{([\pi/3]/\pi - 1)} \tag{12.25}$$

We know that a closed polygon with n sides has n internal angles. The shaded region of Figure 12.17(a) can be regarded as a triangle with two vertices at infinity; it is an example of an **open polygon.** In general an open polygon of n sides with two vertices at infinity will have $(n - 2)$ internal angles; see Figure 12.18(a).

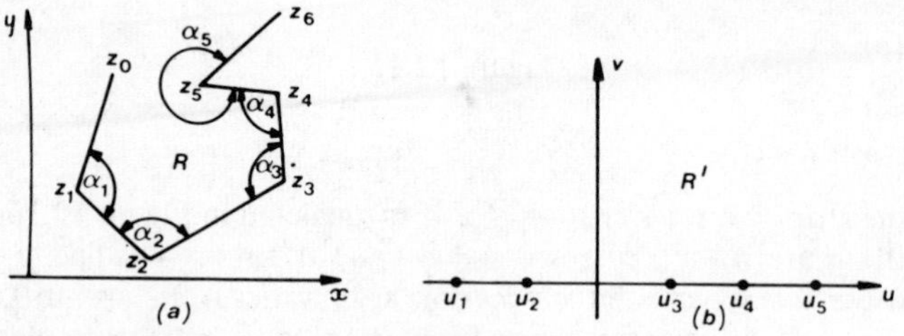

Figure 12.18

We wish to map the region R into the upper half of the w - plane such that the points z_1 to z_5 map to u_1 to u_5 as shown in Figure 12.18(b). It can be shown that this can be accomplished by the transformation given by

$$\frac{dz}{dw} = A(w - u_1)^{\left(\frac{\alpha_1}{\pi} - 1\right)} (w - u_2)^{\left(\frac{\alpha_2}{\pi} - 1\right)} \cdots\cdots (w - u_5)^{\left(\frac{\alpha_5}{\pi} - 1\right)} \tag{12.26}$$

This is the **Schwartz - Christoffel Theorem** applied to this problem. Note the following

(i) The theorem can be modified to cover the case of a closed polygon in the z - plane.

(ii) The (complex) constant A is related to the size and orientation of the polygon.

(iii) One of the vertices of the polygon can be always made to map into $u = \pm\infty$.

(iv) Two of the remaining vertices can be mapped into two specified points on the u - axis.

We do not prove the theorem; we refer you for that to books in the bibliography.

First, note that with the mapping $z = w^{1/3}$, the only interior angle, at B, is $\pi/3$ and that the effect of the mapping is to open out the polygon ABC and place it along the u - axis. A is taken as $u = -\infty$ (or C as $u = \infty$) using (iii); the two other vertices are mapped to $u = 0$ and $u = \infty$.

Finally note that the triangle of Figure 12.19(a) becomes the open polygon of Figure 12.19(b) as the vertex A tends to the point at infinity.

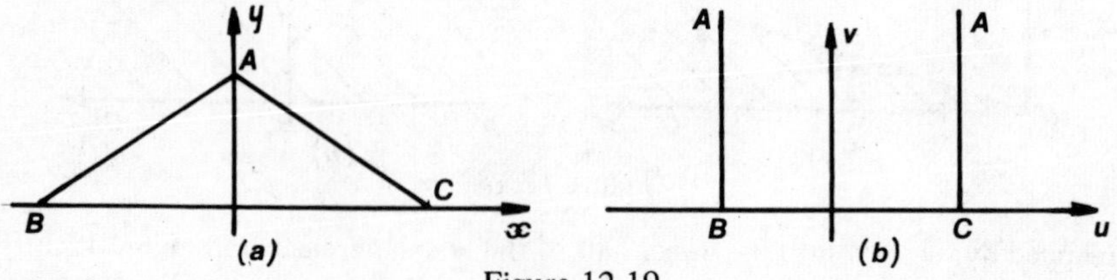

Figure 12.19

Example 1

In order to study the steady-state temperature distribution in the slab shown in

Figure 12.20(a), where A and D are great distances from B and C respectively, we

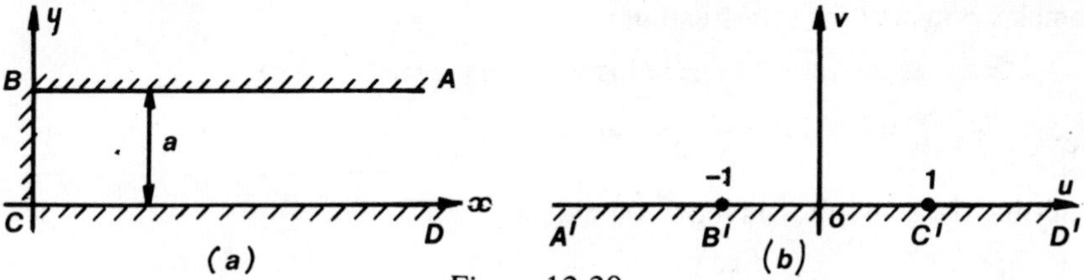

Figure 12.20

shall map the boundary ABCD to the whole real axis in the w - plane as indicated in Figure 12.20(b). Now the only interior angles are those at B and C and each is $\pi/2$. Then (12.26) becomes $\dfrac{dz}{dw} = A\,(w+1)^{-\frac{1}{2}}\,(w-1)^{-\frac{1}{2}} = \dfrac{A}{(w^2-1)^{\frac{1}{2}}}$. Hence $z = A\cosh^{-1} w + B$, B constant. But $z = 0$ corresponds to $w = +1$ and $z = ia$ corresponds to $w = -1$. Since $\cosh^{-1} 1 = 0$, $B = 0$ and $ia = A\cosh^{-1}(-1) = Ai\pi$, giving $A = a/\pi$. (We have used the result that $\cosh^{-1} w = \log[w + \sqrt{w^2-1}]$.) Therefore $w = \cosh(\pi z/a)$.

Suppose now that the edge AB is maintained at a temperature T, the edge BC at 0 and the edge CD at $2T$. Then the temperature distribution for the half w - plane is given by $\Phi = \dfrac{T}{\pi}\tan^{-1}\left(\dfrac{v}{u+1}\right) - \dfrac{2T}{\pi}\tan^{-1}\left(\dfrac{v}{u-1}\right) + 2T$. We can now use the mapping $w = \cosh(\pi z/a)$ i.e. $u = \cosh(\pi x/a)\cos(\pi y/a)$, $v = \sinh(\pi x/a)\sin(\pi y/a)$ to obtain Φ in terms of x and y.

Example 2

Figure 12.21(a) depicts flow into a channel through a narrow slit in one wall. A and D are far to the right and B and C are far to the left. Let πm be the volume/unit

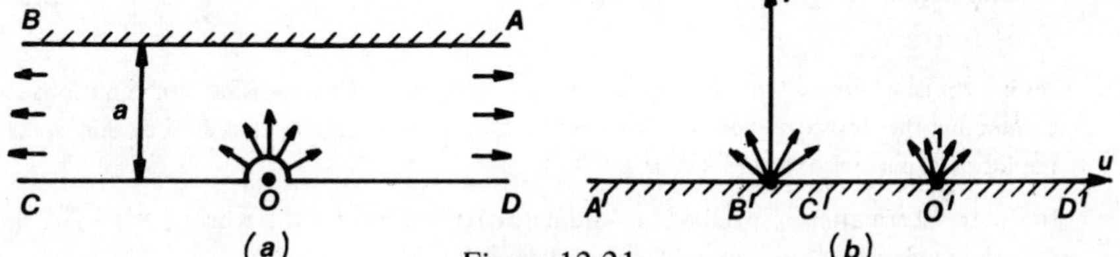

Figure 12.21

thickness flowing into the channel in unit time; this gives flow into the region shown that one would expect from a source of strength m and no wall CD to intervene. By symmetry, the parallel flow at AD or BC will be taken by a sink of strength $\frac{1}{2}m$.

We map the channel region into the upper half w - plane shown in Figure 12.21(b). Here B and C are both mapped to $w = 0$.

The angle at BC is zero and hence $\dfrac{dz}{dw} = Aw^{-1}$ so that $z = A\log w + B$.

Now $z = 0$ maps to $w = 1$ so that $B = 0$; further let $z = ia$ map to $w = -1$ so that $ia = A\log(-1) = A\pi i$. Hence $z = \dfrac{a}{\pi}\log w$ or $w = e^{\pi z/a}$.

Note that we have a line sink of strength $\frac{1}{2}m$ at $w = 0$ and a line source at $w = 1$ of strength m. (Since the end BC absorbed half the output of the source.) The complex potential for this situation is

$$\Phi = -m \log (w - 1) + \tfrac{1}{2} m \log w = -m \log (w^{\frac{1}{2}} - w^{-\frac{1}{2}})$$

Now $w^{\frac{1}{2}} - w^{-\frac{1}{2}} = e^{\pi z/(2a)} - e^{-\pi z/(2a)} = 2 \sinh \dfrac{\pi z}{2a}$

Omitting the constant $-m \log 2$ we obtain

$$\Phi = -m \log \sinh \left(\frac{\pi z}{2a}\right)$$

Since the dividing streamline goes from O perpendicularly to the opposite wall we expect a stagnation point there. Now $\dfrac{d\Phi}{dz} = \dfrac{-m\pi}{2a} \coth \left(\dfrac{\pi z}{2a}\right)$ and this does indeed vanish at $z = ia$.

We pick up two bonuses. If we replace the dividing streamline by a solid wall we have the motion within a semi-infinite channel due to a source at one corner. Alternatively if we use the principle of reflection about the real axis in the z - plane and remove the wall CD we have the flow due to a source midway between two parallel planes.

Similar situations arise in electrostatics and heat flow.

Problems

1. Write down in complex exponential form the equation of the circle $|z| = r$. Hence determine how the circle $|z| = r$, where $r > 1$, maps from the z - plane on to the w - plane under the transformation $w = z + z^{-1}$. (C.E.I.)

2. Show that w describes an ellipse as z describes the circle $|z| = d$, where $d > a$, under the transformation $w = \frac{1}{2}\left[z + \dfrac{a^2}{z}\right]$. (C.E.I.)

3. If $w = f(z)$, where $w = u + jv$ and $z = x + jy$, obtain the Cauchy-Riemann equations connecting the derivatives of u and v with respect to x and y, and deduce that u satisfies Laplace's equation $\partial^2 u/\partial x^2 + \partial^2 u/\partial y^2 = 0$.

   In the transformation to parabolic co-ordinates defined by $z = \zeta^2$, where $\zeta = \xi + j\eta$, show that the Laplacian $(\partial^2 \phi/\partial x^2 + \partial^2 \phi/\partial y^2)$ becomes

   $$\frac{1}{4\rho^2}\left(\frac{\partial^2 \phi}{\partial \xi^2} + \frac{\partial^2 \phi}{\partial \eta^2}\right)$$

   where $\rho^2 = \xi^2 + \eta^2$. Sketch the co-ordinate lines ξ = constant, η = constant. (L.U.)

4. (a) Show, that if a and b are real, the transformation $\dfrac{z - a}{z + a} = j\, e^{jw\pi/b}$, where $w = u + jv$ and $z = x + jy$, transforms the infinite strip bounded by the lines $u = 0$, $u = b$ into the interior of the circle $|z| = a$.

   (b) If $w = \coth (c/z)$ where c is real show that $v = 0$ when $y = 0$ and when $\left|z - \dfrac{jc}{n\pi}\right| = \dfrac{c}{n\pi}$ where n is an integer. (L.U.)

5. Show that if the transformation $w = f(z)$, $z = re^{j\theta}$, regular in $|z| < R$ maps the portion of the circle $|z| = r_0\,(<R)$, $\alpha \leqslant \theta \leqslant \beta$ on to a curve C in the w - plane, then the length of C is

$$\int_{\alpha}^{\beta} |f'(\sigma)|\, r_0 \, d\theta$$

where $\sigma = r_0\, e^{j\theta}$ is a point on the circle.

Show that the transformation $w = 4/(1+z)^2$ transforms the semi-circular arc of unit radius given by $-\frac{1}{2}\pi \leqslant \theta \leqslant \frac{1}{2}\pi$ into an arc of a parabola of length $2\sqrt{2} + 2\log(1+\sqrt{2})$. (L.U.)

6. Use the Schwartz-Christoffel transformation to find a function which maps each of the regions in the z - plane to the upper half of the w - plane

(i)

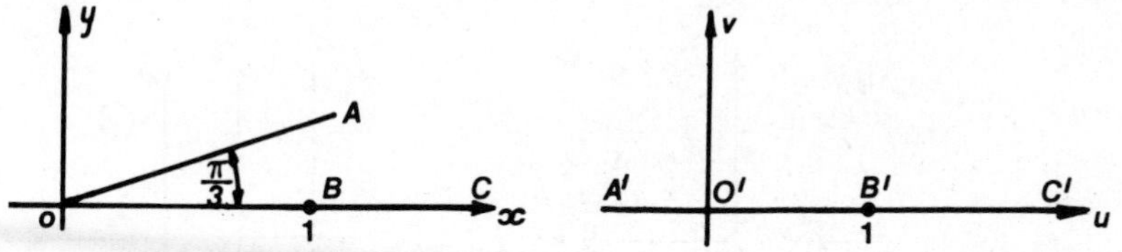

(ii)

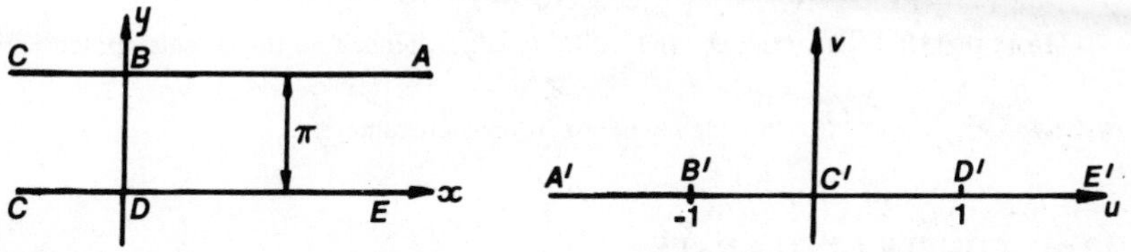

(iii)

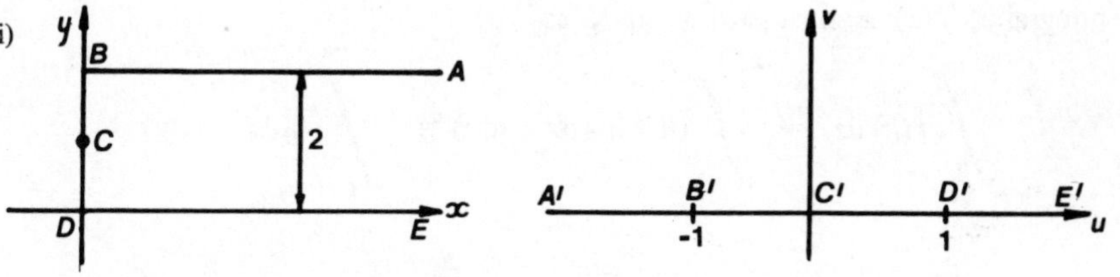

(iv)

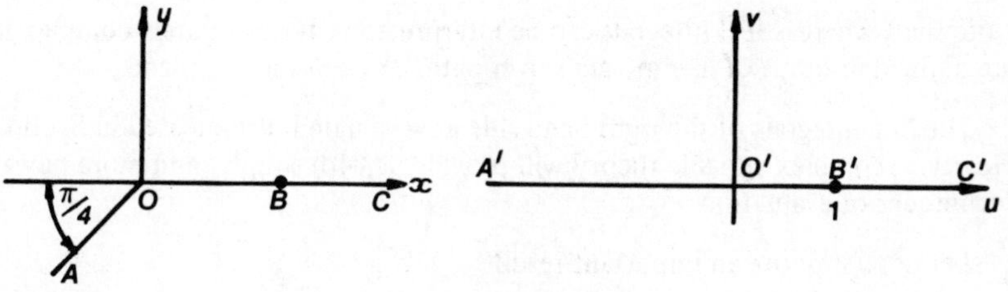

7. A parallel plate capacitor consists of a semi-infinite plate at potential ϕ_1 above a parallel lower plate at potential ϕ_2 a distance h below. Refer to the diagram below.

Consider the semi-infinite plate as two legs of a polygon (BC and CD) the angle at C changing through π radians. The polygon ABCDEA is closed with infinite arcs AB and DE.

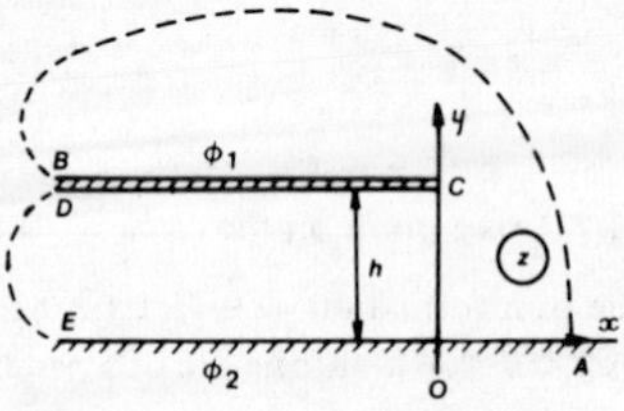

(i) Transform first to the t - plane as shown on the diagram on the left below.

(ii) Transform again to the w - plane as shown in the diagram below to the right.

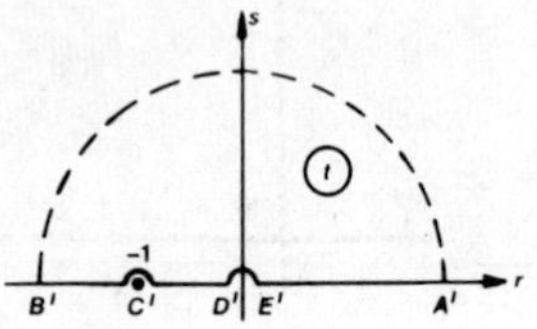

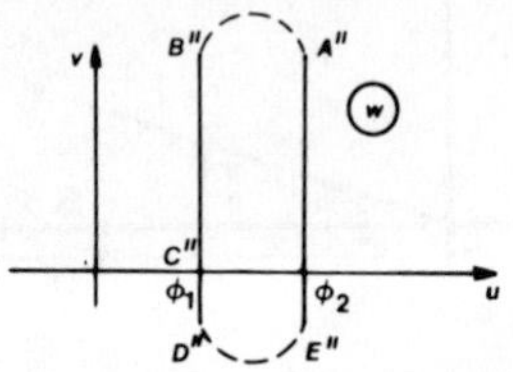

Note that B″C″D″ is at ϕ_1 and A″E″ at ϕ_2. Hence find the complex potential in the form

$$w = \phi_2 + i\,\frac{(\phi_2 - \phi_1)}{\pi} \log t \text{ where } t \text{ is now a parameter.}$$

12.6 COMPLEX INTEGRALS

If $f(z)$ is a single-valued, continuous function in some region R then we define the integral of $f(z)$ along a path C in R as

$$\int_C f(z)\,dz = \int_C (u + iv)(dx + idy) = \int_C (udx - vdy) + i\int_C (vdx + udy) \qquad (12.27)$$

Note that, whereas real integrals can be interpreted in terms of area, complex integrals are defined in terms of line integrals over paths or curves in the plane.

The line integrals of the right-hand side are evaluated as indicated in Section 11.5. However, complex variable theory will provide us with simpler and more powerful techniques of evaluation.

Let us first quote an important result.

$$\left|\int_C f(z)dz\right| \leq \int_C |f(z)| \cdot |dz| \leq M \int_C ds = ML \qquad (12.28)$$

where M is an upper bound of $|f(z)|$ along C and L is the length of the path C.

Example 1

(i) Find $\displaystyle\int_{C_1} z^2\,dz$ where C_1 is that part of the unit circle from $z = 1$ to $z = i$.

(ii) Find $\displaystyle\oint_C z^2\,dz$ where C is the circumference of the unit circle travelled in an anti-clockwise sense.

(i) $$\int_{C_1} z^2\,dz = \int_{C_1} [(x^2 - y^2)\,dx - 2xy\,dy] + i\int_{C_1} [2xy\,dx + (x^2 - y^2)\,dy]$$

On the unit circle we take $x = \cos\theta$, $y = \sin\theta$ and the path of integration is from $\theta = 0$ to $\theta = \pi/2$. We leave you to show that $\displaystyle\int_{C_1} f(z)\,dz = -(1 + i)/3$.

We note that if we merely evaluate $\left[\dfrac{z^3}{3}\right]_1^i$ we again obtain $-\left(\dfrac{1+i}{3}\right)$; in other words we carry out an analogue of real integration. Is this result coincidental or is there a deeper underlying result? Wait and see.

(ii) This time the integral is from $\theta = 0$ to $\theta = 2\pi$. Again we leave you to show $\displaystyle\oint_C z^2\,dz = 0$. Is this also a coincidence?

Example 2

$\displaystyle\oint_C \frac{1}{z}\,dz$ where C is the unit circle may be evaluated by noting that on the unit circle $z = \cos\theta + i\sin\theta = e^{i\theta}$ and $dz/d\theta = ie^{i\theta}$. Hence

$$\oint_C \frac{1}{z}\,dz = \int_0^{2\pi} i\,d\theta = 2\pi i \tag{12.29}$$

Example 3

In a similar manner we find that for integral n

$$\oint_C \frac{dz}{(z - z_0)^n} = \begin{cases} 0 & n > 1 \\ 2\pi i & n = 1 \end{cases} \tag{12.30}$$

where C is the circle centre z_0 and radius r, i.e. $|z - z_0| = r$.

This is one of the most important results in integration. It will be used often later in the chapter. We prove the result as follows. Substituting $z = z_0 + re^{i\theta}$, the integral becomes

$$\oint_C \frac{ire^{i\theta}\,d\theta}{(re^{i\theta})^n} = \int_0^{2\pi} \frac{ie^{(1-n)i\theta}}{r^{n-1}}\,d\theta$$

If $n \neq 1$ we obtain $\frac{i}{r^{n-1}}\left[\frac{e^{(1-n)i\theta}}{(1-n)i}\right]_0^{2\pi} = \frac{[e^{(1-n)2\pi i}-1]}{r^{n-1}}\,\frac{1}{(1-n)} = 0$

If $n = 1$ the integral is $\int_0^{2\pi} i\,d\theta = 2\pi i$

Application. Blasius' Theorem

In Figure 12.22(a) we show a section of cylinder in a steady two-dimensional irrotational flow of an ideal fluid.

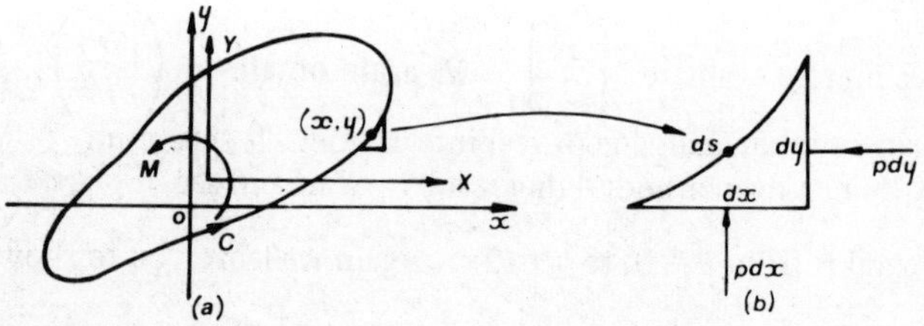

Figure 12.22

The boundary of the cylinder is C. Let X, Y be the components of the force on the cylinder and let M be the moment about the origin, due to fluid pressure. In Figure 12.22(b) we depict a small section of the cylinder surface around a point (x, y). We may write

$$dX = -p\,dy, \quad dY = p\,dx, \quad dM = p\,(x\,dx + y\,dy)$$

Hence $d(X - iY) = dX - i\,dY = -p\,(i\,dx + dy) = -ip\,d(x - iy) = -ip\,d\bar{z}$ and $dM = \mathcal{R}\{p\,(x+iy)(dx - i\,dy)\} = \mathcal{R}\{pz\,d\bar{z}\}$.

Now the pressure at any point is $p = p_0 - \tfrac{1}{2}\rho q^2$ where p_0 is the stagnation pressure, ρ is fluid density and the fluid speed q is given by $q^2 = \left|\frac{dw}{dz}\right|^2 = \frac{dw}{dz}\cdot\frac{d\bar{w}}{d\bar{z}}$. Since p_0 is a constant it has no resultant effect on X, Y or M. Then

$d(X - iY) = +\tfrac{1}{2} i\rho \frac{dw}{dz}\cdot\frac{d\bar{w}}{d\bar{z}}\,d\bar{z} = \tfrac{1}{2} i\rho \frac{dw}{dz}\,d\bar{w}$ and $dM = \mathcal{R}\left\{-\tfrac{1}{2}\rho z \frac{dw}{dz}\,d\bar{w}\right\}$.

Now C is a streamline (on which ψ = constant) and on C, $dw = d\phi$ which is real so that $dw = d\bar{w}$. Using this result and integrating around C to obtain total effects we obtain

$$X - iY = \tfrac{1}{2} i\rho \oint_C \left(\frac{dw}{dz}\right)^2 dz \tag{12.31}$$

$$M = \mathcal{R}\left\{-\tfrac{1}{2}\rho \oint_C z\left(\frac{dw}{dz}\right)^2 dz\right\} \qquad (12.32)$$

These results constitute **Blasius' Theorem.** We shall return shortly to the evaluation of (12.31) and (12.32) in a specific instance.

Example

A circular cylinder $|z| = a$ is in a uniform stream U. It is known that $w = U\left(z + \frac{a^2}{z}\right)$ so that $\left(\frac{dw}{dz}\right)^2 = U^2\left(1 - \frac{2a^2}{z^2} + \frac{a^4}{z^4}\right)$. Hence by (12.30) with $z_0 = 0$ we find from (12.31) that $X = Y = 0$.

Also $z\left(\frac{dw}{dz}\right)^2 = U^2\left(z - \frac{2a^2}{z} + \frac{a^4}{z^3}\right)$. Hence the only term to contribute to (12.32) is $\frac{-2a^2U^2}{z}$ which gives $-4\pi a^2U^2\mathrm{i}$ and this has zero real part. Hence $M = 0$.

These results imply that a cylinder placed in a uniform flow is subject to no net force: this is not borne out in practice. The discrepancy arises from the neglect of viscosity.

Cauchy's Theorem

If $f(z)$ is analytic in a simply-connected bounded region then $\oint_C f(z)\,dz = 0$ for every simple closed path C lying in the region.

[By a simply-connected region we mean that any closed curve in the region may be shrunk to a point without any part of it leaving the region; for example, the interior of a square or of a circle.]

We shall follow Cauchy's proof which made a further condition on $f(z)$: that $f'(z)$ was continuous in this region. Goursat later showed this condition to be unnecessary.

Now $\oint_C f(z)\,dz = \oint_C (u\,dx - v\,dy) + \mathrm{i}\oint_C (v\,dx + u\,dy)$. If $f'(z)$ is continuous then $\frac{\partial u}{\partial x}, \frac{\partial u}{\partial y}, \frac{\partial v}{\partial x}$ and $\frac{\partial v}{\partial y}$ exist and are continuous so that we may apply Green's theorem in the plane. Then if R be the region bounded by C

$$\oint_C f(z)\,dz = \iint_R \left(-\frac{\partial v}{\partial x} - \frac{\partial u}{\partial y}\right) dx\,dy + \mathrm{i}\iint_R \left(\frac{\partial u}{\partial x} - \frac{\partial v}{\partial y}\right) dx\,dy$$

Since $f(z)$ is analytic then the Cauchy-Riemann equations are satisfied and both double integrals vanish automatically. Hence the result follows.

Cauchy's Theorem is often claimed to be the most important one in complex variable theory partly because of its consequences, which we shall examine shortly.

Note that the result $\oint_C z^2 = 0$ where C is the unit circle follows immediately because of the analyticity of z^2. The results of Example 3 for $n \neq 1$ cannot follow from Cauchy's Theorem since the integrand is not analytic at one point, viz. $z = z_0$, of the region contained by C. (Note that $f(z)$ has to be analytic on C as well as inside it.)

Note finally that $\oint_C \frac{1}{z^2}\, dz$, where C is the circle $|z| = 2$ and $\frac{1}{z^2}$ is defined on the region $1 < |z| < 2$ to avoid non-analyticity at $z = 0$, cannot be assumed zero since the region is not simply connected. Check this for yourself.

By analogy with line integrals and reference to Figure 12.23 in which $f(z)$ is assumed analytic inside and on the contour shown

$$\int_{AEB} f(z)\, dz = \int_{ADB} f(z)\, dz$$

In other words we may choose any path between A and B and

$$\int f(z)\, dz$$

will have the same value and hence depends *only* on the points A and B.

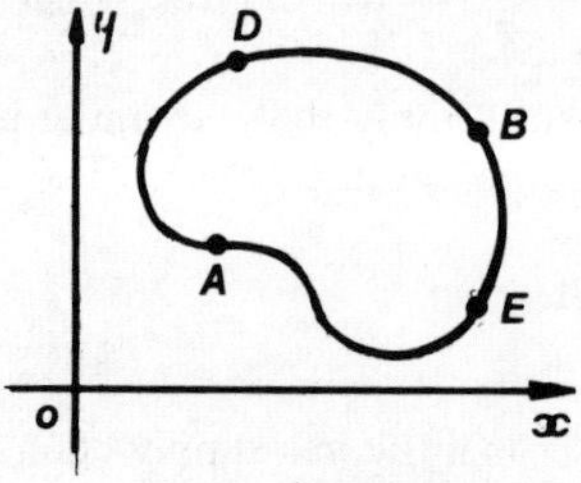

Figure 12.23

This explains the 'coincidence' in Example 1 (i). It allows us to use the real variable integration technique.

Example

$$\int_{i}^{1+2i} \cos z\, dz = [\sin z]_{i}^{1+2i} = \sin(1 + 2i) - \sin i \quad \text{(since } \cos z \text{ is analytic)}$$

Consequences of Cauchy's Theorem

Suppose that $f(z)$ is analytic in a region bounded by the closed curves C_1 and C_2 shown in Figure 12.24. Further suppose that we cut this region by a line joining A and B. Then we claim that

$$\oint_{C_1} f(z)\, dz = \oint_{C_2} f(z)\, dz$$

where both curves are travelled round in the anti-clockwise direction.

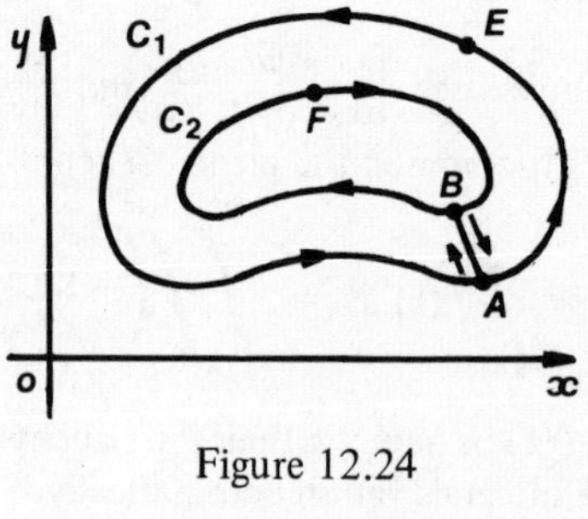

Figure 12.24

To prove this consider the closed curve AEABFBA travelled around in the direction indicated by the arrows. *Now the region between C_1 and C_2 has been cut so that no line can cross the cut and be regarded as remaining inside the region.* Hence the cut region is simply-connected and Cauchy's Theorem applies. Hence

$$\oint_{AEABFBA} f(z)\,dz = 0$$

Now $\displaystyle\int_{AB} f(z)\,dz = -\int_{BA} f(z)\,dz$

Then $\displaystyle\oint_{AEABFBA} = \int_{AEA} + \int_{AB} + \int_{BFB} + \int_{BA} = \int_{AEA} + \int_{BFB} = 0$

Now $\displaystyle\int_{AEA}$ is $\displaystyle\oint_{C_1}$ and $\displaystyle\int_{BFB}$ is $-\displaystyle\oint_{C_2}$ as you can see. The result follows.

In other words to find $\displaystyle\oint_{C_1} f(z)\,dz$ we can replace C_1 by any curve C_2 so long as the region between them contains no singularities of $f(z)$. Quite often a circle is chosen for C_2. (We can also find $\displaystyle\oint_{C_2}$ by replacing C_2 by a suitable C_1.)

Example

Evaluate $\displaystyle\oint_C \frac{3}{z(z-3)}\,dz$ where

C is the curve $|z-3| = 5$; see Figure 12.25.

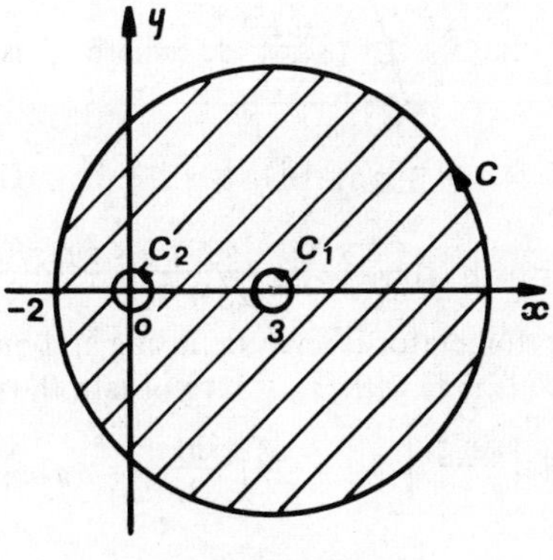

Figure 12.25

Note that $f(z) = \dfrac{3}{z(z-3)}$ is not analytic at $z = 0$ and $z = 3$ but is analytic everywhere else. If we take C_1 as the circle of unit radius centred at $z = 3$ and C_2 as the other unit circle then in the shaded region $f(z)$ is analytic. By an extension of the earlier result, we deduce that

$$\oint_C = \oint_{C_1} + \oint_{C_2}$$

You should prove this by making two suitable cuts in the shaded region. Then

$$\oint_C \frac{3}{z(z-3)}\,dz = \oint_{C_1} \frac{3}{z(z-3)}\,dz + \oint_{C_2} \frac{3}{z(z-3)}\,dz$$

$$= \oint_{C_1} \frac{1}{(z-3)}\,dz - \oint_{C_1} \frac{1}{z}\,dz + \oint_{C_2} \frac{1}{(z-3)}\,dz$$

$$- \oint_{C_2} \frac{1}{z}\,dz = I_1 - I_2 + I_3 - I_4$$

Referring back to (12.30) we can see that $I_1 = 2\pi i$. Similarly from (12.29), $I_4 = 2\pi i$. Since $1/z$ is analytic inside and on C_1, $I_2 = 0$. Since $1/(z-3)$ is analytic inside and on C_2, $I_3 = 0$. Hence $\oint_C \frac{3dz}{z(z-3)} = 2\pi i - 0 + 0 - 2\pi i = 0.$

Cauchy's Integral Formula

If $f(z)$ is analytic within and on the boundary C of a simply-connected region D and if z_0 is any point inside C then

$$\oint_C \frac{f(z)}{(z-z_0)}\,dz = 2\pi i f(z_0) \qquad (12.33)$$

This formula is very useful in saving us work in evaluating some integrals.

Example

Find $\oint_C \frac{\sin z}{z^2+1}\,dz$ where C is the path

(i) $|z - i| = ½$, (ii) $|z + i| = ½$, (iii) $|z| = 2$. Refer to Figure 12.26.

For path (i) we write $\frac{\sin z}{z^2+1} = \frac{\sin z/(z+i)}{z-i}$. Since the numerator is analytic inside and on the path we use (12.33) with $z_0 = i$ to obtain the result

$$2\pi i \left[\frac{\sin z}{z+i}\right]_{z=i} = 2\pi i \frac{\sin i}{2i} = i\pi \sinh 1$$

Likewise, for path (ii) $\frac{\sin z/(z-i)}{z+i}$ is the arrangement which leads us to the result $2\pi i \sin(-i)/(-2i) = i\pi \sinh 1$

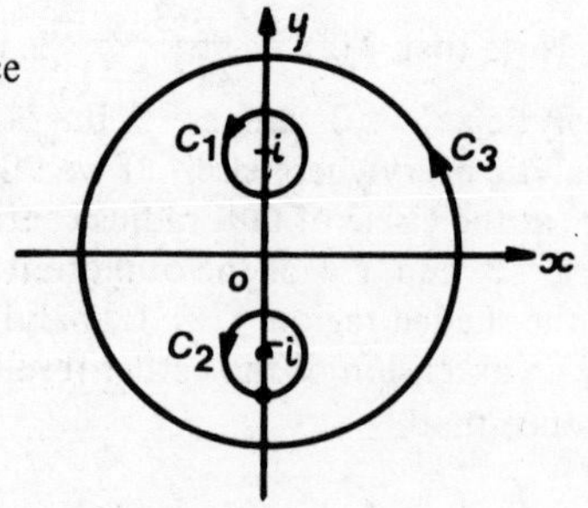

Figure 12.26

For path (iii) we use

$$\oint_{C_3} = \oint_{C_1} + \oint_{C_2} \qquad \text{to obtain the answer } 2\pi\mathrm{i}\sinh 1$$

Derivatives of an analytic function

If $f(z)$ is analytic in a simply-connected domain D, then at any interior point z_0 the derivatives of $f(z)$ of all orders exist and are analytic. (Note now the power of the definition of analyticity.) The derivatives at z_0 are given by

$$f^{(n)}(z_0) = \frac{n!}{2\pi\mathrm{i}} \oint_C \frac{f(z)}{(z - z_0)^{n+1}}\,\mathrm{d}z \tag{12.34}$$

where C is any simple closed curve which encloses z_0 and lies within D.

Problems

1. Evaluate the following integrals along the contours shown

(a) $\displaystyle\int_{(0,1)}^{(2,5)} \{(3x + y)\,\mathrm{d}x + (2y - x)\,\mathrm{d}y\}$ along the straight line joining end points.

(b) $\displaystyle\oint_C |z|^2\,\mathrm{d}z$ around the square with vertices $(0,0)$, $(1,0)$, $(1,1)$, $(0,1)$.

(c) $\displaystyle\int_C (z^2 + 3z)\,\mathrm{d}z$ along the circle $|z| = 2$ from $(2,0)$ to $(0,2)$ anti-clockwise.

(d) $\displaystyle\oint_C \frac{\mathrm{d}z}{z-2}$ (i) around $|z - 2| = 4$ (ii) around $|z - 1| = 5$ (iii) around the square with vertices $\pm 2 \pm 2\mathrm{i}$.

2. Verify Cauchy's theorem for the functions below with C being the square with vertices $\pm 1 \pm \mathrm{i}$

(a) $3z^2 - \mathrm{i}z - 4$ (b) $10 \sin 2z$

3. Evaluate $\displaystyle\int_C f(z)\,\mathrm{d}z$ for functions $f(z)$ and paths C specified below.

(a) $f(z) = z^3 - 1$; $|z - 1| = 1$ (b) $f(z) = z/(z^2 - 1)$; $|z - \pi| = 1$

(c) $f(z) = z^2 + z + 1$; from $z = -\mathrm{i}$ to $z = 2\mathrm{i}$ (d) $f(z) = 1/(z^2 - 2z)$; $|z| = 1$

(e) $f(z) = 2\mathrm{i}/(z^2 + 1)$; $|z - 1| = 6$ (f) $\sin z/(2z - \pi)$; $|z| = 1$

(g) $\sin z/(2z - \pi)$; $|z| = 2$

4. Repeat Problem 3 for the following

(a) $f(z) = e^z/(z-2)$; $|z| = 3$ (b) $f(z) = \sin 3z/(z + \frac{1}{2}\pi)$; $|z| = 5$

(c) $f(z) = e^{3z}/(z - \pi i)$; $|z-1| = 4$ (d) $\sin^6 z/(z - \pi/6)$; $|z| = 1$

(e) $f(z) = 1/\left\{(z+i)z^4\right\}$; $|z - i| = 1.5$ (f) $f(z) = 3/\left\{z^2(z+i)^2\right\}$; $|z| = 5$

(g) $f(z) = 1/(z^4 - 1)$; $|z-1| = 5$

5. Using (12.34) evaluate $\displaystyle\oint_C \frac{e^{2z}}{(z+1)^4}\,dz$ where C is $|z| = 3$

Hint: take $z_0 = -1$ and $n = 3$.

6. (i) Evaluate $\displaystyle\oint_C \frac{z+4}{z^2+2z+5}\,dz$ (a) if C is the circle $|z| = 1$

(b) if C is the circle $|z + 1 - j| = 2$

(ii) By putting $z = e^{j\theta}$ and integrating around the unit circle, show that

$$\int_0^{2\pi} \frac{d\theta}{25 - 16\cos^2\theta} = \frac{2\pi}{15}$$ (C.E.I.)

7. State Cauchy's integral formula for an analytic function and use it to integrate the function $(z^2+1)/(z^2-1)$ along a circle of unit radius with centre at (a) $z = 1$, (b) $z = -1$.

Use the formula to show that if an analytic function takes known values $u + iv$ on the circle $|z| = R$, its value at an interior point z_0 is

$$u_0 + iv_0 = \frac{1}{2\pi}\int_0^{2\pi} \frac{(u+iv)\,z\,d\theta}{z - z_0}, \quad z = R\,e^{i\theta}$$

Prove also that, if z_0 is on the real axis at $z = r$,

$$u_0 = \frac{R}{2\pi}\int_0^{2\pi} \frac{Ru - ru\cos\theta + rv\sin\theta}{R^2 + r^2 - 2Rr\cos\theta}\,d\theta$$ (L.U.)

8. Evaluate $\displaystyle\int_C \frac{(z+2)\,dz}{4z^2+4z+5}$ where C is the circle $|z + \frac{1}{2} - \frac{1}{2}i| = 1$. (L.U.)

9. State Cauchy's Integral formula for the nth derivative. Hence evaluate

(a) $\displaystyle\int_C \frac{dz}{(z-1)^3(z-3)(z-4)}$, where C is the circle $|z-1| = 1$.

(b) $\displaystyle\int_C \frac{\cos^4 z}{z - \pi/6}\,dz$, where C is the circle $|z| = 1$. (C.E.I.)

12.7 TAYLOR AND LAURENT SERIES

Many of the results with which we are familiar from real series extend into complex variables. The idea of **radius of convergence** is extended to a **circle of convergence** $|z - z_0| = \rho$ such that if $|z - z_0| < \rho$ the series under consideration will converge, whereas if $|z - z_0| > \rho$ the series will diverge.

Taylor's Theorem

Let $f(z)$ be analytic at a point z_0. Then there is a power series $\sum_{n=0}^{\infty} a_n (z - z_0)^n$ where the coefficients are given by

$$a_n = f^{(n)}(z_0)/n! \tag{12.35}$$

and which converges to $f(z_1)$ for every z_1 in every neighbourhood of z_0 on which $f(z)$ is analytic so that

$$f(z_1) = \sum_{n=0}^{\infty} \frac{f^{(n)}(z_0)}{n!} (z_1 - z_0)^n$$

Many of the series can be obtained in ways analogous to those for real series and often we can in effect substitute z for x in series expansions we already know. Note that the concept of radius of convergence now means the distance between z_0 and the nearest singularity (if any). Remember that a singularity (or singular point) is a point at which $f(z)$ is not analytic.

Laurent Series

The trouble with Taylor's series is that the circle of convergence is most often merely a part of the region of analyticity of $f(z)$. For example, the series Σz^n converges to $f(z) = 1/(1 - z)$ only *inside* the circle $|z| = 1$ even though $f(z)$ is analytic everywhere *except* at $z = 1$. The **Laurent series** aims to represent $f(z)$ at as many points as possible.

In effect we can expand around a point of singularity right up to, but not including, the singularity itself. This was not possible for Taylor's series.

In Figure 12.27 we show an **annulus of convergence** $r_1 < |z - z_0| < r_2$ within which the series, which is an extension of Taylor's series, converges. The extension is to allow negative powers of $(z - z_0)$ in addition to the positive powers of Taylor's series.

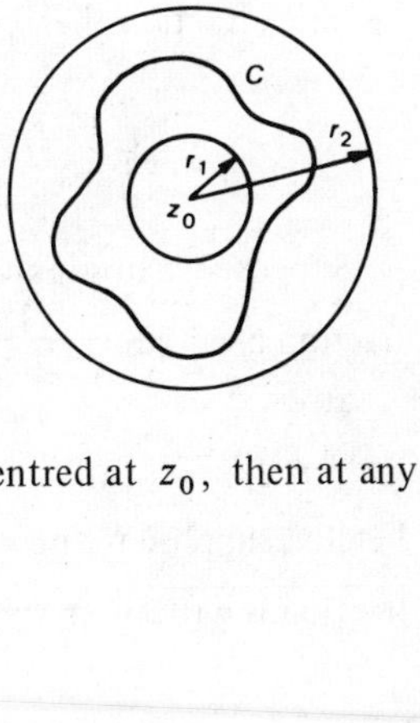

Figure 12.27

Laurent's Theorem

If $f(z)$ is analytic throughout a closed annular region D centred at z_0, then at any point z within the annulus

$$\begin{aligned} f(z) &= a_0 + a_1(z - z_0) + a_2(z - z_0)^2 + \ldots\ldots \\ &\quad + a_{-1}(z - z_0)^{-1} + a_{-2}(z - z_0)^{-2} + \ldots\ldots \end{aligned}$$

where the coefficients are given by

$$a_n = \frac{1}{2\pi i} \oint_C \frac{f(z)}{(z - z_0)^{n+1}}\, dz \qquad n = 0, \pm1, \pm2 \ldots.. \qquad (12.36)$$

each integral being taken round any simple closed path lying in the annulus and encircling the inner boundary; see Figure 12.27.

Example

The function $f(z) = \dfrac{2}{z(z - i)}$ has three distinct series centred at $z_0 = -i$; note the singularities are at $z = 0$ and $z = i$.

(i) A Taylor series converging in $0 < |z + i| < 1$

(ii) A Laurent series converging in $1 < |z + i| < 2$

(iii) A Laurent series converging in $2 < |z + i|$

Sketch the three regions of convergence and mark the positions of the singularities.

Example

Find the Laurent series for $f(z) = 1/\{(z - 1)(z + 1)(z - 2)\}$ about the point $z = 1$ in $1 < |z - 1| < 2$. Now $f(z)$ has singularities at $z = 1$, -1 and 2. Therefore, in the annulus $1 < |z - 1| < 2, f(z)$ is analytic. Make a suitable sketch to verify this.

In partial fractions $f(z) = \dfrac{-1/2}{z - 1} + \dfrac{1/6}{z + 1} + \dfrac{1/3}{z - 2}$. We require an expansion in powers of $(z - 1)$ which we obtain via a binomial expansion, not via (12.36). We therefore 'force' $(z - 1)$ into each denominator thus

$$f(z) = \frac{-1/2}{z - 1} + \frac{1/6}{(z - 1) + 2} + \frac{1/3}{(z - 1) - 1}$$

$$= \frac{-1/2}{(z - 1)} + \frac{1/6}{2(1 + [z - 1]/2)} + \frac{1/3}{(z - 1)[1 - 1/(z - 1)]}$$

By the Binomial expansion, noting that the second fraction requires $\dfrac{|z - 1|}{2} < 1$ and the third fraction that $\dfrac{1}{|z - 1|} < 1$, i.e. $1 < |z - 1| < 2$,

$$f(z) = \frac{1}{12} - \frac{(z - 1)}{24} + \frac{(z - 1)^2}{48} - \ldots - \frac{1/6}{(z - 1)} + \frac{1/3}{(z - 1)^2} + \ldots \qquad (12.37)$$

Suppose we now expand in the annulus $0 < |z - 1| < 1$, then the second partial fraction becomes $\dfrac{1/6}{(z - 1) + 2}$ and the third $\dfrac{-1/3}{1 - (z - 1)}$ and

$$f(z) = -\frac{1}{4} - \frac{3}{8}(z - 1) - \frac{5}{16}(z - 1)^2 - \ldots + \frac{-1/2}{(z - 1)} \qquad (12.38)$$

Finally, suppose we now expand in the annulus $|z - 1| > 2$ so that the second partial fraction is written $\dfrac{1/6}{(z - 1)[1 + 2/(z - 1)]}$ and the third $\dfrac{1/3}{(z - 1)[1 - 1/(z - 1)]}$ then

$$f(z) = \frac{1}{(z-1)^3} - \frac{1}{(z-1)^4} + \frac{3}{(z-1)^5} - \ldots.. \qquad (12.39)$$

Classification of singularities

In the Laurent expansion (12.36) in the annulus $0 < |z - z_0| < \rho$, if the expansion in negative powers of $(z - z_0)$ – the **principal part** of the series – contains an infinite number of terms, then $f(z)$ is said to possess an **essential singularity** at z_0. If the principal part of the series stops short at $a_{-m}(z - z_0)^{-m}$ then $f(z)$ is said to have a **pole of order** m at z_0. A pole of order 1 is called a **simple pole**. Thus the function of the last example has a simple pole at $z = 1$ (as it has at $z = -1$ and at $z = 2$).

Note that $\dfrac{1}{(z^2+1)z^2} = \dfrac{1}{(z-\mathrm{i})(z+\mathrm{i})z^2}$ has simple poles at $z = \mathrm{i}$ and $z = -\mathrm{i}$ and a pole of order two at $z = 0$. Also, $\exp\left\{1/(z-2)\right\}$ has an essential singularity at $z = 2$.

Consider $f(z) = \dfrac{\sin z}{z} = 1 - \dfrac{z^2}{3!} + \dfrac{z^4}{5!} - \ldots.$

There are no negative powers of z in the expansion, but the function is not defined at $z = 0$. However, since $\lim\limits_{z \to 0} f(z) = 1$ we *may* define $f(0) = 1$ and overcome any trouble. We say that $z = 0$ is a **removable singularity**.

Problems

1. Find the Taylor series for each of the following functions with centre of expension c.

   (a) e^z, $c = \mathrm{i}$ (b) $1/(z + 2\mathrm{i})$, $c = 0$ (c) $1/(z^2 - 1)$, $c = 0$

2. Find the series expansion of each of the functions in the given region

   (a) $1/\left\{(z+2)(z-1)\right\}$; $1 < |z - 2| < 4$ (b) as (a) in $4 < |z - 2|$

   (c) $\cos(z-1)/z^3$; $0 < |z| < \infty$ (d) $\sinh z/z^2$; $0 < |z| < \infty$

3. Find all the expansions centred at the origin for $f(z) = 1/\left\{z^2(z-1)(z-2)\right\}$.

4. Expand $\dfrac{1}{z} + \dfrac{1}{z-1} + \dfrac{1}{z-\mathrm{i}}$ in $0 < |z| < 1$.

5. Expand

   (a) $1/z(z-1)$ (b) $(z-1)/\left[z^3(z-2)\right]$ in all Laurent series possible.

6. Expand $1/\left\{(z+1)(z+3)\right\}$ in Laurent series valid for

   (a) $1 < |z| < 3$ (b) $|z| > 3$ (c) $0 < |z + 1| < 2$ (d) $|z| < 1$

7. Classify the singularities of the following functions

   (a) $(z^2 + 1)/z$ (b) $\cos(1/z)$ (c) $1/\left\{(z-1)(z-2)^2\right\}$

   (d) $(\cos z - 1)/z^2$ (e) $(z^2 - 3z + 2)/(z - 2)$ (f) $(e^{2z} - 1)/z^4$

   (g) $(z^2 - 1)/(z^2 + 1)$ (h) $\tan^2 z$

12.8 THE RESIDUE THEOREM

Let $f(z)$ be a function which is analytic inside and on a contour C except for a pole of order m at $z = z_0$ which lies within C. Then expanding $f(z)$ as a Laurent series in powers of $(z - z_0)$, we evaluate $\oint_C f(z)\,dz$ by equating it to $\oint_\Gamma f(z)\,dz$ where Γ is a circle centre z_0, of radius a lying within C. Then it can be shown that the positive powers of the expansion, being analytic, integrate to zero, all the negative powers of the expansion integrate to zero by (12.30) except in the case of $a_{-1}/(z - z_0)$ which gives the result $2\pi i a_{-1}$, again by (12.30). Since a_{-1} is the only coefficient remaining it is called the **residue** of $f(z)$ at z_0. Hence

$$\oint_C f(z)\,dz = 2\pi i a_{-1} \tag{12.40}$$

Finding the residue

Under the conditions above the evaluation of the integral reduces to finding a_{-1}. Obviously we *could* perform a Laurent expansion; we look for simpler techniques. Suppose $f(z)$ has a simple pole at z_0 then

$$f(z) = \frac{a_{-1}}{(z - z_0)} + a_0 + a_1(z - z_0) + a_2(z - z_0)^2 + \ldots..$$

therefore

$$(z - z_0)f(z) = a_{-1} + a_0(z - z_0) + a_1(z - z_0)^2 + a_2(z - z_0)^3 + \ldots..$$

Taking limits as $z \to z_0$

$$\lim_{z \to z_0} \left\{ (z - z_0)f(z) \right\} = a_{-1} \tag{12.41}$$

In general, for a pole of order m at z_0,

$$a_{-1} = \frac{1}{(m-1)!} \lim_{z \to z_0} \left\{ \frac{d^{m-1}}{dz^{m-1}} [(z - z_0)^m f(z)] \right\} \tag{12.42}$$

Example 1

Let $f(z) = \dfrac{z - 3}{(z + 2)^3(z + 4)}$ and we find the residue at $z = -4$. Then $(z + 4)f(z) = \dfrac{(z - 3)}{(z + 2)^3}$ and, taking the limit as $z \to -4$ we find that $a_{-1} = (-4 - 3)/(-4 + 2)^3 = 7/8$.

Example 2

Let $f(z) = (z^2 + 1)/(z + 1)^3$. Then $f(z)$ has a pole of order 3 at $z = -1$. Hence the residue at $z = -1$ is

$$\frac{1}{2!} \lim_{z \to -1} \left\{ \frac{d^2}{dz^2} [(z + 1)^3 f(z)] \right\} = \tfrac{1}{2} \lim_{z \to -1} \left\{ \frac{d^2}{dz^2} (z^2 + 1) \right\} = 1$$

Suppose we seek $\oint_C f(z)\,dz$ where C is $0 < |z+1| < 2$. Then the result is $2\pi i \cdot 1$.

Use of L'Hôpital's Rule

Suppose we wish to find the residue of $f(z) = \dfrac{1}{(z^4+1)(z^2+4)}$ at $z = \dfrac{1+i}{\sqrt{2}}$

We require $\displaystyle\lim_{z \to \frac{1+i}{\sqrt{2}}} \frac{(z-(1+i)/\sqrt{2})}{(z^4+1)(z^2+4)}$

However it is difficult to perform the cancellation needed.

If we apply L'Hôpital's rule, the required limit is equal to

$$\lim_{z \to \frac{1+i}{\sqrt{2}}} \left\{ \frac{1}{4z^3(z^2+4)+(z^4+1)2z} \right\} = \frac{1}{4\left[\frac{1+i}{\sqrt{2}}\right]^3 (i+4)} = \frac{-(5+3i)}{68\sqrt{2}}$$

Extension to the Theorem

We seek $\oint_C f(z)\,dz$ where the only singularities of $f(z)$ on and inside C are poles at $z = z_1$, $z = z_2$, ..., $z = z_r$. Then the required answer is $2\pi i$ x (sum of the residues at the poles inside C).

Example

Find $\oint_C f(z)\,dz$ where C is the unit circle and

$f(z) = (z^2+1)/\left\{(z-2)(2z+1)^2(2z-1)\right\}$.

The poles inside C are at $z = \frac{1}{2}$ and $z = -\frac{1}{2}$. At $z = \frac{1}{2}$, a simple pole, the residue is

$$\lim_{z \to \frac{1}{2}} \left\{ \frac{(z^2+1)}{2(z-2)(2z+1)^2} \right\} = \frac{5/4}{2(-3/2)\,.\,4} = -\frac{5}{48}$$

At $z = -\frac{1}{2}$, a pole of order 2, the residue is

$$\lim_{z \to -\frac{1}{2}} \left\{ \frac{(z-2)(2z-1)\,.\,2z-(z^2+1)\,.\,(4z-5)}{4\ (z-2)^2(2z-1)^2} \right\} = \frac{3}{80}$$

Hence the value of the integral is $2\pi i\left(\frac{-5}{48} + \frac{3}{80}\right) = -2\pi i/15$.

Expansion method for finding residues

Consider $f(z) = (z^8+1)/\left\{z^3(z^2+2)(2z^2+1)\right\}$. This has a pole of order 3 at $z = 0$. To find the residue there it is easiest to expand $f(z)$ in powers of z, i.e.

$$f(z) = \frac{z^8+1}{2z^3(1+\tfrac{1}{2}z^2)(1+2z^2)} = \frac{1}{2z^3}(1+z^8)(1-\tfrac{1}{2}z^2+....)(1-2z^2+....)$$

The coefficient of z^2 in the last three factors is $-5/2$ and so the residue at $z=0$ is $-5/4$.

Problems

1. Find the residue of each of the following functions at each of its singularities.

   (a) $(e^{2z}-1)/z$ (b) $(z^2+1)/(z-1)$ (c) $\sinh z/z^2$

   (d) $(z^2-1)/\left\{(z-2)(z+1)(z-\pi)\right\}$ (e) $\tan z/z^3$ (f) $ze^z/(z^4-z^2)$

   (g) $(z-2)/\left\{z(z-1)\right\}$ (h) $(z-2)/\left\{z^2(z-1)^2\right\}$ (i) $(1-\cos 2z)/z^3$

   (j) $(z+2)/(z^2+9)$ (k) $z^2/(z^4+16)$ (l) $e^{iz}/(z^2+4)$

2. Find the integral of the following functions round the contour $|z|=1$

   (a) $z/(2z-1)$ (b) $(2z^3+4)/(4z-\pi)$ (c) $(z^2+1)/(z^2-2z)$ (d) $\operatorname{cosec} z$

   (e) $(1-e^z)^{-1}$

3. Find the integral of the following functions round the given contour

   (a) $(z^2+1)\,e^z/\left\{(z+i)(z-1)^3\right\}$, $|z|=0.5$ (b) $e^{1/z}$, $|z|=6$ (c) $z/(z^4-1)$, $|z|=4$

12.9 EVALUATION OF REAL INTEGRALS

Several real integrals can be evaluated using contour integration, when other methods fail. We merely list a few common types of integrals.

Type 1

$$\int_0^{2\pi} f(\cos\theta,\ \sin\theta)\,d\theta$$

We let $e^{i\theta}=z$ so that $d\theta = \frac{1}{iz}\,dz$, $\cos\theta = \frac{1}{2}(z+\frac{1}{z})$, $\sin\theta = \frac{1}{2i}(z-\frac{1}{z})$. Then

$$\int_0^{2\pi} f(\cos\theta,\ \sin\theta)\,d\theta \equiv \oint_C g(z)\,dz,$$

where C is the contour $|z|=1$ and $g(z)$ is the expression obtained by substituting for $d\theta$, $\cos\theta$ and $\sin\theta$ in terms of z. The latter integral is $2\pi i$ times the sum of the residues of $g(z)$ at its poles *inside* $|z|=1$.

Example

$$I = \int_0^{2\pi} \frac{d\theta}{3+2\cos\theta}$$

Substituting for $d\theta$ and $\cos\theta$ and expressing the limits of integration in terms of z we obtain

$$I = \frac{1}{i} \oint \frac{dz}{z^2 + 3z + 1}$$

The integrand has simple poles at $z = \frac{-3}{2} \pm \sqrt{\frac{9}{4} - 1}$ i.e. at $z = ½(-3 + \sqrt{5})$ and $z = ½(-3 - \sqrt{5})$. Only the first lies inside the contour $|z| = 1$. The residue there is found to be $\frac{1}{\sqrt{5}}$ and by the general result

$$I = \frac{1}{i} \times 2\pi i \times \frac{1}{\sqrt{5}} = \frac{2\pi}{\sqrt{5}}$$

Type 2

$$\int_{-\infty}^{\infty} f(x)\,dx$$

where $f(x)$ is a real rational function whose denominator is of degree at least 2 greater than the numerator and where the denominator of $f(x)$ is non-zero for all real x. We write

$$\int_{-\infty}^{\infty} f(x)\,dx = \lim_{R\to\infty} \int_{-R}^{+R} f(x)\,dx$$

and convert this latter integral into a *contour* integral as shown in Figure 12.28.

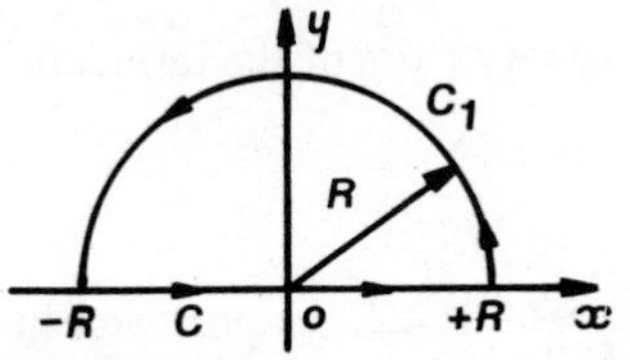

Figure 12.28

The contour C is from $-R$ to $+R$ along the real axis and along the semi-circle C_1. For large enough R the contour will enclose *all* the poles of $f(z)$ in the upper half-plane. Hence $\oint_C f(z)\,dz = 2\pi i$ (Σ residues in upper half-plane.) But

$$\oint_C f(z)\,dz = \int_{-R}^{+R} f(x)\,dx + \int_{C_1} f(z)\,dz$$

and it is possible to show that for the functions satisfying the conditions stated above the integral over the semi-circle tends to zero as the radius $R \to \infty$.

We have in the limit as $R \to \infty$:

$$\int_{-\infty}^{\infty} f(x)\,dx = 2\pi i \text{ (sum of residues of } f(z) \text{ in the upper half-plane)}$$

Example

Evaluate $\displaystyle\int_{-\infty}^{\infty} \frac{dx}{(1+x^2)^3}$. First note that the integrand satisfies the required conditions.

Now $\dfrac{1}{(1+z^2)^3} = \dfrac{1}{(z+i)^3(z-i)^3}$ and has *third* order poles at $z = +i$ and $z = -i$. The pole at $z = +i$ is in the upper half plane so we evaluate the residue there. By the formula for a residue at a third order pole we have: Residue at $z = +i$ is

$$\frac{1}{2!}\left[\frac{d^2}{dz^2}\left(\frac{1}{(z+i)^3}\right)\right]_{z=i}$$

This gives $\dfrac{3}{16i}$ so that the required integral is $2\pi i \cdot \dfrac{3}{16i} = \dfrac{3\pi}{8}$

Type 3

Real integrals of the form

$$\int_{-\infty}^{\infty} f(x)\cos kx\, dx \qquad \text{and} \qquad \int_{-\infty}^{\infty} f(x)\sin kx\, dx$$

where $f(x)$ satisfies conditions similar to those of the $f(x)$ in integrals of Type 2.

We consider the contour integral $\displaystyle\oint_C f(z)\, e^{ikz}\, dz$ over the same contour as in Type 2 integrals. We take real and imaginary parts of the results to obtain the required integrals.

Type 4 A pole on the real axis.

Since the contour of Type 2 would pass through the pole we bend it to by-pass the pole. The contour of Type 2 must be modified by an indentation, formed by a small semi-circle around the pole, which is contracted to the pole. Since we only go 'half-way round the pole' we pick up only πi times the residue there.

Example

$$I = \int_{-\infty}^{\infty} \frac{(1-\cos 2x)}{x^2}\, dx$$

We integrate $f(z) = (1 - e^{2iz})/z^2$ around the contour indicated in Figure 12.29 since there is a double pole at $z = 0$. We let $R \to \infty$ and $r \to 0$.

First note that $\displaystyle\oint_C \to 0$ since it

encloses no poles. We may more easily obtain the result that

$\int_\Gamma \to 0$ by appealing to

Jordan's lemma which states that if $\phi(z)$ is a rational function, i.e. a ratio of two polynomials and if $\psi(z)$ is of the form $e^{imz}g(z)$ where $m > 0$ and $g(z)$ is a rational function then

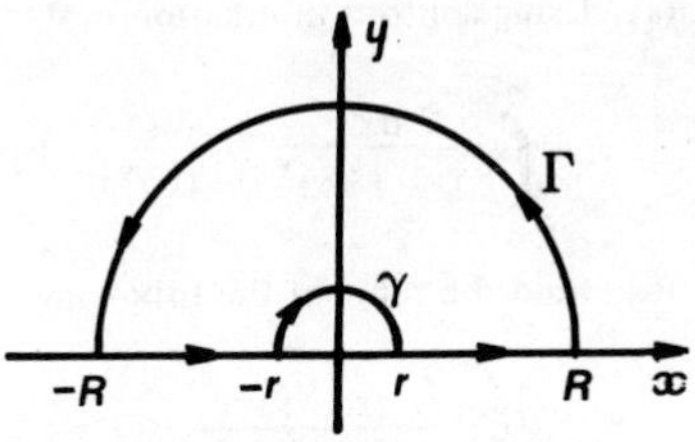

Figure 12.29

(i) if $|z\,\phi(z)| \to 0$ as $R \to \infty$ then $\displaystyle\int_\Gamma \phi(z)\,dz \to 0$

(ii) if $|g(z)| \to 0$ as $R \to \infty$ then $\displaystyle\int_\Gamma \psi(z)\,dz \to 0$

Now $\displaystyle\int_\gamma \to \pi i$ (residue of $f(z)$ at $z = 0$) $= \pi i \times (-2i) = 2\pi$ and $\displaystyle\int_\gamma$ is taken in the clockwise direction. Also $\displaystyle\int_{-R}^{-r} + \int_r^R \to I$. Hence in the limit,

$$0 = \oint_C = \int_{-\infty}^{\infty} + 0 - 2\pi. \quad \text{Therefore } I = 2\pi.$$

Other types of integral and contour will not be discussed here. You are referred to the bibliography for further details.

Problems

1. By contour integration, show that

$$\int_{-\infty}^{\infty} \frac{dx}{(x^2+1)(x^2+4)^2} = \frac{5\pi}{144} \qquad \text{(C.E.I.)}$$

2. Using suitable contours and Cauchy's Residue theorem, evaluate

$$\int_0^{2\pi} \frac{d\theta}{5 - 3\sin\theta} \quad \text{and} \quad \int_{-\infty}^{\infty} \frac{x^2\,dx}{(x^2+1)^2} \qquad \text{(C.E.I.)}$$

3. By using suitable contours show that

$$\text{(i)} \int_0^{2\pi} \frac{d\theta}{5 + 4\cos\theta} = \frac{2\pi}{3} \qquad \text{(ii)} \int_0^{\infty} \frac{dx}{x^6+1} = \frac{\pi}{3} \qquad \text{(C.E.I.)}$$

4. (a) Using contour integration evaluate the following integral and show

$$\int_0^\infty \frac{dx}{(x^2 + a^2)^3} = \frac{3\pi}{16a^5} \qquad (a > 0)$$

(b) Find the value of the following contour integral

$$\oint_C \frac{(z-1)}{(z^2 - 4)(z+1)^4}\,dz$$

where C is (i) the circle $|z| = \frac{1}{2}$ (ii) the circle $|z-1| = \frac{3}{2}$

5. Use the complex integral calculus to show

(a) $\displaystyle\int_0^{2\pi} \frac{\sin^2\theta}{5 + 4\cos\theta}\,d\theta = \frac{\pi}{4}$ (b) $\displaystyle\int_0^\infty \frac{x^2\,dx}{x^4 + a^4} = \frac{\pi\sqrt{2}}{4a}, \quad a > 0$

(c) $\displaystyle\int_0^{2\pi} \frac{\cos 2\theta\,d\theta}{5 - 4\cos\theta} = \frac{\pi}{6}$ (d) $\displaystyle\int_0^\infty \frac{\cos x\,dx}{(1 + x^2)^2} = \frac{\pi}{2e}$ (L.U.)

6. (i) Evaluate $\displaystyle\int \frac{z^2\,dz}{(z-1)(z+2)^2}$, taken round the circle $|z| = 3$

(ii) By integrating round a suitable semi-circle, prove that $\displaystyle\int_0^\infty \frac{dx}{x^4 + 1} = \frac{\pi}{2\sqrt{2}}$

(iii) Show that the function $e^{iz}/(z^2 + \pi^2)$ has only one singularity, a simple pole, above the real axis and find the residue at this pole. By integrating the function round a suitable semi-circle prove that $\displaystyle\int_0^\infty \frac{\cos x}{x^2 + \pi^2}\,dx = \tfrac{1}{2} e^{-\pi}$. (L.U.)

7. By applying contour integration, show that, for $a > b > 0$

(i) $$\int_{-\infty}^\infty \frac{dx}{(a^2 + x^2)(b^2 + x^2)} = \frac{\pi}{ab(a+b)}$$

(ii) $$\int_{-\infty}^\infty \frac{\cos x\,dx}{(a^2 + x^2)(b^2 + x^2)} = \frac{\pi}{(a^2 - b^2)}\left\{\frac{1}{be^b} - \frac{1}{ae^a}\right\}$$ (L.U.)

8. (i) A contour C consists of the part of the real axis between $-R$ and $+R$ together with the semi-circle $|z| = R$, $0 \leqslant \arg z \leqslant \pi$. By integrating $e^{imz}/(1 + z^4)$, where $m > 0$, round C and letting $R \to \infty$, show that

$$\int_0^\infty \frac{\cos mx}{1 + x^4}\,dx = \frac{\pi}{2\sqrt{2}}\,e^{-k}(\cos k + \sin k), \text{ where } k = m/\sqrt{2}.$$

(ii) By integrating e^{iz}/z round the same contour as in (i) above indented at the origin show

that $\int_0^\infty \frac{\sin x}{x}\,dx = \frac{\pi}{2}$. (L.U.)

9. (i) By writing $2\cos\theta = z + \frac{1}{z}$ evaluate $\int_0^{2\pi} \frac{d\theta}{1 - 2x\cos\theta + x^2}$ in the two cases $x<1$ and $x>1$.

(ii) Evaluate $\int_{-\infty}^{\infty} \frac{\cos 2x}{x^2 + a^2}\,dx, (a > 0)$, by contour integration. (L.U.)

12.10 FURTHER APPLICATIONS OF CONTOUR INTEGRATION

In this section, we look at two applications of contour integration: inversion of Laplace Transforms and the Nyquist stability criterion for closed loop systems.

Inversion of Laplace Transforms

The Laplace transform of a function $f(t)$, given by

$$F(s) = \int_0^\infty f(t)\,e^{-st}\,dt \tag{12.43}$$

has an inverse, viz.

$$f(t) = \frac{1}{2\pi i}\int_{a-i\infty}^{a+i\infty} F(s)e^{st}\,ds \tag{12.44}$$

where the integral is taken along a line parallel to the imaginary axis; s being regarded as complex. The real number a is chosen so that all singularities of $F(s)$ lie to the left of the line of integration parallel to the imaginary axis.

In practice, the integral in (12.44) is evaluated by considering the contour integral $\oint_C F(s)\,.\,e^{st}\,ds$, where C is the **Bromwich contour** in the s - plane shown in Figure 12.30. Suppose that the only singularities of $F(s)$ are poles which lie to the left of the line $x = a$. We can apply the residue theorem to show that, if the integral around the circular arc tends to zero as $R \to \infty$, then

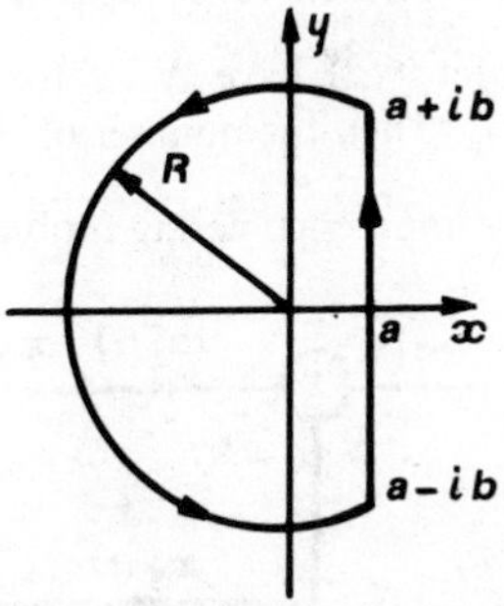

Figure 12.30

$f(t)$ = the sum of the residues of $e^{st}\,F(s)$ at the poles of $F(s)$ (12.45)

Example

Let $q(t)$ represent the charge in a circuit used for charging a capacitor. With the usual notation, the Laplace transform of the charge, viz. $Q(s)$ is given by

$Q(s) = V_0 / \left\{ Rs \left[\frac{1}{RC} + s \right] \right\}$ By our inversion formula (12.44) we have

$$q(t) = \frac{1}{2\pi i} \int_{a-i\infty}^{a+i\infty} \frac{V_0 e^{st}}{Rs(s + 1/RC)} ds$$

Here a can have any positive real value (why?). We need the residues at $s = 0$ and $s = -1/RC$. At $s = 0$ we can show that the residue is $V_0 C$. At $s = -1/RC$ we can show that the residue is $- V_0 C e^{-t/RC}$. Hence, by (12.45), $q(t) = V_0 C(1 - e^{-t/RC})$ as is well known.

Note that if $F(s)$ has branch points then the Bromwich contour must be suitably cut. This is beyond our scope.

Stability criteria

In the analysis of physical systems we often want to know merely whether a system is stable, i.e. a small transient imposed disturbance produces a response which also dies away. If the Laplace transform of the response is of the form $G(s) = P_1(s)/P_2(s)$ where P_1 and P_2 are polynomials, by using the residue theorem, we are enabled to state that:

If all the poles of $G(s)$ lie in the left-hand half-plane then the system is stable.

(We have tacitly assumed that the integral of the appropriate function over a semi-circle tends to zero as the radius of the semi-circle tends to infinity.)

Note that the poles of $G(s)$ are the zeros of $P_2(s)$ which may need to be found by numerical methods.

Nyquist stability criterion

First we quote a result from complex variable theory: Let $f(z)$ be analytic inside and on a closed curve C, apart from a finite number, N_p, of poles within C.

Further let $f(z)$ have N_z zeros inside C. Let C be mapped into a curve C' by $w = f(z)$. Then the number of times C' encircles the origin is equal to $N_z - N_p$.

A common engineering problem is to make the output $x_0(t)$ of a system follow

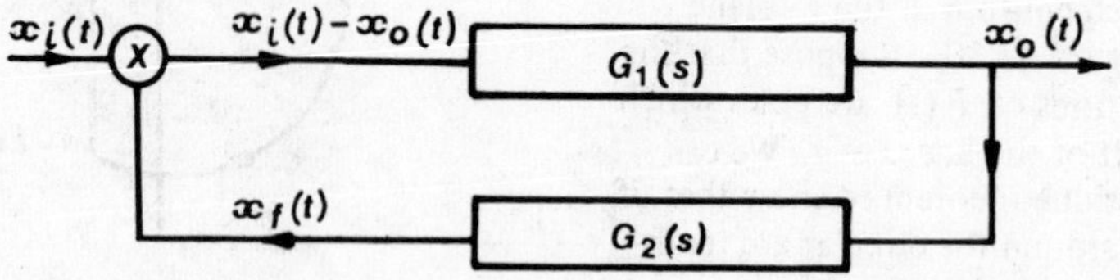

Figure 12.31

quickly and closely the input $x_i(t)$. A feedback loop, such as the one shown in Figure 12.31 samples the output, modifies it to $x_f(t)$ and feeds it back to a differential device which produces an error signal $x_i(t) - x_f(t)$. Remember that the **transfer function** for a system is the ratio of the Laplace transform of the output to the Laplace transform of the input. Let $G_1(s)$ be the transfer function of the original system and $G_2(s)$ be the transfer function of the feedback loop. Then

$$L\{x_0(t)\} = G_1(s)\,[L\{x_i(t)\} - L\{x_f(t)\}]$$

$$L\{x_f(t)\} = G_2(s)\,L\{x_0(t)\}$$

so that on elimination of $L\{x_f(t)\}$ we obtain

$$L\{x_0(t)\} = \frac{G_1(s)}{[1 + G_1(s)\,.\,G_2(s)]}\,L\{x_i(t)\}$$

The stability of the feedback system is important. If the system without the feedback loop is stable for $x_i(t)$ then the product $G_1(s)\,L\{x_i(t)\}$ has no poles in the right-hand half of the s - plane. The stability of the system as a whole depends on the location of the zeros of $[1 + G_1(s)\,.\,G_2(s)]$.

The **Nyquist** criterion, which we state without proof, is that if the locus of $w = G_1(\mathrm{i}\omega)\,.\,G_2(\mathrm{i}\omega)$ is plotted and does not encircle the point $w = -1$ then the system is stable, otherwise the system is unstable.

Problems

1. The Laplace inverse of the transformed function $F(s)$ is given by the contour integral

$$f(t) = \frac{1}{2\pi \mathrm{i}} \oint_C F(z)\, e^{zt}\, \mathrm{d}z$$

where C is a suitably chosen closed contour which encloses the poles of $F(z)$. Use this result to show that the inverses of $(3s + 2)/(s^2 + 4)$ and $1/[s\,(s + 1)^2]$ are $(3 \cos 2t + \sin 2t)$ and $1 - (t + 1)\,e^{-t}$ respectively. (C.E.I.)

2. Invert the following transforms using the method of residues

(a) $\dfrac{1}{(s + 1)(s - 2)^2}$ (b) $\dfrac{s}{(s + 1)^3\,(s - 1)^2}$ (c) $\dfrac{1}{(s^2 + 1)^2}$

3. If $f(z)$ is regular on and inside a simple closed contour C, show, by consideration of the integral $\displaystyle\oint_C \frac{f'(z)}{f(z)}\,\mathrm{d}z$, that the change in $\arg\{f(z)\}$ as the point z moves once round C is $2\pi N$ where N is the number of zeros of $f(z)$ inside C.

Show that the equation $z^4 + z + 1 = 0$ has one root inside each of the four quadrants of the z - plane. Show also that each of these roots lies outside the circle $|z| = 2/3$. (L.U.)

4. Show that the equation $2z^4 + 8z + 7 = 0$ has no real or purely imaginary roots. Show that there is only one root which lies in the first quadrant. (L.U.)

Chapter Thirteen

Statistical Tests of Significance

13.1 INTRODUCTION

We remind ourselves of the main ideas concerning statistical inference that we introduced in *Engineering Mathematics* . The central point is that any inference we draw will be subject to a risk of being wrong: were there sufficient information available to make a definite statement, either about the value of a population parameter or about the validity of a claim under test, then we should not need to resort to statistics.

For samples of size 30 or more, it is a reasonable approximation to say that sample means, sample proportions, differences between sample means and certain other variables are normally distributed. The Central Limit Theorem guarantees that, in the limit as the sample size tends to infinity, the variables mentioned above *are* normally distributed.

The problems we studied fell into two categories: those concerned with estimating a **population parameter** from the observed **sample statistic** (e.g. population mean μ from sample mean $\bar{x}$) and those concerned with testing the validity of a **hypothesis** by examining a suitable sample statistic (e.g. testing the hypothesis that a machine produces articles of a given average length by studying the mean of a sample). In the former case we quote a **confidence interval** for the parameter, an interval about the sample value in which we are reasonably sure that the value of the population parameter lies: the less the risk of being wrong that we are prepared to accept, the wider the interval quoted. In the latter case we state in advance a risk of wrong judgement and base a decision on whether the probability of the observed sample statistic exceeds a value determined by the risk. We have to accept that the confidence interval may not include the true value of the population parameter or that the decision we took may be wrong: in either case our results should be accompanied by a statement of the risk of being wrong. With regard to the test of a hypothesis the errors are categorised as follows. If we reject the hypothesis when it is, in fact, correct we commit a **Type I error**; conversely, if we accept the hypothesis when it is false we commit a **Type II error**. The probability of committing a Type I error is called the **level of significance** for the test employed, or the α **- risk**; the probability of committing a Type II error is called a β **- risk**.

In this chapter we examine some of these ideas in more detail and develop further tests of significance. It is often claimed that it is difficult to know which test should be employed in the solution of a given problem; we hope to make it clear to you the circumstances under which each test is applicable.

13.2 TESTS OF HYPOTHESES

To enable us to illustrate the basic concepts involved we consider an example in which the numbers involved have been so chosen that the arithmetic is particularly simple.

Wire cables are compounded from separate wires; the cables are produced with a mean breaking load of 30kN and the standard deviation of these loads is 0.4kN. The manufacturing process is modified slightly at an increased cost and to examine the effect on the end product, a sample of 100 cables produced by the modified process is tested. The average breaking load for this sample is found to be 30.08kN. What evidence is there that the modification to the process has caused a real change in the mean breaking load of the cables produced?

We begin by noting that the discrepancy between the sample mean of 30.08kN and the supposed population mean* of 30kN can be due (i) to sampling errors or (ii) to the possibility that the population mean is not 30kN or (iii) to both causes. What we shall do is to assume that the population mean is in fact 30kN and that, consequently, the sole reason for the discrepancy is the sampling error. Under this assumption - the **null hypothesis** H_0 - we calculate the probability of a discrepancy as large as the one observed. We have said that any conclusion we reach is subject to a risk of being wrong and we must first decide on a level of risk. We shall, for the moment, arbitrarily choose a risk of 5%. This means that were the experiment (of taking a sample of size 100 and measuring the sample mean) to be repeated a *large* number of times and were the population mean 30kN then in about 5% of experiments the sample mean would be 30.08kN or more (it should be noted that were the population mean less than 30kN, the chances of a sample mean being 30.08kN or more will be smaller than 5%). For a population mean much smaller than 30kN the chances of a sample mean $\geqslant$ 30.08kN become remote.

From the Central Limit Theorem, we know that sample means are approximately normally distributed and therefore we proceed to calculate an acceptance region; see Figure 13.1 and refer to Table AI, page 544.

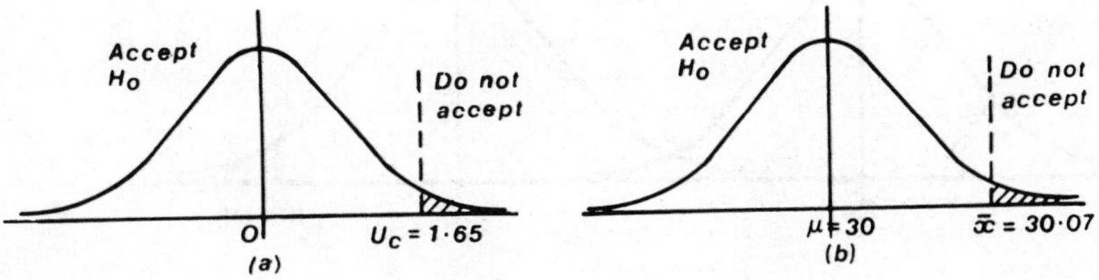

Figure 13.1

The critical value u_c is found from Table AI to be 1.65 (2 d.p.). In order to find the corresponding value for $\bar{x}$ we use the formula

* In this example the population is the set of cables manufactured by the modified process.

$$u_c = \frac{\overline{x} - \mu}{\sigma/\sqrt{n}} \quad \text{which becomes} \quad 1.65 = \frac{\overline{x} - 30}{0.4/10} \quad \text{and this yields}$$

$$\overline{x} = 30 + 0.04 \times 1.65 = 30.066 = 30.07\text{kN} \quad (2 \text{ d.p.})$$

Our decision rule is: accept H_o if the observed $\overline{x} < 30.07$ otherwise do not accept H_o. Note that the α - risk has now decreased because we have rounded up $\overline{x}$ to 2 d.p. The observed value of $\overline{x}$ lies outside the acceptance region and therefore we conclude that we should not accept the hypothesis that the modified process still produces cables of mean breaking load 30kN. We may then say that the modification has caused a real change in the mean breaking load of the cables and that the risk of our judgement being wrong is somewhat smaller than 5%.

Before we proceed further let us just see the effect of reducing the risk of wrong judgement to 1%. From Table AI the critical value u_c is now 2.33 and $\overline{x} = 30 + 0.04 \times 2.33 = 30.0932 = 30.09$ (2 d.p.). The new decision rule is: accept H_o if observed $\overline{x} < 30.09$, otherwise do not accept H_o. From Figure 13.2 we see that the observed value 30.08 now lies within the acceptance region. This time we state that there is no evidence that the modification has caused a change in mean breaking load and the risk of a wrong judgement is 1%.

In the first case we could commit an α - risk or Type I error and the probability of that error is 5%. In the second case we could commit a Type II error or β - risk: we cannot yet state the probability of this happening.

If the population mean were less than 30 kN then the probability of a Type I error would be smaller: see Figure 13.3.

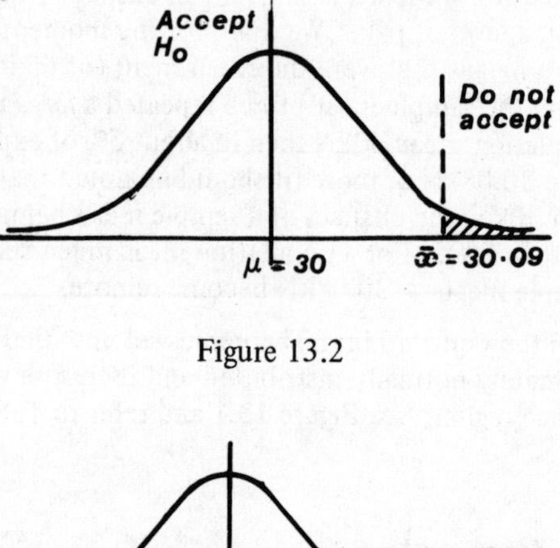

Figure 13.2

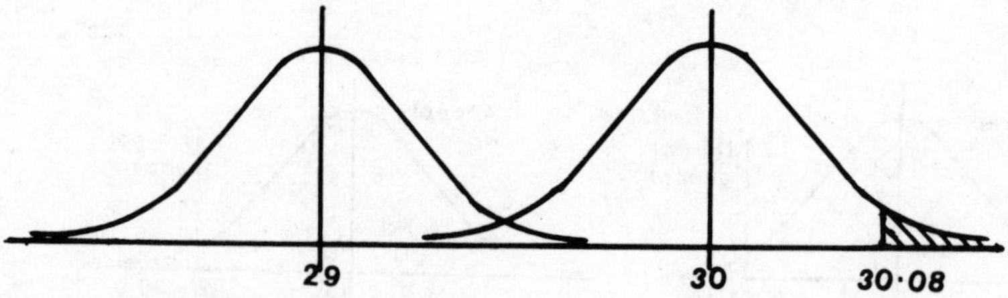

Figure 13.3

Given the decision rule, the α - risk is greatest when $\mu = 30$kN. To fix a value for β we must state a clear alternative hypothesis. As things stand we have a null hypothesis H_o: $\mu = 30$kN and an alternative hypothesis H_1: $\mu > 30$kN. However, we need a more specific alternative, for example $\mu = 30.1$kN; this might correspond to the situation where a mean breaking load of more than 30.1kN would justify the more expensive modified process.

We will assume that we have decided upon an α - risk of 5%. This means that we will accept H_o: μ = 30kN if the observed $\bar{x}$ is less than 30.07kN. If we accept H_o there is a possibility that the alternative hypothesis H_1: μ = 30.1kN is true and we should have committed a Type II error; what is the probability β, of this happening? In other words, what is the probability that we could get a sample mean of less than 30.07 when μ = 30.1? See Figure 13.4. We find a critical value

$$u_c = \frac{30.07 - 30.1}{0.4/10} = \frac{-0.03}{0.04} = -0.75$$

By reference to Table AI we find that the β - risk is 0.2266.

It should be noted that the β - risk is greatest if $\bar{x}$ = 30.07; smaller values of $\bar{x}$ will provide a smaller value of β. See whether you can reason why.

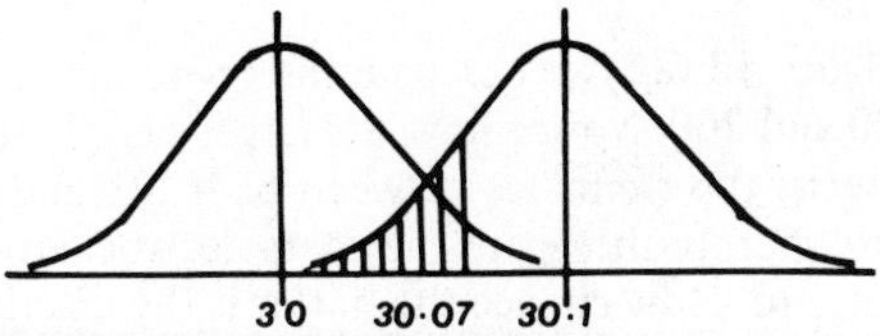

Figure 13.4

We now examine the effect on β of choosing α to be 1%. We remember that this gave a critical $\bar{x}$ of 30.09. The corresponding value $u_c = \dfrac{30.09 - 30.1}{0.04} = -0.25$ and from Table AI we find that β = 0.4013, see Figure 13.5. This means we run a risk of more than 40% of accepting H_o: μ = 30.0kN.

In decreasing the risk of a Type I error we have increased the risk of a Type II error. In general, for a fixed sample size, decreasing one of these errors leads to an increase in the other. The only way to reduce both α - and β - risks simultaneously is to increase the sample size. For example, suppose the sample size n were 200 and that we have the decision rule: accept H_o if $\bar{x} < 30.07$, otherwise do not accept H_o. Then, to find the α - risk we calculate $u_c = \dfrac{30.07 - 30}{0.4/\sqrt{200}} = 2.475$ and this corresponds to an α - risk of 0.0066. The β - risk is found via $u_c = \dfrac{30.07 - 30.1}{0.4/\sqrt{200}} = -1.06$ and this gives β = 0.1446. However, a sample of size 200 may be too costly and we should perhaps consider a better decision rule. Presumably we are concerned to minimise the risk of passing the modification unless the mean breaking load is at least as high as 30.1kN. Given any two of the parameters n, α and β, the third is automatically fixed. Suppose then that we wish to make β at most 0.05 and retain n as 100 then what will the decision rule become and what will α be? For β = 0.05 we shall find $-1.65 = \dfrac{\bar{x} - 30.1}{0.04}$ and therefore

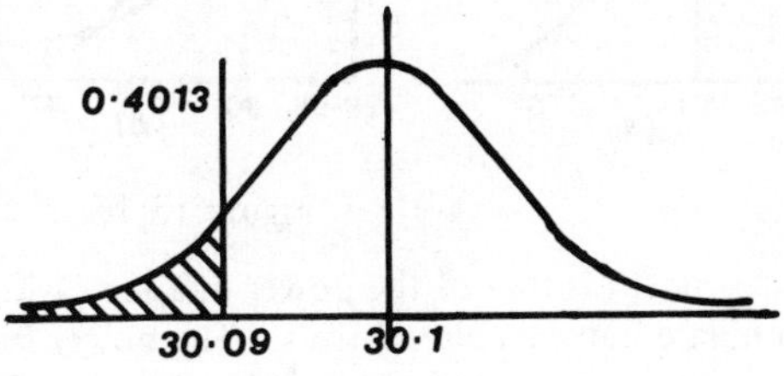

Figure 13.5

$\overline{x} = 30.1 - 0.04 \times 1.65 = 30.03$ (2 d.p.). The decision rule is now: accept H_o if $\overline{x} < 30.03$, otherwise do not accept H_o. Then to determine the α - risk we calculate $u_c = \dfrac{30.03 - 30}{0.04} = 0.75$ and discover that $\alpha = 0.2266$. This may seem too high a risk and it serves to emphasise the point that we need to decide on our priorities before choosing a course of action.

We might ask what is the effect on β of varying H_1. We assume that the decision rule hinges on the critical $\overline{x}$ value of 30.07 and that $n = 100$. It should be clear that the probability of accepting H_o when the true population mean is μ is the area under the normal curve to the left of $u_c = \dfrac{30.07 - \mu}{0.4/10}$. Let us represent this area by $f(\mu)$; then we can plot a graph of $f(\mu)$ against μ. This is called the **Operating Characteristic Curve** for the test criterion, and is shown in Figure 13.6(a).

In Figure 13.6(b) we compare the operating characteristic curves for sample sizes 100 and 200. Notice how the larger sample size has improved the test in that it better discriminates between $\mu = 30$ and any neighbouring value of μ. The more sharply the operating characteristic curve falls from a high chance of acceptance to a low chance the better is the discrimination. The extreme situation would be that we accept H_0 for $\mu \leqslant 30$ and do not accept H_0 for $\mu > 30$. The 'ideal' operating characteristic curve is shown in Figure 13.6(c).

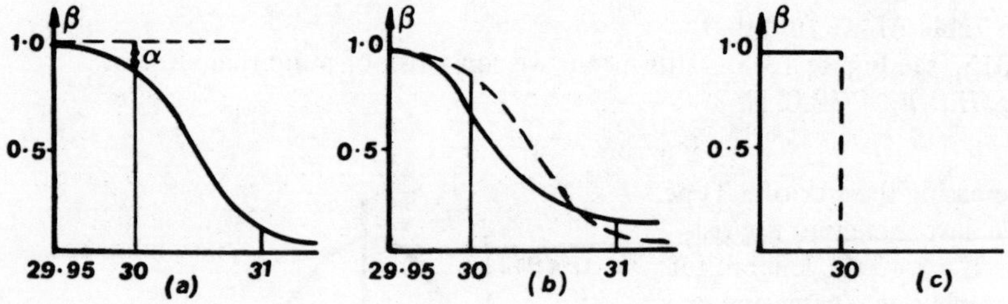

Figure 13.16

A related concept is that of the **power** of a test, which is the ability of the test to discriminate between alternatives. The power function is $(1 - \beta)$ and you should construct its graph for the problem under consideration.

General Strategy

There is a definite strategy that we have adopted in testing the null hypothesis and we summarise the steps involved, illustrating these from the example just considered.

1. *The Null Hypothesis and its alternative are stated*

   In the first case considered, the null hypothesis H_o was $\mu = 30$kN and the alternative hypothesis H_1 was $\mu > 30$kN.

2. *The probability distribution model is selected*

   When considering means of large samples, the normal distribution is appropriate.

3. *The α - risk is chosen*

   In the first case considered this risk was 5%.

4. *Determine the acceptance and rejection regions*

   Since we were employing a one-tail test we found that $u_c > 1.65$ and this leads to rejection of the null hypothesis.

5. *Perform the calculations*

   We obtain $u_c = \dfrac{\bar{x} - \mu}{\sigma/\sqrt{n}} = \dfrac{30.08 - 30}{0.4/10} = 2$

6. *Draw the statistical inference*

   We conclude (with an α - risk less than 5%) that H_o should not be accepted.

7. *Consider further action*

   This will depend on a particular problem and will not be pursued in this chapter. In the present problem, the action might be to adopt the modified process.

   To fix these ideas, we now carry out these steps on a second problem.

Example

Random samples of components manufactured by two machines were taken. From machine A, 100 components were tested and 5 were found to be faulty; whereas 100 components from machine B revealed 9 faulty. Is there evidence that the two machines show different qualities of performance?

1. The null hypothesis H_o is that there is no difference in performance, and the alternative hypothesis H_1 is that there *is* a difference in performance.

2. We shall consider the difference in proportion of defectives; this will be a variable which is approximately normally distributed. If the proportions be p_A and p_B respectively, then the sampling distribution of differences in proportions has mean $\mu_{p_A - p_B} = 0$ and standard deviation

   $\sigma_{p_A - p_B} = \sqrt{p(1-p)\left\{\dfrac{1}{n_A} + \dfrac{1}{n_B}\right\}}$ where p is the population proportion, assumed common.

   We have to estimate p by $\dfrac{n_A p_A + n_B p_B}{n_A + n_B}$ where n_A and n_B are the numbers in the two samples. The standardised variable $u = \dfrac{(p_A - p_B) - 0}{\sigma_{p_A - p_B}}$ follows approximately the distribution $N(0, 1^2)$.

3. We shall select an α - risk of 5%.

4. We have chosen a two-tailed test and therefore we find the acceptance region is $-1.96 < u < 1.96$ as depicted in Figure 13.7.

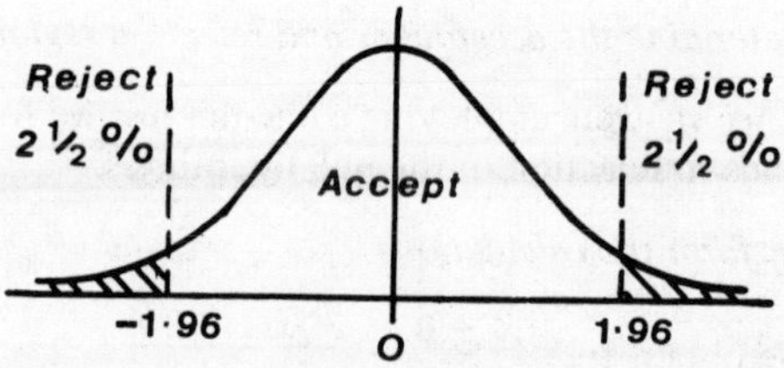

Figure 13.7

5. We first compute $\dfrac{n_A p_A + n_B p_B}{n_A + n_B} = \dfrac{100 \times \frac{5}{100} + 100 \times \frac{9}{100}}{200}$

$$= \frac{14}{200} = 0.07$$

as an estimate of the population proportion p.

Then $\sigma_{p_A - p_B} = \sqrt{(0.07)(0.93)\left(\frac{1}{100} + \frac{1}{100}\right)}$

$$= \sqrt{\frac{0.07 \times 0.93 \times 2}{100}} = 0.036$$

Hence $u = \dfrac{\frac{5}{100} - \frac{9}{100}}{0.036} = \dfrac{-0.04}{0.036} = -1.11$

6. Since u lies in the acceptance region we cannot find sufficient evidence of a real difference in quality, at a 5% α - risk.

Notice that we opted for a two-tailed test in our original statement of the problem. Had we there asked whether there was evidence of machine A giving a higher quality of performance we should have required a one-tailed test. You should show that for this case $u_c = -1.65$ and hence that, with a 5% α - risk the same conclusion holds.

Problems

1. In a chemical plant the acid content of the effluent is monitored. From 400 measurements the acid content in g/litre of effluent had mean 0.142 and standard deviation 0.008. Is the result consistent with a mean acid content of 0.1413 g/litre obtained from previous tests? State any assumptions made.

2. Two observers took separate sets of pressure readings for the same orifice, 100 readings were taken by each and the results were (in mm of mercury)

   First observer: mean reading 3.571, standard deviation 0.203
   Second observer: mean reading 3.637, standard deviation 0.203

   Assuming that each set of readings can be regarded as a random sample of the variations in pressure, determine whether the difference in mean readings is significant at the 5% level.

3. The distribution of daily-output from a particular machine has mean 500 units and standard deviation 50 units. With a new mode of operation a sample of 10 daily runs gave a mean of 550 units. Show that the new method has made a significant change in the daily rate and quote 95% confidence limits for the new mean daily rate.

4. Past experience indicates that wire rods of a particular type have a breaking load which is distributed with standard deviation 60 N. In testing large batches of rods, what sample size should be used to obtain a mean breaking load within 8 N of the mean for the batch with probability 0.99?

5. It is considered satisfactory to know the mean of a measurement within 0.1 of its true value with a probability of 0.95. A batch was measured and found to have standard deviation 0.53. What sample size should we use to obtain a satisfactory estimate of the mean of each batch?

6. A firm manufactures fibres with breaking strengths having mean 400 N and standard deviation 40 N. A new process is claimed to improve the mean breaking strength. What should be the decision rule for rejecting the old process at 1% risk if it is decided to test 100 fibres? What is the probability of accepting the old process if the new process has increased the mean breaking strength to 420 N without altering the standard deviation? Draw an operating characteristic curve and a power curve for values of the new mean of 390, 400, 410, ..., 450.

7. Draw an operating characteristic curve and a power curve for Problem 3.

8. A coin is thrown 100 times. With a 5% risk calculate the probability of accepting the coin as being fair ($p = 0.5$) when $p = 0.1, 0.2, \ldots\ldots, 0.9$. Draw the operating characteristic curve and the power curve.

13.3 MEAN OF A SMALL SAMPLE: THE *t* TEST

In many experiments it is not possible or not economic to take large size samples: also, we may be unaware of the population variance. Under these circumstances the normal model will be inappropriate even though the parent population *is* normally distributed. The statistic $(\overline{x} - \mu)/(\hat{\sigma}/\sqrt{n})$, where $\hat{\sigma}^2$ is the estimate of the variance provided by the sample, will not be normally distributed since not only does $\overline{x}$ vary from sample to sample but so does $\hat{\sigma}$. As well as the sample mean, we need therefore to consider two further variables: the sample variance *and* the sample size. We require a model which will take them both into account.

A suitable model is provided by the *t* **distribution** (sometimes referred to as Student's *t* distribution). It relies on the parent population having a normal distribution $N(\mu, \sigma^2)$. Then if the mean of a random sample of size n is $\overline{x}$ and the sample standard deviation is s, the random variable $t = \dfrac{\overline{x} - \mu}{s/\sqrt{n}}$ follows the t distribution with $\nu = (n - 1)$ degrees of freedom. Note that the sample estimate of σ^2 is calculated from $(n - 1)$ *independent* values of $(x_i - \overline{x})^2$.

Instead of one table for the Normal Distribution $N(0, 1^2)$, we now need a table for each value of ν (starting with $\nu = 1$). In fact, the values provided by t-tables for $\nu > 30$ are extremely close to the corresponding values from

$N(0, 1^2)$ and this ties in with our claim that sample means for sample sizes of about 30 or more can be assumed to be approximately normally distributed. Figure 13.8 shows the graphs of the t distribution for $\nu = 1, 5$ and ∞. The case $\nu \to \infty$ is the normal distribution, though had we also drawn the graph of the case $\nu = 30$ we should not have been able to distinguish them. Note that the t-curves are symmetrical about the vertical axis. The expectation of the distribution is 0 and its variance is $\dfrac{\nu}{\nu - 2}$ (for $\nu > 2$).

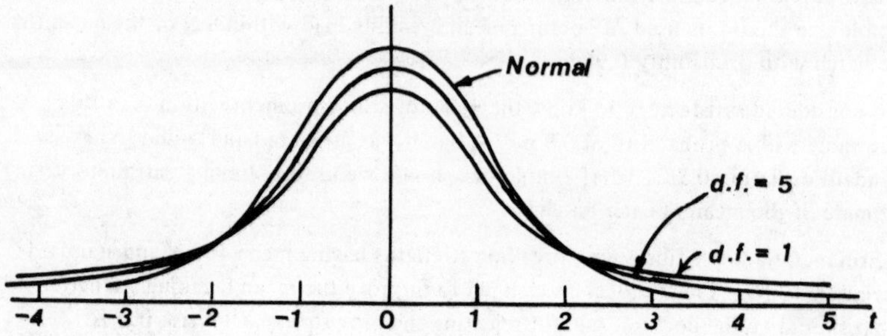

Figure 13.8

Table AII shows selected percentage points of the t distribution. Note that the interpretation is similar to that for the normal distribution. For example, the probability that with 4 degrees of freedom $t > 2.13$ is 5%, that $t < -2.13$ is (by symmetry) 5% and that $|t| > 2.78$ is 5%. You should try Problems 1 and 2 at the end of this section to ensure that you are familiar with the use of this table.

It may be argued that the t distribution is more general than the normal distribution because we do not require knowledge of σ ; however, the t distribution model *does* require a normal parent population. In practice, provided this population is not too heavily skewed and follows roughly a Gaussian 'bell-shaped' distribution, the application of the t test should give satisfactory results.

Confidence Intervals

We now show the application of the t test in estimating a confidence interval for a population mean. The principles are exactly the same as for a normal distribution.

Example

A random sample of 5 steel beams produced by a mill was tested for ultimate compressive strength. The results were, in MN/m^2.

301500 329600 318700 307200 322400

We require to find the 95% confidence interval for the mean ultimate compressive strengths of steel beams produced at the mill.

Calculations produce $\bar{x} = 315880$ $s^2 = 130210 \times 10^3$

Hence the estimated standard error of the mean $\frac{s}{\sqrt{5}} = 5103$.

From Table AII we see that for 4 degrees of freedom there is a 5% chance that $|t| > 2.78$ and hence a 95% chance that $|t| \leq 2.78$. Therefore the 95% confidence interval for the mean ultimate compressive strength is

$315880 \pm 2.78 \times 5103$ i.e. 315880 ± 14186

It is now probably reasonable to round off these results to give

315900 ± 14200 and hence the interval is 301700 to 330100

You should now show that the 99% confidence interval is 292400 to 339400.

Tests of Hypothesis

We demonstrate the application of the t test to tests of hypothesis.

Example 1

A machine is set to pack 10 kg of cement in a bag. A sample of 9 bags is taken and for this sample the mean weight of cement in a bag is found to be 9.5 kg and the sample standard deviation is 0.5 kg. Test the hypothesis that the machine is in proper working order using a risk of wrong rejection of (a) 5% (b) 1%.

(a) Let us follow the strategy laid down earlier.

1. The null hypothesis is H_o: $\mu = 10$ and the alternative hypothesis is H_1: $\mu \neq 10$.
2. The risk of wrong rejection is 5%.
3. The t distribution is chosen (why?) with 8 degrees of freedom.
4. From Table AII we find that we should reject H_0 if $|t| > 2.314$ We use a two-tailed test because of the nature of H_1.
5. From the sample of size 9 we find $t = \frac{\bar{x} - \mu}{s/\sqrt{9}} = \frac{3(9.5 - 10)}{0.5} = -3$
6. We therefore do not accept H_o, with a risk of wrong judgement of less than 5%.

(b) The only changes are

2. The risk of wrong rejection is 1%
4. From Table AII we find that we should reject H_o if $|t| > 3.36$
6. We therefore do not reject it.

Example 2

An electrical component is claimed to have a mean life of 4000 hours. Laboratory tests are conducted on a random sample of 5 components and the recorded lifetimes were 3960, 4010, 3940, 3950 and 3990 hours. Is the claim justified?

1. The null hypothesis is H_o: $\mu = 4000$ and the alternative hypothesis is $H_1: \mu < 4000$ (Why?)

2. We shall choose a risk of wrong judgement of 5%
3. We select the t distribution with 4 degrees of freedom
4. From Table AII we see that we should reject H_o if $t < -2.13$ (See Figure 13.9). This time we use a one-tailed test.

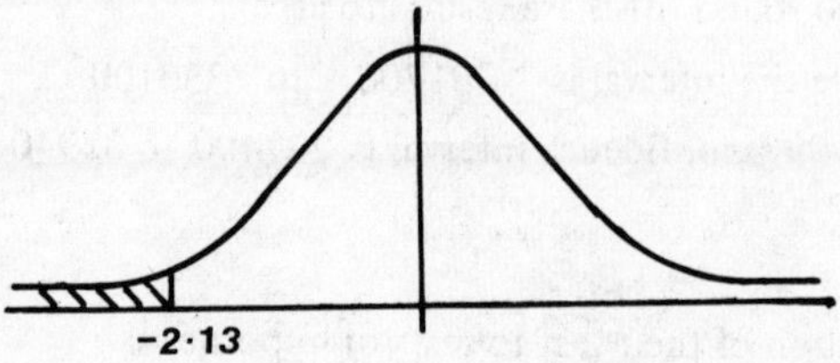

Figure 13.9

5. The sample mean $\bar{x} = 3970$ and $\sum_{i=1}^{5} (x_i - \bar{x})^2 = 3400$

$$\therefore \hat{\sigma}^2 = s^2 = \frac{3400}{4} = 850$$

$$\therefore s = 29.2 \text{ (3sf)}$$

$$\text{Hence } t = \frac{\bar{x} - \mu}{s/\sqrt{n}} = \frac{3970 - 4000}{29.2/\sqrt{5}} = -2.29$$

6. We therefore reject the claim.

Problems

1. Using Table AII, calculate the probabilities that

(a) $|t| > 3.01$, $\nu = 13$ (b) $|t| > 2.08$, $\nu = 21$ (c) $|t| > 1.76$, $\nu = 14$
(d) $|t| \leqslant 2.84$, $\nu = 7$ (e) $|t| \leqslant 4.77$, $\nu = 5$ (f) $|t > 7.45$, $\nu = 3$
(g) $t < -2.39$, $\nu = 24$ (h) $t \leqslant 11.24$, $\nu = 8$

2. Find 5% critical values of t for $\nu = 15, 28, 35, 250$.

3. A machine is designed to produce bolts of diameter 0.14 cm. To find whether the machine is in working order a sample of 10 bolts is taken. Recorded thicknesses are 0.1325, 0.14, 0.135, 0.1425, 0.135, 0.1375, 0.1375, 0.1275, 0.14, 0.1325. With a level of significance of 5%, determine whether the machine is in working order.

4. Paper is produced with a supposed mean thickness of 0.004 cm. 9 sample measurements are taken and their mean is 0.0037 cm. with a standard deviation of 0.0003 cm. With a 5% risk, decide whether the output differs significantly from the supposed standard.

5. A random sample of 10 items has a mean of 224 and a standard deviation of 13.2. Could the sample have come from a population with a mean of 208?

6. A machine produces components to a nominal length of 4.800 cm. It is reset every morning. The first hour's product was sampled and the following lengths measured.

4.802, 4.801, 4.804, 4.801, 4.798, 4.802, 4.800, 4.801, 4.801, 4.802, 4.803, 4.801, 4.799, 4.800, 4.801, 4.805. Is there evidence of an incorrect setting?

7. Measurements of the diameters of a random sample of 200 bearings made by a firm during one week showed a mean of 1.224 cm and a standard deviation of 0.042 cm. Find 95% and 99% confidence limits for the mean diameter of all the bearings.

8. Boilers at a factory used a mean weight of fuel per day of 30.10 units. A new design of boiler was tested over 10 days and the following weights of fuel were recorded: 30.15, 29.84, 28.32, 27.50, 30.32, 29.41, 29.57, 28.54, 32.43, 29.82. Is there evidence of an economy in fuel with the new design?

9. A sample of 10 items gave the following results: 33.5, 34.8, 34.0, 33.4, 34.6, 35.1, 33.6, 34.2, 34.2, 34.6. Find 95% and 99% confidence intervals for the population mean. State any assumptions made.

10. A sample of 5 items has masses 6.7, 5.9, 6.4, 5.8 and 6.2 kg. What are the 95% confidence limits for the mean of the population?

11. Ten measurements of a physical constant gave a mean of 123.77 units and the sum of their squared deviations from this value was 0.9261 squared units. Find 95% confidence limits for the true value of the constant.

12. A random sample of size 12 had a mean of 14.3 and variance of 2.1. Test at the 5% level of significance the claim that the population mean is 15.0 as opposed to the alternative hypothesis that the mean is less than 15.0.

13.4 TEST OF SAMPLE VARIANCE: THE χ^2 DISTRIBUTION

Often, articles need to be manufactured with precision; if components are to interlock it is important that certain parameters, for example diameter, do not vary much from a quoted mean. It is often necessary to check the variance of a sample against a quoted population variance to see whether the machine is now producing articles with greater variability.

Now we know that the sample variance, $s^2 = \dfrac{1}{n-1}\sum_{i=1}^{n}(x_i - \bar{x})^2$, is an unbiased estimator of the population variance σ^2.

It follows that $\dfrac{\Sigma(x_i - \bar{x})^2}{\sigma^2}$, i.e. $\dfrac{(n-1)s^2}{\sigma^2}$ is an unbiased estimator of $(n-1)$. In fact, if the sample is drawn from a normal population $N(\mu, \sigma^2)$ the value of $\dfrac{(n-1)s^2}{\sigma^2}$ is a random variable which follows the chi-square distribution having $\nu = (n-1)$ degrees of freedom. Chi-square is written χ^2.

Note, as for the t test, the assumption of a normal population. If the population is 'nearly' normal, the χ^2 distribution is 'nearly' followed.

Different values of ν give rise to different shapes of χ^2 distribution.

In Figure 13.10 we graph the χ^2 distribution for three values of ν. Notice that the curves are markedly skew. In the special case $\nu = 2$ the curve is different from those with other values of ν since it possesses no local maximum.

Table AIII gives selected percentage points for various χ^2 distributions. For example, the probability that, with $\nu = 1$, $\chi^2 > 6.635$ is 1%, that $\chi^2 > 3.841$ is 5% and with $\nu = 4$ the probability that $\chi^2 < 0.2971$ is 1%. You should try Problem 1 at the end of this section to gain familiarity with the table.

For non-tabulated distributions we may use interpolation or the two results:

(a) for large ν, χ^2 is approximately distributed $N(\nu, 2\nu)$ ($\chi^2 \simeq \nu + u\sqrt{2\nu}$ where u is the corresponding percentage point for the normal distribution $N(0, 1^2)$).

(b) for $\nu > 30$ the variable $(\sqrt{2\chi^2} - \sqrt{2\nu - 1})$ is approximately normally distributed $N(0, 1^2)$, so that $\chi^2 \simeq \frac{1}{2}(\sqrt{2\nu - 1} + u)^2$.

We have effectively examined the differences between s^2 and σ^2 by forming the ratio s^2/σ^2 for which we have found a sampling distribution; we cannot easily find a distribution for $s^2 - \sigma^2$.

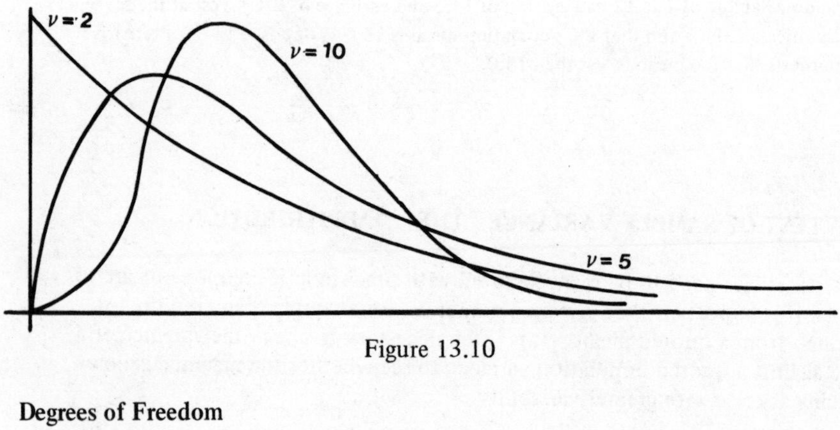

Figure 13.10

Degrees of Freedom

It should be pointed out at this stage that the number of degrees of freedom ν for a statistic is not always one less than the number of items n in the sample chosen. In general, the number of degrees of freedom for a statistic is given by $\nu = n - \gamma$ where γ is the number of population parameters that must be estimated from a sample. For the one-sample t test we needed only to estimate μ, for χ^2 only σ^2; in both these cases, then, $\gamma = 1$. We shall see when we come to analyse regression lines that $\gamma = 2$ there since we use the sample of observations to estimate the slope and the intercept of the line.

Confidence Intervals

It is easiest to introduce the ideas by means of an example. The principles are exactly the same as for other distributions that we have examined.

Example 1

A sample of 10 batteries were tested for useful life. These lives, in months, were recorded to 1 d.p. and are as follows: 15.0, 14.2, 15.9, 14.8, 14.9, 15.1, 16.0, 14.8, 15.1, 15.3. We now seek a 95% confidence interval for the standard deviation of the lives of batteries produced by the process under consideration.

From the sample readings we find $s^2 = 0.281$. There are 9 degrees of freedom.

In Figure 13.11 we see that the critical χ^2 values are those corresponding to percentage points 0.025 and 0.975; these, from Table AIII are found to be 2.70 and 19.0 respectively.

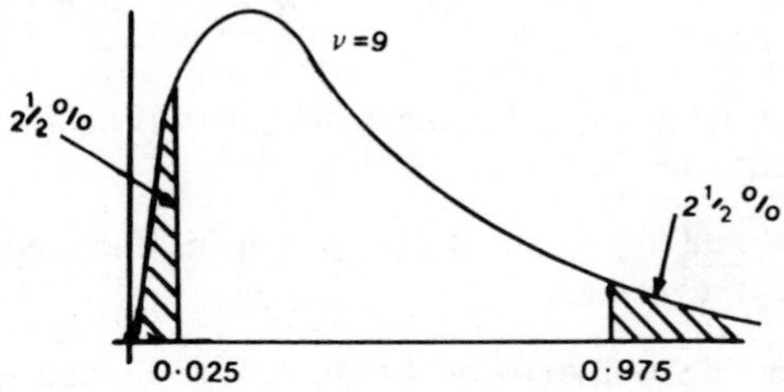

Figure 13.11

Now σ^2 is estimated by $\frac{(n-1)s^2}{\chi^2}$. Corresponding, therefore, to the value $\chi^2 = 2.70$ is the estimated value $\hat{\sigma}^2$ of $\frac{9 \times 0.281}{2.70}$, i.e. 0.937 (3 d.p.)

In the same way, corresponding to $\chi^2 = 19.0$ is the value $\frac{9 \times 0.281}{19.0} = 0.133.$

Therefore the 95% confidence interval for σ^2 is $0.133 < \sigma^2 < 0.937.$ Notice that this interval is *not* symmetrical about the observed sample value of 0.281 .

We can therefore obtain a 95% interval for σ by taking square roots to obtain $0.365 < \sigma < 0.968.$

If we now widen the limits to obtain a 99% confidence interval we find that $0.107 < \sigma^2 < 1.458$ and hence $0.327 < \sigma < 1.207.$

Let us check the accuracy of the statements made about approximations to the χ^2 distribution for $\nu > 30.$

Example 2

The standard deviation of the lifetimes of a sample of 101 light bulbs is measured to be 50 hours. Find 95% confidence limits for the variance of the population of light bulbs, (a) using χ^2 tables directly, (b) using the normal distribution approximation.

(a) For 100 degrees of freedom the critical values of χ^2 are found to be 74.2 and 129.6 and hence the confidence interval for σ^2 is
$\frac{100 \times 2500}{129.6} < \sigma^2 < \frac{100 \times 2500}{74.2}$ i.e. $1929.0 < \sigma^2 < 3369.3$ and hence $43.9 < \sigma < 58.1.$

(b) The values of χ^2 are approximately $100 \pm 1.96 \times \sqrt{200}$ i.e. 72.29 and 127.71. Calculations similar to (a) give the interval as $44.2 < \sigma < 58.8.$

Test of Hypothesis

The strategy that we developed in Section 13.2 still holds good. The only difference is in the distribution employed.

Example

A manufacturer claims that certain machine parts he produces will have mean diameter 4 cm with standard deviation 0.01 mm. The diameters of 5 parts are measured and found to be (in mm.): 39.98, 40.01, 39.96, 40.03, 40.025. Is the claim valid?

First, we test whether the sample mean is significantly different from the population mean.

1. The null hypothesis is H_o: $\mu = 4$ cm and the alternative hypothesis is H_1: $\mu \neq 4$ cm.

2. The t distribution with 4 degrees of freedom is selected since we have a small sample.

3. The α - risk is chosen to be 5%.

4. The acceptance region is found from Table AII to be $|t| \leqslant 2.78$.

5. We calculate $\bar{x} = 40.001$, $s = 0.0301$ and hence $t = \dfrac{40.001 - 40.0}{0.0301/\sqrt{5}}$ $= 0.074$.

6. Clearly, there is no evidence at 5% risk of $\mu \neq 4$ cm.

Second, we test whether the sample standard deviation is significantly higher that 0.01 mm (or, equivalently, whether the sample variance is significantly higher than 0.0001 $(\text{mm})^2$).

1. The null hypothesis is H_0: $\sigma^2 = 0.0001$ and the alternative hypothesis is H_1: $\sigma^2 > 0.0001$.

2. We select the χ^2 distribution with 4 degrees of freedom.

3. The α - risk is chosen to be 5%,

4. The acceptance region is found from Table AIII to be $\chi^2 < 9.488$.

5. $\chi^2 = \dfrac{(n-1)s^2}{\sigma^2} = \dfrac{4 \times (0.0301)^2}{0.0001} = \dfrac{0.003625}{0.0001} = 36.25$

6. We conclude, with much less than 5% risk, that the variability in diameter is greater than was claimed.

Problems

1. (i) Find the values of χ^2 which are exceeded with a probability of 5% for $\nu = 12, 7, 3, 80, 200$.

(ii) Repeat the above for a probability of 1%.

(iii) Repeat for a probability of 50%.

(iv) What are the approximate probabilities that χ^2 will be greater than 20 for $\nu = 12, 30$?

(v) What is the probability that (a) $\chi^2 \leqslant 26.30$, (b) $\chi^2 \leqslant 23.54$, (c) $7.962 \leqslant \chi^2 \leqslant 15.34$ with $\nu = 16$?

2. The variance of a group of 8 observations is 3.52. What are the 95% confidence limits for the population variance? State any assumptions made.

3. A sample of 20 observations gave a standard deviation 3.72. Is this compatible with the hypothesis that the sample is from a normal distribution with variance 4.35?

4. The following data are from a normal population. With a 5% risk, test the claim that $\sigma = 2.5$ as opposed to the alternative claim that $\sigma \neq 2.5$: 20.2, 40.2, 37.6, 31.2, 26.0, 36.6, 26.4, 21.6, 32.0.

5. Repeat 4 for the alternative claim that $\sigma > 2.5$.

6. A company manufactures components with a certain life. A random sample of 25 components has a standard deviation of 140 hours. What are the 95% confidence limits for the standard deviation of all components?

7. The standard deviation of wires manufactured by a firm is 533.8 N. A new process is tested and the standard deviation of a sample of 10 wires is 667.2N. Is the decrease in variability significant?

8. Find a 95% confidence interval for the variance of the normal population from which a random sample of 6 items gave $\Sigma (x_i - \bar{x})^2 = 40$.

9. A sample of size 20 gave an estimate for σ^2 of 16.4. Find a 99% confidence interval for the population standard deviation.

10. Find a 99% confidence interval for the variance of the population from which the following sample was taken: 7.4, 12.5, 9.7, 9.0, 10.6, 11.3.

13.5 FURTHER APPLICATIONS OF χ^2: INFERENCES ABOUT FREQUENCIES

We mentioned in the last section that $\dfrac{\sum_{i=1}^{n} (x_i - \bar{x})^2}{\sigma^2}$ is an unbiased estimator of $(n-1)$ and if the sample $\{x_1, x_2, \ldots, x_n\}$ is drawn from a Normal population, $\sum_{i=1}^{n} \dfrac{(x_i - \bar{x})^2}{\sigma^2}$ follows a χ^2 distribution with $(n-1)$ degrees of freedom.

We now approach the distribution from a different viewpoint. There are two results we quote.

(a) If x is normally distributed $N(0, 1)$ then x^2 follows a χ^2 distribution with 1 degree of freedom.

(b) If $x_1, x_2, x_3, \ldots\ldots, x_r$ are r independent random variables each distributed $N(0, 1)$ then the random variable $x_1{}^2 + x_2{}^2 + x_3{}^2 + \ldots\ldots + x_r{}^2$ follows a χ^2 distribution with r degrees of freedom.

(Note that if x_1 is a random variable distributed normally then $\dfrac{(x_1 - \bar{x})}{\sigma}$ follows approximately a $N(0, 1)$ distribution and therefore

$$\sum_{i=1}^{n} \frac{(x_i - \bar{x})^2}{\sigma^2}$$ will be the sum of squares of $(n - 1)$ independent random variables distributed $N(0, 1)$. We say $(n - 1)$ since if $x_1, x_2, \ldots, x_{n-1}$ are known and $\bar{x}$ is given, x_n is already determined.)

Tests on a single proportion

The Normal distribution may sometimes be used as an approximation to the Binomial distribution in testing hypotheses about proportions. We now work through an example using first the Normal distribution and then the χ^2 distribution.

Example 1

(See Figure 13.12.) It is claimed that during a week day mid-morning the proportion of traffic south of a junction which either goes onto the minor road or has emerged from it is at least $\frac{1}{3}$.

To test this claim a count lasting two hours was taken. This found that of 1436 vehicles south of the junction 434 either then proceeded along the minor road or had just emerged from it. Is the claim valid?

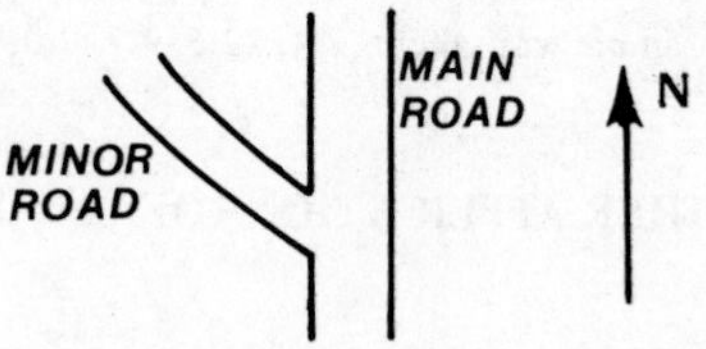

Figure 13.12

(a) *Normal distribution*

1. The null hypothesis is

   H_o: $p = \frac{1}{3}$ and the

   alternative hypothesis is

   H_1: $p < \frac{1}{3}$. (One tail test.)

2. The α - risk is taken to be 1%.
3. We examine the sampling distribution of proportions; the standard error is $\sqrt{\dfrac{p(1-p)}{n}}$ where $p = \frac{1}{3}$; $n = 1436$.
4. The acceptance region will be $u > -2.33$ (Why?)

5. We calculate $u = \dfrac{\dfrac{434}{1436} - \dfrac{1}{3}}{\sqrt{\dfrac{\frac{1}{3} \times \frac{2}{3}}{1436}}} = \dfrac{.3023 - .3333}{0.0124} = -2.49$

6. We must reject the claim at 1% risk.

(Note that we could have calculated u as $\dfrac{434 - \left(\frac{1}{3} \times 1436\right)}{\sqrt{1436 \times \frac{1}{3} \times \frac{2}{3}}}$

i.e. $\dfrac{x - np}{\sqrt{n\ p(1-p)}}$ and we are then dealing with frequency instead of proportion.)

(b) χ^2 *distribution*

We draw up the Table 13.1 .

Table 13.1

	On minor road	Not on minor road	Total
Observed Frequency	$x_1 = 434$	$x_2 = 1002$	1436
Expected Frequency	$\frac{1}{3} \times 1436 \simeq 479$	$\frac{2}{3} \times 1436 \simeq 957$	1436

1. The null hypothesis is still H_o: $p = \frac{1}{3}$ and the alternative hypothesis is now $H_1: p \neq \frac{1}{3}$. (This is because χ^2 is constructed to cope with discrepancies.)
2. The α - risk is 1%.
3. We shall calculate $\dfrac{(x_1 - np)^2}{np} + \dfrac{(x_2 - n(1-p))^2}{n(1-p)}$ where $x_1 + x_2 = n.$

   We see that for the two cases - 'on minor road' - and - 'not on minor road' - we are calculating $\dfrac{(\text{observed frequency} - \text{expected frequency})^2}{\text{expected frequency}}$ and adding the results.

   Let $x_1 = n\pi$ and $x_2 = n(1-\pi)$.

   Then $\dfrac{(x_1 - np)^2}{np} + \dfrac{([n - x_1] - n(1-p))^2}{n(1-p)}$

$$= \frac{n^2(\pi - p)^2}{np} + \frac{n^2(p-\pi)^2}{n(1-p)} = \frac{n^2(\pi-p)^2}{np(1-p)}[1-p+p]$$

$$= \frac{(\pi-p)^2}{p(1-p)/n} = \left[\frac{\pi - p}{\sqrt{p(1-p)/n}}\right]^2$$

and we see that this is just the square of the u score that we calculated in part (a) and is in fact a variable which follows a χ^2 distribution. There would appear to be two degrees of freedom since there are two discrepancies between observed and expected frequencies. However, once x_1 is decided upon, x_2 is fixed by the constraint $x_1 + x_2 = n$ and hence there is only one degree of freedom. We shall therefore note that we can approximate the problem by a χ^2 distribution with one degree of freedom.

4. For the χ^2 distribution with 1 degree of freedom Table AIII indicates that the critical score is 6.635. We shall therefore accept H_o if the calculated score < 6.635 and reject H_o if it is greater than 6.635. (See Figure 13.13.)

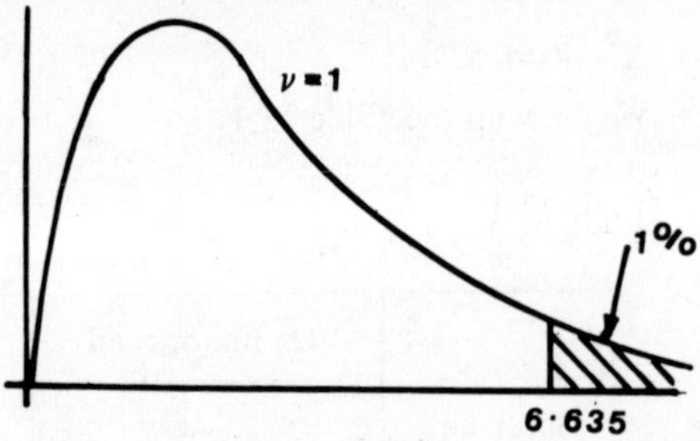

Figure 13.13

5. We calculate $\dfrac{(x_1 - np)^2}{np} + \dfrac{(x_2 - n(1-p))^2}{n(1-p)}$

$$= \frac{(434-479)^2}{479} + \frac{(1002-957)^2}{957}$$

$$= 4.228 + 2.116 = 6.344$$

6. We therefore do not reject H_o.

Now it may well be argued that we have a conflict here, since the two tests give different conclusions. However, it must be remembered that when we square $\dfrac{x - np}{\sqrt{n\,p(1-p)}}$ we must, to be consistent, recognise that the range $-1.96 < u < 1.96$ is equivalent to $\chi^2 < (1.96)^2$. In other words, the χ^2 test is equivalent to a two-tailed test using the normal distribution. Therefore, to compare like with like, we should repeat the test using the normal distribution applied in a two-tailed test where, for 1% risk, the acceptance region will be $-2.58 < u < 2.58$. The conclusion on this occasion is not to reject H_o; note that $(2.58)^2 \triangleq 6.65$ (the slight discrepancy is because the value of u corresponding to 1% on a two-tailed test is somewhat less than 2.58).

You should note that the normal distribution test is more flexible since it permits us to use either a one-tail or a two-tail test, whereas the χ^2 test can only operate on a two-tail basis.

Example 2

A firm keeps a record of absenteeism from one of its factories, attributing the cause of days lost either to illness or to some other reason. Over a given period there were 3000 days lost by illness, of which 1900 were lost by male employees. In the factory, out of a work force of 1500, 900 were men. Is there evidence to show that there is a difference in rate of absenteeism due to illness between male and female employees?

1. If there is no difference between men and women with regard to illness-caused absenteeism then the ratio of male to total days so lost should be the same as the ratio of male to total employees (apart from sampling errors); this proportion should be 900:1500 i.e. 3:5. The null hypothesis is therefore H_o: $p = 3{:}5$ and the alternative hypothesis is $H_1 : p \neq 3{:}5$.
2. We take an α - risk of 5%.
3. We use a χ^2 distribution with 1 degree of freedom.
4. From Table AIII we find that the critical score is 3.841 and we shall accept H_o if the calculated score for χ^2 is < 3.841.
5. We first construct the Table 13.2

Table 13.2

	Male	Female	Total
Observed Frequency	1900	1100	3000
Expected Frequency	1800	1200	3000
Difference	+100	−100	

Then we calculate $\dfrac{(+100)^2}{1800} + \dfrac{(-100)^2}{1200} = 13.89$ (2 d.p.)

6. We reject H_o with a risk of somewhat less than 5%. (In fact if the α - risk had been 1% we should still have rejected H_o.)

Additive Property of χ^2

We now make use of a result, hinted at earlier, that if x_1^2 has a χ^2 distribution with ν_1 degrees of freedom and x_2^2 has a χ^2 distribution with ν_2 degrees of freedom then the random variable $x_1^2 + x_2^2$ has a χ^2 distribution with $\nu_1 + \nu_2$ degrees of freedom. This result clearly extends to more than two variables. This allows us to pool information from a repeated experiment.

Example

A null hypothesis H_o is tested by three experiments. The experiments are performed under the same conditions and the value of χ^2 calculated in each case. The results are 3.6, 1.9 and 2.7 respectively and for each experiment the number of degrees of freedom is 1. Comparison with Table AIII which gives a χ^2 value for 5% risk of 3.841, shows that the null hypothesis cannot be rejected from the results of any one experiment. However, if we use the additive property of χ^2, we find that the pooled value is 8.2 with 3 degrees of freedom. The critical value is 7.815 and this means that, with a 5% risk, we can reject H_o on the basis of all three experiments.

Test on several proportions

If we observe several sample proportions of an attribute, we can use the χ^2 test to examine whether there is a significant difference between the proportions. Suppose we have k such samples. We need to compare the observed proportion p_i of the attribute in each sample with the expected proportion; for the purpose of the test we assume that the expected proportions of the attribute are the same for each sample. This is our null hypothesis. We then estimate p by $\hat{p} = \dfrac{p_1 + p_2 + \ldots + p_k}{N}$ where N is the total number in all the samples.

Next, we calculate for each sample the expected number of successes (i.e. the number in the sample which possess the attribute).

Sample	1	2	....	k
Proportion of successes	p_1	p_2	....	p_k
Number of successes	n_1p_1	n_2p_2	....	n_kp_k
Number of failures	$n_1(1-p_1)$	$n_2(1-p_2)$	....	$n_k(1-p_k)$
Total number	n_1	n_2	....	n_k

For sample i this is $s_i = \left[\dfrac{\sum_{j=1}^{k} n_j p_j}{N} \cdot n_i\right]$

We also calculate for each sample the expected number of failures $f_i = n_i - s_i$

Then we calculate $\sum_{i=1}^{k} \dfrac{(n_i p_i - s_i)^2}{s_i} + \sum_{i=1}^{k} \dfrac{(n_i(1-p_i) - f_i)^2}{f_i}$

This is really the sum of $2k$ terms, corresponding to the k frequencies of success and k frequencies of failure we have observed; for each cell we calculate (observed frequency−expected frequency)^2/expected frequency and add the results. This number is a statistic which follows the χ^2 distribution; in a

method analagous to that in the subsection on testing a single proportion, we can show that the number produced is simply $\sum_{i=1}^{k} \frac{(x_i - n_i p)^2}{n_i p(1-p)}$ where x_i is the observed number of successes; p is the pooled estimate $\frac{x_1 + x_2 + \ldots + x_k}{n_1 + n_2 + \ldots + n_k}$.
From the general results for the χ^2 variates we see that the number we calculate is a χ^2 variate with $(k-1)$ degrees of freedom. (We can decide arbitrarily on $(k-1)$ proportions, then the rest are fixed by the given totals in each sample and the total number of successes.)

Example

Tests are made on the proportion of defective castings produced by 4 different moulds. For mould A there were 15 failures out of 155 castings, for mould B there were 16 out of 100 castings, for mould C there were 13 out of 125 and for D there were 18 defectives out of 160 castings. With a 5% level of significance, test the hypothesis that the proportion of failures is the same for each mould.

We first tabulate observed frequencies as in Table 13.3.

Table 13.3

	Mould A	Mould B	Mould C	Mould D	Total
Failures	15	16	13	18	62
Non-failures	140	84	112	142	478
Total	155	100	125	160	540

From this table we see that the overall proportion of failures is $\frac{62}{540} = 0.115$ (3 d.p.).

1. The null hypothesis H_o is that the proportion of failures for each mould is 0.115. The alternative hypothesis H_1 is that the proportions are not all 0.115.
2. The α - risk is 5%.
3. We use a χ^2 distribution. To determine the number of degrees of freedom note that fixing any three frequencies, each in a different column, is necessary and sufficient to determine all the other five entries for observed frequencies, since the row and column totals determine these. For example, of the eight frequencies, fixing 15, 84 and 18 will allow the others to be calculated from the row and column totals. We have therefore three degrees of freedom. Note that the number of degrees of freedom is the product (number of rows -1)(number of columns -1).
4. The acceptance region is found from Table AIII to be $\chi^2 < 7.815$

5. We first compute a table of frequencies expected under H_o. For mould A we expect 0.115 x 155 failures and 155–(0.115 x 155) non-failures.

	Mould A	Mould B	Mould C	Mould D	
Failures	17.80	11.48	14.35	18.37	62
Non-failures	137.20	88.52	110.65	141.63	478
Total	155	100	125	160	540

For each of the central eight cells we calculate the quantity (observed frequency–expected frequency)^2/expected frequency and add the results. We now produce a composite table; in each of the central cells we arrange the numbers thus:

observed frequency

$$\frac{(\text{o.f.} - \text{e.f.})^2}{\text{e.f.}}$$

expected frequency

and obtain Table 13.4

Table 13.4

	Mould A		Mould B		Mould C		Mould D		Totals
Failures	15 17.80	0.440	16 11.48	1.780	13 14.35	0.127	18 18.37	0.007	62
Non-Failures	140 137.20	0.057	84 88.52	0.231	112 110.65	0.016	142 141.63	0.001	478
Totals	155		100		125		160		540

Then $\chi^2 = 0.440 + 1.780 + \ldots.. + 0.016 + 0.001 = 2.659$

6. Since the calculated values of χ^2 lies within the acceptance region we must accept H_o.

Yates' Correction

When the calculated value of χ^2 is close to the edge of the acceptance region it is important to determine this calculated value as accurately as possible. Since we are applying a continuous distribution to discrete data we should really apply a correction for continuity. The form of correction we use is known as *Yates' correction.* For each cell we calculate

(|observed frequency − expected frequency| − 0.5)^2 / (expected frequency). Note that Yates' correction always decreases the calculated χ^2. We now work through an example to see the effect of this correction. Unless the calculated value of χ^2 is just in excess of the value at the edge of the acceptance region the correction is seldom applied to situations with more than 1 degree of freedom.

Minimum Class Frequency

So far, all the examples we have chosen have yielded expected frequencies $\geqslant 5$. If this condition is not observed the χ^2 distribution becomes a bad approximation. To overcome this, we have to combine classes and an example of this is given in the section on Goodness of Fit.

Example

Two new experimental methods of manufacturing concrete slabs were compared by testing a sample of 24 slabs from each method. Of the slabs from method **A** 15 had a low workability compared with 7 of those from method **B**. With a 5% risk of wrong rejection test whether there is a difference in the proportion of slabs with low workability.

We draw up the Table 13.5, calculating the expected frequencies on the basis of H_o: the proportions of slabs of low workability (and hence of not low workability) are the same for the output population of each method; this proportion is (15 + 7)/(24 + 24) i.e. 22/48.

Table 13.5

	Method A		Method B		Total
Low	15		7		
Workability		1.455		1.455	22
	11		11		
Not	9		17		
Low		1.231		1.231	26
	13		13		
Total	24		24		48

$$\chi^2 = 1.455 + 1.455 + 1.231 + 1.231 = 5.372$$

There is one degree of freedom in this example (why?) and since Table AIII gives the critical value $\chi^2_{5\%}$ as 3.841 we have evidence strong enough to reject H_o. It is instructive however to see the effect of using Yates' correction.

$$\chi^2 = \frac{(|15-11|-0.5)^2}{11} + \frac{(|7-11|-0.5)^2}{11} + \frac{(|9-13|-0.5)^2}{13} + \frac{(|17-13|-0.5)^2}{13}$$

$$= 1.114 + 1.114 + 0.942 + 0.942 = 4.112$$

The value is now considerably reduced, but still in excess of $\chi^2_{5\%}$ and we must still reject H_o.

Contingency Tables

Where observed frequencies can be categorised in a table with r rows and k columns we have an r - by - k **contingency table.** Often we test these observed frequencies against a null hypothesis that the two attributes, which categorise the frequencies into rows and columns, are independent: we are then carrying out a **contingency analysis.** We have, of course, encountered several such contingency tables already; what we now seek to do is to extend ideas.

Contingency tables can arise in two ways:

(a) Samples are taken from k different populations and the observations from each sample are classified under more than two outcomes, which may be qualitative in nature (e.g. major damage, minor damage or no damage).

(b) A sample is taken from one population and the observations are categorised in two ways in a qualitative fashion (e.g. colour and hardness).

In the first case, each column total (sample size) is fixed; however, in the second case, only the grand total of observations is fixed. We shall now work through two examples to tie all the concepts together.

Example 1

Three kinds of protection against corrosion are tested by taking samples of metal alloy and applying one of the protection treatments on each sample. The samples were then subjected to corroding agents under identical conditions. The following results were observed:

	Method A	Method B	Method C	Totals
Severe corrosion	9	8	14	31
Minor corrosion	27	17	30	74
No corrosion	50	63	47	160
Totals	86	88	91	265

It is required to establish whether there is a difference in effectiveness of treatment by the three methods. (In our context we want to establish whether the relative frequencies of each category of corrosion are independent of the method of treatment.) This is an example of case (a).

1. The null hypothesis H_o is that the proportion of each category is the same, viz. 31 : 74 : 160.

2. We take an α - risk of 5%.

3. We use a χ^2 distribution. To determine the number of degrees of freedom, note that fixing 6 of the 9 central frequencies is sufficient; the column totals and the grand total will determine the other three. (The number of degrees of freedom is $(r-1) \times (k-1)$).

4. The acceptance region is found to be $\chi^2 \leqslant 16.92$.

5. We first compute a composite table of frequencies so that each central cell contains three numbers arranged as described on page 468. (See Table 13.6).

Table 13.6

	Method A		Method B		Method C		Totals
	9		8		14		
Severe corrosion		0.11		0.51		1.05	31
	10.06		10.29		10.65		
	27		17		30		
Minor corrosion		0.37		2.33		0.83	74
	24.02		24.57		25.41		
	50		63		47		
No corrosion		0.07		1.83		1.15	160
	51.92		53.14		54.94		
Totals	86		88		91		265

Calculations are shown to 2 d.p. Notice, for example, that the expected frequency for severe corrosion under Method B is calculated as $\frac{31}{265} \times 88$.

Then we calculate $\chi^2 = 0.11 + 0.51 + ... + 1.15 = 8.25$.

6. The value calculated lies well within the acceptance region and we cannot reject H_o.

Criticism of χ^2 test

It is often stated that whenever numerical variables appear, as opposed to qualitative attributes, they should be analysed in a way which exploits their numerical nature. All the χ^2 test can do is to indicate whether a significant difference exists: it cannot give a quantitative measure of the difference.

Example 2

The following table relates to classification of 600 metal specimens. We require to test the possibility of an association between hardness and colour.

Colour \ Hardness	Soft	Fairly Hard	Hard	Very Hard	Totals
Light	143	118	91	48	400
Medium	65	55	24	6	150
Dark	32	7	5	6	50
Totals	240	180	120	60	600

1. The null hypothesis H_o is that there is no association between hardness and colour (i.e. that the two attributes are independent). For example the probability of a light-coloured, hard specimen is $\frac{400}{600} \times \frac{120}{600}$ and, hence, the expected frequency of such samples is $\frac{400 \times 120}{600} = 80$.
2. We work with an α - risk of 1%.
3. We use a χ^2 distribution with 6 degrees of freedom (why?).
4. The acceptance region is $\chi^2 \leqslant 16.81$.
5. The Table 13.7 is computed in a similar way to the Table 13.6 of the last example.
   We obtain a calculated χ^2 value of 28.36.
6. Since this value is far in excess of the critical value of 16.81 we conclude that we must reject H_o.

Table 13.7

Hardness / Colour	Soft		Fairly Hard		Hard		Very Hard		Totals
	143		118		91		48		
Light		1.81		0.03		1.51		1.60	400
	160		120		80		40		
	65		55		24		6		
Medium		0.42		2.22		1.20		5.40	150
	60		45		30		15		
	32		7		5		6		
Dark		7.20		4.27		2.50		0.20	50
	20		15		10		5		
Totals	240		180		120		60		600

This implies that there is a significant association between the colour and the hardness of the metal specimens.

Goodness of Fit

We can use the χ^2 distribution to test goodness of fit between a set of observed frequencies and a set of frequencies obtained from a theoretical model. We must beware of three factors.

(a) The determination of the frequency classes must ensure that no observed frequency is $\leqslant 5$.

(b) We must beware of too good a fit. If the calculated χ^2 value is very small, then we can see whether the chances of obtaining such a low value are less than say, 1%. If so, we might become suspicious and wonder whether the results have been faked.

(c) We must be careful to calculate correctly the number of degrees of freedom. If there are k frequency classes, then the number of degrees of freedom is $k - 1 - m$, where m is the number of parameters of the theoretical distribution which have to be estimated from the observations. For a Poisson distribution we need only the parameter λ and for the binomial distribution we need p, but for the normal distribution we require both μ and σ. We illustrate the principles by two examples.

Example 1

Background radioactivity being usually random, the number of pulses observed in a small time interval should follow a Poisson model. In a laboratory experiment 100 observations, each lasting 10 seconds, produced the following results:

Number of pulses recorded	0	1	2	3	4	5	6	7	8
Number of intervals	5	15	21	23	18	10	5	2	1

If the theoretical distribution is a Poisson one we need to estimate λ.

This we can do from $\overline{x} = \dfrac{\Sigma x_i f_i}{\Sigma f_i} = \dfrac{(0 \times 5 + 1 \times 15 + \ldots\ldots + 7 \times 2 + 8 \times 1)}{100}$

$$\therefore \lambda = \overline{x} = \frac{300}{100} = 3$$

We now calculate the theoretical frequencies. The probability of 0 pulses is $e^{-3} = 0.0498$ (4 d.p.) and hence the expected frequency of 0 pulses is $100 \times e^{-3} = 4.98$ (2 d.p.). The other expected frequencies are calculated similarly. The calculations for χ^2 are shown below:

Number of pulses Recorded	Observed Frequencies	Expected Frequencies	$\dfrac{(O_i - E_i)^2}{E_i}$ (4 d.p.)
0	5	4.98	0.0001
1	15	14.94	0.0002
2	21	22.41	0.0887
3.	23	22.41	0.0155
4	18	16.81	0.0842
5	10	10.08	0.0006
6	5 }	5.04 }	
7	2 } 8	2.16 } 8.01	0.0000
8	1 }	0.81 }	
		Total	0.1893

You should notice that we have grouped the last three frequency classes so that no class contains fewer than 5 members. We have now to set these results in the framework of our strategy.

1. The null hypothesis, H_o, is that there is no difference between the observed and expected frequencies.
2. We choose an α - risk of 5%.
3. We employ a χ^2 distribution with $7 - 1 - 1 = 5$ degrees of freedom (there are 7 classes for the calculation of χ^2 and we need to estimate 1 parameter from the observations).
4. We find that the critical value for χ^2 is 11.07; we also check the value of χ^2 which is exceeded on 99% of occasions: this is 0.55 (2 d.p.).
5. The calculations are as above.
6. The calculated value of χ^2, 0.19 (2 d.p.) is well below the 5% value of 11.07 and we should therefore accept H_o. However, the calculated value is considerably less than the 99% value of 0.55; this causes us at least to examine the experimental observations more critically: the fit is almost too good to be true.

Example 2

The lengths of a sample of 250 bolts produced by a company were measured. The results are shown below.

Length (cm)	Frequency f_i	Mid-value x_i	u_i	$f_i u_i$	$f_i u_i^2$
2.5347 – 2.5349	2	2.5348	–5	–10	50
2.5350 – 2.5352	6	2.5351	–4	–24	96
2.5353 – 2.5355	8	2.5354	–3	–24	72
2.5356 – 2.5358	15	2.5357	–2	–30	60
2.5359 – 2.5361	42	2.5360	–1	–42	42
2.5362 – 2.5364	68	2.5363	0	0	0
2.5365 – 2.5367	49	2.5366	1	49	49
2.5368 – 2.5370	25	2.5369	2	50	100
2.5371 – 2.5373	18	2.5372	3	54	162
2.5374 – 2.5376	12	2.5375	4	48	192
2.5377 – 2.5379	4	2.5378	5	20	100
2.5380 – 2.5382	1	2.5381	6	6	36
Totals	250			97	959

We have to determine whether the observed frequencies follow a normal distribution; we work to a 5% level of significance.

The first calculations are those of the mean and standard deviation and for this we code the variables so that $u_i = (x_i - 2.5363)/0.0003$.

Then $\bar{u} = (\Sigma f_i u_i)/(\Sigma f_i) = \dfrac{97}{250} = 0.388$

and hence $\hat{\mu} = 0.0003\, u + 2.5363 = 2.5364$ cm (4 d.p.)

Further $\hat{\sigma}^2 = \dfrac{1}{n} \Sigma f_i x_i^2 - (\hat{\mu})^2 = V(x)$

Now $V(u) = \dfrac{1}{250} \Sigma f_i u_i^2 - \bar{u}^2 = \dfrac{959}{250} - (0.388)^2$

$= 3.836 - 0.151 = 3.685$

Hence $\hat{\sigma}^2 = V(x) = (0.0003)^2 \; V(u)$

and therefore $\hat{\sigma} = 0.0003 \times \sqrt{3.685} = 0.000577$ cm (3 s.f.)

We are now in a position to use our theoretical model of a normal distribution with mean 2.5364 and standard deviation 0.000577 cm to calculate expected frequencies. We must first calculate the probabilities of respectively obtaining a value in the ranges (2.53465, 2.53495), (2.53495, 2.53525), ..., (2.53795, 2.53825) and then converting these to expected frequencies by multiplying the probabilities by 250. The results are shown below, with the expected frequencies quoted to 1 d.p.

Observed Frequencies O_i	Expected Frequencies E_i	$\dfrac{(O_i - E_i)^2}{E_i}$ (2 d.p.)
2	1.1	0.74
6	4.0	1.00
8	11.1	0.87
15	23.9	3.31
42	39.5	0.16
68	50.2	6.31
49	49.0	0
25	36.6	3.68
18	21.1	0.46
12	9.4	0.72
4	3.1	0.26
1	1.0	0
Totals 250	250.0	17.51

1. As before, the null hypothesis is that the observations exactly fit the quoted normal distribution.

2. The α - risk is 5%.
3. We use a χ^2 distribution with $12 - 1 - 2 = 9$ degrees of freedom. (Why?)
4. We find that the acceptance region is $\chi^2 \leqslant 16.92$.
5. The calculations are as on the previous page.
6. We reject the null hypothesis and state that the fit is poor.

Problems

1. A supposedly fair coin is thrown 200 times; 85 heads were observed. With a risk of (a) 5%, (b) 1%, test the claim.

2. It is claimed that 3 out of 10 cars do not meet certain safety requirements. Test the claim at a 1% level of significance if among 400 cars tested, 73 failed to meet the requirements.

3. An experiment was repeated 4 times. The first experiment yielded $\chi^2 = 7.90$ with $\nu = 4$ and the corresponding values for the other experiments were 15.01, 8; 17.32, 11; 12.11, 6. What conclusions can be drawn? Use 5% risk.

4. Four experiments with $\nu = 10$, 21, 3 and 7 respectively gave χ^2 values of 20, 33.5, 6.4 and 15 respectively. What conclusions can be drawn at a 1% risk?

5. Over a long period of time the output of a kiln includes on average 5% defectives. A load of 3000 items has 200 defectives. Does this suggest unusual conditions during the manufacture of this load?

6. A random number table of 706 digits showed the following distribution. Does it differ significantly from that expected?

Digit	0	1	2	3	4	5	6	7	8	9
Frequency	75	63	58	50	95	73	102	63	46	81

7. The results below give the number of failures on four types of components. Is there any evidence that failure is more likely on some components than others?

Type	A	B	C	D
Number of failures	46	31	38	49

8. (i) A die is tossed 100 times with the results as shown below. Is the die fair?

Score	1	2	3	4	5	6
Frequency	11	18	20	18	19	14

(ii) If a further 240 throws yield frequencies 33, 48, 40, 41, 29, 49, is the conclusion altered?

9. Four machines were examined for faults over a long period. The numbers of faults that developed were 15, 9, 14 and 12 respectively. Is there evidence of differences in performance? Apply Yates' correction; does the conclusion change?

10. Armatures used in particular machines come from two manufacturers. After 10000 hours, 45 of the 60 armatures from manufacturer A and 66 of the 100 from B are still in use. Is this evidence of a real difference between the manufacturing processes?

11. Four machines A, B, C and D produce output which is categorised as Grade I, II or III. From the following table test the hypothesis that there is no difference between the machine outputs. Use a 1% risk.

Machine	A	B	C	D
Grade I	620	750	400	530
Grade II	130	200	140	130
Grade III	50	50	60	40

What is the effect of Yates' correction?

12. Articles are tested for finish and length. The results of testing a day's output are shown below. Test the hypothesis that the quality of finish is independent of the length.

	Good Finish	Average Finish	Poor Finish
Satisfactory length	324	225	127
Unsatisfactory length	57	50	37

What is the effect of Yates' correction?

13. A garage recorded the following faults on four makes of cars.

Make \ Faults	Mechanical	Electrical	Chassis
A	32	26	12
B	60	54	22
C	52	38	16
D	28	25	12

Is the type of fault independent of the make of car?

14. Samples of three kinds of product were subjected to severe pressure changes. The results of the experiments are shown below. With a 5% risk, test whether there is any association between types of product and nature of pressure-induced defect.

Defect \ Material	A	B	C
Strong	15	9	8
Slight	31	18	27
None	48	62	51

15. In three firms the numbers of products graded as perfect, satisfactory and unsatisfactory are as shown. Is there a significant difference between the proportions of the various kinds of products?

Firm	A	B	C
Perfect	28	42	50
Satisfactory	33	76	71
Unsatisfactory	39	82	79

16. A skilled operative has the following record of rejects over a period of 300 working days. It is believed the rejects are randomly distributed.

Rejects/day	0	1	2	3	4	5	6
Frequency	147	95	46	15	11	4	1

Explain why you would expect the rejects to follow a Poisson distribution and test for goodness of fit.

17. The following data relates to tensile strength of screws. Could the sample have come from a normal population?

Value	1350–1950	1950–2550	2550–3150	3150–3750	3750–4350
Frequency	19	66	114	207	250
Value	4350–4950	4950–5550	5550–6150		
Frequency	236	101	7		

18. The number of accidents reported daily in a factory over 300 working days is tabulated below. Test for agreement with a Poisson distribution of the same mean and total frequency. Comment.

No. of accidents/day	0	1	2	3	4	5	6	7	8
Frequency	17	43	69	68	50	28	13	8	4

19. The frequencies of daily failures of a machine are given in the following table:

No. of failures/day	0	1	2	3	4	5	6
Frequency	51	111	37	37	8	4	2

On the assumption of a constant failure rate these results should fit a Poisson distribution. Find this distribution and use a χ^2 - test to examine the assumption that the failure rate is constant. (L.U.)

20. Of 100 articles produced by factory A, 21 are found to be defective while of 200 produced by factory B the number defective is 20. Assuming the samples to be taken at random at each factory, can it be concluded that the production from B is more satisfactory? (L.U.)

13.6 SAMPLE VARIANCES: THE F TEST

If you have glanced at the next chapter you may have noticed that we are at some stage involved with comparing variances (often set out as a variance analysis table). Variances can be regarded as measures of error within a system and and we are often able to look for significant sources of error by this means.

We have not yet compared the means of two samples and this problem is tackled in the next section. We require first to decide whether the variances of the two samples s_1^2 and s_2^2 can be regarded as equal: if so we are able to 'pool' the variances so that we get a total estimate of the variance. This decision can be made by reference to the so-called F **test.**

Other circumstances in which the comparison of variances is important are when we wish to establish whether two samples could have come from the same population. We should test whether their means are significantly different and then whether their variances differ significantly. Notice that we look at the *difference* between two means and at the *ratio* of their variances.

F distribution

If samples of two independent random variables have sample variances s_1^2 and s_2^2 based on ν_1 and ν_2 degrees of freedom respectively, where the corresponding population variances are σ_1^2 and σ_2^2, then the ratio

$F = \dfrac{s_1^2\,\sigma_2^2}{s_2^2\,\sigma_1^2}$ follows an F distribution with ν_1 and ν_2 degrees of freedom.

Note that there are two independent χ^2 distributions, $\chi_1^2 = \dfrac{\nu_1 s_1^2}{\sigma_1^2}$

and $\chi_2^2 = \dfrac{\nu_2 s_2^2}{\sigma_2^2}$ lurking around, so that we could define $F = \dfrac{\nu_2 \chi_1^2}{\nu_1 \chi_2^2}$

In many cases we are concerned with working under a null hypothesis that σ_1^2 and σ_2^2 are equal and then we test the ratio $\dfrac{s_1^2}{s_2^2}$ against the F distribution.

In this book we always arrange that $s_1^2 \geqslant s_2^2$ so that the calculated ratio $\geqslant 1$.

We may then employ the result that if s_1^2 and s_2^2 are the sample variances of two independent random variables, of sizes n_1 and n_2 respectively, both taken from normal populations with the same variance then the ratio $\dfrac{s_1^2}{s_2^2}$ follows an F distribution with $\nu_1 = n_1 - 1$ and $\nu_2 = n_2 - 1$ degrees of freedom; this is written $F_{\nu_2}^{\nu_1}$

Again, note the underlying assumption of a parent normal population. However, this time we have the two parameters ν_1 and ν_2 to cope with. Figure 13.14 shows the graphs of some F distributions for various combinations of ν_1 and ν_2.

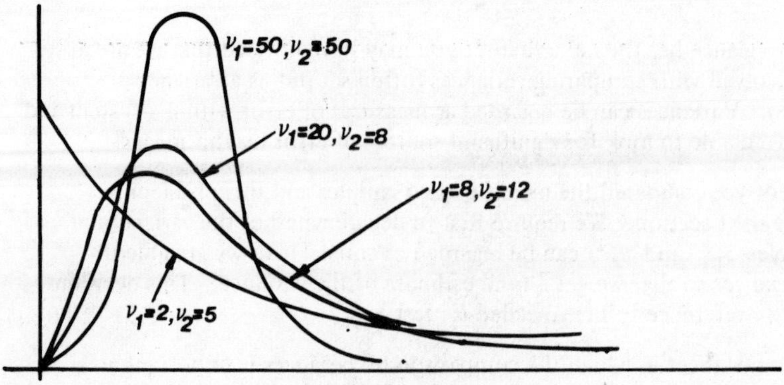

Figure 13.14

From this diagram we can get a qualitative idea of how the shape of the distribution is affected by changing the values of ν_1 and ν_2. Notice that the basic shape is the same. Table AIV shows selected percentage points of several F distributions.

The tabulated value, f_α, is the value of F above which there is an area of α (i.e. the probability of obtaining a value of $F > f_\alpha$ is α) see Figure 13.15; of course, we must first specify ν_1 and ν_2. Sometimes we write $f_\alpha(\nu_1, \nu_2)$, e.g. $f_{0.05}$ (2, 3), to emphasise all three parameters. Table AIV is used as follows.

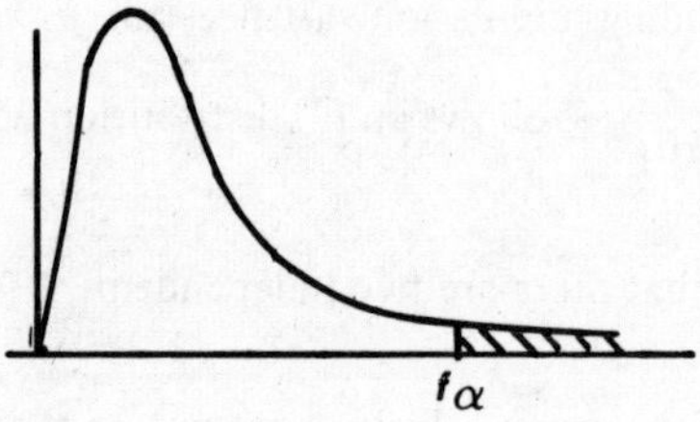

Figure 13.15

1. Select the column corresponding to ν_1
2. Select the row corresponding to ν_2
3. Observe that there are 3 entries corresponding to ν_1 and ν_2. Of these the top one will be relevant to a 5% risk, the middle one to a 2½ % risk and the bottom one to a 1% risk.

Examples

(i) We seek $f_{0.05}$ (2, 3). The three numbers in the table are 9.55, 16.0 and 30.8 and of these we require 9.55.

(ii) We seek $f_{0.01}$ (8, 60). The three numbers are 2.10, 2.41 and 2.82: of these we require 2.82.

Notice that with 8 and 60 degrees of freedom respectively we are saying that the probability of an F score > 2.82 is 0.01 or 1%. Observe also that for a fixed value of α, the values of $F \searrow 1$ as ν_1 and ν_2 get larger. To gain familiarity with the table you should try Problem 1 at the end of this section.

Where the particular combination of ν_1 and ν_2 is not present you must either use interpolation or consult more comprehensive tables.

When testing null hypotheses $\sigma_1^2 = \sigma_2^2$ we shall either have the alternative hypothesis $\sigma_1^2 > \sigma_2^2$ or $\sigma_1^2 < \sigma_2^2$ or $\sigma_1^2 \neq \sigma_2^2$. In the second case we can express the alternative hypothesis as $\sigma_2^2 > \sigma_1^2$ and calculate s_2^2/s_1^2. In the third case we note that we work with a risk of $\alpha/2$, and use a two-tailed test.

Properties of the F distribution

(i) $\nu_2 \to \infty;\ F(\nu_1, \infty) = \dfrac{\chi_1^2}{\nu_1}$ (Hence the χ^2 distribution is a special case of the F distribution.)

(ii) $\nu_1 = 1;\ F(1, \nu_2) = t^2(\nu_2)$ (In some books the variance analysis tables will have a column headed 'F or t^2'. This implies that the t distribution is a special case of the F distribution.)

(iii) $f_\alpha(\nu_1, \nu_2) = \dfrac{1}{f_{1-\alpha}(\nu_1, \nu_2)}$

For example, $f_{0.99}(2, 3) = \dfrac{1}{f_{0.01}(2, 3)} \triangleq \dfrac{1}{30.8} \triangleq 0.0325$

Tests of hypothesis about two variances

We have now set up all the necessary background for testing a hypothesis concerning two variances. We illustrate the basic ideas via examples.

Example 1

Observations of water hardness were taken from the outputs of two separate water treatments. The first sample comprised 11 observations and the second comprised 15 observations. The results were recorded in parts per million of calcium carbonate and it was found that the variance of the first sample was (in coded form) 94.6 and the variance of the second sample was 36.8. Is there a significant difference in the variances?

Although $s_1^2 > s_2^2$ it does not follow that $\sigma_1^2 > \sigma_2^2$, however, since $s_1^2 >> s_2^2$ we are led to suspect that σ_1^2 may be greater than σ_2^2. It is now a question of how much greater than 1 the ratio s_1^2/s_2^2 can be.

1. The Null Hypothesis, H_o is that $\sigma_1^2 = \sigma_2^2$ and the alternative hypothesis, H_1 is that $\sigma_1^2 > \sigma_2^2$. We adhere to our convention that the larger variance is divided by the smaller and so the variance ratio is always $\geqslant 1$.
2. The α - risk is chosen to be 5%. (Note that this must be split equally.)
3. We choose an F distribution with 10 and 14 degrees of freedom respectively.
4. The acceptance and rejection regions are shown in Figure 13.16. Note that the upper limit of the acceptance region is given by $f_{0.05}(10, 14)$

The upper value is found to be 2.60. Therefore we cannot accept H_0 at the 5% level of significance if $F > 2.60$.

5. We calculate $F = 94.6/36.8 = 2.57$ (3 s.f.).

6. Since this lies in the acceptance region we cannot reject H_0.

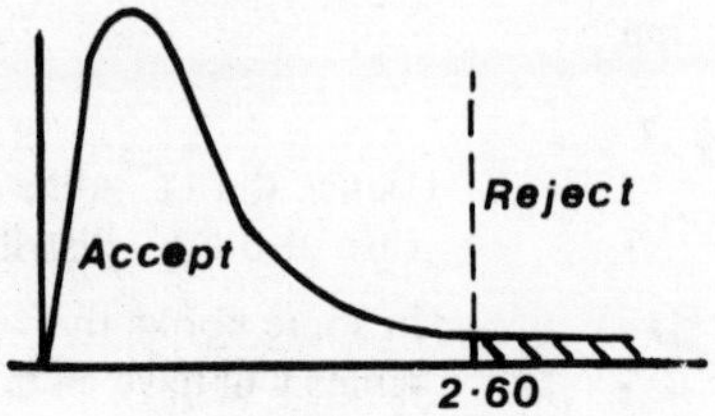

Figure 13.6

Example 2

The effectiveness of diluting a poisonous substance was determined by taking 6 samples from the container after 5 minutes of mixing and a further 6 samples after 5 more minutes. The following results were obtained for the percentage of substance in each sample.

after 5 minutes:	21.3	21.6	19.3	20.8	21.2	21.8
after 10 minutes:	19.6	20.6	20.8	21.8	22.0	21.2

We require to determine whether there is a significant increase in the homogeneity of the dilution.

1. The null hypothesis, H_o is that there is no change in the homogeneity i.e. $\sigma_1{}^2 = \sigma_2{}^2$. The alternative hypothesis, H_1 is that there is an *increase* in the homogeneity, i.e. $\sigma_1{}^2 > \sigma_2{}^2$.
2. We work with an α - risk of 0.05.
3. We employ an F distribution with 5 and 5 degrees of freedom and use a one-tailed test.
4. We find that $f_{0.05}\ (5, 5) = 5.05$; notice that we have concentrated the 5% at one end: see Figure 13.17.
5. We calculate $s_1{}^2$ (relating to 5 minutes) as 0.812 and s_2^2 (relating to 10 minutes) as 0.768. This gives $F = 0.812/0.768 = 1.06$.
6. Clearly we must accept H_o and state that at the 5% level of significance the extra 5 minutes mixing has had no marked effect on the homogeneity of the mixing.

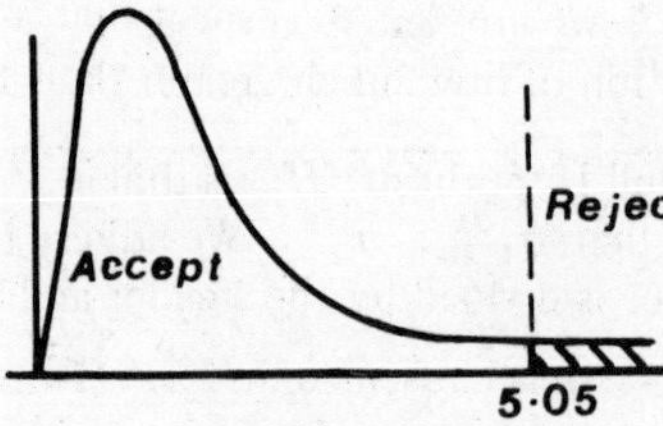

Figure 13.17

Pooling of variances

If two sample variances, s_1^2 and s_2^2 when tested with an F distribution do not suggest that there is a significant difference between them we may wish to obtain an overall (or pooled) estimate of the common variance. What we do is to take a weighted average of the sample variance, where the weights are the respective degrees of freedom. The larger sample - the amalgamation of the two original samples - gives a better estimate of the variance. The pooled estimate is, with the usual notation, $\dfrac{(\nu_1 s_1^2 + \nu_2 s_2^2)}{\nu_1 + \nu_2}$ based on $\nu = n_1 + n_2 - 2$ $= \nu_1 + \nu_2$ degrees of freedom.

Example

In Example 1 above we may pool the variances and obtain an estimate of

$$\frac{10 \times 94.6 + 14 \times 36.8}{10 + 14} = \frac{1461.2}{24} = 60.9$$

Problems

1. Find (a) $f_{0\cdot05}(4, 6)$, $f_{0\cdot05}(10, 7)$, $f_{0\cdot05}(12, 32)$, $f_{0\cdot05}(12, 34)$, $f_{0\cdot05}(12, 35)$, $f_{0\cdot05}(11, 35)$

   (b) $f_{0\cdot025}(3, 3)$, $f_{0\cdot025}(10, 120)$

   (c) $f_{0.01}(10, 8)$, $f_{0.01}(8, 10)$

   (d) $f_{0.99}(3, 10)$, $f_{0.95}(3, 10)$

2. A sample of 8 observations has variance 2.59 and a sample of 16 has variance 4.92. Can these samples be reasonably regarded as coming from the same normal population?

3. For each of the cases below we show the sample size and sample variance for two samples. Find whether they can be reasonably regarded as coming from the same normal population and if so obtain a pooled estimated of the population. Work with 5% risk.

   (a) size 4, variance 0.537; size 16, variance 0.155

   (b) 10, 760/9; 8, 587.5/7

   (c) 16, 6.2; 21, 3.9

   (d) 21, 42.9; 13, 146.0

4. Two voltmeters are tested by comparing ten independent readings from each with the voltage of a standard cell (whose voltage is known to be 2.10 volts). The results are shown below. Is there evidence that one meter is more consistent than the other?

First meter	2.11	2.15	2.14	2.10	2.09	2.11	2.12	2.15	2.13	2.14
Second meter	2.12	2.06	2.02	2.08	2.11	2.05	2.06	2.03	2.05	2.08

5. The following data for two samples is given. Assuming the observations from each sample to be normally distributed test the hypothesis that the samples could have come from the same population.

First sample $n = 11$, $\Sigma x_i = 66$ $\Sigma x_i^2 = 402.4$

Second sample $n = 6$, $\Sigma x_i = 37.6$ $\Sigma x_i^2 = 195.36$

6. The data for two samples of a test for water hardness are shown below. The hypothesis is that there is no significant difference in the variances. Is it justified?

First sample $n = 9$, $\Sigma x_i = 252$, $\Sigma x_i^2 = 7275$

Second sample $n = 14$, $\Sigma x_i = 434$, $\Sigma x_i^2 = 13550$

7. A machine was modified to try to improve its precision. 14 measurements of a quantity were made before modification with results: 2.20, 2.36, 2.22, 2.39, 2.41, 2.25, 2.38, 2.19, 2.57, 2.52, 2.49, 2.33, 2.70, 2.30.

   After modification 10 measurements of the same quantity were made with results: 2.22, 2.27, 2.50, 2.52, 2.33, 2.35, 2.40, 2.44, 2.51, 2.32.

   Is there evidence that modification has improved the precision of the machine?

8. The outputs of two machines were measured for length. From machine A the variance of 21 measurements was 42.9 and from machine B the variance of 13 measurements was 146.0. Is it likely that the variances of the two outputs are equal?

13.7 COMPARISON OF SAMPLE MEANS

When population variances are known we may compare sample means using the normal distribution. We concentrate in this section on situations where one or both of the following conditions obtain

(i) the population variance is unknown
(ii) the sample sizes are small.

By comparison with Section 13.3 we expect an application of the t test. There are two basic cases to study: where the samples may sensibly be *paired* and where they may not. An example of the former case is where two treatments are compared by trying each under several sets of conditions so that in each condition we can compare the true difference of effects. We take up this case later in the section. For the moment we look at examples where pairing is not possible or not meaningful.

Independent Samples

We consider the situation where we wish to compare the outputs from two machines or two factories where it is not necessarily true that the numbers in each sample are equal. Even if the numbers in each sample are the same, the conditions under which the samples were taken may vary and we are entitled to assume that the samples are independent.

We assume that the variances of the populations from which the samples were taken are, in fact, *equal*. We further *assume* that the two populations are *normally* distributed. For reference we state that the two populations will be distributed as $N(\mu_1, \sigma_1^2)$ and $N(\mu_2, \sigma_2^2)$, that samples are drawn of size n_1 with sample mean $\bar{x}_1$ and sample variance s_1^2 from the first population and of size n_2 with sample mean $\bar{x}_2$ and sample variance s_2^2 from the second.

The technique will be to set up a null hypothesis that the two sample means are equal and perform a t test on the difference in the sample means $(\bar{x}_1 - \bar{x}_2)$. In other words, we shall assume that $\mu_1 = \mu_2$ and then calculate

$$\frac{(\bar{x}_1 - \bar{x}_2) - (\mu_1 - \mu_2)}{\text{s.e.}} = \frac{(\bar{x}_1 - \bar{x}_2)}{\text{s.e.}}$$

where s.e. (the standard error of the difference in the means) is equal to $\sqrt{\dfrac{\sigma_1^2}{n_1} + \dfrac{\sigma_2^2}{n_2}}$. However, we have assumed that $\sigma_1^2 = \sigma_2^2 = \sigma^2$ and hence we take the standard error to be

$$\sqrt{\frac{\sigma^2}{n_1} + \frac{\sigma^2}{n_2}} = \sigma\sqrt{\frac{1}{n_1} + \frac{1}{n_2}}.$$

Then $\dfrac{(\bar{x}_1 - \bar{x}_2)}{\sigma\sqrt{\dfrac{1}{n_1} + \dfrac{1}{n_2}}}$ will follow the distribution $N(0, 1)$. For each sample the number of degrees of freedom is one less than the number of items in the sample since the determination of the mean reduces the degrees of freedom by 1; hence the total number of degrees of freedom in the system is $(n_1 - 1) + (n_2 - 1)$. But how valid is the assumption that the variances are equal? We shall first have to check the validity by means of an F test. As regards the assumption of normally distributed populations, the same remarks apply here as applied in Section 13.3.

Example 1

Two makes of steel cable are tested for strength. The breaking loads for samples of each make were measured and the results recorded below in kN. Is there a significant difference in the mean breaking loads? Work with a 5% risk of wrong judgement.

Make I	30.38	30.25	30.34	30.27		
Make II	30.37	30.36	30.35	30.30	30.32	30.39

For ease in following the argument we shall perform some calculations before proceeding.

For Make I $\quad \nu_1 = 3, \ \bar{x}_1 = 30.31$ and $s_1^2 = \dfrac{110 \times 10^{-4}}{3}$

For Make II $\quad \nu_2 = 5, \ \bar{x}_2 = 30.35$ and $s_2^2 = \dfrac{56 \times 10^{-4}}{5}$

(a) *Are the sample variances significantly different?*

1. We set up the null hypothesis that the population variances are equal; the alternative hypothesis is that they are not equal.
2. We work with an α - risk of 5%.
3. We use an F distribution with 3 and 5 degrees of freedom.
4. We find that $f_{0.95}(3, 5) = 5.41$. The rejection region is therefore $F > 5.41$.
5. We calculate $F = \dfrac{s_1^2}{s_2^2} = \dfrac{110 \times 10^{-4}}{3} \times \dfrac{5}{56 \times 10^{-4}} = 3.27$ (3 s.f.)
6. Hence we cannot conclude (with a 5% risk) that the variances are significantly different, though we should have been happier with a value of F closer to 1.

(b) *Pooling of variances*

We now estimate σ^2 as $\dfrac{3 \times \left(\dfrac{110}{3} \times 10^{-4}\right) + 5\left(\dfrac{56}{5} \times 10^{-4}\right)}{3 + 5}$

$= 2.075 \times 10^{-3}$

(c) *Are the sample means significantly different?*

1. The null hypothesis is that $\mu_1 = \mu_2$ or, equivalently, that $\mu_1 - \mu_2 = 0$. The alternative hypothesis is that $\mu_1 \neq \mu_2$.
2. The α - risk is 5%.
3. We use a t distribution with 8 degrees of freedom since we estimate σ from the sample data and employ a two-tailed test.
4. The critical values of t are ± 2.75 and hence the rejection region is $|t| > 2.75$.
5. We calcualte $t = \dfrac{(30.35 - 30.31) - 0}{\sqrt{2.075 \times 10^{-3}\left[\dfrac{1}{4} + \dfrac{1}{6}\right]}} = 1.36$ (3 s.f.)
6. Since the value calculated lies in the acceptance region, we conclude that the mean breaking loads are not significantly different at the 5% level.

Example 2

A standard cell, the voltage of which was known to be 1.10 volts was used to test the accuracy of two voltmeters A and B. Nine independent readings of the voltages from each meter give the following values.

Meter	Variate	Measured Voltages								
A	x_1	1.12	1.16	1.15	1.11	1.10	1.13	1.16	1.14	1.10
B	x_2	1.12	1.09	1.04	1.09	1.09	1.05	1.06	1.03	1.08

Using a maximum risk of 5% of finding spurious differences, determine whether there is a significant difference in the mean readings of the voltmeters. Give a 99% confidence interval for the difference in mean readings.

First we perform some calculations.

$\bar{x}_1 = 1.13, \quad \bar{x}_2 = 1.07$

$s_1{}^2 = \text{var}(\bar{x}_1) = 5.75 \times 10^{-4}; \quad s_2{}^2 = \text{var}(x_2) = 8.44 \times 10^{-4}$

(a) *Are the sample variances significantly different?*

We list merely the differences from the previous examples. The α - risk is 1% and this together with an F distribution with 8 and 8 degrees of freedom produces an acceptance region of $F < 3.44$. In fact,

$F = \dfrac{s_2{}^2}{s_1{}^2} = \dfrac{8.44 \times 10^{-4}}{5.75 \times 10^{-4}} = 1.47$ and this lies in the range appropriate to

an α - risk of 5%. Hence the sample variances are not significantly different.

(b) *Pooling of variances*

We estimate σ^2 as $\dfrac{8 \times (8.44 \times 10^{-4}) + 8 \times (5.75 \times 10^{-4})}{8 + 8} = 7.10 \times 10^{-4}$

(3 s.f.)

(c) *Are the means significantly different?*

Again we merely outline the steps to be taken. With a null hypothesis that $\mu_1 = \mu_2$ and an α - risk of 5%, we employ a t-distribution with 16 degrees of freedom and operate a two-tailed test. Table AII yields the critical value of 2.12. We calculate

$$t = \frac{(1.13 - 1.07) - 0}{\sqrt{7.10 \times 10^{-4}\left[\frac{1}{9} + \frac{1}{9}\right]}} = 4.78$$

Now we must reject the null hypothesis and conclude that the means are significantly different.

(d) *Confidence Interval for difference in means*

Now that we have established a significant difference in the means let us determine a 95% confidence interval for the difference in the means. The observed difference was 0.06 and therefore the confidence interval required will be

0.06 ± 2.12 s.e., i.e. $0.06 \pm 2.12 \times \sqrt{7.10 \times 10^{-4}\left[\frac{1}{9} + \frac{1}{9}\right]}$

i.e. 0.0600 ± 0.0266 (4 d.p.)

This gives a 95% confidence interval of (0.0334, 0.0866) which would seem to be quite large.

(NOTE: If one end of the interval were positive and the other end were negative we should expect the corresponding null hypothesis $\mu_1 = \mu_2$ to have been accepted at an α - risk of 5%.

If on applying the F test, we are forced to reject the null hypothesis $\sigma_1^2 = \sigma_2^2$ then the procedures above will be inapplicable.)

Paired samples

Suppose we are testing whether a particular factor in some experimental situation is causing the mean of a parameter to be significantly different from that value which was expected. How can we be sure that no other factor may be contributing to the difference being so large? One solution, which is extensively used, is to set up a controlled experiment in which for each item in the test sample there is one item in the control sample which is subject to the same conditions.

Alternatively we may wish to compare the effects of two processes on several items which are paired off (in the sense of possessing identical or similar characteristics or in the sense of being subjected to identical conditions). We can then examine the different effects of the processes and be reasonably sure that we have eliminated or at least reduced the effects of other experimental factors. Previously, with independent samples we had ignored these other effects.

We then work with a set of differences and the null hypothesis will be that the mean of such differences is zero. Two examples will clarify the ideas.

Example 1

Ten mixes of concrete were sampled after 2 and 4 minutes of mixing. The samples were tested for porosity. The results, in suitable units are shown below. Determine whether there is evidence that time in mixing affects the porosity of the concrete.

MIX	1	2	3	4	5	6	7	8	9	10
After 2 mins	2.0	3.1	2.5	2.3	3.0	2.7	2.1	2.4	2.8	2.2
After 4 mins	2.2	2.7	2.8	2.4	2.6	2.5	2.4	2.6	2.9	2.5

We shall first effect a solution by the method of independent samples.

Let the variate x_1 be the porosity after 2 minutes and the variate x_2 be the porosity after 4 minutes.

Calculations yield $\overline{x}_1 = 2.51$; $\overline{x}_2 = 2.56$; $s_1{}^2 = 0.143$; $s_2{}^2 = 0.043$.

(A cursory glance at the ratio $s_1{}^2/s_2{}^2$ shows that we may pool the variances to estimate σ^2 as $\dfrac{(9 \times 0.143) + (9 \times 0.043)}{18} = 0.093$

1. The null hypothesis is that $\mu_1 = \mu_2$ and the alternative hypothesis is that $\mu_1 \neq \mu_2$.
2. We work with an α - risk of 5%.
3. We use a t distribution with 18 degrees of freedom and employ a two-tailed test.
4. The acceptance region is $|t| < 2.10$.
5. We calculate $t = \dfrac{(2.56 - 2.51) - 0}{\sqrt{0.093\left[\dfrac{1}{10} + \dfrac{1}{10}\right]}} = \dfrac{0.05}{\sqrt{0.0186}} = 0.367$
6. We conclude that mixing time does not affect porosity.

If however, we proceed via the paired sample approach then we first form a set of differences, being careful to subtract consistently the first reading in each pair from the second.

MIX	1	2	3	4	5	6	7	8	9	10
Differences (4 min – 2 min)	0.2	−0.4	0.3	0.1	−0.4	−0.2	0.3	0.2	0.1	0.3

From this set of 10 readings we calculate

$$\overline{d} = 0.05, \quad \text{var}(d) \equiv s_d^2 = 0.0783$$

1. We set up a null hypothesis, that the expected difference $\mu_d = 0$. The alternative hypothesis is that $\mu_d \neq 0$.

2. We work with an α - risk of 5%.

3. We use a t distribution with 9 degrees of freedom (we have effectively reduced our problem to a study of 10 numbers) and apply a two-tailed test.

4. The acceptance region is found to be $|t| < 2.26$.

5. We calculate $t = \dfrac{0.05 - 0}{\sqrt{\dfrac{0.0783}{10}}} = 0.565$

6. Hence the mean difference is not significant at the 5% level and we have not sufficient evidence that the time in mixing affects porosity.

It can be seen that although both methods yield the same conclusion the values of t obtained are different. What then would be the interpretation if the t-values had been such that in one test the t-value lay in the acceptance region but in the other test the t-value did not. What must be borne in mind is that although we may not be able to detect a significant difference by one of our tests, there may be an underlying real difference in population means. Consequently, any indication of such a difference must be heeded and the fact that one test does give a significant result is sufficient. Note too, that the pairing approach does eliminate factors other than the true difference in effects of the two treatments and we would expect it to be more sensitive. Whether the unpaired test will also give a significant result depends on the amount to which the other factors obscure this essential difference in means.

Example 2

An experiment was conducted to compare the degree of corrosion on pipes which were coated with a protecting agent and on pipes which were not treated. Twenty specimens were tested in pairs. Each pair comprised one treated pipe and one untreated pipe and each pair was subjected to a different set of conditions, so that the items in each pair endured the same test environment. After a year the samples were examined for effects of corrosion. The effects were measured by finding the deepest hollow caused by corrosion and determining its depth. This depth is denoted as x_1 for the uncoated specimens and x_2 for the coated specimens. The results are recorded below in mm. Has the treatment had a significant effect? Work with a risk of 5%.

Pair	1	2	3	4	5	6	7	8	9	10
x_1	1.07	0.94	1.55	1.88	1.40	1.45	1.12	1.78	1.32	1.40
x_2	0.99	1.09	1.09	1.32	1.32	1.50	1.02	1.57	1.02	0.69

We first form a table of differences $d = (x_1 - x_2) \times 100$; the coding is for ease of arithmetic.

Pair	1	2	3	4	5	6	7	8	9	10
d	8	−15	46	56	8	−5	10	21	30	71
d^2	64	225	2116	3136	64	25	100	441	900	5041

Hence $\Sigma\, d = 230$ and $\Sigma\, d^{\,2} = 12112.$

Therefore $\bar{d} = 23$ and $s_d^{\,2} = 758$ (3 s.f.)

1. The null hypothesis is that $\mu_d = 0$, i.e. the protecting agent has no effect on corrosion. The alternative hypothesis is that $\mu_d \neq 0$.
2. The α - risk has already been selected to be 5%.
3. We use the t distribution with 9 degrees of freedom on a two-tailed test.
4. The acceptance region is $|t| < 2.26$.
5. The t score is $\dfrac{23 - 0}{s_d/\sqrt{n}} = \dfrac{23}{\sqrt{758/10}} = 2.64$
6. We therefore reject the null hypothesis at the 5% level.

   NOTE that with an α - risk of 1%, the acceptance region would have been $|t| < 3.25$ and the null hypothesis would not have been rejected.

Finally we calculate 95% and 99% confidence intervals for the difference in means.

The 95% interval is $23 \pm 2.26 \times 8.706 = 23 \pm 19.68$ i.e. (3.32, 42.68).

The 99% interval is $23 \pm 3.25 \times 8.706 = 23 \pm 28.29$ i.e. (−5.29, 51.29).

Notice that the latter interval does include the possibility $\mu_d = 0$ and hence ties in with the acceptance of the null hypothesis at this level. If we convert these intervals back in terms of the original units we have the results (quoted to 2 d.p.).

95% interval (0.03, 0.43); 99% interval (−0.05, 0.51)

Tests of Hypothesis or Confidence Intervals?

It may be argued that a confidence interval may be more meaningful under some circumstances than the test of a hypothesis. A confidence interval can provide a measure of reliability on a sample estimate, whereas a test of hypothesis will only yield rejection or non-rejection. We have seen that the two concepts are linked. That is, for example, a 95% confidence interval on the mean of a sample may include the proposed mean which forms the basis of the null hypothesis. However, since the confidence interval also gives an index of reliability, it is often preferred.

Yet tests of hypothesis do have a role to play: in statistical quality control where the output from a process is continually monitored. It should be borne in mind that economic considerations must also be taken into account and these may modify a decision reached on purely statistical grounds.

Problems

1. Tests were made on the output of two machines. The number of hours of useful life of the items from the two machines are tabulated below. Is there a significant difference in mean useful life?

A	300,	310,	290,	300,	290,	300,	280,	300,	300,	310
B	290,	300,	310,	290,	280,	290,	300,	290		

2. In a determination of the percentage content of a metallic element in an alloy produced by two manufacturers the results were as follows. Are the differences significant?

A	19.2,	15.8,	18.7,	16.3,	12.4
B	13.2,	11.9,	15.2,	11.7,	12.9

3. Six operators made tests on each of 2 machines. The times taken to perform the tasks are shown below. Compare the results if

(a) the difference between operators is ignored.
(b) the difference between operators is eliminated.

	I	II	III	IV	V	VI
A	8.2	7.1	7.4	7.2	7.5	7.0
B	7.5	7.4	7.6	7.0	7.7	6.9

4. The lifetimes of two kinds of abrasive stones are compared. For 10 stones of the first kind the average is 29 units with standard deviation 3 units; for 12 stones of the second kind the average is 33 units with a standard deviation of 2 units. Is the average of the second kind significantly higher than the first?

5. Two makes of component, one plastic and one metal were tested by finding how many operations they could perform before surface imperfections rendered them inoperable. 4 plastic components gave results 2120, 2300, 2245, 2095, whilst 4 metal components gave results 3350, 3270, 3100 and 3360. Is there sufficient evidence that the metal components last 1000 operations more than plastic ones? Use a 1% level of significance.

6. Two balances A and B are tested for comparability. 8 items were weighed on each balance. Do the results suggest that there is a significant discrepancy between the balances?

A	1.003	2.537	2.538	1.528	4.533	4.518	4.025	5.032
B	1.000	2.535	2.535	1.530	4.531	4.516	4.026	5.032

7. The tensile strengths of two alloys were compared by taking samples of 10 specimens of each alloy. The first alloy has mean strength 15700 with standard deviation 2983; the second alloy has mean strength 13550 with standard deviation 2032. Test the hypothesis that the mean strengths of the two alloys do not differ; work with a 1% risk.

8. Nine pieces of fibre are treated with two chemicals A and B, one half of each piece is treated with A and the other half with B. From the results for breaking strengths given, can you conclude that there is a significant difference between the mean breaking strengths of those parts treated with A and those treated with B?

	I	II	III	IV	V	VI	VII	VIII	IX
A	331	381	384	367	365	403	409	346	378
B	355	376	365	367	411	368	395	336	373

9. Pressures at two stations A and B are recorded at random times and the results in millibars are shown below.

A	1007	1001	1003	1032	1036	1020	996
B	977	989	992	1004	984	970	

Test the hypothesis that the mean pressure at A is the same as at B. Is there a more suitable way of testing this hypothesis?

10. The percentages of a substance found in six random samples taken from a large quantity of metal were

4.21 4.57 4.60 4.55 4.32 4.60

Find 95% confidence limits for the mean percentage of the substance in the metal. Subsequently a further seven random samples were taken from a similar quantity of a second metal giving percentages

4.40 4.60 4.80 4.85 4.70 4.45 4.55

Use the variance ratio test to compare the variability of the two metals and the t test to compare the mean percentages of the substance in the metals. Comment on the results of the tests. (L.U.)

13.8 INTRODUCTION TO ANALYSIS OF VARIANCE

Three brands of lubricating oil were tested on three cars of the same make. Each car was test-driven over the same routes on the same days and carried the same load. At the end of the test period, measurements were made of the maximum wear in each cylinder. The results are shown in Table 13.8 ; the wear is recorded in hundredths of a millimetre.

Table 13.8

Lubricant	Cylinder wear			
A	22	21	25	18
B	16	19	21	18
C	10	17	12	15

We have taken readings for each of the four cylinders in the cars, and, if we take the sample mean reading for each car, we may assume that their difference reflects the difference in the performance of the lubricating oil. At first glance it would seem that oil A is least effective and that C is most effective, but it remains to be seen whether the differences are significant.

We can attempt a simple-minded analysis of the data. The row means are, respectively 21½, 18½, 13½. To eliminate the effects of differences between lubricants we shall subtract 3½ from each observation in the first row, ½ from the observations in the second row and add 4½ to those in the third row; this will make all row means equal to 18. The results will then be as in Table 13.9

Table 13.9

					Means
A	18½	17½	21½	14½	18
B	15½	18½	20½	17½	18
C	14½	21½	16½	19½	18

What then are the remaining variations due to? We attribute them to *experimental* error; this could be pure observational error or its combination with unmentioned factors. We can see that there is still a variation *within* each sample against which must be weighed the variation *between sample means.*

Assumptions for model of one - factor analysis of variance

1. The observations are formed into k rows which are effectively random samples from normal distributions.
2. The normal populations have the same variance, viz. σ^2.

We have not specified that the sample sizes are equal. The analysis can easily be modified for unequal sizes, but we shall work with k samples each of size s.

Table 13.10 illustrates the general case.

Table 13.10

Sample	Observations	Totals	Means
1	$x_{11}, x_{12}, \ldots\ldots, x_{1s}$	T_1	$\bar{x}_1$
2	$x_{21}, x_{22}, \ldots\ldots, x_{2s}$	T_2	$\bar{x}_2$
.	. . .	.	.
.	. . .	.	.
.	. . .	.	.
k	$x_{k1}, x_{k2}, \ldots\ldots, x_{ks}$	T_k	$\bar{x}_k$

The mean of sample i is $\bar{x}_i = T_i/s$

The grand total $T = \sum_{i=1}^{k} T_i$; the overall mean $\bar{x} = T/ks$

The total number of observations, $N = ks$

We have effectively assumed that the i th sample comes from a normal population $N(\mu_i, \sigma^2)$.

Partition of total sums of squares

We consider the total sum of squares $\sum_{i=1}^{k}\sum_{j=1}^{s}(x_{ij}-\bar{x})^2$. (The double Σ notation $\sum_{i=1}^{k}\sum_{j=1}^{s}$ means that we first sum the quantities $(x_{ij}-\bar{x})^2$ over each sample and then add all these sample totals).

Theorem

$$\sum_{i=1}^{k}\sum_{j=1}^{s}(x_{ij}-\bar{x})^2 = \sum_{i=1}^{k} s(\bar{x}_i-\bar{x})^2 + \sum_{i=1}^{k}\sum_{j=1}^{s}(x_{ij}-\bar{x}_i)^2 \quad (13.1)$$

or $$\text{Total sum of squares} = \begin{matrix}\text{Between samples}\\ \text{sum of squares}\end{matrix} + \begin{matrix}\text{Within samples}\\ \text{sum of squares}\end{matrix}$$

This result shows that we can split the total variation of the observations about the overall mean into a component representing the variation between sample means and a component representing the variation within each sample. The first term on the right-hand side of (13.1) represents the departures of sample means from the overall mean. The presence of the factor s indicates the size of each sample and suggests that one of the summations has been done.

Estimates of σ^2

(i) Since the k samples came from populations with assumed equal variances σ^2, we are entitled to pool the variances of each of the k samples to obtain a better estimate of σ^2.

The variance of sample i is $s^2{}_i = \frac{1}{(s-1)}\sum_{j=1}^{s}(x_{ij}-\bar{x}_i)^2$.

Since we are assuming that the sample sizes are equal, the pooled estimate of σ^2 is

$$s^2{}_p = \frac{1}{k}\sum_{i=1}^{k} s_i^2 = \frac{1}{k(s-1)}\sum_{i=1}^{k}\sum_{j=1}^{s}(x_{ij}-\bar{x}_i)^2 \quad (13.2)$$

based on $k(s-1)$ degrees of freedom. This is equal to $\frac{1}{k(s-1)} \times \begin{pmatrix}\text{within samples}\\ \text{sum of squares}\end{pmatrix}$.

(ii) We now set up a null hypothesis that the k samples come from populations with equal means, i.e. $H_0: \mu_1 = \mu_2 = ... = \mu_k = \mu$ (13.3). The alternative hypothesis H_1 is that at least one of the population means is not equal to the others.

Under this null hypothesis we can make a second and independent estimate of σ^2. We may now regard the original N observations as comprised of k samples from the same population $N(\mu, \sigma^2)$.

In this case we can consider the k sample means as k observations from the same population. We know that the variance of the means of samples of size s from a population with variance σ^2 is the standard error of sample means, viz. $\dfrac{\sigma^2}{s}$.

If we consider the sample means, $\bar{x}_1, \bar{x}_2, \ldots\ldots, \bar{x}_k$, the variance of this set of k observations is $\dfrac{1}{(k-1)} \sum_{i=1}^{k} (\bar{x}_i - \bar{x})^2$. If this is an estimate of $\dfrac{\sigma^2}{s}$ it follows that we obtain a second estimate of σ^2 as

$$s_m^2 = \frac{s}{(k-1)} \sum_{i=1}^{k} (\bar{x}_i - \bar{x})^2 \qquad (13.4)$$

based on $(k-1)$ degrees of freedom.

This is equal to $\dfrac{1}{(k-1)}$ x (between samples sum of squares).

Notice that since $sk - 1 \equiv (k-1) + k(s-1)$ we have partitioned the total degrees of freedom in the system as well as the total sum of squares. We illustrate the partition of degrees of freedom by Figure 13.18.

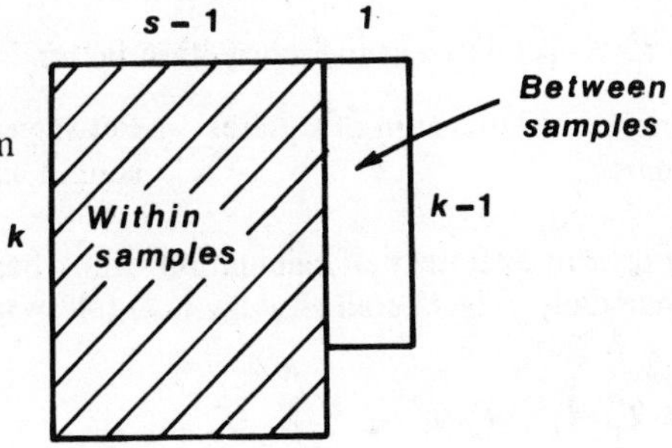

Figure 13.18

The cross-shaded area represents the 'unexplained' part of sum of squares. We shall examine whether the ratio $\dfrac{s_m^2}{s_p^2}$ is significantly larger than 1 by means of an F test.

We can hence construct a theoretical variance analysis table. (see Table 13.11)

Simplification of Calculations

The following simplifications ease the calculations.

(i) Total sum of squares: $\sum_{i=1}^{k} \sum_{j=1}^{s} (x_{ij} - \bar{x})^2 = \sum_{i=1}^{k} \sum_{j=1}^{s} x_{ij}^2 - \dfrac{T^2}{N}$ (13.5a)

Table 13.11

Source of variation	Components of sum of squares	Degrees of Freedom	Components of Variance	F ratio
Between samples	$s\sum_{i=1}^{k}(\bar{x}_i-\bar{x})^2$	$k-1$	$\frac{s}{(k-1)}\sum_{i=1}^{k}(\bar{x}_i-\bar{x})^2=s_m^2$	$\frac{s_m^2}{s_p^2}$
Within samples	$\sum_{i=1}^{k}\sum_{j=1}^{s}(x_{ij}-\bar{x}_i)^2$	$k(s-1)$	$\frac{1}{k(s-1)}\sum_{i=1}^{k}\sum_{j=1}^{s}(x_{ij}-\bar{x}_i)^2=s_p^2$	
Total	$\sum_{i=1}^{k}\sum_{j=1}^{s}(x_{ij}-\bar{x})^2$	$ks-1$		

(ii) Between samples sum of squares: $s\sum_{i=1}^{k}(\bar{x}_i-\bar{x})^2=\frac{1}{s}\sum_{i=1}^{k}T_i^2-\frac{T^2}{N}$ (13.5b)

The term T^2/N is known as the **correction factor.**

(iii) Within samples sum of squares = Total sum of squares − Between samples sum of squares (13.5c)

We may now present a strategy of calculation using these simplifications for a new analysis of variance table. The overall strategy is as follows.

(i) Calculate T_i, T_i^2, T, T^2/N; $\sum_{i=1}^{k}T_i^2$.

(ii) Calculate $\sum_{i=1}^{k}\sum_{j=1}^{s}x_{ij}^2$.

(iii) Calculate total sum of squares via (13.5a), between samples sum of squares via (13.5b) and within samples sum of squares via (13.5c).

(iv) Complete ANOVA (analysis of variance) table.

(v) Carry out decision-making strategy.

Example

Let us now apply the technique to the example on page 492.

We see that $s = 4$, $k = 3$ and hence that $N = ks = 12$. First we re-present the data and provide an extended table (Table 13.12).

Table 13.12

i	x_{ij}				T_i	T_i^2
1	22	21	25	18	86	7396
2	16	19	21	18	74	5476
3	10	17	12	15	54	2916
				Total	214	15788

(i) The correction factor $T^2/N = (214)^2/12 = 3816.\dot{3}$.

(ii) $\sum_{i=1}^{k} \sum_{j=1}^{s} x_{ij}^2 = 22^2 + 21^2 + \ldots + 15^2 = 4014$

(iii) Total sum of squares $= \sum_{i=1}^{k} \sum_{j=1}^{s} x_{ij}^2 - \frac{T^2}{N} = 4014 - 3816.\dot{3} = 197.\dot{6}$

Between samples sum of squares $= \frac{1}{s} \sum_{i=1}^{k} T_i^2 - \frac{T^2}{N} = \left(\frac{1}{4} \text{ x } 15788\right) - 3816.\dot{3} = 130.\dot{6}$

Within samples sum of squares $= 197.\dot{6} - 130.\dot{6} = 67$

(iv) The variance analysis table is as follows .

Source of variation	Components of sum of squares	Degrees of freedom	Components of variance	F ratio	Critical value of F
Between samples	$130.\dot{6}$	2	$65.\dot{3}$	8.78	1%
Within samples	67	9	$7.\dot{4}$		$F_9^2 = 8.02$
Total	$197.\dot{6}$	11			

(v) The decision-making strategy is

1. $H_0 : \mu_1 = \mu_2 = \mu_3$.
2. We work with an α - risk of 1%.
3. The appropriate F distribution has 2 and 9 degrees of freedom.

4. The rejection region is $\frac{s_m^2}{s_p^2} > 8.02$

5. The calculated F value is 8.78.

6. There is a significant difference between sample means, and hence we conclude with slightly less than 1% risk of wrong judgement that there is a difference in effectiveness of lubricating oils.

Two - factor analysis of variance

Consider the following example.

The life of a car tyre may depend upon the make and on the driver. In an experiment drivers were asked to test each of four makes of tyre. The lives of the tyres shown below are the means of four tyres of each make measured in hundreds of miles over 12000 miles; the same make of car is used throughout the twelve experiments. We require to test for significance between tyres and between drivers.

Driver \ Tyre	1	2	3	4
A	–2	5	–3	–4
B	2	9	0	–3
C	–1	5	–2	1

We may well question the experimental design. Did the drivers test each tyre make under identical conditions of road and weather? For this example we may assume that this precaution was taken as far as possible.

We need to examine the effects of two factors on the lives of tyres: the make of tyre and the driver. We shall partition the total sum of squares into *three* components: one due to differences between tyre makes (row differences), one due to differences between drivers (column differences) and one due to experimental error. The procedure is similar to that used for one-factor analysis of variance. Notice that we might have been able to study whether a particular position of cylinder was more subject to wear , had the data of the previous example been presented so as to record which cylinder was which; this might have given us an indication as to the effectiveness of the method of applying the lubricating oil. (We can never be sure that we have included *all* factors in the design of an experiment, but we may hope to include all *significant* factors.) Each factor we study takes a part of the residual sum of squares and we want to ensure that the factors do make *independent* contributions as far as possible.

Model for two - factor ANOVA

We shall assume that each of the s levels of the first factor is combined with each of the r levels of the second factor so that we have a set of $N = rs$ observations x_{ij}, where $j = 1, \ldots\ldots, s$ and $i = 1, \ldots\ldots, r$.

Assumptions

1. The observations are independent random variables.
2. The variable x_{ij} comes from a population which is normally distributed $N(\mu_{ij}, \sigma^2)$.

Partition of total sum of squares

For the purposes of this subsection we shall denote the mean of row i as $\overline{x}_i$ and the mean of column j as $\overline{x}_j$.

We quote the result

$$\sum_{i=1}^{r}\sum_{j=1}^{s}(x_{ij}-\overline{x})^2 = s\sum_{i=1}^{r}(\overline{x}_i-\overline{x})^2 + r\sum_{j=1}^{s}(\overline{x}_j-\overline{x})^2 + \sum_{i=1}^{r}\sum_{j=1}^{s}(x_{ij}-\overline{x}_i-\overline{x}_j+\overline{x})^2 \qquad (13.6)$$

or Total sum of squares = Between rows sum of squares + Between columns sum of squares + Residual sum of squares

Null hypothesis

We set up two null hypotheses which we can test via ANOVA.

(i) There is no difference between row means.

(ii) There is no difference between column means.

Estimates of σ^2

(i) $s_r^{\,2} = \dfrac{1}{(r-1)}\displaystyle\sum_{i=1}^{r}(x_i-\overline{x})^2$ is an unbiased estimator of σ^2/s based on $(r-1)$ degrees of freedom when hypothesis (i) is true.

(ii) Similarly $s_c^{\,2} = \dfrac{1}{(s-1)}\displaystyle\sum_{j=1}^{s}(\overline{x}_j-\overline{x})^2$ is an unbiased estimator of $\dfrac{\sigma^2}{r}$ based on $(s-1)$ degrees of freedom if there are no real differences between columns. In both cases we are using the result that the standard error of sample means of size n from a normal population $N(\mu, \sigma^2)$ is σ^2/n.

(iii) We quote the result that $s_p^{\,2} = \dfrac{1}{(r-1)(s-1)}$ x (Residual sum of squares) is an unbiased estimator of σ^2.

We associate this variance with $(r-1)(s-1)$ degrees of freedom. Figure 13.19 illustrates the partitioning of the degrees of freedom. This illustrates the identity $rs-1 \equiv (r-1)+(s-1)+(r-1)(s-1)$.

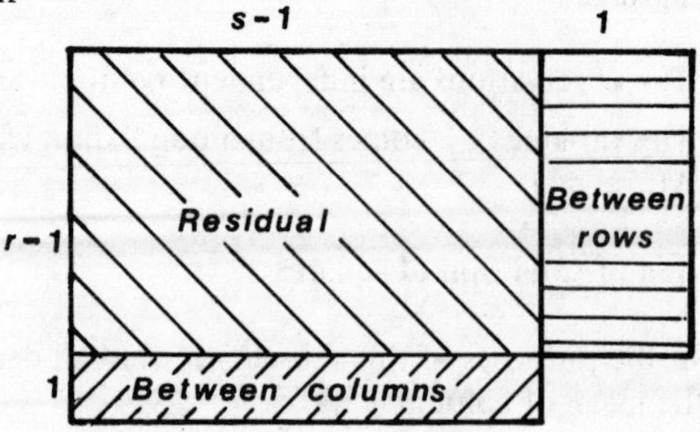

Figure 13.19

We shall aim to test whether the variance of the row differences, $s_r^{\,2}$ or the variance of the column differences, $s_c^{\,2}$, is significant relative to the residual variance, $s_p^{\,2}$.

Simplification of calculations

Before applying these results to the data we note that the calculations can be simplified.

(i) Total sum of squares:

$$\sum_{i=1}^{r}\sum_{j=1}^{s}(x_{ij}-\bar{x})^2 = \sum_{i=1}^{r}\sum_{j=1}^{s}x_{ij}^{\,2} - \frac{T^2}{N} \qquad (13.7a)$$

(ii) Between rows sum of squares:

$$s\sum_{i=1}^{r}(\bar{x}_i-\bar{x})^2 = \frac{1}{s}\sum_{i=1}^{r}T_i^{\,2} - \frac{T^2}{N} \qquad (13.7b)$$

(iii) Between columns sum of squares:

$$r\sum_{j=1}^{s}(\bar{x}_j-\bar{x})^2 = \frac{1}{r}\sum_{j=1}^{s}T_j^{\,2} - \frac{T^2}{N} \qquad (13.7c)$$

(iv) Residual sum of squares = Total sum of squares − Between rows sum of squares − Between columns sum of squares (13.7d)

Strategy

(i) Calculate T_i, T_j; $T_i^{\,2}$, $T_j^{\,2}$; $\sum_{i=1}^{r} T_i^{\,2}$, $\sum_{j=1}^{s} T_j^{\,2}$

(ii) Calculate T, T^2/N

(iii) Calculate $\sum_{i=1}^{r}\sum_{j=1}^{s} x_{ij}^{\,2}$

(iv) Calculate total sum of squares, between rows sum of squares, between columns sum of squares and residual sum of squares.

(v) Complete ANOVA table.

(vi) Carry out decision-making strategy for differences in row means.

(vii) Carry out decision-making strategy for differences in column means.

Example

We first re-present the data for the example on page 498 in Table 13.13.

We may use the result $\sum_{i=1}^{r} T_i = T = \sum_{j=1}^{s} T_j$ as a check.

Table 13.13

Driver \ Tyre	1	2	3	4	T_i	T_i^2
A	−2	5	−3	−4	−4	16
B	2	9	0	−3	8	64
C	−1	5	−2	1	3	9
T_j	−1	19	−5	−6	7	89 Σ
T_j^2	1	361	25	36	423	
					Σ	

Note that $r = 3,\ s = 4,\ N = 12$

(i) $\sum_{i=1}^{r} T_i^2 = 89, \quad \sum_{j=1}^{s} T_j^2 = 423$

(ii) $T = 7$; Correction Factor, $T^2/N = 49/12 = 4.083$

(iii) $\sum_{i=1}^{r} \sum_{j=1}^{s} x_{ij}^2 = (-2)^2 + 5^2 + \ldots.. + 1^2 = 179$

(iv) Total sum of squares $= \sum_{i=1}^{r} \sum_{j=1}^{s} x_{ij}^2 - T^2/N = 179 - 4.083 = 174.917$

Between rows sum of squares $= \frac{1}{s} \sum_{i=1}^{r} T_i^2 - T^2/N$

$= (¼ \times 89) - 4.083 = 18.167$

Between columns sum of squares $= \frac{1}{r} \sum_{j=1}^{s} T_j^2 - T^2/N$

$$= \left(\frac{1}{3} \times 423\right) - 4.083 = 136.917$$

Residual sum of squares = 174.917 −18.167 −136.917 = 19.833

(v) We shall round off these sums of squares to 1 d.p. for the ANOVA table.

Source of variation	Components of sum of squares	Degrees of freedom	Components of variance	F Ratio	Critical values of F
Between rows	18.2	2	9.1	2.76	1% $F_6^2 = 10.92$ 5% $F_6^2 = 5.14$
Between columns	136.9	3	45.6	13.82	1% $F_6^3 = 9.78$ 5% $F_6^3 = 4.76$
Residual	19.8	6	3.3		
Total	174.9	11			

We leave you to draw conclusions from the above.

Problems

1. Five mixes of the same alloy were made and for each mix, five determinations of density are made. Is there any evidence, on the basis of the results below, that certain mixes have a higher mean density than others?

Mix	Density (g.ml^{-1})				
A	5.6	5.5	5.7	5.1	5.2
B	5.3	5.5	5.4	5.2	5.4
C	5.5	5.3	5.4	5.4	5.2
D	5.5	5.4	5.0	5.3	5.8
E	5.7	5.4	5.6	5.5	5.4

2. For measuring strains in members, thin wire gauges are fixed to the surface of the strained member with an adhesive. The fixing of the adhesive requires that it should be cured at a high temperature. In order to analyse the effect of temperature on the sensitivity of the gauge, five different curing temperatures were used each on six gauges. The gauges were applied to suitable surfaces and these surfaces were subjected to known strains. The gauge response was expressed as a percentage of the known applied strain. From the results given does it seem that changes in curing temperature affect the ability of the gauges?

Curing Temperature °C	Percentage gauge response					
80	76	79	68	74	81	78
90	95	82	90	100	92	87
105	103	91	98	90	93	69
120	77	78	65	69	80	69
130	70	56	69	68	61	60

3. From the following measurements made in a precipitator determine whether the flow rate affects the exit dust loading.

Total flow (ft^3/h)	*Exit dust loading (grains per cubic yard in flue gas)*		
200	1.2	1.0	1.6
300	2.0	1.8	2.5
400	2.4	3.0	3.5
500	3.1	3.8	4.4

(C.E.I.)

4. The resistance of wire from four sources was tested by taking a sample of five from each source. From the data below is there a significant difference between the resistances of wires from the four sources?

Source	Resistance (Ω)				
A	15.4	15.0	14.4	14.6	15.8
B	17.2	17.2	17.6	17.6	18.0
C	18.8	18.8	18.4	19.0	19.2
D	11.6	11.0	11.4	11.6	11.0

5. Five materials are tested for water absorption. Samples of four items of each material are tested and the absorption of water (percentage of dry weight) is measured. From the results below, can you conclude that there is a significant difference in absorption?

Type	I	II	III	IV	V
	3.9	6.2	6.0	6.2	4.6
Absorption	4.4	6.7	5.3	5.3	4.2
	4.1	6.3	4.2	5.3	4.6
	4.0	5.8	5.4	5.0	4.7

6. An experiment was carried out to test the hardness of an alloy. Three operators were each asked to make the measurement on the five machines available. The results, in coded form, are below. Test for differences between machines and between operators.

Operator \ Machine	I	II	III	IV	V
A	15	3	4	6	7
B	25	16	5	17	6
C	13	24	13	17	9

7. Samples of four makes of tyre were tested under identical conditions for braking distances, with the following results:

Trial	1	2	3	4
Brand A	27	30	25	26
Brand B	25	20	22	21
Brand C	27	31	30	32
Brand D	26	26	25	23

Test at a level of significance of 1% whether the differences between the mean braking distances obtained for the four makes of tyre are significant. (C.E.I.)

8. The following table gives the productivity (measured in suitable units against some standard) of three shifts in a given industrial plant over a period of one week.

Shift	Mon.	Tues.	Wed.	Thurs.	Fri.
First shift	6	11	9	15	8
Second shift	−2	15	10	8	0
Third shift	−12	7	−9	8	1

Are there significant differences in productivity (a) from shift to shift, (b) from day to day? (C.E.I.)

9. Test the following data for differences in rows and columns.

6	4	6	7
8	5	9	9
6	8	2	8
6	3	4	6

Chapter Fourteen

Linear Regression

14.1 INTRODUCTION TO SIMPLE REGRESSION

In this chapter we shall be concerned mainly with studying the relationship between several variables which appear from a cursory glance at the data to be related. We first tackle these problems from a *regression* approach, that is, we regard one of the variables as the dependent variable and the others as independent variables; the second line of attack is to study *correlation* effects which show to what degree the variables are related. We shall see that correlation is a less powerful technique than regression; however, there is a connection between the two ideas.

We now look at two examples which will be used to illustrate the main ideas of *simple regression* and correlation (i.e. only two variables are involved in each case). An introduction to *Multiple regression* which involves more than two variables is given at the end of the chapter.

Example 1

The following data gives the percentage of sand in soil at different depths.

depth (cm)	0	15	30	45	60	75	90	105	120
% sand	75.6	58.0	59.3	57.5	52.5	54.2	35.8	41.9	32.6

We make a start by labelling the observations of depth as x_i and the observations of percentage of sand as y_i. The data suggests that the percentage of sand decreases with depth from a first glance. What we shall seek to do is to answer the following questions (though not necessarily in the order shown).

(i) Is there a significant relationship between depth of measurement and percentage of sand?

(ii) Is there a simple relationship between depth and percentage of sand which will enable us to predict percentage of sand given the depth?

(iii) Could we find a simple relationship to estimate depth, having measured the percentage of sand in a soil sample?

(iv) In particular, is a linear relationship a good model?

(v) If we are able to obtain a linear relationship, then how reliable will its predictions be? Can we perhaps place confidence intervals on the predictions?

Note that this situation is typical of many laboratory experiments. Often a straight line is fitted either to the observed data or to data derived from the observations in such a way as to make a straight line reasonable. For example, if we have a theoretical model equation $p\,v^{\gamma} = c$ where c and γ are constants and both pairs of readings for p and v are experimentally observed then we can rearrange the model equation as

$$\log p + \gamma \log v = \log c \quad \text{or} \quad P + \gamma V = D \quad \text{or} \quad P = D - \gamma V$$

which suggests a linear relationship between $P(= \log p)$ and $V(= \log v)$. Usually we would fit a straight line to the derived data and then estimate D and γ from this line. By a simple calculation we can deduce c, the original constant. Regression analysis puts this procedure on a rigorous footing by deriving the equation of the straight line algebraically (hence removing subjective fitting of the straight line) and also by measuring the goodness of fit of the line, placing confidence limits on the line slope and intercept and placing a confidence interval on any prediction from the line.

In this example, a straight line fit is attempted in the hope that it is a reasonable fit and that it will provide a reasonable prediction model.

Example 2

The maintenance costs of a particular kind of machine are expected to increase with the age of the machine. In addition, because of inflation the cost of labour and parts will increase. To determine the effect of age on cost of maintenance a survey of 12 machines in a factory was undertaken and the results tabulated as below.

Age of machine (years) x_i	1	1.5	2	3	3.5	4	5	6	7
Cost of last year's maintenance (£100's) y_i	22	20	22	21 23	25	27 24	25 27	26	29

Notice that, here, some readings are duplicated (e.g. for $x = 3$, $y = 21$ and 23) and this leads us to wonder whether we should have taken 2 or 3 readings of percentages of sand for each depth in the previous example.

The questions we might ask are

(i) Is there a significant relationship between age of machine and annual maintenance cost? (We were careful to take all observations in the same year so that effects of inflation were reduced, since inflation has almost monthly increases it may well be that an operation carried out in August may well cost significantly more than the same operation carried out in February.)

(ii) Will a linear relationship provide a good fit?

(iii) How reliable will predictions from a linear model be?

(iv) Can we suggest a useful life for the machine, such that after a certain number of years it is no longer economical to maintain the machine?

Our approach will be to work through the development of the techniques with the first example and then, with the techniques at our disposal to attack the second example, developing what further techniques appear to be required.

14.2 SCATTER DIAGRAMS AND THE LEAST SQUARES MODEL

Provided the observations are relatively few in number, it is worthwhile plotting each observation (x_i, y_i) on a graph with Cartesian axes. Once again, we emphasise that our observations are but a sample of all possible observations. The observations from Example 1 are plotted in Figure 14.1. The points show a general downwards trend from left to right, but it is by no means clear that the fitting of a straight line will give reasonable predictions. In Figure 14.2 we have drawn in a straight line which qualitatively seems to fit the plotted data. It has the special feature that it passes through the point with coordinates $(\bar{x}, \bar{y})$; the reason for this will become apparent later. We shall fit the line

$$y = \alpha + \beta x \tag{14.1}$$

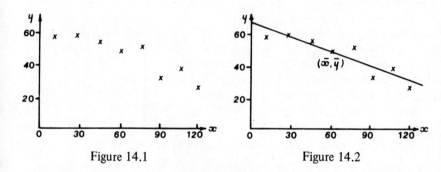

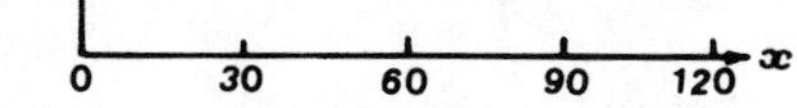

Figure 14.1

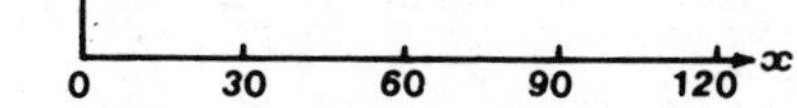

Figure 14.2

The following points are to be noted

(i) For each observed value of x_i it may be that the observed value y_i does not equal the value of y obtained when substituting $x = x_i$ in (14.1). We may write $y_i = \alpha + \beta x_i + \xi_i$, where $i = 1, \ldots\ldots, n.$ (14.2) Equation (14.2) is the **simple regression model.** The term ξ_i is an error term and the ξ_i are independent random variables having mean 0 and variance σ^2.

(ii) We have to obtain estimates of α and β from the observations.

(iii) If $\beta = 0$ there is no regression model and we must test whether $\hat{\beta} = 0$ is a possible estimate.

(iv) We *assume* that the values x_i are not subject to error but that y_i alone is subject to the errors in the model.

(v) The model itself may not be an ideal relationship for the true population of x and y, but even if it were, the samples of observations may not fit any such model exactly because of observational errors in y. In other words, we will obtain an *estimated regression line*

$$\hat{y}_i = \hat{\alpha} + \hat{\beta} x_i$$

Note that the 'hats' on α, β and y_i indicate that they are estimates.

14.3 ESTIMATION OF THE PARAMETERS OF THE REGRESSION LINE

We shall first estimate the parameters α and β and then place confidence intervals on them.

From (14.2) we have an expression for the error at each observed point

$$\xi_i = y_i - \alpha - \beta x_i \tag{14.3}$$

The method of least squares minimises

$$\sum_{i=1}^{n} \xi_i^2 = \sum_{i=1}^{n} (y_i - \alpha - \beta x_i)^2 = S(\alpha, \beta)$$

We solve the equations

$$\frac{\partial S}{\partial \alpha} = 0 = \frac{\partial S}{\partial \beta}$$

These lead to the *normal equations*

$$\left.\begin{aligned} \sum_{i=1}^{n} y_i &= n\alpha + \beta \sum_{i=1}^{n} x_i \\ \text{and} \quad \sum_{i=1}^{n} x_i y_i &= \alpha \sum_{i=1}^{n} x_i + \beta \sum_{i=1}^{n} x_i^2 \end{aligned}\right\} \tag{14.4}$$

We shall now introduce a notation which will be useful in subsequent analysis.

$$\left.\begin{aligned} \text{Let } S_{xy} &= \sum_{i=1}^{n} (x_i - \bar{x})(y_i - \bar{y}) \equiv \sum_{i=1}^{n} x_i y_i - \bar{x} \sum_{i=1}^{n} y_i \\ &\equiv \sum_{i=1}^{n} x_i y_i - \bar{y} \sum_{i=1}^{n} x_i \\ S_{xx} &= \sum_{i=1}^{n} (x_i - \bar{x})^2 \equiv \sum_{i=1}^{n} x_i^2 - \bar{x} \sum_{i=1}^{n} x_i \\ S_{yy} &= \sum_{i=1}^{n} (y_i - \bar{y})^2 \equiv \sum_{i=1}^{n} y_i^2 - \bar{y} \sum_{i=1}^{n} y_i \end{aligned}\right\} \tag{14.5}$$

Eliminating α from equations (14.4) we find that the estimate of β is

$$\hat{\beta} = \frac{n \sum_{i=1}^{n} x_i y_i - (\sum_{i=1}^{n} x_i)(\sum_{i=1}^{n} y_i)}{n \sum_{i=1}^{n} x_i^2 - (\sum_{i=1}^{n} x_i)^2}$$

$$= \frac{\sum_{i=1}^{n} x_i y_i - \frac{1}{n}(\sum_{i=1}^{n} x_i)(\sum_{i=1}^{n} y_i)}{\sum_{i=1}^{n} x_i^2 - \frac{1}{n}(\sum_{i=1}^{n} x_i)(\sum_{i=1}^{n} x_i)}$$

$$= \frac{\sum_{i=1}^{n} x_i y_i - \bar{x} \sum_{i=1}^{n} y_i}{\sum_{i=1}^{n} x_i^2 - \bar{x} \sum_{i=1}^{n} x_i} = \frac{S_{xy}}{S_{xx}} \qquad (14.6)$$

From the first equation of (14.4), division by n and rearrangement produces

$$\hat{\alpha} = \bar{y} - \hat{\beta}\bar{x} \qquad (14.7)$$

Equations (14.6) and (14.7) produce least squares estimates of the parameters α and β. The estimated regression is $\hat{y} = \hat{\alpha} + \hat{\beta}x$

From (14.7) we obtain this equation in the form

$$\hat{y} - \bar{y} = \hat{\beta}(x - \bar{x}) \qquad (14.8)$$

and we see immediately that the point with coordinates $(\bar{x}, \bar{y})$ lies on the line as hinted at earlier.

Example

For our first example we shall calculate α and β. We must first calculate S_{xx} and S_{xy}. We shall also calculate S_{yy} since this will be required in future analysis. We shall obtain these values by coding the data given on page 505.

First the data is coded by the equations $X = (x - 60)/15$, $Y = y - 50$. Hence $\bar{X} = 0$, $\bar{Y} = 1.9\dot{3}$ from Table 14.1. Also

$$S_{XX} = \sum_{i=1}^{9} X_i^2 - \bar{X} \sum_{i=1}^{9} X_i = 60 - 0$$

$$S_{YY} = \sum_{i=1}^{9} Y_i^2 - \bar{Y} \sum_{i=1}^{9} Y_i = 1456 - 1.9\dot{3} \times 17.4 = 1456 - 33.64 = 1422.36$$

Table 14.1

x	y	X	Y	X^2	Y^2	XY
0	75.6	−4	25.6	16	655.36	−102.4
15	58.0	−3	8.0	9	64.00	−24.0
30	59.3	−2	9.3	4	86.49	−18.6
45	57.5	−1	7.5	1	56.25	−7.5
60	52.5	0	2.5	0	6.25	0
75	54.2	1	4.2	1	17.64	4.2
90	35.8	2	−14.2	4	201.64	−28.4
105	41.9	3	−8.1	9	65.61	−24.3
120	32.6	4	−17.4	16	302.76	−69.6
	Σ	0	17.4	60	1456.00	−270.6

$$S_{XY} = \sum_{i=1}^{9} X_i Y_i - \bar{Y} \sum_{i=1}^{9} X_i = -270.6 - 0$$

Decoding, $\bar{x} = 15\bar{X} + 60$ i.e. $\bar{x} = 60$

$\bar{y} = \bar{Y} + 50$ i.e. $\bar{y} = 51.93$

$S_{xx} = 225\ S_{XX} = 13500$

$S_{yy} = S_{YY} = 1422.36$

$S_{xy} = 15\,S_{XY} = -4059$

Hence, from (14.6), $\hat{\beta} = \dfrac{-4059}{13500} = -0.301$ (3 s.f.)

and from (14.7), $\hat{\alpha} = 51.93 + 0.301 \times 60 = 69.99$ (2 d.p.)

This gives the required regression line as $y = 69.99 - 0.301x$

Alternatively, we could have used (14.8) to obtain $y - 51.93 = -0.301(x - 60)$.

Properties of the estimators $\hat{\alpha}$ and $\hat{\beta}$: confidence intervals

Notice that $\hat{\beta} = \dfrac{\sum_{i=1}^{n} x_i y_i - \bar{x} \sum_{i=1}^{n} y_i}{\sum_{i=1}^{n} x_i^2 - \bar{x} \sum_{i=1}^{n} x_i}$ is a linear function of the

values y_i. Further, $\hat{\alpha} = \bar{y} - \hat{\beta}\bar{x} = \frac{1}{n} \Sigma\, y_i - \hat{\beta}\bar{x}$ is also a linear function of the values y_i. (We are treating x_i values as given constants.) It can be shown that these estimators are unbiased., i.e. $E(\hat{\alpha}) = \alpha$ and $E(\hat{\beta}) = \beta$.

Before we proceed further, let us take stock of the situation. We have assumed that the y_i values are statistically independent and that for each x_i the corresponding y_i has expected value lying on the regression line. The error term may be compounded from a measurement error and an error due to the inherent irreproducible nature of many phenomena.

Figure 14.3 attempts to show the relationship between the actual regression line and the consequent errors in y_i values on the one hand and the estimated regression line and the sample observations y_i on the other.

The blobs are the observations (x_i, y_i) and the dotted line is the estimated line. The true line is shown as a solid line.

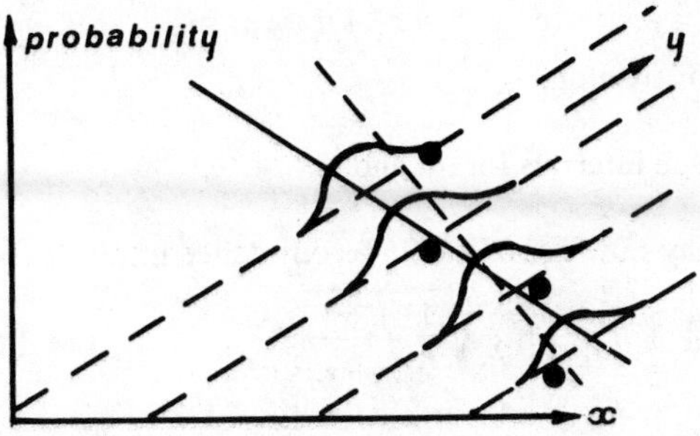

Figure 14.3

Assumption of normal errors

We now make the assumption that the errors ξ_i are distributed as $N(0, \sigma^2)$. It follows that $\hat{\beta}$ and $\hat{\alpha}$ are also distributed normally. It can be shown that

$$\hat{\beta} \sim N(\beta, \sigma^2/S_{xx}) \quad \text{and} \quad \hat{\alpha} \sim N(\alpha, [\frac{1}{n} + \frac{\bar{x}^2}{S_{xx}}]\sigma^2).$$

If the experiment was poorly designed, and all the x_i values were relatively close together then the deviations $(x_i - \bar{x})$ would be small and hence S_{xx} would be small. This would imply that the variance of $\hat{\beta}$ would be comparatively large and hence the estimate of the slope of the regression line of y on x would be poor. If the values x_i were more spread out then the estimate of the slope would be more reliable. Draw some scatter diagrams and convince yourself of this.

Distributions of $\hat{\alpha}$ and $\hat{\beta}$

If the values ξ_i, $\hat{\beta}$ and $\hat{\alpha}$ are distributed normally then so too are the y_i values. We could say that the random variable $z = \dfrac{\beta - \hat{\beta}}{\sqrt{\sigma^2/S_{xx}}}$ is distributed $N(0, 1)$. However, since we do not know σ^2 for the variable ξ_i we must estimate it by means of the sample statistic s^2 calculated from the observations:

$$s^2 = \frac{1}{n-2} \sum_{i=1}^{n} (y_i - \hat{\alpha} - \hat{\beta} x_i)^2 = \frac{1}{n-2} \sum_{i=1}^{n} (y_i - \hat{y}_i)^2 \qquad (14.9)$$

where $\hat{y}_i = \hat{\alpha} + \hat{\beta} x_i$.

Best results for an unbiased estimator for σ^2 are obtained with $(n-2)$. Alternatively, we may argue that the number of degrees of freedom is $(n-2)$ since we had to estimate two parameters α and β from the sample of observations.

Therefore the statistic $\dfrac{\hat{\beta} - \beta}{\sqrt{s^2/S_{xx}}}$ follows a t distribution with $(n-2)$ degrees of freedom.

Similarly, $\dfrac{\alpha - \hat{\alpha}}{s\sqrt{(S_{xx} + n\bar{x}^2)/n\, S_{xx}}}$ follows a t distribution with $(n-2)$ degrees of freedom.

Confidence intervals for α and β

We may now quote the 95% confidence intervals for α and β as

$$\hat{\alpha} \pm t_{0.025}\, s \sqrt{\frac{1}{n} + \frac{\bar{x}^2}{S_{xx}}} \qquad (14.10a)$$

and
$$\hat{\beta} \pm t_{0.025} \frac{s}{\sqrt{S_{xx}}} \qquad (14.10b)$$

Example

We return to our example to first calculate s^2. Although we shall see later how this can be obtained by a back-door approach, we shall, for the moment have to calculate each ξ_i directly. The following calculations (find $\hat{y}_i$ first from the equation $\hat{y}_i = \hat{\alpha} + \hat{\beta} x_i$) produce $s^2 = \dfrac{201.98}{7} = 28.85$ and hence $s = 5.372$ (3 d.p.). [See Table 14.2]

A 95% confidence interval is first found by taking from Table AII the value of $t_{0.025}$ with 7 degrees of freedom to be 2.365.

For α the 95% confidence interval is $69.99 \pm 2.365 \times 5.372 \sqrt{\dfrac{1}{9} + \dfrac{3600}{13500}}$

i.e. 69.99 ± 7.81 (3 s.f.) i.e. (62.18, 77.80)

For β the 95% confidence interval is $-0.301 \pm 2.365 \times \dfrac{5.372}{\sqrt{13500}}$

i.e. -0.301 ± 0.109 (3 s.f.) i.e. (–0.410, –0.192)

Now, since the confidence interval for β does not include the possibility $\beta = 0$ we assume that we can reject (at the 5% level) the hypothesis $\beta = 0$ and this suggests that the regression equation is a significant one.

Table 14.2

x_i	y_i	$\hat{y}_i$	$y_i - \hat{y}_i$	$(y_i - \hat{y}_i)^2$
0	75.6	69.99	5.61	31.47
15	58.0	65.47	−7.47	55.80
30	59.3	60.96	−1.66	2.76
45	57.5	56.44	1.06	1.12
60	52.5	51.93	0.57	0.32
75	54.2	47.41	6.79	46.10
90	35.8	42.90	−7.10	50.41
105	41.9	38.38	3.52	12.39
120	32.6	33.87	−1.27	1.61
			Σ	201.98

Problems

For each of the sets of data in Problems 1 to 10, estimate the parameters of the regression line of y on x, $y = \alpha + \beta x$ and find 95% and 99% confidence intervals for the parameters α and β.

1.

x	100	200	300	400	500
y	4.1	8.0	12.6	16.3	19.4

2.

x	1	2	3	4	5	6
y	15	35	41	63	77	84

3.

x	0	10	20	30	40	50
y	52	60	64	73	76	81

4.

x	26.8	25.4	28.9	23.6	27.7	23.9	24.7	28.1	26.9	27.4	22.6	25.6
y	26.5	27.3	24.2	27.1	23.6	25.9	26.3	22.5	21.7	21.4	25.8	24.9

5.

x	1.0	1.1	1.2	1.3	1.4	1.5	1.6	1.7	1.8	1.9	2.0
y	8.1	7.8	8.5	9.8	9.5	8.9	8.6	10.2	9.3	9.2	10.5

6.

x	65	63	67	64	68	62	70	66	68	67	69	71
y	68	66	68	65	69	66	68	65	71	67	68	70

7.

x	0.01	0.02	0.03	0.04	0.05	0.06	0.07
y	1.1	1.3	1.3	1.6	1.5	1.9	1.8

8.

x	56	42	72	36	63	47	55	49	38	42	68	50
y	147	125	160	118	149	128	150	145	115	140	152	150

9.

x	43	35	51	47	46	62	32	36	41	39	53	48
y	12	8	14	9	11	16	7	9	12	10	13	11

10.

x	43	35	51	47	46	62
y	12	8	14	9	11	16

11. Write a computer program to carry out the tasks required from Problems 1 to 10.

12. An experiment was conducted to evaluate a new method for determining the carbon content of clays used in the manufacture of sewer pipes. The carbon contents of 37 clays were determined both by the new method, represented by y, and by a standard method known to be very accurate, represented by x. Estimate the regression equation which assesses the results y obtained by the new method in terms of the true carbon contents, x, using the crude sums of squares and products given below.

$$n = 37, \quad \sum_i^n x_i = 74, \quad \sum_i^n y_i = 111, \quad \sum_i^n x_i^2 = 348, \quad \sum_i^n y_i^2 = 958,$$

$$\sum_i^n x_i y_i = 522.$$

13. (i) Given a set of observations (x_1, y_1), (x_2, y_2), ..., (x_n, y_n), derive the Normal equations which are used for the least-squares estimation of the parameters a, b in the linear relation $y = a + bx$ connecting the variates x and y.

(ii) Tensile tests on a steel specimen yielded the following results:

Tensile force	x	1	2	3	4	5	6
Elongation	y	15	35	41	63	77	84

Assuming the regression of y on x to be linear, estimate the parameters of the regression line and determine 95% confidence limits for its slope. (C.E.I.)

14. The following data is believed to follow a relationship of the form $y = a + b/x$. Obtain the regression equation

x	2	3	4	5	6	7	8	9
y	14.3	14.8	15.1	15.3	15.5	15.7	15.8	16.0

15. Fit a regression curve $y = ax + bx^2$ to the data

x	1	2	3	4	5
y	2.2	5.7	9.4	14.7	20.4

14.4 CONFIDENCE INTERVALS FOR PREDICTIONS

We now want to conclude our study of confidence intervals by finding a confidence interval for a value y_0, corresponding to x_0. For example, suppose we want to predict the percentage of sand at a depth of 80 cm. We must tread very carefully, since the substitution of x_0 into the regression equation gives the *average* of the distribution of the y - values that might correspond to x_0. Remember that the regression line we have obtained is only an estimate of the population regression line because we only have a sample of observations at our disposal. We shall first find a confidence interval for the mean of the y - values for a given x and then extend our results to the case of a single y - observation. We would expect the estimation of a simple observation to be less precise and, consequently the confidence interval to be wider. Then we shall see the dangers of extrapolation.

For the mean of the distribution of values of y for a given value of x, x_0 say, we need the variance of the mean which is equal to

$$\frac{s^2}{n} + \frac{s^2}{S_{xx}}(x_0 - \bar{x})^2$$ (We omit the proof of this result.)

Hence the 95% confidence interval for the mean value is

$$\hat{\alpha} + \hat{\beta} x_0 \pm t_{0.025}\, s\sqrt{\frac{1}{n} + \frac{(x_0 - \bar{x})^2}{S_{xx}}} \qquad (14.11)$$

When we come to put a confidence interval around the estimate of a *single* observation, we have another error to take into account, the error ξ which is built into the model equation (14.2). This adds a term σ^2 to the variance which now is estimated as

$$\frac{s^2}{n} + \frac{s^2}{S_{xx}}(x_0 - \bar{x})^2 + s^2$$ (Notice that we estimate the error variance of ξ as s^2.)

Hence the 95% confidence interval for a single observation y corresponding to the value x_0 is

$$\hat{\alpha} + \hat{\beta} x_0 \pm t_{0.025}\, s\sqrt{\frac{1}{n} + \frac{(x_0 - \bar{x})^2}{S_{xx}} + 1} \qquad (14.12)$$

In each case, the t value is from the distribution with $(n - 2)$ degrees of freedom.

Example

Referring to our current example we shall now calculate 95% confidence intervals for the mean y - value and a single observation corresponding to $x = 80$. Then we shall find the 95% confidence interval corresponding to $x = 180$.

First, we collect the numbers we need for the calculations $\hat{\alpha} = 69.99$, $\hat{\beta} = -0.301$, $x_0 = 80$, $n = 9$, $s = 5.372$, $\bar{x} = 60$, $S_{xx} = 13500$, $t_{0.025}(7) = 2.365$.

From equation (14.11) we find the 95% confidence intervals for the mean of the y - values corresponding to $x = 80$. It is

$$69.99 - 0.301 \times 80 \pm 2.365 \times 5.372\sqrt{\frac{1}{9} + \frac{(80 - 60)^2}{13500}}$$

i.e. 45.91 ± 4.77 or $(41.14, 50.68)$

This interval seems quite large: if we examine equation (14.11) we can see that the further x_0 is from $\bar{x}$ then the larger the quantity $(x_0 - \bar{x})^2$ and hence the wider the confidence interval.

Second, we find the 95% confidence interval for a single observation is, from equation (14.12)

$$69.99 - 0.301 \times 80 \pm 2.365 \times 5.372 \sqrt{\frac{1}{9} + \frac{(80-60)^2}{13500} + 1}$$

i.e. 45.91 ± 13.57 or $(32.34, 59.48)$

This interval is clearly so large as to be of little comfort with regard to the precision of an estimate.

Finally, we find the 95% confidence interval for a single observation corresponding to $x = 180$. Using equation (14.12) we obtain

$$69.99 - 0.301 \times 180 \pm 2.365 \times 5.372 \sqrt{\frac{1}{9} + \frac{(180-60)^2}{13500} + 1}$$

i.e. 15.81 ± 18.75 or $(-2.94, 34.56)$

We have clearly got a useless result here. This pinpoints the dangers of *extrapolation*. In the first place, the value of $(x_0 - \bar{x})^2$ is so large that the confidence interval is very wide indeed. In the second place, we cannot be sure that a linear model is applicable outside the range of observations; see Figure 14.4. We should therefore be very wary of extrapolation and avoid it wherever possible.

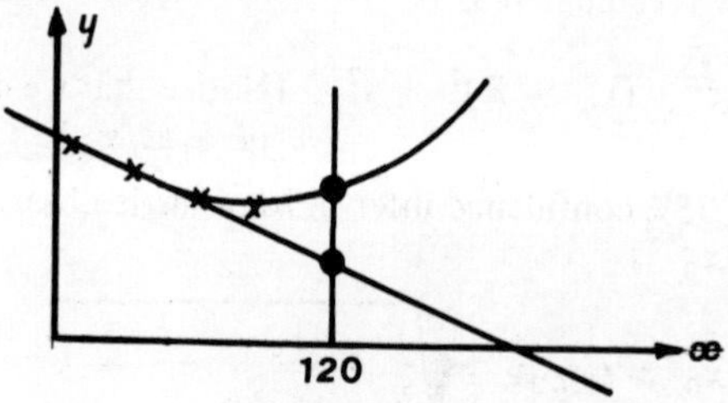

Figure 14.4

Further conclusions

It is worth pursuing this example a little further. First we shall see what is the minimum width of confidence interval by taking $x_0 = 60$. Then equation (14.11) gives the 95% confidence interval for the mean of the y - values as

$$69.99 - 0.301 \times 60 \pm 2.365 \times 5.372 \sqrt{\frac{1}{9} + 0}$$

i.e. 51.93 ± 4.23 or $(47.70, 56.16)$

Even this interval is quite wide - about 16% of the middle value. Second, we remark that for a single observation even at $x_0 = \bar{x}$ and even if $n \to \infty$, the confidence interval is still not zero.

The 95% confidence interval has a limiting minimum value of

$$69.99 - 0.301 \times 60 \pm 2.365 \times 5.372 \sqrt{0 + 0 + 1}$$

i.e. 51.93 ± 12.70 or $(39.23, 64.63)$

This width is due to the inherent variability in the data.

Finally, we illustrate diagrammatically the effect on the estimates of the mean y - value and on the single observation of varying certain parameters.

Figure 14.5(a) shows the effect on the interval estimate of the mean value y

assuming that only the slope β is subject to error; Figure 14.5(b) shows the effect if only the intercept α is subject to error; Figure 14.5(c) shows the combined effects on the mean value and on a single observation.

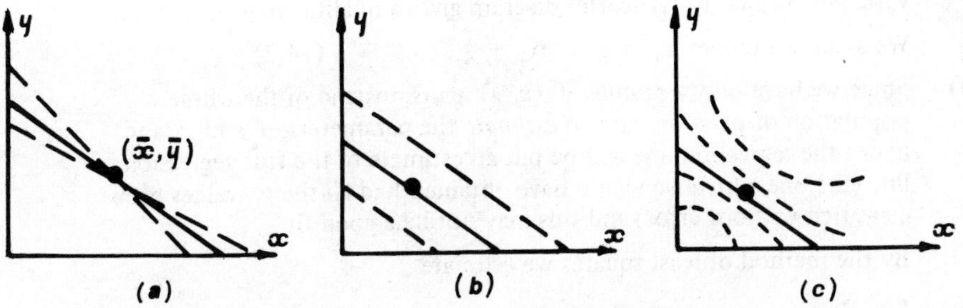

Figure 14.5

It should be noted that the effects increase as we move away from $\bar{x}$.

Problems

1. In a study of machine maintenance costs in a factory, data on 12 machines was collected, consisting of, for each machine, its age (x) in months and its maintenance cost rate (y) in pence per hour.

   From this data, calculations gave:

   $\bar{x} = 24, \quad \bar{y} = 18, \quad \Sigma (x - \bar{x})^2 = 1950, \quad \Sigma (y - \bar{y})^2 = 472,$

   $\Sigma (x - \bar{x})(y - \bar{y}) = 780.$

   The regression of cost rate on age can be assumed to be linear.

   On this basis, obtain an estimate of the mean maintenance cost rate of machines which are 29 months old and find the 95% confidence limits for the maintenance cost rate of one such machine.

2. Find 95% confidence intervals for particular values of y predicted by the regression equations of Problems 1 to 10 of Section 14.3. The corresponding x - values are indicated below;

   ((i) refers to problem 1 (ii) to problem 2 and so on)

   (i) $x = 250$ (ii) $x = 5.2$ (iii) $x = 15$ (iv) $x = 25$ (v) $x = 1.43$
   (vi) $x = 63.8$ (vii) $x = 0.058$ (viii) $x = 60$ (ix) $x = 50$ (x) $x = 50$

3. The following data is given

x	1.0	1.1	1.2	1.3	1.4	1.5	1.6	1.7	1.8	1.9	2.0
y	5.6	5.3	6.0	7.3	7.1	6.4	6.1	7.7	6.8	6.7	8.0

   Estimate the linear regression line $y = \alpha + \beta x$ and estimate the value of y when $x = 1.75$. Find 95% and 99% confidence intervals for α and β. Sketch the regression lines and the 90% confidence limits for the mean value of y for $x = 1.6$ and for the value of y when $x = 1.6$.

14.5 REVIEW EXAMPLE

It would be as well to review progress to date and consolidate ideas with an artificially contrived example.

(i) We assume that there is a relationship $y = \alpha + \beta x$ connecting two variables x and y. A scatter diagram gives a qualitative justification.

(ii) We assume a model $y_i = \alpha + \beta x_i + \xi_i$. (14.2)

(iii) Since we have only a *sample* of (x, y) pairs, instead of the whole population of pairs we have to *estimate* the parameters α and β and hence the regression line will be but an estimate of the true regression line (the one which we would have obtained had all the y - values been measured without error) and this may not be a good fit.

(iv) By the method of least squares we estimate

$$\hat{\beta} = \frac{S_{xy}}{S_{xx}} \qquad (14.6)$$

where $S_{xy} = \sum_{i=1}^{n} x_i y_i - \bar{x} \sum_{i=1}^{n} y_i$

(14.5)

and $S_{xx} = \sum_{i=1}^{n} x_i^2 - \bar{x} \sum_{i=1}^{n} x_i$

We also estimate $\hat{\alpha} = \bar{y} - \hat{\beta}\bar{x}$ (14.7)

(v) We can calculate certain confidence intervals. With a 95% confidence level and $(n - 2)$ degrees of freedom we find

(a) for α: $\hat{\alpha} \pm t_{0.025}\, s \sqrt{\frac{1}{n} + \frac{\bar{x}^2}{S_{xx}}}$ (14.10a)

(b) for β: $\hat{\beta} \pm t_{0.025}\, s \sqrt{\frac{1}{S_{xx}}}$ (14.10b)

(c) for mean y: $\hat{\alpha} + \hat{\beta}x_0 \pm t_{0.025}\, s \sqrt{\frac{1}{n} + \frac{(x_0 - \bar{x})^2}{S_{xx}}}$ (14.11)

(d) for single y: $\hat{\alpha} + \hat{\beta}x_0 \pm t_{0.025}\, s \sqrt{\frac{1}{n} + \frac{(x_0 - \bar{x})^2}{S_{xx}} + 1}$ (14.12)

where x_0 is the value of x corresponding to the particular value of y we require to predict and

$$s^2 = \frac{1}{n-2} \sum_{i=1}^{n} (y_i - \hat{\alpha} - \hat{\beta}x_i)^2 \qquad (14.9)$$

(vi) The further the value of x is from $\bar{x}$, the worse the precision of estimates of the mean y and the single observation y. Extrapolation is dangerous.

(vii) There are several sources of variability built into the system which can be measured as variances. We shall see later in the chapter how to use a technique of *variance analysis* to compare their effects and make judgements about the goodness of fit of the estimated regression line.

Example

For the data shown below, fit a simple regression equation and obtain 99% confidence intervals for α, β and for a single observation of y corresponding to $x = 0.365$.

x	0.31	0.32	0.33	0.34	0.35	0.36	0.37
y	2.1	2.3	2.3	2.6	2.5	2.9	2.8

(i) First we code the data by the relations $X = 100(x - 0.3)$, $Y = 10(y - 2.0)$. The calculations then proceed as in Table 14.3.

Table 14.3

x	y	X	Y	X^2	Y^2	XY
0.31	2.1	1	1	1	1	1
0.32	2.3	2	3	4	9	6
0.33	2.3	3	3	9	9	9
0.34	2.6	4	6	16	36	24
0.35	2.5	5	5	25	25	25
0.36	2.9	6	9	36	81	54
0.37	2.8	7	8	49	64	56
	Σ	28	35	140	225	175

Then $\bar{X} = 4$, $\bar{Y} = 5$ and $S_{XX} = 140 - 4 \times 28 = 28$;
$S_{XY} = 175 - 4 \times 35 = 35$; $S_{YY} = 225 - 5 \times 35 = 50$.

(We have included ΣY^2 because we shall require its value later in the chapter.)

Decoding, we find that $\bar{x} = 0.01\bar{X} + 0.3 = 0.34$,

$\bar{y} = 0.1\bar{Y} + 2.0 = 2.5$

$S_{xx} = 10^{-4} S_{XX} = 0.0028$ and $S_{xy} = 10^{-1} 10^{-2} S_{XY} = 0.035$

$S_{yy} = 10^{-2} S_{YY} = 0.5$

(ii) We estimate $\hat{\beta} = \dfrac{S_{xy}}{S_{xx}} = \dfrac{0.035}{0.0028} = 12.5$

and $\hat{\alpha} = \bar{y} - \hat{\beta}\bar{x} = 2.5 - 12.5 \times 0.34 = -1.75$.

Hence the estimated regression equation is $y = -1.75 + 12.5x$

(iii) In the estimation of the confidence intervals we require to estimate σ^2 by s^2, as given by (14.9).

We now demonstrate a method for determining s^2, based on the values we have so far calculated. If you wish, you may omit the derivation and proceed straight to the result (14.13)

$$(n-2)s^2 = \sum_{i=1}^{n} (y_i - \hat{\alpha} - \hat{\beta}x_i)^2$$

$$= \sum_{i=1}^{n} (y_i - \bar{y} + \hat{\beta}\bar{x} - \hat{\beta}x_i)^2 \qquad \text{from (14.7)}$$

$$= \sum_{i=1}^{n} [(y_i - \bar{y}) - \hat{\beta}(x_i - \bar{x})]^2$$

$$= \sum_{i=1}^{n} (y_i - \bar{y})^2 + \hat{\beta}^2 \sum_{i=1}^{n} (x_i - \bar{x})^2 - 2\hat{\beta} \sum_{i=1}^{n} (x_i - \bar{x})(y_i - \bar{y})$$

$$= S_{yy} + \hat{\beta}^2 S_{xx} - 2\hat{\beta}S_{xy} \qquad \text{from (14.5)}$$

$$= S_{yy} + \frac{(S_{xy})^2}{S_{xx}} - 2\frac{(S_{xy})^2}{S_{xx}} \qquad \text{from (14.6)}$$

$$= S_{yy} - \frac{(S_{xy})^2}{S_{xx}}$$

Hence $$s^2 = \frac{1}{n-2}\left[S_{yy} - \frac{(S_{xy})^2}{S_{xx}}\right] \qquad (14.13)$$

In this example $$s^2 = \frac{1}{5}\left[0.5 - \frac{(0.035)^2}{0.0028}\right] = 0.0125$$

(iv) The 99% confidence intervals require the value of $t_{0.005}$ for 5 degrees of freedom; this is 4.032.

For α: $-1.75 \pm 4.032\sqrt{0.0125}\sqrt{\frac{1}{7} + \frac{(0.34)^2}{0.0028}}$

i.e. $-1.75 \pm 0.4508 \times 6.437$ or $(-4.65, 1.15)$

For β: $12.5 \pm 4.032\sqrt{0.0125}/\sqrt{0.0028}$

i.e. 12.5 ± 8.51 or $(3.99, 21.01)$

For mean y: $-1.75 + 12.5 \times 0.365 \pm 4.032\sqrt{0.0125} \times \sqrt{\frac{1}{7} + \frac{(0.365 - 0.34)^2}{0.0028}}$

i.e. 2.81 ± 0.27 or $(2.54, 3.08)$

$$\textit{For single } y\text{:}\quad -1.75 + 12.5 \times 0.365 \pm 4.032\sqrt{0.0125} \times \sqrt{\frac{1}{7} + \frac{(0.365 - 0.34)^2}{0.0028} + 1}$$

i.e. 2.81 ± 0.52 or $(2.29,\ 3.33)$

14.6 A SIMPLE INTRODUCTION TO CORRELATION

What we have done so far is to assume that only the y - values are subject to error and to fit a straight line $y = \alpha + \beta x$ to the data using a least squares criterion. In the example on percentage y of sand in soil at a given depth x we obtained on page 510 the estimated regression line for y on x:

$$y = 69.99 - 0.301x \qquad (14.14)$$

Suppose we were interested in predicting the depth knowing the percentage of sand. We should then be fitting a relationship $x = a + by$ (the regression line of x on y) to the data and we would repeat the analysis on pages 518 and 519 interchanging the roles of y and x.

Thus $\hat{b} = \dfrac{S_{xy}}{S_{yy}} = \dfrac{-4059}{1422.36} = -2.85$ (3 s.f.)

and $\hat{a} = \bar{x} - \hat{b}\bar{y} = 60 + 2.85 \times 51.93 = 207.9$

Therefore the regression equation of x on y is

$$x = 207.9 - 2.85y \qquad (14.15a)$$

Now if we re-express this equation with y as the left-hand side we obtain

$$y = \frac{207.9}{2.85} - \frac{x}{2.85} \quad \text{i.e. } y = 72.95 - 0.351x \qquad (14.15b)$$

This is not the same as equation (14.14). However, the lines given by equations (14.14) and (14.15a) pass through the point with coordinates $(\bar{x},\ \bar{y})$; see Figure 14.6 (which is not to scale).

If the data points lay exactly on a straight line we would expect the regression line of y on x and the regression line of x on y to coincide as the line on which the data points lay. The worse the fit of either regression line the greater the difference in their slopes. Notice that since both lines pass through $(\bar{x},\ \bar{y})$ we could transfer the origin of axes to this point. This would introduce new variables $(x_i - \bar{x})$ and $(y_i - \bar{y})$ which explains the presence of these terms in S_{xx}, S_{xy} and S_{yy}.

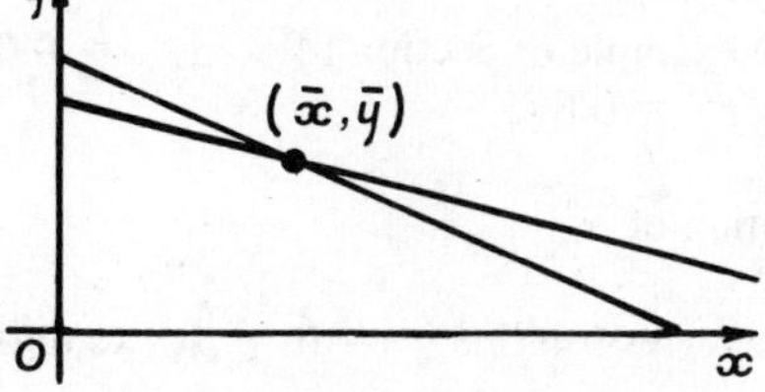

Figure 14.6

What fitting a regression line does not tell us is the degree of linear relationship between y and x, i.e. how closely they are related linearly. We shall seek to measure this closeness (and hence the goodness of fit of either regression line). This technique will measure the degree of *correlation* between x and y; neither variable is assumed to be independent as is the assumption with regression.

For two variables x and y from populations with means μ_x and μ_y and variances $\sigma_x{}^2$ and $\sigma_y{}^2$ respectively, we define the **population correlation coefficient** $\rho = E\left\{\left(\frac{x_i - \mu_x}{\sigma_x}\right)\left(\frac{y_i - \mu_y}{\sigma_y}\right)\right\}$ where $E\{Z\}$ is the expectation of Z.

We will estimate ρ by means of the **sample correlation coefficient**

$$r = \frac{1}{n-1} \sum_{i=1}^{n} \left(\frac{x_i - \bar{x}}{s_x}\right)\left(\frac{y_i - \bar{y}}{s_y}\right).$$

Since $s_x{}^2 = \frac{1}{n-1} \sum_{i=1}^{n} (x_i - \bar{x})^2 = \frac{1}{n-1} S_{xx}$ and

$s_y{}^2 = \frac{1}{n-1} \sum_{i=1}^{n} (y_i - \bar{y})^2 = \frac{1}{n-1} S_{yy}$ we may rewrite $r = \frac{S_{xy}}{\sqrt{S_{xx}}\sqrt{S_{yy}}}$

More usually we work in terms of $r^2 = \frac{(S_{xy})^2}{S_{xx}\, S_{yy}}$ (14.16)

It can be shown that $r^2 \leqslant 1$; since, clearly, $r^2 \geqslant 0$ it follows that $-1 \leqslant r \leqslant +1$

Example 1

For the problem on percentage of sand in soil at various depths, we had $S_{xx} = 13500$, $S_{yy} = 1422.36$, $S_{xy} = -4059$ and hence $r^2 = 0.858$ (3.d.p.).

Example 2

For the example of Section 14.5 $S_{xx} = 0.0028$, $S_{yy} = 0.5$, $S_{xy} = 0.035$ and hence $r^2 = 0.875$.

Interpretation of r^2

For the regression line $y = \hat{\alpha} + \hat{\beta}x$ we defined its slope $\hat{\beta} = \frac{S_{xy}}{S_{xx}}$, whilst

for the regression line $x = \hat{a} + \hat{b}y$ we defined $\hat{b} = \frac{S_{xy}}{S_{yy}}$.

If we re-express the second line as $y = -\frac{\hat{a}}{\hat{b}} + \frac{1}{\hat{b}}x$ its slope is $\frac{1}{\hat{b}} = \frac{S_{yy}}{S_{xy}}$.

The ratio of these two slopes is $\hat{\beta}/(\frac{1}{\hat{b}}) = \frac{S_{xy}}{S_{xx}} \Big/ \frac{S_{yy}}{S_{xy}} = \frac{(S_{xy})^2}{S_{xx}\ S_{yy}} = r^2$

If the two lines had coincided then $r^2 = 1$ and we have perfect linear correlation. If the slope of either regression line is positive we take r positive and in the case of $r^2 = 1$ we see that $r = 1$ gives perfect positive linear correlation; similarly $r = -1$ gives perfect negative linear correlation. It should be clear that the closer r^2 is to 1 then the better the fit of the regression line of y on x (and of the regression line of x on y) and the more closely the variables x and y are correlated.

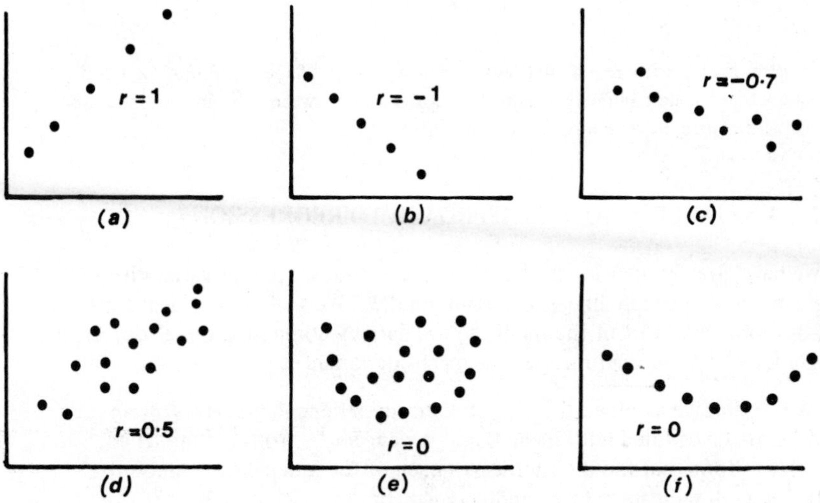

Figure 14.7

In Figure 14.7 we see schematic representations of various values of r. In (a) we have a case of perfect positive correlation whilst in (b) we have perfect negative correlation. In (c) and (d) we see the lowering of $|r|$ as the scatter increases. The last two diagrams have the same value of r but represent different situations altogether; whereas (e) represents random scatter, (f) has definite indications of a relationship between x and y which, because it is not linear, produces a zero r score. As long as we accept the reasonableness of a linear model, then the higher the value of $|r|$ the better we are pleased.

It is not possible to give definite guidelines as to what values of r^2 correspond to good fits. In some scientific experiments a value of r^2 of 0.96 would perhaps be considered only as fair; on the other hand in social science studies an r^2 value of 0.70 is considered very good. It all depends on the data and how it can be collected.

Problems

1. Evaluate the coefficient r^2 for each of the Problems 1 to 10 of Section 14.3.

2. Find r^2 for the data below. Comment.

x	4	5	9	14	18	22	24
y	1.6	2.2	1.1	1.6	0.7	0.3	1.7

3. Repeat Problem 2 for the data

x	70	92	80	74	65	83
y	70	80	59	83	74	86

4. The extension y cm of a spring under a load of x newtons is measured for eight values of x and the results are given in the following table:

x	1	2	3	4	5	6	7	8
y	1.9	3.8	5.1	6.0	7.0	7.8	8.4	8.8

It is thought that there is a relation between x and y of the form $y = ax + bx^3$ Show that the data indicate a significant correlation between y/x and x^2, and use this correlation to estimate values of a and b. (L.U.)

14.7 VARIANCE ANALYSIS APPROACH TO REGRESSION

We have already mentioned that there are several sources of variability associated with a simple linear regression model. We shall now endeavour to test the goodness of fit of the model to the data by comparing these sources of variability which we shall measure by means of variances.

Notice that we assumed that the statistically independent errors ξ_i in our model were distributed with mean 0 and variance σ^2; from the model equation (14.2) it follows that the distribution of each of the independent random variables y_i has mean $\alpha + \beta x_i$ and variance σ^2.

We can show that

$$\sum_{i=1}^{n} \hat{\xi}_i^2 = \sum_{i=1}^{n} (y_i - \bar{y})^2 - \sum_{i=1}^{n} \hat{\beta}^2 (x_i - \bar{x})^2 = \sum_{i=1}^{n} (y_i - \bar{y})^2 - \sum_{i=1}^{n} (\hat{y}_i - \bar{y})^2$$

where $\hat{y}_i$ is the estimated mean value of y_i from the regression equation [see equation (14.8)].

Hence we may rearrange the relationship to give

$$S_{yy} = \sum_{i=1}^{n} (y_i - \bar{y})^2 = \sum_{i=1}^{n} (\hat{y}_i - \bar{y})^2 + \sum_{i=1}^{n} \hat{\xi}_i^2 \qquad (14.17)$$

If we refer to Figure 14.8, we find that $(y_i - \bar{y})$ is the difference between the observation y_i and the mean of observations $\bar{y}$ and that $(\hat{y}_i - \bar{y})$ is the difference between the estimated value $\hat{y}_i$ and the mean $\bar{y}$. Hence we may write equation (14.17) in words:

$$\begin{pmatrix}\text{Total sum of squares due to}\\ \text{observed values } y_i\end{pmatrix} = \begin{pmatrix}\text{Sum of squares due to}\\ \text{estimated values } \hat{y}_i\end{pmatrix} + \begin{pmatrix}\text{Sum of squares}\\ \text{due to errors}\end{pmatrix} \qquad (14.18)$$

We have therefore partitioned the total sum of squared deviations of the original observations about the mean into

(a) a component due to values estimated by the regression equation and
(b) a component due to errors about the regression line.

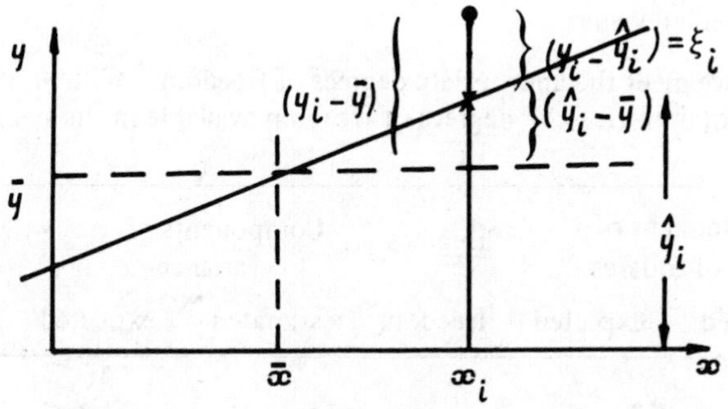

Figure 14.8

Now $(\hat{y}_i - \bar{y})^2 = \hat{\beta}^2 (x_i - \bar{x})^2$ and hence the component

$$\sum_{i=1}^{n} (\hat{y}_i - \bar{y})^2 = \hat{\beta}^2 \sum_{i=1}^{n} (x_i - \bar{x})^2 = \hat{\beta}^2 S_{xx} = \frac{(S_{xy})^2}{S_{xx}} = \frac{(S_{xy})^2 S_{yy}}{S_{yy} S_{xx}}$$

Also $\quad r^2 = \dfrac{(S_{xy})^2}{S_{xx} S_{yy}}$ from equation (14.16).

Therefore $$\sum_{i=1}^{n} (\hat{y}_i - \bar{y})^2 = r^2 S_{yy} \qquad (14.19a)$$

Hence $$\sum_{i=1}^{n} \hat{\xi}_i^2 = (1 - r^2) S_{yy} \qquad (14.19b)$$

Second interpretation of r^2

The quantity $r^2 S_{yy}$ may be regarded as that part of the total sum of squares which is 'explained' by the regression line. The quantity $(1 - r^2) S_{yy}$ is the sum of squares due to errors about the regression line. Hence r^2 measures the proportion of the total sum of squares explained by the regression equation. Therefore we see that the closer r^2 is to 1, the less the sum of squares due to errors. In the ideal case, $r^2 = 1$, there is no sum of squares term due to errors (all the sum of squares being explained by the regression line) and the line fits perfectly, since if $\sum_{i=1}^{n} \hat{\xi}_i^2 = 0$ it follows that *all* the individual errors $\hat{\xi}_i$ are zero.

Analysis of variance

The technique will be to form the ratio $\left(\dfrac{\text{variance due to regression}}{\text{variance due to errors}}\right)$ and to examine its value with an F test. Each variance is obtained by dividing the appropriate sum of squares by the appropriate number of degrees of freedom. We now refer to the theoretical analysis of variance table below.

In the second column we have written down the estimated sums of squares that we derived earlier; in the third column we have produced the corresponding theoretically expected values.

In column 4 we meet the appropriate degrees of freedom. With n sample items there is a total of $(n-1)$ degrees of freedom available in the system.

Source of variability	Component of sums of squares: estimated	Component of sums of squares: expected	Degrees of freedom	Components of variance: estimated	Components of variance: expected	F
Due to regression line	$r^2 S_{yy}$	$\sigma^2 + \beta^2 S_{xx}$	1	$r^2 S_{yy}$	$\sigma^2 + \beta^2 S_{xx}$	$F^1_{(n-2)}$
About regression line	$(1-r)^2 S_{yy}$	$(n-2)\sigma^2$	$(n-2)$	$\dfrac{(1-r^2)S_{yy}}{(n-2)}$	σ^2	$\dfrac{r^2(n-2)}{1-r^2}$
Total	S_{yy}	$(n-1)\sigma^2 + \beta^2 S_{xx}$	$(n-1)$			

Since 2 data points are sufficient to determine any line it follows that $(n-2)$ points are competing to share the total error about the regression line. This leaves one degree of freedom to be associated with the sum of squares due to the regression.

Alternatively, we notice that the number of degrees of freedom is the coefficient of σ^2 in the appropriate expected sum of squares. We will then obtain unbiased estimates of the variance.

In columns 5 and 6 we produce the variance components by dividing a sum of squares by its associated number of degrees of freedom.

Finally, we divide the two variances to obtain the ratio $\dfrac{r^2(n-2)}{(1-r^2)}$.

Example 1

For the example of Section 14.5 we complete the variance analysis table as follows.

We first recall that $S_{yy} = 0.5$, $n = 7$, and $r^2 = 0.875$, hence $r^2 S_{yy} = 0.875 \times 0.5 = 0.4375$. The critical values of $F_5{}^1$ were obtained from Table A IV.

Source of variability	Component of sum of squares	Degrees of freedom	Component of variance	F	Critical values of F
Due to regression line	$0.4375 = r^2 S_{yy}$	1	0.4375	35	5% 6.61
Errors about line	0.0625	5	0.0125		1% 16.26
Total	$S_{yy} = 0.5000$	6			

Since the calculated F value is much greater than the 1% value we conclude, with considerably less than a 1% risk, that the regression is significant.

Example 2

For the sandy soil example, we had $S_{yy} = 1422.36$, $n = 9$ and $r^2 = 0.858$. The variance analysis table is as below.

Source of variability	Component of sum of squares	Degrees of freedom	Component of variance	F	Critical values of F
Due to regression line	1220.38	1	1220.38	42.2	5% 5.59
Errors about line	201.98	7	28.91		1% 12.25
Total	1422.36	8			

Again we find that the calculated F value is much greater than the 1% value and we therefore conclude, with a much less than 1% risk, that the regression is a significant one.

Error variance

It is customary to speak of the second entry in the component of variance column of the VA table, $(1 - r^2)S_{yy}/(n - 2)$, as the **error variance** associated with the linear regression model. Hence we might quote the estimated regression equation together with the error variance. Then for the example of Section 14.5 we have $\hat{y} = -1.75 + 12.5x$, with error variance 0.0125. For the sandy soil example the equation is $\hat{y} = 69.99 - 0.301x$ with error variance 28.91.

Review example

In Section 14.1 we introduced an example concerning the cost of maintaining a machine. We now collect together all the techniques we have developed so far and apply them to this example.

We now repeat the data

Age of machine in years, x_i	1	1.5	2	3	3	3.5	4	4	5	5	6	7
Cost of previous year's maintenance (£100's) , y_i	22	20	22	21	23	25	24	27	25	27	26	29

The steps we shall perform are as follows

(i) By coding the data, calculate $\overline{x}$, $\overline{y}$, S_{xx}, S_{xy}, S_{yy}, using Table 14.4

(ii) Calculate the percentage overall fit; $100\,r^2$.

(iii) Validate the regression equation by a variance analysis.

(iv) Estimate the regression equation.

(v) Place confidence intervals on $\hat{\alpha}$ and $\hat{\beta}$.

(vi) Place confidence intervals on the estimated cost $\hat{y}$ and on an observed cost y_i for a given value x_0.

(vii) Draw appropriate conclusions.

Table 14.4

Raw data x_i	y_i	X	Y	X^2	XY	Y^2
1	22	−6	−3	36	18	9
1.5	20	−5	−5	25	25	25
2	22	−4	−3	16	12	9
3	21	−2	−4	4	8	16
3	23	−2	−2	4	4	4
3.5	25	−1	0	1	0	0
4	24	0	−1	0	0	1
4	27	0	2	0	0	4
5	25	2	0	4	0	0
5	27	2	2	4	4	4
6	26	4	1	16	4	1
7	29	6	4	36	24	16
		−6	−9	146	99	89

We assume a regression line of the form $y = \alpha + \beta x$

(i) We code the data by writing $X = 2(x - 4)$, $Y = (y - 25)$
Then $\bar{X} = -0.5$, $\bar{Y} = -0.75$, $S_{XX} = \Sigma X^2 - \bar{X}\,\Sigma X = 143$
$S_{XY} = \Sigma XY - \bar{X}\,\Sigma Y = 94.5$, $S_{YY} = \Sigma Y^2 - \bar{Y}\,\Sigma Y = 82.25$

Decoding produces $\bar{x} = 3.75$, $\bar{y} = 24.25$; $S_{xx} = 35.75$, $S_{xy} = 47.25$, $S_{yy} = 82.25$

(ii) Since $r^2 = \dfrac{(S_{xy})^2}{S_{xx}\,S_{yy}} = \dfrac{(47.25)^2}{35.75 \times 82.25} = 0.7593$ (4 s.f.)
the percentage overall fit is 75.93%.

(iii) We now form a variance analysis table: note that the component of sum of squares due to regression is $r^2 S_{yy} = 62.45$ and that due to errors is $(1 - r^2)\,S_{yy} = 82.25 - 62.45 = 19.80$.

Source of variability	Component of sum of squares	Degrees of freedom	Component of variance	F	Critical values of F
Due to regression line	62.45	1	62.45	31.54	5% F^1_{10} = 4.96
Errors about line	19.80	10	1.98		1% F^1_{10} = 10.04
Total	82.25	11			

Hence we conclude, with considerably less than 1% risk of being wrong, that the regression effect is significant.

(iv) To find the regression equation we first calculate

$$\hat{\beta} = \frac{S_{xy}}{S_{xx}} = \frac{47.25}{35.75} = 1.322$$

Then we use the equation for the estimated regression line in the form
$\hat{y} - \bar{y} = \hat{\beta}(x - \bar{x})$
i.e. $\hat{y} - 24.25 = 1.322(x - 3.75)$

We obtain the estimated regression equation as $\hat{y} = 1.32x + 19.29$ with error variance 1.98.

(v) To estimate the 95% confidence intervals we note that there are 10 degrees of freedom, and the appropriate t - score is found to be 2.228. We also require $s = \sqrt{1.98} = 1.407$. Hence the 95% confidence interval for β is

$$\hat{\beta} \pm 2.228\,\frac{s}{\sqrt{S_{xx}}} \quad \text{i.e. } 1.322 \pm 2.228 \times \frac{1.407}{5.979} \quad \text{i.e. } 1.322 \pm 0.524$$

or (0.798, 1.846).

The 95% confidence interval for α is

$$\hat{\alpha} \pm 2.228\, s\sqrt{\frac{1}{n} + \frac{\bar{x}^2}{S_{xx}}} \quad \text{i.e. } 19.29 \pm 2.228 \times 1.407\sqrt{\frac{1}{12} + \frac{(3.75)^2}{35.75}}$$

i.e. 19.29 ± 2.16 or (17.13, 21.45)

(vi) The 95% confidence interval for the expected value $\hat{y}$ corresponding to the value x_0 is $\hat{\alpha} + \hat{\beta}x_0 \pm 2.228 s\sqrt{\frac{1}{n} + \frac{(x_0 - \bar{x})^2}{S_{xx}}}$ and that for a single observation

is $\hat{\alpha} + \hat{\beta}x_0 \pm 2.228\ s\sqrt{\frac{1}{n} + \frac{(x_0 - \bar{x})^2}{S_{xx}} + 1}$

(vii) Our conclusions are that, with much less than a 1% error of wrong judgement, there is a significant increase in the annual cost of maintaining the machines with age of the machine. The relationship is represented by the regression equation $\hat{y} = 1.32x + 19.29$ with error variance 1.98, $\hat{y}$ is the estimated annual maintenance cost at age x years. In the sample of machines examined whose ages ranged between 1 and 7 years, 75.93% of the variability observed in annual maintenance cost could be attributed to age effects. Since the 95% confidence interval for α does not include the value zero we can conclude (with less than 5% risk) that the expected cost of the first year's maintenance of a machine of the type studied is not zero. (Perhaps this may represent some basic fixed annual charge.)

The marginal increase in annual maintenance cost is estimated as $1.32(\hat{\beta})$.

Problems

1. For each of Problems 1 to 10 of Section14.3, calculate the percentage overall fit ($100r^2$) and validate (or otherwise) the regression equation by a variance analysis.

2. The following data was recorded from an experiment.

x	0	5	10	15	20	25
y	22	36	70	92	125	143

Find the regression equation $y = \alpha + \beta x$; estimate the value of y for $x = 17$; test the hypothesis that $\alpha = 6$ as opposed to the alternative $\alpha \neq 6$, using a 1% level of significance; decide whether a linear model is appropriate.

3. Write a computer program to carry out the tasks required for Problem 1.

14.8 CONCLUDING REMARKS

(i) One question with reference to Example 2 of Section14.1 remains to be answered: what effect, if any, do the repeated observations have on the variance analysis table? Is there a possibility of subdividing further the sum of squares due to errors about the regression line?

Suppose, in general that there are k distinct values of x: $x_1, x_2, \ldots, x_k$ out of the n observations. Let n_1 be the number of occasions on which x_1 occurs, n_2 the number of occasions on which x_2 occurs etc. (any of these values n_i can be 1). Then, of course, $n = n_1 + n_2 + \ldots + n_k$. We denote by y_{i1}, y_{i2} etc. those observed values of y which correspond to the value x_i and denote by $\bar{y}_i$ the average of these values y_{i1}, y_{i2} etc. Now we can partition the error sum of squares into two halves: that part which is due to the variability of y_{i1}, y_{i2} etc. which correspond to each given x_i and a second part which really represents lack of fit of the model. In other words, the first part reflects experimental or observational error, whilst the second part is the error due to the simplicity of a linear model. Perhaps we can apply this reasoning to Example 2.

First we will compute the component of sum of squares due solely to experimental error. We note that this has $(n - k)$ associated degrees of freedom (k different values of x_i). Then we subtract this from the error sum of squares and produce a variance analysis table which contains three sub-components.

Applying this procedure to Example 2, we had three instances of repeated measurements in 12 observations i.e. $k = 9$, For $x_4 = 3$ we had $y_{41} = 21$, $y_{42} = 23$ and therefore $\bar{y}_4 = 22$.

Similarly, for $x_6 = 4$ we had $y_{61} = 27$, $y_{62} = 24$ and $\bar{y}_6 = 25.5$; for $x_7 = 5$ we had $y_{71} = 25$, $y_{72} = 27$ and $\bar{y}_7 = 26$.

If we form the quantity

$$\sum_{i=1}^{n_1} (y_{1i} - \bar{y}_1)^2 + \sum_{i=1}^{n_2} (y_{2i} - \bar{y}_2)^2 + \ldots + \sum_{i=1}^{n_k} (y_{ki} - \bar{y}_k)^2$$

we notice that, since most of the observations are not replicated, all but three of these terms will vanish and we shall be left with

$$\sum_{i=1}^{2} (y_{4i} - \bar{y}_4)^2 + \sum_{i=1}^{2} (y_{6i} - \bar{y}_6)^2 + \sum_{i=1}^{2} (y_{7i} - \bar{y}_7)^2$$

$$= (21 - 22)^2 + (23 - 22)^2 + (27 - 25.5)^2 + (24 - 25.5)^2 + (25 - 26)^2 + (27 - 26)^2$$

$$= 1 + 1 + 2.25 + 2.25 + 1 + 1 = 8.5$$

This sum of squares is associated with $n - k = 12 - 9$ i.e. 3 degrees of freedom. If we refer back to page 529, we find that the sum of squares due to errors about the line was 19.80 with 10 associated degrees of freedom. Since replication of observations has accounted for 8.5 of this total and has 3 associated degrees of freedom, it follows that we still have to account for $19.80 - 8.5 = 11.30$ of the sum of squares and $10 - 3 = 7$ degrees of freedom; this vâriability is attributable to lack of fit of the line to the data. The extended variance analysis table (compare with the table on page 529) follows.

Source of variability	Component of sum of squares	Degrees of freedom	Component of variance	F	Critical values of F
Due to regression line	62.45	1	62.45	22.07	5% F_7^1 = 5.59 1% F_7^1 = 12.25
Errors about line:					
Lack of fit	11.3	7	1.61	0.57	5% F_3^7 = 8.85
Experimental error	8.5	3	2.83		1% F_3^7 = 27.7
Total	82.25	11			

We can now see that the regression is still clearly significant at the 1% level (22.07 > 12.25). Furthermore, the contribution of the lack of fit is not significant at even the 5% level (0.57 << 8.85). This latter result implies that we need not consider a model containing higher order powers of x, since a linear model seems to fit the data quite well. Note that in this example the presence of replications has shifted the emphasis onto the experimental error and the value 2.83 is now the unbiased estimate of σ^2. It might well be argued that it is not worth partitioning the error sum of squares unless all observations are replicated, certainly only when the majority are; this is a conclusion we should support as only 3 terms contributed to the value 2.83 and the estimate is unlikely to be a good one.

(ii) We ought to say a little more about the connection between regression and correlation. Correlation states the degree to which two normal variates are related, whereas regression (and it need not be linear regression) tells us how the variables are related. Regression answers more interesting questions than does correlation and we have, in fact, relegated correlation to the role of an aid to regression analysis. It must be borne in mind that a high value of the correlation coefficient r *does not necessarily indicate a cause and effect relationship.* If two variables are highly correlated it suggests further investigation of the situation. It would be foolish to assume that a high correlation coefficient between output of steel and production of sulphuric acid implied that the one had a marked effect on the other; the truth is more likely that both are influenced by a third factor - growing demand for industrial products. Yet even a linear regression analysis might give a value of $\hat{\beta}$ significantly different from zero. More investigation is clearly called for before any definite conclusions can be reached.

14.9 INTRODUCTION TO MULTIPLE REGRESSION

So far we have been considering regression models involving one independent variable; now we extend the models to include several independent variables. For example, with three independent variables (often called **regressor variables**) the model will be

$$y = \alpha + \beta_1 x_1 + \beta_2 x_2 + \beta_3 x_3 \quad \text{or alternatively} \quad x_0 = \alpha + \beta_1 x_1 + \beta_2 x_2 + \beta_3 x_3$$

The parameters will be estimated by the method of least squares, as was the case for simple regression. The variable y or x_0 is sometimes called the **response variable.** In many areas of mathematics it is the case that the sub-class of problems which involves the least possible number of variables has certain special properties which no longer hold when a further variable is introduced. This new sub-class of problems then usually can point the way to a general solution which will cater for sub-classes with many variables. (Of course, the number of variables may be so large as to cause difficulties in practical computation, but that is a separate issue.) Occasionally, the introduction of a second or a third extra variable into mathematical models does raise other points to be studied.

In the case of regression problems we are now able to adopt a new attitude of thinking. We are now able to cast our net wide to include in our regression model all those regressor variables we think may influence the response variable, *provided that we can measure them.* But we should be aware of the questions we hope to answer by our regression model. Can we ask all the same questions we asked of the simple linear regression model? What other questions can we ask? In the first place, having constructed the model we should need to determine the percentage fit and validate the whole model equation by using a variance analysis table. Then we should estimate the parameters in the model by the method of least squares. We must decide whether all the variables in the model deserve their places; those that cannot justify their inclusion (in the sense that their omission does not produce a statistically significant change) will be put to one side: the method of **backwards elimination.** We should then examine the simplified model and test, one by one, the discarded variables to see whether they can justify their inclusion at this stage: the method of **forward selection.** Other questions which might be asked are whether each of the regressor variables in the model is independent from the others and what implications interdependence may have. We may find further questions which require answering as the analysis is developed.

Example

It is believed that the current gain x_0 of a transistorised electrical component is related to the two processing variables sheet resistance x_1 and diffusion time x_2. The following data was collected and the proposed model is $x_0 = \alpha + \beta_1 x_1 + \beta_2 x_2$.

x_0	6.2	8.7	8.3	10.7	11.7	10.0	9.0	8.1	7.4	13.5
x_1 (ohm cm)	1.7	2.7	0.7	1.4	2.8	0.5	2.6	2.2	0.9	1.8
x_2 (hours)	5.5	7.6	5.8	13.0	8.2	9.3	10.0	6.7	5.5	11.2

This is a fairly straightforward problem and the only worries we have at this stage are whether the model has included all significant regressor variables and whether a linear model is suitable.

Least Squares Model: normal equations and their solution

We may characterise the linear multiple regression model by the following properties. There is a single response variable, x_0, which is to be determined by several independent regressor variables x_1, x_2, ..., x_m. These regressor variables are measured accurately (except for small observational errors which may be ignored with regard to other errors in the system). The response variable is subject to errors ξ which have a mean of zero and a variance of σ^2; these errors are present because the regression model may not faithfully represent the behaviour of x_0. (This can arise because the model has not included *all* the variables which affect x_0 and because the linearity of the model may not be apposite to the problem.) The governing model equation is

$$x_0 = \alpha + \beta_1 x_1 + \beta_2 x_2 + \dots\dots\dots + \beta_m x_m + \xi \qquad (14.20)$$

where the constants α, β_1, β_2, ..., β_m are the parameters of the model.

The sample of n observations that we take consists of n sets each of $(m + 1)$ values: one value is for the response variable and one value for each of the m regressor variables. We shall find it convenient to represent these values in the form of a **data matrix** which has $(m + 1)$ rows and n columns; each row comprises the n values of one of the variables x_0, x_1, ..., x_m and each column contains one set of observations.

A typical data matrix for three regressor variables is shown below

$$\begin{bmatrix} x_{01} & x_{02} & x_{03} & x_{04} & x_{05} \\ x_{11} & x_{12} & x_{13} & x_{14} & x_{15} \\ x_{21} & x_{22} & x_{23} & x_{24} & x_{25} \\ x_{31} & x_{32} & x_{33} & x_{34} & x_{35} \end{bmatrix}$$

Estimation

We shall find the best estimates of α, β_1, ..., β_m and the error variance σ^2 by the least squares method. This necessitates the minimisation of

$$S = \sum_{j=1}^{n} \xi_j^2 = \sum_{j=1}^{n} (x_{0j} - \alpha - \sum_{i=1}^{m} \beta_i x_{ij})^2 \qquad (14.21)$$

and will produce an estimated regression equation $x_0 = \hat{\alpha} + \hat{\beta}_1 x_1 + \dots\dots + \hat{\beta}_m x_m$ (14.22)

The normal equations

The estimation of the parameters α, β_1, β_2, ..., β_m requires the minimisation of S given by equation (14.21). Notice that for a fixed set of data $\{(x_{ij}\,; i = 0, m),\ j = 1, n\}$, S depends upon the values of the parameters.

We may therefore write $S = S(\alpha, \beta_1, \beta_2, \ldots\ldots, \beta_m)$. We require $\frac{\partial S}{\partial \alpha} = \frac{\partial S}{\partial \beta_1} = \frac{\partial S}{\partial \beta_2} = \ldots\ldots\ldots\ldots = \frac{\partial S}{\partial \beta_m} = 0$. This set of conditions provides $(m+1)$ equations. For example, with two regressor variables

$S = \sum_{j=1}^{n} (x_{0j} - \alpha - \beta_1 x_{1j} - \beta_2 x_{2j})^2$ we obtain the **normal equations**

$$\left.\begin{aligned}
\sum_{j=1}^{n} x_{0j} &= n\hat{\alpha} + \hat{\beta}_1 \sum_{j=1}^{n} x_{1j} + \hat{\beta}_2 \sum_{j=1}^{n} x_{2j} \\
\sum_{j=1}^{n} x_{0j}x_{1j} &= \hat{\alpha} \sum_{j=1}^{n} x_{1j} + \hat{\beta}_1 \sum_{j=1}^{n} x^2_{1j} + \hat{\beta}_2 \sum_{j=1}^{n} x_{2j}x_{1j} \\
\sum_{j=1}^{n} x_{0j}x_{2j} &= \hat{\alpha} \sum_{j=1}^{n} x_{2j} + \hat{\beta}_1 \sum_{j=1}^{n} x_{1j}x_{2j} + \hat{\beta}_2 \sum_{j=1}^{n} x^2_{2j}
\end{aligned}\right\} \qquad (14.23)$$

In matrix form the set can be written

$$\begin{bmatrix} \sum_{j=1}^{n} x_{0j} \\ \sum_{j=1}^{n} x_{0j}x_{1j} \\ \sum_{j=1}^{n} x_{0j}x_{2j} \end{bmatrix} = \begin{bmatrix} n & \sum_{j=1}^{n} x_{1j} & \sum_{j=1}^{n} x_{2j} \\ \sum_{j=1}^{n} x_{1j} & \sum_{j=1}^{n} x^2_{1j} & \sum_{j=1}^{n} x_{2j}x_{1j} \\ \sum_{j=1}^{n} x_{2j} & \sum_{j=1}^{n} x_{1j}x_{2j} & \sum_{j=1}^{n} x^2_{2j} \end{bmatrix} \begin{bmatrix} \hat{\alpha} \\ \hat{\beta}_1 \\ \hat{\beta}_2 \end{bmatrix} \qquad (14.24)$$

Notice that the first matrix on the right-hand side of (14.24) is symmetric.

Example

For the example currently under study equations (14.24) become

$$\begin{bmatrix} 93.6 \\ 164.76 \\ 812.57 \end{bmatrix} = \begin{bmatrix} 10 & 17.3 & 82.8 \\ 17.3 & 36.37 & 145.59 \\ 82.8 & 145.59 & 744.96 \end{bmatrix} \begin{bmatrix} \hat{\alpha} \\ \hat{\beta}_1 \\ \hat{\beta}_2 \end{bmatrix} \qquad (14.25)$$

It can be shown by Gauss Elimination that these equations have the solution $\hat{\alpha} = 3.822$, $\hat{\beta}_1 = 0.213$, $\hat{\beta}_2 = 0.624$ and hence the estimated regression equation is $\hat{x}_0 = 3.822 + 0.213x_1 + 0.624x_2$.

Conditions required for a solution

The first condition that a solution to the normal equations shall exist is that the number of sets of observations, n, must be greater than $(m + 1)$, the number of parameters to be estimated. This is an extension of the result that in fitting a straight line there must be at least two data points. An infinite number of lines will be able to pass through one given data point, reflected by the fact that the normal equations will be singular in such a case. Two data points will determine exactly the straight line that passes through them and more than two data points will produce a regression line which will not necessarily fit the points exactly. In the same way, for multiple regression to operate there must be at least one more data point than parameters in the model.

If we are going to solve the linear normal equations, we shall have to contend with the problems that arise in the solution of linear equations. If the regressor variables are related so that at least one is a linear combination of some of the others then the matrix will be *singular* and the normal equations will have no solution. We can overcome this condition if we can omit the relevant regressor variables (and the same number of normal equations) from the model. If we cannot find which of the regressor variables should be removed we shall be in trouble, the model is said to be **overspecified** because at least one of its regressor variables is redundant.

Since we are carrying out practical numerical calculations we must expect round-off to interfere and we may not get an exactly singular matrix. Instead of a division by zero, we should be dividing by a near-zero value. This is the problem of **ill-conditioning** and we can attempt to avoid these problems by rejecting all regressor variables which lead to a division by a small value (less than a specified value); such a check can be built into a computer program.

Confidence intervals on the parameters

It is possible to obtain confidence intervals for the estimates of the parameters, of the mean value of x_0, given x_1 and x_2 and for an individual observation of x_0. The procedure is analogous to that for simple regression, However, the form of approach we shall adopt is based on the variance analysis method. The reader who wishes to pursue the technique of confidence intervals can consult the appropriate books mentioned in the bibliography.

Validation analysis

Before we turn to the consideration of whether each of the regressor variables is entitled to its place in the model we should first measure the goodness of fit of the whole model.

Even though we obtain a regression equation which has a high percentage fit, it does not follow that we have the best regression model available. In order to attain this model we shall perform two kinds of operation: a **backwards elimination process** and a **forward selection process.**

In backwards elimination we take the model comprising all the regressor variables x_i and study the effects on the model of omitting in turn only the variable x_i. The variable x_i is replaced and x_{i+1} omitted and so on. If this is unrelated to the other regressor variables, then its omission will be a simple affair and we can judge whether leaving it out from the model makes a significant difference. If, however, x_i is related to the other regressor variables (the variance analysis table will give us an indication of the degree of correlation between the regressor variables) then its omission needs more study.

The rejection of a regressor variable during the backwards elimination stage does not mean that it is necessarily lost from the model. The second stage of the analysis is the forward selection phase for the rejected variables. In this stage the separate effects of adding each of the rejected variables one at a time to the model is studied. The implication is that a new set of calculations must be undertaken, starting from the new information matrix, each time a new model is proposed. If the regressor variables are correlated among themselves, backwards elimination and forward selection would give different results if each were used as the first stage.

When many regressor variables are involved, the two-phase cycle of backwards elimination and forward selection must be repeated until a stage is reached where the omission of any of the current regressor variables would markedly reduce the percentage overall fit and the addition of another regressor variable would not significantly improve the percentage fit.

Forward selection phase to censor rejected variables

Particularly where a model shows a high degree of correlation, it is not wise to reject variables completely. As a first step, we must certainly re-calculate $\hat{\alpha}$ and those $\hat{\beta}_i$ associated with the variables retained in the model.

As we have already remarked, if two variables, x_3 and x_4, say, have been rejected in the backwards elimination analysis it may be that they are correlated among themselves and it may be possible to obtain a better regression model by including one of them (but not both). In general, a subset of regressor variables would be rejected because the component of sum of squares due to them was shared among the inter-related variables so that no one of them could make a significant contribution to the model.

In the forward selection phase, the rejected regressor variables are added one at a time to see whether anyone of them by itself could make a significant contribution. When one is found to do so, it is added to the model, all the variables in the improved model are validated by backwards elimination and those rejected are subjected to forward selection. The process continues until a 'best' model is obtained: no variable can be omitted and no other variable will significantly improve the situation.

Since every new intermediate model requires a substantial amount of re-calculation, it is wise to use the computer for such analysis and standard packages exist in some computer centres.

The order of adding regressor variables to an intermediate model is important if the variables are inter-related. Consider Figure 14.9.

We shall suppose that the three inter-related regressor variables x_2, x_5 and x_6 have been rejected during the backwards elimination phase and are now candidates during the forward selection phase. If we first fit back x_2 it will also take with it the shared components which we can symbolise by set notation as $x_2 \cap x_5$, $x_2 \cap x_6$ and $x_2 \cap x_5 \cap x_6$; the shaded areas in Figure 14.9. Fitting next the variable x_5 will take in also $x_5 \cap x_6$. When x_6 comes to the selection panel it will have no shared elements to accompany it, it will have only the component due to it alone and it may not be able to produce a significant contribution to justify admission to the model. How unfair, you may think, for if the variables had been presented in the order x_5, x_6, x_2 the state of affairs would have been different. Accompanying x_5 would be the shared elements $x_5 \cap x_6$, $x_5 \cap x_2$ and $x_5 \cap x_6 \cap x_2$; accompanying x_6 would be the shared element $x_6 \cap x_2$, whereas x_2 would be on its own and this time *it* would most likely not gain admission. Yet, given the inter-relationships between these variables, we must expect the best regression model not to include all three and, since they are inter-related, it really does not matter which are in the model so long as the significant contributions are included.

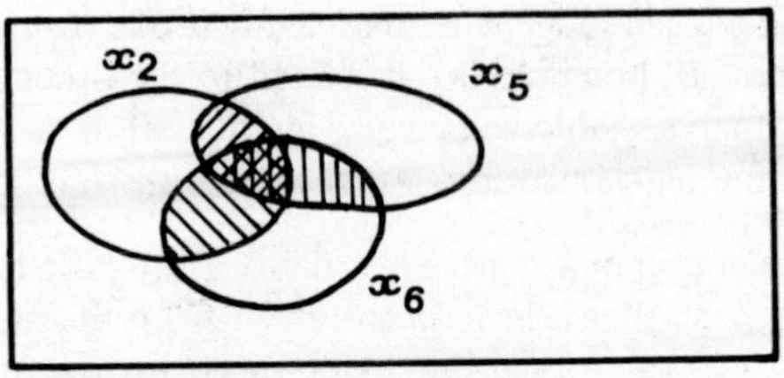

Figure 14.9

Problems

1. Estimate the multiple linear regression equation for the following sets of data.

(i)

(i)	y	2	5	7	8	5
	x_1	4	4	2	3	−1
	x_2	1	2	2	4	5

(ii)

(ii)	x_1	24	24.5	25	27	27.5	28	29	30	31	32
	x_2	24.5	26.4	13.2	35.2	15.5	17.5	35.0	36.5	27.3	38.0
	y	51.5	51.2	22.0	42.5	21.5	12.5	43.0	46.8	24.8	42.7

(iii)

(iii)	x_1	3	5	6	8	12	14
	x_2	16	10	7	4	3	2
	y	90	72	54	42	30	12

2. Write a computer program to find the regression equation for y on x_1 and x_2 given n sets of data.

3. The following data gives information on carbon content x_1, silicon content x_2 and tensile strength y for 25 samples of cylindrical steel rod.

$$\Sigma x_1 = 15, \quad \Sigma x_2 = 20, \quad \Sigma y = 25, \quad \Sigma x_1 y = 75, \quad \Sigma x_2 y = 100,$$

$$\Sigma x_1 x_2 = 55, \quad \Sigma y^2 = 180, \quad \Sigma x_1{}^2 = 45, \quad \Sigma x_2{}^2 = 80.$$

Estimate the best linear equation $\hat{y} = a + b_1 x_1 + b_2 x_2$ for predicting y from x_1

and x_2, and determine the proportion of the sum of squares in tensile strength readings 'explained' by the fitted regression equation. Carry out a goodness-of-fit test.

4. Find the linear regression equation of x on y and z for the following data, stating clearly the method used.

x	64	71	53	67	55	58
y	57	59	49	62	51	50
z	8	10	6	11	8	7

(Make use of the further information $\Sigma xy = 20293$, $\Sigma y^2 = 18076$). (C.E.I.)

5. A civil engineering contractor wishes to estimate the amount of road that his company can construct in a certain time. This depends upon the proportion of time (number of tenths) for which rain is falling and the number of men working on the road. The available information is:

Amount of road constructed in a fixed time - z	12	19	9	8	5	13
Number of tenths of time in which rain was falling - x	9	6	3	1	4	7
Number of men working on the road - y	7	8	6	3	4	8

Find the multiple regression line of z on x and y and estimate the number of men needed to ensure progress (namely $z > 0$) under the worst possible conditions. (C.E.I.)

6. Large bolts are to be used on structures exposed to the atmosphere. They are given a corrosion-resistant coating. The process of coating involves passing the bolts through a bath of coating fluid, spinning the coated bolts in a centrifuge and finally heat drying them. We form a model $y = \alpha + \beta_1 x_1 + \beta_2 x_2 + \beta_3 x_3$ where y is the resulting thickness of the coating, x_1 is the viscosity of the coating fluid, x_2 is the time of spinning and x_3 is the drying temperature. Observations of the thickness y and of values of x_1, x_2 and x_3 were made on a sample of 14 bolts. The results are as follows. Obtain the estimated model equation.

x_1	15	24	21	19	13	16	25	22	9	8	14	17	23	18
x_2	29	30	35	31	47	45	31	34	37	28	39	33	38	36
x_3	35	31	20	25	37	30	34	29	22	27	33	30	28	23
y	92	105	101	97	93	85	87	102	94	86	84	109	110	103

Appendix

DIFFERENTIATION UNDER THE INTEGRAL SIGN

Suppose $f(x, y)$ is a continuous function of x and y. Consider the definite integral $\int_a^b f(x, y)\,\mathrm{d}x$; this will be a function of y, $F(y)$. During the integration with respect to x, y is effectively constant. Let

$$\delta F = \int_a^b f(x, y + \delta y)\,\mathrm{d}x - \int_a^b f(x, y)\,\mathrm{d}x$$

$$= \int_a^b \left\{ f(x, y + \delta y) - f(x, y) \right\} \mathrm{d}x$$

$$= \int_a^b \delta y\, f_y(x, y + \xi \delta y)\,\mathrm{d}x \qquad \text{by the Mean Value Theorem}$$

where $0 < \xi < 1$.

Since y is a constant with respect to the integration we may write

$$\frac{\delta F}{\delta y} = \int_a^b f_y(x, y + \xi \delta y)\,\mathrm{d}x$$

In the limit as $\delta y \to 0$ we obtain

$$\frac{\mathrm{d}F}{\mathrm{d}y} = \int_a^b \frac{\partial f}{\partial y}\,\mathrm{d}x \tag{A.1}$$

This result assumes that $\frac{\partial f}{\partial y}$ is a continuous function of x and y.

Example

Let $F(y) = \displaystyle\int_0^{\pi/2} \frac{\sin(y\,x)}{x}\,dx$

Then $\dfrac{\partial F}{\partial y} = \displaystyle\int_0^{\pi/2} \frac{x\cos(y\,x)}{x}\,dx = \int_0^{\pi/2} \cos(y\,x)\,dx$

$$= \left[\frac{1}{y}\sin(y\,x)\right]_0^{\pi/2} = \frac{1}{y}\sin\left(\frac{\pi y}{2}\right).$$

Infinite Integrals

If the range of the defining integral is infinite (or semi-infinite) the conditions under which it is possible to differentiate under the integral sign for $y_1 \leqslant y \leqslant y_2$ are

(i) $\dfrac{d}{dy}\displaystyle\int_a^b f(x, y)\,dx = \int_a^b \frac{\partial f}{\partial y}\,dx$

(ii) $\displaystyle\int_a^\infty f(x, y)\,dx$ converges

(iii) $\displaystyle\int_a^\infty \frac{\partial f}{\partial y}\,dx$ converges uniformly for $y_1 \leqslant y \leqslant y_2$

If these conditions are met then $\dfrac{d}{dy}\displaystyle\int_a^\infty f(x, y)\,dx = \int_a^\infty \frac{\partial f}{\partial y}\,dx$

Example

Let $I(y) = \displaystyle\int_0^\infty e^{-xy}\frac{\sin x}{x}\,dx$

Then $\dfrac{\partial I}{\partial y} = -\displaystyle\int_0^\infty x\,e^{-xy}\frac{\sin x}{x}\,dx = -\int_0^\infty e^{-xy}\sin x\,dx$

$$= \left[\frac{e^{-xy}(y\sin x + \cos x)}{1 + y^2}\right]_0^\infty \text{, as you can verify}$$

$$= -1/(1 + y^2)$$

Hence $I(y) = C - \tan^{-1} y$ for C constant. When $y \to \infty$, $I(y) \to 0$ and hence $C = \pi/2$. Therefore $I(y) = \pi/2 - \tan^{-1} y$.

(Note that we have still to check that the required conditions hold.)

Finally, if we let $y = 0$ we can actually evaluate

$$\int_0^\infty \frac{\sin x}{x}\, \mathrm{d}x = \pi/2 \tag{A.2}$$

By taking $y = \alpha x$, $\alpha \neq 0$ we can show that

$$\int_0^\infty \frac{\sin \alpha x}{x}\, \mathrm{d}x = \pi/2 \tag{A.3}$$

Variable limits of integration

Consider $F(\alpha) = \displaystyle\int_a^b f(x, \alpha)\, \mathrm{d}x$ where a and b are functions of α. Under suitable conditions it can be shown that

$$\frac{\mathrm{d}F}{\mathrm{d}\alpha} = \int_a^b \frac{\partial f}{\partial \alpha}\, \mathrm{d}x + f(b, \alpha)\frac{\mathrm{d}b}{\mathrm{d}\alpha} - f(a, \alpha)\frac{\mathrm{d}a}{\mathrm{d}\alpha} \tag{A.4}$$

Example

Let $F(\alpha) = \displaystyle\int_{\pi/(6\alpha)}^{\pi/(2\alpha)} \frac{\sin \alpha x}{x}\, \mathrm{d}x$

Then $F'(\alpha) = \displaystyle\int_{\pi/(6\alpha)}^{\pi/(2\alpha)} \frac{x \cos \alpha x}{x}\, \mathrm{d}x + \frac{\sin(\pi/2)}{\pi/(2\alpha)}\left(\frac{-\pi}{2\alpha^2}\right) - \frac{\sin(\pi/6)}{\pi/(6\alpha)}\left(\frac{-\pi}{6\alpha^2}\right)$

$$= \left[\frac{\sin \alpha x}{\alpha}\right]_{\pi/(6\alpha)}^{\pi/(2\alpha)} - \frac{1}{\alpha} + \frac{1}{2\alpha} = \frac{1}{\alpha} - \frac{1}{2\alpha} - \frac{1}{\alpha} + \frac{1}{2\alpha} = 0$$

Then $F(\alpha) =$ constant and the original integral is independent of α; by a simple substitution, try to establish this.

Integration under the integral sign

We work via an example, viz. $I = \displaystyle\int_0^\infty e^{-x^2} \cos 2\alpha x\, \mathrm{d}x$. Now it can be shown that

$$\frac{dI}{d\alpha} = -2 \int_0^\infty e^{-x^2} x \sin 2\alpha x \, dx$$

$$= \left[e^{-x^2} \sin 2\alpha x\right]_0^\infty - 2\alpha \int_0^\infty e^{-x^2} \cos 2\alpha x \, dx = 0 - 2\alpha I$$

Therefore $I = A\, e^{-\alpha^2}$

But when $\alpha = 0$, $A = \int_0^\infty e^{-x^2}\, dx = \frac{\sqrt{\pi}}{2}$ as can be shown via an appeal to double integrals. Hence

$$I = \frac{\sqrt{\pi}}{2} e^{-\alpha^2} \tag{A.5}$$

If we integrate both sides of (A.5), with respect to α, from 0 to α we obtain

$$\text{L.H.S.} = \int_0^\alpha I \, d\alpha = \int_0^\infty e^{-x^2} \left\{ \int_0^\alpha \cos 2\alpha x \, d\alpha \right\} dx$$

$$= \int_0^\infty e^{-x^2} \frac{\sin 2\alpha x}{2x} \, dx$$

$$\text{R.H.S.} = \frac{\sqrt{\pi}}{2} \int_0^\alpha e^{-\alpha^2} \, d\alpha$$

Remembering the definition $\operatorname{erf} \alpha = \frac{2}{\sqrt{\pi}} \int_0^\alpha e^{-u^2} du$ and cancelling a factor of 2,

$$\int_0^\infty e^{-x^2} \frac{\sin 2\alpha x}{x} \, dx = \frac{\pi}{2} \operatorname{erf} \alpha \tag{A.6}$$

Table A.I. The Normal Probability Integral

$\frac{x-\mu}{\sigma}$	0	1	2	3	4	5	6	7	8	9
0	0000	0040	0080	0120	0160	0199	0239	0279	0319	0359
.1	0398	0438	0478	0517	0557	0596	0636	0675	0714	0753
.2	0793	0832	0871	0909	0948	0987	1026	1064	1103	1141
.3	1179	1217	1255	1293	1331	1368	1406	1443	1480	1517
.4	1555	1591	1628	1664	1700	1736	1772	1808	1844	1879
.5	1915	1950	1985	2019	2054	2088	2123	2157	2190	2224
.6	2257	2291	2324	2357	2389	2422	2454	2486	2517	2549
.7	2580	2611	2642	2673	2703	2734	2764	2794	2822	2852
.8	2881	2910	2939	2967	2995	3023	3051	3078	3106	3133
.9	3159	3186	3212	3238	3264	3289	3315	3340	3365	3389
1.0	3413	3438	3461	3485	3508	3531	3554	3577	3599	3621
1.1	3643	3665	3686	3708	3729	3749	3770	3790	3810	3830
1.2	3849	3869	3888	3907	3925	3944	3962	3980	3997	4015
1.3	4032	4049	4066	4082	4099	4115	4131	4147	4162	4177
1.4	4192	4207	4222	4236	4251	4265	4279	4292	4306	4319
1.5	4332	4345	4357	4370	4382	4394	4406	4418	4429	4441
1.6	4452	4463	4474	4484	4495	4505	4515	4525	4535	4545
1.7	4554	4564	4573	4582	4591	4599	4608	4616	4625	4633
1.8	4641	4649	4656	4664	4671	4678	4686	4693	4699	4706
1.9	4713	4719	4726	4732	4738	4744	4750	4756	4761	4767
2.0	4772	4778	4783	4788	4793	4798	4803	4808	4812	4817
2.1	4821	4826	4830	4834	4838	4842	4846	4850	4854	4857
2.2	4861	4865	4868	4871	4875	4878	4881	4884	4887	4890
2.3	4893	4896	4898	4901	4904	4906	4909	4911	4913	4916
2.4	4918	4920	4922	4925	4927	4929	4931	4932	4934	4936
2.5	4938	4940	4941	4943	4945	4946	4948	4949	4951	4952
2.6	4953	4955	4956	4957	4959	4960	4961	4962	4963	4964
2.7	4965	4966	4967	4968	4969	4970	4971	4972	4973	4974
2.8	4974	4975	4976	4977	4977	4978	4979	4979	4980	4981
2.9	4981	4982	4982	4983	4984	4984	4985	4985	4986	4986

3.0	3.1	3.2	3.3	3.4	3.5	3.6	3.7	3.8	3.9
4987	4990	4993	4995	4997	4998	4998	4999	4999	4999

Table A.II. Percentage points of the t distribution

ν \ P%	50	25	10	5	2.5	1	0.5	0.1
1	1.00	2.41	6.31	12.7	25.5	63.7	127.	637.
2	.816	1.60	2.92	4.30	6.21	9.92	14.1	31.6
3	.765	1.42	2.35	3.18	4.18	5.84	7.45	12.9
4	.741	1.34	2.13	2.78	3.50	4.60	5.60	8.61
5	.727	1.30	2.01	2.57	3.16	4.03	4.77	6.86
6	.718	1.27	1.94	2.45	2.97	3.71	4.32	5.96
7	.711	1.25	1.89	2.36	2.84	3.50	4.03	5.40
8	.706	1.24	1.86	2.31	2.75	3.36	3.83	5.04
9	.703	1.23	1.83	2.26	2.68	3.25	3.69	4.78
10	.700	1.22	1.81	2.23	2.63	3.17	3.58	4.59
11	.698	1.21	1.80	2.20	2.59	3.11	3.50	4.44
12	.695	1.21	1.78	2.18	2.56	3.05	3.43	4.32
13	.694	1.20	1.77	2.16	2.53	3.01	3.37	4.22
14	.692	1.20	1.76	2.14	2.51	2.98	3.33	4.14
15	.691	1.20	1.75	2.13	2.49	2.95	3.29	4.07
16	.690	1.19	1.75	2.12	2.47	2.92	3.25	4.01
17	.689	1.19	1.74	2.11	2.46	2.90	3.22	3.96
18	.688	1.19	1.73	2.10	2.44	2.88	3.20	3.92
19	.688	1.19	1.73	2.09	2.43	2.86	3.17	3.88
20	.687	1.18	1.72	2.09	2.42	2.85	3.15	3.85
21	.686	1.18	1.72	2.08	2.41	2.83	3.14	3.82
22	.686	1.18	1.72	2.07	2.41	2.82	3.12	3.79
23	.685	1.18	1.71	2.07	2.40	2.81	3.10	3.77
24	.685	1.18	1.71	2.06	2.39	2.80	3.09	3.74
25	.684	1.18	1.71	2.06	2.38	2.79	3.08	3.72
26	.684	1.18	1.71	2.06	2.38	2.78	3.07	3.71
27	.684	1.18	1.70	2.05	2.37	2.77	3.06	3.69
28	.683	1.17	1.70	2.05	2.37	2.76	3.05	3.67
29	.683	1.17	1.70	2.05	2.36	2.76	3.04	3.66
30	.683	1.17	1.70	2.04	2.36	2.75	3.03	3.65
40	.681	1.17	1.68	2.02	2.33	2.70	2.97	3.55
60	.679	1.16	1.67	2.00	2.30	2.66	2.91	3.46
∞	.674	1.15	1.64	1.96	2.24	2.58	2.81	3.29

Table A.III. **Percentage points of the χ^2 distribution**

	P	0.990	0.950	0.900	0.500	0.100	0.050	0.010
ν	1	1571×10^{-7}	3932×10^{-6}	0.01579	0.4549	2.705	3.841	6.635
	2	0.0201	0.1026	0.2107	1.386	4.605	5.991	9.210
	3	0.1148	0.3518	0.5844	2.366	6.251	7.815	11.34
	4	0.2971	0.711	1.064	3.357	7.779	9.488	13.28
	5	0.5543	1.145	1.610	4.351	9.236	11.07	15.09
	6	0.8721	1.635	2.204	5.348	10.64	12.59	16.81
	7	1.239	2.167	2.833	6.346	12.02	14.07	18.48
	8	1.646	2.733	3.490	7.344	13.36	15.51	20.09
	9	2.088	3.325	4.168	8.343	14.68	16.92	21.67
	10	2.558	3.940	4.865	9.342	15.99	18.31	23.21
	11	3.053	4.575	5.578	10.34	17.28	19.68	24.73
	12	3.571	5.226	6.304	11.34	18.55	21.03	26.22
	13	4.107	5.892	7.042	12.34	19.81	22.36	27.69
	14	4.660	6.571	7.790	13.34	21.06	23.68	29.14
	15	5.229	7.261	8.547	14.34	22.31	25.00	30.58
	16	5.812	7.962	9.312	15.34	23.54	26.30	32.00
	17	6.408	8.672	10.09	16.34	24.77	27.59	33.41
	18	7.015	9.390	10.86	17.34	25.99	28.87	34.81
	19	7.633	10.12	11.65	18.34	27.20	30.14	36.19
	20	8.260	10.85	12.44	19.34	28.41	31.41	37.57
	21	8.897	11.59	13.24	20.34	29.62	32.67	38.93
	22	9.542	12.34	14.04	21.34	30.81	33.92	40.29
	23	10.20	13.09	14.85	22.34	32.01	35.17	41.64
	24	10.86	13.85	15.66	23.34	33.20	36.42	42.98
	25	11.52	14.61	16.47	24.34	34.38	37.65	44.31
	26	12.20	15.38	17.29	25.34	35.56	38.89	45.64
	27	12.88	16.15	18.11	26.34	36.74	40.11	46.96
	28	13.56	16.93	18.94	27.34	37.92	41.34	48.28
	29	14.26	17.71	19.77	28.34	39.09	42.56	49.59
	30	14.95	18.49	20.60	29.34	40.26	43.77	50.89
	40	22.16	26.51	29.05	39.34	51.81	55.76	63.69
	50	29.71	34.76	37.69	49.33	63.17	67.50	76.15
	60	37.48	43.19	46.46	59.33	74.40	79.08	88.38
	70	45.44	51.74	55.33	69.33	85.53	90.53	100.4
	80	53.54	60.39	64.28	79.33	96.58	101.9	112.3
	90	61.75	69.13	73.29	89.33	107.6	113.1	124.1
	100	70.06	77.93	82.36	99.33	118.5	124.3	135.8

Table A.IV. *F* distribution - 5% 2½% and 1% points

ν_1 / ν_2	1	2	3	4	5	6	7	8	10	12	24	∞
	161	200	216	225	230	234	237	239	242	244	249	254
1	648	800	864	900	922	937	948	957	969	977	997	1018
	4052	5000	5403	5625	5764	5859	5928	5981	6056	6106	6235	6366
	18.5	19.0	19.2	19.2	19.3	19.3	19.4	19.4	19.4	19.4	19.5	19.5
2	38.5	39.0	39.2	39.2	39.3	39.3	39.4	39.4	39.4	39.4	39.5	39.5
	98.5	99.0	99.2	99.2	99.3	99.3	99.4	99.4	99.4	99.4	99.5	99.5
	10.13	9.55	9.28	9.12	9.01	8.94	8.89	8.85	8.79	8.74	8.64	8.53
3	17.4	16.0	15.4	15.1	14.9	14.7	14.6	14.5	14.4	14.3	14.1	13.9
	34.1	30.8	29.5	28.7	28.2	27.9	27.7	27.5	27.2	27.1	26.6	26.1
	7.71	6.94	6.59	6.39	6.26	6.16	6.09	6.04	5.96	5.91	5.77	5.63
4	12.22	10.65	9.98	9.60	9.36	9.20	9.07	8.98	8.84	8.75	8.51	8.26
	21.2	18.0	16.7	16.0	15.5	15.2	15.0	14.8	14.5	14.4	13.9	13.5
	6.61	5.79	5.41	5.19	5.05	4.95	4.88	4.82	4.74	4.68	4.53	4.36
5	10.01	8.43	7.76	7.39	7.15	6.98	6.85	6.76	6.62	6.52	6.28	6.02
	16.26	13.27	12.06	11.39	10.97	10.67	10.46	10.29	10.05	9.89	9.47	9.02
	5.99	5.14	4.75	4.53	4.39	4.28	4.21	4.15	4.06	4.00	3.84	3.67
6	8.81	7.26	6.60	6.23	5.99	5.82	5.70	5.60	5.46	5.37	5.12	4.85
	13.74	10.92	9.78	9.15	8.75	8.47	8.26	8.10	7.87	7.72	7.31	6.88
	5.59	4.74	4.35	4.12	3.97	3.87	3.79	3.73	3.64	3.57	3.41	3.23
7	8.07	6.54	5.89	5.52	5.29	5.12	4.99	4.90	4.76	4.67	4.42	4.14
	12.25	9.55	8.45	7.85	7.46	7.19	6.99	6.84	6.62	6.47	6.07	5.65
	5.32	4.46	4.07	3.84	3.69	3.58	3.50	3.44	3.35	3.28	3.12	2.93
8	7.57	6.06	5.42	5.05	4.82	4.65	4.53	4.43	4.30	4.20	3.95	3.67
	11.26	8.65	7.59	7.01	6.63	6.37	6.18	6.03	5.81	5.67	5.28	4.86
	5.12	4.26	3.86	3.63	3.48	3.37	3.29	3.23	3.14	3.07	2.90	2.71
9	7.21	5.71	5.08	4.72	4.48	4.32	4.20	4.10	3.96	3.87	3.61	3.33
	10.56	8.02	6.99	6.42	6.06	5.80	5.61	5.47	5.26	5.11	4.73	4.31
	4.96	4.10	3.71	3.48	3.33	3.22	3.14	3.07	2.98	2.91	2.74	2.54
10	6.94	5.46	4.83	4.47	4.24	4.07	3.95	3.85	3.72	3.62	3.37	3.08
	10.04	7.56	6.55	5.99	5.64	5.39	5.20	5.06	4.85	4.71	4.33	3.91
	4.84	3.98	3.59	3.36	3.20	3.09	3.01	2.95	2.85	2.79	2.61	2.40
11	6.72	5.26	4.63	4.28	4.04	3.88	3.76	3.66	3.53	3.43	3.17	2.88
	9.65	7.21	6.22	5.67	5.32	5.07	4.89	4.74	4.54	4.40	4.02	3.60
	4.75	3.89	3.49	3.26	3.11	3.00	2.91	2.85	2.75	2.69	2.51	2.30
12	6.55	5.10	4.47	4.12	3.89	3.73	3.61	3.51	3.37	3.28	3.02	2.72
	9.33	6.93	5.95	5.41	5.06	4.82	4.64	4.50	4.30	4.16	3.78	3.36

ν_2 \ ν_1	1	2	3	4	5	6	7	8	10	12	24	∞
	4.60	3.74	3.34	3.11	2.96	2.85	2.76	2.70	2.60	2.53	2.35	2.13
14	6.30	4.86	4.24	3.89	3.66	3.50	3.38	3.29	3.15	3.05	2.79	2.49
	8.86	6.51	5.56	5.04	4.70	4.46	4.28	4.14	3.94	3.80	3.43	3.00
	4.49	3.63	3.24	3.01	2.85	2.74	2.66	2.59	2.49	2.42	2.24	2.01
16	6.12	4.69	4.08	3.73	3.50	3.34	3.22	3.12	2.99	2.89	2.63	2.32
	8.53	6.23	5.29	4.77	4.44	4.20	4.03	3.89	3.69	3.55	3.18	2.75
	4.41	3.55	3.16	2.93	2.77	2.66	2.58	2.51	2.41	2.34	2.15	1.92
18	5.98	4.56	3.95	3.61	3.38	3.22	3.10	3.01	2.87	2.77	2.50	2.19
	8.29	6.01	5.09	4.58	4.25	4.01	3.84	3.71	3.51	3.37	3.00	2.57
	4.35	3.49	3.10	2.87	2.71	2.60	2.51	2.45	2.35	2.28	2.08	1.84
20	5.87	4.46	3.86	3.51	3.29	3.13	3.01	2.91	2.77	2.68	2.41	2.09
	8.10	5.85	4.94	4.43	4.10	3.87	3.70	3.56	3.37	3.23	2.86	2.42
	4.26	3.40	3.01	2.78	2.62	2.51	2.42	2.36	2.25	2.18	1.98	1.73
24	5.72	4.32	3.72	3.38	3.15	2.99	2.87	2.78	2.64	2.54	2.27	1.94
	7.82	5.61	4.72	4.22	3.90	3.67	3.50	3.36	3.17	3.03	2.66	2.21
	4.20	3.34	2.95	2.71	2.56	2.45	2.36	2.29	2.19	2.12	1.91	1.65
28	5.61	4.22	3.63	3.29	3.06	2.90	2.78	2.69	2.55	2.45	2.17	1.83
	7.64	5.45	4.57	4.07	3.75	3.53	3.36	3.23	3.03	2.90	2.52	2.06
	4.15	3.29	2.90	2.67	2.51	2.40	2.31	2.24	2.14	2.07	1.86	1.59
32	5.53	4.15	3.56	3.22	3.00	2.84	2.72	2.62	2.48	2.38	2.10	1.75
	7.50	5.34	4.46	3.97	3.65	3.43	3.26	3.13	2.93	2.80	2.42	1.96
	4.11	3.26	2.87	2.63	2.48	2.36	2.28	2.21	2.11	2.03	1.82	1.55
36	5.47	4.09	3.51	3.17	2.94	2.79	2.66	2.57	2.43	2.33	2.05	1.69
	7.40	5.25	4.38	3.89	3.58	3.35	3.18	3.05	2.86	2.72	2.35	1.87
	4.08	3.23	2.84	2.61	2.45	2.34	2.25	2.18	2.08	2.00	1.79	1.51
40	5.42	4.05	3.46	3.13	2.90	2.74	2.62	2.53	2.39	2.29	2.01	1.64
	7.31	5.18	4.31	3.83	3.51	3.29	3.12	2.99	2.80	2.66	2.29	1.80
	4.00	3.15	2.76	2.53	2.37	2.25	2.17	2.10	1.99	1.92	1.70	1.39
60	5.29	3.93	3.34	3.01	2.79	2.63	2.51	2.41	2.27	2.17	1.88	1.48
	7.08	4.98	4.13	3.65	3.34	3.12	2.95	2.82	2.63	2.50	2.12	1.60
	3.92	3.07	2.68	2.45	2.29	2.18	2.09	2.02	1.91	1.83	1.61	1.25
120	5.15	3.80	3.23	2.89	2.67	2.52	2.39	2.30	2.16	2.05	1.76	1.31
	6.85	4.79	3.95	3.48	3.17	2.96	2.79	2.66	2.47	2.34	1.95	1.38
	3.84	3.00	2.60	2.37	2.21	2.10	2.01	1.94	1.83	1.75	1.52	1.00
∞	5.02	3.69	3.12	2.79	2.57	2.41	2.29	2.19	2.05	1.94	1.64	1.00
	6.63	4.61	3.78	3.32	3.02	2.80	2.64	2.51	2.32	2.18	1.79	1.00

For each value of $F_{\nu_2}^{\nu_1}$, the upper number is the 5% point, the middle one is the 2½% point and the lowest one is the 1% point.

Bibliography

The following is a selected list of books related to the subject matter of this text.

1. BAJPAI, A.C., CALUS, I.M., FAIRLEY, J.A. and WALKER, D., *Mathematics for Engineers and Scientists, Vol. 2.* Wiley, London, 1973.

2. BAJPAI, A.C., MUSTOE, L.R. and WALKER, D., *Engineering Mathematics.* Wiley, London, 1974.

3. BRACEWELL, R., *The Fourier Transform and its Applications.* McGraw-Hill, New York, 1965.

4. CHURCHILL, R.V., *Operational Mathematics.* McGraw-Hill, New York, 1958.

5. COHEN, A.M. et al., *Numerical Analysis.* McGraw-Hill, London, 1973.

6. CRANK, J., *The Mathematics of Diffusion.* O.U.P., London, 1976.

7. DIXON, L.C.W., *Nonlinear Optimisation.* E.U.P., London, 1972.

8. HALD, A., *Statistical Theory with Engineering Applications.* Wiley, New York, 1960.

9. LLEWELLYN, R.W., *Linear Programming.* Holt, Rinehart and Winston, New York, 1964.

10. RAINVILLE, E.D., *Special Functions.* Macmillan, New York, 1960.

11. SMITH, G.D., *Numerical Solution of Partial Differential Equations.* O.U.P., London, 1965.

12. SNEDECOR, G.W. and COCHRAN, W.G., *Statistical Methods.* Iowa State University Press, Ames, Iowa, U.S.A., 1967.

Answers

Chapter 1

Page 7

1. Any two of the first set; any pair of the second set except for the last pair.
3. Two, in each case.
4. (i), (iii) and (v) are subspaces.
5. (i) $(-2\lambda, -2\lambda, \lambda)$; (ii) $(\mu, 0, -\mu)$; (iii) $(3\nu, -2\nu, \nu)$
6. (i) $\left\{\left(\frac{2}{\sqrt{14}}, \frac{1}{\sqrt{14}}, \frac{3}{\sqrt{14}}\right), \left(\frac{4}{\sqrt{42}}, \frac{5}{\sqrt{42}}, \frac{1}{\sqrt{42}}\right), \left(\frac{1}{\sqrt{3}}, \frac{1}{\sqrt{3}}, -\frac{1}{\sqrt{3}}\right)\right\}$

   (ii) $\left\{\left(\frac{\sqrt{2}}{2}, \frac{-\sqrt{2}}{2}, 0\right), \left(\frac{\sqrt{2}}{6}, \frac{\sqrt{2}}{6}, \frac{-2\sqrt{2}}{3}\right), \left(\frac{-2}{3}, \frac{-2}{3}, \frac{-1}{3}\right)\right\}$

   (iii) $\left\{\left(\frac{\sqrt{2}}{2}, 0, \frac{\sqrt{2}}{2}\right), (0, 1, 0), \left(\frac{\sqrt{2}}{2}, 0, \frac{-\sqrt{2}}{2}\right)\right\}$

   (iv) $\left\{\left(\frac{2\sqrt{5}}{5}, \frac{-\sqrt{5}}{5}, 0\right), \left(\frac{\sqrt{5}}{5}, \frac{2\sqrt{5}}{5}, 0\right), (0, 0, 1)\right\}$
7. (i) $\left\{\frac{1}{\sqrt{3}}(1, 1, -1), \frac{1}{\sqrt{2}}(1, 0, 1), \frac{1}{\sqrt{6}}(2, -1, 1)\right\}$

   (ii) $\left\{\frac{1}{\sqrt{51}}(7, -1, -1), \frac{1}{\sqrt{2}}(0, 1, -1), \frac{\sqrt{102}}{7}\left(\frac{2}{7}, 1, 1\right)\right\}$
8. $\frac{1}{\sqrt{2}}(1, 0, 1, 0), \frac{1}{\sqrt{6}}(1, 2, -1, 0), \frac{1}{\sqrt{21}}(-2, 2, 2, 3), \frac{1}{\sqrt{7}}(-1, 1, 1, 2)$
9. $\left\{\frac{1}{\sqrt{2}}(0, 1, 1, 0), \frac{1}{3}(0, 2, -2, -1), \frac{1}{9}(-3, -2, 2, -8)\right\}$

Page 13

1. $\begin{bmatrix} 1 & 3 & 2 \\ 2 & 1 & 1 \\ 3 & 2 & 3 \end{bmatrix}$ 2. (6, 4, 8), (8, 9, 19), (14, 13, 27)
3. (i), (iii) and (iv); (1, 2, −2), (−1, 3, 1) and (1, 1, 1)
4. (1, −1, 1); (2, 1, −1); (2, −1, 0) and (3, 0, −1)
5. (i) 2, 1 (ii) 2, 1 (iii) 2, 2 (iv) 2, 1 (v) 2, 1 (vi) 2, 2
6. $(-x_2, -x_1)$ and (x_2, x_1) 7. Space with basis (1, 1, 1, 0) and (2, 1, 0, 1)

8. (i) $\left(4 - \frac{7}{3}a, \ \frac{4}{3}a - 1, \ a\right)$ (ii) $(a, \ -a, \ -a)$

9. (i) 2, $\{(1, 0, 1), (0, 1, 1)\}$ (ii) 1, $\{(1, -1, 1)\}$ (iii) 1, $\{(2, 3, 5)\}$
(iv) 0, no basis (v) 0, no basis (vi) 2, $\{(1, -1, 0), (1, 0, -1)\}$

10. (i) 2, $\{(1, 0, 2), (3, -2, 0)\}$, $(0, 0, -1)$, $\alpha\begin{bmatrix}1\\0\\2\end{bmatrix} + \beta\begin{bmatrix}3\\-2\\0\end{bmatrix} + \begin{bmatrix}0\\0\\-1\end{bmatrix}$

(ii) 1, $\{(1, 0, -2)\}$, $(0, 0, 1)$, $\begin{bmatrix}0\\0\\1\end{bmatrix} + \alpha\begin{bmatrix}1\\0\\-2\end{bmatrix}$

(iii) 1, $\{(5, -2, 13)\}$, $(1, 0, 3)$ $\begin{bmatrix}1\\0\\3\end{bmatrix} + \alpha\begin{bmatrix}5\\-2\\13\end{bmatrix}$

(iv) 0, no basis, gen. sol. is $\begin{bmatrix}3\\2\\0\end{bmatrix}$

Page 18

1. 2 2. 2, 2, 2, 2, 3 8. $q = \dfrac{K\Delta T}{l} f_1(\rho C_p vL/K) f_2(\mu C_p/K)$

Page 26

5. $L = \begin{bmatrix}2 & 0 & 0\\1 & -3 & 0\\-1 & 0 & 7/2\end{bmatrix}$, $U = \begin{bmatrix}1 & -1 & 3/2\\0 & 1 & -11/6\\0 & 0 & 1\end{bmatrix}$, Det = −21

6. (i) 1, −2, 3 (ii) −6/125, 47/125, 99/125 (iii) 77/18, 32/9, 28/9, −7/3
(iv) 4, 2, −3, 3 (v) 5, 16, −6

7. 1, 1, 1, $A^{-1} = \dfrac{1}{64}\begin{bmatrix}39 & 14 & -22\\14 & -36 & 20\\-22 & 20 & -4\end{bmatrix}$

8. (a) 7/16, 1/2, 1/2, 1/2

9. $x^{(2)} = (2.052, \ 4.791, \ 8.147)$

Page 30

2. (i) $\begin{bmatrix}23 & 20\\55 & 48\\87 & 76\end{bmatrix}$ (ii) $\begin{bmatrix}47 & 50\\54 & 62\\65 & 70\\72 & 82\end{bmatrix}$ (iii) $\begin{bmatrix}10\\28\\46\end{bmatrix}$ (iv) $\begin{bmatrix}ax + by\\cx + dy\end{bmatrix}$

(v) $\begin{bmatrix}a^2 + h^2 + g^2 & ah + hb + gf & ag + hf + gc\\ha + bh + fg & h^2 + b^2 + f^2 & hg + bf + fc\\ga + fh + cg & gh + fb + cf & g^2 + f^2 + c^2\end{bmatrix}$

(vi) $\begin{bmatrix}A_1 & A_2\\A_3 & A_4\end{bmatrix}$ (vii) $\begin{bmatrix}A_1 & 0\\A_3 & A_4 - A_3A_1^{-1}A_2\end{bmatrix}$

5. (i) $\begin{bmatrix} -23 & 29 & -64/5 & -18/5 \\ 10 & -12 & 26/5 & 7/5 \\ 1 & -2 & 6/5 & 2/5 \\ 2 & -2 & 3/5 & 1/5 \end{bmatrix}$ (ii) and (iii) are inverses of each other

(iv) $\frac{1}{18}\begin{bmatrix} 2 & 5 & -7 & 1 \\ 5 & -1 & 5 & -2 \\ -7 & 5 & 11 & 10 \\ 1 & -2 & 10 & 5 \end{bmatrix}$

(v) $\begin{bmatrix} \cos\alpha & -\sin\alpha & -\cos(\alpha-\beta+\gamma) & \sin(\alpha-\beta+\gamma) \\ \sin\alpha & \cos\alpha & -\sin(\alpha-\beta+\gamma) & -\cos(\alpha-\beta+\gamma) \\ 0 & 0 & \cos\gamma & -\sin\gamma \\ 0 & 0 & \sin\gamma & \cos\gamma \end{bmatrix}$

Chapter 2

Page 36

1. $y = \cos t \begin{bmatrix} 2 \\ 4 \end{bmatrix} + \begin{bmatrix} -4 \\ 2 \end{bmatrix} \sin\sqrt{6}\,t$ 3. $y = \begin{bmatrix} 0.5 \\ 0.5 \end{bmatrix} \cos t + \begin{bmatrix} 0.5 \\ -0.5 \end{bmatrix} \cos\sqrt{3}\,t$

4. (i) $\frac{6\sqrt{2}}{5}\begin{bmatrix} 1 \\ 2 \end{bmatrix} \sin \frac{1}{\sqrt{2}}\,t - \frac{7}{5\sqrt{3}}\begin{bmatrix} -2 \\ 1 \end{bmatrix} \sin\sqrt{3}\,t$

(ii) $1.2916\begin{bmatrix} 1 \\ 2 \end{bmatrix} \cos\sqrt{3}\,t - 0.2916\begin{bmatrix} -2 \\ 1 \end{bmatrix} \cos 3\sqrt{2}\,t$

Page 41

1. (i) 2(−1, 1), 3(−2, 1) (ii) 1(−1, 1, 2), 2(0, 1, 2), 3(0, 0, 1)
(iii) 1(−1, 0, 1), 1(−2, 1, 0) are possible solutions, 5(1, 1, 1)
(iv) 1(1, −1, 0); 2(2, −1, −2), 3(1, −1, −2) (v) 0(1, −1, 0); 1(0, 0, 1); 4(1, 1, 0)
(vi) 1(1, 0, −1) & (0, 1, −1); 3(1, 1, 0)
(vii) −1(0, 1, −1); i (1+i, 1, 1); −i (1−i, 1, 1)
(viii) 2(2, −1, 0); 0(4, −1, 0); 1(4, 0, −1) (ix) −1(1, 0, 1); 2(1, 3, 1); 1(3, 2, 1)
(x) 1(1, 0, −1, 0) & (1, −1, 0, 0); 2(−2, 4, 1, 2); 3(0, 3, 1, 2)

2. (ii) $\begin{bmatrix} -3 & 0 & 3 \\ -2 & 1 & 2 \\ -6 & 0 & 6 \end{bmatrix}$

3. 1(−1, 1, 1); −2(11, 1, −14); 3(1, 1, 1)

Page 46

1. (i) −2(1, −1); 7(4, 5) (ii) 3+2i (1, i); 3−2i (1, −i)
(iii) 4(1, 0, 0); −2(3, −2, 0); 7(24, 8, 9) (iv) 9(4, 1, −1); −9(1, −4, 0) & (1, 0, 4)

2. 0(1, −1, 0), 1(0, 0, 1), 4(1, 1, 0). Real, symmetric matrix.

3. $0(1, 1, -1), (3+\sqrt{3})(1,\ 1+\sqrt{3},\ 2+\sqrt{3}), (3-\sqrt{3})(1,\ 1-\sqrt{3},\ 2-\sqrt{3})$

4. $2(1, 0, -1), (2+\sqrt{2})(1,\ \sqrt{2},\ 1), (2-\sqrt{2})(1, -\sqrt{2}, 1)$

5. $-2,\ 2(1,1,-1),\ 4$

6. $(2,\ 1,\ 2)\quad \begin{bmatrix} 3 & -2 & 4 \\ -2 & -2 & 6 \\ 4 & 6 & -1 \end{bmatrix}$

8. $(0,0,1),\ (1,-1,0),\ (1,0,-1)$ are possible answers.

11. (i) $\begin{bmatrix} 7 & 10 \\ 5 & 7 \end{bmatrix},\ \begin{bmatrix} 17 & 24 \\ 12 & 17 \end{bmatrix},\ \begin{bmatrix} -1 & 2 \\ 1 & -1 \end{bmatrix}$

(ii) $\begin{bmatrix} 42 & 31 & 29 \\ 45 & 39 & 31 \\ 53 & 45 & 42 \end{bmatrix},\ \begin{bmatrix} 193 & 160 & 144 \\ 224 & 177 & 160 \\ 272 & 224 & 193 \end{bmatrix},\ \frac{1}{11}\begin{bmatrix} -2 & 5 & -1 \\ -1 & -3 & 5 \\ 7 & -1 & -2 \end{bmatrix}$

(iii) $\begin{bmatrix} 15 & 19 & -26 \\ 14 & 46 & -52 \\ 14 & 19 & -25 \end{bmatrix},\ \begin{bmatrix} 31 & 65 & -80 \\ 30 & 146 & -160 \\ 30 & 65 & -79 \end{bmatrix},\ \frac{1}{6}\begin{bmatrix} 0 & -1 & 4 \\ -6 & 1 & 8 \\ -6 & -1 & 10 \end{bmatrix}$

12. $\omega^2 = 2, 3, 6$ with corresponding eigenvectors $(0,1,-1),\ (1,-1,-1),\ (2,1,1)$

13. (i) $2\begin{bmatrix} -1 & -2 \\ 1 & 2 \end{bmatrix} + 3\begin{bmatrix} 2 & 2 \\ -1 & -1 \end{bmatrix}$

(iv) $\frac{1}{2}\begin{bmatrix} 0 & -2 & 1 \\ 0 & 2 & -1 \\ 0 & 0 & 0 \end{bmatrix} + 2\begin{bmatrix} 2 & 2 & 0 \\ -1 & -1 & 0 \\ -2 & -2 & 0 \end{bmatrix} + 3\tfrac{1}{2}\begin{bmatrix} -2 & -2 & -1 \\ 2 & 2 & 1 \\ 4 & 4 & 2 \end{bmatrix}$

(v) $1\begin{bmatrix} 0 & 0 & 0 \\ 0 & 0 & 0 \\ 0 & 0 & 1 \end{bmatrix} + 4\tfrac{1}{2}\begin{bmatrix} 1 & 1 & 0 \\ 1 & 1 & 0 \\ 0 & 0 & 0 \end{bmatrix}$

Page 54

1. (i) $2x_1^2 + 2x_2^2 - 5x_3^2 - 6x_1x_2 + 2x_1x_3 + 8x_2x_3$
(ii) $3x_1^2 + 4x_2^2 + 6x_3^2 + 4x_1x_3 - 10x_2x_3$
(iii) $x_1^2 + 2x_3^2 - 4x_1x_2 + 8x_1x_3 + 6x_2x_3$

2. (i) $\begin{bmatrix} 2 & -3 & 0 \\ -3 & 0 & 0 \\ 0 & 0 & 1 \end{bmatrix}$ (ii) $\begin{bmatrix} 3 & -4 & 5/2 \\ -4 & 4 & 0 \\ 5/2 & 0 & 0 \end{bmatrix}$ (iii) $\begin{bmatrix} 1 & 2 & 3 \\ 2 & -2 & -4 \\ 3 & -4 & -3 \end{bmatrix}$

3. (i) $y_1^2 + y_2^2 + y_3^2$ (ii) $y_1^2 - y_2^2 + y_3^2$ (iii) $4y_1^2 - 16y_2^2 + 16y_3^2$
(iv) $y_1^2 + 128y_2^2 - 128y_3^2$ (v) $4y_1^2 - 16y_3^2 + 12y_4^2$ (vi) $y_1^2 - 8y_2^2$

4. $4, 6,\ \mathbf{H} = \begin{bmatrix} 0 & 1 & 1 \\ 1 & 1 & 0 \\ -1 & 0 & -1 \end{bmatrix}$

5. $\mathbf{P} = \begin{bmatrix} 1 & 1 \\ 1 & -1 \end{bmatrix},\ \mathbf{B}^{-1} = \frac{1}{7}\begin{bmatrix} 4 & 3 \\ 3 & 4 \end{bmatrix}$

6. $3(0,1,-1),\ 3(1,2,0),\ -3(-2,1,1)$ are possible answers, indefinite.

8. $\mathbf{H} = \begin{bmatrix} 0 & 1 & 2 \\ 1 & -1 & 1 \\ -1 & -1 & 1 \end{bmatrix}$ 9. $\mathbf{P} = \begin{bmatrix} 2 & 1 & 2 \\ -2 & 2 & 1 \\ 1 & 2 & -2 \end{bmatrix}$

10. Positive definite

Page 58

1. $2.436EI/L^2$ [0.076, 0.293, 0.617, 1]
2. 0[0, 0, 1, 0, 1, 1] is smallest, −24.208[0, 0.526, 0, −0.526, −1, 1] is largest
3. $0.6176\sqrt{\delta_{11}}$ [1, .5325, .1573], $3.755\sqrt{\delta_{11}}$ [−0.7502, 1, 0.9223], $9.689\sqrt{\delta_{11}}$ [−0.3210, 1, −0.8971]

Page 65

2. 7.873[0.225, 0.549, 1] 3. 12.289[0.274, 1, 0.871]
4. [1, −1.492, 1.376, −0.784] 5. 12.59[0.77, 1, 0.32]
6. 19.29[20, 13, 9] 7. $\frac{1}{18}\begin{bmatrix} 2 & 2 & 4 \\ 2 & 5 & -2 \\ 4 & -2 & 2 \end{bmatrix}$

Page 70

2. 6.425[.731, .233, 1]
3. (i) 2[1, 0, −1], $(2+\sqrt{2})[1, -\sqrt{2}, 1]$, $(2-\sqrt{2})[1, -\sqrt{2}, 1]$ (ii) 26.305[0.070, 0.233, 0.530, 1] 2.203[−0.828, −1, −0.612, 0.615], 0.454[1, −0.207, −0.676, 0.337], 0.038[0.427, −1, 0.822, −0.233] 4. 5.851[0.106, 1, 0.663] 5. 19.286[1, 0.668, 0.450]
6. 1[1, 1, −1], 2.38742[1, −.38742, .07504], .27924[.58112, 1, .86037]
7. 19.2[1, 0.7, 0.4], −7.2[−1, 0.9, 0.9] 8. 12.265[1, −1.485, 1.322]

Chapter 3

Page 80

1. (a) 5(3, 2), (b) 7(3,2); 9, any positive coordinates on $x + 3y = 9$; 8(4, 0)
   (c) 138(30, 26), (d) 6(0, 6) is max, unbounded solution, (e) no solution
   (f) max. 23(4, 15), min 10

Page 86

2. 22(4, 1)

Page 89

1. (a) $17\frac{2}{3}$(17/3, 2/3, 0), (b) $7\frac{2}{3}$(5/3, 4/3, 0), (c) 40 (0, 12, 2) (d) 15(3, 0, 0)
   (e) 244(24, 24, 28)
2. 7*X*, 4*Y*, 2*Z* 3. (a) 10(0, 0, $1\frac{2}{3}$, 0), (b) 6.5 at (2.5, 0, 2, 0)
4. Lead, 34 tons; tin, 6 tons; B, 60 tons. 5. 0A, 600B, 0C

Page 94

1. (a) max is 20(1¼, 4¼), (b) min is −12(2, 0)

2. (a) 5(0.130, 0.087), (b) 7(0.565, 0.044); 9(1, 0); 8.8(0, 0.2), (c) 138(1.274, 0.091), (d) min. 6(1, 0, 0, 0), unbounded above, (e) no solution, (f) 23(0.2, 0.4); 34(1.6, 0.2)

4. (i) min 4(0, 4, 0, 0, 0), not unique
(ii) (a) yes (b) no (c) yes 5. 7.5(1.5, 0); unbounded; unbounded; 5(1, 0)

Chapter 4

Page 98

2. (i) $f(x)$ is not differentiable at its minimum (ii) $f'(x)$ never zero for $x \geqslant 2$

Page 105

2. (b) $\tau = \frac{1}{2}(1 + \sqrt{5})$

4. (a) 0.54 at $x = -1$ or 1, (b) 0.416 at $x = 2$, (c) 0.94492 at $x = 0.104042$
(d) -2.7290 at $x = 1.32472$, (e) 0 at $x = 1$

Page 111

1. -1.205 at $(1, 1.35)$ 2. -108 at $(8, 6)$

3. (a) 0 at $(0, 0)$, (b) 0.2222 at $(1.1111, 1.1111)$, (c) 0 at $(1, 1)$,
(d) 0 at $(1, 1)$, (e) 0 at $(3, 0.148)$, (f) four minimum values of 0 at $(\pm 1, 0)$, $(0, \pm 1)$

Page 117

1. (i) minimum $z = x^2 + xy + y^2$, (ii) minimum $z = (x + y)^2$

3. $\pm 52/\sqrt{104}$ at $(\pm 10/\sqrt{104}, \pm 1/\sqrt{104})$

4. minimum at $(0, 0)$

5. Saddle $f = 0$ at $(0, 0)$, minimum $f = -108a^3$ at $(6a, 18a)$

7. 48.3 at $(1, -1, 1)$ 8. $\left(\dfrac{-1}{\sqrt{14}}, \dfrac{-3}{\sqrt{14}}, \dfrac{-2}{\sqrt{14}}\right)$

12. [.2265, .3727, .8999] 13. (i) $4\pi k^3/3$, (ii) e^{-1} 15. (ii) $6\sqrt{6}$

17. $r = 5.419$, $h = 10.838$ 18. 17/23 at $\left(\dfrac{10}{23}, \dfrac{1}{23}, -\dfrac{1}{23}, \dfrac{17}{23}\right)$

Page 126

1. $(2\mathbf{i} + \mathbf{j})/\sqrt{5}$, $(4\mathbf{i} - 5\mathbf{j})/\sqrt{41}$, $(-2\mathbf{i} + 5\mathbf{j})/\sqrt{26}$ 2. 10.05 at point $(-0.44, 0.24)$

4. $f_{min} = \frac{1}{2}$ at $(1\frac{1}{2}, 2)$ 5. $\{58, 12, 152\}/362 \sim \{0.160, 0.033, 0.420\}$

6. 2.73

8. (a) $f = 0$ at $(5, 5)$, (b) 34.1

9. $x^{(3)} = x^{(4)} = [2.37, 1.84]$

Page 130

1. $x_{n+1} = \sin(x_n y_n) - \dfrac{y_n}{2\pi}$, $y_{n+1} = 2\pi x_n - (\pi - \frac{1}{4})(e^{2x_n - 1} - 1)$, $|2\pi - (2\pi - \frac{1}{2})e^{2x-1}| < 1$

2. Roots at $(0.896, -0.444)$ and $(0, -1)$

3. $(-\frac{1}{4}, \frac{1}{4})$ 4. $x = -2.3498$, $y = 1.6430$ and $x = -3.1617$, $y = -2.0793$

5. [0.775, –3.786]

Page 132

1. Min $z = 18$ when $x = y = 3$

Chapter 5

Page 137

2. $x(0.2) = 0.1627$
3. $y(0.2) = 1.248$, $y(0.4) = 1.670$; by Taylor $y(0.2) = 1.253$, $y(0.4) = 1.675$
4. (i) $y(1) = 12.4451$ (ii) $y(1) = 1.41421$ (iii) $y(1) = 2.37797$ (iv) $y(1) = 0.25720$

Page 145

1. $y(-0.1) = 0.895$, $y(0.1) = 1.094$, $y(0.2) = 1.176$, $y(0.3) = 1.241$
2. 1.0100, 1.0405, 1.0923, 1.1668 3. 0.684 4. 1.015
5. 0.6536 6. 0.30934 7. 0.851, 0.780

Page 150

1. $3^n[A\cos\frac{n\pi}{3} + B\sin\frac{n\pi}{3}] + 3^{n-2} + \frac{13}{49}(7n+1)$
2. (a) $A(-7)^n + B + 3^n/20$, (b) $\{1+(-7)^{n+1}\}/\{1+(-7)^n\} - 2$
3. (a) $(An+B)2^n + 3^n$, (b) $x_n = 4^n + 2 + n^2/2 - n/2$, $y_n = 2.4^n - 2 - n$
4. (a) $(An+B)3^n + 2^n + \frac{1}{4}(n+1)$, (b) $n(A\,4^n + B)$
5. (a) $13^{n/2}[A\cos 0.588n + B\sin 0.588n] + 2^n/5$, (b) $(An+B)3^n + (n+4)2^n + n^2 3^n/18$
8. Frequencies are $\omega_i = \frac{1}{\pi}\sqrt{\frac{k}{m}}\sin[\pi(i+\frac{1}{2})/21]$, eigenvalues $\lambda_i = 4\sin^2[\pi(i+\frac{1}{2})/21]$

Page 152

1. $y = x+1$, $y = x+1+0.1e^x$, $y = x+1-0.1e^x$, $y(5) = 20.84$ or -8.84
4. See problem 2
5. Analytical solution $y = \frac{1}{2}x$

Page 161

1. $y(0.5) = 1.32$, $y(1) = 1.41$
2. $y'_1 = y_2$, $y'_2 = y_3$, $y'_3 = -\frac{1}{2}y_1y_3$, $y_1(1) = 0.2$, $y_2(1) = 0.397$, $y_3(1) = 0.387$
4. As an example $y(0.5) = -0.241$

Page 164

1. First iteration gives 0.0704, 0.1213, 0.1392, 0.1213, 0.0704
2. 0.1002, 0.2013, 0.3046 3. Analytical solution is $y = \sin x$
4. For example $y(0.5) \simeq 0.30$ 5. For example $y(0.5) = 0.363$
6. For example $y(0.5) = -0.694$

Chapter 6

Page 176

1. (a) $A[1+4x+8x^2/3+.....]+Bx^{3/4}[1+x+4x^2/15+.....]$
 (b) $Ax[1+x^2/10+x^4/360+.....]+B\sqrt{x}[1+x^2/6+x^4/168+.....]$
 (c) $A[1+2x/3+x^2/3]+Bx^4[1+2x^5+3x^6+.....]$
 (d) $A[1-x/3+5x^2/42-.....]+Bx^{1/4}[1-2x/5+2x^2/15-.....]$
 (e) $A\sqrt{x}[1-x+x^2/2-.....]+Bx[1-2x/3+4x^2/15-.....]$
 (f) $Ax^{-1}[1-x^2/2!+x^4/4!-.....]+B[1-x^2/3!+x^4/5!-.....]$
 (g) $A[1+x+2x^2+14x^3/3+.....]+Bx^{7/3}[1+12x/5+\frac{396}{65}x^2+.....]$
 (h) $y=au+bv$ where $u=A[1-2x+3x^2/2-2x^3/3+.....]$,
 $v=u\log x+[3x-13x^2/4+\frac{31}{18}x^3+.....]$

3. $y=A(1-x+x^2/6-x^3/90+.....)+B\sqrt{x}(1-x/3+x^2/30-x^3/630+.....)$; $A=1$, $B=0$

4. $A\left[1+\frac{2x}{3}+\frac{x^2}{6}\right]+Bx^4\left[1+\frac{2!\,x}{5.1}+\frac{3!x^2}{6.5.2.1}+\frac{4!\,x^3}{7.6.5.3.2.1}+.....\right]$

6. (i) $x=0$, $x=1$ are singular points, $x=1$ is regular.

7. $A[1+x+\frac{x^2}{(2!)^2}+\frac{x^3}{(3!)^2}+.....]=u(x)$, $Bu(x)\log x-2B[\frac{x}{(1!)^3}+\frac{3x^2}{(2!)^3}+\frac{11x^3}{(3!)^3}+.....]$

8. (a) $a_0x^{1-\lambda}\left[1-\left(\frac{2+\lambda}{2-\lambda}\right)x+\left(\frac{1+\lambda}{3-\lambda}\right)\left(\frac{2+\lambda}{2-\lambda}\right)x^2+.....\right]$
 (b) $ku(x)\log x+kx^{-1}\left[5x-\frac{37x^2}{2}+\frac{53x^3}{6}+.....\right]$

9. $A[1-nx^2/2!+n(n-2)x^4/4!-.....]+Bx[1-(n-1)x^2/3!+(n-1)(n-3)x^4/5-.....]$,
 $y=\exp(-x^2/4)[1-2x^2+x^4/3]$

10. $T=A[1+\frac{14000}{3}x^{3/2}+14{,}000{,}000x^{5/2}+.....]+B[x+3750x^2+\frac{2800}{3}x^{5/2}+.....]$, yes

Page 179

2. $1.66\Gamma(1.6)$, $1.6\Gamma(1.6)$, $1.9644\times10^8\,\Gamma(1.6)$, $-6.6672\times10^{-17}\,\Gamma(1.6)$

3. 5.24

4. 3.50

5. $\sqrt{\pi}\,a^{(n/2+1)}\Gamma\left(\frac{1}{n}\right)\Big/2n\Gamma\left(\frac{3}{2}+\frac{1}{n}\right)$

6. $\frac{1}{5}\left[\Gamma\left(\frac{1}{3}\right)\right]^2$

Page 183

5. (i) $(384/x^4-72/x^2+1)J_1(x)+(12/x-192/x^3)J_0(x)$
 (ii) $-\sqrt{\frac{2}{\pi x}}\left[\frac{\cos x}{x}+\sin x\right]$
 (iii) $2xJ_3(2x)+2x^2\left[\left\{1-\frac{3}{x^2}\right\}J_2(2x)+\frac{3}{2x}J_1(2x)\right]$, $J_0(x^2)-2x^2J_1(x^2)$,
 $\left[1-\frac{4}{x^2}\right]J_1(x)+\frac{2}{x}J_0(x)$

6. $\tfrac{1}{2}J_0(x)(\sin x+\cos x)$, $\tfrac{1}{2}J_0(x)(\sin x-\cos x)$, $-\tfrac{1}{2}J_0(x)(\sin x+\cos x)$,
 $\tfrac{1}{2}J_0(x)(\cos x-\sin x)$, $\frac{x}{2}J_0(x)(\sin x+\cos x)-\tfrac{1}{2}J_0(x)\sin x$,

$\frac{x}{2} J_0(x)(\sin x - \cos x) + \frac{1}{2} J_0(x) \cos x,$

$\frac{x}{2} J_0(x)(\sin x + \cos x) - \frac{1}{2} J_0(x) \sin x - xJ_1(x) \cos x,$

$xJ_1(x) \sin x - \frac{x}{2} J_0(x)(\sin x - \cos x) - \frac{1}{2} J_0(x) \cos x, \quad -xJ_1(x),$

$x^2 J_1(x) + xJ_0(x) - \int J_0(x)\,dx, \quad x^3 J_1(x) - 2x^2 J_2(x), \quad -J_0(x),$

$-xJ_0(x) + \int J_0(x)\,dx$

10. 0.9385, 0.2423, 0.0307

Page 186

1. (i) $AJ_0(x) + BY_0(x)$, (ii) $AJ_3(x) + BJ_{-3}(x)$, (iii) $x(AJ_1(x) + BY_1(x))$,
(iv) $AJ_3(3x) + BJ_{-3}(3x)$, (v) $\frac{1}{x}[AJ_1(x) + BY_1(x)]$,
(vi) $AJ_0\left(\frac{3\sqrt{x}}{2}\right) + BY_0\left(\frac{3\sqrt{x}}{2}\right)$, (vii) $AJ_{n/k}(ke^x) + BJ_{-n/k}(ke^x)$

5. From flanges $-\frac{1}{2}\pi k(\partial T/\partial r)_{r=a}$. From rim $-2\pi k(\partial T/\partial r)_{r=b}$, $T = T_1 + AI_0(\lambda r) + BK_0(\lambda r)$,
$\lambda = \sqrt{2h/k}$ and $AI_0(\lambda a) + BK_0(\lambda a) = T_0 - T_1$, $AI_1(\lambda b) - BK_1(\lambda b) = -\frac{1}{2}\lambda[AI_0(\lambda b) + BK_0(\lambda b)]$

Page 188

2. $T = 2A \sum_\alpha J_0\left(\frac{\alpha r}{a}\right) \sinh\left(\frac{\alpha Z}{a}\right) \Big/ \alpha J_1(\alpha) \sinh\left(\frac{\alpha l}{a}\right)$

4. $z = J_n\left(\frac{\omega r}{c}\right) \cos n\theta \cos(\omega t - \epsilon)$, $n = 0, 1, 2, \ldots..$ and $J_n\left(\frac{\omega a}{c}\right) = 0$

Page 193

1. $P_4(x) = \frac{5.7}{2.4}x^4 - 2 \cdot \frac{3.5}{2.4}x^2 + \frac{1.3}{2.4}$, $P_5(x) = \frac{7.9}{2.4}x^5 - \frac{2.5.7}{2.4}x^3 + \frac{3.5}{2.4}x$,
$P_6(x) = \frac{7.9.11}{2.4.6}x^6 - \frac{3.5.7.9}{2.4.6}x^4 + \frac{3.3.5.7}{2.4.6}x^2 - \frac{1.3.5}{2.4.6}$

3. $P_3(x) = \frac{3.5}{2.3}x^3 - \frac{3.1.3}{2.3}x$, $P_4(x)$ as in question 1

4. $1 = P_0(x)$, $x = P_1(x)$, $x^2 = \frac{1}{3}P_0(x) + \frac{2}{3}P_2(x)$, $x^3 = \frac{2}{5}P_3(x) + \frac{3}{5}P_1(x)$,
$x^4 = \frac{8}{35}P_4(x) + \frac{4}{7}P_2(x) + \frac{7}{5}P_0(x)$

8. $T = \frac{1}{2}(T_1 + T_2) + \frac{3}{4}(T_1 - T_2)\left\{\frac{r}{a}\right\} P_1(\cos\theta) + \frac{7}{16}(T_2 - T_1)\left\{\frac{r}{a}\right\}^3 P_3(\cos\theta) + \ldots..$

Chapter 7

Page 202

3. $0.9453x - 0.125x^3$

4. $-0.01204 + 1.039x - 0.297x^2 + 0.0417x^3 - 0.719x^4 + 0.7x^5$

5. (i) $1 + 3.75x + 3x^2$, 0.25; (ii) x^2, 0; (iii) $0.812 - 2.812x + 3x^2$, 1/32

8. $1 + 0.75x + x^2$ 9. $1 + x$, $1.25 + x$, approx. ½ and 5/12

Page 213

1. (a) $\frac{3}{2}-\frac{2}{\pi}\,[\sin x + \frac{1}{3}\sin 3x + \frac{1}{5}\sin 5x + ...]$

(b) $\frac{\pi}{4}-\frac{2}{\pi}\,[\cos x + \frac{1}{3^2}\cos 3x + \frac{1}{5^2}\cos 5x + ...] + [\sin x - \frac{1}{2}\sin 2x + \frac{1}{3}\sin 3x - \;....]$

(c) $-\pi^2 - 8(\cos x + \frac{1}{3^2}\cos 3x + ...) + \frac{2}{\pi}\left[\left\{\frac{3\pi^2}{1} - \frac{4}{13}\right\}\sin x + \frac{\pi^2}{2}\sin 2x + ...\right]$

(d) $\frac{2}{\pi}\left[1 - \sum_{n=1}^{\infty}\frac{\cos 2nx}{(4n^2-1)}\right]$

(e) $\frac{2}{\pi}\left[\sin x + \frac{1}{3}\sin 3x + ...\right]$

(f) $\frac{1}{4}(\pi+2) - \frac{2}{\pi}\sum_{n=1}^{\infty}\frac{\cos(2n-1)x}{(2n-1)^2} - \frac{1}{\pi}\sum_{n=1}^{\infty}\frac{(\pi-1)(-1)^n}{n}\sin nx$

(g) $\frac{2}{\pi}\,(\cos x - \frac{1}{3}\cos 3x + ... + \sin 2x + \frac{1}{3}\sin 6x + \frac{1}{5}\sin 10x + ...)$

2. $\pi/4 - \frac{1}{2}$ 3. $\frac{4}{\pi\sqrt{2}}\,[\sin x + \frac{\sin 3x}{3^2} - \frac{\sin 5x}{5^2} - .....]$

5. $\frac{2}{\pi}\,[(\pi^2-4)\sin x - \frac{\pi^2}{2}\sin 2x + \left[\frac{\pi^2}{3} - \frac{4^2}{3^3}\right]\sin 3x + ...]$

Page 217

1. (a) even (b) neither (c) neither (d) odd (e) even (f) neither

2. $\frac{1}{\pi} + \frac{1}{2}\cos x - \frac{2}{\pi}\sum_{n=1}^{\infty}\frac{(-1)^n\cos 2nx}{(4n^2-1)}$

3. (a) $\frac{4}{\pi}\,(\sin x + \frac{1}{3}\sin 3x + ...)$; 2

(b) $2\left[\left\{\frac{\pi}{1} - \frac{4}{1^3\pi}\right\}\sin x - \frac{\pi}{2}\sin 2x + \left\{\frac{\pi}{3} - \frac{4}{3^3\pi}\right\}\sin 3x - .....\right]$;

$\frac{\pi^3}{3} - 4\,(\cos x - \frac{1}{4}\cos 2x + \frac{1}{9}\cos 3x - .....)$

(c) $\frac{2}{\pi}\,[(\pi+2)\sin x - \frac{\pi}{2}\sin 2x + \frac{1}{3}(\pi+2)\sin 3x - \frac{\pi}{4}\sin 4x + ...]$

4. $\frac{4}{\pi}\left[\sin x + \frac{2}{3}\sin 2x + \frac{1}{3}\sin 3x + \frac{4}{15}\sin 4x + \frac{1}{5}\sin 5x + ...\right]$

5. $\frac{\pi}{4} - \frac{1}{2}$ 6. $2[\sin x - \frac{1}{2}\sin 2x + \frac{1}{3}\sin 3x - .....]$

7. $\frac{\pi}{9} + \frac{2}{\pi}\left[\left\{\frac{\pi}{2\sqrt{3}} - 1\right\}\cos x + \frac{\pi}{4\sqrt{3}}\cos 2x - \frac{4}{9}\cos 3x + ...\right]$

Page 227

1. (a) $1 + \frac{2}{\pi}\sum_{n=1}^{\infty}\frac{1}{n}\,[1 - 2(-1)^n]\sin nx$

(b) $\sum_{n=1}^{\infty} b_n \sin\frac{n\pi x}{2}$ where $b_n = \frac{-2}{n\pi}$ for n even and $\frac{2}{n\pi}\left[1 + \frac{1}{n\pi}(-1)^{(n-1)/2}\right]$ for n odd

(c) $\frac{\sqrt{3}}{\pi}\left[\frac{\sin x}{1^2}+\frac{\sin 2x}{2^2}-\frac{\sin 4x}{4^2}-\frac{\sin 5x}{5^2}+\ldots..\right]$

(d) $\frac{8a}{\pi^2}\left[\sin\frac{2\pi x}{l}-\frac{1}{3^2}\sin\frac{6\pi x}{l}+\frac{1}{5^2}\sin\frac{10\pi x}{l}-\ldots..\right]$

(e) $\frac{2al^2}{\pi^2 b\,(l-b)}\sum_{n=1}^{\infty}\frac{1}{n^2}\sin\frac{n\pi b}{l}\sin\frac{n\pi x}{l}$

2. (a) $\frac{3a}{\pi}\left[\sin 2x+\frac{1}{2}\sin 4x+\frac{1}{3}\sin 6x+\frac{1}{5}\sin 10x+\frac{1}{7}\sin 14x+\ldots..\right]$

(b) $\frac{8\sqrt{2}\,a}{\pi^2}\left[\cos\frac{\pi x}{4b}-\frac{1}{3^2}\cos\frac{3\pi x}{4b}-\frac{1}{5^2}\cos\frac{5\pi x}{4b}+\ldots..\right]$

3. (a) $\frac{16}{\pi}\sum_{n=1}^{\infty}\frac{(1-\cos n\pi)}{n}\sin\frac{n\pi x}{2}$

(b) $2-\frac{8}{\pi^2}\sum_{n=1}^{\infty}\frac{(1-\cos n\pi)}{n^2}\cos\frac{n\pi x}{4}$

(c) $20-\frac{40}{\pi}\sum_{n=1}^{\infty}\frac{1}{n}\sin\frac{n\pi x}{5}$

(d) $\frac{3}{2}+\sum_{n=1}^{\infty}\left[\frac{6(\cos n\pi-1)}{n^2\pi^2}\cos\frac{n\pi x}{3}-\frac{6\cos n\pi}{n\pi}\sin\frac{n\pi x}{3}\right]$

Page 232

1. $0.28 - 13.18\cos x + 5.08\cos 2x + 0.66\cos 3x + 25.12\sin x + 6.14\sin 2x + 1.74\sin 3x$

2. Coefficients in same order
   (a) 3.82, −1.02, 0, 0.86; 2.83, 0.03, −0.63
   (b) 1.63, −1.27, −0.18, −0.07; 1.00, −0.17, −0.07
   (c) 1.37, 0.38, −0.65, −0.28; 1.30, −0.49, −0.12

3. (a) 1.21, 1.00, −0.01, 0.23, 0.25; 2.73, −0.21, 0.27, −0.12
   (b) $a_i = 0$; 3.95, −1.27, −0.76, 0.69, 0

Page 237

2. The cubics will be:
   (i) $8.427\times10^{-2}(x-0.15)^3-0.218(x-0.15)^2-5.511\times10^{-2}(x-0.15)+0.3945$ for $x\in[0.15,\ 0.89]$
   (ii) $0.1155(x-0.89)^3-3.091\times10^{-2}(x-0.89)^2-0.2393(x-0.89)+0.2685$ for $x\in[0.89,\ 1.07]$
   (iii) $2.912\times10^{-2}(x-1.07)^3+3.144\times10^{-2}(x-1.07)^2-0.2392(x-1.07)+0.2251$ for $x\in[1.07,\ 2.11]$

4. (i) $0.3090x^3+0.3765x^2+1.033x+1.000$ for $x\in[0,\ 1]$
   (ii) $0.6550(x-1)^3+1.304(x-1)^2+2.713(x-1)+2.718$ for $x\in[1,\ 2]$

5. $f(x)$ is an even function so we only need to fit cubics for $x\in[0,\ 1]$
   (i) $n=4,\ h=0.5;\ p(x)=4.973x^3-5.935x^2+1$ for $x\in[-1,\ 1]$
   For $n=20$ ($h=0.1$), some of the cubics are:
   If $x\in[0,\ 0.2]$ the cubic is $44.68x^3-18.40x^2-0.6064x+1$
   If $x\in[0.4,\ 0.5]$ the cubic polynomial is
   $-3.772(x-0.4)^3+2.226(x-0.4)^2-0.8056[x-0.4]+0.2000$
   If $x\in[0.8,\ 1.0]$ the cubic polynomial is
   $-0.2501(x-0.8)^3+0.2334(x-0.8)^2-0.1385(x-0.8)+.005882$

6. For $x \in [0,\ 250]$ the fitted polynomial is
$$-1.624 \times 10^{-7}x^3 + 1.052 \times 10^{-3}x^2 + 5.924 \times 10^{-2}x + 60$$
For $x \in [250,\ 365]$ the fitted polynomial is
$$-1.408 \times 10^{-4}(x-250)^3 + 9.298 \times 10^{-4}(x-250)^2 + 0.5546(x-250) + 138$$

Chapter 8

Page 248

1. (a) hyperbolic, (b) hyperbolic, (c) elliptic, (d) elliptic, (e) hyperbolic
(f) hyperbolic, (g) parabolic

Page 252

4. $\theta = e^{-\pi^2 y}\sin \pi x + e^{-4\pi^2 y}\sin 2\pi x$ 5. $V = 4\cosh 15x \sin 5t$

Page 264

1. (a) $\theta = -200 \sum_1^\infty \frac{(-1)^n}{n\pi} \sin \frac{n\pi x}{50}\, e^{\frac{-n^2\pi^2 Kt}{50}}$

(b) $\theta = 100 - 200 \sum_1^\infty \frac{(2(-1)^n - 1)}{n\pi} \sin \frac{n\pi x}{50}\, e^{\frac{-n^2\pi^2 Kt}{50}}$

(c) $\theta = x - 100 \sum_1^\infty \frac{(-1)^n}{n\pi} \sin \frac{n\pi x}{50}\, e^{\frac{-n^2\pi^2 Kt}{50}}$

(d) $\theta = 25 + x - 50 \sum_1^\infty \frac{((-1)^n + 1)}{n\pi} \sin \frac{n\pi x}{50}\, e^{\frac{-n^2\pi^2 Kt}{50}}$

(e) $\theta = 50 + 2x + 100 \sum \frac{((-1)^n + 1)}{n\pi} \sin \frac{n\pi x}{50}\, e^{\frac{-n^2\pi^2 Kt}{50}}$

2. $u = \frac{4}{\pi}\left(e^{-kt}\sin x - \frac{1}{9}e^{-9kt}\sin 3x + \frac{1}{25}e^{-25kt}\sin 5x - \ldots\right)$

3. $u = u_0 + \frac{8e^{-kt}}{\pi^3} \sum \frac{1}{(2s-1)^3} \sin (2s-1)\pi x \,.\, e^{-\{(2s-1)h\pi\}^2 t}$,

$u = u_0 + e^{-kt} \sum B_r \sin r\pi x \,.\, e^{-(r\pi h)^2 t}$

6. $\bar{U} = 1 - \frac{8}{\pi^2} \sum_{1,3,5,\ldots}^{\infty} \frac{1}{n^2} e^{-\left\{\frac{n\pi^2}{4H^2} c_v t\right\}}$

13. (ii) $\underline{F} = 3e^{-2y}\sin 3x$

14. (i) $y = A\sin 2\pi x \cos 2\pi ct$, (ii) $y = \sum_1^\infty \frac{12A}{n^3\pi^3}(-1)^{n+1} \sin n\pi x \cos n\pi ct$

(iii) $y = \frac{2}{\pi^2}\left\{\frac{(\sqrt{2}-1)}{1^2}\sin \pi x \cos \pi ct + \frac{2\sin 2\pi x \cos 2\pi ct}{2^2} + \frac{(\sqrt{2}+1)}{3^2}\sin 3\pi x \cos 3\pi ct - \frac{(\sqrt{2}+1)}{5^2}\sin 5\pi x \cos 5\pi ct - \frac{2}{6^2}\sin 6\pi x \cos 6\pi ct - \ldots\right\}$

(iv) $y = \sum_1^\infty \frac{-4A}{n^3\pi^3}\left\{1 + 5(-1)^n + \frac{12}{n^2\pi^2}(1 - (-1)^n)\right\} \sin n\pi x \cos n\pi ct$

15. $y = \sum(\alpha \cosh ax + \beta \sinh ax + \gamma \cos ax + \delta \sin ax)\sin(ca^2 t + \epsilon)$, $c^2 = EI/\rho A$

16. $2\pi/\omega$ where $\rho A\omega^2 = EI\pi^4/l^4$

20. $a^2 = \frac{1}{2}[(RG - LC\omega^2) + K]$, $b^2 = \frac{1}{2}[-(RG - LC\omega^2) + K]$,
$K = \sqrt{(RG - LC\omega^2)^2 + (RC + GL)^2\omega^2}$

21. $f(\theta) = A_1 \cos n\theta + B_1 \sin n\theta$, $V = \dfrac{1}{r^2} \sin 2\theta$

22. (i) $V = a^2 - r^2 \cos 2\theta$, (ii) $f = Cx^2 e^{k(x^3/3-t)}$, $f = x^2 e^{(x^3/3-1/3-t)}$

Page 276

1. $u(10, 0.4) = 10$ (nearest integer), $u(10, 0.8) = 14$ (nearest integer)
2. $r = 0.1$, $u(10, 0.4) = 9$ (nearest integer), $u(10, 0.5) = 10$ (nearest integer)
   $r = 0.4$, $u(10, 0.4) = 12$ (nearest integer), $u(10, 0.8) = 14$ (nearest integer)
   $r = 0.8$, $u(10, 0.8) = 24$ (nearest integer)
   $r = 1.0$, $u(10, 1) = 30$ (nearest integer)
3. $u(12, 0.4) \simeq 110$, $u(12, 0.8) \simeq 90$, $u(10, 0.4) \simeq 16$, $u(10, 0.8) \simeq 29$
4. $u(10, 0.4) \simeq 6$, $u(10, 0.8) \simeq 10$
5. $u(10, 0.4) \simeq 0$, $u(10, 0.8) \simeq 1$
6. $h = 2$ gives (1) $r = 0.8$, $u(10, 3.2) \simeq 30$, $u(10, 16) \simeq 75$
   $r = 2$, $u(10, 8) \simeq 55$, $u(10, 16) \simeq 72$
   $r = 5$, $u(10, 20) \simeq 99$, $u(10, 40) \simeq 123$
   (5) $r = 0.8$, $u(10, 3.2) \simeq 10$, $u(10, 16) \simeq 29$
   $r = 2$, $u(10, 8) \simeq 24$, $u(10, 16) \simeq 34$

Page 282

1. Temperatures at A, B, C, D are 24, 23, 25, 27 to nearest integer
2. Values at A, B, C, D are 0.7, 1.4, 0.4, 1.0
3. Values at A, B, C, D are 64, 71, 37, 43 to nearest integer
4. Values at A, B, C, D are 15, 17, 15, 13 to nearest integer
5. $h = 1$ gives for example $V(1, 2) = 27$, $V(2, 2) = 8$, $V(3, 3) = -10$ all to nearest integer
6. $h = 1$ gives $\theta(0, 0) \simeq 30$, $\theta(1, 0) = \theta(-1, 0) \simeq 20$, $\theta(1, 1) = \theta(-1, 1) \simeq -32$,
   $\theta(1, -1) = \theta(-1, -1) \simeq 82$, $\theta(0, 1) \simeq -28$, $\theta(0, -1) \simeq 108$
7. $h = 1$ gives with origin at centre of pipe section, $\theta(0, 3) = 133$, $\theta(-2, 3) = \theta(2, 3) = 114$,
   $\theta(3, 3) = \theta(-3, 3) = 122$, $\theta(4, 4) = \theta(-4, 4) = 95$, $\theta(-3, 0) = \theta(3, 0) = 128$

Page 285

1. ϕ takes value 1 at all internal mesh points
2.

		38	83
		42	90
38	42	65	106
83	90	106	137

3. $\phi(1, 1) = 4$, $\phi(2, 1) = 4$, $\phi(1, 2) = 8$, $\phi(2, 2) = 8$, $\phi(1, 3) = 10$, $\phi(2, 3) = 6$

Page 289

1. Elliptic, parabolic or hyperbolic for
   (a) $x > 0, y > 0$ and $x < 0, y < 0$; $x = 0, y = 0$; $x > 0, y < 0$ and $x < 0, y > 0$

(b) no region;	all x and y;	no region
(c) $x^2 + y^2 < 1$;	$x^2 + y^2 = 1$;	$x^2 + y^2 > 1$
(d) $x^2 < 4y$;	$x^2 = 4y$;	$x^2 > 4y$

3. (i) $\xi = 3x + y,\ \eta = 2x + y,\ \dfrac{\partial^2 u}{\partial\xi\partial\eta} = 0$

   (ii) $\xi = 4x + y,\ \eta = -2x + y,\ 2\dfrac{\partial^2 u}{\partial\xi\partial\eta} = 2\dfrac{\partial u}{\partial\xi} - \dfrac{\partial u}{\partial\eta}$

   (iii) $\xi = 2x + y,\ \eta = y,\ \dfrac{\partial^2 u}{\partial y^2} = 0$

   (iv) $\xi = y - x,\ \eta = y,\ \dfrac{\partial^2 u}{\partial\eta^2} = \dfrac{\partial u}{\partial\xi} - 9u$

4. $a = 1,\ b = 2\ ,\ z = f(x + y) + g(x + 2y)$
5. See answer to Problem 3 of Page 264
6. $-0.59,\ -1.90,\ -1.12$

Chapter 9

Page 294

1. $s/\ (s+1)(s^2 + 2s + 5)\ ,\ (2s^5 - 4s^3 - 6s)/(s^2+1)^4,\ k/(s^2 + 4k^2),$
   $2\omega(s+2)/[(s+2)^2 + \omega^2]^2,\ (s+4)/(s^2 + 8s + 12),\ (6s^2 - 29s + 36)/(s-2)^3,$
   $\frac{1}{2}\log[(s+1)/(s-1)]$

2. $N_1(t) = \dfrac{a}{\lambda_1}(1 - e^{-\lambda_1 t}),\quad N_2(t) = \dfrac{a}{(\lambda_2 + b)} - \dfrac{ae^{-\lambda_1 t}}{(\lambda_2 - \lambda_1 - b)} + \dfrac{a\lambda_1 e^{-(\lambda_2 + b)t}}{(\lambda_2 + b)(\lambda_2 - \lambda_1 + b)}$

   After shut down $N_1(t) = \dfrac{a}{\lambda_1} e^{-\lambda_1 t}$,

   $$N_2(t) = \frac{a}{(\lambda_2 + b)}\left[e^{-\lambda_2 t} + \frac{(\lambda_2 + b)}{(\lambda_2 - \lambda_1)}(e^{-\lambda_1 t} - e^{-\lambda_2 t})\right]$$

Page 301

1. $y = \cos 2t + 3U(t-2)\ [\ \frac{1}{4} - \frac{1}{4}\cos 2(t-2)]$
2. (i) $f(t) = n + 1$ for $n\pi < t < (n+1)\pi$
   (ii) $f(t) = \cos t,\ 0 < t < \pi,\ = 0,\ \pi < t < 2\pi$, solution $f(t) =$ (i) – (ii)
3. $be^{-sc}/(s^2 + b^2)(1 - e^{-s\pi/b})$
6. $te^{-4t}(e^4 - 5)/16 - e^{-4t}(e^4 + 3)/32,\ 0 < t < 1,$
   $-5te^{-4t}/16 - 3e^{-4t}/32 - t/16 + 3/32,\ t > 1$
7. (i) e^{-bs}/s, (ii) e^{-bs}/s^2,
   (iii) $\dfrac{x^2}{2} + \dfrac{1}{6a}(x - a/2)^3\ U(x - a/2)$
8. $53\omega l^4/30720EI$
9. $$\left.\begin{array}{l} CE_0(t - \frac{1}{n}\sin nt)/T,\ 0 < t < T \\ CE_0[T\cos n(t-T) + \frac{1}{n}\sin n(t-T) - \frac{1}{n}\sin nt]/T,\ t > T \end{array}\right\}\ n = \frac{1}{\sqrt{LC}}$$

Page 305

3. (a) $\frac{1}{2}(e^{-t} - \cos t + \sin t)$, (b) $\frac{1}{9}(e^{2t} - e^{-t} - 3te^{-t})$

Page 316

1. (i) $\frac{\sin \lambda b}{2\lambda a}$ $(\lambda \neq 0)$, $\frac{b}{a}$ $(\lambda = 0)$, (ii) $\frac{\lambda}{\pi^2 + \lambda^2}$, $\frac{\pi}{2} e^{-\pi m}$

2. (i) $4\lambda/(1 + \lambda^2)^2$, (ii) $[(2\lambda^2 - 1)\sin 2\lambda + 2\lambda \cos 2\lambda]/\lambda^3$

3. (i) $\sin \lambda\pi/(1 - \lambda^2)$, (ii) $1/(1 + \lambda^2)$

4. $\sin \lambda a/\lambda$, $f(x) = \begin{cases} 0, & x < 0 \\ x, & 0 < x < a \\ 0, & x > a \end{cases}$

Chapter 10

Page 327

1. Analytical results are (a) 2/3, (b) 1/4, (c) 0.946083, (d) 19.579498 (e) 0.001543

2. 3.8730 3. Four ordinates 3.7605, five ordinates 3.9973, 3/8 rule 4.0418

4. Five ordinates −64/45

Page 330

4. Analytically 0.693147 5. Analytically 6.389056

6. Analytically 4.047190 7. (a) 1.5707963, (b) 1.3506439

8. Analytically 1

Page 335

7. Analytically $\pi/2$ 8. Analytically 1

9. Analytically 1.570796, 2-point 1.5, 4-point 1.5686, 6-point 1.57073

10. Use transformation $z = 4x/\pi - 1$

Page 346

1. (a) 18, (b) 32/3, (c) 2/9, (d) 4/15, (e) 9/20

2. (a) 1/15, (b) 32/3, (c) 1/12

3. (a) 84/5, (b) 11/3, (c) 17/2, (d) $\pi(\sqrt{3} - 1)/4$

4. (a) 4, (b) $\frac{2}{3}\pi abc$, (c) $\frac{1}{6}abc$, (d) 1/10, (e) 64/35

5. (a) 1, (b) 64/3, (c) $\frac{\pi}{2} - \frac{1}{3}$, (d) 16/5, (e) 8π

Page 354

1. (a) $5/12$, (b) $2 - \pi/2$, (c) $4/3 + \pi/2$
2. (a) $16/3$, (b) $625\pi/2$, (c) 2π
3. (a) 0, (b) $\pi/2$
4. (a) $128/15$, (b) $kac^3/12$, (c) $3/80$
5. (a) $(4/5, 9/20)$, (b) $(e - 2, (e^2 + 1)/8)$
6. (a) $\pi(1 - 1/e)$, (b) $\pi^2(\pi^2 + 12)/48$, (c) $1/10$
7. (a) $\pi\sqrt{3}$, (b) $\pi(\tan \frac{1}{2} - \tan \frac{1}{4})$, (c) $32/9$, (d) $8\pi/3 - 3\sqrt{3}$
8. (a) $16\pi/3 + 8\sqrt{3}$, (b) $64/3$, (c) $\pi - 9\sqrt{3}/16$, (d) $\pi/2 - 2/3$
9. (a) $-8uv$, (b) $4(v^2 - v - u)$, (c) $-\sin t \cos t$
11. (a) 6, (b) $\sqrt{6}$, (c) $\pi(2\sqrt{2} - 1)/6$, (d) $a^3(\pi - 2)/3$, (e) $28\pi/3$, (f) $4\pi a\left[a - \sqrt{a^2 - b^2}\right]$
15. M. of I. $= 656\lambda a^6/45$
16. New torque $= 2\mu wa/3$, old $= \mu wa/2$
18. $2\pi\gamma$
19. $5\pi a^4/8$
20. 48

Page 360

1. (a) $\pi/8$, (b) $9\pi/2$, (c) 24π, (d) 96
5. $4\pi(a^2 - b^2)^{3/2}/3$
6. $\pi a^3(\tan\alpha - \tan\beta)$
7. $2\pi a^3(1 - \cos^4\alpha)/3$
8. $\pi a^3/6$
9. $\pi a^3(2 - \sqrt{2})/3$
10. $3\pi a^4/2c$
11. $\pi a^4(p + q)/pq$

Page 364

1. (a) $-10/3$, (b) $71/420$,
2. (a) $a^2b^2c^2/8$, (b) $4\pi/15$, (c) $\dfrac{40\pi}{3} - \dfrac{256}{9}$, (d) $7\pi/2$
3. (a) $\pi a^4/4$, (b) $2(e - 1)/3$
4. (a) $3\pi a^5/10$, (b) $\pi(2 \log 2 - 1)$
5. (a) $16a^3/3$, (b) $\pi a^3(2 - \sqrt{2})/3$, (c) $45\pi/2$, (d) $3\pi a^3/4$
6. (a) $8k\pi a^5/15$, (b) $4k\pi abc(a^2 + b^2)/15$
7. (a) $4\pi a^3bc/15$, (b) $a^8/64$,
8. $3h/8$

Chapter 11

Page 367

1. (i) Parallel planes, (ii) hyperboloids
2. Radial field
3. Directions of (i) x axis, (ii) x axis ($x \neq \pm 1$), (iii) $-2\mathbf{i} \pm 4x^2\mathbf{j}$, (iv) all

Page 370

1. $-e^{-t}\mathbf{i} - 12\cos 3t\mathbf{j} + 2\cos t\mathbf{k}$, $e^{-t}\mathbf{i} + 36\sin 3t\mathbf{j} - 2\sin t\mathbf{k}$
2. $(4xy - 4x^3)\,\mathbf{i} + y\cos x\mathbf{j}$, $2x^2\mathbf{i} + \sin x\mathbf{j} - \sin y\mathbf{k}$
5. $m = -n(1 \pm i)$, $\mathbf{r} = e^{-nt}(\mathbf{A}\cos nt + \mathbf{B}\sin nt)$, $\mathbf{r} = \mathbf{v}/n\, e^{-nt}\sin nt$

Page 373

1. (a) $\mathbf{i}+\mathbf{j}+\mathbf{k}$, $\sqrt{3}$, (b) $2(x\mathbf{i}+y\mathbf{j}-z\mathbf{k})$, $\sqrt{x^2+y^2+z^2}$,
   (c) $(z^4 - xy)\,\mathbf{i} - x^2\mathbf{j} + 8xz^3\mathbf{k}$, $\sqrt{4(z^4-xy)^2 + x^4 + 64x^2z^6}$
   (d) $-3x^2y\mathbf{i} - x^3\mathbf{j} + 2\mathbf{k}$, $\sqrt{9x^4y^2 + x^6 + 4}$
2. (i) 2, (ii) 28, (iii) -3
3. (i) $\dfrac{x^2+y^2}{2}$, (ii) $(3x^2 - 2y^2 + z^2)/2$
4. (i) $(\mathbf{i}+\mathbf{j}+\mathbf{k})/\sqrt{3}$, (ii) $(\mathbf{i}+\mathbf{j})/\sqrt{2}$
5. $\dfrac{7}{\sqrt{131}}, \dfrac{9}{\sqrt{131}}, \dfrac{-1}{\sqrt{131}}$ 6. $-7/\sqrt{6}$
7. Sphere $x^2+y^2+z^2 = 49$, $\pm(6\mathbf{i}+2\mathbf{j}+3\mathbf{k})/7$, $\dfrac{-4}{2401}(6\mathbf{i}+2\mathbf{j}+3\mathbf{k})$

Page 375

1. (i) $3\mathbf{i}+\mathbf{j}+2\mathbf{k}$, (ii) $2x\mathbf{i}+2y\mathbf{j}+2z\mathbf{k}$, (iii) $y\mathbf{i}+z\mathbf{j}+x\mathbf{k}$, (iv) $3\left(\dfrac{1}{r^{3/2}} - \dfrac{1}{r^{1/2}}\right)$

Page 382

1. (i) $4a^2\tan^{-1}\dfrac{b}{a} + 4b^2\tan^{-1}\dfrac{a}{b}$, (ii) $3k^2\pi/2$, (iii) $c^2\pi/4$
2. $a^3(k-4)/6$, $a^3(k-3)/3$ 3. $1 - 2/\pi - \pi/4$
4. (b) $a^2(2 + 4/\pi)$ 5. (i) 40, (ii) 4, (iii) $-152/3$
7. 120 9. $\phi = x^3+y^3+z^3 - 3xyz + x^2z + z^2y$, -30, 0

Page 386

1. (i) 0, (ii) 0, (iii) $-y\mathbf{i} - z\mathbf{j} - x\mathbf{k}$, (iv) 0
2. (i) $5y\mathbf{k}$, (ii) 0, (iii) 0
3. $2x^2y - xz^3$

Page 388

6. r^{-4}

Page 391

1. (a) 81/4, (b) -72, (c) 36π, (d) $64\pi/3$
2. $2\sqrt{2}\,\pi a$ 3. $-3\pi a^4/4$ 5. -4π 6. $-2\pi b^2$

Chapter 12

Page 403

1. (a) $u = x^3 - 3xy^2$, $v = 3x^2y - y^3$, (b) $(x^2 + y^2)^{1/2}$, 0, (c) x, 0
   (d) $e^x \cos y$, $e^x \sin y$, (e) $x/(x^2 + y^2)$, $-y/(x^2 + y^2)$
   (f) $x(x^2 + y^2 + 1)/[(x^2 - y^2 + 1)^2 + 4x^2y^2]$, $y(1 - x^2 - y^2)/[(x^2 - y^2 + 1)^2 + 4x^2y^2]$
   (g) $2(x^2 - y^2) + 7x + 6$, $y(4x + 7)$

2. (a) Translation by 1 unit, (b) Rotation by $\pi/2$
   (c) Reflection in the x-axis and rotation by $\pi/2$

3. (a) $r^3 \cos 3\theta + ir^3 \sin 3\theta$, (c) $z = 0$ and $\alpha \leqslant \arg z < \alpha + 2\pi/3$
   (d) Equal moduli; arguments differ by $2\pi/3$

4. (a) $1 - \frac{1}{2}i$, (b) $-(2 + i)/5$, (c) $5(7 - 4i)/66$, (d) $(1 + i)/\sqrt{2}$, (e) $\frac{1}{4}$

5. (a) $1 - i$, (b) $4/5 - 4i$, (c) 0. Limit as $z \to 0$ does not exist.

7. 3i, no, yes

8. (a) $-1 \pm i$, (b) $\pm 2 \pm 2i$ (c) nowhere in that region

9. (a) $2z + 3$, (b) $-z^{-2}$

10. (a) $-i$, $i/(z + i)^2$, (b) $-1 \pm 2i$, $(19 + 4z - 3z^2)/(z^2 + 2z + 5)^2$

11. (a) analytic, (b) not, (c) not, (d) analytic

14. (a) y, z (b) $e^x \sin y$, e^z (c) $2 \tan^{-1}(y/x)$, $\log z^2$
    (d) $ye^x \cos y + xe^x \sin y$, ze^z
    (e) $x^2 - y^2 + 2xy - 3x - 2y$, $(1 + i)z^2 - (2 + 3i)z$
    (f) $4xy - x^3 + 3xy^2$, $2z^2 - iz^3$

15. $v = 2y^2 - 2x^2 - 3x^2y + y^3$, $f(z) = -z^3 - 2iz^2$

16. (a) not harmonic, (b) harmonic, $v = x^2 - y^2 + 2y$, $f(z) = 2z + iz^2$

Page 408

3. (i) $2k\pi i/3$, (ii) $\frac{1}{8}\pi i + \frac{1}{2}k\pi i$, (iii) $\frac{1}{2}\log 2 + (2k\pi - \frac{\pi}{4})i$

4. $\cos z$, $-\sin z$

6. (i) $\cosh 1$, (ii) $\sin 1 \cosh 1 + i \cos 1 \sinh 1$, (iii) $-i \tanh \pi$

8. (i) $\pm i \log(2 + \sqrt{3}) + \frac{1}{2}\pi + 2k\pi$

   (ii) $-i \log(\sqrt{2} + 1) + \frac{1}{2}\pi + 2k\pi$, $-i \log(\sqrt{2} - 1) + \frac{3}{2}\pi + 2k\pi$

11. (i) $i\sqrt{3}$, (ii) i

12. $\log(\sqrt{2} + 1) + \pi i/2 + 2k\pi i$, $\log(\sqrt{2} - 1) + 3\pi i/2 + 2k\pi i$

13. (i) $(2k\pi - \pi/2)i$, (ii) $\log 5 + (\tan^{-1} 4/3 + 2k\pi)i$, (iii) $\frac{1}{2}\log 5 + (\tan^{-1}(-\frac{1}{2}) + 2k\pi)i$

14. -2

17. (i) $e^{-\pi/4 + 2k\pi}\left\{\cos(\frac{1}{2}\log 2) + i \sin(\frac{1}{2}\log 2)\right\}$
    (ii) $\sqrt{5}\, e^{(\tan^{-1} 2 - 2k\pi)}$ [cis $(\log\sqrt{5} - \tan^{-1} 2 + 2k\pi)$]

(iii) $e^{-4k\pi}$ cis (2 log 2) ; cis $\theta \equiv \cos\theta + i \sin\theta$

18. arg w = ½ arg (z − i) + ½ arg (z + i)

Page 415

1. (i) Circular region, centre (−2 + i), radius 4
(ii) Annulus, centre 2i, radii 1 and 2, (iii) Right half plane
(iv) Circle centre $-\dfrac{5}{3}$ i, radius 4/3
(v) Region between branches of hyperbola $x^2 - y^2 = 2$ (vi) Region inside ellipse foci (±3, 0)

2. (i) $6x^2y^2 - x^4 - y^4$ = const., (ii) $y^2 - x^2 - 2e^{-x} \sin y$ = const.,
(iii) $r^2 \sin 2\theta$ = const.

3. (i) Yes, streamlines are rectangular hyperbolas, (ii) No, (iii) Yes, straight line flow.

5. w_1 – line charge at point x_1, w_2 – opposite line charge at point x_2, $w_1 + w_2$ is combination of these giving circular lines of force.

7. Vortex of strength K at the origin; sink at the origin.

9.

	arg $z = \pi/6$	$\lvert z\rvert = 2$	$R(z) = -2$	$I(z) = 1$
(i)	$v + u\sqrt{3} = 0$	$u^2 + v^2 = 4$	$v = -2$	$u = -1$
(ii)	$v + u\sqrt{3} = 2$	$u^2 + (v-2)^2 = 4$	$v = 0$	$u = -1$
(iii)	$v + u\sqrt{3} = 2(1 + \sqrt{3})$	$\left[\dfrac{u}{2} - 1\right]^2 + \left[\dfrac{v}{2} - 1\right]^2 = 4$	$v = -2$	$u = 0$

10. $u^2 = 4\left[1 - \dfrac{v}{2}\right]$

11. (i) $u^2 + 2v = 1$, (ii) $\lvert w\rvert > 4$, (iii) Region between $v^2 = 4(1-u)$ and $v^2 = 16(4-u)$,
(iv) $1 < \lvert z\rvert < 4$ and $\lvert \arg w\rvert < \pi/2$

12. (i) $u^2 + v^2 - u - v = 0$, (ii) $v = -½$, (iii) $u = 0$,
(iv) $u^2 + v^2 + 2u + 2v + 1 = 0$, (v) $u = -½$

13. (0, 0) 15. Translation + inversion 16. $w = (10z - 12)/(z - 1)$

17. (i) $w = 6z + 4$, (ii) $w = \dfrac{1}{z}$, (iii) $w = iz$, (iv) $w = \dfrac{(z_2 - z_3)(z - z_1)}{(z_2 - z_1)(z - z_3)}$

20. (b) Region bonded by ellipse $u^2/\cosh^2 \dfrac{N\pi}{a} + v^2/\sinh^2 \dfrac{N\pi}{a} = 1$ and real axis.

21. $w = \left[iz + \dfrac{1}{\sqrt{2}}(1+i)\right] \Big/ \left[z + \dfrac{1}{\sqrt{2}}(1+i)\right]$, $v = \dfrac{4u - 2 - u^2}{4u + 2 + 3u^2}$

22. $u = (x^2 + y^2 - 1)/[(x+1)^2 + y^2]$, $v = 2y/[(x+1)^2 + y^2]$

23. (i) major arc of circle $u^2 + v^2 = \dfrac{1}{e^4}$ between $\left(0, -\dfrac{1}{e^2}\right)$ and $\left(-\dfrac{1}{e^2}, 0\right)$,
(ii) semicircle $u^2 + v^2 = e^2$, upper half.

24. (i) straight line $u = 3 \log 3$, (ii) straight line $v = \dfrac{\pi}{4}$

25. (i) $v \geqslant 0$, $-\infty < u < \infty$, (ii) $v \geqslant 0$, $u > 0$

Page 422

1. $z = re^{i\theta}$, ellipse $u^2/(r + \frac{1}{r})^2 + v^2/(r - \frac{1}{r})^2 = 1$

6. (i) $w = z^3$, (ii) $w = e^z$, (iii) $w = \cosh(\pi z/2)$, (iv) $w = z^{4/5}$

Page 431

1. (a) 32, (b) $-1 + i$, (c) $-44/3 - 8i/3$, (d) (i) $2\pi i$, (ii) $2\pi i$, (iii) $2\pi i$
3. (a) 0, (b) 0, (c) $-3/2$, (d) $-\pi i$, (e) 0, (f) 0, (g) πi
4. (a) $2\pi i e^2$, (b) $2\pi i$, (c) $-2\pi i$, (d) $\pi i/32$, (e) $-2\pi i$, (f) 0, (g) 0
5. $\dfrac{8\pi i}{3e^2}$
6. (i) (a) 0, (b) $\dfrac{\pi}{2}(3 + 2i)$
7. (a) $2\pi i$, (b) $-2\pi i$
8. $\pi\left[\dfrac{3}{2} + i\right]$
9. (a) $19\pi i/108$, (b) $9\pi i/8$

Page 435

1. (a) $e^i \sum_{n=0}^{\infty} \dfrac{(z-i)^n}{n!}$ (b) $\sum_{n=0}^{\infty} \dfrac{(-1)^n}{(2i)(n+1)} z^n$ (c) $-\sum_{n=0}^{\infty} z^{2n}$
2. (a) $\dfrac{1}{3} \sum_{n=0}^{\infty} (-1)^n \left[\dfrac{1}{(z-2)^{n+1}} - \dfrac{(z-2)^n}{4^{n+1}}\right]$ (b) $\dfrac{1}{3} \sum_{n=0}^{\infty} (-1)^n \dfrac{1-4^n}{(z-2)^{n+1}}$

   (c) $\sum_{n=0}^{\infty} \dfrac{(-1)^n}{(2n)!} (z-1)^{2n-1}$ (d) $\sum_{n=0}^{\infty} z^{2n-1}/(2n+1)!$
3. $\dfrac{1}{2z^2} + \dfrac{3}{4z} + \sum_{n=0}^{\infty} \left[1 - \dfrac{1}{2^{n+3}}\right] z^n$ $\quad |z| < 1$

   $\dfrac{1}{2z^2} + \dfrac{3}{4z} + \sum_{n=0}^{\infty} \dfrac{1}{z^{n+1}} - \sum_{n=0}^{\infty} \dfrac{1}{2^{n+3}} z^n$ $\quad 1 < |z| < 2$

   $\dfrac{1}{2z^2} + \dfrac{3}{4z} - \sum_{n=0}^{\infty} \dfrac{1}{z^{n+1}} - \sum_{n=0}^{\infty} \dfrac{2^{n-2}}{z^{n+1}}$ $\quad 2 < |z| < \infty$
4. $\dfrac{1}{z} + \sum_{n=0}^{\infty} \left[\dfrac{1}{i^{n-1}} - 1\right] z^n$
5. (a) $-\dfrac{1}{z} - 1 - z - z^2$, $0 < |z| < 1$; $z^{-2} + z^{-3} + z^{-4} + \ldots..$ $|z| > 1$

   (b) $-\dfrac{1}{4}z^3 - \dfrac{1}{2}z^4 - \dfrac{11}{16}z^5 - \dfrac{13}{16}z^6 - \ldots..$ $|z| < 1$

   $1 + z + z^2 + \tfrac{3}{4}z^3 + \ldots.. + \dfrac{1}{z} + \dfrac{1}{z^2} + \dfrac{1}{z^3} + \ldots..$ $\quad 1 < |z| < 2$

   $1 + 5/z + 49/z^3 + 129/z^4$ $\quad |z| > 2$
6. (a) $-\dfrac{1}{2z^4} + \dfrac{1}{2z^3} - \dfrac{1}{2z^2} + \dfrac{1}{2z} - \dfrac{1}{6} + \dfrac{z}{18} - \dfrac{z^2}{56} + \dfrac{z^3}{162}$

   (b) $\dfrac{1}{z^2} - \dfrac{4}{z^3} + \dfrac{13}{z^4} - \dfrac{40}{z^5}$

(c) $\dfrac{1}{2(z+1)} - \dfrac{1}{4} + \dfrac{1}{8}(z+1) - \dfrac{1}{16}(z+1)^2 + \ldots..$

(d) $\dfrac{1}{3} - \dfrac{4}{9}z + \dfrac{13}{27}z^2 - \dfrac{40}{81}z^3$

7. (a) Pole of order 1 at $z = 0$ (b) Essentl. at $z = 0$ (c) Pole of order 1 at $z = 1$, Pole of order 2 at $z = 2$ (d) Removable singularity at $z = 0$, $f(0) = -½$
(e) Removable singularity at $z = 2$, $f(2) = 1$ (f) Pole of order 3 at $z = 0$
(g) Poles of order 1 at $z \pm i$ (h) Pole of order 2 at $z = (2k+1)\pi/2$

Page 438

1. (a) 0, (b) 2, (c) 1, (d) $1/(2-\pi)$, 0, $(\pi-1)/(\pi-2)$, (e) 0
(f) 0, −1; −1, $e^{-1}/2$; 1, e/2, (g) 0, 2; 1, −1, (h) 0, −3; 1, 3
(i) $-\dfrac{1}{16} + \dfrac{1}{24}i$, (j) $\dfrac{3-2i}{6}$; $-3i$, $\dfrac{3+2i}{6}$
(k) $2e^{\pm i\pi/4}$, $\dfrac{(1\mp i)}{8\sqrt{2}}$; $2e^{\pm 3i\pi/4}$, $\dfrac{(-1\mp i)}{8\sqrt{2}}$, (l) $\pm 2i$, $\mp ie^2/4$
2. (a) $½\pi i$, (b) $(2-\pi^3/64)\pi i$, (c) $-\pi i$, (d) $2\pi i$, (e) $-2\pi i$
3. (a) 0, (b) 0, (c) 0

Page 441

2. $\pi/2$, $\pi/2$
4. (b) (i) 0, (ii) $\pi i/162$
6. (i) $2\pi i$
9. (i) $2\pi/(1-x^2)$, $0 < x < 1$, $2\pi/(x^2-1)$, $x > 1$, (ii) $\pi e^{-2a}/a$

Page 445

2(a) $[e^{-t} + e^{2t}(3t-1)]/9$, (b) $[e^{-t} + e^{t}(2t-1)]/4$, (c) $(\sin t - t\cos t)/2$

Chapter 13

Page 452

1. Yes 2. No 3. (519, 581) 4. 375 5. 108
6. Reject old process if mean breaking strength of sample exceeds 409.4 kN; 0.004

Page 456

1. (a) 1%, (b) 5%, (c) 10%, (d) 97.5%, (e) 99.5%, (f) 0.25%, (g) 1.25%, (h) 87.5%
2. 2.13, 2.05, 2.03, 1.96 3. No; highly significant
4. $t = -2.655$; significant, 5. No 6. $t = 2.88$; yes
7. 1.224 ± 0.006 cm; 1.024 ± 0.008 cm. 8. No; $t = 1.203$
9. 34.2 ± 0.4, 34.2 ± 0.6 10. 6.24 ± 0.45

11. 123.77 ± 0.23 12. Cannot reject claim

Page 460

1. (i) 21.03, 14.07, 7.815, 101.9, 124.3
   (ii) 26.22, 18.48, 11.34, 112.3, 135.8
   (iii) 11.34, 6.346, 2.366, 79.33, 99.33, (iv) 91%
   (v) (a) 95%, (b) 90%, (c) 45%

2. (1.54, 14.6) 3. No, $\chi^2 = 10.46$ 4. Cannot reject first claim

5. Reject first claim 6. $111.6 < \sigma < 198.8$ 7. No

8. $2.77 \leqslant \sigma^2 \leqslant 32.25$ 9. (2.84, 6.75) 10. (0.962, 39.1)

Page 476

1. (a) Reject claim, (b) cannot reject claim
2. z - value is -5.1, reject claim 3. Cannot reject separately, reject combined
4. Can reject combined value 5. Significant
6. Yes, $\chi^2 = 42.75$ 7. $\chi^2 = 4.8$, No
8. Not sufficient evidence to reject; no 9. Not significant
10. Not significant 11. Very significant
12. Not significant 13. Not significant
14. $\chi^2 = 7.69$; cannot assume a connection
15. Not significant 16. Reject claim of randomness
17. No 18. Too good a fit
19. Reject claim of constant failure rate at 0.1% significance level
20. Reject claim that factories have similar standards of production, with 1% risk

Page 483

1. (a) 4.53, 3.64, 2.07, 2.05, 2.04, 2.08, (b) 15.4, 2.16,
   (c) 5.81, 5.06, (d) 0.153, 0.270
2. Yes, $F = 1.90$
3. (a) No (b) Yes, $\hat{\sigma}^2 = 84.2$ (c) Yes, $\hat{\sigma}^2 = 4.9$ (d) No
4. No, $F = 1.51$ 5. Not significant at 5% level
6. Significant at 5% level, not at 1% level. 7. No, $F = 2.04$ 8. No

Page 491

1. $t = 0.976$, No 2. Doubtful, $t = 2.56$
3. Not significant in either case 4. Yes 5. No
6. No significant difference between balances.

7. No significant difference in means
8. No
9. Reject; take readings at same times
10. (i) 95% confidence between variances (almost too small an F value to be true) and no significant difference between means.

Page 502

1. No
2. Yes
3. Reject hypothesis that mean dust loadings are the same for different flow rates
4. Yes
5. Yes
6. No significant difference between operators or machines
7. Reject at 1% level
8. Significant differences between shifts, but not day to day.
9. Neither difference significant

Chapter 14

Page 513

1. 95% C.I. for α: 0.41 ± 1.58 — 95% C.I. for β: 0.039 ± .005
   99% C.I. for α: 0.41 ± 2.89 — 99% C.I. for β: 0.039 ± .008
2. 95% C.I. for α: 3.20 ± 11.22 — 95% C.I. for β: 14.09 ± 2.88
   99% C.I. for α: 3.20 ± 18.58 — 99% C.I. for β: 14.09 ± 4.78
3. 95% C.I. for α: 53.24 ± 3.25 — 95% C.I. for β: 0.58 ± 0.11
   99% C.I. for α: 53.24 ± 5.39 — 99% C.I. for β: 0.58 ± 0.18
4. 95% C.I. for α: 42.60 ± 14.49 — 95% C.I. for β: −0.69 ± 0.56
   99% C.I. for α: 42.60 ± 20.61 — 99% C.I. for β: −0.69 ± 0.79
5. 95% C.I. for α: 6.41 ± 2.09 — 95% C.I. for β: 1.81 ± 1.36
   99% C.I. for α: 6.41 ± 2.17 — 99% C.I. for β: 1.81 ± 1.42
6. 95% C.I. for α: 35.82 ± 22.68 — 95% C.I. for β: 0.48 ± 0.34
   99% C.I. for α: 35.82 ± 32.25 — 99% C.I. for β: 0.48 ± 0.48
7. 95% C.I. for α: 1.00 ± 0.24 — 95% C.I. for β: 12.50 ± 5.42
   99% C.I. for α: 1.00 ± 0.38 — 99% C.I. for β: 12.50 ± 8.51
8. 95% C.I. for α: 83.70 ± 23.62 — 95% C.I. for β: 1.09 ± 0.45
   99% C.I. for α: 83.70 ± 33.59 — 99% C.I. for β: 1.09 ± 0.64
9. 95% C.I. for α: −0.86 ± 4.57 — 95% C.I. for β: 0.27 ± 0.10
   99% C.I. for α: −0.86 ± 6.49 — 99% C.I. for β: 0.27 ± 0.14
10. 95% C.I. for α: −2.20 ± 11.00 — 95% C.I. for β: 0.29 ± 0.23
    99% C.I. for α: −2.20 ± 18.20 — 99% C.I. for β: 0.29 ± 0.38
12. $y = 1.65x - 0.30$
13. 3.2, 14.1
14. $y = 16.27 - 4.21/x$
15. $y = 0.43x - 0.01x^2$

Page 517

1. 20p/hour 95% C.I. 20 ± 8.287

2. (i) 10.14 ± 1.66 (ii) 76.45 ± 13.90 (iii) 61.90 ± 4.96
   (iv) 25.43 ± 3.84 (v) 9.00 ± 1.50 (vi) 66.22 ± 3.40
   (vii) 1.73 ± 0.32 (viii) 149.20 ± 18.40 (ix) 12.49 ± 3.04
   (x) 12.45 ± 1.97

3. 3.94, 1.80

   95% C.I. For α, β 3.94 ± 2.13, 1.8 ± 1.39
   99% C.I. For α, β 3.94 ± 3.05, 1.8 ± 2.00
   $y(1.6) = 6.82 \pm 1.24$ [90% C.I.]
   Mean value of $y(1.6) = 6.82 \pm 0.73$ [90% C.I.]

Page 523

1. 0.996, 0.98, 0.98, 0.43, 0.50, 0.49, 0.87, 0.75, 0.78, 0.76

2. 0.14 3. 0.058 4. $\hat{a} = 1.83$, $\hat{b} = -1.3 \times 10^{-2}$, $r^2 = 0.91$

Page 530

1. $\underline{F^1_3}$ 680.6, $F^1_4 = 184.3$, $F^1_4 = 223.8$, $F^1_{10} = 7.54$, $F^1_9 = 9.00$, $F^1_{10} = 9.75$, $F^1_5 = 35.0$,
   $F^1_{10} = 29.46$, $F^1_{10} = 34.65$, $F^1_4 = 12.67$

2. 17.48, 5.11; 104.3. Reject hypothesis that $\alpha = 6$ at 5% level.

Page 538

1. (i) $\hat{y} = 0.86 + 0.51x_1 + 1.19x_2$, (ii) $\hat{y} = 95.67 - 3.78x_1 + 1.68x_2$
   (iii) $\hat{y} = 61.40 - 3.65x_1 + 2.54x_2$

3. $\hat{y} = -0.055 + 0.88x_1 + 0.66x_2$

4. $\hat{x} = -5.02 + 1.22y - 0.003z$

5. $\hat{z} = -1.22 - 0.49x + 2.45y$ $(\geqslant 3)$

6. $\hat{y} = 104.83 + 0.71x_1 - 0.003x_2 - 0.71x_3$

Index